EMBRYOGENESIS EXPLAINED

EMBRYOGENESIS EXPLAINED

Natalie K. Gordon
University of Manitoba, Canada (retired)

Richard Gordon
Gulf Specimen Aquarium & Marine Laboratory, USA
Wayne State University, USA

World Scientific

NEW JERSEY • LONDON • SINGAPORE • BEIJING • SHANGHAI • HONG KONG • TAIPEI • CHENNAI • TOKYO

Published by

World Scientific Publishing Co. Pte. Ltd.
5 Toh Tuck Link, Singapore 596224
USA office: 27 Warren Street, Suite 401-402, Hackensack, NJ 07601
UK office: 57 Shelton Street, Covent Garden, London WC2H 9HE

Library of Congress Cataloging-in-Publication Data
Names: Gordon, Natalie K., author. | Gordon, Richard, 1943– , author.
Title: Embryogenesis explained / Natalie K Gordon, retired (University of Manitoba, Canada), Richard Gordon (Gulf Specimen Aquarium & Marine Laboratory, USA & Wayne State University, USA).
Description: Hackensack, New Jersey : World Scientific, 2016. | Includes bibliographical references and index.
Identifiers: LCCN 2015046966 | ISBN 9789814350488 (hardback : alk. paper)
Subjects: LCSH: Embryology.
Classification: LCC QL955 .G67 2016 | DDC 612.6/4--dc23
LC record available at http://lccn.loc.gov/2015046966

British Library Cataloguing-in-Publication Data
A catalogue record for this book is available from the British Library.

Printed in Singapore

Dedication

A true scientist is not doing science for the money, the prestige, or the possibility of fame. Rather, the true scientist collects data, stands in awe of nature and then shares the results of research freely and completely with others, all in the pursuit of truth. The heavy lifting of science falls mostly on the student who is diligently tackling one piece of the larger puzzle their mentor is unraveling. Students often work long hours and make great personal sacrifices completing their small piece of the larger puzzle. The most coveted prize, and often the only reward, of this hard work is the right to put their name on a real scientific publication alongside their mentor. The citations we give to published papers and figures often represent the contributions of such students. It is to the undergraduate project student, the summer research assistants, the graduate students and the postdoctoral fellows whose names are listed here that we dedicate this work. We know how much of your soul was poured into this and we deeply appreciate the privilege you have conferred on us of our benefiting from it.

Preface

Toad Hall Toy Store Tale (Winnipeg, ~1991):

> "May I help you?"
>
> "I am looking for something that could be used to model functioning of the cell nucleus during embryogenesis."
>
> "I think I'll leave you to it," the young clerk responded, backing away from Natalie as if she were insane.

It is a rare privilege in life to have predicted a phenomenon fundamental to our understanding of the universe, and then to find it as predicted. In our case, Dick (Richard Gordon) predicted differentiation waves on embryos in 1985, and Natalie Gordon (then Natalie Björklund) found the first one in 1989. Our work has not yet shaken the foundations of biology, despite our numerous publications, including Dick's massive tome[1]. Dick was advised[2] to write a more concise treatment of the topic. Natalie wanted to turn her experience explaining the waves to graduate students, especially those lacking a biology background, into written form. Thus, the idea for this book was proposed and accepted by World Scientific.

The title of this book *Embryogenesis Explained* is based on our own possibly arrogant belief that we have it right. Our model fits. For any theory of embryogenesis to be the "right" theory, it must have certain key elements. First and foremost, the right theory must be universal.

We can expect some organisms to evolve specializations in development. An example of this is the early development of the fruit fly *Drosophila melanogaster*. *Drosophila* became a "model organism" when embryologist Thomas Hunt Morgan[3] turned his attention to genetics around 1908[4]. The fruit fly's egg is packed full of several types of messenger RNA (mRNA) in a spatial gradient[5]. mRNA is RNA templated from DNA and then carried as a message to the protein making machinery and used as a code for stringing together amino acids in the correct order to get a functioning protein. It is presumed by many that some of this mRNA (or its protein counterpart[6]) acts as a morphogen, a molecule that somehow "controls" the shaping of the organism[7].

Many people have agreed among themselves that "morphogen gradients" combined with "inductive interactions" by which one cell "tells" another what to do, constitute the fundamental mechanism of embryogenesis[8,9]. However, there is another insect of similar shape to *Drosophila*, the polyembryonic parasitic wasp (Figure 0.1), in whose eggs there are no such initial gradients[10]. In fact, the polyembryonic wasp develops literally thousands of little wasps, all genetic clones, from one single egg. Even vertebrates from fish[11] to armadillos[12] and humans[13] can be polyembryonic. If the gradients of *Drosophila* were the fundamental underlying mechanism in which the body plan of the insect is laid out, how do we explain the parasitic wasp? The answer Natalie heard in one professor's class: "We are talking about *Drosophila*, not wasps", will not do, especially as it is looking more and more like the gradient explanation is insufficient of itself even to the task of laying out the anterior and posterior (head and tail) of the fruit fly[14].

Figure 0.1. **Left**: One fruit fly (*Drosophila melanogaster*) egg yields one adult fly. From[15] with permission under the Creative Commons Attribution-Share Alike 2.5 Generic license. **Right**: One polyembryonic wasp egg yields up to 2000 adult wasps[16]. From a web page by Scott Justis[17] with his kind permission. Yet the overall structure of both is quite similar. A beetle is also an insect, yet it develops by cell division at all stages[18], whereas *Drosophila* starts by nuclear divisions until there are 6000 nuclei in one "syncytial" cell (the fertilized egg), before cell division splits them apart[19]. So three quite

different appearing embryonic mechanisms produce adult insects[20]. Are they really unrelated mechanisms? Syncytial sponges start out as cellular embryos, suggesting "that the original metazoans [multicellular animals] were cellular, not syncytial"[21], also suggesting the developmental mechanism in *Drosophila* is modified from its ancestors[22].

It cannot be accepted that mRNA gradients are the universal mechanism for laying out a body plan unless it can be shown that nature made many paths to final development, including a unique one for fruit flies and one that is entirely different for wasps, despite their obvious similarity as adults. Such a conclusion would require ignoring the evidence of our eyes and the general observation that nature is conservative. If the wasp and the fruit fly share their basic biochemistry, protein making machinery and metabolism, form, genetic structure and anatomy, can we really conclude they evolved two entirely independent and different methods of embryogenesis? Must we not instead conclude that if gradients of mRNA as found in the *Drosophila* egg do not guide wasp development, then they cannot be the universal underlying mechanism of embryogenesis? The wasp takes two weeks to develop, whereas *Drosophila* does so in hours. Is it possible that the gradients only represent a clever shortcut for that particular stage and only in that particular organism because it allows the embryo to develop so quickly?

Furthermore, the gradients, if they were the underlying mechanism, must be used over and over again throughout fly development. Yet they are not. The individual cells in the eyes of developing fruit flies, the ommatidia, do not have thousands of tiny gradients in them as they develop. The same goes for any other later steps of fly development such as the imaginal discs that form into components like legs and wings.

We think a mechanism based on morphogen gradients evolved later. Gradients evolved as an innovative specialization of fruit flies which allows them the speedy early development that so distinguishes them from other insects. The fruit fly takes a shortcut around the slower fundamental mechanism by prepackaging the egg with mRNA. The wasp does not prepackage the egg[23] so it must use the older, standard, slower mechanism.

Discovering the short cut mechanism of the fruit fly egg is no small scientific accomplishment[24]. Yet, it tells us more about hindrances in the conventions of the field than it does about the fundamental mechanism of embryogenesis. Some other mechanism, besides maternal mRNA gradients, is involved and has awaited discovery. We must insist that any model of embryogenesis that has valid explanatory power must apply to all organisms, and the model mechanism must be demonstrably repeated over and over again in multiple steps of development across multiple phyla and all the way back to the common ancestor to all life on Earth, or at least to LUCA: the last universal common ancestor[25,26]. This must be the case, in spite of the occasional clever shortcut, or the model simply cannot be the correct one.

What does the maternal mRNA gradients theory tell us about embryologists? It is rather unfortunate that throughout the history of embryology scientists have consistently chosen as model organisms those organisms that develop very quickly. It is convenient for a scientist to use a fruit fly that produces new offspring once every three days instead of once every several weeks or once a year. Regrettably this often means scientists have ended up spending their entire careers studying the clever shortcuts, such as gradients of mRNA, instead of the fundamental mechanism. In the life and death race to get grants and to be “right”, embryologists with a pet model tend to ignore or even denigrate the work of others. It is therefore no coincidence that Nobel prizes in embryology are so rarely awarded[27-30].

Embryology also attracts those student biologists most interested in the fundamental processes of life. These problems are tough nuts to crack. The embryo repeats its miracle in front of us over and over again yet rarely yields its secrets. Before a student can understand this stuff, huge quantities of time-changing anatomy must be memorized. The most complex and demanding tools of molecular biology are needed to unravel the inner workings. And, as we shall see, physics and engineering are needed to bridge the gaps between molecules, cells and embryos, placing a huge demand on the modern embryologist to be skilled in many disciplines. It can therefore feel difficult for a non-embryologist to assess a theory of embryogenesis critically. Embryologists are not subjected to the same kind of intense critiquing by their fellow scientists because so few of our fellow scientists are qualified to do so.

The result has been that there is almost a mysticism that goes along with being an embryologist. There are probably more deeply religious people among embryologists than among any other subset of scientists[31].

Embryology is also one of the few places in science where one can, in effect, embrace vitalism. After all, what other group of scientists could say "inductive interactions" and claim the phrase has any real explanatory power? And it is only developmental biologists who can claim that cellular components, like genes, can "foresee", "recognize", or "organize" life without being mocked. The logicians and pragmatists among students who find themselves wondering about this tend to quietly leave embryology after their first exposure to answers like "auto-inductive interactions mediated by neuralizing factor" and similar epicycles because they prefer to study something easier, like particle physics or brain surgery. They then avoid embryology as much as possible in their future. The embryologist who has developed a wonderful model about the shortcut used by one organism is not likely to be exposed to data from another organism, much less have their theory challenged by it. And so our requirement for universality of a model across phyla is largely our personal bugaboo and not a requirement of the field as whole.

In addition to being universal, the second thing that we think we must demand of the embryogenesis mechanism is that it must be traceable in some form back in time to before embryogenesis was required. Nature is conservative. She would not employ the mechanism of embryogenesis if the components had not already been tried and tested over multiple generations of single celled organisms. True, when we examine the single celled organisms of Earth today we have to keep in mind that they are not precise copies of early ancestors. They too have no doubt developed their own shortcuts and bypasses since our common ancestors parted ways. Yet, if we stick to the universal basis of embryogenesis we can certainly expect to learn about how we work by examining these fellow single-celled inhabitants of our little planet.

Finally, whether we embryologists would like to admit it or not, because life began in such a simple form, the underlying mechanism of embryogenesis must also be simple. All the major breakthroughs of biology often look ridiculously simple to the student looking back after the problem was solved. How many of us both admired the greats of our scientific past while picking up the basics of their entire life's work in a class or two as part of a lecture series in introductory biology? As long as we must invoke patronizing exhortations like "It's just too complex for you to understand at this level of your education", we don't have an explanation worth keeping. Each of the fine details may take a PhD thesis to be fully understood. Still, once we have figured out embryogenesis, we really should be able to explain it in a form that an intelligent elementary school fifth grader can grasp.

We will present clear evidence that the cytoskeleton has just such a mechanism for embryogenesis. It is firmly linked to the same machinery responsible for the earliest forms of motility and perception. It is universally present in that it can be traced from the most primitive forms of life existing in the hostile primordial environments of Earth and yet still be functioning right up to the present in our very own human cells. We will later introduce the concept of a differentiation tree that portrays the higher hierarchical structures we seek to unravel in embryogenesis Evolution moves forward in tiny increments of novelty but occasionally also takes a big leap forward[32]. We will reenter this debate by introducing a new way to look at microevolution vs. macroevolution that unifies and explains both. Our model of embryogenesis is both universal and explanatory. It's also simple. This is why we think we are right.

The problem of embryogenesis is literally the problem of "which came first, the chicken or the egg?" All life on Earth seems to trace back to a common origin, in an ultimate sense. The origin of life is a problem we have yet to solve[33-35]. We therefore make the assumption that the explanation for the origin of life lies in pre-life physics, and furthermore that we may speak of an embryo physics that somehow creates each organism and will check if our present understanding of physics is up to the task. Some of us will always prefer the finger of God stirring a primordial pool no matter what physics tells us. *Embryogenesis Explained* will merely show the religious how God did what he did when he stirred the pool. For those who prefer explanations without God, we will offer suggestions on how physics might adapt and grow to embrace the whole organism.

This shorter but conceptually expanded work (cf.[1]) is written in the expectation that the reader already has some general interest in scientific knowledge. It should be useful as a jumping off point for the curious layman or the scientist or professional encountering the problem of embryogenesis without the formal biology background. There will also be material useful for the seasoned biologist caught up in the new rush of information about the role of mechanics in developmental biology and cellular level mechanics in medicine. We build our argument anew, starting from the origin of life (Chapter 1). We quickly review important concepts such as developmental anatomy (Chapter 2), genetics (Chapter 3) and epigenetics (Chapter 4). Then we go into the cell delving into the cytoskeleton (Chapter 5). Hopefully these chapters provide enough background for solid understanding without miring the reader in endless details. We then introduce the cytoskeleton-based cell state splitter in each cell and the differentiation waves they support (Chapter 6), which we predicted way back in 1985 and found in 1990. In Chapter 7 we bring in the concept of the differentiation tree as a way of organizing and understanding the development of an embryo in terms of its differentiation waves, relating them to developmental anatomy via fate maps. Afterwards we explain how signals from the cell state splitter reach the nucleus (Chapter 8) and how the nucleus responds by changes in gene expression (Chapter 9). Here, with advances made by scientists from around the world, we are able to pull together a far more comprehensive story than we could two decades ago. In Chapter 10 we put differentiation waves in the context of the many other waves that have been found in embryos. At the end we turn to a bit of speculation, showing how the differentiation tree might explain the progress of evolution over geological time (Chapter 11), and delve into questions in the philosophy of science that embryos force us to confront, ending with the concept of the cybernetic embryo (Chapter 12).

For all the supposed complexity of life and depiction of development as a disjoint set of problems[36], we have arrived at the conclusion that embryogenesis is fundamentally simple, and therefore simple to explain.

This text was tried out on the wonderful, eclectic mixture of participants from around the globe[37] in the Embryo Physics Course, held since January 2009 (Figure 0.2). We have been hosting them in the virtual world known as Second Life®[38]. In fact, they were an integral part of the writing itself, as each draft chapter was also presented as a talk, starting January 2011. We value their feedback and those who have read various partial drafts: Mohsen Abdel-Hadi, Patricia F. Anderson[39], Dorothea Blostein, William R. Buckley, John W.M. Bush, James A. Coffman, Susan Crawford-Young, Bruce Damer, Michael V. Danilchik, Ille C. Gebeshuber, Kyle Harrington, Nabiha H. Husain, Vishal Jindal, Stephen M. Levin, Iavor Kostov, Lu Kai, Stephen P. McGrew, Stuart A. Newman, Evgenii Rudnyi[40], Luigia Santella, Robert Stone, Jack A.

Figure 0.2. A virtual classroom for the Embryo Physics Course in Second Life®[37].

Tuszynski, Tanya Wolff, Borys Wróbel. Bradly Alicea and George E. Mikhailovsky kindly read and commented on the whole penultimate draft. Thanks also go to Cheryl Millan Violon (née Buors), Nicole Boudreau and Candice Lewis, Neil John Maclean Health Sciences Library, and Evelyn Bruneau and Kristen Kruse of the Library Electronic Resources Unit, University of Manitoba, for retrieving many references, the present and former staff of World Scientific Publishing Company who worked with us: Jihan Abdat, Rajesh Babu, Yolande Koh, V K Sanjeed, Ian Seldrup and Catherine Yeo, and Jack Rudloe, the late Anne E. Rudloe, and Cypress Rudloe, who hosted us at their Gulf Specimen Marine Laboratory during much of the writing. William R. Buckley[41], Evgenii Rudnyi[42] and Lu Kai[43,44] also served as webmasters for the Embryo Physics Course. Starting in 2014, under the guidance of Stephen A. Krawetz[45], graduate credit was offered by Wayne State University. Most of the slides from the talks are available online[42]. Thanks are due to our colleagues who also work on our favorite animal, the axolotl, especially the late John B. Armstrong, the late Pieter D. Nieuwkoop, and S. Randal Voss, Chris Muzinic and Laura Muzinic of the Ambystoma Genetic Stock Center at the University of Kentucky, who have carried on the work of George Malacinski and Susan Duhon in providing the world with axolotls and their embryos[46]. We thank Richard J. Feldmann for paying us a visit at our quarter section in Manitoba, Silver Bog, where we discussed his concept of connectrons[47]. We think his connectrons are part of the hierarchical control of the genome and a component of the DNA sequences used by the nuclear state splitter we describe in Chapter 9. Our discussion during his visit clarified some of our thinking about how the nuclear state splitter could work.

Our book is profusely illustrated, thanks to hundreds of people whose work is acknowledged and/or cited in the figure captions. (We've also added a few of our own.) Those from nonarchival, eventually cobweb Internet pages are preserved here.

Finally we thank the toy makers, for in the course of this work we accumulated a whole box of toys that provided models for various aspects of embryogenesis. We can now put our toys aside, distributing them to our grandchildren, especially the Wurfel, which gave us a "toy model"[48] of how the nucleus might work in development (Figures 0.3 and 0.4).

Scientists build on one another's work, and should be acknowledged for their contributions. Their work also justifies our assertions, although we

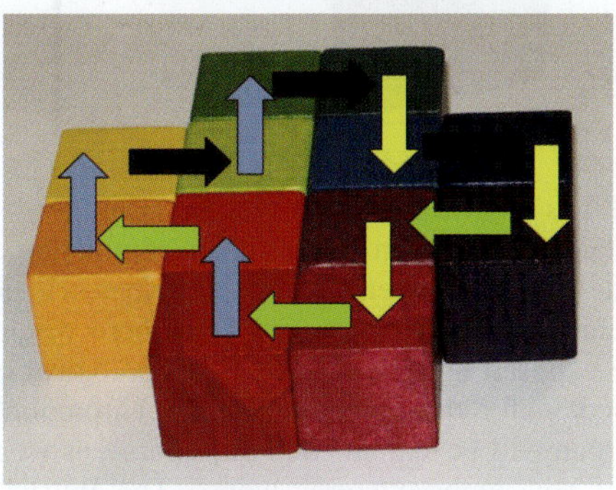

Figure 0.3. We bought our first Wurfel at Toad Hall Toys (Winnipeg). It provided a toy model for the behavior of the nucleus during cell differentiation that sustained us for decades until we recently figured out what we think is the real mechanism. The idea was, for instance, that each block in a Wurfel could represent a chromosome domain. Where two domains touched on their inner, hidden surfaces, genes would be exposed and be available for transcription. Those on the outside surfaces would be sequestered. See §10.13 *A Clockwork Universe Within: Nuclear Tensegrity Mechanics (Wurfels) as a Foundation for the Nuclear State Splitter* in *Hierarchical Genome*[1].

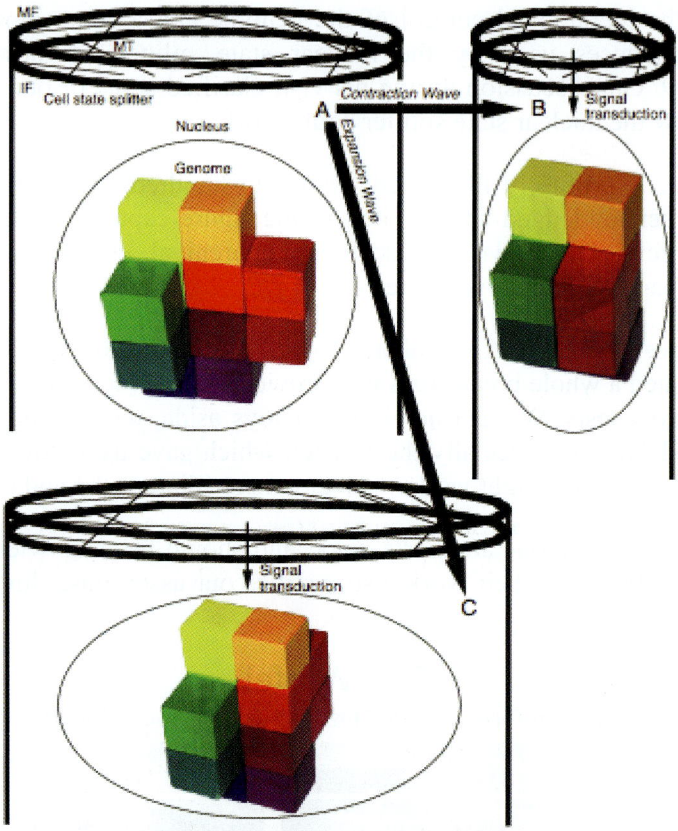

Figure 0.4. A Wurfel has many possible configurations[49], so we thought that when cell A, via its cell state splitter, participates in a contraction wave, the current configuration changes one way (B), whereas if it participates in an expansion wave, the Wurfel-like genome changes another way (C). Thus there would be a change in which subsets of genes were exposed or sequestered. The topological circularity of the Wurfel also conveniently stood in for the potential circular chromolinker organization of the genome. All of this jargon will be explained in due course, as we introduce our richer and more realistic ideas of how embryogenesis works. The Wurfel is now an antique, no longer manufactured. RIP.

often interpret their findings in new ways they may not like. Reference to our own work establishes our credentials. To keep the references convenient, yet so they do not interfere with the reading, they are all placed in chapter endnotes, where we hope their titles will entice further reading. Sample references to most topics are given. We often chose to refer to review articles as they make a good jumping off point to further explanation. Most of the papers listed are thus samples from large bodies of hard work by many people possibly not named. Free copies of some can be downloaded via the wonderful free databases Google Scholar[50] and PubMed[51] and preprint archives[52,53] which are turning the scientific literature into a single, accessible resource[54], and our own work is available on request. Just email us. Work citing a particular paper can be found using Google Scholar or Web of Science[55], which help bring one right up to date in any field. PubMed shows "Related citations". We frequently cite those articles in Wikipedia[56] that we think are reasonably well written, kept up to date by dedicated volunteers and aid quick understanding. Our previous work[1] had nearly 7000 references, which would be overwhelming here in this simpler format. Science cannot proceed without the hard work and intense debate of honest scholars, all of whom contribute to the common pool of knowledge. Even if an idea turns out to be wrong, it must still be respected. It is based on such contributions and with such respect that our own ideas are hereby offered.

For this work we would like to especially thank the embryologist Thomas Hunt Morgan whose 1897 book[3] *The Development of the Frog's Egg, An Introduction to Experimental Embryology* first brought us together on April 16, 1988 at the Fort Whyte Nature Centre Library in Winnipeg (although it was years before either of us realized the significance of this meeting), to Antone G. Jacobson who presenting his work with then graduate student Beth Burnside[57] woke up Dick at the Woods Hole Embryology Course in the summer of 1969, to the late Hans Laale[58], Natalie's first mentor in embryology who encouraged her to accept a job with Dick because, even though he disagreed with Dick's ideas, he expected she would learn a lot from him, and to Lorne E. Seargeant, one of Natalie's PhD committee members[59-62] who strongly encouraged her to publish her hypotheses and not just her results.

Natalie K. Gordon (nee Björklund), PhD, Human Genetics,
NKBGordon@gmail.com

Richard Gordon, PhD, Chemical Physics,
DickGordonCan@gmail.com

Gulf Specimen Marine Laboratory & Aquarium, Panacea, Florida USA

C.S. Mott Center for Human Growth and Development,
Wayne State University, Detroit, Michigan USA

1. Gordon, R. (1999). *The Hierarchical Genome and Differentiation Waves: Novel Unification of Development, Genetics and Evolution*. Singapore & London, World Scientific & Imperial College Press: [http://www.worldscientific.com/worldscibooks/10.1142/2755].
2. Papageorgiou, S. (2001). The Hierarchical Genome and Differentiation Waves: Novel Unification of Development, Genetics and Evolution (Vol. I, II) (1999) Richard Gordon. World Scientific and Imperial College Press, London, 1836pp. ISBN: 981-02-2268-8 [book review]. *BioEssays* **23**(6), 559.
3. Morgan, T.H. (1897). *The Development of the Frog's Egg, An Introduction to Experimental Embryology*. New York, Macmillan Co.
4. Kohler, R.E. (1994). *Lords of the Fly: Drosophila Genetics and the Experimental Life*. Chicago, University of Chicago Press.
5. Gaspar, I. (2011). Microtubule-based motor-mediated mRNA localization in *Drosophila* oocytes and embryos. *Biochemical Society Transactions* **39**, 1197-1201.
6. Gurdon, J.B., A. Mitchell & K. Ryan (1996). An experimental system for analyzing response to a morphogen gradient. *Proceedings of the National Academy of Sciences USA* **93**(18), 9334-9338.
7. Stathopoulos, A. & M. Levine (2002). Dorsal gradient networks in the *Drosophila* embryo. *Developmental Biology* **246**(1), 57-67.
8. Neumann, C. & S. Cohen (1997). Morphogens and pattern formation. *BioEssays* **19**(8), 721-729.
9. Chan, T.M., W. Longabaugh, H. Bolouri, H.L. Chen, W.F. Tseng, C.H. Chao, T.H. Jang, Y.I. Lin, S.C. Hung, H.D. Wang & C.H. Yuh (2009). Developmental gene regulatory networks in the zebrafish embryo. *Biochimica et Biophysica Acta-Gene Regulatory Mechanisms* **1789**(4), 279-298.
10. Grbić, M. (2003). Polyembryony in parasitic wasps: evolution of a novel mode of development. *International Journal of Developmental Biology* **47**(7-8), 633-642.
11. Laale, H.W. (1984). Polyembryony in teleostean fishes: double monstrosities and triplets. *Journal of Fish Biology* **24**, 711-719.
12. Enders, A.C. (2002). Implantation in the nine-banded armadillo: how does a single blastocyst form four embryos? *Placenta* **23**(1), 71-85.
13. Blickstein, I. & L.G. Keith (2007). On the possible cause of monozygotic twinning: lessons from the 9-banded armadillo and from assisted reproduction. *Twin Research and Human Genetics* **10**(2), 394-399.
14. Lucchetta, E.M., J.H. Lee, L.A. Fu, N.H. Patel & R.F. Ismagilov (2005). Dynamics of *Drosophila* embryonic patterning network perturbed in space and time using microfluidics. *Nature* **434**(7037), 1134-1138.
15. Karwath, A. (2005). *Drosophila melanogaster* - side. http://en.wikipedia.org/wiki/File:Drosophila_melanogaster_-_side_%28aka%29.jpg
16. Grbić, M., L.M. Nagy & M.R. Strand (1998). Development of polyembryonic insects: a major departure from typical insect embryogenesis. *Development Genes and Evolution* **208**(2), 69-81.
17. Justis, S. (2010). Wasp ID - *Copidosoma*. http://bugguide.net/node/view/466062/bgimage

18. da Fonseca, R.N., J.A. Lynch & S. Roth (2009). Evolution of axis formation: mRNA localization, regulatory circuits and posterior specification in non-model arthropods. *Current Opinion in Genetics & Development* **19**(4), 404-411.
19. Mavrakis, M., R. Rikhy & J. Lippincott-Schwartz (2009). Plasma membrane polarity and compartmentalization are established before cellularization in the fly embryo. *Developmental Cell* **16**(1), 93-104.
20. Lynch, J.A., E. El-Sherif & S.J. Brown (2012). Comparisons of the embryonic development of *Drosophila*, *Nasonia*, and *Tribolium*. *WIREs Developmental Biology* **1**(1), 16-39.
21. Leys, S.P., E. Cheung & N. Boury-Esnault (2006). Embryogenesis in the glass sponge *Oopsacas minuta*: Formation of syncytia by fusion of blastomeres. *Integrative and Comparative Biology* **46**(2), 104-117.
22. Wiegmann, B.M., M.D. Trautwein, I.S. Winkler, N.B. Barr, J.-W. Kim, C. Lambkin, M.A. Bertone, B.K. Cassel, K.M. Bayless, A.M. Heimberg, B.M. Wheeler, K.J. Peterson, T. Pape, B.J. Sinclair, J.H. Skevington, V. Blagoderov, J. Caravas, S.N. Kutty, U. Schmidt-Ott, G.E. Kampmeier, F.C. Thompson, D.A. Grimaldi, A.T. Beckenbach, G.W. Courtney, M. Friedrich, R. Meier & D.K. Yeates (2011). Episodic radiations in the fly tree of life. *Proceedings of the National Academy of Sciences USA* **108**(14), 5690-5695.
23. Zhurov, V., T. Terzin & M. Grbić (2007). (In)discrete charm of the polyembryony: evolution of embryo cloning. *Cellular and Molecular Life Sciences* **64**(21), 2790-2798.
24. Nobelprize.org (1995). The Nobel Prize in Physiology or Medicine 1995 jointly to Edward B. Lewis, Christiane Nüsslein-Volhard and Eric F. Wieschaus for their discoveries concerning "the genetic control of early embryonic development". http://www.nobelprize.org/nobel_prizes/medicine/laureates/1995/press.html
25. Di Giulio, M. (2011). The Last Universal Common Ancestor (LUCA) and the ancestors of archaea and bacteria were progenotes. *Journal of Molecular Evolution* **72**(1), 119-126.
26. Morange, M. (2011). Some considerations on the nature of LUCA, and the nature of life. *Research in Microbiology* **162**(1), 5-9.
27. Hamburger, V. (1985). Hans Spemann, Nobel Laureate 1935. *Trends in Neurosciences* **8**(9), 385-387.
28. Sulston, J.E. (2003). *Caenorhabditis elegans*: The cell lineage and beyond (Nobel lecture). *Chembiochem* **4**(8), 688-696.
29. Marine Biological Laboratory (2007). The Jean & Katsuma Dan Lectureship in Embryology/Physiology. http://hermes.mbl.edu/events/special/07dan.html
30. Johnson, M.H. & J. Cohen (2012). Reprogramming rewarded: the 2012 Nobel prize for Physiology or Medicine awarded to John Gurdon and Shinya Yamanaka. *Reproductive Biomedicine Online* **25**(6), 549-550.
31. Goodenough, U. (1998). *The Sacred Depths of Nature*. New York, Oxford University Press.
32. Reid, R.G.B. (2007). *Biological Emergences: Evolution by Natural Experiment*. Cambridge, MIT Press.
33. Gordon, R. & R.B. Hoover (2007). Could there have been a single origin of life in a Big Bang universe? *Proceedings of the SPIE* **6694**, doi:10.1117/1112.737041.

34. Gordon, R. (2008). Hoyle's tornado origin of artificial life, a computer programming challenge. In: *Divine Action and Natural Selection: Science, Faith and Evolution*. Eds.: J. Seckbach & R. Gordon. Singapore, World Scientific: 354-367.
35. Sharov, A.A. & R. Gordon (2013). Life Before Earth. http://arxiv.org/abs/1304.3381
36. Love, A.C. (2014). The erotetic organization of developmental biology. In: *Towards a Theory of Development*. Eds.: A. Minelli & T. Pradeu, Oxford University Press: 33-55.
37. Gordon, R. (2013). Conception and development of the Second Life® Embryo Physics Course [invited]. *Systems Biology in Reproductive Medicine* **59**, 131-139.
38. Linden Lab. (2013). Second Life®. http://secondlife.com/
39. Anderson, P.F. (2011). Cool Toys Pic of the day - About Writing Books in New and Different Ways. https://cooltoysu.wordpress.com/2011/01/14/cool-toys-pic-of-the-day-about-writing-books-in-new-and-different-ways/
40. Rudnyi, E. (2011). Perception, Feedback, and Qualia. http://blog.rudnyi.ru/2011/01/perception-feedback-and-qualia.html
41. Buckley, W.R. (2008). Computational ontogeny. *Biological Theory* **3**(1), 1-6.
42. Rudnyi, E. (2014). Embryo Physics Course – Embryogenesis Explained. http://embryogenesisexplained.com/author-index
43. Lu, K., T. Cao & R. Gordon (2012). A cell state splitter and differentiation wave working-model for embryonic stem cell development and somatic cell epigenetic reprogramming. *BioSystems* **109**(3, Special Issue on Biological Morphogenesis), 390-396.
44. Lu, K., R. Gordon & T. Cao (2013). Reverse engineering the mechanical and molecular pathways in stem cell morphogenesis. *Journal of Tissue Engineering and Regenerative Medicine*, doi: 10.1002/term.1672.
45. Jodar, M., S. Selvaraju, E. Sendler, M.P. Diamond & S.A. Krawetz (2013). The presence, role and clinical use of spermatozoal RNAs. *Human Reproduction Update* **19**(6), 604-624.
46. Smith, J.J., S. Putta, J.A. Walker, D.K. Kump, A.K. Samuels, J.R. Monaghan, D.W. Weisrock, C. Staben & S.R. Voss (2005). Sal-Site: integrating new and existing ambystomatid salamander research and informational resources. *BMC Genomics* **6**, 181 and http://www.ambystoma.org.
47. Feldmann, R.J. (2001). *Algorithmic determination of connectrons [US Patent WO2001094542 A2]*. Washington, DC, USA, US Patent Office.
48. Alicea, B. & R. Gordon (2014). Toy models for macroevolutionary patterns and trends. *BioSystems* **122**(Special Issue: Patterns of Evolution), 25-37.
49. Tromp, J.T. & R. Gordon (2006). The number of 3D configurations of a labeled size 2*n "Wurfel" [A117613: The On-Line Encyclopedia of Integer Sequences]. http://oeis.org/A117613
50. Google (2013). Google Scholar. http://scholar.google.ca/
51. U.S. National Library of Medicine (2013). PubMed. http://www.ncbi.nlm.nih.gov/pubmed
52. Cold Spring Harbor Laboratory (2015). bioRχiv, The Preprint Server for Biology. http://www.biorxiv.org
53. Cornell University Library (2015). arXiv.org. http://arxiv.org

54. Gordon, R. & B.J. Poulin (2008). There is but one journal: the scientific literature. *PLoS Medicine* **5**(10), e201.
55. Thomson Reuters (2013). Web of Science. http://wokinfo.com/products_tools/multidisciplinary/
56. Wikipedia (2014). Wikipedia, the free encyclopedia that anyone can edit., https://en.wikipedia.org/wiki/Main_Page
57. Burnside, M.B. & A.G. Jacobson (1968). Analysis of morphogenetic movements in the neural plate of the newt *Taricha torosa*. *Developmental Biology* **18**, 537-552.
58. Laale, H.W. (1987). *An Illustrated Guide to the Embryology of the Amphibians*. Winnipeg, University of Manitoba.
59. Greenberg, C.R., A.N. Prasad, L.A. Dilling, J.R. Thompson, J.C. Haworth, B. Martin, P. Wood-Steiman, L.E. Seargeant, B. Seifert, F.A. Booth & C. Prasad (2002). Outcome of the first 3-years of a DNA-based neonatal screening program for glutaric acidemia type 1 in Manitoba and northwestern Ontario, Canada. *Molecular Genetics and Metabolism* **75**(1), 70-78.
60. Björklund, N.K., B.N. Chodirker, C.R. Greenberg, L.E. Seargeant & J.A. Evans (2003). C677T MTHFR frequencies in different Canadian Provinces [http://jmg.bmj.com/content/40/8/619.full/reply - jmedgenet_el_46]. *Journal of Medical Genetics* **40**(8).
61. Björklund, N.K., J.A. Evans, C.R. Greenberg, L.E. Seargeant, C.E. Schneider & B.N. Chodirker (2004). The C677T methylenetetrahydrofolate reductase variant and third trimester obstetrical complications in women with unexplained elevations of maternal serum alpha-fetoprotein. *Reproductive Biology and Endocrinology* **2**(1), doi: 10.1186/1477-7827-1182-1165.
62. Björklund, N.K. (2005). *The Association of the C677T5,10 methylenetetrahydrofolate reductase Variant with Elevated Maternal Serum a-fetoprotein and Complications of Pregnancy [Ph.D. Thesis]*. Winnipeg, Department of Biochemistry and Medical Genetics, University of Manitoba.

Contents

Contents

Chapter 1

How Embryogenesis Began in Evolution

We define embryogenesis as the process of producing the right kind of new cell, in the right place at the right time. This phrase "the right cell, in the right place, at the right time" is an expression attributed to Hans Driesch around 1890[1,2]. All other aspects of embryology like morphogenesis, differentiation, and regeneration arise from embryogenesis of a first simplest proto-life form. Where and how did this first life originate?

There is a delightful scene from the Star Trek Next Generation television series where a powerful, immortal being, Q, takes Captain Jean Luc Picard back in time and space to stand looking down at a tiny pool on a hostile volcanic planet[3]. Q tells Picard this is Earth eons ago and Picard should look in the pool. Q then laughs. Q says that THE miracle is about to occur. Right there in that puddle, Picard's far distant first ancestor is about to come into existence. Q comments that humans are the mere product of a pond of goo[4]. In their ongoing debate about the merit of humans Picard quotes Shakespeare saying:

> "I know Hamlet. And what he might say with irony I say with conviction. 'What a piece of work is man! How noble in reason! How infinite in faculty! In form, in moving, how express and admirable. In action, how like an angel. In apprehension, how like a god…'"[5].

Our God-like form is assumed to have begun in a primordial soup from some primitive bit of RNA (or proteins[6]) that was able to self-replicate[7]. There are many hypotheses about the conditions that allowed first life to form and many attempts to simulate the event by chemistry[8-14] or computer[15-17]. The two most common places that this remarkable feat is thought to have been accomplished inside the simplest of membranes are in a pool of water or near a hot hydrothermal vent[18,9].

Our universe started with the Big Bang 13.75 billion years ago[21] (Figure 1.1). It started hot and cooled down, forming galaxies and stars and perhaps planets as early as 200 million years after the Big Bang[22]. Earth formed much later, 4.54 billion years ago[23] (Figure 1.2). Our records of life on Earth begin a mere 4.1 billion years ago with what looks like fossil prokaryotes[24,25] and chemical signatures[26-28] presumed to be unique to life. While prokaryotes are simple compared to us, they are organisms that have evolved to levels of complexity far beyond the proto-life forms scientists envision at the origin of life. So where did life begin? Did first life start here on Earth at some time before 4.1 billion years? Did it

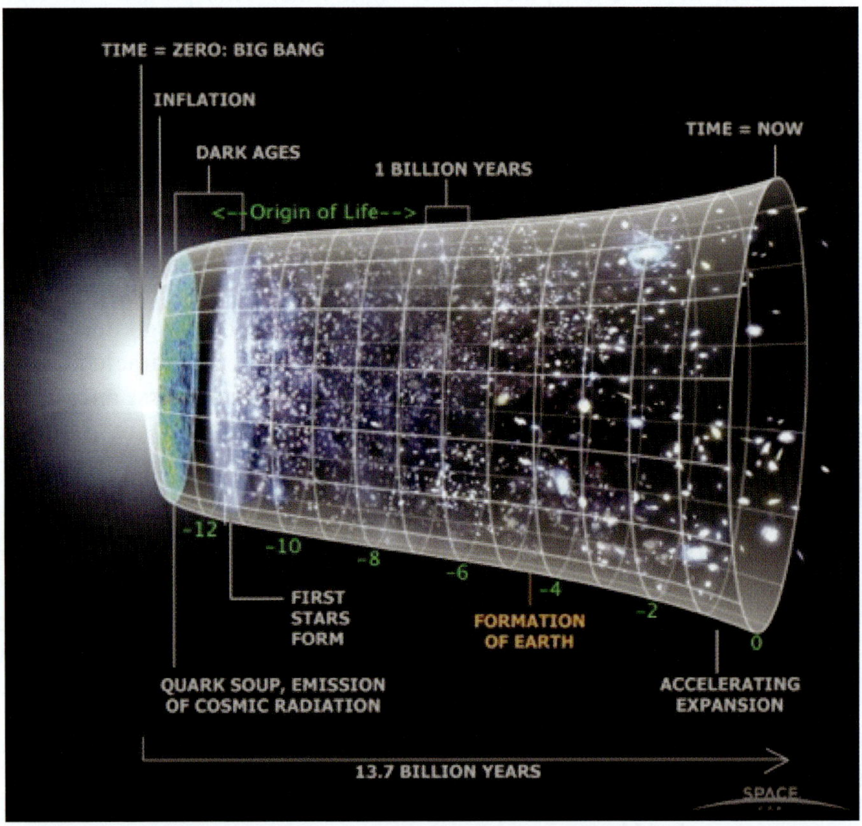

Figure 1.1. Development of the universe since the Big Bang, courtesy of the Hubble Space Telescope Science Institute, on which we have superimposed Alexei Sharov's estimate for the origin of life, 9.7 ± 2.5 billion years ago. For details see "Life Before Earth" (2013, 2016)[19,20].

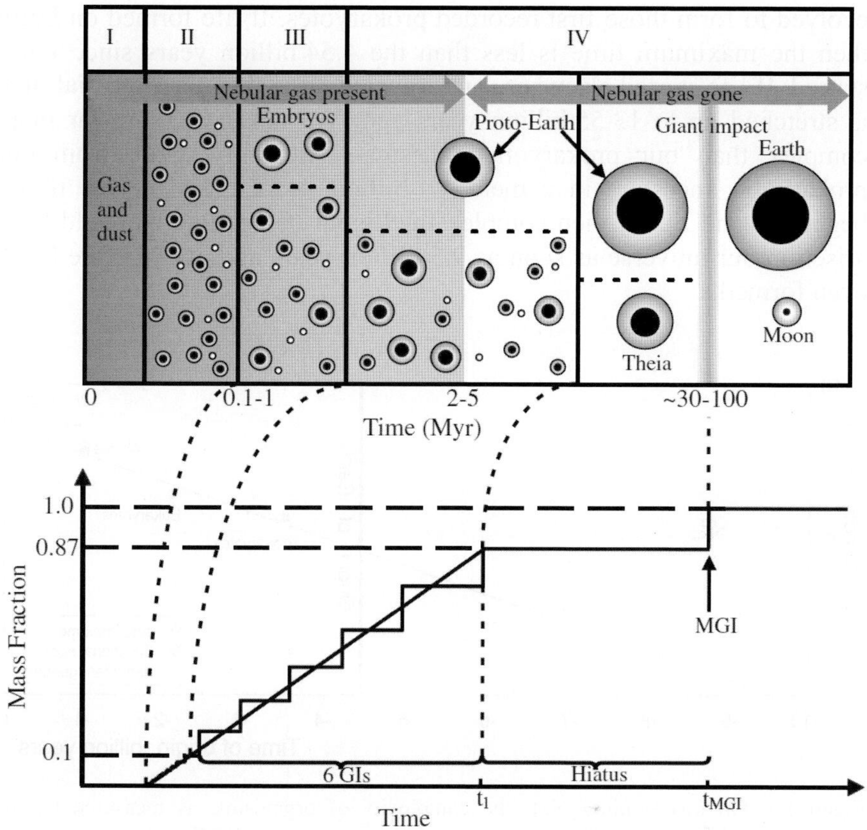

Figure 1.2. In this diagram the Earth formed 4.54 billion years ago from gas and dust during a period of 3 to 6 million years. The "embryos" here at stage III are formed from gas, dust and planetesimals of stages I and II. The denser core of the Earth and the other bodies is shown in black, and the outer part in grey. Collision of the Earth with another planet the size of Mars, called Theia, melted much or all of the surface, in this scenario. This figure is used with the kind permission of Gang Yu[32].

begin somewhere else[29] and then was transported here[30,31]? As mankind gets ready to travel beyond our moon this is becoming a more urgent question and many hypotheses have been brought forward to try to answer it.

If life started elsewhere in our turbulent Milky Way galaxy[33] and traveled to Earth, this increases the possible environments that life may have started in. It also increases the total time available for life to have evolved to form those first recorded prokaryotes. If life formed on Earth then the maximum time is less than the 4.54 billion years since Earth started. If life started elsewhere and traveled here, then the potential time is stretched up to 13.55 billion years ago. We eukaryotes are far more complex than our prokaryotic ancestors. Humanity arose from our prokaryotic ancestors in a mere 4.1 billion years. This means, if life began beyond Earth, then complex intelligent beings like us could have arisen in our universe as soon as 9.5 billion years ago, well before Earth even formed.

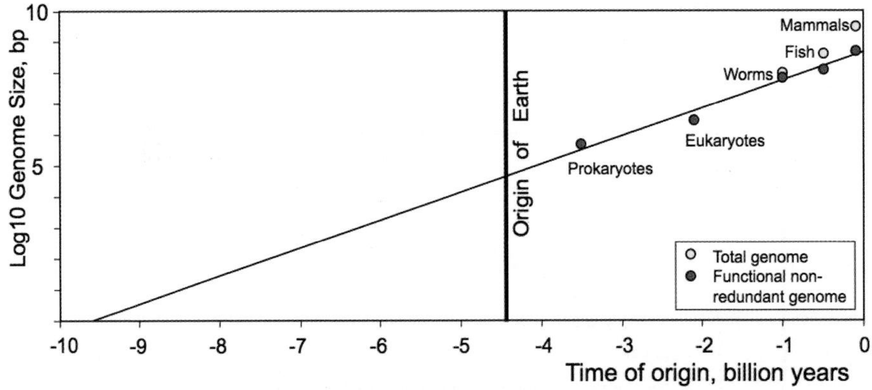

Figure 1.3. On this semilog plot, the complexity of organisms, as measured by the logarithm of the amount of coding nucleotide base pairs in DNA[20], is linear with time (i.e., exponentially increasing). Time is counted backwards in billions of years before the present (time 0). This method of plotting allows extrapolation back to just one base pair between 9 and 10 billion years ago, albeit with a large margin of error (not shown). The extrapolation to "one" (log10 = 0) molecule is justifiable in one model for the origin of life[34], or as the result of an abiotic process[35]. This plot suggests: 1) there was no intelligent life in our universe prior to the origin of Earth; 2) life took a long time, about 5 billion years, to reach the complexity of bacteria (prokaryotes); 3) Earth was seeded by a primitive form of life, rather than anything more complex than prokaryotes, and the panspermia hypothesis is correct[29]. However this model eliminates the idea that former intelligent life seeded Earth[36]; 4) life did not start on Earth; 5) attempts to synthesize the origin of life from scratch may have to emulate many cumulative rare events that occurred during that 5 billion year period before bacteria; 6) the Drake equation for guesstimating the number of civilizations in the universe[37] may be wrong, as intelligent life like us has just begun appearing in our universe. It has been hypothesized that there could be many planets with intelligent life in the universe. This model would explain why we have not found any sign of it yet (Fermi's paradox[38]). The time scale is roughly compatible with the idea that life may have originated shortly after parts of the universe cooled down after the Big Bang[39]. This figure is from "Life before Earth"[19].

Maybe life on Earth started with nothing more complicated than prokaryotes because it took 9.5 billion years for life to evolve that far. Extrapolated data suggests this might have been the case (Figure 1.3). If that idea holds up, then life indeed did not originate on Earth but life did find the right conditions for the evolution of complexity beyond prokaryotes to produce us here on this obscure little blue planet[19,40] (Figure 1.1). In our own galaxy there may be 15 to 30 billion habitable planets[41] (exoplanets) and even more exomoons that are habitable[42]. If that is true then the prokaryotes that arrived here on Earth may have also been deposited on other suitable planets as well and may have been evolving along at the same pace as us (Figure 1.4). Perhaps we are not alone and all complex life like us is approximately the same age as we are.

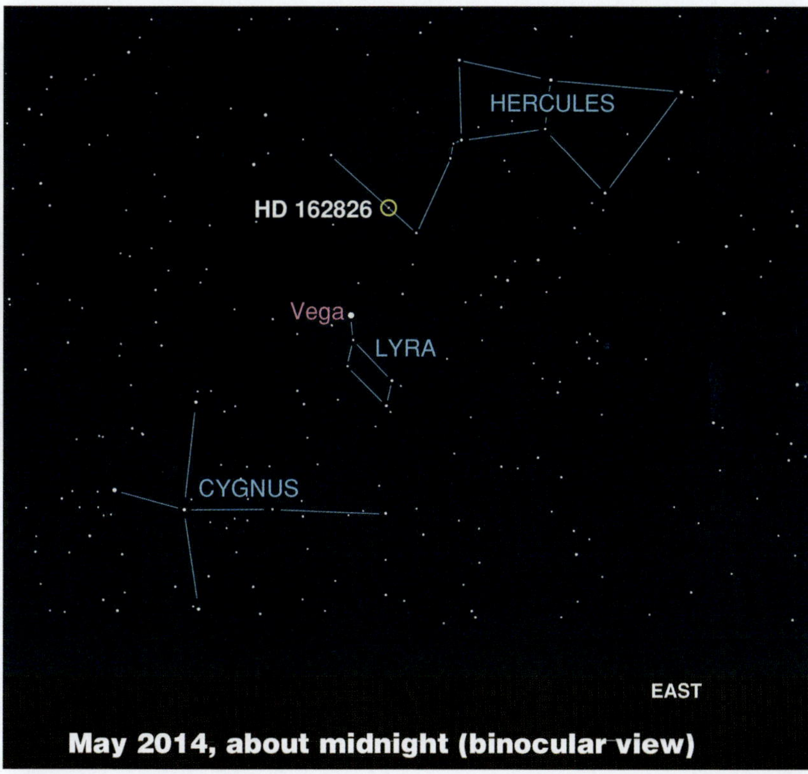

Figure 1.4. A nearby star, just 110 light years away, visible with binoculars, that probably formed at the same time as our Sun and its solar system, from the same galactic cloud. From "Solar Sibling HD"[43] with permission of McDonald Observatory. Cf.[44].

The universe is expanding much faster than primitive life could have travelled about in space, even if borne on the winds of supernovae explosions[45] (Figure 1.5), hypervelocity stars[46] (Figure 1.6), galactic ejecta, or transferred during frequent galaxy collisions[47]. Whole "rogue" planets are occasionally ejected from their solar systems[48-50] and bounced out, sometimes towards capture by other stars, like celestial billiard balls[51]. Taking into account all of these phenomena, we estimated that the "speed of life" without technological assistance is limited to 1000 kilometers per second[39]. This is 1/300th of the speed of light, which is 300,000 kilometers per second (299,792.458 kilometers per second in vacuum, to be more precise[52]). Thus despite the vast distances in space (Figure 1.7), there are many means for life to get around, even between galaxies. But life can only get so far.

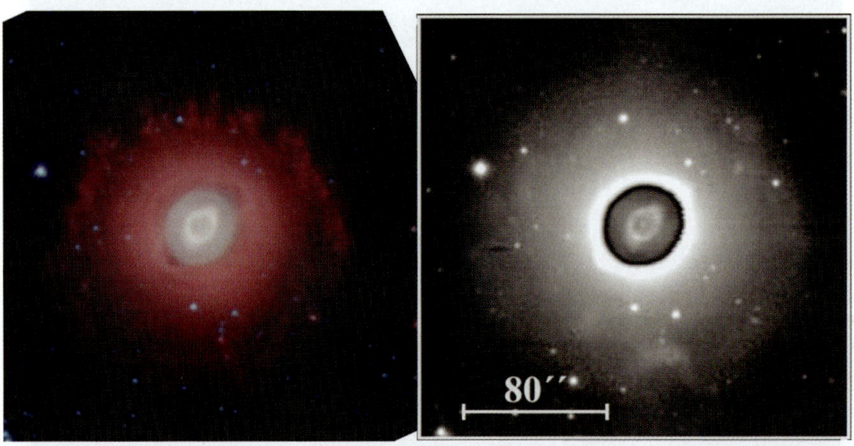

Figure 1.5. Dying stars eject significant fractions of their mass at speeds of up to 4500 kilometers per second[53]. This can result in the ejection of whole planets[48]. If we take 4500 km/s as the speed of life, instead of our previous estimate[39] of 1000 km/s, then the number of independent origins of life needed to fill the visible universe is reduced from 50,000 to 10,000. Here we show the Ghost of Jupiter Nebula NGC 3242, which has undergone multiple ejections, resulting in an onion-like structure around the star, expanding "only" 25 to 30 kilometers per second[54]. **Left**: showing inner rings only, from NASA Mission pages[55] with permission of NASA, rotated to match the image on the right. **Right**: showing visible extent of the ejections, from the Royal Astronomical Society[56] with permission of Oxford University Press (photographic reverse of their image). The visible radius is about 100 arc seconds, which at a distance of 1.6 kiloparsecs comes to 8 trillion kilometers. The distance of the Earth from our sun is 150 million kilometers. Were we near NGC 3242 at that distance, we would be inside the innermost ring.

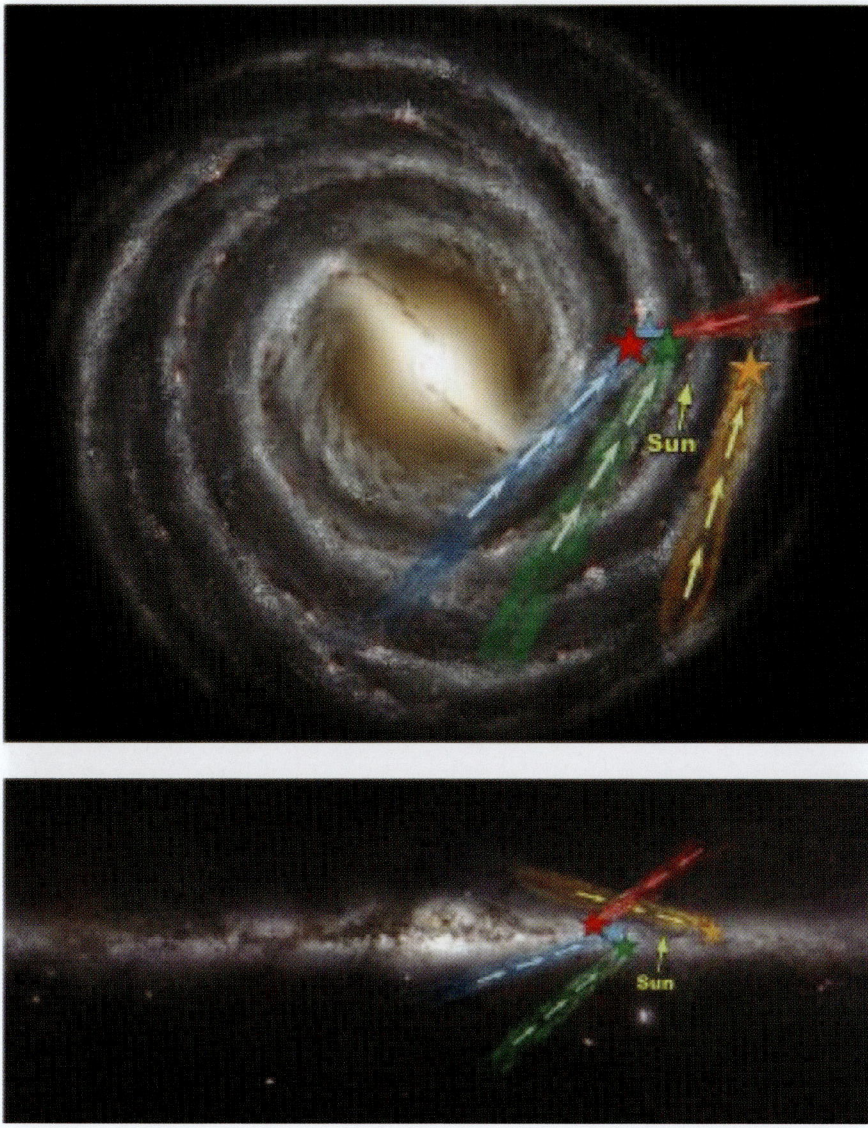

Figure 1.6. "Top and side views of the Milky Way galaxy show the location of four of the new class of hypervelocity stars. These are sun-like stars that are moving at speeds of more than a million miles per hour [450 kilometers per second] relative to the galaxy: fast enough to escape its gravitational grasp. The general directions from which the stars have come are shown by the colored bands." From Hypervelocity Star Illustration[57] with blanket permission. Such stars could carry along with them a still hypothetical intergalactic panspermia (any living thing that has been transferred in space).

Figure 1.7. The difficulty of detecting life on other planets is emphasized by this photograph of Earth as seen from "close-by" Saturn. From NASA Mission Pages[58] courtesy of NASA/JPL-Caltech.

Making the generous assumption that life is somewhat evenly distributed across the universe, Dick, writing with Richard Hoover, calculated that life must have started independently at least 50,000 times, if not many more[39]. New results suggest that the speed of life could approach the speed of light, as may happen with stars and their solar systems flung out of colliding galaxies by close encounters with the pair of their black holes[59-62], possibly carrying along their inhabited planets[63,64,60]. Even close to the speed of light, there would still have to have been 170 independent starts to life, to fill our universe.

If we assume first life began here on Earth and did not arrive from elsewhere, then the organic molecules required for life to start must be from somewhere. There is more than one hypothesis about that. Perhaps organic molecules were made fresh here on Earth from nothing more than simple gases by fierce activity of lightning[65] and/or volcanoes[66]. Perhaps they were delivered by crashing comets[67] or meteorites[68] on our young planet before it had an atmosphere[69]. Our galaxy is certainly rich in organic molecules we know are synthesized around stars[70,71].

The religious among us who nonetheless embrace the overwhelming evidence for evolution often see the finger of God as stirring that primordial goo: yes, life did start out as goo somewhere in the universe, maybe more than once, but it could only have done so because God did it. With 170 to 50,000 plus start-ups, we could suggest that He's still busy doing it. As none of us have a Q to take us back to the very first primordial goo, these questions may never be satisfactorily resolved. The best we can aim for now is a re-creation "from scratch" of something we would call life[81,82]. And if one must invoke God, then we can at the very least, expect to figure out exactly how He does it.

Whatever the details of the debate on how life began, it is accepted by most scientists and all but the most literal of creationists, that life started as a simple, single celled organism. We are here after all. We therefore begin our explanation of the origin of embryogenesis at the first point in time when a self-replicating primitive form of life first arose, whether it formed here or elsewhere in the universe and then traveled here.

We see probable descendants, and thus perhaps examples of the earliest life forms that survived our planet's pre-oxygen atmosphere, in the bacteria in Yellowstone Park[72] and hot deep-sea vents[73] (Figure 1.8). These high temperatures and toxic environments would kill most other life present on Earth today. A new method of "resurrecting" ancient thermostable proteins shows that microorganisms started on Earth at temperatures near boiling, and this is corroborated by geological evidence of the temperatures of ancient oceans[74]. Conditions were also

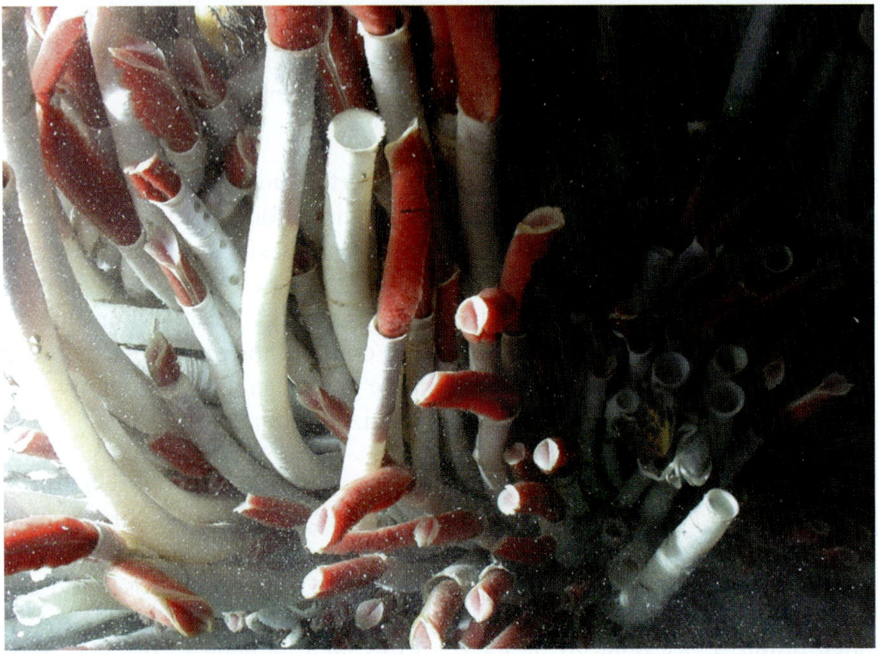

Figure 1.8. "*Riftia pachyptila*, a siboglinid polychaete, … is a giant tube worm (up to 1.5 m long) and very common on the East-Pacific Rise where it lives in clusters in the immediate vicinity of the vents. Inside the tube, the worm's body is colorless, and holds a large sack [sac] called a trophosome. This sack contains symbiotic bacteria that provide nutrition to the worm. The worm has no mouth, eyes, or stomach"[73]. From Ferrini & Lilley (2004)[78] with kind permission of Marvin Lilley and Vicki Ferrini and the Lamont-Doherty Earth Observatory of Columbia University under a Creative Commons Attribution-Noncommercial-Share Alike 3.0 License.

acidic[75]. In other words, much of the whole "cooled" surface of Earth may have been pockmarked with acidic hot springs. Some of these organisms may have been fossilized, a process that occurs rapidly today[76] and so probably did then and we might be lucky enough to be able to find them (Figure 1.9). Given speculations that high pressure was also involved in the origin of life[77], then it may have occurred near deep sea vents, or perhaps in deep, hot, wet rock of the Earth's crust. Those conditions would make finding any fossils to give us a clear answer unlikely. Since we can't answer that question now, we will simply accept we exist and move on to how we got here.

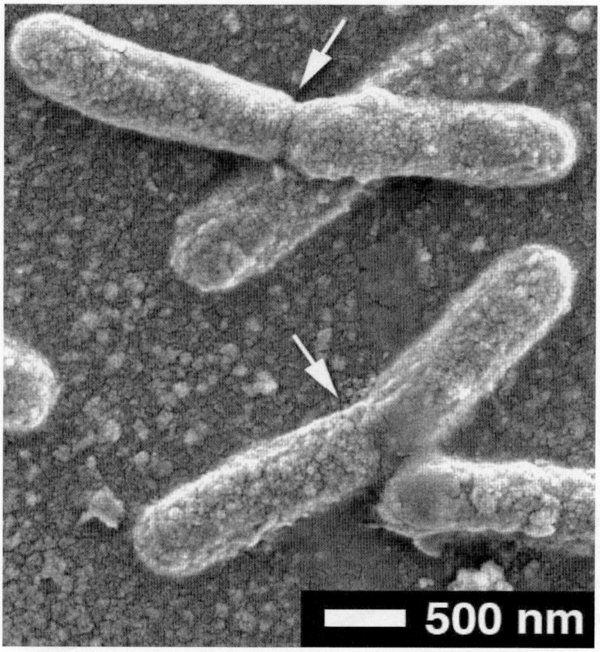

Figure 1.9. These are dividing bacteria from a hot, acidic spring in New Zealand. This genus, *Alicyclobacillus*, is also found in Yellowstone Park[79]. These bacteria silicified in two days, and thus are rapidly made fossils. From Jones & Renaut (2006)[80] with kind permission of Howard E. Harper, Jr. of the Society for Sedimentary Geology. Silicification of mammalian cells has recently been achieved[81,82].

The most important determinant of life is the ability to self-replicate. One cell becomes two, becomes four and so on. Once self-reproduction of cells was achieved (Figure 1.10) the environment was no longer "simply" a goo of organic molecules. The environment rapidly became crowded with other cells filling the space (Figure 1.11). Competition was born.

Competition means not all can survive (Figure 1.12). Differences between the cells assured survival and reproduction for some but death for others. Early life may have eaten their neighbors directly causing their death or simply scavenged their remains for additional food after their death. Competition means selection for winners. The losers were removed from the new life pool and did not get to reproduce. As soon as there was competition and differential survival, evolution began.

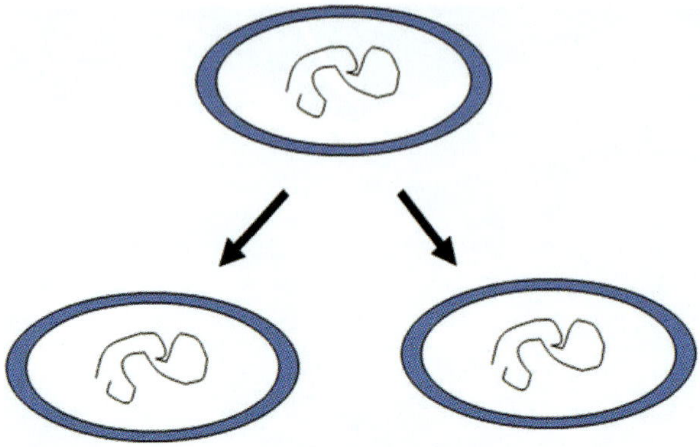

Figure 1.10. The ability of one cell to become two similar cells (self-reproduction) is presumed to be the major first event in the origin of life.

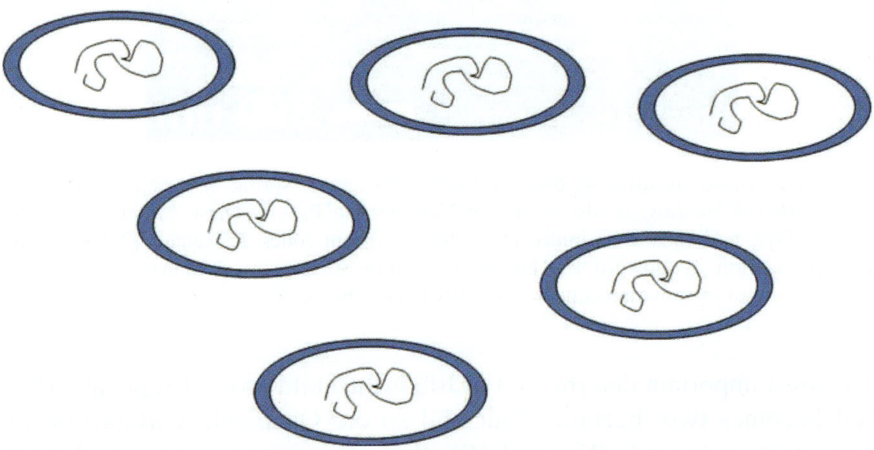

Figure 1.11. Once self-replication was achieved, the world became crowded. Competition begins. This would be the second major event of the origin of life after self-reproduction.

Even if cells did not at first eat each other, or the remains of the already dead losers, the crowding would soon mean there was not enough nonliving primordial "goo food" to go around. We postulate the second ability to evolve after replication would be movement. At first such movement would have been random (as in a by-product of a need to excrete insoluble compounds, described in detail in Chapter 5) and resulted in random benefit to those organisms that simply moved away

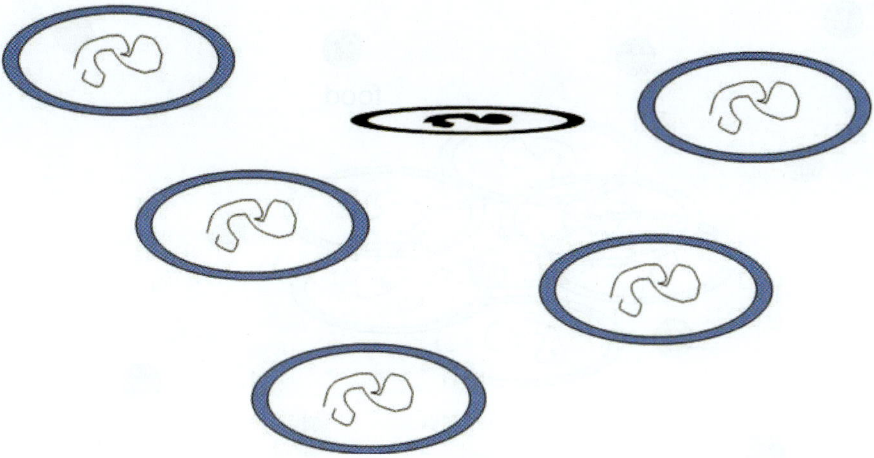

Figure 1.12. Competition means not all can survive. Evolution, the third major event of the origin of life, begins with differential survival arising from competition.

from others to an area where competition was reduced (Figure 1.13). Later, this random movement might have including stopping when the cells reached food and the correct pH and temperature for functioning. Once such a spot was located, and could be recognized, the organism that could deliberately stop movement and stay put in such a favorable place would have the advantage.

First movement was likely accidental and random but once movement became directed in response to an external stimulus, perception was born[83] (Figure 1.14). The most primitive perception was likely: "to stop moving now, as this spot is a good one". Another important directed movement based on perception would be to start moving randomly again when food runs out or if conditions have changed to unfavorable. Directed movement towards that which was desirable for survival would shortly follow the combination of random movement with deliberate stops and starts. So it was replication, movement and then perception that acted as the first critical elements upon which future complexity would be built.

This means any cells that evolved a way to perceive where they were relative to food or food sources, and/or danger, would have a powerful advantage over their neighbors. Those cells that didn't succeed in

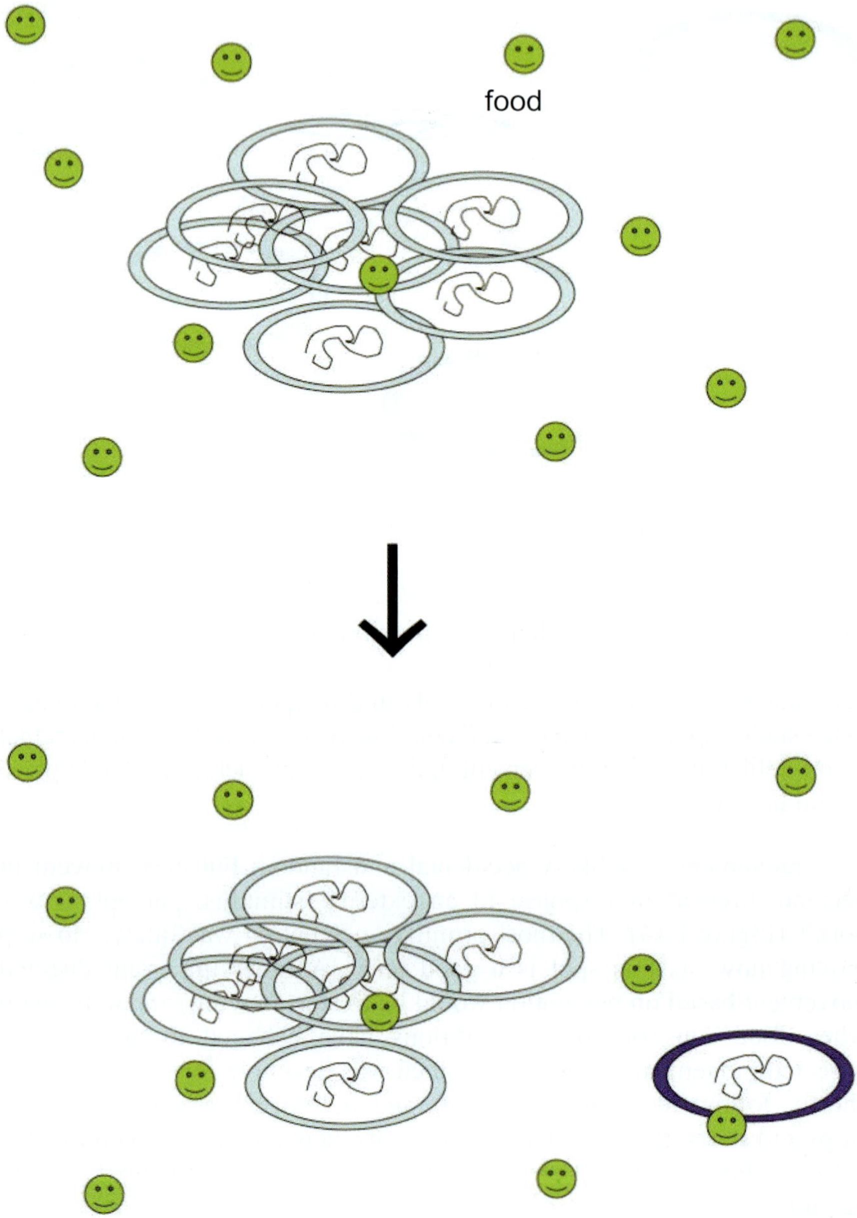

Figure 1.13. The first new novelty to arise from evolutionary duplication and divergence was likely ability to move randomly away from the crowd, where there might be more food. A model for such movement at the origin of life based on motile oil droplets has been proposed[85].

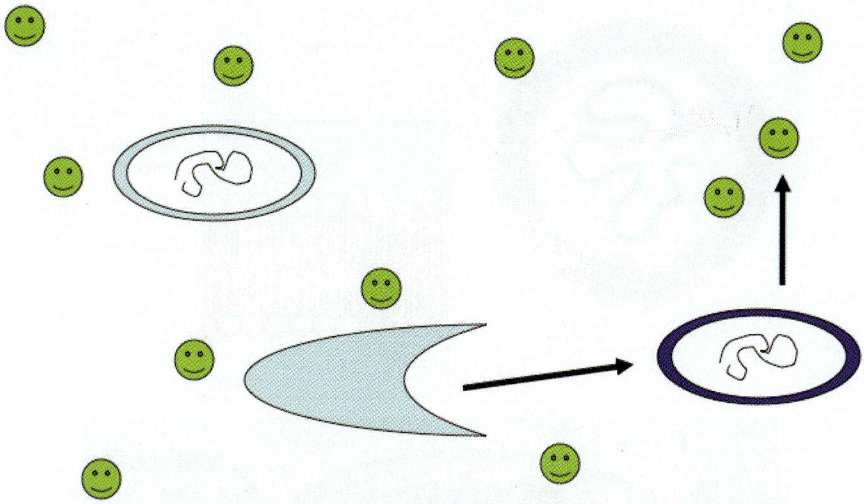

Figure 1.14. An important step in early evolution would have been the ability to perceive food (smiley face) or a predator (pointed shape) and move towards or away from it. The cell that can perceive and move would have a supreme advantage over the cell that cannot. Therefore perception and motility likely evolved rapidly and early after the origin of life.

becoming motile in some way died or were eaten[84]. Directed movement, towards food or away from a predator or deleterious condition, would confer an enormous advantage. Therefore, there would be strong evolutionary selection for directed movement and improved perception. We can therefore speculate that the first cells rapidly developed sensors to perceive what was around them, mechanisms to move as required in response to what was sensed, and a primitive internal logic that determined which way to move[86]. This first act of perception may have been quite simple, no more complex than the "perception" of an automatic door of a moving person or a thermostat of a change in temperature. If we examine some of the species of creatures that exist here on Earth today and yet seem to have the characteristics we guess for first life, most have the ability to sense and to move, no matter how primitive (Figure 1.15).

What would come next? An organism that could sense conditions were not optimum for reproduction and could hold off dividing until conditions improved would then have the advantage. It would not

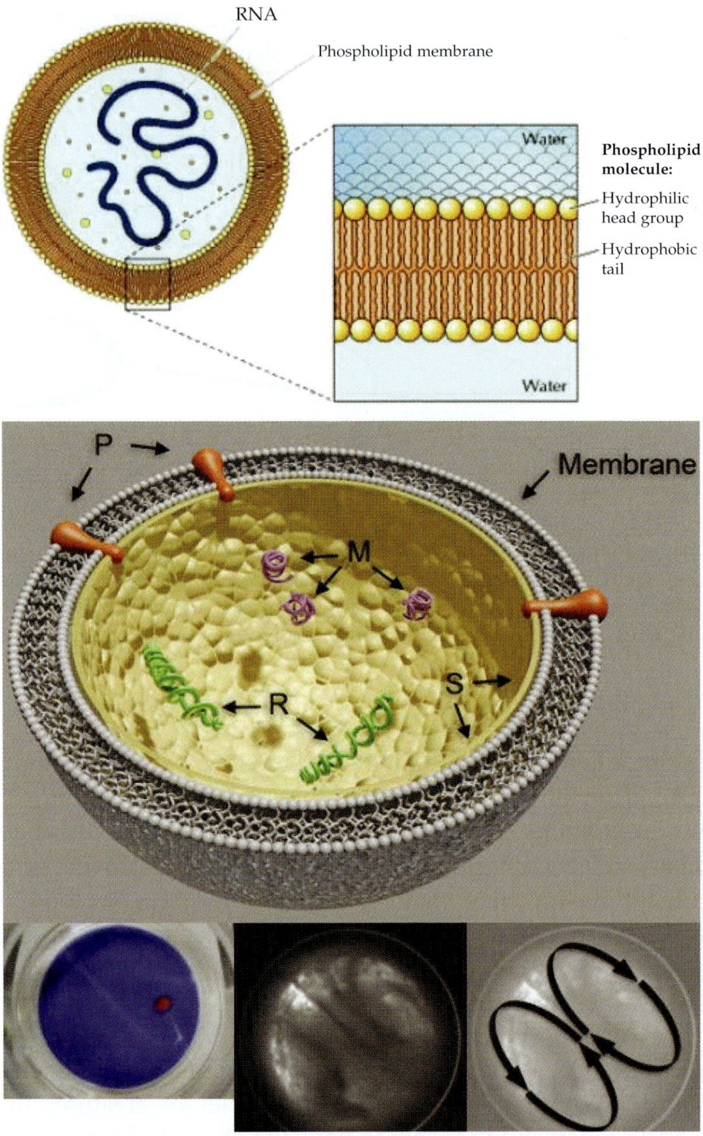

Figure 1.15. **Top**: Rough guess at the structure of the first living organism. From Cooper (2000)[87], in public domain. **Middle**: Present guess at the minimal components for a living protocell: "An S-polymer arises by chance that stabilizes protocell membranes allowing them to survive to return their contents to the anhydrous phase. Protocells then evolve the P-polymer, which gives them access to nutrients through transmembrane pores. Access to nutrients supports the emergence of metabolism catalyzed by M-polymers. Metabolism will generate products that support replication (R-polymers)"[13]

Note that the hypothesized first component, the S-polymer, is in the place of future cortical cytoplasm, suggesting that a cytoskeletal component was the first step after the cell membrane in the origin of life. This model is guiding the search for the simplest molecules that can have the four functions. With kind permission of Bruce F. Damer and David W. Deamer. **Bottom**: An alternative model for the first protocell is based on an oil droplet that absorbs a tar-like complex mixture of abiotic compounds, has no membrane, yet is capable of movement in various ways (**left**)[88,89] via internal oil flow (**middle, right**)[85] (100 μm droplet of 1:1 oleic acid: nitrobenzene placed in a solution of 10 mM oleate at pH 12 on a glass slide). From Martin Hanczyc with his kind permission. Such oil droplets can hydrodynamically split themselves, imitating cell division without specific polymers being involved[90]. See also two additional references cited below[91,92]. Thus there are many competing conceptions of the first protocells and which functions they required and acquired, in what order, on their way to producing cells.

increase the chance of survival if, while moving away from a predator, the cell paused to divide. Nor would there be much point in replicating oneself when there was already a shortage of food. Reproduction needed to be linked to perception of the environment. The early organisms that succeeded in this thrived. Those that did not succeed in controlling the timing of their reproduction likely vanished in short order. The twin machinery linking reproduction and perception/motile response to the environment is therefore the oldest machinery life has.

When did simple replication change to embryogenesis? Embryogenesis is the production of a new offspring, simpler than an adult, with the potential to grow and develop into a copy of the parent. We know that some organisms make the simplest of spores in response to a harsh environmental condition. The ability to form a spore allows survival until the danger has passed (sometimes to 20 to 40 million years[93,94], and possibly even longer while traveling through space). The spore is different from its parent vegetative cell when it is created. Yet it can change back into the parental form once conditions are right again. Therefore, we regard the creation of the first primitive spore as the moment marking the beginning of embryogenesis (Figure 1.16).

What evidence is there that sporulation is the first form of embryogenesis? Actually there is very little evidence. One reason is that the formation of full spores is limited to the phylogenetic class of organisms called bacteria. The archaea phylum is probably older than bacteria[95,96] and yet they do not form spores. Both are presumed to have arisen from early protobacteria life on Earth. This means a very long time

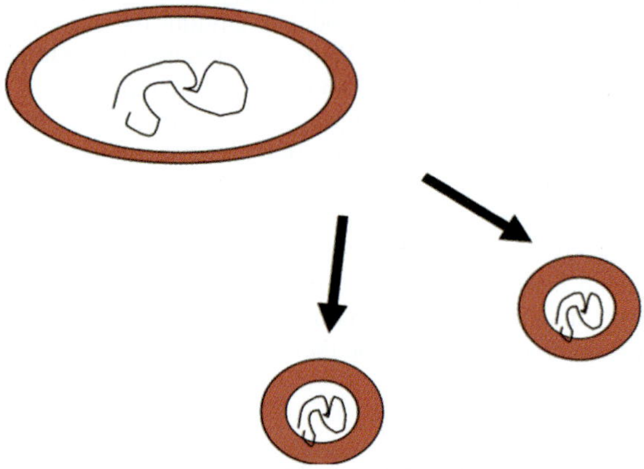

Figure 1.16. The environment is hostile and ever changing, especially on the scale of a single cell. The next novelty in evolution after movement and perception may have been spore production. This would allow the cell to go into a quiescent form to survive adverse conditions. Determining adverse conditions and changing replication to spore formation require both perception and the ability to access a different subset of the genome and therefore spore formation is linked to both self-reproduction and perception.

in evolution passed between when life first evolved and life began to make spores. The ability to form spores as a much later evolutionary step may have been the key to the success of bacteria over sporeless archaea. This may be why archaea is limited largely to certain specific harsh environments in today's world. They are the sporeless remnant class of the original protobacteria.

But if Archaea were sporeless, then how could life have traveled through space to here on Earth if we assume life began elsewhere? Alternatively, there may have been a dormant but not full spore form of sporulation present in both archaea and in bacteria arising from a protobacterial ancestor. Only the bacterial phylum perfected the spore as we know it today, over evolutionary time. So-called "persisters", cells that become metabolically dormant under adverse conditions[97], may represent the original form that spores of today evolved from and the form that traveled through space. This form of spore is not morphologically different and therefore does not represent a true spore. There is evidence that bacteria with true spores may have all evolved from a single ancient *Firmicutes*[98]. Some archaea contain genes that are involved in sporulation in other prokaryotes[99]. Thus we have a few candidates for this first event of true spore formation that appears to have happened here on Earth.

However, a more important reason it is hard to find the evidence that spores are the first form of embryogenesis is because of lateral gene transfer (also called horizontal gene transfer)[100,101]. Prokaryotes are notoriously promiscuous with their genes, transferring over entire stretches of DNA complete with multiple gene sets shared among unrelated bacteria and archaea[102]. Lateral gene transfer makes the study of the evolution of sporulation (and pretty much anything else using DNA sequences) fraught with difficulties in interpretation. We hypothesize that the first form of embryogenesis is some form of sporulation but it will take some time to sort out when and in what kind of organism this occurred. This is especially so given that the relationships between the presumed original three kingdoms (Archaea, Bacteria, Eukaryotes) are still unsettled[103].

Once embryogenesis began, what came next? Let's return to the primordial mud puddle of goo where life may have begun. If you have ever visited a hot spring you have likely been struck by the lovely patterns of brilliant colors that exist at the edges of the spring as heated water runs off. Narrow bands form with each band representing a specific zone of temperature and pH. Each band also has its unique bacterial denizens with different pigment coloration that allows us to see the different zones (Figure 1.17). The result is spectacular bands of colors, some only centimeters wide. These can be seen today in places like Yellowstone Park[72].

Walking around Yellowstone Park one is also struck by something else (Figure 1.18). That little world of a single colored band is so very fragile! Hot springs regularly change their course. They also stop and die, or a new one begins spewing in a new location. This must represent a major catastrophe for the tiny colored critters in those narrow environmental zones. How many times did life begin in a small pool somewhere only to be wiped out by some random act of a volcanic welling or unexpected crumpling of the edge of a hot spring from erosion? Yet in hot springs all over the world these delightfully colored rainbow bugs survive and thrive. The first form of embryogenesis, as we define it, is required to explain this.

Figure 1.17. At least 5 bands of colored, thermophilic bacteria in a hot spring in Yellowstone Park can be seen. The green band is the coolest, probably cyanobacteria and/or eukaryotic algae. Scale: about 1 meter across. From Blum (2010)[104] with kind permission of Bruce Blum. Cf.[105]. At least some Yellowstone bacteria form spores[106]. Smithsonian Travel magazine also has wonderful images[107]. Hot springs change course or vanish altogether with astonishing frequency. The environment of these single cells organisms is fragile and ever changing. Spore formation is an adaptive response to this.

Imagine then that you are a little bacterium living in such a narrow band. You are motile, traveling forward if conditions are good and backing away where you perceive the edges of your narrow zone of welcoming conditions is ending. Suppose you started dropping a spore every time you were at the edge of the zone and not just when you were facing imminent collapse of your environment? Each time you move to the edge

Figure 1.18. Natalie looking into a hot spring in Yellowstone National Park. This is one of the best places in the world to contemplate how volatile and changing our planet is. Here we saw examples of primordial soups teeming with bacterial life· We could feel, as we never did before, the volcanic depths of our planet right under our feet. Hot springs frequently shift their location and when that happens, trees die, as did this twisted "skeleton" in the foreground. Yellowstone represents a rich microcosm of life and death in our universe.

of the hospitable zone and perceive that one end is in less desirable conditions, you go through an unequal cell division (Figure 1.19). Instead of producing two identical daughter cells, you produce one spore that is dropped and left in place and one "vegetative" cell that then moves backward, away from the new spore and the hostile zone, back into the conditions you love (Figure 1.20). Your species would have a decided advantage if the hot spring suddenly caves in and reforms somewhere else. The edges that you found so inhospitable that you dropped a spore, might suddenly be in the center of the zone your species thrives in. The spore is sitting there waiting for just such an event.

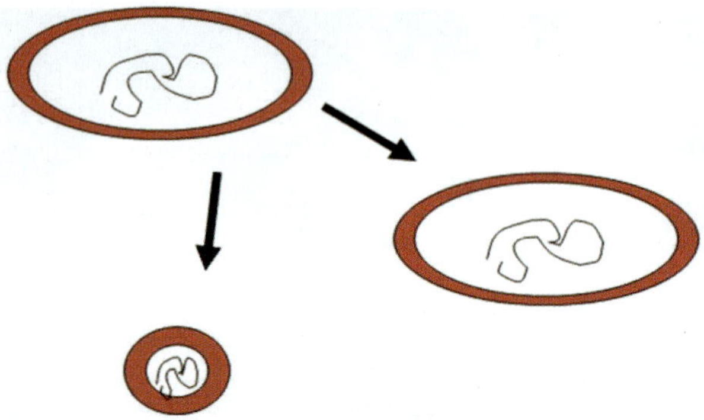

Figure 1.19. The next novelty in evolution might have been unequal cell division, a cell replicating into one self-reproducing cell like itself (a "vegetative" cell) and one spore. This would have been the start of asymmetric cell division. No cell division is precisely symmetric, and small advantages of one daughter cell over another[108,109] may be the evolutionary origin of asymmetric cell division. Simple physical mechanisms may have prevailed in early cell evolution[110,111], including asymmetric division[112].

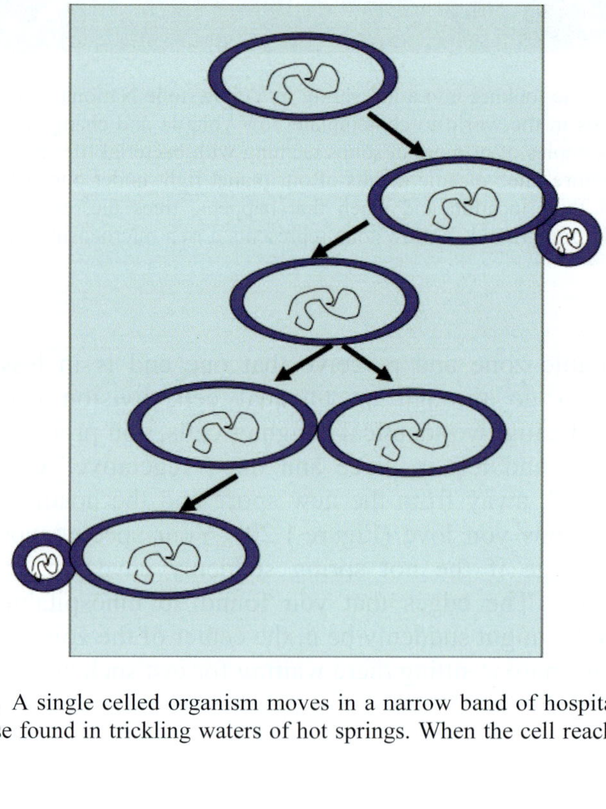

Figure 1.20. A single celled organism moves in a narrow band of hospitable conditions such as those found in trickling waters of hot springs. When the cell reaches the edge of

the hospitable zone, it undergoes an unequal cell division dropping a spore at the edge. The vegetative daughter cell moves back away into the hospitable zone. When in an area where conditions are right, the cell undergoes instead an equal cell division into two identical vegetative daughter cells. If the cell reaches the other side and again encounters a hostile zone then it again drops a spore. The essence of embryogenesis is the getting of a right new cell of the right kind in the right place at the right time. This is the first instance of embryogenesis and means the right cell (spore or vegetative), in the right place (central or edge of the hospitable zone), of the right kind (vegetative division or spore) at the right time (when cell movement has the spore dropping in the inhospitable edge of the zone) with vegetative division occurring when the cell is entirely in the vegetative zone.

This is why Earth humans must have tetanus shots and use pressure to can low acid food. The Earth is absolutely full of spores, many airborne. Botulism is just one example. If we create an anaerobic low acid environment by accidentally putting a *Clostridium botulinum* spore in a sealed low acid can of beans, the spore springs to life and grows. It replicates, eventually filling the can with toxins that stop all muscle function if we eat the toxins with the beans. Likewise if we accidentally insert one of the *Clostridium tetani* spores that ordinary soil is chock full of into our tissues via a deep cut, we can get tetanus unless our immune system as been primed to prevent it by previous immunization.

The spore system works. Such a system evolved early to allow the survival of bacterial life even when the environment changed abruptly. Many hot springs bacteria produce spores[113] and some bacteria become motile and move as they do so[114]. Not surprisingly then, the genes involved in motility appear to be derived from the same set of original genes involved in spore formation[115].

This system would require some form of sensing the environment and the ability to move forward or away from what was sensed. Thermotaxis or chemotaxis would require continuous communication of the environmental state on the outside to the internal machinery of the cell. It would require being able to sense that one side was friendly and one side was less friendly. This first form of early embryogenesis would also, by necessity, result in a direct linking of the machinery of both perception and the motility response to cell replication and spore formation. We should therefore be able to evolutionarily trace all the genetics of embryogenesis to one of these three basic fundamental processes: replication, movement and perception (Figure 1.21). Since the processes

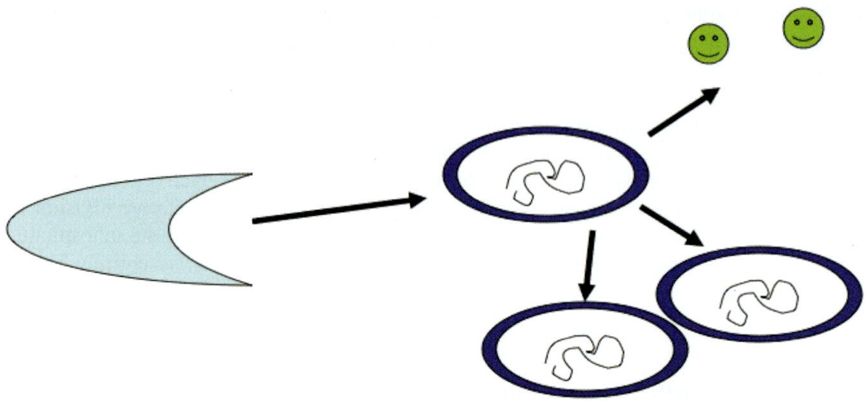

Figure 1.21. Three fundamental first life processes would be replication, movement and perception. We should be able to evolutionarily trace all cell functions to one of these three basic fundamental processes if our explanation of the origin of embryogenesis is correct.

are so fundamental to first life, everything else cells do should also be genetically traceable back to these functions in some fashion.

Perhaps initially, there would have been rather very few species. Perhaps because of inaccuracies in the replication machinery, there would be so many variations in size, composition, and other aspects of life, that each individual was unique. This collection of unique individuals was only reduced to sets of species later[116]. Horizontal gene transfer, in which RNA or DNA is transferred between individuals of nominally different "species", occurs in three distinct ways[117,118]: uptake of loose molecules from the environment, transfer via bacterial viruses called phage, and by conjugation, i.e., sex (Figure 1.22). All three are promiscuous, not only in regards to individuals involved, but often also in regards to species. Now RNA or DNA are themselves a form of food, but once ingested in a cell, instead of being broken down into nucleotides, there is the alternate possibility that they replicate. They may even become incorporated into the main genome of the cell. Thus a side effect of an eat or be eaten world is that you may become and therefore be what you eat. And thus food led to sex. Despite many other mechanisms for sex that have arisen, horizontal gene transfer continues to be an important mechanism in extant microorganisms, especially in symbiotic consortia of bacteria. Eukaryotic diatoms, for instance, are chock full of incorporated bacterial genes[119], as is also an extremophilic red alga[120]. In eukaryotes, viruses[121] and perhaps occasional successful matings between sperm and egg of different species[122] contribute to the horizontal transfer of genes and genomes. By the sharing of genes, novelties such as new proteins with new functions became widely distributed and this may account for the common, rich biochemistry shared by most extant organisms.

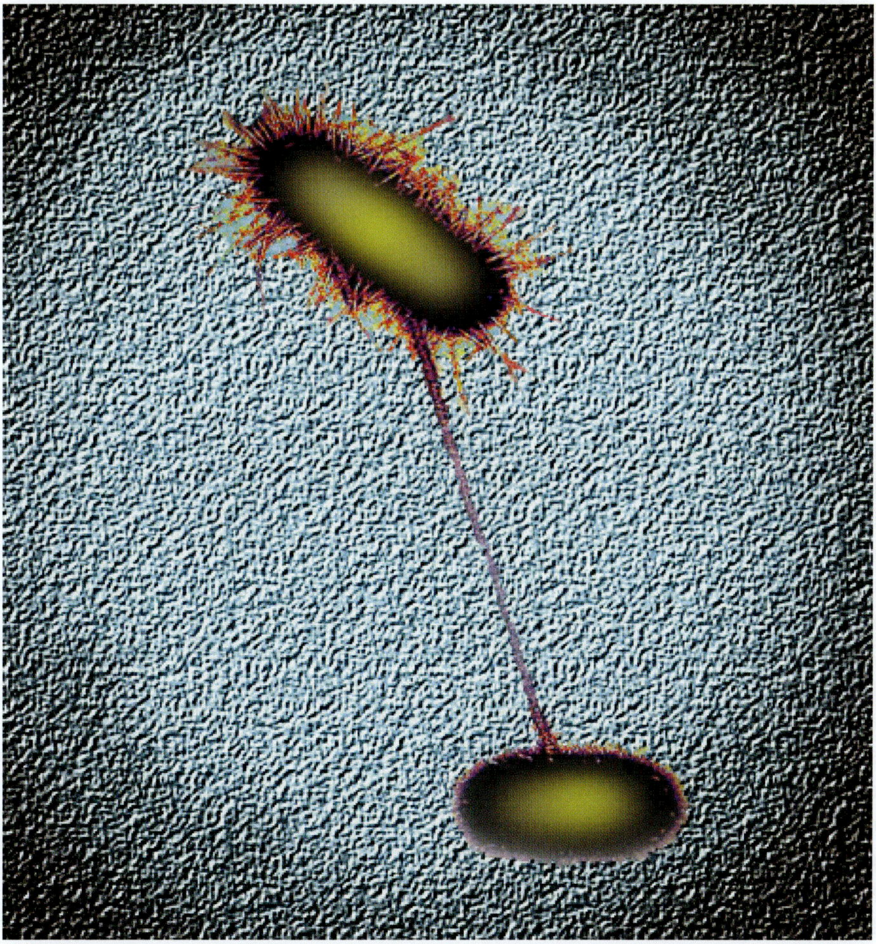

Figure 1.22. Transfer of DNA between bacteria. This is the prokaryote version of sex, called "bacterial conjugation". From Cann's "Bacterial Conjugation"[123] with permission under a Creative Commons Attribution Share-Alike license.

We can also expect to find evidence of a single universal starting motif for each of these critical first steps connecting replication, motility, perception, directed movement, controlled reproduction, and embryogenesis. This is because of the nature of evolution itself. Nature rarely begins things anew. Each novelty that evolved tended to reuse time-tested mechanisms already on hand, a process that has been called tinkering[124], as if evolution were accomplished by a not so intelligent being without a grand plan. Rather, evolution stumbles about hitting all kinds of obstacles and then moves back and forward again in yet another attempt. But once something new works, it tends to be retained and repeated, and shared, over and over again. Repetition can result in multiple variations on a theme, whether at the species level, or within individuals. We know for example that certain small DNA sequences code for portions of protein structures referred to as motifs. Motifs are repeated over and over and over again with very little change in the sequence of the motif even though it is used in many different proteins, because the motif works[125]. So we can expect to find that there will be motifs for embryogenesis that can be traced right back to the first successful replication event.

The paths of evolution are hardly straight ones. Nature has regularly gone off on other tracks with great flourishing of what appear to be bizarre offshoots of the complex branching of the tree of life in Earth's long past[126] (Figure 1.23). These branches were occasionally severely trimmed (Figure 1.24) and then a new branch would take off. However, we can usually look about us and see some survivor of the earlier form, possibly changed somewhat, but nonetheless present[127]. Scientists assume there are no phyla known to be totally extinct if they exist in the fossil record[128]. Admittedly, though, there may have been some soft bodied or tiny phyla that are no longer with us that escaped fossilization[129], but there is no evidence of them. The horsetail is an example of the persistence of some survivors of all phyla. The modest sized horsetail we see along a creek edge or in a wet meadow is a leftover survivor of a phylum that used to dominate the Earth with tall forests, of which the largest extant species still gives a hint[130] (Figure 1.25). Earth once had big insects with a 70 cm wingspan commanding the skies[131], possibly terrifying our proto-mammal therapsid ancestors[132]. Today the comparatively little dragonfly is a charming summer note (unless you are a mosquito). And so it is useful to look around at organisms present today or preserved in fossil form to understand life.

Figure 1.23. *Hallucigenia*, an animal from the Burgess Shale formation in Canada, with kind permission of Karen Carr Studio[133]. It is probably an ancestor of velvet worms[134].

Figure 1.24. An asteroid wipes out many of the dinosaurs[135], with kind permission of Mark P. Witton.

Figure 1.25. A colony of the horsetail *Equisetum giganteum* in the valley of the Río Lluta, northern Chile[130], with kind permission of Chad E. Husby. It is generally accepted that all phyla here on Earth have a representative species here even though they are no longer the dominant species. Earth was once covered with forests of towering horsetails. These are the largest ones left.

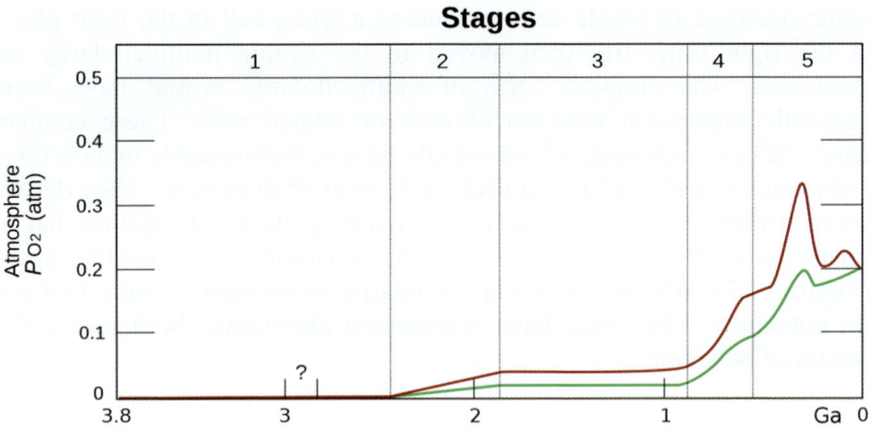

Figure 1.26. Upper and lower estimates of the atmospheric oxygen concentration over the past 3.8 billion years (Ga = Giga-annum = 1 billion years). The present atmospheric level of O_2 is 20% or a partial pressure of 0.2 atmospheres on the left scale. The Great Oxidation Event is placed at 2.1 to 2.5 Ga, preceded by a few "whiffs" of oxygen between 2.5 to 3.1 Ga[144] which may have been due to earlier photosynthetic prokaryotes that went extinct due to the tail end of the period of relatively frequent collisions of asteroids with the Earth[145]. The fluctuations in O_2 over the past 0.5 Ga may correlate with fluctuations in CO_2 (carbon dioxide) over the same time[146]. From Wikipedia[147] with permission under the terms of a GNU Free Documentation License.

In the distant Archaean past of Earth, a new creature evolved that used light for energy and expelled the potentially toxic waste product, oxygen[136], perhaps as long as 2.8 to 3 billion years ago[137,138]. In doing so it eventually changed our planet to a 20% oxygen atmosphere, starting the Great Oxidation Event of 2.4 billion years ago[139] (Figure 1.26). Some scientists think that life on Earth could not move onto land until after the Great Oxidation event because it took oxygen to create the ozone that protects us from UV damage[140,141] although others disagree[142,143]. Regardless of competing hypotheses, a little photosynthetic prokaryote that harvested light and produced oxygen profoundly changed Earth.

Anabaena is a possible representative example of the prokaryote Great Oxidation Event. *Anabaena* was called a blue-green alga by early botanists, or nowadays a cyanobacterium. Species of this genus consist of a long string of sister "vegetative" cells. In other words, each string is a clone. These vegetative cells use a simple form of photosynthesis converting carbon dioxide and water into sugars and oxygen. From the embryogenesis of single cells producing a spore cell in the right place at the right time, life then moved to the simple multicellularity of *Anabaena*. The simplest form of multicellularity would have been colonial cooperation between clusters of related cells. These cellular cooperatives, each made of individuals fully self-sustainable in their own right, stick together after cell division instead of dispersing. They derive some advantage or some benefit by remaining clustered together that is greater than the earlier advantage of separating and moving apart (Figure 1.27). Whether such multicellularity or dispersion came first we do not know. They may have represented alternative, both successful modes of behavior.

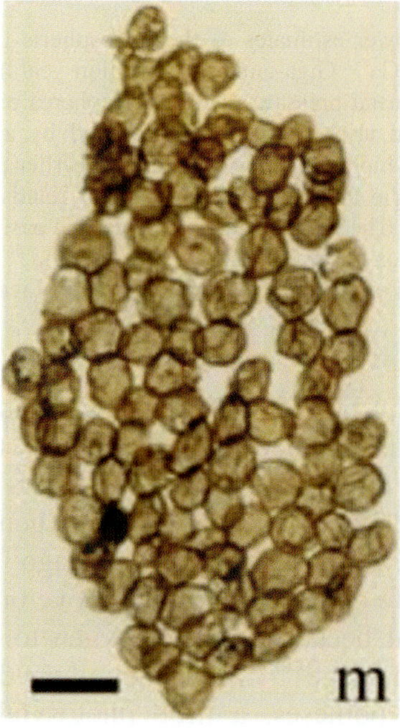

Figure 1.27. A fossilized colonial cyanobacterium *Ostiana microcystis* consisting of irregularly attached spherical cells, which nevertheless form a monolayer. It is possible that the first cells at the origin of life were "attached" neighbors in clones like this, rather than independent, separate cells. Thus contrary to general presumptions[148], multicellularity may have preceded solitary cell life. This contradicts[149] the presumption that life ratcheted from single cells to multicellular organisms[150]. Bar = 50 μm. From *Precambrian Research*[151] with kind permission of Andrew K. Knoll and permission under the Harvard University Open Access Policy http://nrs.harvard.edu/urn-3:HUL.InstRepos:3934555.

Under stress conditions *Anabaena* forms spores called akinetes, which generally remain attached to the string of cells until the string dies. When fall arrives in a northern pond, it is spotted with *Anabaena* akinetes dropped onto the mud bottom to wait for spring[152] (Figure 1.28). Fossil akinetes have been found as old as 2.1 billion years[153], and the origin of cyanobacterial akinetes likely coincided with the Great Oxidation Event[154,155].

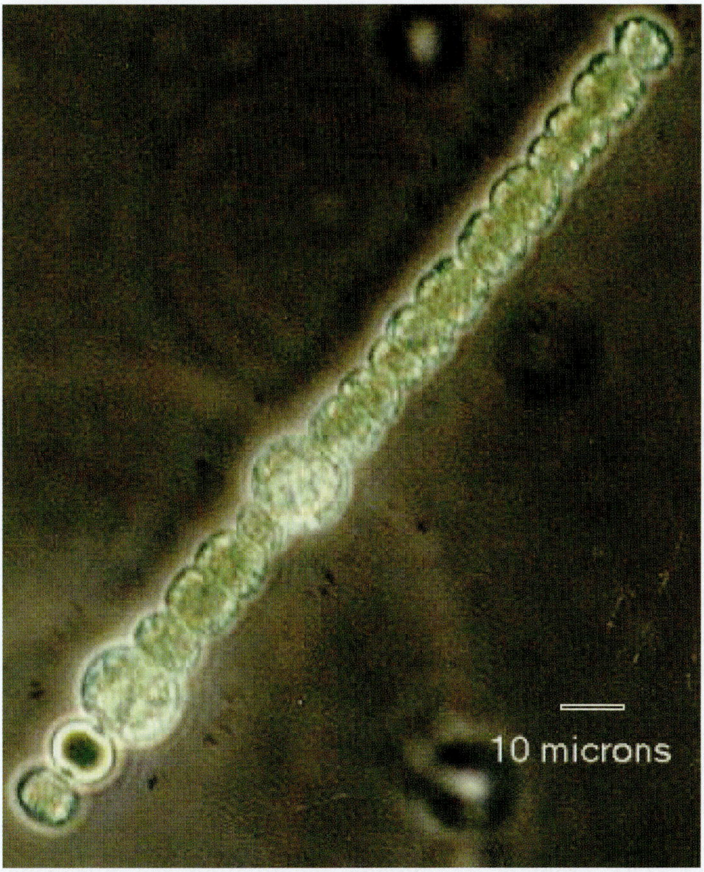

Figure 1.28. A clonal colony of the cyanobacterium *Anabaena scheremetievi* with two spores called akinetes and another kind of differentiated cell, second from the left end, called a heterocyst. The rest of the cells are called "vegetative" cells. Akinetes and heterocysts differentiate from vegetative cells. From Purdue Images[156] with kind permission of Roger Burks (University of California at Riverside), Mark Schneegurt (Wichita State University), and Cyanosite[157].

Anabaena's akinetes are not by themselves a surprise or novelty. Sporulation was already well known by the time they appeared. What is so fascinating about *Anabaena* is that it also has a third type of cell on the string. After the early steps of formation of spores, the next step in embryogenesis should have been an active but different type of cells that arises from a sporulation-like change but which continues to exist in an active but different form, side by side with the parent cell. The third cell type of *Anabaena*, the heterocyst, does exactly that (Figure 1.28).

Cyanobacteria are presently able to live anywhere there is sunshine and at least some water[158]. This is because they can fix nitrogen from the air and turn it into a form that is required for functions like DNA replication. It is presumed that earlier cyanobacteria were able to fix nitrogen, but as O_2 levels rose, the nitrogen fixing enzymes stopped working. This presented an evolutionary conundrum for these cyanobacteria. On the one hand sunlight and water could easily be converted into sugars by producing oxygen. On the other hand, this ready source of sugar meant a byproduct, oxygen, that prevented the bacteria from producing their own fixed nitrogen. They could not do both at once. Some bacteria solved the problem by simply avoiding photosynthesis altogether[159]. Other present-day single-cell cyanobacteria solve this conundrum by alternating between photosynthesis during the day and only doing nitrogen fixation at night when photosynthesis is not possible[160].

As always, evolution meant some organisms adapted, sometimes in different ways. *Anabaena* is one such adaptor. At regular intervals along the multicellular *Anabaena* strands, one of the vegetative cells will form a new cell that at first glance looks like a spore. It isn't. Instead, it is a cell with a thick spore-like outer layer impervious to oxygen, called a heterocyst. Heterocysts, considered a specialized form of spore cell[161], although they rarely germinate[162], arose as a novelty shortly after the Great Oxidation Event[155]. With the emergence of heterocysts, the nitrogen fixing enzymes were protected from external oxygen, by being housed within a thick walled spore-like cell from which oxygen could be excluded. The problem of oxygen inactivating the nitrogenase was solved by having two different cell types in a single strand. Doing this required a new adaptation. Cells had to cooperate. Cooperation itself was not new, as we can assume that consortia[163] of different, metabolically cooperating prokaryotes preceded the formation of heterocysts. Horizontal gene transfer is another form of cooperation between cells.

What was new was that incompatible cell functions became housed in two separate kinds of cells in the same strand and that the resources they produced were shared via the pore connecting a vegetative cell to a heterocyst[164].

Because the vegetative cells in cyanobacteria continue to divide and usually produce a pair of vegetative cells when they do so, the distance between the nondividing heterocysts increases. Vegetative cells that are most distant from the heterocysts will not just divide and reproduce vegetative cells. Instead, they form proheterocysts, a near spore state, and then just one of the proheterocysts continues on to form a heterocyst, while the other goes back to being a vegetative cell. The patterning mechanism is still under investigation[165,166], despite decades since the first modeling[167,168]. (The mechanism of transport of nitrogen through the pores from the heterocysts to and through the adjoining vegetative cells is also uncertain[169].) It is presumed that heterocyst formation is related to declining nitrogen levels in the vegetative cells farthest from heterocysts. When sister vegetative cells prepare to divide, they somehow assess the amount of nitrogen present and if the nitrogen levels are too low they respond by producing one vegetative cell and one heterocyst instead of the usual two vegetative cells. This kind of simple cooperation between two sister-cells is colonial. Colonial cooperation formed the entire basis for multicellularity of complex organisms as evolution continued. Colonial organisms brought the simple spore formation of embryology much closer to the embryology of complex organisms. The first colonial organism represented a huge advance in the task of producing the right cell, in the right place, at the right time.

Thus, at somewhat regular intervals along the strand[170], one of the *Anabaena* cells will form a thick spore-like heterocyst. The heterocyst cannot use sunshine to make food like the rest of the cells in the chain, because that would produce oxygen. Instead, it fixes nitrogen, which cannot be fixed in the presence of oxygen. It converts atmospheric nitrogen, N_2, into a usable fixed form of the chemical. The nitrogen-fixing heterocyst shares its bounty with nearby vegetative cells and they in turn share their food acquired by photosynthesis converting carbon dioxide and water into sugars and oxygen.

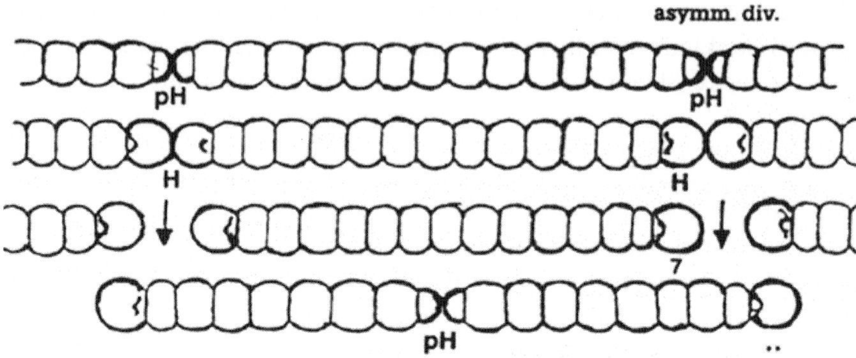

Figure 1.29. In the colonial cyanobacterium *Anabaenopsis* two proheterocysts (pH) are formed at even intervals along the chain of vegetative cells. When these mature to heterocysts (H), the heterocysts come apart. As a result, the colony size does not vary by more than roughly a factor of 2 in number of cells. This may be the first example of size regulation in a multicellular organism. From Komárek & Anagnostidis (1989)[171] with kind permission of Waltraud Bosch of Schweizerbart und Borntraeger (http://www.schweizerbart.de).

Among species of filamentous cyanobacteria are some that form pairs of heterocysts instead of single ones. The strand length of a colony is limited because the paired heterocysts disconnect at the juncture between each pair (Figure 1.29). The result is small short filaments of consistent range of lengths with one heterocyst at the end of each filament. This may represent the previous stage of evolution of the algal strand studded with evenly spaced heterocysts. This may also represent a mechanism for controlling the size of a colony to allow it to better adapt to a turbulent environment. A turbulent environment tears apart long strands and this might have been a mechanism to create small short strands that nonetheless each have a heterocyst. The control of a cellular collective's size is a critically important ability required in later embryogenesis. We shall see that embryos of more complex organisms require a form of cellular collective size control even though they do not use this particular mechanism.

Our Earth spent a long period dominated by these blue green algae/ cyanobacterial life forms. Pattern formation and differentiation were born in their era. One pattern form is regular spacing of the heterocysts on a chain. Another pattern formed is the branching pattern of the vegetative cells present in other cyanobacteria, going beyond one-dimensional chains (Figure 1.30). It is curious to note that even in this simple system, differentiation and its associated pattern formation are part and parcel of the same process (Figure 1.31). Much of biological research separates these two concepts and presumes they are somehow independent processes. On the contrary, it seems that they are almost always tightly coupled. The pattern of equally spaced heterocysts within a chain of vegetative cells is differentiation intrinsically linked to evolution of more complex organisms such as ourselves.

There is one more critically important step required for this prokaryotic-based simple style of prokaryote differentiation to become the kind of full blown organismal process that can generate complex differentiation. This step is the evolution of the eukaryotic cell. Eukaryotic cells arose about 2 billion years ago[175]. The process is not entirely understood but essentially some smaller organisms were folded into larger ones or engulfed by them. The recently discovered Lokiarchaeota in an Arctic

Figure 1.30. A filamentous cyanobacterium, *Hapalosiphon* that exhibits what is called "true" branching, i.e., where the cells at a branch point seem to be attached and not simply bound together by an external sheath. From Tsukii (2005)[172] with kind permission of Yuuji Tsukii.

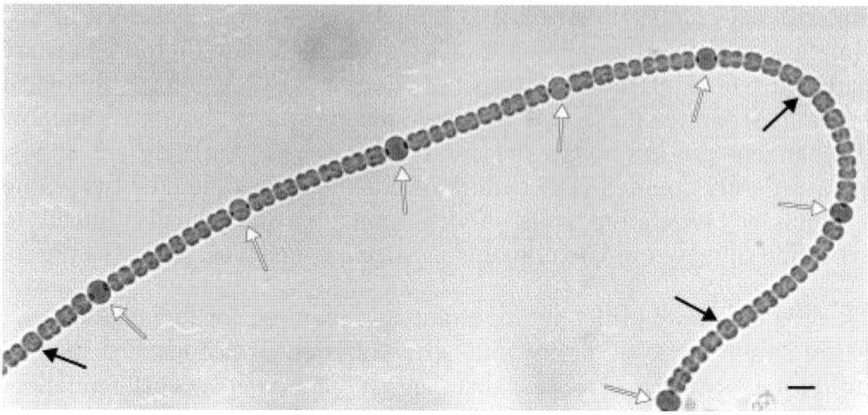

Figure 1.31. One can actually watch the formation of new heterocysts, called proheterocysts (black arrows), at positions about halfway between existing heterocysts (white arrows), in cyanobacteria. From Adams (2000)[173] with kind permission from Elsevier and from[174] with kind permission of Cambridge University Press.

Ocean vent community have been touted as a missing link between archaea and eukaryotes[176]. They have what appear to be original membrane proteins previously thought to be present only in eukaryotes. Only the eukaryotic cell, the type of cells that make up our own human bodies, has the kind of membrane required for the complexity and organization that has sustained more advanced forms of embryogenesis.

These smaller organisms did not simply die when enfolded in that membrane but instead adapted and grew and reproduced inside the larger one. Over time the little life forms were more and more adapted to life on the inside instead of outside. They exchanged independence for security and a stable environment and earned their keep by providing critical functions for the larger cell. It was as if the heterocyst moved from being at regular places along a chain of cells to the inside of the sister vegetative cells. Such a step would certainly make the distribution of sugars and nitrogen more efficient and this kind of advantage may be what drove the evolution of eukaryotes.

The best example is the mitochondrion. At some point in the past a small oxygen-using "purple" bacterium evolved. This little bacterium could produce enormous amounts of energy in the form of "ATP" (adenosine triphosphate) with a then recently evolved proton pump using carbohydrates and oxygen. This smaller bacterium was integrated into a larger non-oxygen using cell[177,178]. The smaller cell was not consumed but instead provided ATP based energy for the larger one. Most eukaryotic cells that live in oxygenated environments today have some form of mitochondria and without them cannot produce energy. The mitochondria have a double membrane with an inner prokaryotic-like membrane and an outer eukaryotic membrane supporting the notion that this bacterium was originally acquired by being engulfed in an infolding.

Mitochondria retain their own prokaryotic style single loop of DNA. Many of the genes on their DNA have moved to the central eukaryotic nucleus[179], which is so much more efficient at producing required proteins. The more genes that could be taken up by the larger nucleus, the more the smaller mitochondria could concentrate resources on energy production. The genes that remain are almost all genes for complex membrane embedded proteins that are inserted into the membrane as they are translated. Such proteins would be very difficult to move into the mitochondria after being synthesized elsewhere and their genes remained in the mitochondria. The extreme of the process is seen in some eukaryotes which contain hydrogenosomes instead of mitochondria, for which the genes all migrated to the nucleus (*Proposition 123* in *Hierarchical Genome*[180] and Zimmer (2009)[175]).

A later major though very similar event was the addition of the chloroplast to eukaryotic cells[181], perhaps from filamentous cyanobacteria containing heterocysts[182]. A small cyanobacterium with the ability to photosynthesize was incorporated alongside the mitochondria in some ancestral eukaryotic organism that eventually gave rise to plants on Earth. The same partial transfer of the genes to the nucleus also occurred. This event came at a later time and so only plants and eukaryotic algae (with few exceptions[292]) have chloroplasts.

Arguably, the most important single requirement for the evolution of eukaryotes was the creation of inner membrane bound regions that are divided from the rest of the cell. Primarily, these enclosed areas are the Golgi and endoplasmic reticulum, which appear to have formed entirely from infolding. These structures are essential for the creation and trafficking of new proteins to the appropriate location. The most specialized area of the cell is the area where the genetic material is sequestered, the nucleus. The nucleus has a complex membrane underlaid and overlaid with structural elements of the cell's cytoskeleton and studded with openings called nuclear pores. The nucleus contains archaea and bacterial like structures and may itself be a symbiont[183]. We will look at these details of the eukaryotic cell, the true master of embryogenesis, in far greater depth in later chapters.

And so, by the time simple single cell eukaryotes appeared on the Earth, all life forms had already achieved the following critical abilities. Life could reproduce. It could perceive its environment. It could move toward or away from whatever it perceived or it could stop moving. It could produce a different cell type, a spore, in response to its perceptions. It could produce colonial living cells that each had different functions and those different cells could cooperate and share bounty to their mutual benefit. The eukaryotic cell was formed when an amoeboid type ancestral cell engulfed a prokaryotic cell. That prokaryotic cell then became a symbiont instead of a meal. It took more than half the time that life has been on our planet to achieve these milestones. Once achieved though, the complexity of life could now explode and cover the Earth. And so it did. And yet complexity still remains nothing more than the ability to produce the right cell, in the right place, at the right time.

However, we should not be too hasty in dismissing the morphogenetic capabilities of prokaryotes as evolutionarily dead ends. Before we move along to look more closely at eukaryotic embryogenesis we should pause and look at what prokaryotes have accomplished. Molecular structures that are known to be highly relevant to embryogenesis, such as the components of the cytoskeleton, are explained in greater detail in later chapters. For now let us lay the ground by presenting a summary of the above and other remarkable morphological and behavioral abilities of various prokaryotes that presage properties of eukaryotes[184]:

1. Some prokaryotes have gone beyond spores (Figure 1.32) to differentiate into at least four kinds of cells using the same DNA, such as vegetative cells, akinetes, heterocysts (Figure 1.28) and hormogonia (motile vegetative cells in motile filaments)[185] in cyanobacteria. Asymmetric cell divisions producing two kinds of cells may be occurring in these prokaryotic cell differentiations, such as occur in eukaryotes. The parallel is enhanced by the discovery that bacteria have a mechanism similar to the spindle apparatus in eukaryotes that is involved in separation of duplicated

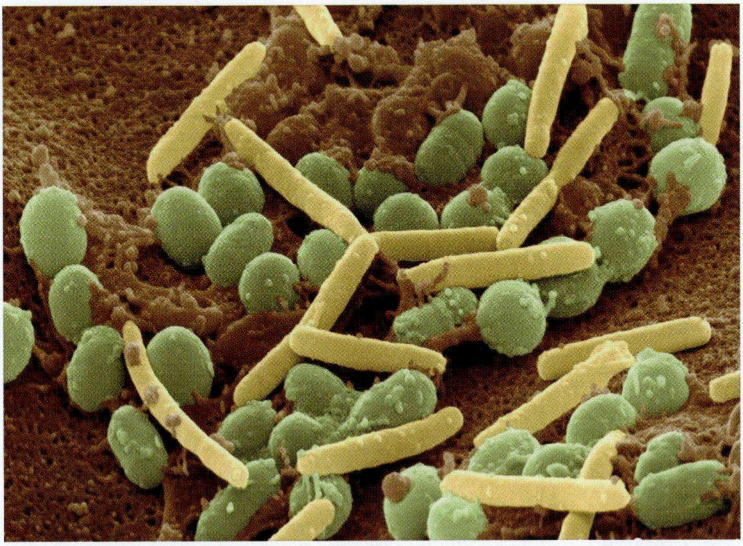

Figure 1.32. SEM (scanning electron microscopy) of "rods" and round spores of the bacterium *Myxococcus xanthus*. From Velicer *et al.* (2010)[189] with kind permission of Gregory J. Velicer and courtesy of Indiana University.

DNA[186]. Vertebrate embryos go much further in producing perhaps as many as 7000 cell types (*Permissions and Note Added in Proof* in *Hierarchical Genome*[180] and Bard *et al.*, 1998[187]).

2. Prokaryotes can mate, via the one-way transmission of DNA[188] (Figure 1.22).

3. Some individual prokaryotes are spiral in shape, and therefore exhibit chirality (Figure 1.33). This chirality can extend to whole colonies (Figures 1.34 and 1.35). Eukaryotic nematode worms also exhibit whole body chirality[190].

4. Some prokaryotes form colonies that are square (Figure 1.36) or cubic (Figure 1.37), demonstrating that they can perpendicularly alter the directions of successive cell divisions. These oriented cell divisions may predate the evolution of eukaryotes[191], in which spatially oriented cell division is common in plants[192] and animals[193]. Some eukaryotic algae also form square colonies (Figure 1.38).

Family	Species	Shape
γ-proteo	*Escherichia coli*	
γ-proteo	*Vibrio cholerae*	
γ-proteo	*Francisella tularensis*	
β-proteo	*Neisseria gonorrhoeae*	
β-proteo	*Bordetella pertussis*	
α-proteo	*Sinorhizobium meliloti*	
α-proteo	*Caulobacter crescentus*	
α-proteo	*Rickettsia typhi*	
ε-proteo	*Helicobacter pylori*	
Firmicutes	*Bacillus subtilis*	
Firmicutes	*Staphylococcus aureus*	
Firmicutes	*Streptococcus pyogenes*	
Mollicutes	*Mycoplasma pneumoniae*	
Actino-bacteria	*Corynebacterium glutamicum*	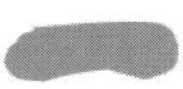
Spirochetes	*Borrelia burgdorferi*	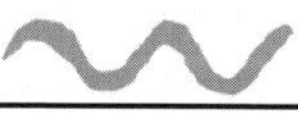

Figure 1.33. Individual bacteria have a wide repertoire of shapes, including spiral, chiral forms. From Margolin (2009)[195] with permission of Elsevier. We are working on the role of shape in the origin of life and its relationship to the origin and evolution of the cytoskeleton[293].

Figure 1.34. A colony of the cyanobacterium *Anabaena spiralis* with a heterocyst in the middle. Its shape indicates a handedness or chirality to the whole colony. From the Kernow Microscopical Society[196] with kind permission of Pete Terry, Hon. Sec.

5. Some single celled prokaryotes are square in shape[194] (Figure 1.39). Amongst eukaryotes, some plant tissues form checkerboards (Figure 1.40) and some single cell eukaryotic diatoms (Figure 1.41) are square in "valve" view.

6. Gliding motility is motion of a microorganism over a surface with no obvious moving parts, and the mechanism(s) is/are still uncertain, both for prokaryotes[197] and eukaryotes[198-204]. Single cell myxobacteria exhibit gliding motility that may involve helical cytoskeletal components[205]. Myxobacteria reverse directions periodically and collisions of these gliding cells can synchronize their motion forming waves and patterns in groups of moving cells[206]. This is analogous to the motions of the colonial, eukaryotic diatom *Bacillaria paradoxa*[203,204] (Figure 1.42), and indeed the same model for gliding has been proposed for diatoms, cyanobacteria and myxobacteria[207]. The motion of sheets of eukaryotic cells, such as the melanocytes over the yolk sac in embryonic zebrafish[208], often exhibit propagating waves (a form of "collective behavior") of various motility, shape and orientation parameters[209-212].

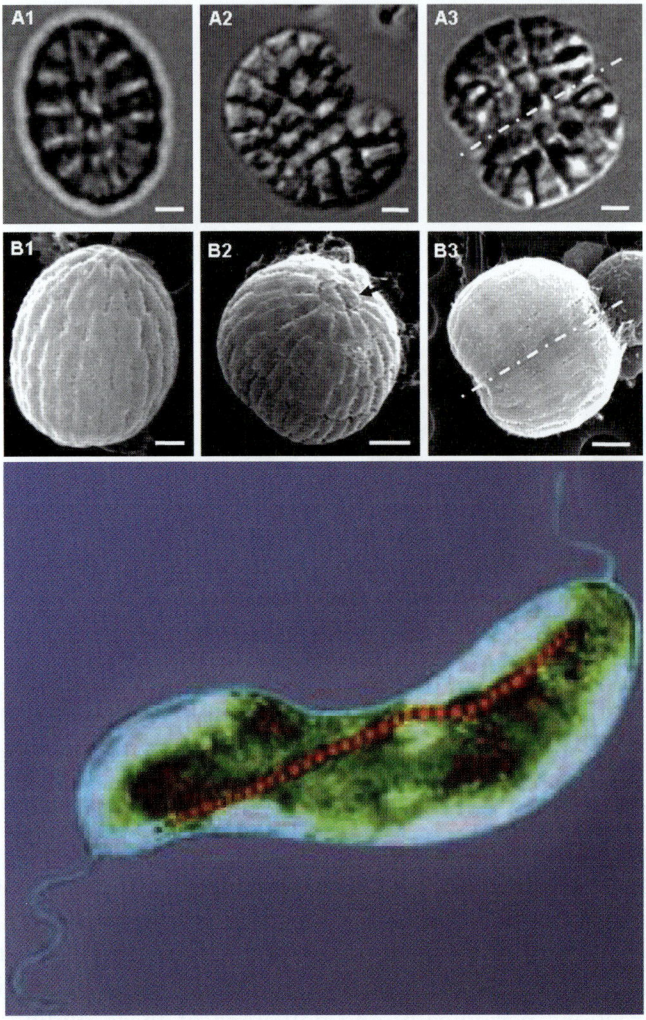

Figure 1.35. **Top**: These newly discovered magnetotactic prokaryotes (see also Abreu *et al.*, 2007[215]) are colonies that are ellipsoidal in shape and act somewhat like a single cell, doubling in size and then splitting in two. The top row shows light micrographs and the bottom row shows SEMs (scanning electron micrographs). The colony has spiral symmetry. Each cell has a flagellum, and since the colony swims as a unit, it has been concluded that their individual corkscrew rotations must be somehow coordinated. These properties are shared with many eukaryotic embryos at their blastula stage. The bars are 2 μm long. From Zhou *et al.* (2012)[216] with permission of John Wiley and Sons. **Bottom**: A flagellated bacterium containing a string of magnetosomes, which are tiny magnets synthesized by the bacterium. From Wikipedia[217] with permission under a Creative Commons Attribution-Share Alike 3.0 Unported license.

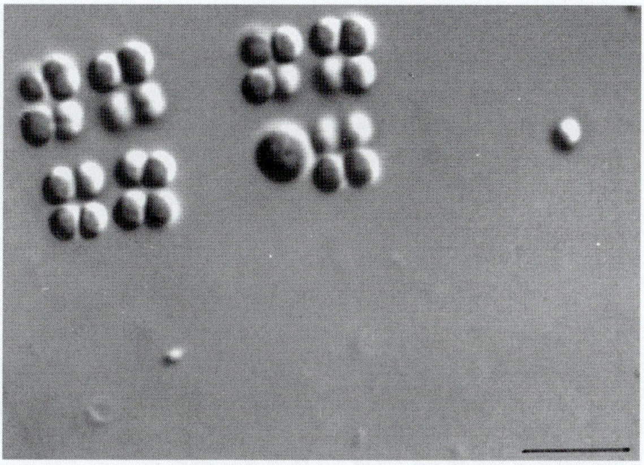

Figure 1.36. Cells of the cyanobacterium *Merismopedia* divide in consecutively perpendicular directions in a plane, resulting in square colonies. In culture, sometimes larger cells are formed. Bar = 25 μm. Image used with kind permission of Krzysztof Raciborski, Finnish Zoological and Botanical Publishing Board[218].

Figure 1.37. A cyanobacterium *Eucapsis alpina* that forms cubic colonies. Four views of the same colony. From National University of Ireland[219] with kind permission of Christopher F. Carter.

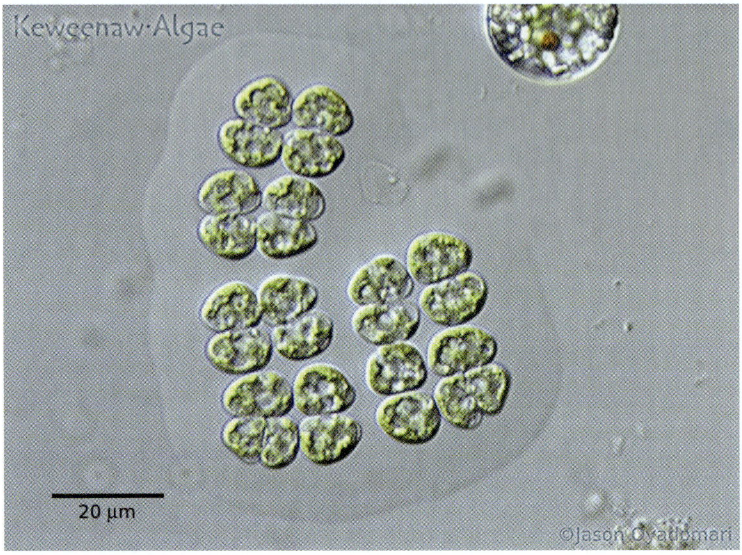

Figure 1.38. A eukaryotic alga that forms square colonies. From Chlorophycean Gallery #2[220] with kind permission of Jason Oyadomari.

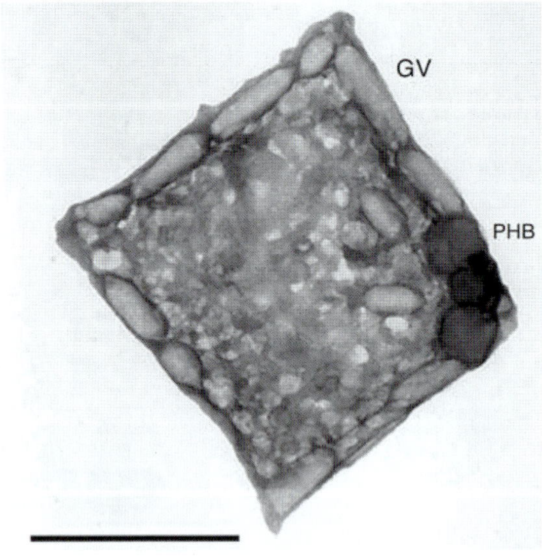

Figure 1.39. Square bacteria live in high salt environments[221]. This is an electron micrograph of one cell. GV = gas vesicles, PHB = poly-ß-hydroxybutyrate granules. Bar = 1 μm. From Burns *et al.* (2004)[222] with permission of John Wiley and Sons.

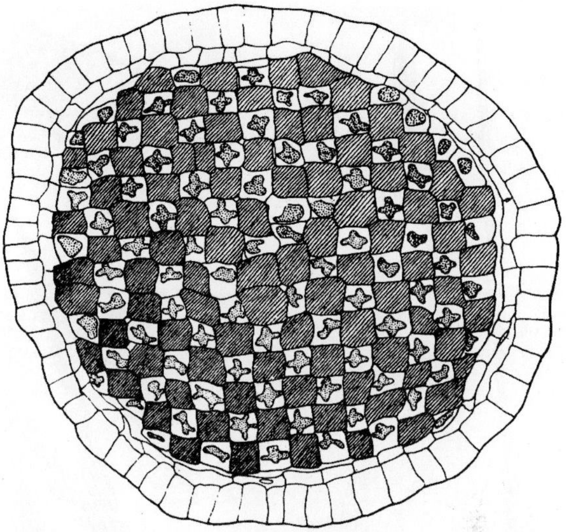

Figure 1.40. A eukaryote plant with square cells in a checkerboard pattern: "Section through developing liverwort sporangium showing differentiation of alternating spores and elater cells. (From Goebel.)"[192].

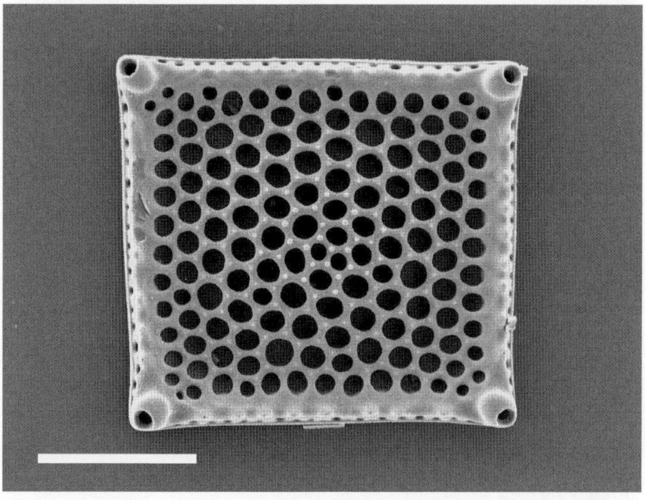

Figure 1.41. The silica shell of a single cell eukaryotic diatom *Triceratium*, with a square shape, found in a fossil deposit at Oamaru, New Zealand. Scale bar = 50 μm. From Mary Ann Tiffany with her kind permission. Some diatoms achieve an uncanny perfection of symmetry[223]. See Gschmeissner (2013)[224] for a prism shaped diatom with two triangular faces and three square faces.

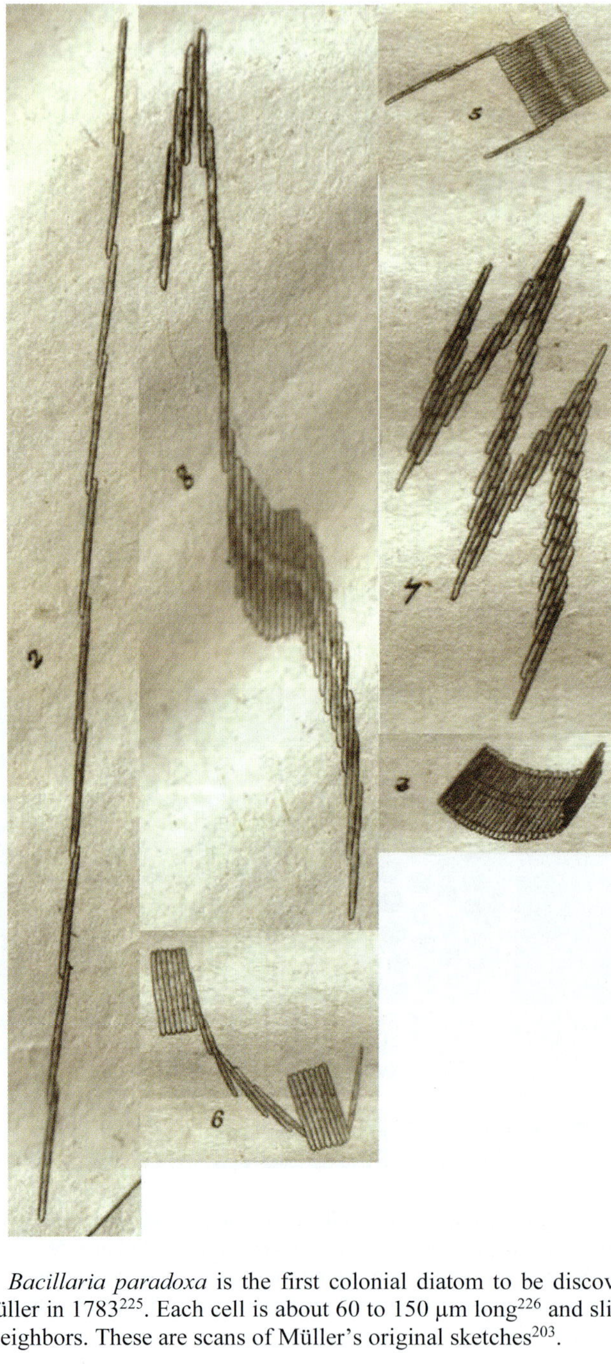

Figure 1.42. *Bacillaria paradoxa* is the first colonial diatom to be discovered, by Otto Friedrich Müller in 1783[225]. Each cell is about 60 to 150 μm long[226] and slides against its one or two neighbors. These are scans of Müller's original sketches[203].

7. Some cyanobacteria filamentous colonies move[213], indicating behavioral coordination between cells in gliding motility. Another filamentous cyanobacterial colony, *Synechococcus*, swims with no obvious moving parts[214], again suggesting coordination between cells. Other prokaryote colonies swim with coordinated flagella[227] (Figure 1.35). Coordinated cilia beating occurs in most eukaryotes such as single cell ciliates, the colonial swimming ball of *Volvox* cells[228], early animal embryos, and in our lungs and fallopian tubes, though the mechanism may not involve direct cell-cell communication, but rather hydrodynamic interactions between adjacent cilia[229], i.e., mechanical cell-cell interactions.

8. Some prokaryotes change their behavior when the cell density reaches a threshold, a phenomenon called "quorum sensing"[230]. Quorum sensing is also found in eukaryotic fungi, such as yeast[231], and is speculated to be involved in size control of tissues in animals[232]. Some researchers consider that bacteria have achieved a form of "social intelligence" via these cooperative behaviors[233,234]. If we think of quorum sensing as long distance communication between cells, then the ability of some bacteria to conduct electrical signals over long distances[235] might be considered analogous to nervous conduction in animals (Figure 1.43).

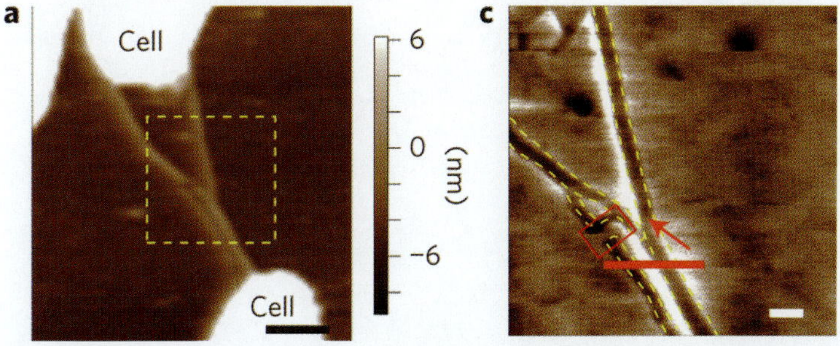

Figure 1.43. Many bacteria have protein rods called pili protruding from them[236]. In *Geobacter sulfurreducens*, of length 2 μm (left), the pili conduct electricity, which can be visualized by electrostatic force microscopy as the white charge cloud around the pili (right). From Malvankar *et al.* (2014)[237] with permission of Nature Publishing Group. Scale bars are left: 400 nm, right: 100 nm. "The pilus is only 3 nm wide and propagates charges in a manner similar to a carbon nanotube" (Nikhil S. Malvankar, University of Massachusetts Amherst).

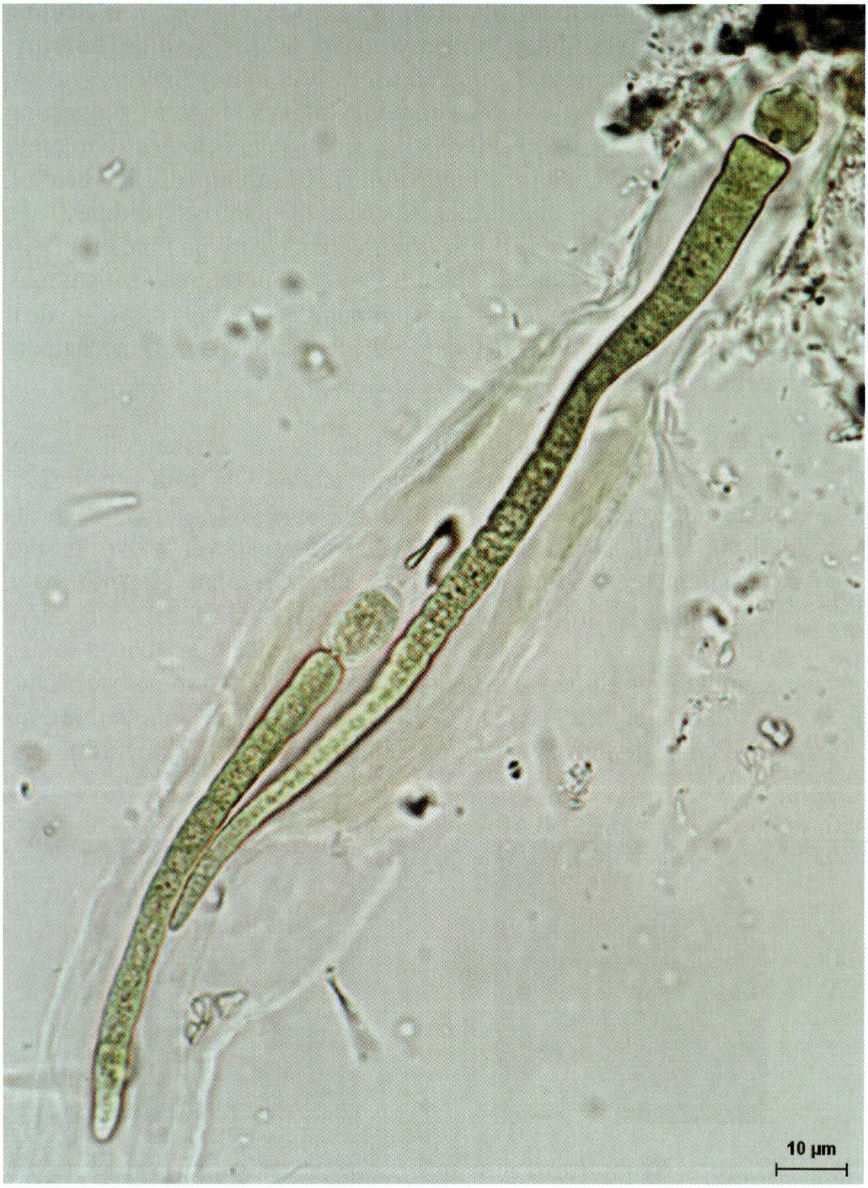

Figure 1.44. A tapered filamentous cyanobacterium, *Dichothrix gypsophila*. Note the heterocyst at the opposite end from the taper and the extracellular material forming a sheath around both filaments. From Sheath *et al.* (2011)[238] with kind permission of Robert Sheath.

9. Some filamentous colonies have tapered cells, so that each cell is different, at least in shape (Figure 1.44).

10. Some prokaryotes form branched filaments (Figure 1.30).

11. Some swimming prokaryotes form wonderful patterns when grown in thin layers of medium[239] (Figure 1.45). Such "swarming" behavior can result in whole groups of bacteria moving together in a common direction (Figure 1.46). Swarm behavior of self-propelled, nonliving particles which can lead to phase transitions, may explain some of this behavior[240].

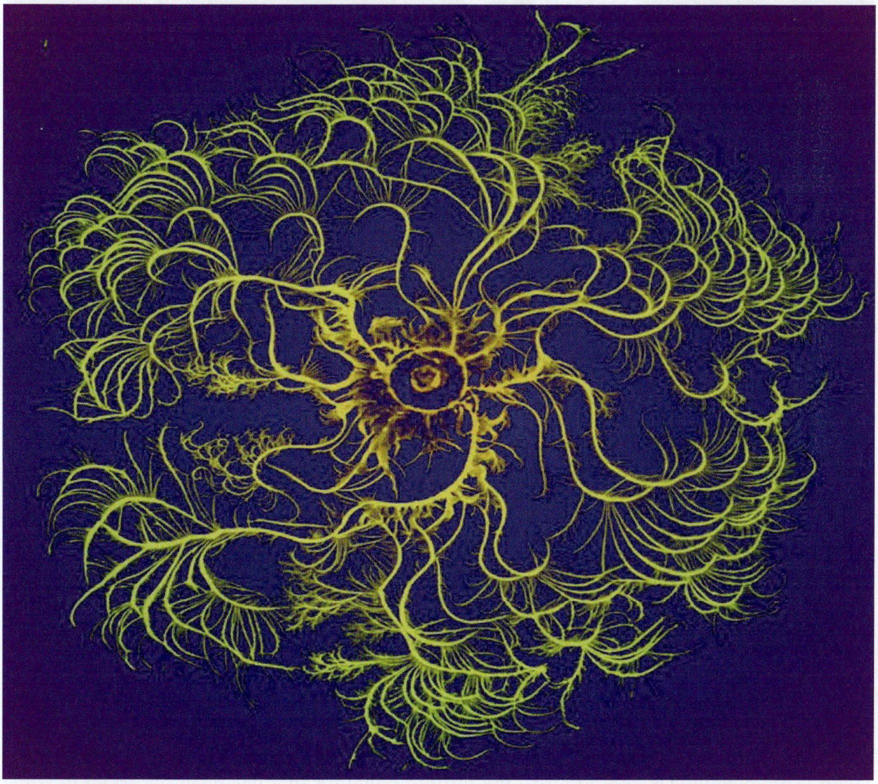

Figure 1.45. A pattern formed by swimming bacteria in a shallow medium in an 8 cm diameter Petri dish. From Alber (2005)[242] with kind permission of the late Eshel Ben-Jacob. Motile cyanobacteria also form swirling[290] and reticulate[291] patterns.

Figure 1.46. Each bright droplet in this 8 cm diameter colony consists of up to millions of the bacterium *Paenibacillus vortex* swimming in circles around the center of the drop[243]. From Wassenaar *et al.* (2011)[244] with kind permission of the late Eshel Ben-Jacob. This behavior is analogous to the milling behavior that has been observed in some ants and schooling fish[245] (Charles M. Breder, personal communication, 1963).

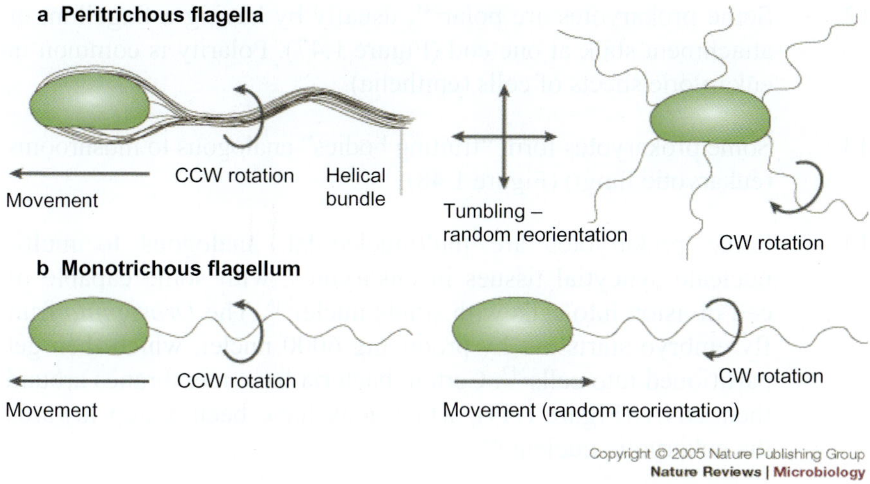

Figure 1.47. Bacteria that have flagella may have one or more. Some have them just at one end, making the cell essentially polar, i.e., having distinguishable ends. CW = clockwise, CCW = counter clockwise. From Butler & Camilli (2005)[249] with permission of Nature Publishing Group.

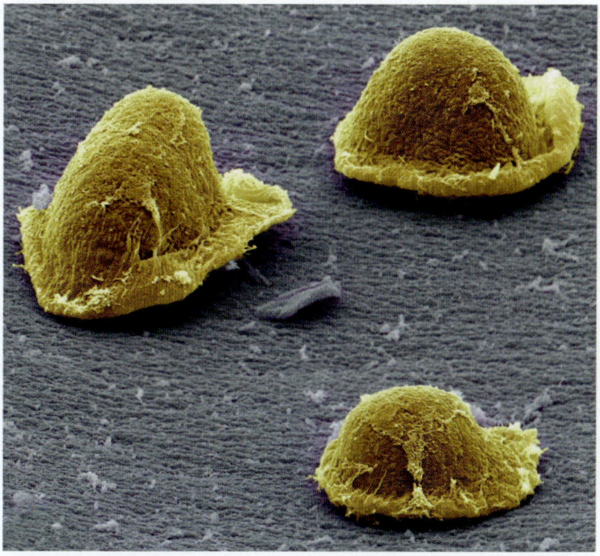

Figure 1.48. SEM of fruiting bodies of the bacterium *Myxococcus xanthus*, in which the bacteria aggregate to make a three dimensional structure. From Velicer (2010)[250] with kind permission of Gregory J. Velicer and courtesy of Indiana University.

12. Some prokaryotes are polar[241], usually by having a flagellum or attachment stalk at one end (Figure 1.47). Polarity is common in eukaryotic sheets of cells (epithelia).

13. Some prokaryotes form "fruiting bodies" analogous to mushrooms (eukaryotic fungi) (Figure 1.48).

14. Some prokaryotes are multinucleoidal, analogous to multinucleate syncytial tissues in eukaryotes, with some capable of cell division into cells with single nuclei[246]. The *Drosophila* fruit fly embryo starts out by producing 6000 nuclei, which then get partitioned into cells[247]. Certain bacteria have membranes around their DNA (Figure 1.49), which may have been a step towards the eukaryotic nucleus[248].

15. Some prokaryotes contain crystals of magnetite and therefore can swim oriented to the Earth's magnetic field[252] (Figure 1.35). The same has now been found in fish[253]. It has generally been assumed that the tiny magnets were sufficient in magnetic strength to turn a bacterium in the Earth's magnetic field. Of course, a whole fish needs to somehow amplify the twist applied to these subcellular magnets. This amplification required by the physics has also been found in colonial magnetotactic bacteria[254].

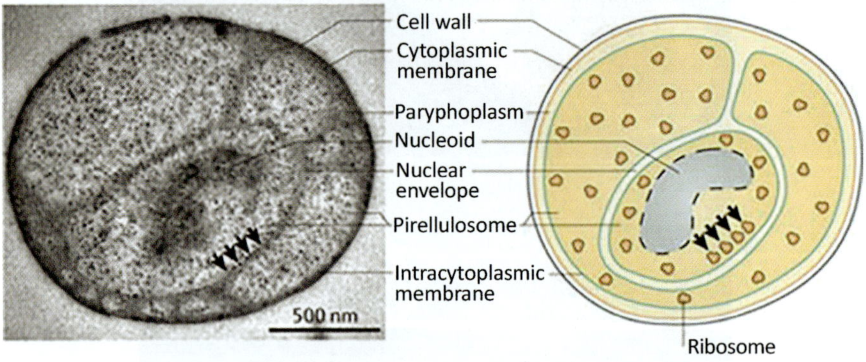

Figure 1.49. The planctomycetes bacterium *Gemmata obscuriglobus* has a membrane (nuclear envelope) around its DNA (nucleoid). Arrows indicate membrane associated ribosomes. Most prokaryotes do not have a nuclear membrane. From Fuerst & Sagulenko (2011)[251] with permission of Nature Publishing Group.

16. Some prokaryotes control the size of their colonies by doubling the number of cells and then splitting the colony in two (Figures 1.29 and 1.35). Such fission of the whole organism[255], which apparently avoids going through a single cell stage in its life cycle[256], also occurs in the eukaryote turbellian flatworm *Stenostomum incaudatum* (*Proposition 123* in *Hierarchical Genome*[180]). The "simplest" multicellular animal, the placozoan *Trichoplax*[257], also usually reproduces by splitting its one flat sheet of cells into two or three parts[258].

17. Some prokaryotes are huge, larger than many eukaryotic cells[259], and have an electric potential across their membrane[260].

18. Some prokaryotes have been shown to have membrane surface waves during cell division[261,262]. Cf. Chapter 10.

19. Some prokaryotes have a cytoskeleton containing microfilaments and microtubules, made of forms of actin and tubulin, respectively, and intermediate filaments, giving them the full complement of cytoskeleton found in eukaryotes[263]. The cytoskeleton is involved in spore formation[264]. See Chapter 5.

20. Cell death, which is common in the later stages of development of eukaryotic embryos, for example separating the fingers in us humans (Figure 1.50), (but not digits in amphibians[265]), is also found in some colonial prokaryotes[266,267]. Mechanisms of aging of cells in prokaryotes resemble those of eukaryotic cells[268].

21. The mechanical properties of bacteria include an enormous pressure inside of some, up to 50 atmospheres[269], down to none at all for the square bacteria that can live in high salt conditions (Figure 1.39). Internal pressure therefore plays a role in morphology in bacteria, as it does in embryos[270].

22. While bacteria generally do not have a membrane bound nucleus, some have been demonstrated to contain membrane bound vacuoles[271]. In eukaryotic plants vacuoles play a major role in cell shape, turgor and thus whole plant morphogenesis[272,273].

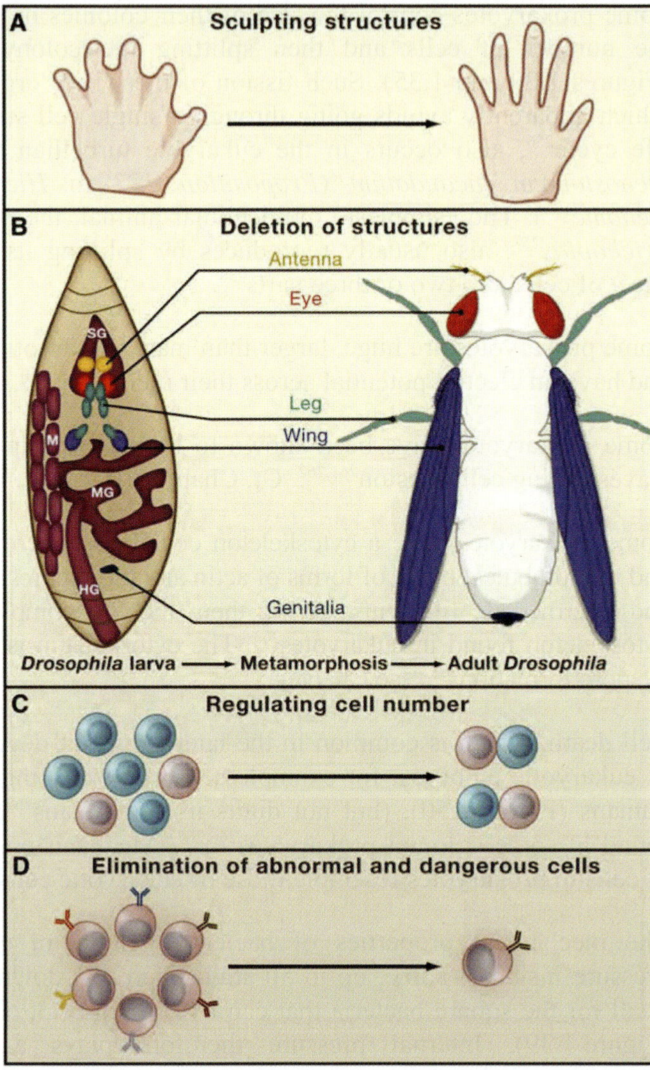

Figure 1.50. Programmed cell death is part of normal embryogenesis in many organisms, including plants[281]. For example: (A) in humans the webbing between fingers and toes comes about by death of cells in the webs between digits. In insects that go through a grub (larval) stage, (B) most of the tissues die except the brain and small regions called "imaginal discs", when the grub or caterpillar metamorphoses to an adult. In some tissues, such as our brains, there is an "overproduction" of cells, and many die off (C). Finally, the immune system destroys "defective" cells (D). From Fuchs & Steller (2011)[282] with permission from Elsevier.

23. The extracellular matrix, outside our cells, but in between cell layers, is an important component of embryos. Bacteria also secrete a matrix of extracellular polymers, known colloquially as slime, that holds them together in organized biofilms[274].

24. Bacteria of different species can complement one another's capabilities in "consortia"[275], which often includes horizontal gene transfer between them[276]. This is analogous to the division of labor of differentiated cell types in embryos.

25. In some cases bacteria undergo mirror symmetric cell divisions[277]. Such divisions are also found in eukaryotic cells in cell culture[278], in the single cell ciliates[279], and in certain cells at the midline in intact embryos[280].

We should not be too hasty in dismissing the morphogenetic capabilities of single eukaryotic cells either. For example, certain diatoms can have two shells ("valves") that are so different that, separated, they would be presumed to be from two different species (Figure 1.51). Syncytial (also known as coenocytic) organisms, including marine algae meters long, can consist of one huge cell with many nuclei with "differentiated" parts (Figure 1.52). In this book we will focus on embryogenesis of multicellular eukaryotic organisms that generally have one nucleus per cell.

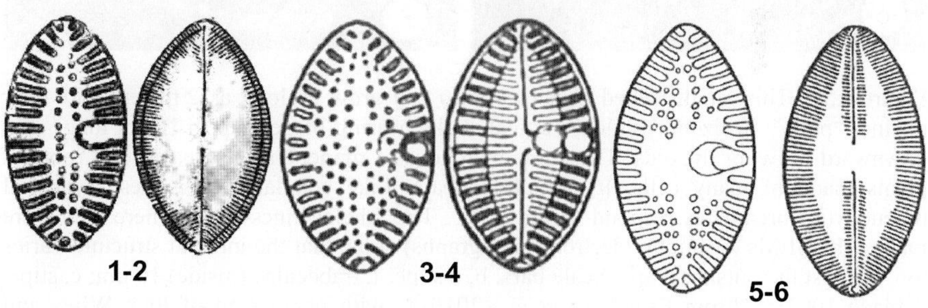

Figure 1.51. Three diatom species in which opposite shells (valves) are substantially different from one another. As one and the same cell is inside each pair of valves, they are formed from the same genome in the same cell, an unsolved conundrum. The vertical lines are slits in the shell called raphes that are involved in diatom motility. Note that they only appear on one valve and not the other in these species. From[283] with kind permission of Luc Ector and the Editors of *Phytotaxa*, Carlos Eduardo Wetzel and Saúl Blanco. For further discussion see our "Diatoms and nanotechnology"[284].

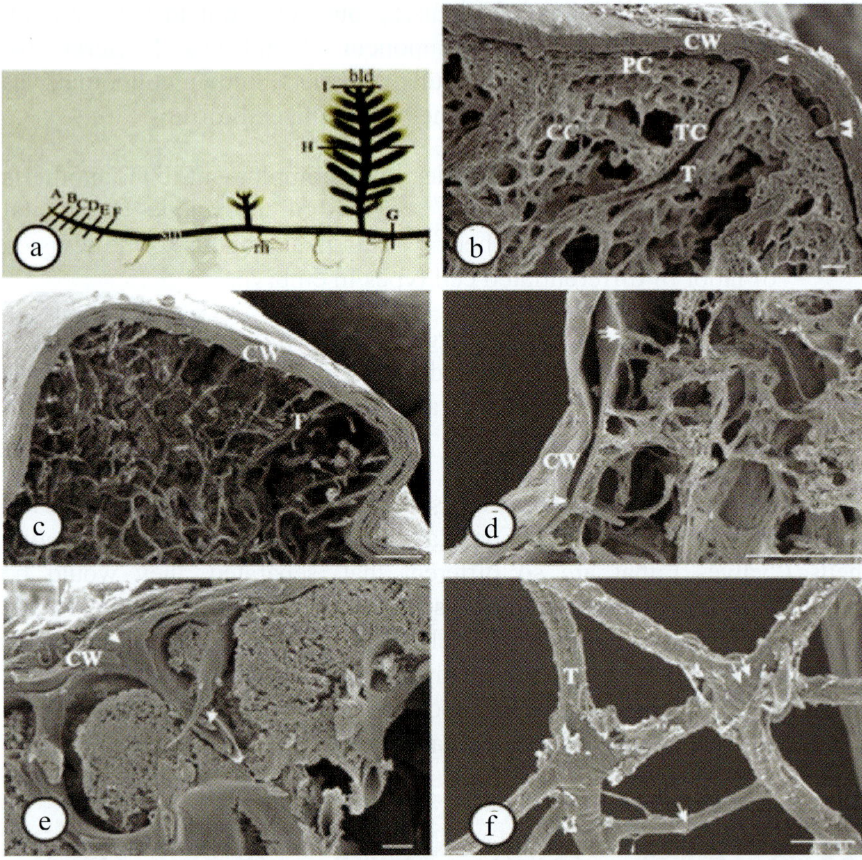

Figure 1.52. This single celled alga *Caulerpa* can grow as long as 2 meters with four distinct "parts": horizontal stolon, vertical stalks bearing blades (up to 10 cm long), and downward growing rhizoids. These are analogous to runners, stems, leaves and roots of plants made of many cells. It is multinucleate. Like cellular plants, these syncytial (coenocytic) organisms respond to gravity[285]. The labeled lines show where cuts were made. The SEMs (scanning electron micrographs) show that the internal structure varies too, not just the outside shape. Scale bars: b, e stipe, f, trabeculae (inside) 10 μm; c, stipe, d blade 100 μm. From Fagerberg *et al.* (2010)[286] with permission of John Wiley and Sons. See also Ranjan *et al.* (2015)[287]. Evidence from syncytial fungi suggests that flow patterns in the cytoplasm may isolate nuclei into groups that then differentiate in terms of gene expression, even though no cell membranes separate them[288,289].

1. Reid, R.G.B. (1985). *Evolutionary Theory: The Unfinished Synthesis*. London, Croom Helm.
2. Witkowski, J.A. (1985). The hunting of the organizer: an episode in biochemical embryology. *Trends in Biochemical Sciences* **10**(10), 379-381.
3. Moore, R.D. & B. Braga (1994). "See this? This is you." Star Trek: The Next Generation. Season 7, Episode 25: All Good Things... http://www.ovguide.com/tv_episode/star-trek-the-next-generation-season-7-episode-25-all-good-things-1-46684
4. Follmann, H. & C. Brownson (2009). Darwin's warm little pond revisited: from molecules to the origin of life. *Naturwissenschaften* **96**(11), 1265-1292.
5. Sherwin, J. (1999). *Quotable Star Trek*, Simon & Schuster.
6. Ikehara, K. (2012). [GADV]-protein world hypothesis on the origin of life. In: *Genesis - In the Beginning: Precursors of Life, Chemical Models and Early Biological Evolution*. Ed.: J. Seckbach. Dordrecht, Springer: 107-122.
7. Atkins, J.F., R.F. Gesteland & T.R. Cexh, Eds. (2011). *RNA Worlds: From Life's Origins To Diversity In Gene Regulation*. Cold Spring Harbor, New York, USA, Cold Spring Harbor Laboratory Press.
8. Rasmussen, S., M.A. Bedau, L. Chen, D. Deamer, D.C. Krakauer, N.H. Packard & P.F. Stadler, Eds. (2008). *Protocells: Bridging Nonliving and Living Matter*. Cambridge, MIT Press.
9. Deamer, D. (2011). *First Life: Discovering the Connections Between Stars, Cells, and How Life Began*, University of California Press.
10. Pascal, R., A. Pross & J.D. Sutherland (2013). Towards an evolutionary theory of the origin of life based on kinetics and thermodynamics. *Open Biology* **3**(11), #130156.
11. Barge, L.M., T.P. Kee, I.J. Doloboff, J.M.P. Hampton, M. Ismail, M. Pourkashanian, J. Zeytounian, M.M. Baum, J.A. Moss, C.-K. Lin, R.D. Kidd & I. Kanik (2014). The fuel cell model of abiogenesis: A new approach to origin-of-life simulations. *Astrobiology* **14**(3), 254-270.
12. Deacon, T.W., A. Srivastava & J.A. Bacigalupi (2014). The transition from constraint to regulation at the origin of life. *Frontiers in Bioscience-Landmark* **19**, 945-957.
13. Damer, B. & D. Deamer (2015). Coupled phases and combinatorial selection in fluctuating hydrothermal pools: A scenario to guide experimental approaches to the origin of cellular life. *Life* **5**, 872-887.
14. Luisi, P.L. (2015). Chemistry constraints on the origin of life. *Israel Journal of Chemistry* **55**(8), 906-918.
15. Markoff, J. (2009). Wanted: Home Computers to Join in Research on Artificial Life. *New York Times*(September 29), D3.
16. Damer, B., P. Newman, R. Gordon, T. Barbalet, D.W. Deamer & R. Norkus (2010). The EvoGrid: a framework for distributed artificial chemistry cameo simulations supporting origins of life endeavors. In: *Artificial Life XII. Proceedings of the Twelfth International Conference on the Synthesis and Simulation of Living Systems*. Eds.: H. Fellermann, M. Dorr, M. Hanczyc, L.L. Laursen, S. Maurer, D. Merkle, P.-A. Monnard, K. Sty & S. Rasmussen. Cambridge, MIT Press: 73-79.
17. Damer, B., P. Newman, R. Norkus, R. Gordon & T. Barbalet (2012). Cyberbiogenesis and the EvoGrid: a Twenty-First Century grand challenge. In:

Genesis - In the Beginning: Precursors of Life, Chemical Models and Early Biological Evolution. Ed.: J. Seckbach. Dordrecht, Springer: 267-288.

18. Martin, W., J. Baross, D. Kelley & M.J. Russell (2008). Hydrothermal vents and the origin of life. *Nature Reviews Microbiology* **6**(11), 805-814.
19. Sharov, A.A. & R. Gordon (2013). Life Before Earth. http://arxiv.org/abs/1304.3381
20. Sharov, A.A. & R. Gordon (2016). Life before Earth. In: *Habitability of the Universe Before Earth [in series: Astrobiology: Exploring Life on Earth and Beyond, eds. Pabulo Henrique Rampelott, Joseph Seckbach & Richard Gordon]*. Eds.: R. Gordon & A.A. Sharov, Elsevier. In preparation.
21. Jarosik, N., C.L. Bennett, J. Dunkley, B. Gold, M.R. Greason, M. Halpern, R.S. Hill, G. Hinshaw, A. Kogut, E. Komatsu, D. Larson, M. Limon, S.S. Meyer, M.R. Nolta, N. Odegard, L. Page, K.M. Smith, D.N. Spergel, G.S. Tucker, J.L. Weiland, E. Wollack & E.L. Wright (2011). Seven-year Wilkinson Microwave Anisotropy Probe (WMAP) observations: sky maps, systematic errors, and basic results. *Astrophysical Journal Supplement Series* **192**(2), 1-15.
22. Zheng, W., M. Postman, A. Zitrin, J. Moustakas, X. Shu, S. Jouvel, O. Høst, A. Molino, L. Bradley, D. Coe, L.A. Moustakas, M. Carrasco, H. Ford, N. Benítez, T.R. Lauer, S. Seitz, R. Bouwens, A. Koekemoer, E. Medezinski, M. Bartelmann, T. Broadhurst, M. Donahue, C. Grillo, L. Infante, S.W. Jha, D.D. Kelson, O. Lahav, D. Lemze, P. Melchior, M. Meneghetti, J. Merten, M. Nonino, S. Ogaz, P. Rosati, K. Umetsu & A. van der Wel (2012). A magnified young galaxy from about 500 million years after the Big Bang. *Nature* **489**(7416), 406-408.
23. Wikipedia (2015). Age of the Earth. https://en.wikipedia.org/wiki/Age_of_the_Earth
24. Brasier, M., N. McLoughlin, O. Green & D. Wacey (2006). A fresh look at the fossil evidence for early Archaean cellular life. *Philosophical Transactions of the Royal Society B-Biological Sciences* **361**(1470), 887-902.
25. Brasierf, M. (2015). Deep questions about the nature of early-life signals: a commentary on Lister (1673) 'A description of certain stones figured like plants'. *Philosophical Transactions of the Royal Society A-Mathematical Physical and Engineering Sciences* **373**(2039), #20140254.
26. Rasmussen, B., I.R. Fletcher, J.J. Brocks & M.R. Kilburn (2008). Reassessing the first appearance of eukaryotes and cyanobacteria. *Nature* **455**(7216), 1101-1104.
27. Xie, S. & H. Yin (2014). Progress and perspective on frontiers of geobiology. *Science China-Earth Sciences* **57**(5), 855-868.
28. Bell, E.A., P. Boehnke, T.M. Harrison & W.L. Mao (2015). Potentially biogenic carbon preserved in a 4.1 billion-year-old zircon. *Proceedings of the National Academy of Sciences*, doi: 10.1073/pnas.1517557112.
29. McNichol, J. & R. Gordon (2012). Are we from outer space? A critical review of the panspermia hypothesis. In: *Genesis - In the Beginning: Precursors of Life, Chemical Models and Early Biological Evolution*. Ed.: J. Seckbach. Dordrecht, Springer: 591-620.
30. Bada, J.L. (2004). How life began on Earth: a status report. *Earth and Planetary Science Letters* **226**(1-2), 1-15.

31. Gordon, R. & J. McNichol (2012). Recurrent dreams of life in meteorites. In: *Genesis - In the Beginning: Precursors of Life, Chemical Models and Early Biological Evolution*. Ed.: J. Seckbach. Dordrecht, Springer: 549-590.
32. Yu, G. & S.B. Jacobsen (2011). Fast accretion of the Earth with a late Moon-forming giant impact. *Proceedings of the National Academy of Sciences USA* **108**(43), 17604-17609.
33. Finkbeiner, A. (2012). The new Milky Way. A fresh look at our galaxy points to a chaotic past and a violent end. *Nature* **490**, 24-27.
34. Sharov, A.A. (2009). Coenzyme autocatalytic network on the surface of oil microspheres as a model for the origin of life. *International Journal of Molecular Sciences* **10**(4), 1838-1852.
35. Hanczyc, M.M. & D. Merkle (2014). Hydrogen cyanide and the computational origin of life [slides]. http://embryogenesisexplained.com/2014/03/computational-origin-of-life.html
36. Crick, F.H.C. & L.E. Orgel (1973). Directed panspermia. *Icarus* **19**(3), 341-346.
37. Wikipedia (2012). Drake equation. http://en.wikipedia.org/w/index.php?title=Drake_equation&oldid=486661162
38. Wikipedia (2012). Fermi paradox. http://en.wikipedia.org/wiki/Fermi_paradox
39. Gordon, R. & R.B. Hoover (2007). Could there have been a single origin of life in a Big Bang universe? *Proceedings of the SPIE* **6694**, doi:10.1117/1112.737041.
40. Sharov, A.A. (2012). *A Short Course on Biosemiotics: 2. Evolution of Natural Agents: Preservation, Development, and Emergence of Functional Information [PowerPoint presentation]*. Second Life®, Embryo Physics Course: http://embryogenesisexplained.com/2012/04/a-short-course-on-biosemiotics-2.html.
41. Becker, A., M. Campbell & P. Aldhous (2013). First interactive map of galaxy's habitable planets, https://www.newscientist.com/article/dn24269-first-interactive-map-of-galaxys-habitable-planets/
42. Heller, R., D. Williams, D. Kipping, M.A. Limbach, E. Turner, R. Greenberg, T. Sasaki, E. Bolmont, O. Grasset, K. Lewis, R. Barnes & J.I. Zuluaga (2014). Formation, habitability, and detection of extrasolar moons. *Astrobiology* **14**(9), 798-835.
43. Ramirez, I. & T. Jones (2014). Solar Sibling HD 162826. https://mcdonaldobservatory.org/news/gallery/solar-sibling-hd-162826
44. de Vries, K. (2014). Solar siblings? Astronomers discover sun's long-lost brother. http://www.foxnews.com/science/2014/05/09/solar-siblings-astronomers-discover-sun-long-lost-brother/?cmpid=NL_fntop
45. Postnov, K.A. & M.E. Prokhorov (1992). Planet survival? *Astronomy & Astrophysics* **258**(2), L17-L18.
46. Palladino, L.E., K.J. Schlesinger, K. Holley-Bockelmann, C. Allende Prieto, T.C. Beers, Y.S. Lee & D.P. Schneider (2014). Hypervelocity star candidates in the SEGUE G and K dwarf sample. *The Astrophysical Journal* **780**(1), #7.
47. Iono, D., M.S. Yun & P.T.P. Ho (2005). Atomic and molecular gas in colliding galaxy systems. I. The data. *Astrophysical Journal Supplement Series* **158**(1), 1-37.

48. Veras, D., M.C. Wyatt, A.J. Mustill, A. Bonsor & J.J. Eldridge (2011). The great escape: how exoplanets and smaller bodies desert dying stars. *Monthly Notices of the Royal Astronomical Society* **417**(3), 2104-2123.
49. Veras, D. & C.A. Tout (2012). The great escape - II. Exoplanet ejection from dying multiple-star systems. *Monthly Notices of the Royal Astronomical Society* **422**(2), 1648-1664.
50. Veras, D., N.W. Evans, M.C. Wyatt & C.A. Tout (2014). The great escape - III. Placing post-main-sequence evolution of planetary and binary systems in a Galactic context. *Monthly Notices of the Royal Astronomical Society* **437**(2), 1127-1140.
51. Kratter, K.M. & H.B. Perets (2012). Star hoppers: Planet instability and capture in evolving binary systems. *Astrophysical Journal* **753**(1), #91.
52. Wikipedia (2015). Speed of light. https://en.wikipedia.org/wiki/Speed_of_light
53. Smith, N., W. Li, R.J. Foley, J.C. Wheeler, D. Pooley, R. Chornock, A.V. Filippenko, J.M. Silverman, R. Quimby & J.S. Bloom (2007). SN 2006gy: discovery of the most luminous supernova ever recorded, powered by the death of an extremely massive star like η Carinae. *The Astrophysical Journal* **666**(2), 1116-1128.
54. Ruiz, N., M.A. Guerrero, Y.-H. Chu & R.A. Gruendl (2011). Physical structure of the planetary nebula NGC 3242 from the hot bubble to the nebular envelope. *The Astronomical Journal* **142**(3), #91.
55. Clavin, W. (2013). A Ghostly Trio from NASA's Spitzer Space Telescope. http://www.nasa.gov/mission_pages/spitzer/news/spitzer20131028.html - .UnaDPCQeJ9R
56. Corradi, R.L.M., D. Schönberner, M. Steffen & M. Perinotto (2003). Ionized haloes in planetary nebulae: new discoveries, literature compilation and basic statistical properties. *Monthly Notices of the Royal Astronomical Society* **340**(2), 417-446.
57. Turner, J. & D. Salisbury (2014). Hypervelocity Star Illustration. http://www.eurekalert.org/multimedia/pub/66984.php?from=257672
58. NASA/JPL-Caltech/Space Science Institute (2013). The Day the Earth Smiled. http://www.nasa.gov/mission_pages/cassini/multimedia/pia17171.html - .UjZ5Kbx1FfX
59. Loeb, A. & J. Guillochon (2014). Observational cosmology with semi-relativistic stars. http://arxiv.org/abs/1411.5030
60. Gordon, R. (2015). (Ultimately) Alone in the Universe. http://embryogenesisexplained.org/2015/09/16/ultimately-alone-in-the-universe-by-richard-dick-gordon/
61. Guillochon, J. & A. Loeb (2015). The fastest unbound stars in the universe. *Astrophysical Journal* **806**(1), #124.
62. Loeb, A. (2015). *From the First Star to Milkomeda*, Amazon Digital Services, Kindle.
63. Lemonick, M.D. (2014). You've Heard of Shooting Stars, but This is Ridiculous. http://time.com/3602047/hypervelocity-stars/
64. Woo, M. (2014). The Fastest Stars in the Universe May Approach Light Speed. http://www.wired.com/2014/12/superfast-stars-black-holes/
65. Miller, S.L. (1953). A production of amino acids under possible primitive earth conditions. *Science* **117**(3046), 528-529.

66. Voronkov, M.G., E.N. Sukhomazova, E.P. Levanova, V.A. Grabel'nykh, L.K. Papernaya, N.V. Russavskaya & N.A. Korchevin (2007). High-temperature formation of heteroaromatic structures during volcanic eruptions as the basis of the origin of life on the Earth. *Doklady Chemistry* **415**, 170-171.
67. Svettsov, V.V. (2002). Evaluating cometary delivery of organics to the early Earth. *Solar System Research* **36**(1), 50-61.
68. Wirick, S., G.J. Flynn, L.P. Keller, K. Nakamura-Messenger, C. Peltzer, C. Jacobsen, S. Sandford & M. Zolensky (2009). Organic matter from comet 81P/Wild 2, IDPs, and carbonaceous meteorites; similarities and differences. *Meteoritics & Planetary Science* **44**(10), 1611-1626.
69. Shaw, G.H. (2008). Earth's atmosphere - Hadean to early Proterozoic. *Chemie der Erde-Geochemistry* **68**(3), 235-264.
70. Kwok, S. (2012). Synthesis of organic matter by stars and its effect on the origin of life on Earth. *Ninth Pacific Rim Conference on Stellar Astrophysics* **451**, 3-8.
71. Kwok, S. (2015). Organic compounds in circumstellar and interstellar environments. *Origins of Life and Evolution of Biospheres* **45**(1-2), 113-121.
72. Sheehan, K.B., D.J. Patterson, B.L. Dicks & J.M. Henson (2005). *Seen and Unseen: Discovering the Microbes of Yellowstone*. Helena, Globe Pequot Press.
73. Minic, Z. (2009). Organisms of deep sea hydrothermal vents as a source for studying adaptation and evolution. *Symbiosis* **47**(3), 121-132.
74. Gaucher, E.A., S. Govindarajan & O.K. Ganesh (2008). Palaeotemperature trend for Precambrian life inferred from resurrected proteins. *Nature* **451**(7179), 704-707.
75. Perez-Jimenez, R., A. Inglés-Prieto, Z.M. Zhao, I. Sanchez-Romero, J. Alegre-Cebollada, P. Kosuri, S. Garcia-Manyes, T.J. Kappock, M. Tanokura, A. Holmgren, J.M. Sanchez-Ruiz, E.A. Gaucher & J.M. Fernandez (2011). Single-molecule paleoenzymology probes the chemistry of resurrected enzymes. *Nature Structural & Molecular Biology* **18**(5), 592-596.
76. Drum, R.W. (1968). Petrifaction of plant tissue in the laboratory. *Nature* **218**, 784-785.
77. Hazen, R.M., N. Boctor, J.A. Brandes, G.D. Cody, R.J. Hemley, A. Sharma & H.S. Yoder (2002). High pressure and the origin of life. *Journal of Physics-Condensed Matter* **14**(44), 11489-11494.
78. Ferrini, V. & M. Lilley (2004). Tube worms near hydrothermal vents at EPR (2004). http://media.marine-geo.org/image/tube-worms-near-hydrothermal-vents-epr-2004
79. Johnson, D.B., N. Okibe & F.F. Roberto (2003). Novel thermo-acidophilic bacteria isolated from geothermal sites in Yellowstone National Park: physiological and phylogenetic characteristics. *Archives of Microbiology* **180**(1), 60-68.
80. Jones, B. & R.W. Renaut (2006). Growth of siliceous spicules in acidic hot springs, Waiotapu geothermal area, North Island, New Zealand. *Palaios* **21**(5), 406-423.
81. Kaehr, B., J.L. Townson, R.M. Kalinich, Y.H. Awad, B.S. Swartzentruber, D.R. Dunphy & C.J. Brinker (2012). Cellular complexity captured in durable silica biocomposites. *Proceedings of the National Academy of Sciences of the United States of America* **109**(43), 17336-17341.

82. Ying, J.Y. (2012). Biocomposites: Cells made of silica. *Nature Nanotechnology* **7**(12), 777-778.
83. Martin, C.C. & R. Gordon (2001). The evolution of perception. *Cybernetics & Systems* **32**(3-4), 393-409.
84. de Nooijer, S., B.R. Holland & D. Penny (2009). The emergence of predators in early life: there was no Garden of Eden. *PLoS One* **4**(6), e5507.
85. Hanczyc, M.M. (2011). Metabolism and motility in prebiotic structures. *Philosophical Transactions of the Royal Society B-Biological Sciences* **366**(1580), 2885-2893.
86. Faguy, D.M. & K.F. Jarrell (1999). A twisted tale: the origin and evolution of motility and chemotaxis in prokaryotes. *Microbiology* **145**(Pt 2), 279-281.
87. Cooper, G.M. (2000). *The Cell: A Molecular Approach* http://www.ncbi.nlm.nih.gov/books/NBK9841/. Sunderland, MA, Sinauer Associates, 2nd.
88. Hanczyc, M.M. & T. Ikegami (2010). Chemical basis for minimal cognition. *Artificial Life* **16**(3), 233-243.
89. Horibe, N., M.M. Hanczyc & T. Ikegami (2011). Mode switching and collective behavior in chemical oil droplets. *Entropy* **13**(3), 709-719.
90. Caschera, F., S. Rasmussen & M.M. Hanczyc (2013). An oil droplet division-fusion cycle. *Chempluschem* **78**(1), 52-54.
91. Banno, T., R. Kuroha, S. Miura & T. Toyota (2015). Multiple-division of self-propelled oil droplets through acetal formation. *Soft Matter* **11**(8), 1459-1463.
92. Turk-MacLeod, R., P. Nghe, G. Woronoff, D. Schnettler, E. Szathmáry & A.D. Griffiths (2015). Compartmentalization of the formose reaction to test metabolism-first theories on the origin of life. In: *Astrobiology Science Conference 2015, Habitability, Habitable Worlds, and Life, June 15–19, 2015, Chicago, Illinois*. Ed.: P. Doran. Houston, Lunar and Planetary Institute: #7162.
93. Cano, R.J. & M.K. Borucki (1995). Revival and identification of bacterial-spores in 25-million-year-old to 40-million-year-old Dominican amber. *Science* **268**(5213), 1060-1064.
94. Martin-González, A., J. Wierzchos, J.C. Gutiérrez, J. Alonso & C. Ascaso (2009). Microbial Cretaceous park: biodiversity of microbial fossils entrapped in amber. *Naturwissenschaften* **96**(5), 551-564.
95. Wang, M., L.S. Yafremava, D. Caetano-Anolles, J.E. Mittenthal & G. Caetano-Anolles (2007). Reductive evolution of architectural repertoires in proteomes and the birth of the tripartite world. *Genome Research* **17**(11), 1572-1585.
96. Wikipedia (2014). Archaea. http://en.wikipedia.org/wiki/Archaea
97. Feng, J., D.A. Kessler, E. Ben-Jacob & H. Levine (2013). Growth feedback as a basis for persister bistability. *Proceedings of the National Academy of Sciences USA* **111**(1), 544-549.
98. Vollmer, W. (2012). Bacterial outer membrane evolution via sporulation? *Nature Chemical Biology* **8**(1), 14-18.
99. Onyenwoke, R.U., J.A. Brill, K. Farahi & J. Wiegel (2004). Sporulation genes in members of the low G+C Gram-type-positive phylogenetic branch (Firmicutes). *Archives of Microbiology* **182**(2-3), 182-192.
100. Sapp, J. (2012). Horizontal gene transfer and the tree of life. In: *Genesis - In the Beginning: Precursors of Life, Chemical Models and Early Biological Evolution*. Ed.: J. Seckbach. Dordrecht, Springer: 743-755.

101. Syvanen, M. (2012). Evolutionary implications of horizontal gene transfer. *Annual Review of Genetics* **46**, 341-358.
102. Nelson, K.E., R.A. Clayton, S.R. Gill, M.L. Gwinn, R.J. Dodson, D.H. Haft, E.K. Hickey, J.D. Peterson, W.C. Nelson, K.A. Ketchum, L. McDonald, T.R. Utterback, J.A. Malek, K.D. Linher, M.M. Garrett, A.M. Stewart, M.D. Cotton, M.S. Pratt, C.A. Phillips, D. Richardson, J. Heidelberg, G.G. Sutton, R.D. Fleischmann, J.A. Eisen & C.M. Fraser (1999). Evidence for lateral gene transfer between Archaea and Bacteria from genome sequence of *Thermotoga maritima*. *Nature* **399**(6734), 323-329.
103. Reynaud, E.G. & D.P. Devos (2011). Transitional forms between the three domains of life and evolutionary implications. *Proceedings of the Royal Society B-Biological Sciences* **278**(1723), 3321-3328.
104. Blum, B. (2010). Thermal Bacteria, Yellowstone NP, WY. http://www.bandwmag.com/galleries/color/contests/7/categories/7/photographs/26064
105. Nakagawa, T. & M. Fukui (2002). Phylogenetic characterization of microbial mats and streamers from a Japanese alkaline hot spring with a thermal gradient. *Journal of General and Applied Microbiology* **48**(4), 211-222.
106. Darland, G. & T. Brock (1971). *Bacillus acidocaldarius* sp.nov., an acidophilic thermophilic spore-forming bacterium. *Journal of General Microbiology* **67**(July), 9-15.
107. Geiling, N. (2014). The Science Behind Yellowstone's Rainbow Hot Spring: The Grand Prismatic Spring might seem photoshopped, but there's very real science beneath its vibrant colors. http://www.smithsonianmag.com/travel/science-behind-yellowstones-rainbow-hot-spring-180950483/?utm_source=smithsoniantopic&utm_medium=email&utm_campaign=20140511-Weekender
108. Ackermann, M., L. Chao, C.T. Bergstrom & M. Doebeli (2007). On the evolutionary origin of aging. *Aging Cell* **6**(2), 235-244.
109. Laney, S.R., R.J. Olson & H.M. Sosik (2012). Diatoms favor their younger daughters. *Limnology and Oceanography* **57**(5), 1572-1578.
110. Koonin, E.V. & A.Y. Mulkidjanian (2013). Evolution of cell division: From shear mechanics to complex molecular machineries. *Cell* **152**(5), 942-944.
111. Mercier, R., Y. Kawai & J. Errington (2013). Excess membrane synthesis drives a primitive mode of cell proliferation. *Cell* **152**(5), 997-1007.
112. Sato, Y., K. Yasuhara, J.-i. Kikuchi & T.N. Sato (2013). Synthetic cell division system: Controlling equal vs. unequal divisions by design. *Scientific Reports* **3**, #3475.
113. Wagner, I.D. & J. Wiegel (2008). Diversity of thermophilic anaerobes. *Annals of the New York Academy of Sciences* **1125**, 1-43.
114. Li, Y.H., L. Mandelco & J. Wiegel (1993). Isolation and characterization of a moderately thermophilic anaerobic alkaliphile, *Clostridium paradoxum* sp. nov. *International Journal of Systematic Bacteriology* **43**(3), 450-460.
115. Luciano, J., R. Agrebi, A.V. Le Gall, M. Wartel, F. Fiegna, A. Ducret, C. Brochier-Armanet & T. Mignot (2011). Emergence and modular evolution of a novel motility machinery in bacteria. *PLoS Genetics* **7**(9), e1002268.
116. Shirt-Ediss, B., K. Ruiz-Mirazo, F. Mavelli & R.V. Solé (2014). Modelling lipid competition dynamics in heterogeneous protocell populations. *Scientific Reports* **4**, #5675.

117. Ochman, H., J.G. Lawrence & E.A. Groisman (2000). Lateral gene transfer and the nature of bacterial innovation. *Nature* **405**(6784), 299-304.
118. Davis, B.M. & M.K. Waldor (2013). Horizontal gene transfer: Linking sex and cell fate. *Current Biology* **23**(3), R118-R119.
119. Bowler, C., A.E. Allen, J.H. Badger, J. Grimwood, K. Jabbari, A. Kuo, U. Maheswari, C. Martens, F. Maumus, R.P. Otillar, E. Rayko, A. Salamov, K. Vandepoele, B. Beszteri, A. Gruber, M. Heijde, M. Katinka, T. Mock, K. Valentin, F. Verret, J.A. Berges, C. Brownlee, J.P. Cadoret, A. Chiovitti, C.J. Choi, S. Coesel, A. De Martino, J.C. Detter, C. Durkin, A. Falciatore, J. Fournet, M. Haruta, M.J. Huysman, B.D. Jenkins, K. Jiroutova, R.E. Jorgensen, Y. Joubert, A. Kaplan, N. Kröger, P.G. Kroth, J. La Roche, E. Lindquist, M. Lommer, V. Martin-Jezequel, P.J. Lopez, S. Lucas, M. Mangogna, K. McGinnis, L.K. Medlin, A. Montsant, M.P. Secq, C. Napoli, M. Obornik, M.S. Parker, J.L. Petit, B.M. Porcel, N. Poulsen, M. Robison, L. Rychlewski, T.A. Rynearson, J. Schmutz, H. Shapiro, M. Siaut, M. Stanley, M.R. Sussman, A.R. Taylor, A. Vardi, P. von Dassow, W. Vyverman, A. Willis, L.S. Wyrwicz, D.S. Rokhsar, J. Weissenbach, E.V. Armbrust, B.R. Green, Y. Van de Peer & I.V. Grigoriev (2008). The *Phaeodactylum* genome reveals the evolutionary history of diatom genomes. *Nature* **456**(7219), 239-244.
120. Schonknecht, G., W.H. Chen, C.M. Ternes, G.G. Barbier, R.P. Shrestha, M. Stanke, A. Brautigam, B.J. Baker, J.F. Banfield, R.M. Garavito, K. Carr, C. Wilkerson, S.A. Rensing, D. Gagneul, N.E. Dickenson, C. Oesterhelt, M.J. Lercher & A.P. Weber (2013). Gene transfer from bacteria and archaea facilitated evolution of an extremophilic eukaryote. *Science* **339**(6124), 1207-1210.
121. Villarreal, L.P. & G. Witzany (2010). Viruses are essential agents within the roots and stem of the tree of life. *Journal of Theoretical Biology* **262**(4), 698-710.
122. Vickers, S.E. & D.I. Williamson (2011). Interspecies hybrids. In: *Chimeras and Consciousness: Evolution of the Sensory Self.* Eds.: L. Margulis, C.A. Asikainen & W.E. Krumbein. Cambridge, MIT Press: 183-197.
123. Cann, A.J. (2007). Bacterial conjugation. http://www.flickr.com/photos/ajc1/1103490291/
124. Turner, J.S. (2007). *The Tinkerer's Accomplice: How Design Emerges from Life Itself.* Cambridge, Harvard University Press.
125. Wikipedia (2015). Protein domain. https://en.wikipedia.org/wiki/Protein_domain
126. Monge-Nájera, J. & X.G. Hou (2000). Disparity, decimation and the Cambrian "explosion": comparison of early Cambrian and Present faunal communities with emphasis on velvet worms (Onychophora). *Revista de Biología Tropical* **48**(2-3), 333-351.
127. Wikipedia (2012). Phylum. http://en.wikipedia.org/wiki/Phylum
128. Margulis, L. & M.J. Chapman (2009). *Kingdoms & Domains: An Illustrated Guide to the Phyla of Life on Earth.* Amsterdam, Academic Press/Elsevier, 4th.
129. Valentine, J.W. (2004). *On the Origin of Phyla.* Chicago, University of Chicago Press.
130. Husby, C.E. (2007). Colony of *Equisetum giganteum.* http://w3.fiu.edu/husby/

131. Clapham, M.E. & J.A. Karr (2012). Environmental and biotic controls on the evolutionary history of insect body size. *Proceedings of the National Academy of Sciences USA* **109**(27), 10927-10930.
132. Wikipedia (2015). Mammal. https://en.wikipedia.org/wiki/Mammal
133. Carr, K. (2011). *Hallucigenia*. http://www.karencarr.com
134. Smith, M.R. & J. Ortega-Hernández (2014). *Hallucigenia*'s onychophoran-like claws and the case for Tactopoda. *Nature*, doi:10.1038/nature13576.
135. Witton, M. (2010). Dinosaurs 'came to rule world after mass extinction'. http://www.telegraph.co.uk/science/science-news/7499113/Dinosaurs-came-to-rule-world-after-mass-extinction.html
136. Kendall, B., C.T. Reinhard, T.W. Lyons, A.J. Kaufman, S.W. Poulton & A.D. Anbar (2010). Pervasive oxygenation along late Archaean ocean margins. *Nature Geoscience* **3**(9), 647-652.
137. Farquhar, J., A.L. Zerkle & A. Bekker (2011). Geological constraints on the origin of oxygenic photosynthesis. *Photosynthesis Research* **107**(1), 11-36.
138. Crowe, S.A., L.N. Døssing, N.J. Beukes, M. Bau, S.J. Kruger, R. Frei & D.E. Canfield (2013). Atmospheric oxygenation three billion years ago. *Nature* **501**(7468), 535-538.
139. Lyons, T.W. & C.T. Reinhard (2011). Earth science sea change for the rise of oxygen. *Nature* **478**(7368), 194-195.
140. Lowry, B., C. Hebant & D. Lee (1980). The origin of land plants: A new look at an old problem. *Taxon* **29**(2-3), 183-197.
141. Scott, A.C. (1984). The early history of life on land. *Journal of Biological Education* **18**(3), 207-219.
142. Wolf, E.T. & O.B. Toon (2010). Fractal organic hazes provided an ultraviolet shield for early Earth. *Science* **328**(5983), 1266-1268.
143. Cockell, C.S. & J.A. Raven (2007). Ozone and life on the Archaean earth. *Philosophical Transactions of the Royal Society A-Mathematical Physical and Engineering Sciences* **365**(1856), 1889-1901.
144. Lyons, T.W., C.T. Reinhard & N.J. Planavsky (2014). The rise of oxygen in Earth's early ocean and atmosphere. *Nature* **506**(7488), 307-315.
145. Bottke, W.F. (2015). Early solar system bombardment and Earth's habitability. In: *Astrobiology Science Conference 2015, Habitability, Habitable Worlds, and Life, June 15–19, 2015, Chicago, Illinois*. Ed.: P. Doran. Houston, Lunar and Planetary Institute: #7700.
146. Gordon, R. & B.J. Poulin (2012). Quitting cold turkey: rapid oil independence for the USA. In: *The Science of Algal Fuels: Phycology, Geology, Biophotonics, Genomics and Nanotechnology*. Eds.: R. Gordon & J. Seckbach. Dordrecht, Springer: 3-20.
147. Holland, H.D. (2011). Estimated evolution of atmospheric P_{O2}. http://en.wikipedia.org/wiki/File:Oxygenation-atm-2.svg
148. Ratcliff, W.C., M.D. Herron, K. Howell, J.T. Pentz, F. Rosenzweig & M. Travisano (2013). Experimental evolution of an alternating uni- and multicellular life cycle in *Chlamydomonas reinhardtii*. *Nature Communications* **4**, #2742.
149. Alicea, B. & R. Gordon (2014). Ratchets, Constructions, Games, and Borg in the Reading Queue. http://syntheticdaisies.blogspot.com/2014/11/ratchets-constructions-games-and-borg.html?showComment=1416817271771 - c5447421839988830006

150. Libby, E. & W.C. Ratcliff (2014). Ratcheting the evolution of multicellularity. *Science* **346**(6208), 426-427.

151. Vorob'eva, N.G., V.N. Sergeev & A.H. Knoll (2009). Neoproterozoic microfossils from the margin of the East European Platform and the search for a biostratigraphic model of lower Ediacaran rocks. *Precambrian Research* **173**(1-4), 163-169.

152. Kaplan-Levy, R.N., O. Hadas, M.L. Summers, J. Ruecker & A. Sukenik (2010). Akinetes: dormant cells of cyanobacteria. *Topics in Current Genetics* **21**, 5-27.

153. Tomitani, A., A.H. Knoll, C.M. Cavanaugh & T. Ohno (2006). The evolutionary diversification of cyanobacteria: molecular–phylogenetic and paleontological perspectives. *Proceedings of the National Academy of Sciences USA* **103**(14), 5442-5447.

154. Schirrmeister, B.E., A. Antonelli & H.C. Bagheri (2011). The origin of multicellularity in cyanobacteria. *BMC Evolutionary Biology* **11**, #45.

155. Schirrmeister, B.E., J.M. de Vos, A. Antonelli & H.C. Bagheri (2013). Evolution of multicellularity coincided with increased diversification of cyanobacteria and the Great Oxidation Event. *Proceedings of the National Academy of Sciences USA* **110**(5), 1791-1796.

156. Burks, R. (1998). *Anabaena scheremetievi* - 400x phase contrast. http://www-cyanosite.bio.purdue.edu/images/lgimages/SCHERE1

157. Schneegurt, M.A. (2015). Cyanosite: For Research on Cyanobacteria. http://www-cyanosite.bio.purdue.edu/

158. Bergman, B., W.W. Zheng, J. Klint & L. Ran (2008). On the origin of plants and relations to contemporary cyanobacterial-plant symbioses. *Plant Biotechnology* **25**(3), 213-220.

159. Bothe, H., H.J. Tripp & J.P. Zehr (2010). Unicellular cyanobacteria with a new mode of life: the lack of photosynthetic oxygen evolution allows nitrogen fixation to proceed. *Archives of Microbiology* **192**(10), 783-790.

160. Stal, L.J. & J.R. Zehr (2008). Cyanobacterial nitrogen fixation in the ocean: Diversity, regulation, and ecology. In: *Cyanobacteria: Molecular Biology, Genomics and Evolution*. Eds.: A. Herrero & E. Flores. Wymondham, UK, Caister Academic Press: 423-446, A414, A415.

161. Rodrigues, J.F.M., D.J. Rankin, V. Rossetti, A. Wagner & H.C. Bagheri (2012). Differences in cell division rates drive the evolution of terminal differentiation in microbes. *PLoS Computational Biology* **8**(4), e1002468.

162. Wolk, C.P. (1965). Heterocyst germination under defined conditions. *Nature* **205**, 201-202.

163. Subashchandrabose, S.R., B. Ramakrishnan, M. Megharaj, K. Venkateswarlu & R. Naidu (2011). Consortia of cyanobacteria/microalgae and bacteria: Biotechnological potential. *Biotechnology Advances* **29**(6), 896-907.

164. Walsby, A.E. (2007). Cyanobacterial heterocysts: terminal pores proposed as sites of gas exchange. *Trends in Microbiology* **15**(8), 340-349.

165. Meeks, J.C. & J. Elhai (2002). Regulation of cellular differentiation in filamentous cyanobacteria in free-living and plant-associated symbiotic growth states. *Microbiology and Molecular Biology Reviews* **66**(1), 94-121.

166. Gerdtzen, Z.P., J.C. Salgado, A. Osses, J.A. Asenjo, I. Rapaport & B.A. Andrews (2009). Modeling heterocyst pattern formation in cyanobacteria. *BMC Bioinformatics* **10**(Suppl. 6), S16.

167. Baker, R.W. & G.T. Herman (1970). CELIA - a cellular linear iterative array simulator. In: *Proceedings of the Fourth Annual Conference on Applications of Simulation*. Eds.: P.J. Kiviat & M. Araten, Winter Simulation Conference: 64-73

168. Gordon, R. & A.E. Walsby (1970). Development of the heterocyst spacing problem in *Anabaena*. In: *Serbelloni Symposium on Theoretical Biology, June, 1970*.

169. Brown, A.I. & A.D. Rutenberg (2012). Reconciling cyanobacterial fixed-nitrogen distributions and transport experiments with quantitative modelling. *Physical Biology* **9**(1), #016007.

170. Allard, J.F., A.L. Hill & A.D. Rutenberg (2007). Heterocyst patterns without patterning proteins in cyanobacterial filaments. *Developmental Biology* **312**(1), 427-434.

171. Komárek, J. & K.K. Anagnostidis (1989). Modern approach to the classification system of Cyanophytes 4 - Nostocales. *Archiv für Hydrobiologie* **82**(3), 247-345.

172. Tsukii, Y. (2005). Cyanophyceae: Stigonematales: Stigonemataceae: *Hapalosiphon* sp., http://protist.i.hosei.ac.jp/pdb/images/Prokaryotes/Stigonemataceae/sp_08.html

173. Adams, D.G. (2000). Heterocyst formation in cyanobacteria. *Current Opinion in Microbiology* **3**(6), 618-624.

174. Adams, D.G. (1992). Multicellularity in cyanobacteria. *Symposia of the Society for General Microbiology* **47**, 341-384.

175. Zimmer, C. (2009). On the origin of eukaryotes. *Science* **325**(5941), 666-668.

176. Spang, A., J.H. Saw, S.L. Jorgensen, K. Zaremba-Niedzwiedzka, J. Martijn, A.E. Lind, R. van Eijk, C. Schleper, L. Guy & T.J.G. Ettema (2015). Complex archaea that bridge the gap between prokaryotes and eukaryotes. *Nature* **521**(7551), 173-179.

177. Margulis, L. (1993). *Symbiosis in Cell Evolution: Microbial Communities in the Archean and Proterozoic Eons*. New York, W.H. Freeman and Co., 2nd.

178. Cavalier-Smith, T. (2006). Origin of mitochondria by intracellular enslavement of a photosynthetic purple bacterium. *Proceedings of the Royal Society B-Biological Sciences* **273**(1596), 1943-1952.

179. Ueda, M. & K.I. Kadowaki (2012). Gene content and gene transfer from mitochondria to the nucleus during evolution. *Advances in Botanical Research* **63**, 21-40.

180. Gordon, R. (1999). *The Hierarchical Genome and Differentiation Waves: Novel Unification of Development, Genetics and Evolution*. Singapore & London, World Scientific & Imperial College Press.

181. Howe, C.J., A.C. Barbrook, R.E.R. Nisbet, P.J. Lockhart & A.W.D. Larkum (2008). The origin of plastids. *Philosophical Transactions of the Royal Society B-Biological Sciences* **363**(1504), 2675-2685.

182. Deusch, O., G. Landan, M. Roettger, N. Gruenheit, K.V. Kowallik, J.F. Allen, W. Martin & T. Dagan (2008). Genes of cyanobacterial origin in plant nuclear genomes point to a heterocyst-forming plastid ancestor. *Molecular Biology and Evolution* **25**(4), 748-761.

183. Shinozawa, T., T. Horiike & K. Hamada (2002). Nucleus symbiosis hypothesis: formation of eukaryotic cell nuclei by the symbiosis of Archaea in bacteria. In:

Symbiosis: Mechanisms and Model Systems. Ed.: J. Seckbach. Dordrecht, Kluwer Academic Publishers: 231-235.

184. Shapiro, J.A. & M. Dworkin, Eds. (1997). *Bacteria as Multicellular Organisms*. New York, Oxford University Press.

185. Meeks, J.C., E.L. Campbell, M.L. Summers & F.C. Wong (2002). Cellular differentiation in the cyanobacterium *Nostoc punctiforme*. *Archives of Microbiology* **178**(6), 395-403.

186. Ptacin, J.L., S.F. Lee, E.C. Garner, E. Toro, M. Eckart, L.R. Comolli, W. Moerner & L. Shapiro (2010). A spindle-like apparatus guides bacterial chromosome segregation. *Nature Cell Biology* **12**(8), 791-798.

187. Bard, J.B.L., R.A. Baldock & D.R. Davidson (1998). Elucidating the genetic networks of development: a bioinformatics approach. *Genome Research* **8**(9), 859-863.

188. Llosa, M., F.X. Gomis-Rüth, M. Coll & F. de la Cruz (2002). Bacterial conjugation: a two-step mechanism for DNA transport. *Molecular Microbiology* **45**(1), 1-8.

189. Velicer, G.J., J. Berger & S. Kadam (2010). *Myxococcus xanthus* rods and spores. http://newsinfo.iu.edu/web/page/normal/assets/9155.html

190. Bergmann, D.C., J.R. Crew, J.M. Kramer & W.B. Wood (1998). Cuticle chirality and body handedness in *Caenorhabditis elegans*. *Developmental Genetics* **23**(3), 164-174.

191. Licari, G.R., P.E. Cloud & W.D. Smith (1969). A new chroococcacean alga from the Proterozoic of Queensland. *Proceedings of the National Academy of Sciences USA* **62**(1), 56-62.

192. Sinnott, E.W. (1960). *Plant Morphogenesis*. New York, McGraw-Hill Book Co.

193. Castanon, I. & M. Gonzalez-Gaitan (2011). Oriented cell division in vertebrate embryogenesis. *Current Opinion in Cell Biology* **23**(6), 697-704.

194. Walsby, A.E. (2005). Archaea with square cells. *Trends in Microbiology* **13**(5), 193-195.

195. Margolin, W. (2009). Sculpting the bacterial cell. *Current Biology* **19**(17), R812-822.

196. Kernow Microscopical Society (2005). *Anabæna spiralis* showing a large heterocyst. http://kernowmicro.org.uk/mediac/400_0/media/Ana

197. McBride, M.J. (2001). Bacterial gliding motility: multiple mechanisms for cell movement over surfaces. *Annual Review of Microbiology* **55**, 49-75.

198. Gordon, R. & R.W. Drum (1970). A capillarity mechanism for diatom gliding locomotion. *Proceedings of the National Academy of Sciences USA* **67**(1), 338-344.

199. Drum, R.W., R. Gordon, R. Bender & N.S. Goel (1971). On weakly coupled diatomic oscillators: *Bacillaria*'s paradox resolved. *Journal of Phycology* **7**(Suppl.), 13-14.

200. Gordon, R. (1987). A retaliatory role for algal projectiles, with implications for the mechanochemistry of diatom gliding motility. *Journal of Theoretical Biology* **126**, 419-436.

201. Kapinga, M.R.M. & R. Gordon (1992). Cell attachment in the motile colonial diatom *Bacillaria paxillifer*. *Diatom Research* **7**(2), 215-220.

202. Kapinga, M.R.M. & R. Gordon (1992). Cell motility rhythms in *Bacillaria paxillifer*. *Diatom Research* **7**(2), 221-225.

203. Ussing, A.P., R. Gordon, L. Ector, K. Buczkó, A.G. Desnitskiy & S.L. VanLandingham (2005). *The Colonial Diatom "Bacillaria paradoxa": Chaotic Gliding Motility, Lindenmeyer Model of Colonial Morphogenesis, and Bibliography, with Translation of O.F. Müller (1783), "About a peculiar being in the beach-water"*. Ruggell, A.R.G. Ganter Verlag Kommanditgesellschaft. **5**.
204. Gordon, R. (2016). Partial synchronization of the colonial diatom *Bacillaria "paradoxa"*. *Research Ideas and Outcomes (RIO) 2*, e7869.
205. Nan, B. & D.R. Zusman (2011). Uncovering the mystery of gliding motility in the myxobacteria. *Annual Review Genetics* **45**, 21-39.
206. Igoshin, O.A., J. Neu & G. Oster (2004). Developmental waves in myxobacteria: A distinctive pattern formation mechanism. *Physical Review E* **70**(4), #041911.
207. Drum, R.W. & R. Gordon (2003). Star Trek replicators and diatom nanotechnology. *Trends in Biotechnology* **21**(8), 325-328.
208. Choudhri, S. (1984). Electron microscopy of cell-cell contacts between migrating yolk sac melanocytes of the zebrafish, *Brachydanio rerio* [Supervisor: Richard Gordon]. In: *UM Program and Proceedings, B.Sc. (Med) Presentations*. Winnipeg, Faculty of Medicine, University of Manitoba: 71-78.
209. Cox, B. (2010). A multi-scale, discrete-cell simulation of organogenesis: Application to the effects of strain stimulus on collective cell behavior during ameloblast migration. *Journal of Theoretical Biology* **262**(1), 58-72.
210. Baker, R.E. & M.J. Simpson (2012). Models of collective cell motion for cell populations with different aspect ratio: Diffusion, proliferation and travelling waves. *Physica A-Statistical Mechanics and Its Applications* **391**(14), 3729-3750.
211. Markham, D.C., R.E. Baker & P.K. Maini (2014). Modelling collective cell behaviour. *Discrete and Continuous Dynamical Systems* **34**(12), 5123-5133.
212. Zaritsky, A., D. Kaplan, I. Hecht, S. Natan, L. Wolf, N.S. Gov, E. Ben-Jacob & I. Tsarfaty (2014). Propagating waves of directionality and coordination orchestrate collective cell migration. *PLoS Computational Biology* **10**(7), e1003747.
213. Read, N., S. Connell & D.G. Adams (2007). Nanoscale visualization of a fibrillar array in the cell wall of filamentous cyanobacteria and its implications for gliding motility. *Journal of Bacteriology* **189**(20), 7361-7366.
214. Koiller, J., K.M. Ehlers & F. Chalub (2010). Acoustic streaming, the "small invention" of cyanobacteria? *Arbor-Ciencia Pensamiento y Cultura* **186**(746), 1089-1115.
215. Abreu, F., J. Lopes Martins, T. Souza Silveira, C. Neumann Keim, H.G.P. Lins De Barros, F.J. Gueiros & U. Lins (2007). '*Candidatus* Magnetoglobus multicellularis', a multicellular, magnetotactic prokaryote from a hypersallne environment. *International Journal of Systematic and Evolutionary Microbiology* **57**, 1318-1322.
216. Zhou, K., W.Y. Zhang, Y.Z. Kui, H.M. Pan, S.D. Zhang, W.J. Zhang, H.D. Yue, Y. Li, T. Xiao & L.F. Wu (2012). A novel genus of multicellular magnetotactic prokaryotes from the Yellow Sea. *Environmental Microbiology* **14**(2), 405-413.
217. Wikimedia Commons (2011). File:Small magnetosome.jpg. https://commons.wikimedia.org/wiki/File:Small_magnetosome.jpg

218. Palińska, K.A. & W.E. Krumbein (1998). Patterns of growth in coccoid, aggregate forming cyanobacteria. *Annales Botanici Fennici* **35**(3), 219-227.
219. Guiry, M.D., G.M. Guiry & C.F. Carter (2012). AlgaeBase. World-wide electronic publication, National University of Ireland, Galway: *Eucapsis alpina* Clements & Schantz. http://www.algaebase.org/search/species/detail/?species_id=ba27e292474b1c306&sk=0&from=results
220. Oyadomari, J. (2010). Keweenaw Algae, Chlorophycean Gallery #2, *Dispora*, Perrault Bog, floating algal mass, 6/17/10, pH 5.0, 23°C, 41 uS http://www.keweenawalgae.mtu.edu/gallery_pages/chlorophyceans2.htm
221. Walsby, A.E. (1980). A square bacterium. *Nature* **283**, 69-71.
222. Burns, D.G., H.M. Camakaris, P.H. Janssen & M.L. Dyall-Smith (2004). Cultivation of Walsby's square haloarchaeon. *FEMS Microbiology Letters* **238**(2), 469-473.
223. Sterrenburg, F.A.S., R. Gordon, M.A. Tiffany & S.S. Nagy (2007). Diatoms: living in a constructal environment. In: *Algae and Cyanobacteria in Extreme Environments. Series: Cellular Origin, Life in Extreme Habitats and Astrobiology, Vol. 11*. Ed.: J. Seckbach. Dordrecht, The Netherlands, Springer: 141-172.
224. Gschmeissner, S. (2013). Diatom, SEM. http://fineartamerica.com/featured/26-diatom-sem-steve-gschmeissner.html
225. Wikipedia (2015). Otto Friedrich Müller. https://en.wikipedia.org/wiki/Otto_Friedrich_M%C3%BCller
226. Hustedt, F. (1985). *The Pennate Diatoms*. Koenigstein, Koeltz Scientific Books.
227. Acosta-Avalos, D., L.M. dos Santos Azevedo, T.S. Andrade & H.L. de Barros (2012). Magnetic configuration model for the multicellular magnetotactic prokaryote *Candidatus* Magnetoglobus multicellularis. *European Biophysics Journal with Biophysics Letters* **41**(5), 405-413.
228. Ueki, N., S. Matsunaga, I. Inouye & A. Hallmann (2010). How 5000 independent rowers coordinate their strokes in order to row into the sunlight: phototaxis in the multicellular green alga *Volvox*. *BMC Biology* **8**, #103.
229. Niedermayer, T., B. Eckhardt & P. Lenz (2008). Synchronization, phase locking, and metachronal wave formation in ciliary chains. *Chaos* **18**(3), #037128.
230. Goldstone, R.J., R. Popat, M.P. Fletcher, S.A. Crusz & S.P. Diggle (2012). Quorum sensing and social interactions in microbial biofilms. In: *Microbial Biofilms: Current Research and Applications*. Eds.: G. Lear & G.D. Lewis, Caister Academic Press: 1-24.
231. Lim, C.S., R. Rosli, H.F. Seow & P.P. Chong (2012). *Candida* and invasive candidiasis: back to basics. *European Journal of Clinical Microbiology & Infectious Diseases* **31**(1), 21-31.
232. Gomer, R.H., W. Jang & D. Brazill (2011). Cell density sensing and size determination. *Development, Growth & Differentiation* **53**(4), 482-494.
233. Armitage, J.P., C.J. Dorman, K. Hellingwerf, R. Schmitt, D. Summers & B. Holland (2003). Thinking and decision making, bacterial style: Bacterial Neural Networks, Obernai, France, 7th-12th June 2002. *Molecular Microbiology* **47**(2), 583-593.

234. Ben-Jacob, E., I. Becker, Y. Shapira & H. Levine (2004). Bacterial linguistic communication and social intelligence. *Trends in Microbiology* **12**(8), 366-372.
235. El-Naggar, M.Y. & S.E. Finkel (2013). Live wires: Discoveries of microbial communities that transfer electrons between cells and across relatively long distances are launching a new field of microbiology. http://www.the-scientist.com//?articles.view/articleNo/35299/title/Live-Wires/
236. Van Gerven, N., G. Waksman & H. Remaut (2011). Pili and flagella: Biology, structure, and biotechnological applications. *Progress in Molecular Biology and Translational Science* **103**, 21-72.
237. Malvankar, N.S., S.E. Yalcin, M.T. Tuominen & D.R. Lovley (2014). Visualization of charge propagation along individual pili proteins using ambient electrostatic force microscopy. *Nature Nanotechnology*, doi: 10.1038/nnano.2014.1236.
238. Sheath, R., P. Kociolek, B. Fetscher & R. Stancheva (2012). Soft-Bodied Stream Algae of California: *Dichothrix gypsophila* (Kütz.) Bornet et Flahault. http://dbmuseblade.colorado.edu/DiatomTwo/sbsac_site/species.php?g=Dichothrix&s=gypsophila
239. Wu, Y.L., Y. Jiang, A.D. Kaiser & M. Alber (2011). Self-organization in bacterial swarming: lessons from myxobacteria. *Physical Biology* **8**(5), #055003.
240. Redner, G.S., A. Baskaran & M.F. Hagan (2013). Reentrant phase behavior in active colloids with attraction. *Physical Review E* **88**(1), #012305.
241. Gitai, Z., N. Dye & L. Shapiro (2004). An actin-like gene can determine cell polarity in bacteria. *Proceedings of the National Academy of Sciences USA* **101**(23), 8643-8648.
242. Alber, M. (2005). Seminar on Interdisciplinary Biological Research: Mathematical and Computational Modeling in Biology. http://www.nd.edu/~malber/sem_un/sem_und.html
243. Cohen, I., I. Golding, I.G. Ron & E. Ben-Jacob (2001). Biofluiddynamics of lubricating bacteria. *Mathematical Methods in the Applied Sciences* **24**(17-18), 1429-1468.
244. Wassenaar, T.M., K. Farkas & E. Ben-Jacob (2011). *Bacteria: The Benign, the Bad, and the Beautiful*, John Wiley & Sons.
245. Lukeman, R., Y.X. Li & L. Edelstein-Keshet (2009). A conceptual model for milling formations in biological aggregates. *Bulletin of Mathematical Biology* **71**(2), 352-382.
246. Schwedock, J., J.R. McCormick, E.R. Angert, J.R. Nodwell & R. Losick (1997). Assembly of the cell division protein FtsZ into ladder-like structures in the aerial hyphae of *Streptomyces coelicolor*. *Molecular Microbiology* **25**(5), 847-858.
247. Mazumdar, A. & M. Mazumdar (2002). How one becomes many: blastoderm cellularization in *Drosophila melanogaster*. *BioEssays* **24**(11), 1012-1022.
248. Fuerst, J.A. & E. Sagulenko (2012). Keys to eukaryality: planctomycetes and ancestral evolution of cellular complexity. *Frontiers in Microbiology* **3**, #167.
249. Butler, S.M. & A. Camilli (2005). Going against the grain: chemotaxis and infection in *Vibrio cholerae*. *Nature Reviews Microbiology* **3**(8), 611-620.
250. Velicer, G.J., J. Berger & S. Kadam (2010). An sRNA controls a bacterium's social life. http://newsinfo.iu.edu/news-archive/14412.html

251. Fuerst, J.A. & E. Sagulenko (2011). Beyond the bacterium: planctomycetes challenge our concepts of microbial structure and function. *Nature Reviews Microbiology* **9**(6), 403-413.
252. Martel, S., M. Mohammadi, O. Felfoul, Z. Lu & P. Pouponneau (2009). Flagellated magnetotactic bacteria as controlled MRI-trackable propulsion and steering systems for medical nanorobots operating in the human microvasculature. *International Journal of Robotics Research* **28**(4), 571-582.
253. Eder, S.H.K., H. Cadiou, A. Muhamad, P.A. McNaughton, J.L. Kirschvink & M. Winklhofer (2012). Magnetic characterization of isolated candidate vertebrate magnetoreceptor cells. *Proceedings of the National Academy of Sciences USA* **109**(30), 12022-12027.
254. Greenberg, M., K. Canter, I. Mahler & A. Tornheim (2005). Observation of magnetoreceptive behavior in a multicellular magnetotactic prokaryote in higher than geomagnetic fields. *Biophysical Journal* **88**(2), 1496-1499.
255. Lefevre, C.T., F. Abreu, U. Lins & D.A. Bazylinski (2010). Nonmagnetotactic multicellular prokaryotes from low-saline, nonmarine aquatic environments and their unusual negative phototactic behavior. *Applied and Environmental Microbiology* **76**(10), 3220-3227.
256. Keim, C.N., J.L. Martins, F. Abreu, A.S. Rosado, H.L. de Barros, R. Borojevic, U. Lins & M. Farina (2004). Multicellular life cycle of magnetotactic prokaryotes. *FEMS Microbiology Letters* **240**(2), 203-208.
257. Gordon, R. (2001). Making waves: the paradigms of developmental biology and their impact on artificial life and embryonics. *Cybernetics & Systems* **32**(3-4), 443-458.
258. Srivastava, M., E. Begovic, J. Chapman, N.H. Putnam, U. Hellsten, T. Kawashima, A. Kuo, T. Mitros, A. Salamov, M.L. Carpenter, A.Y. Signorovitch, M.A. Moreno, K. Kamm, J. Grimwood, J. Schmutz, H. Shapiro, I.V. Grigoriev, L.W. Buss, B. Schierwater, S.L. Dellaporta & D.S. Rokhsar (2008). The *Trichoplax* genome and the nature of placozoans. *Nature* **454**(7207), 955-960.
259. Gallardo, V.A. & C. Espinoza (2007). Large multicellular filamentous bacteria under the oxygen minimum zone of the eastern South Pacific: a forgotten biosphere. *Proceedings of the SPIE* **6694**, #6941H.
260. Slayman, C. & H. Felle (1980). Direct measurement of membrane potential and resistance in giant bacteria using coated microelectrodes. *Federation Proceedings* **39**(6), 2130.
261. Loose, M., E. Fischer-Friedrich, J. Ries, K. Kruse & P. Schwille (2008). Spatial regulators for bacterial cell division self-organize into surface waves in vitro. *Science* **320**(5877), 789-792.
262. Ivanov, V. & K. Mizuuchi (2010). Multiple modes of interconverting dynamic pattern formation by bacterial cell division proteins. *Proceedings of the National Academy of Sciences USA* **107**(18), 8071-8078.
263. Møller-Jensen, J. & J. Löwe (2005). Increasing complexity of the bacterial cytoskeleton. *Current Opinion in Cell Biology* **17**(1), 75-81.
264. Müller, F.D., C.W. Schink, E. Hoiczyk, E. Cserti & P.I. Higgs (2012). Spore formation in *Myxococcus xanthus* is tied to cytoskeleton functions and polysaccharide spore coat deposition. *Molecular Microbiology* **83**(3), 486-505.

265. Cameron, J.A. & J.F. Fallon (1977). The absence of cell death during development of free digits in amphibians. *Developmental Biology* **55**(2), 331-338.
266. Miguélez, E.M., C. Hardisson & M.B. Manzanal (1999). Hyphal death during colony development in Streptomyces antibioticus: morphological evidence for the existence of a process of cell deletion in a multicellular prokaryote. *Journal of Cell Biology* **145**(3), 515-525.
267. Nariya, H. & M. Inouye (2008). MazF, an mRNA interferase, mediates programmed cell death during multicellular *Myxococcus* development. *Cell* **132**(1), 55-66.
268. Coelho, M., A. Dereli, A. Haese, S. Kühn, L. Malinovska, M.E. DeSantis, J. Shorter, S. Alberti, T. Gross & I.M. Tolić-Nørrelykke (2013). Fission yeast does not age under favorable conditions, but does so after stress. *Current Biology* **23**(19), 1844-1852.
269. Arnoldi, M., M. Fritz, E. Bäuerlein, M. Radmacher, E. Sackmann & A. Boulbitch (2000). Bacterial turgor pressure can be measured by atomic force microscopy. *Physical Review. E, Statistical Physics, Plasmas, Fluids, and Related Interdisciplinary Topics* **62**(1 Pt B), 1034-1044.
270. Beloussov, L.V. (1998). *The Dynamic Architecture of a Developing Organism: An Interdisciplinary Approach to the Development of Organisms*. Dordrecht, Kluwer Academic Publishers.
271. Guo, H. (2008). Vacuole-like structures in prokaryote - A personal account. In: *Plant Cell Compartments - Selected*. Ed.: B. Schoefs, Research Signpost: 385-391.
272. Obroucheva, N.V. (2008). Cell elongation as an inseparable component of growth in terrestrial plants. *Russian Journal of Developmental Biology* **39**(1), 13-24.
273. Oda, Y., T. Higaki, S. Hasezawa & N. Kutsuna (2009). New insights into plant vacuolar structure and dynamics. *International Review of Cell and Molecular Biology* **277**, 103-135.
274. Flemming, H.-C. (2011). The perfect slime. *Colloids and Surfaces B-Biointerfaces* **86**(2), 251-259.
275. Kato, S. & K. Watanabe (2010). Ecological and evolutionary interactions in syntrophic methanogenic consortia. *Microbes and Environments* **25**(3), 145-151.
276. Rawlings, D.E. (2005). Characteristics and adaptability of iron- and sulfur-oxidizing microorganisms used for the recovery of metals from minerals and their concentrates. *Microbial Cell Factories* **4**, #13.
277. Toro, E. & L. Shapiro (2010). Bacterial chromosome organization and segregation. *Cold Spring Harbor Perspectives in Biology* **2**(2), a000349.
278. Albrecht-Buehler, G. (1977). Daughter 3T3 cells. Are they mirror images of each other? *J. Cell Biol.* **72**(3), 595-603.
279. Jerka-Dziadosz, M. (1983). The origin of mirror-image symmetry doublet cells in the hypotrich ciliate *Paraurostyla weissei. Wilhelm Roux's Archives of Developmental Biology* **192**(3-4), 179-188.
280. Tawk, M., C. Araya, D.A. Lyons, A.M. Reugels, G.C. Girdler, P.R. Bayley, D.R. Hyde, M. Tada & J.D.W. Clarke (2007). A mirror-symmetric cell division that orchestrates neuroepithelial morphogenesis. *Nature* **446**(7137), 797-800.

281. Lord, C.E. & A.H. Gunawardena (2012). The lace plant: a novel model system to study plant proteases during developmental programmed cell death in vivo. *Physiologia Plantarum* **145**(1), 114-120.
282. Fuchs, Y. & H. Steller (2011). Programmed cell death in animal development and disease. *Cell* **147**(4), 742-758.
283. Wetzel, C.E. & L.U.C. Ector (2014). *Planothidium lagerheimii comb. nov.* (Bacillariophyta, Achnanthales) a forgotten diatom from South America. *Phytotaxa* **188**(5), 261-267.
284. Gordon, R. (2010). Diatoms and nanotechnology: early history and imagined future as seen through patents. In: *The Diatoms: Applications for the Environmental and Earth Sciences*. Eds.: J.P. Smol & E.F. Stoermer. Cambridge, Cambridge University Press. **2nd**: 585-602.
285. Jacobs, W.P. (1993). Rhizome gravitropism precedes gravimorphogenesis after inversion of the green algal coenocyte *Caulerpa prolifera* (Caulerpales). *American Journal of Botany* **80**(11), 1273-1275.
286. Fagerberg, W.R., E. Hodges & C.J. Dawes (2010). The development and potential roles of cell-wall trabeculae in *Caulerpa mexicana* (Chlorophyta). *Journal of Phycology* **46**(2), 309-315.
287. Ranjan, A., B.T. Townsley, Y. Ichihashi, N.R. Sinha & D.H. Chitwood (2015). An intracellular transcriptomic atlas of the giant coenocyte *Caulerpa taxifolia*. *PLoS Genetics* **11**(1), e1004900.
288. Alberti, S. (2015). Don't go with the cytoplasmic flow. *Developmental Cell* **34**(4), 381-382.
289. Pieuchot, L., J. Lai, R.A. Loh, F.Y. Leong, K.H. Chiam, J. Stajich & G. Jedd (2015). Cellular subcompartments through cytoplasmic streaming. *Developmental Cell* **34**(4), 410-420.
290. Sato, N., Y. Katsumata, K. Sato & N. Tajima (2014). Cellular dynamics drives the emergence of supracellular structure in the cyanobacterium, *Phormidium* sp KS. *Life-Basel* **4**(4, Sp. Iss. SI), 819-836.
291. Tamulonis, C. & J. Kaandorp (2014). A model of filamentous cyanobacteria leading to reticulate pattern formation. *Life-Basel* **4**(3, Sp. Iss. SI), 433-456.
292. Serôdio, J., S. Cruz, P. Cartaxana & R. Calado (2014). Photophysiology of kleptoplasts: photosynthetic use of light by chloroplasts living in animal cells. *Philosophical Transactions of the Royal Society B-Biological Sciences* **369**(1640), #20130242.
293. Gordon, R., N.D. Denkov, M.A. Tiffany, N.K. Gordon, M.M. Hanczyc & S.K. Smoukov (2016). A tensegrity origin of life? Protocells, flat archaea, desmids and diatoms as polygonally shaped droplets. *Advances in Colloid and Interface Science*, in preparation.

Chapter 2

Developmental Anatomy of the Axolotl

In order to understand how embryogenesis works, you must understand what embryogenesis is. This obvious statement evades many people who begin and end their studies with molecular developmental biology. It is as if understanding architecture required only a deep appreciation of bricks and mortar, i.e., in our case, molecules and their interactions. Learning the basic anatomy of the embryo is no longer required in many university-level biology curricula. This was partly the fault of the classical embryologists who often required students to regurgitate letter perfect details written in arcane vocabularies. Today, there is an almost exclusive emphasis on the latest new techniques in molecular biology. Biochemical tools are not seen as valuable adjuncts to classical embryology: they have replaced them. Finding detailed lists of genes and proteins and their presumed interactions for this book was easy. In contrast, we had difficulty finding good anatomical figures of developing embryos.

Embryogenesis encompasses anatomy versus time. This includes molecular events that may be causes. However, these events may also be consequences of other processes at size scales bigger than molecules. How much bigger? The ratio of the volume of the human body to the volume of a single amino acid is about[1] 5×10^{26} to 1. For comparison, the ratio of the mass of the Earth to the average mass of a human is "only" 10^{24} to 1, i.e., 500 times lower. How can someone expect to interpret the results of molecular biology without knowing the anatomy? It is simply not possible to do so.

In this chapter we will introduce the anatomy of an early developing axolotl embryo and explain why we like it as a model organism. We will also introduce important concepts of classical embryology that are critical to understanding the embryo. As you will see, it is the unique properties of the axolotl embryo that allowed us to accidentally stumble upon what we now think is the fundamental underlying mechanism of all embryogenesis. We will overlay and expand this basic anatomy and introduce embryological concepts as we progress in subsequent chapters. If you already know the basics of descriptive embryology you can safely skim ahead. If you don't, then you should pay close attention here. The detail is thick. One thing we will not do is hit you with all the heavy jargon that tends to make the average person doze off as they study descriptive embryology. The other thing to keep in mind is that most students of biology who had have the good fortune to be exposed to classical embryology will memorize this process in minutia, often without really understanding it, and then promptly forget it after the final exam. If you are in that category, we hope this chapter will enrich your understanding and refresh your memory and maybe, just maybe, make you wonder if you made a mistake when you ran away from embryology.

Figure 2.1. An axolotl with freshly laid eggs (2 mm diameter), by Susan Crawford-Young, with her kind permission. Note that its genus and species names are now *Ambystoma mexicanum*, but the genus was earlier called *Amblystoma*.

Figure 2.2. Dick with grandson Nick, delivering axolotls for the display at the Royal Tyrrell Museum, Drumheller, Alberta, Canada in 2010. For a movie of them see YouTube[2].

We begin by introducing our favorite model animal, a salamander[3], called the axolotl[4] (*Ambystoma mexicanum*) (Figure 2.1). The axolotl is a big salamander, typically 20 centimeters as an adult, although they can often be a lot larger (Figure 2.2). Most of us have watched water dwelling frog tadpoles grow up, lose their tails and gills, and become air breathing adults. Most salamanders do the same thing. The axolotl is a neotenic salamander. Neotenic means it never goes through this metamorphosis to become a land dweller. The axolotl remains in the larval state, complete with external gills, even as a breeding adult. This makes them convenient to use in a laboratory because they can be kept in tanks like tropical fish through their whole life cycle. Neoteny is a recent adaptation in evolutionary time for the axolotl[5], and with a little artificial hormonal prodding in the laboratory, the axolotl can still transform into a regular looking terrestrial salamander, although some do not survive forced metamorphosis[6]. The axolotl can also crossbreed and produce fertile offspring with its closest salamander relative, the tiger salamander (*Ambystoma tigrinum*) which does metamorphose naturally[7]. The axolotl is otherwise a typical salamander, despite its fully aquatic life cycle.

In the laboratory, the axolotl has been used as an experimental embryo model since 1865[8-10]. It is easy to raise[11,12] and get spawnings of 400 to 4000 eggs. The axolotl is native to the lakes that once existed where Mexico City is now[13]. Unfortunately, it is endangered in the wild[14] due to this habitat destruction. The regulations required for keeping and sharing endangered species have had the ironic effect of suppressing international cooperation between researchers[15], many of whom maintain substantial breeding colonies of laboratory raised animals. Such regulations affect even animals that are the product of hundreds of generations in laboratories. This is one reason why this once popular model organism has fallen from favor in embryology.

Another reason is that the axolotl is a relatively slow developer compared to other favorite organisms of embryologists. The axolotl takes two years of tending to go from an embryo to a fertile adult. Given the average grant cycle of 1-3 years of funding, this slow development makes the axolotl a major liability in the grant system. Funding comes one year at a time if you are lucky and you must have results to get more. At a time when virtually all embryologists have become a subspecies of molecular biologists, waiting two years to get offspring from a planned mutation or after the introduction of special marker genes is often a risk to one's career. Hard problems sometimes require a lifetime or more to solve[16]. The current system of doing science simply does not permit that kind of extended devotion.

Yet another reason axolotls have fallen out of favor as animal models is their huge genome. The axolotl genome has only 2% more genes than humans. That 2% more is spread amongst 64 billion nucleotide bases of double helix DNA[17], compared to 6.406 billion nucleotide bases for human females and 6.294 billion for human males[18], i.e., axolotls have about 10 times as much DNA per cell as we have. The extra DNA also represents a major barrier to molecular biology. That is a lot of DNA to work around.

Full sequencing of the axolotl's DNA is likely in the near future[19,20] but it is not yet amongst the many sequenced organisms[21,22]. Why they have so much more DNA without having lots more genes is a puzzle[5].

The huge size of the axolotl genome actually makes it much easier for investigating embryogenesis morphologically (by watching shape change) because in general the size of cells increases with the amount of DNA they contain[23]. The axolotls have more DNA and therefore have bigger cells. Bigger cells means easier microscopy, micromanipulation, and computer simulation of cell movements. An example of the size advantage is that neural plate cells, the cells that later form the brain and spinal cord, are 15 μm wide and 50 μm tall in axolotl embryos, whereas chick embryo neural plate cells at a comparable stage are about 8 μm wide and 20 μm tall[24], and human neural plate cells are about 5 μm wide and 20 μm tall (Figure 2.3). The ratio of volumes of axolotl to human neural plate cells is almost 600 to 1. As is so often the case, the amount of DNA doesn't correlate to complexity. In terms of complexity, an axolotl has pretty much the same tissues as we do. The axolotl has a relatively huge egg. The egg is 2 mm in diameter, or 2000 μm, compared to 70 μm for a human egg (Figure 2.4).

Axolotl embryos have variegated pigmentation. This means that the amount of surface pigment varies greatly from one cell to the next (Figure 2.5). Axolotl pigmentation granules are embedded in the egg cortex and are membrane bound[25], which makes them handy markers of the movements of the individual cells on an embryo's external surface. While much larger chicken eggs are often studied[26], the chick embryo itself develops as an initially tiny structure on the surface of the huge yolk, with small cells that are not pigmented. Human and chick early embryos have no pigmentation, and frog embryos have dark, uniform dorsal (top side) pigmentation. All the cells look the same making tracking any one cell difficult at best. With the axolotl embryo's large cells and variegated pigment it has proven easy to watch every surface cell with a modest microscope[27]. This is one reason that use of axolotls allowed us to discover differentiation waves. A secondary reason is that these waves involve contraction or expansion of the apical surfaces of the cells, which makes the pigment granules come closer together or move farther apart. With practice, one can actually see a contraction wave on the surface of an axolotl embryo because it shows up as a transient darker band of cells. In time-lapse movies it is quite apparent[27].

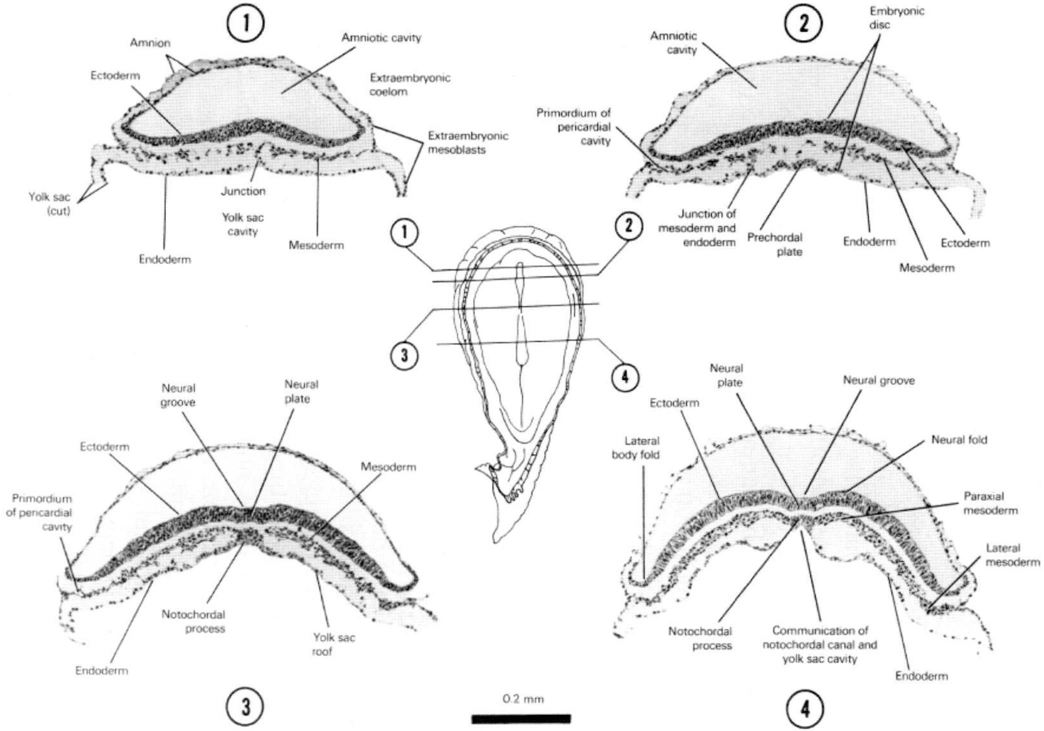

Figure 2.3. The open neural plate Carnegie Stage 8 of human embryos. Note that the cross sections show that the neural plate is one cell thick, as in axolotl. From the Atlas of Human Embryos[28] with kind permission of Raymond Gasser. 3D animated reconstructions of human embryos are available online[29].

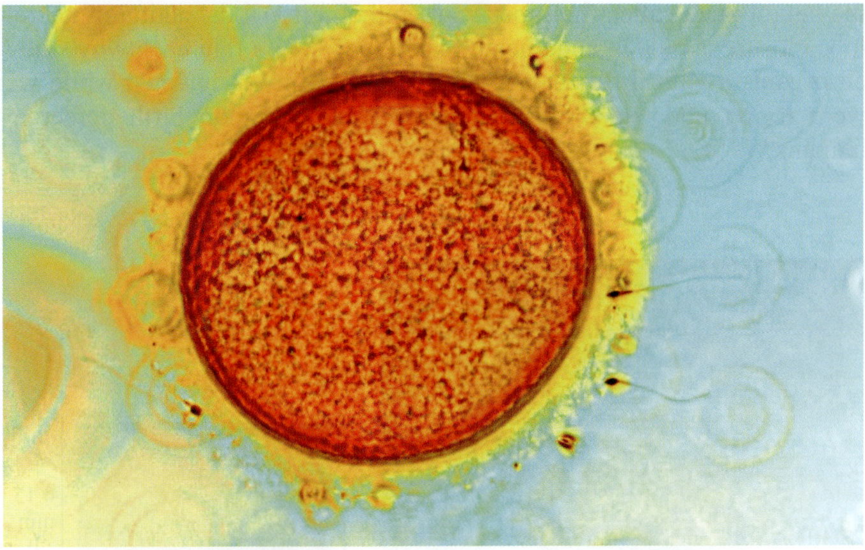

Figure 2.4. "A light micrograph of a human egg, which is surrounded by a matrix of glycoproteins, called the zona pellucida (brown rim), and a layer of follicle cells, called the corona radiata (yellow)." Intact sperm are also seen. The ones that look empty headed are merely out of focus. Diameter of the egg[30] is about 70 μm. This is from Spike Walker with his kind permission[31].

From our perspective, the axolotl has additional important positive features, besides cell size and pigmentation. These other advantages made it an ideal model organism for us to work with. First, unlike highly specialized model animals, such as the fruit fly or the zebrafish, the axolotl is not bizarre or unusual or in a rush to achieve adulthood via special mechanisms. The axolotl appears to contain none of the evolved clever shortcuts that confuse the picture of what is going on. In particular, the axolotl grows and closes the neural tube that eventually forms the spinal cord and brain in much the same way that the human embryo does (Figure 2.3). Indeed, we stumbled onto evidence of our explanation of embryogenesis while we were actually looking for an answer to a common human developmental disability that occurs when neural tube closure fails[32,33] (Figure 2.6). This birth defect is called spina bifida (Figure 2.7), and occurs worldwide in about 6 out of 1000 births. The neural plate in axolotls[34,35] and humans is a layer just one cell thick, whereas the most commonly used, fast developing amphibian, the South African clawed toad *Xenopus laevis*, has an extra layer of cells covering the neural plate called the superficial epithelium. All the action of neural tube closure is hidden underneath that layer[36,37]. We figured[33]: why study an organism with different neural tube closure from our own while we were trying understand this debilitating and common birth defect in humans?

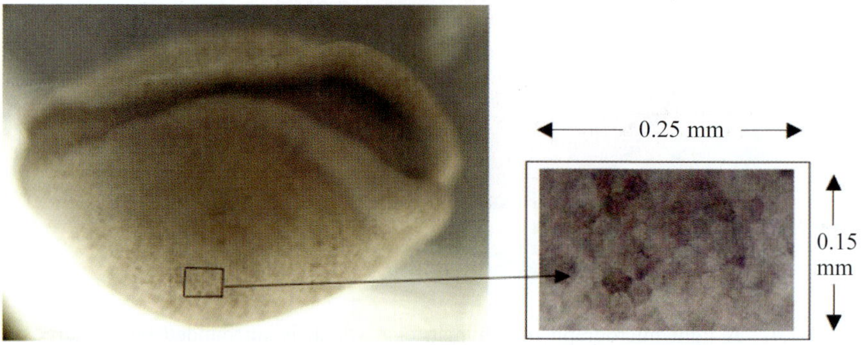

(a)

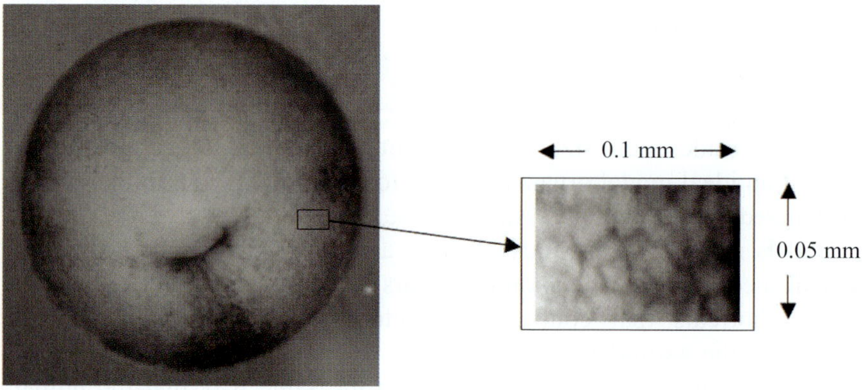

(b)

Figure 2.5. Axolotl embryos vary significantly in surface pigmentation from one cell to the next[38-40]. (a) An embryo at neural tube closure, Stage 16. The magnified portion shows individual cells with their variegated pigmentation. (b) Even at the bottom side, where pigmentation is light, individual cells are easy to see (Stage 10.5). From her Masters thesis[41] with kind permission of Susan Crawford-Young.

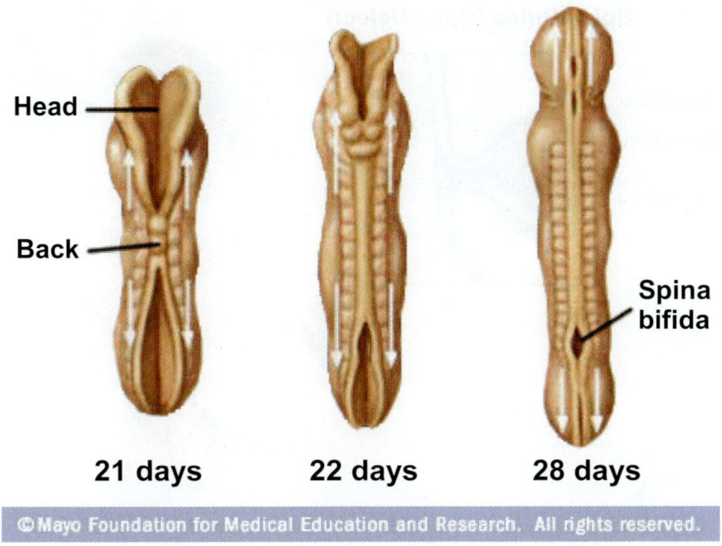

Figure 2.6. In human embryos, the neural plate starts to roll into the neural tube in the middle and zips up towards both ends. If it fails to seal at any point, the result is a neural tube defect, such as spina bifida. From the Mayo Foundation[42] with permission of Mayo Clinic Health Information. Note how for humans a time clock, days, is used to describe developmental whereas in the axolotl a structural appearance clock is used. This is because the speed of axolotl development depends on temperature with lower temperatures resulting in slower development and higher temperatures resulting in faster development. The human embryo, sheltered in the womb, develops at a constant temperature.

This egg also has the benefit of preferring to grow in a simple solution of slightly salted water while sitting on a desk in a lab at the same temperatures scientists like to sit and work at. It will even grow in double distilled water. Scientists as a group are quite willing to put up with hot or cold conditions to get their all important data but it is always nicer not to have to be freezing in an oversized refrigerator or cooking in an incubator room. It's also nice to not have to fuss with complicated serums, and sterile conditions.

Another reason for studying axolotls is that, in terms of evolution, axolotls are reasonably close to our common ancestor with them. Axolotls, like all vertebrates including us, have a notochord. The notochord is a stiff rod of tissue that forms very early in embryogenesis and acts as a support structure underlying the region that will become the

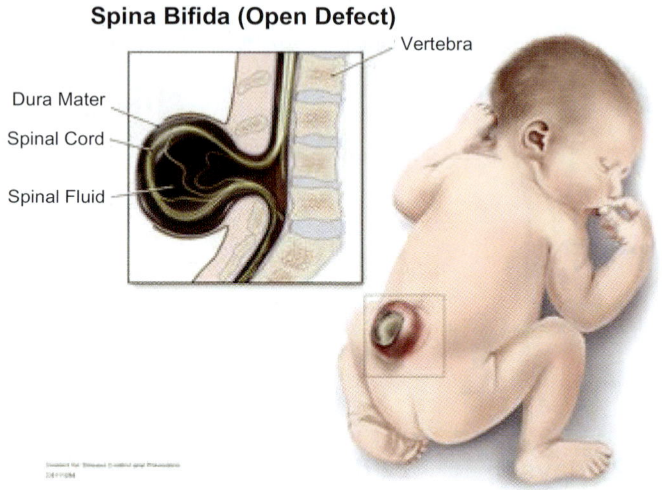

Figure 2.7. The birth defect spina bifida can lead to paralysis from the point on the spine where it occurs on down. From the CDC[43] and in Public Domain. The opening is caused very early in development if the neural plate fails to seal into a complete tube, during a process called neural tube closure.

spine. The notochord preceded the spine in evolution. In more phylogenetically basal (more primitive) organisms, the notochord persists through the adult stage and it is the primary axial support running from the base of the brain all the way to the end of the tail, such as amphioxus (Figure 2.8). In most vertebrates, including us, the spine forms above the notochord and the spine takes over the function of main axial support. The notochord forms the *nucleus pulosus* where it acts as a shock absorbing material and separates the vertebrae (Figure 2.9). The axolotl is intermediate between amphioxus and humans with a proper spine but also a persistent notochord.

During the evolution of higher species on earth, animals split into two broad groups, the invertebrates like insects and sea urchins and the vertebrates. All four legged vertebrates passed through, i.e., branched off from, an axolotl-like ancestor (Figure 2.10). Frogs went off on their own separate branch. Axolotls look very much like the earliest salamander fossils that have been found (Figure 2.11), so they haven't changed much in appearance. Our common ancestor is likely to have had the many wonderful properties that are still available for study in axolotls.

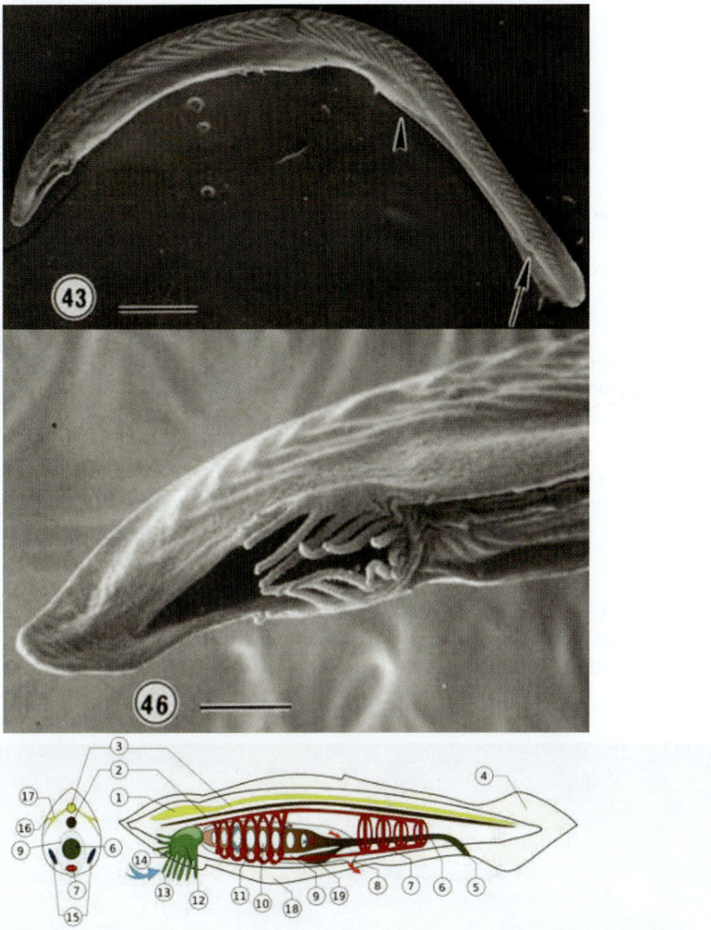

Figure 2.8. The most primitive extant fish, amphioxus (*Branchiostoma lanceolatum*), is a filter feeder. It retains the notochord as an adult, and has an eyespot at the head end of its nerve cord. Length of an adult is about 6 centimeters. **Top**: left side of a late larva (they are not completely bilaterally symmetric, with the anus, at the arrow in the lower right hand corner, on the left side), scale bar 300 μm. **Middle**: a juvenile, showing development of oral cirri around the mouth, also on the left side[44], scale bar 100 μm. Scanning electron micrographs (SEMs) from Stokes & Holland (1995)[45] with permission of John Wiley and Sons. **Bottom**: Anatomy of amphioxus. "1. brain-like blister 2. notochord 3. dorsal nerve cord 4. post-anal tail 5. anus 6. food canal 7. blood system 8. abdominal porus 9. overpharynx lacuna 10. gill's slit 11. pharynx 12. mouth lacuna 13. mimosa 14. mouth gap 15. gonads (ovary/testicle) 16. light sensor 17. nerves 18. abdominal ply 19. hepatic caecum 20. swim bladder 21. lateral line." From Wikimedia[46] with permission under a Creative Commons Attribution-Share Alike 3.0 Unported license.

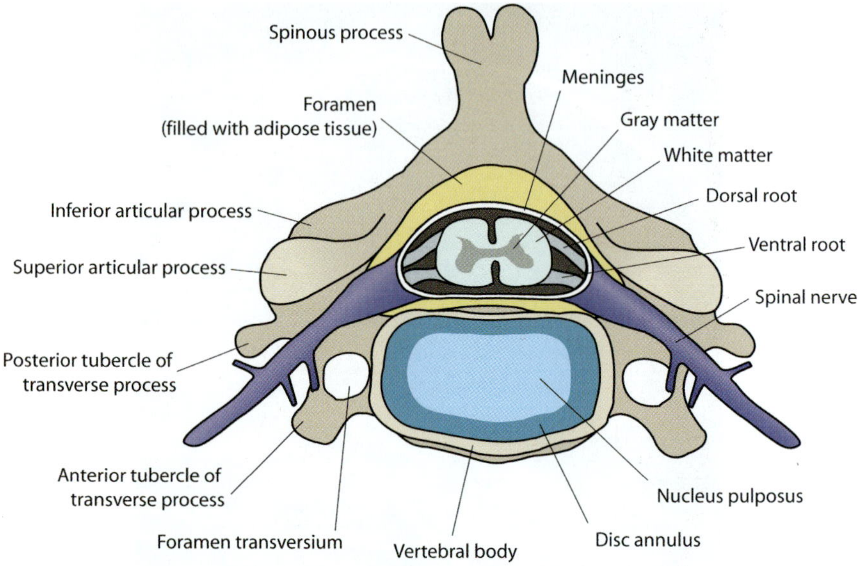

Figure 2.9. Cross section of a cervical (neck) vertebra, showing the nucleus pulposus, which is a tissue that differentiates from notochord. From Wikimedia[47] with permission under a Creative Commons Attribution-Share Alike 3.0 Unported license.

Figure 2.10. This Temnospondyl, of the genus *Capetus*, estimated at 1.5 meters in length[48], is from a group that may have included ancestors of both amphibians and mammals. They lived as long as 300 million years ago and had a variety of ecological

niches[49]. From Wikipedia[50] under a GNU Free Documentation License. The biggest Temnospondyl genera found so far are: *Metoposaurus*, up to 3 meters in length[51] dated from 220 million years ago[52], and *Prionosuchus* at 4.5 meters, 270 million years ago[53,54]. Temnospondyls are "stem salamanders", whereas axolotls are considered "crown group salamanders"[55]. For how these phylogenetic distinctions are defined, see Wikipedia[56].

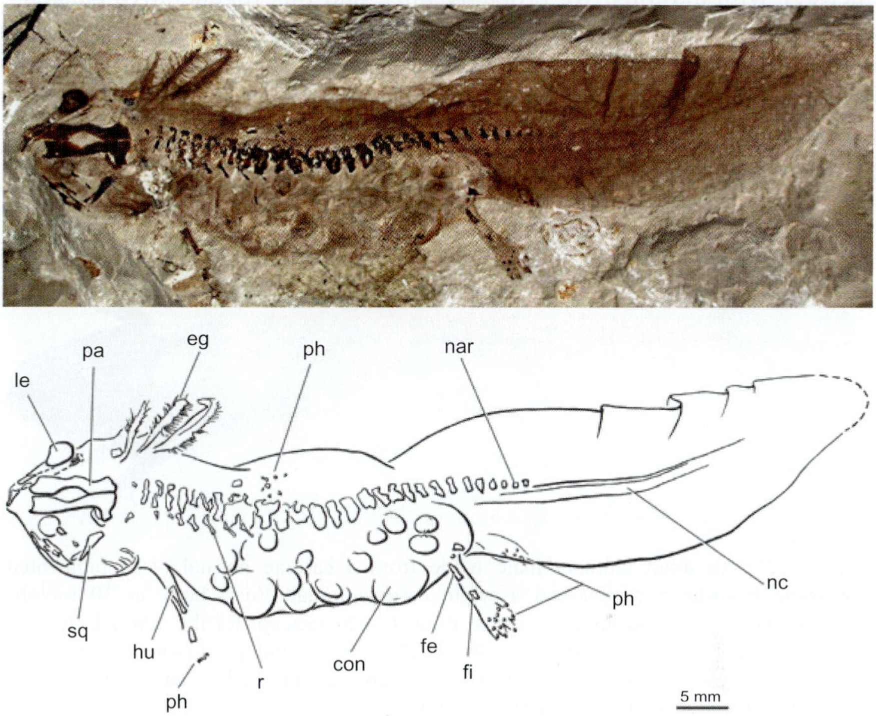

Figure 2.11. A fossilized small larval "crown group salamander" about 165 million years old (oldest crown salamander now known) showing exceptional preservation of soft tissues. Note the external gills (eg) just like axolotls. From Gao & Shubin (2003)[57] with permission of Nature Publishing Group. For further discoveries of salamander fossils see PNAS[58,59].

Another one of their marvelous properties is that axolotls have a tremendous capacity to regenerate[60]. They are aggressive towards one another, and a large axolotl will bite off the leg of a small one. While the latter is slowed down and less of a competitor, it nevertheless regenerates its leg in 6-8 weeks[61]. Jaws regenerate too[62]. Skin wounds heal without scarring[63]. If they lose their tail[64] or a even part of their heart[65-67] they will likewise regenerate, as does a severed spinal cord[68]. A chunk of brain may be successfully transplanted[69] and in a closely related salamander even memory has been transplanted[70]. These are all wonderful capacities we wish we had, under the current promise of buzzwords related to "regenerative medicine"[71]. Given our common ancestry with axolotls, we probably once had this regenerative capacity and then we lost these abilities somewhere along the way. If we understood better how the axolotl does them[72], perhaps we could get our own bodies to repeat our ancestors' feats of regeneration[73]. If you cut two axolotl larvae in half and put the halves together it heals nicely and grows normally[74,75] (Figure 2.12).

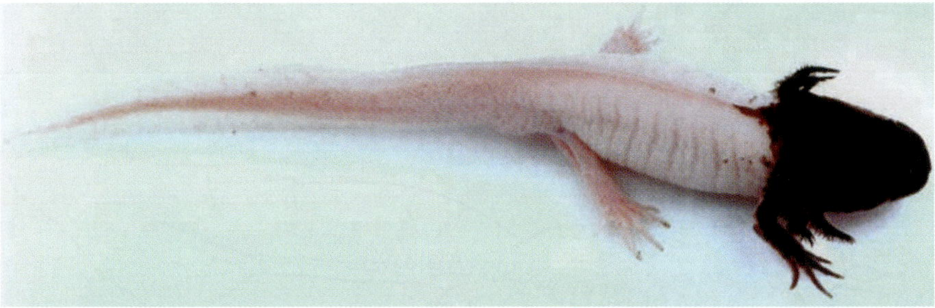

Figure 2.12. "An adult chimera male made from a cardiac normal (+/+) pigmented embryonic head-heart region and a white cardiac mutant (*c/c*) body at 30 months after surgery. The axolotl appears normal in every way (except for the unusual pigmentation pattern)." From Lemanski *et al.* (2001)[75] with kind permission of Springer Science+Business Media. Embryos at Stages 21-24 were cut in half and front and rear pieces fused to achieve this "magician's" result. Cf.[310].

The axolotl is thus what is called a "model organism"[76]. One question that often needs to be asked is: "Model for what?" Since we have spent most of our scientific lives in the medical research community, insofar as embryogenesis is relevant to human medical problems we favor the axolotl. It is our model organism. There are many alternative model organisms for development. They have advantages and disadvantages as well. We list here the most commonly used but this is not an exhaustive list by any means:

- Amphioxus, which retains its notochord as an adult, perhaps the most primitive fish[77,78] (Figure 2.8).

Figure 2.13. *Arabidopsis thaliana* is a plant with a small genome widely used as a model organism. From what is now a cobweb page[79] with permission of the European Commission in conformation with their reuse policy. Flowers are 4 mm wide[80].

- *Anabaena*, the colonial cyanobacterium with a spacing pattern of heterocysts we met in Chapter 1[81] (Figures 1.28, 1.29 and 1.34).
- *Arabidopsis*, a mustard plant[82] (Figure 2.13) noted for the fact that it has very little DNA that is not used for coding genes as compared to other plants.

- *Caenorhabditis elegans*, a nematode worm, perhaps the most complicated mosaic organism[83]. All embryos come in one of two classes, mosaic or regulating. If you cut an embryo in half at a very early stage, a regulating embryo will produce perfect twins. A mosaic embryo will produce two complimentary half embryos that eventually die (Figures 2.14 and 2.15).

Figure 2.14. Two intertwined nematodes (*Caenorhabditis elegans*) widely used as a model organism. From National Institute of Standards and Technology[84] with kind permission of Bryant C. Nelson. Length is about 1 millimeter.

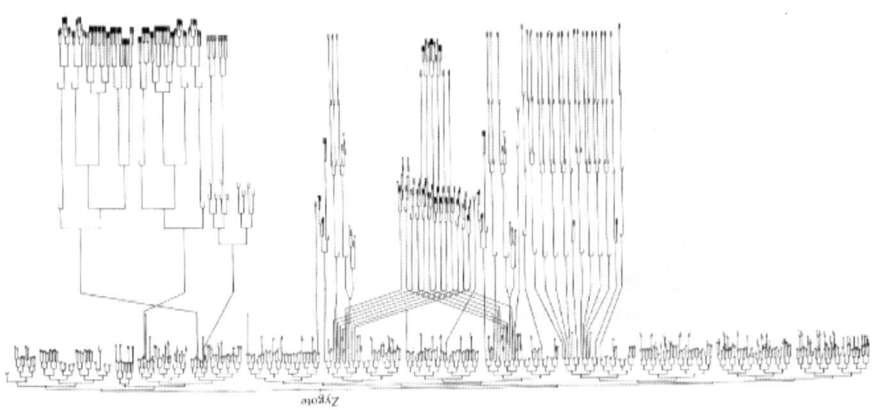

Figure 2.15. The lineage tree of the nematode *Caenorhabditis elegans* (Figure 2.14), which is invariant between genetically identical individuals. The time scale is vertical ranging from 0 to 800 minutes (hatching). At each cell division, indicated by a bifurcating branch point, the cell closest to the head end (anterior) is placed to the right. This is a montage of figures from Sulston *et al.* (1983)[85] made by John E. Sulston[86] and reproduced with his kind permission. This could be turned into a differentiation tree[87] (defined in Chapter 6) by rearranging the branches so that for each asymmetric division the small cell is on the left (single cell contraction wave) and the large cell is on the right (single cell expansion wave).

- Chicken[88], a very well known model organism and more recently studied as the living representative of the dinosaurs.
- *Drosophila melanogaster*, the fruit fly[89] (Preface Figure 0.1).
- Grasses, including our grain crops which are so important to agriculture[90].
- Mouse, as the common representative of mammals[91].
- *Paramecium*, a single-cell ciliate, with a patterned cortex that allows us to study early eukaryotes as well[92] (Figure 2.16).
- Parasitic wasps as a way to study how one egg can produce many individuals[93] (Preface Figure 0.1).
- Sea squirts, also known as tunicates or ascidians, whose larvae have a notochord and whose development is similar to but simpler than our own[94,95] (Figure 2.17).
- Sea urchins[96], especially studied for their early development, are also one of the first model organisms many students learn embryology from in high school (Figure 2.18).
- Sponges[97] are arguably the simplest of multicellular organisms with embryos.
- *Volvox*, a colonial green alga that forms swimming balls that look like animal blastulae[98] (Figure 2.19).
- *Xenopus laevis*, the South African clawed toad, long the favorite of amphibians with which to study embryogenesis and in which the molecular biology is widely explored[99].
- Zebrafish, *Danio rerio*, a small freshwater aquarium fish with a 3 month life cycle[100] used as representative of that phylum. Dick has studied their pigment cell patterns[101-104].

While model organisms tend to be similar across fields of study, anyone's favorite organism can be declared by them to be a "model organism". The reader should note that, because scientists' efforts are dispersed over many organisms, data for some questions is not yet available for each species. This forces us to work by analogies between species, called homologies when they can be justified via evolutionary evidence of common ancestry, and hope that when the detailed work is done we will not be shown to have been far off the mark. Rather than explain this each time, if you note that a specific reference is to a different organism other than the axolotl, it is because we think the homology is close enough to warrant transfer of a concept from that species to our preferred model animal.

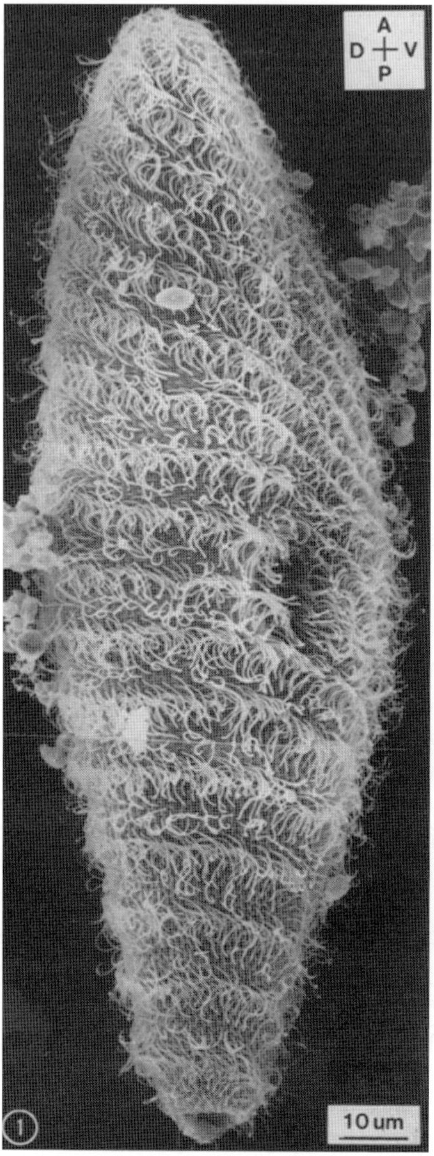

Figure 2.16. *Paramecium* is a single celled animal with an ornate structure covered with cilia, with which it swims. The centrosomes from which the cilia grow are formed in waves that may be homologous to differentiation waves and gastrulation (Proposition 113 in *Hierarchical Genome*[105]). A-P = anterior-posterior, D-V = dorsal-ventral. From Tamm (1972)[106] with kind permission of Suzanne O'Donnell, Rockefeller University Press Permissions Department, ©1972.

Figure 2.17. The tunicate (alias sea squirt or ascidian) starts life as a tadpole that adheres to a surface and becomes a sessile adult. The notochord (straight shaft head through tail) is lost during this metamorphosis. From the British Museum[107]: in the public domain ©. The adult (below) is a sessile filter feeder. From Wikimedia[108] with permission under the Creative Commons Attribution 2.0 Generic license. Adults shown are 4 cm tall. Tunicate larvae range in length[308] from 1 to 4.5 mm.

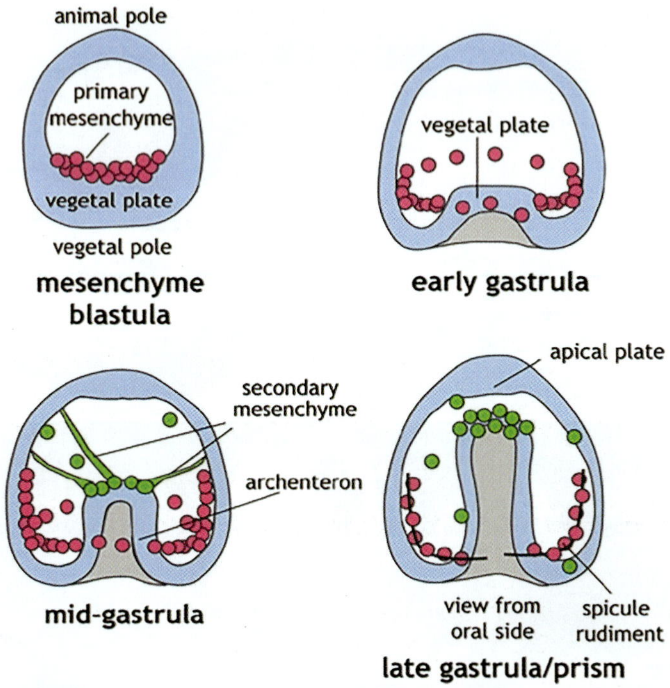

Figure 2.18. Echinoderms such as the sea urchin illustrate a form of simple two-layer gastrulation in an evolutionarily primitive organism. Echinoderms are members of the Bilateria phylum, meaning they are bilateral organisms, though imperfectly so. Bilateral means that the left and right sides are mirror images of one another. Their body plan both radiates outward from a central point and has a left and right side. They also have a top and bottom orientation. Gastrulation begins with expansion of the presumptive ectoderm (animal pole) followed by "ingression" of the endoderm (vegetal plate). Some cells (red and green circles in the diagram) lose their affinity for the adjacent cells. These cells come from the boundary zone between the animal and vegetal pole. They undergo a dramatic rearrangement of their cytoskeleton and change shape. They also move inwards, breaking free of their neighboring cells. They migrate as individual cells along the inside of the blastocoel forming a layer that will become the mesoderm (secondary mesenchyme). (This breaking free of some cells from a sheet of cells, shape change and migration occurs over and over again in embryogenesis across multiple phyla up to and including ourselves.) The cells in the boundary region between the animal and vegetal poles which don't break free also change shape and elongate, undergoing extreme apical constriction as bottle cells. In doing so, they change the shape of the entire embryo, which becomes flat on the bottom. This flat bottom moves inwards starting from the same point at which the mesodermal cells left the layer elongating until it touches the inside at the top. Once the cells touch, they adhere. Gastrulation is now complete. The essential body plan is set: ectoderm on the outside, mesoderm in a middle layer and endoderm in the center. The rest of embryogenesis involves refinements on that plan. From Hardin (2005)[109] with kind permission of Jeff Hardin.

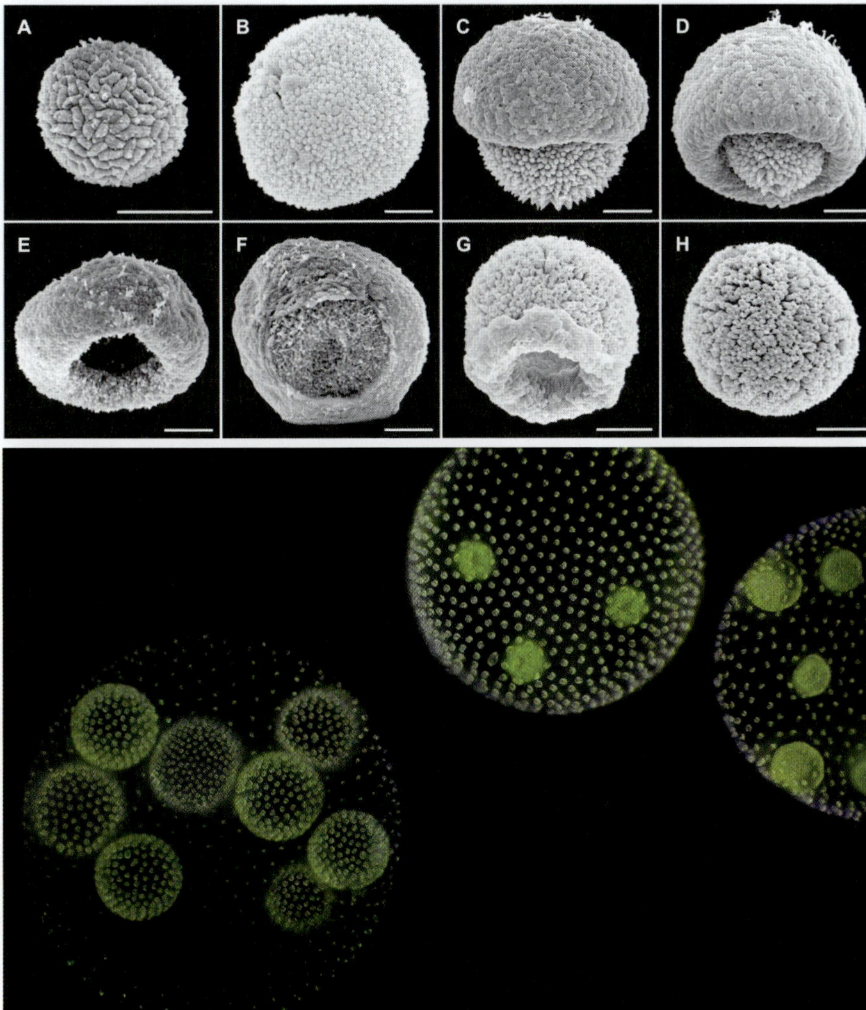

Figure 2.19. **Top**: The multicellular alga *Volvox* shares two properties with animal embryos: the embryo undergoes an inversion analogous to gastrulation, illustrated here, and the cells then have externally oriented cilia that beat in a coordinated fashion. From Höhn & Hallmann (2001)[110] with permission under a Creative Commons Attribution License. Original SEMs kindly supplied by Armin Hallmann. Scale bars: 20 μm. **Bottom**: Live *Volvox*, including internal colonies formed from their "mother" colony, which undergo inversion after formation. From Wikimedia[111] with permission under a Creative Commons Attribution-Share Alike 3.0 Germany license.

Figure 2.20. Unfertilized axolotl eggs (oocytes) are visibly opaque due to their high yolk content. Each has a surface layer on one hemisphere that is darkly pigmented that will rotate to the top after fertilization, as the white hemisphere has more yolk, of higher density than the cytoplasm. Diameter = 2 mm. From Brooke (2011)[112] with kind permission of Andrew D. Johnson.

The freshly laid axolotl egg itself is a massively yolked bottom heavy sphere 2 millimeters in diameter[113], making the embryo translucent and highly light scattering (Figure 2.20), unlike many clear marine embryos. The axolotl embryo cannot eat between being laid as an egg and becoming a larva that is able to go hunt down its prey. It has no placenta attached from which to derive nourishment. It must carry this substantial yolk supply with it for its two-week journey (at 20°C) from egg to a hatched larva with a functional mouth and brain[114]. For comparison, *Drosophila* hatches as a grub in under 24 hours[115] and zebrafish larvae within 3 days[116]. *Xenopus* hatches in 2 days[117].

If embryogenesis is a physics phenomenon, then the standard approach in physics would be to define the initial conditions of the embryo, apply the appropriate dynamics, and calculate stepwise over time[118-120]. Unfortunately, there has been no concerted effort to define the beginning state of any embryo in a physically precise manner. Since the life cycle of an organism returns approximately to the same place (egg + sperm > embryo > larva > adult > egg or sperm), from a physics point of view, we could try to define the initial conditions at any arbitrary time point in the cycle. We have chosen the just fertilized egg (Figure 2.4) as our starting point, and will impart what sparse information we have about it. Thus we leave out of our account oogenesis[121] (formation of the egg in the female), spermatogenesis (formation of sperm in the male), both of which have their own fascinating and substantial literature[122,123], and most of the formation of the polar bodies, which are remnants of successful preparation of the egg's DNA for joining the sperm's DNA. Considered from our chosen starting point, they are at the end of the life cycle.

There is a widespread assumption that the environment somehow determines which cells in an embryo become which kinds. At every step of development, each cell is presumed to react to its local environment within the embryo in order to "decide" what to do next, and this concept of cells making decisions permeates much thinking about embryogenesis. The major problem with such wishful thinking is that each cell contains the same DNA, and passes identical copies on to its daughter cells (barring mutation and a few rare exceptions), so that there is no obvious reason or genetic mechanism for cells to create new environments for one another. That all of these interactions between genetically identical cells should somehow work themselves out in the creation of many distinct microenvironments, all in the right place at the right time, is about as plausible as having a musically untrained crowd of chattering people suddenly switch their cacophony to four part harmony. Thus, notions of environmental control of development have to be reexamined. As one author put it "...the embryo is provided with little more spatial information than the simple instruction: 'This side up'"[124]. Given that inverted embryos[118] or those in zero gravity[125] can develop pretty normally, even "this side up" may not be "provided".

The top half of the fertilized axolotl egg is pigmented a dark color while the bottom half is white (Figure 2.20). In *Xenopus* fertilized eggs these two hemispheres differ in surface viscosity as well[126] and we presume the same is true of the axolotl. The two hemispheres in *Xenopus* also have two different kinds of "cortical granules" attached to them[127,128]. There are no cortical granules in axolotl[25] but surface protrusions and folds on eggs of the newt *Triturus alpestris* differ in surface density between the two hemispheres[129]. Since these newts are a close relative of the axolotl we can assume cortical granules of the frog have some kind of functional equivalent to surface protrusions and folds and the same differences are true for the axolotl. There are measurable differences between the top and bottom hemisphere in the amphibian egg. We will get into the mechanical importance of this in subsequent chapters.

Figure 2.21. On rare occasions an axolotl is found that is pigmented on one side. **Top**: from Hess (2009)[130] with kind permission of Manuela Hess. **Bottom**: with kind permission of Jacob Bidinger[131]. This can be due to either the fertilized egg being rotated and stuck rotated prior to cortical rotation, so that the pigmented hemisphere (Figures 2.20 and 2.28) is not on top, or to the fusion of two early embryos sharing the same jelly capsule and forming a chimeric embryo, i.e. an embryo that has cells from two genetically different siblings.

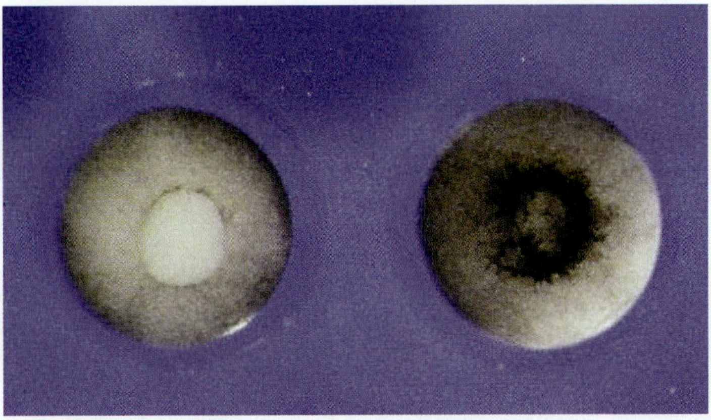

Figure 2.22. The 1.2 mm diameter *Xenopus* embryo on the left is viewed from the bottom and is normal, shown at the stage where the blastopore is starting to close over the unpigmented endoderm cells. The one on the right was centrifuged upside down at 20 g before cortical rotation, and is also seen from the bottom. The blastopore is now pigmented and contains most of the yolk. But development is still normal. Kindly provided by Steven Black (cf.[132]).

By the end of embryogenesis the larva ends up white on the belly and dark on the top. This makes it camouflaged in both directions, from predators looking up towards the sky, who will see white, or down to a dark lake bottom where the dark pigmented surface will be disguised. We have seen a few cases where the egg got stuck on its side so that the upper pigment layer was rotated by about 90° relative to the interior of the cell (Figure 2.21). The result was an axolotl larva that was white on one side and dark on the other either down the length of it or with either the front half or the back half lacking pigment. If an egg is deliberately turned upside down and held in place, it has a 50% chance of developing into an adult with white on top and black on the bottom[133,118]. The other 50% will simply die. This indicates that the "top" and "bottom" orientation of the egg, while present from the beginning, does not exist in a fixed way until some time after it is fertilized and laid. The top versus bottom demarcation of the egg is thus not essential to embryogenesis (Figure 2.22).

There are seven layers of noncellular material referred to as jelly on the outside of a freshly laid axolotl egg. Each layer has a distinct chemistry[134,135]. The egg's jelly layers are barely visible when freshly laid

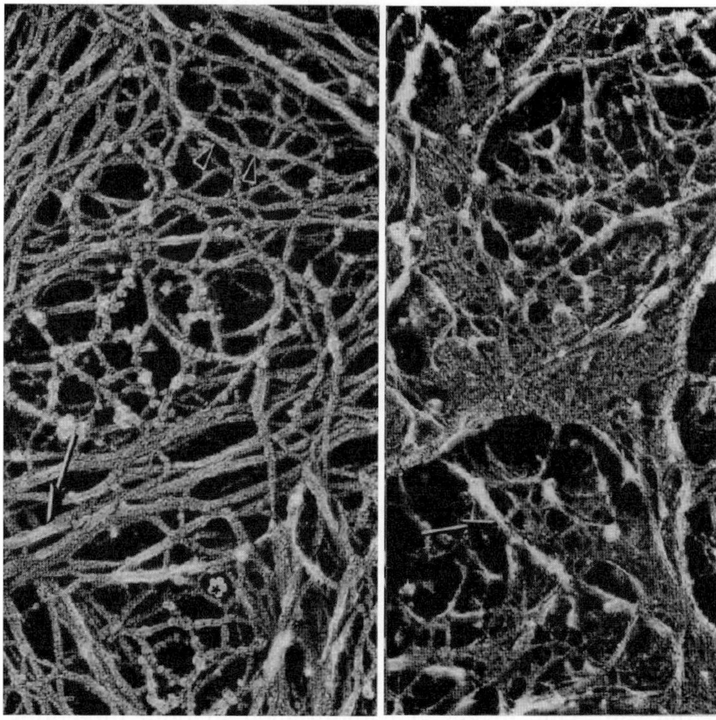

Figure 2.23. Frog vitelline membrane before (**left**) and after (**right**) fertilization. In frogs the altered vitelline membrane is a barrier to further "hopeful" sperm. For scale, the wider fibrils on the left are 19 nm in diameter. From Larabell *et al.* (1989)[136] with permission of John Wiley and Sons.

but swell up as the layers absorb water to varying degrees over a few hours. Fully swollen, the total diameter of the jelly reaches 13 mm around the 2 mm diameter one-cell embryo. The outer jelly layer of one embryo may be fused to another as the female lays a string of fertilized eggs[137] (Figure 2.1). These layers are deposited in the mother's oviduct after internal fertilization, and the jelly layers consist mostly of mucopolysaccharides, i.e., polymers of sugars, and a bit of protein[138-140]. The axolotl jelly layers are:

1. The innermost vitelline membrane, which at early stages wraps the embryo tightly, and which is removable for experimentation. In axolotls, the vitelline membrane tears and handles like cellophane (Figure 2.23). It is 1 μm = 0.001 mm thick in *Xenopus*, 2 μm after

fertilization[136] and is of similar thickness in the axolotl. The embryo is maintained as approximately spherical and has a vertical diameter that is 92% of the horizontal diameter. If the vitelline membrane is removed, the embryo pancakes into a flattened blob that nevertheless develops normally (Figure 2.24). At each cell division, the otherwise flattened cell changes shape and elongates upward against the force of gravity and then falls back to its flattened shape after cell division. The 2 mm diameter newt embryo *Triturus alpestris* (Figure 2.24) has a density slightly higher than water, so the flattening is probably caused by gravity and resisted by the elasticity of the cells' cortex[141]. The reason behind the dynamic shape change is not known and could be due to changes in the membrane and/or forces of the cytoskeleton[142].

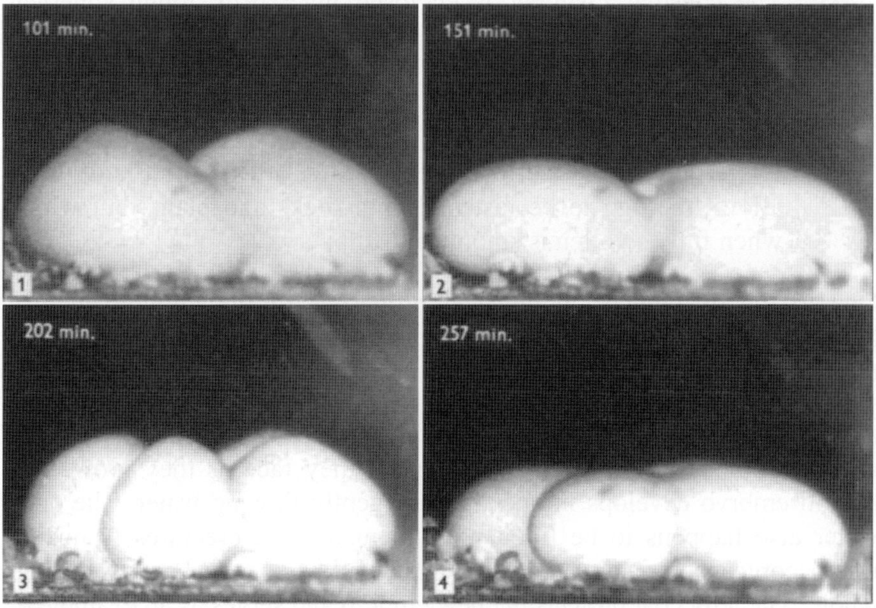

Figure 2.24. "Illustrations from a film sequence of an egg cleaving on a flat surface after the vitelline membrane had been removed. 1, a 'rounded-up' position of the egg at first cleavage. 2, a 'relaxed' position of the egg between first and second cleavage. 3, a 'rounded-up' position of the egg at second cleavage; note that one of the blastomeres is of greater height than breadth. 4, the subsequent' relaxed' position of the egg between second and third cleavage." From Selman & Waddington (1955)[142] with kind permission of the Company of Biologists. This newt embryo *Triturus alpestris* is a 2 mm diameter sphere when the vitelline membrane is intact.

Later in development the whole embryo develops it's own internal pressure and even without the vitelline membrane rounds up again. Therefore the vitelline membrane is not essential to normal development.

2. There is a shell of mucus that swells and liquefies[143] after the egg is laid, called the capsular chamber. When first laid eggs are laying at all angles. Once swollen, this chamber allows the bottom heavy embryo to rotate and right itself so it is always floating dark side up[113,144]. This chamber has higher internal pressure compared to the external pressure of the environment. It is when this chamber is disrupted or more than one egg gets crowded inside one such chamber that you can naturally end up with an embryo that has the pigmentation pattern of dark on top and light underneath altered.
3. The capsular chamber is enclosed by a firm, tough spherical shell of jelly.
4. A second firm layer of jelly.
5. A soft jelly layer.
6. A firm jelly layer.
7. A soft jelly layer that is extremely sticky when first laid which allows the egg to be fastened onto rocks and plants but which when fully swollen is smooth and slippery.

Due to the differences in physical properties of the jelly layers, we have been able to observe two of them so far using optical coherence tomography (Figure 2.25).

All embryos have special adaptations that have evolved over time to deal with the unique environmental challenges they face as they grow. The axolotl embryo develops in open, still or gently flowing water. The open water also happens to be filled with all manner of creatures from tiny microscopic bacteria to predators who would be quite happy to eat it whole. The jelly coat protects the new embryo and serves as an example of how evolution produces complex adaptations for successful reproduction.

The egg must be joined with sperm. Fertilization in the axolotl is internal The male deposits a packet of sperm (Figure 2.26), and is followed along by a female, who picks it up with her cloaca[145]. This may sound like a rather boring and uninteresting way to transfer sperm to egg but the axolotls don't appear to agree. In leucistic (white) axolotls we have seen

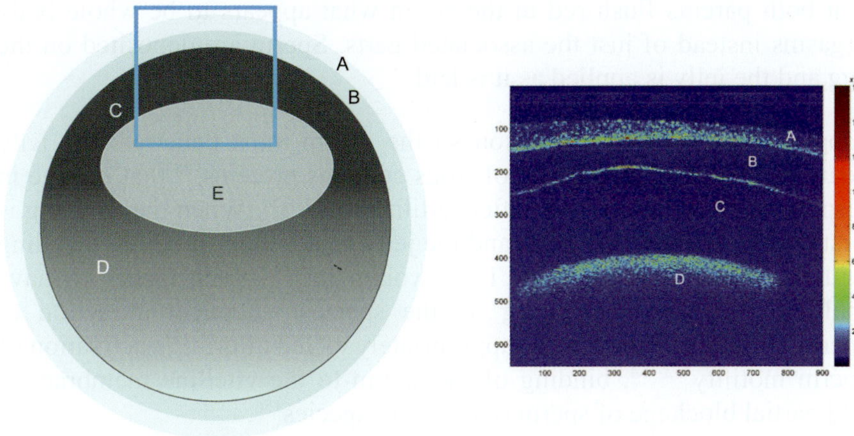

Figure 2.25. This is one of the first images of a Stage 8 to 9 axolotl blastula imaged by optical coherence tomography (OCT). **Left**: Blastula showing the portion of the embryo scanned. **Right**: OCT scan of the top of the blastula. A: outer jelly, B: inner jelly, C: animal half of the egg/embryo, D: Vegetal half of the egg/embryo, E: Blastocoel. From Crawford-Young *et al.* (2016)[146] where current embryo imaging methods are compared, with kind consent of Susan Crawford-Young and Sherif S. Sherif.

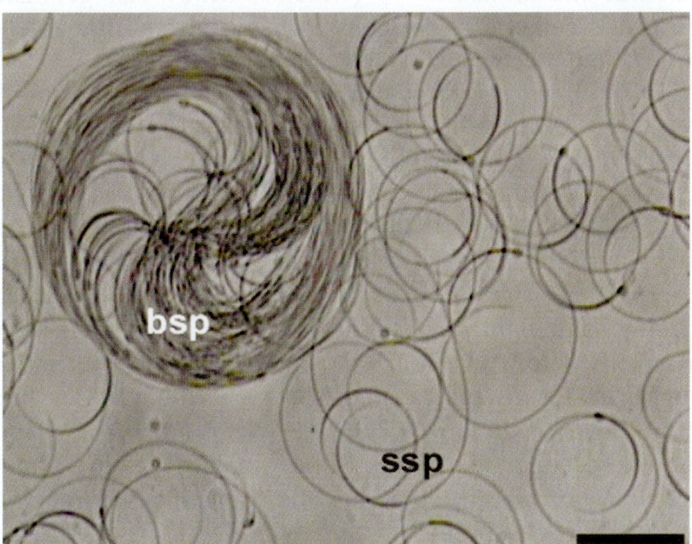

Figure 2.26. Single and bundled sperm (spermatophores) of the axolotl. Bar = 10 μm. From Mansour *et al.* (2011)[147] with permission of Elsevier.

that both parents flush red in the act in what appears to be whole body orgasms instead of just the associated parts. Sperm are deposited on the egg and the jelly is applied as it is laid.

Frogs have external fertilization so the sperm must penetrate the jelly coat. The vitelline membrane of frogs contains proteins[148] that change in response to the sperm[148] and after fertilization[149,150]. When the frog egg is first laid, it is typically sticky and the jelly coat is thin. It begins swelling with water upon being laid. The swollen jelly layers appear to have multiple functions depending on the species: chemical attraction of sperm by release of a protein appropriately called allurin[151]; activation of sperm motility[152-154], binding of the sperm to the vitelline membrane[155] and partial blockage of sperm of different species[156].

The jelly coat of amphibians also controls retention of heat which protects from freezing[157], perhaps in part due to the light focusing effect of the jelly, acting as a spherical "fish eye" gradient index lens[158]. The jelly coat lets oxygen in and carbon dioxide out[159]. It keeps embryos physically far enough apart, so they all have access to oxygen[160]. It also harbors symbiotic eukaryotic algae that produce oxygen and use up carbon dioxide and ammonia[161-163] (which somehow also get inside some of the embryo's cells[164]). It provides protection from ultraviolet light[165], desiccation[166] and mechanical damage[166].

We have observed that the parasitic protozoan *Costia* eats dejellied axolotl embryos[167], but dies if entangled in egg jelly. Many harmless microorganisms can also make their home in the jelly[164]. Fungal molds are thwarted by jelly[168]. Furthermore, the jelly may protect the embryos from predation by larger beasts, directly[169,170], by being unpalatable[171]. By offering a clear view of the embryo, seen dark against a dark bottom from the top, and light against the sky when seen from below, the transparent jelly helps camouflage the embryo. The jelly is not complete protection. Fish, for instance, cannot always be fooled[172] and frequently feast on the eggs, which is why one rarely finds amphibians laying eggs in places where larger fish swim. Some amphibians have cannibalistic brothers and sisters who hatch earlier and can eat right through the jelly and consume nearby embryos[173].

In the newt family, the growing embryo "hatches" from its vitelline membrane early in development while staying inside the capsular chamber until it can swim freely. The capsular chamber is either big enough to hold the embryo through to hatching, or swells further as development proceeds[137]. Frog tadpoles do all their development to the free swimming stage inside the vitelline membrane which swells as development proceeds[174].

Interestingly, mammals also have to hatch from their own version of the vitelline membrane called the "zona pellucida" via a process called zona hatching. Mammalian hatching occurs prior to implantation when the embryo is a ball of only a few cells. Infertility specialists often assist *in vitro* fertilized human eggs to hatch because this increases the rate of successful pregnancy in some types of infertility (Figure 2.27). This is an example of how organisms that have evolved from a common ancestor will later diverge so that different species share how development proceeds but have different timing for individual steps of development.

There is a second rotation of the amphibian one-cell embryo that takes place after laying (Figure 2.20). The whole outer cortex[175], both the pigmented top and the lightly or unpigmented bottom, rotates slightly (about 30°) between fertilization and first cleavage. One then has a window into the upper, relatively yolk free cytoplasm, that is in the shape of a crescent, and is thus called the grey crescent (Figure 2.28). The motion that forms the grey crescent is called cortical rotation. This motion is visible because the pigment granules are embedded in the cortex, mostly in the upper hemisphere. We will address theories for what causes cortical rotation, and its consequences, in Chapter 5, but simply note here that the cortex is a complex of cytoskeletal and organellar components bound together just under the cell surface[176]. It has a lower viscosity than the cytoplasm contents of the one-cell embryo[175]. The result is much like a toy "Martian eye" which always looks up at you no matter how you turn it (Figure 2.29). Before first cleavage, many events occur besides cortical rotation. These will be considered in Chapter 10.

By the time that the first cell division begins, the axolotl will have an approximate top and bottom, a future area that will form the front/mouth end of the embryo, and a back or cloacal end. (The anus, urethra and genital openings are combined into one opening in amphibians, birds and reptiles, called the cloaca.) The visible evidence of this process of head/genital orientation is the formation of the grey crescent on one side

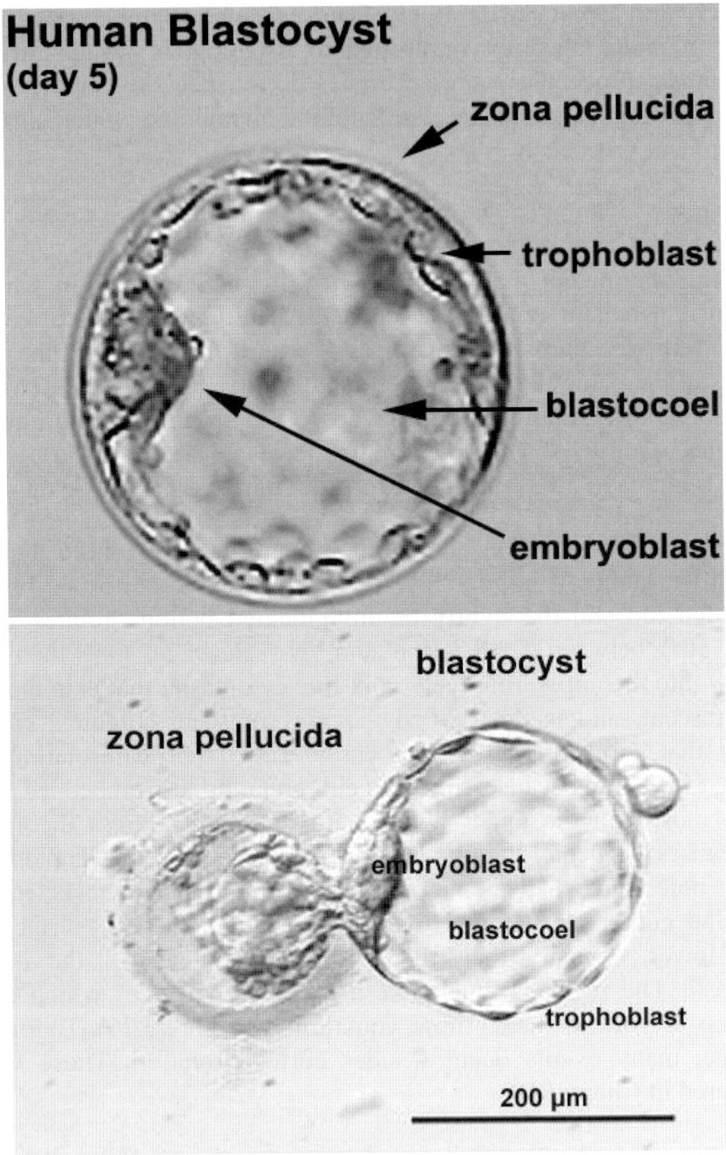

Figure 2.27. **Top**: Human embryo at the blastocyst stage. The trophoblast later becomes the placenta, and the fetus develops from the cells of the embryoblast (also called the inner cell mass). From Zhang *et al.* (2009)[177] with permission under a Creative Commons Attribution (CC BY) license. **Bottom**: Human blastocyst hatching from the zona pellucida. Labelled in Hill (2015)[178]. From Klimanskaya *et al.* (2006)[179] with permission of Nature Publishing Group.

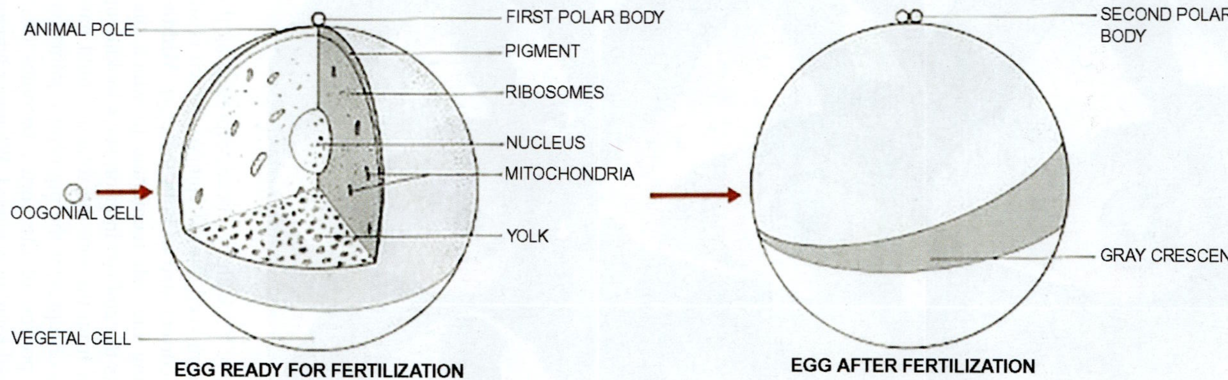

Figure 2.28. Rotation of the cortex by about 30° results in visualization of the "grey crescent". Basically this is a window into the less yolky cytoplasm of the upper hemisphere. The darkly pigmented top hemisphere of the cortex is tilted up in a process called cortical rotation so that a portion of the unpigmented bottom hemisphere now lies above the equator. The yolk inside, mostly in the lower hemisphere, sloshes back[180,118], remaining at the bottom. From Kimball (2012)[181] with permission under a Creative Commons Attribution 3.0 Unported (CC BY 3.0) license. In *Xenopus*, where the pigmentation occurs in both hemispheres, the grey crescent is usually not visible[182].

Figure 2.29. **Top row**: A "Martian eye"[183] with two beads representing the polar bodies attached at the gravitational top of a one-cell axolotl embryo (which is sitting in an old seashell that was handy and of the correct size to hold the egg). **Middle row**: During cortical rotation the cortex, including the cell membrane to which the polar bodies adhere, rotates about 30°. The eye, as the embryo interior, stays looking up because it is bottom heavy. **Bottom row**: as seen from the top, the rotation could have been in any direction. Red lines on top hemisphere represent the pigmented animal cortex. **Arrows** indicate the grey crescent. Which portion of the cortex becomes the grey crescent depends on which direction the cortical rotation occurred. Whatever happens, the equator of rotation later becomes the midline of the animal, defining the left and right sides. Thus left/right is not determined until cortical rotation occurs. In *Xenopus* only one sperm gets into the egg. While this biases the direction of cortical rotation to be within 8° of the

sperm entry point, it does not determine that direction. In fact, if you grab the surface of a one-cell *Xenopus* embryo and rotate it manually, you determine the midline. In axolotl, which allows many sperm in, though only one becomes functional, there is no observable bias but the site of the first sperm that is fully inside the egg and whose pronucleus fuses with the egg nucleus marks the future midline.

of the embryo shortly after fertilization (Figure 2.28). The grey crescent normally corresponds to the cloacal end of the salamander[184] along with birds, reptiles, fish and mammals.

In insects and other invertebrates the mouth generally but not always[185,186] forms at the grey crescent side[187,188]. This "reversal" between invertebrates and vertebrates was originally noted by Étienne Geoffroy Saint-Hilaire in 1822[189-191] (Figure 2.30). This idea of a single evolutionary reversal has been supported by molecular evidence[192-195], although uncertainties remain[196,44]. Nevertheless, in terms of the developmental anatomy, vertebrates, with a notochord, develop their mouth opposite the grey crescent and invertebrates, without a notochord, develop their mouth on the same side as the grey crescent.

Physicians commonly use weeks when talking about human development. All eight-week human embryos are pretty much the same because all humans develop at a consistent temperature and thus speed. When two physicians refer to an eight-week embryo they both know what stage of development that means. However, in the axolotl the speed of development is temperature dependent. If the egg is warmer it develops more quickly. If it is cooler, development slows. Because of this, when two embryologists are discussing axolotl embryogenesis, they don't talk about development in terms of time for those early stages. These stages are defined by shape, numbered and illustrated in a publication[197], such as that for axolotls[114] (Figure 2.31), which then becomes the standard nomenclature for subsequent work although some work refines[198] or extends[199] the original staging. Similar stage numbers are used for *Xenopus* (Figure 2.32). Because the stage numbers are based on morphology, the time differences between them (at a given temperature) are irregular. One stage can be very short while another can be very long.

The axolotl egg pretty much runs on maternal RNA until[205,114] about the end of Stage 8 (Figure 2.33) or about 13 cycles of cell division. The first eight stages therefore consist of rapid and repeated cell division[206]

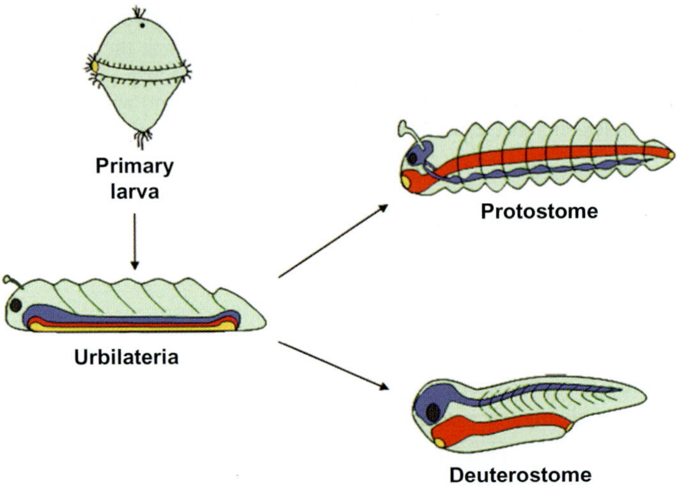

Figure 2.30. "Protostomes (proto, first; stomo, mouth) develop the mouth near the initial site of the blastopore. Deuterostomes… develop the anus close to the blastopore and the mouth is perforated secondarily. Urbilateria is the hypothetical last common ancestor of the two bilaterian branches…. Ectoderm is shown in green, CNS [central nervous system] in blue, eye in black, and endoderm in red, with its openings in yellow." From De Robertis (2008)[200] with kind permission of Diane Sullenberger. Copyright (2008) National Academy of Sciences USA. Note that in deuterostomes, which includes the vertebrates, the nervous system is on top (dorsal), while in protostomes it is on the bottom (ventral), except in the head region.

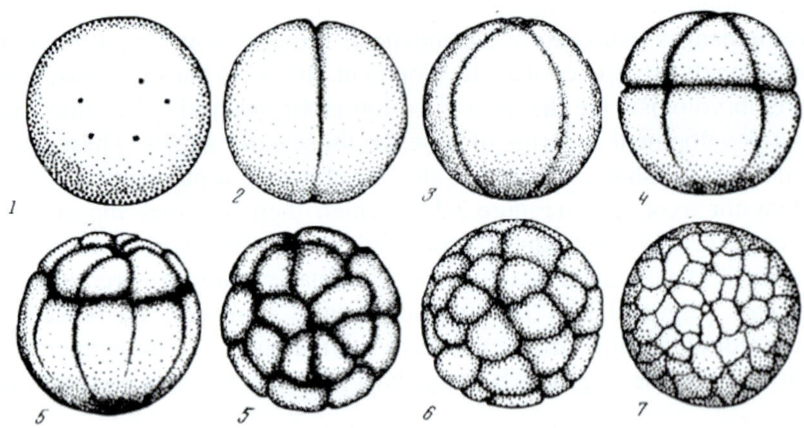

Figure 2.31. The axolotl egg and early embryo are 2 mm in diameter. Stages 1-5 are shown from the side, and Stages 5-7 from the top. The cleavages in Stages 4-5 may reflect either the higher density and perhaps viscosity of the yolky cytoplasm at the

bottom or the different surface viscosities of the top (animal) and bottom (vegetal) hemispheres of the cortex[126,201]. Stages 4 and 5 are caught here with two cleavage furrows still in progress. Cleavage furrows occur about half an hour after contraction waves that start from the animal pole and probably always propagate to the vegetal pole[202]. Cell division is close to synchronous at these stages, so that the number of cells is a power of 2: 2^{n-1} at Stage n, i.e., 1, 2, 4, 8, 16, 32, 64 cells, respectively. Note that the size of this embryo remains nearly constant, while the cells get smaller and smaller. From Sal-Site[114,203] with kind permission of Laura Muzinic, Chris Muzinic, and Randal S. Voss. For an earlier definition of axolotl stages see Schreckenberg & Jacobson (1975)[204].

(Figure 2.31) with almost no new gene expression. Each round of cell division doubles the number of cells. Each step requires approximately halving each cell in size but doubling the DNA in the embryo, since each cell contains the same amount of DNA.

The amount of DNA created by replication required to accomplish this feat is astonishing and no other period of normal development matches it. We speculate that the mRNA packed into the egg allows the replication to proceed at full pace without pausing to express genes along the way. We do know that until the midblastula transition the DNA is tightly coiled and inactivated. DNA replication in each nucleus starts from many replication origins at random. The problem of how this "parallel processing" nevertheless results in exactly one duplication of the whole genome in the short time for these cell divisions was only recently solved[207]. (We will cover the replication of DNA in greater detail in Chapter 3.)

Cells divide into approximately equal volumes at first but then into larger and smaller cells. Axolotls follow what is called the "standard" cleavage pattern in amphibian embryos. They split twice, top to bottom, forming four cells and then they have one horizontal split to make eight cells. The latitudinal position of this third cleavage in the standard cleavage pattern of axolotls depends on the gravitational force (Figure 2.34). On Earth the result is smaller cells on the top layer and larger cells on the bottom. Synchrony with respect to size of the cells ends with the third cleavage. Four cleavage pattern classes have been recognized in amphibians, defined in terms of both the number of synchronous cell divisions, and the directions of the first three cleavages[215] (Figure 2.35). The boundaries separating the early cells as they divide (i.e., the cleavage furrows) reside in a coordinate system similar to latitudes and longitudes on a globe of the Earth. Longitudes run from the animal pole towards the

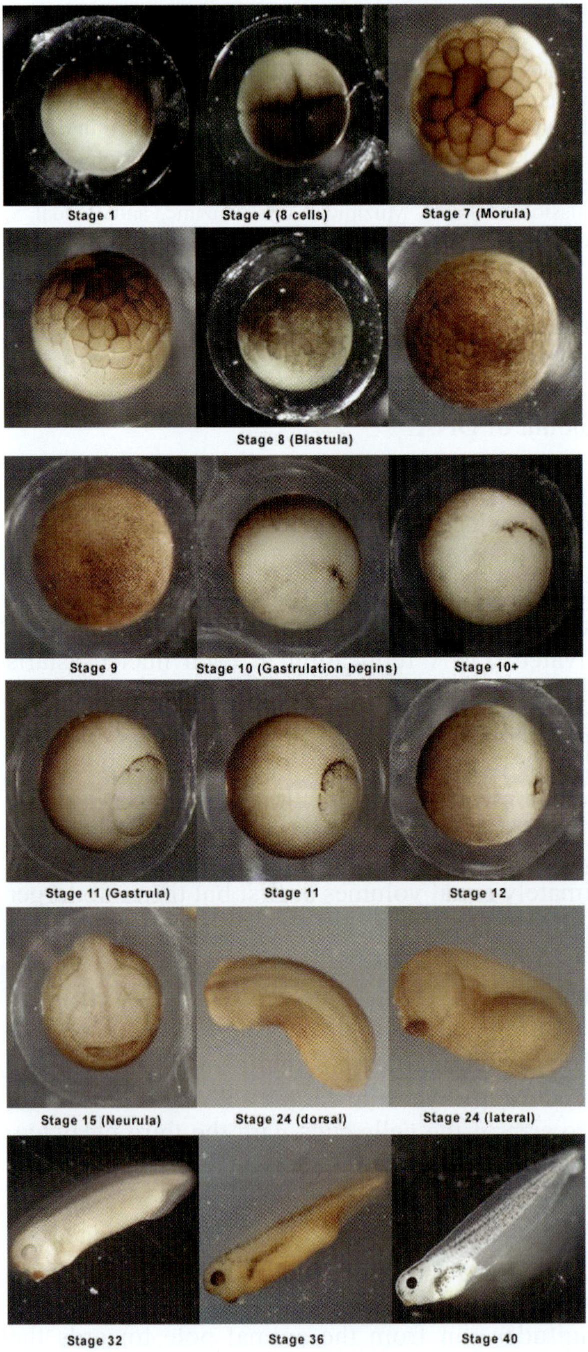

Figure 2.32. The staging numbers[208,209] for *Xenopus laevis* correspond quite well to those in axolotl, even though this embryo is a bit smaller (1.2 mm diameter), develops much faster, and has a superficial epithelium at early stages. From "*Xenopus* embryo staging"[210] with kind permission of Judith A. Cebra-Thomas.

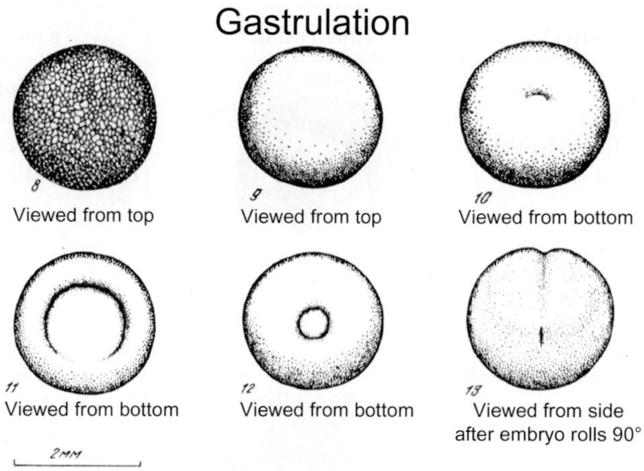

Figure 2.33. After axolotl Stage 8 the cells become so numerous and so small that the artist stopped drawing them, and the surface becomes taut and smooth, making them even harder to distinguish. Beginners often can't distinguish Stage 9 from Stage 1 (Figure 2.31). Note how in the axolotl, because speed of development varies with temperature, stages are marked by specific external structures being visible. Therefore the time from one Stage to the next will vary considerably, and will be temperature dependent. The arc in Stage 10 is called the dorsal lip of the blastopore. It extends (Stage 11) to a full circle (Stage 12) which shrinks down to a point (Stage 13). By Stages 11 to 13 the cleavage waves begin losing their synchrony[211] and a shift has occurred in where the waves are initiated and end[212]. From Sal-Site[114,203] with kind permission of Laura Muzinic, Chris Muzinic, and Randal S. Voss.

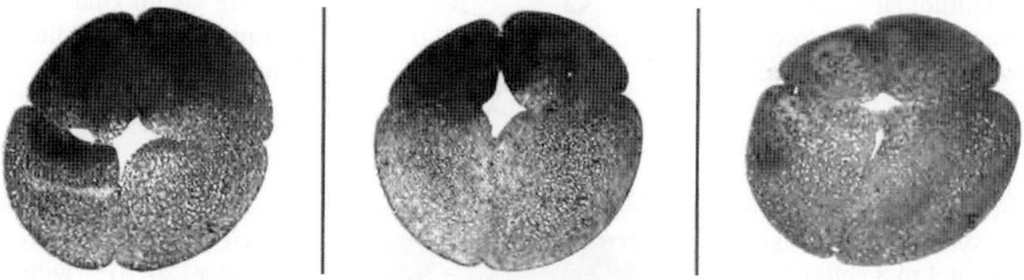

Figure 2.34. The latitude of the third cleavage in *Xenopus* rises as the force of gravity increases: 0G, 1G and 3G left to right (microgravity simulated via a clinostat, normal gravity, and increased gravity simulated via a centrifuge, respectively). The force of gravity was down the page in these embryo slices. It is zero in space or zero on average when the embryo is rotated in a clinostat, which rotates it upside down at regular intervals. Although the number of 3G embryos surviving to the tadpole stage was significantly reduced, surviving embryos at 0G and 3G were nearly normal, with only slight head and eye dysmorphologies (see Crawford-Young (2006)[213] for a review). From Yokota *et al.* (1992)[214] with kind permission of Amaia Etxebarria of University of the Basque Country Press. Embryo diameter = 1.2 mm.

Cleavage pattern	Illustration	Reference
"Standard"[215] 10-12 synchronous divisions (synchrony lost at 1024 to 4096 cell stage) #1 longitudinal #2 longitudinal #3 latitudinal Examples: axolotl (*Ambystoma mexicanum*), *Xenopus laevis*	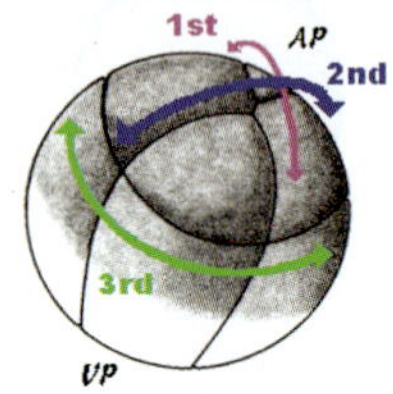	From Developmental Biology Online[216] with kind permission of Colin DeMill.
5 or more synchronous divisions (lost at 32 cells or more) #1 longitudinal #2 longitudinal #3 longitudinal Sole example: *Eleutherodactylus coqui*[217,218]	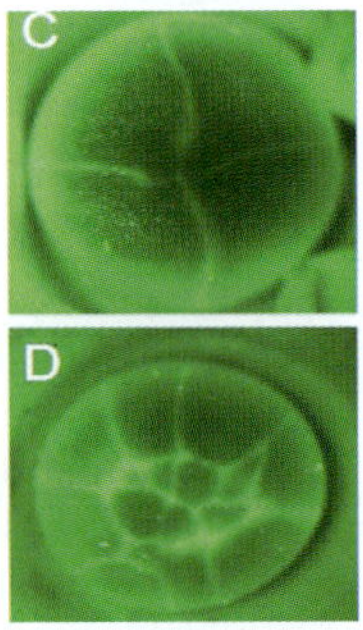	C: VP view, D: AP view. Diameter: 35 mm. From Elinson *et al.* (2011)[219] with permission under a Creative Commons Attribution License.
3-4 synchronous divisions (lost at 8 to 16 cells) #1 longitudinal #2 longitudinal #3 latitudinal Examples: *Myobatrachus gouldii*[220], *Heleioporus eyrei*[221]	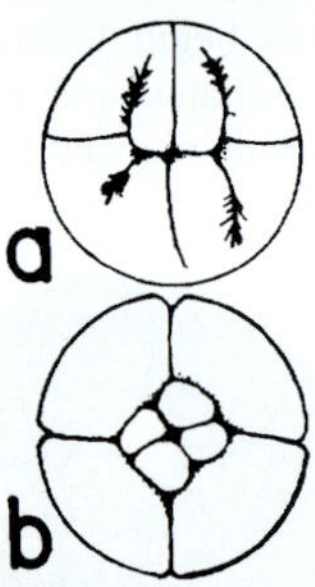	AP view. a: 3rd cleavage in progress, b: completed. Diameter: 2.6-2.9 mm. From Packer (1966)[221] with kind permission of the Editor, Christopher Beachy.
2-3 synchronous divisions (lost at 4 to 8 cells) #1 longitudinal #2 longitudinal #3 longitudinal Examples: *Gastrotheca riobambae*[222], *Ensatina eschscholtzii*[223]	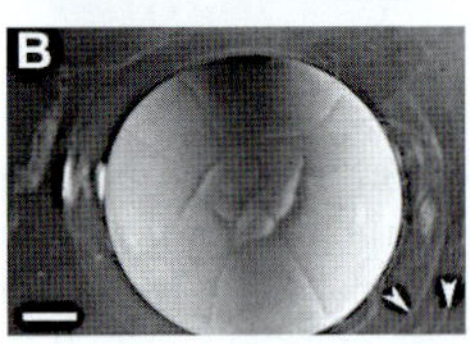	AP view. From Collazo & Keller (2010)[223] with permission under a Creative Commons Attribution License. Bar = 1 mm. Arrows point to jelly layers.

Figure 2.35. Amphibian embryos exhibit four different kinds of cell cleavage[215], reverse ordered here by the number of synchronous cell divisions. AP = animal pole, VP = vegetal pole.

vegetal pole, i.e., from the North Pole towards the South Pole. Latitudes are parallel to the equator. These four classes cross anuran/urodele categories, and there is no strict correlation with egg size.

This cell division seems to be roughly coordinated by waves of contraction of the cortex that start at the top of the embryo and then move downwards[202,224]. These waves also start out synchronously but as cleavage progresses the wave moves more quickly on the left side than on the right side (Figure 2.36). We speculate this is due to asymmetries of the cytoskeleton that causes these waves, something we will discuss in Chapter 5.

In terms of evolution, large, yolky eggs with only a few synchronous divisions may have come first[225]. If one looks at the evolutionary tree leading up to mammals and compares yolk size of eggs relative to the rest of the embryo, the yolk of the amphibian embryo is the largest yolk

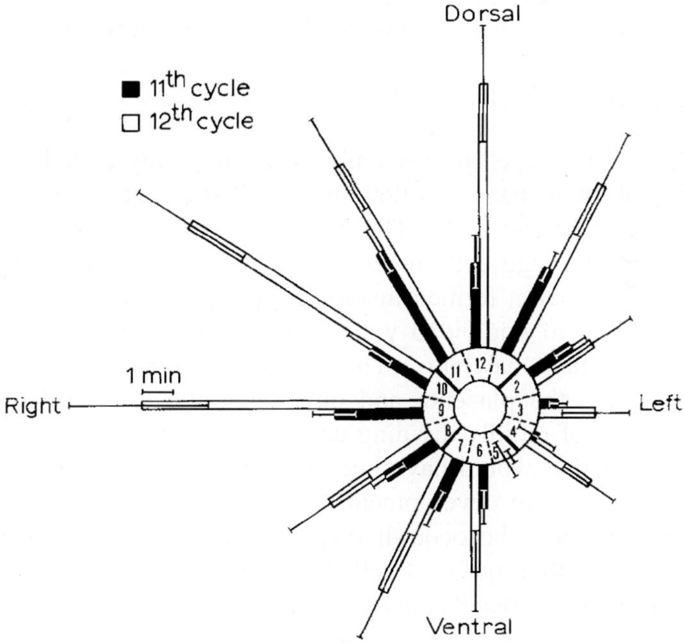

Figure 2.36. In *Xenopus laevis*, which has "standard" cleavage waves, there is an as yet unexplained left/right asymmetry in the durations of the final synchronous waves of cell division. From Boterenbrood *et al.* (1983)[212] with permission of Springer.

mass that is moved to the inside of the embryo mostly within the endoderm layer although each cell has some stored yolk. In reptiles, birds and marsupial mammals, the embryo develops on top of the much larger yolk. At no time does the yolk move to the interior. The difficulty in getting the yolk inside becomes apparent towards the end of gastrulation. The move to development with the embryo sitting on top the yolk mass was a major step in evolution. Mammals have cleavage that is pretty well even during the early stages but mammals have special early adaptations to make the placenta and support tissue which we will examine later. Their separate yolk sac varies greatly between mammalian species[309].

By Stage 8 of the axolotl, the ball of cells now has a hollow top called the blastocoel under a loose dome of small cells (Figure 2.37). The lower half is solid cells with no open space between them. The cells on the bottom are much larger than the cells on the top. They are heavy with yolk. The loose heavy yolk-filled cells on the inside do not normally contribute to the eventual construction of the embryo. They form nutritive cells that sustain and nourish the embryo and the new hatchling (Figure 2.38) until about a week after hatching. Once the larvae can start to hunt and eat, these cells vanish.

The "ball of cells stage" of any embryo is commonly called a blastula. Amphibian blastulas have a hollow, water filled chamber under a cap of cells (Figures 2.39 and 2.40). This hollow is called the blastocoel with the yolky nutritive egg cells inside on the bottom. In fish, reptiles and birds, the blastocoel is formed inside the clump of cells perched on top of the huge blob of undivided yolk. In mammals, the blastocoel forms before mammalian hatching. The outer layer of the mammalian blastocyst forms the placenta and outer amniotic sac. Tucked on the interior, a mass of rapidly dividing cells which form the embryo proper sits in the spot that would have been nutritive yolk cells in the axolotl (Figure 2.41). Later in development, the inner cell mass of the embryo will reform a second blastocoel-like space as part of normal mammalian development but that space is called the exocoelom to differentiate it from the original blastocoel (Figure 2.42).

In mammals, the blastocoel fluid also contains nutrients[226] and many proteins presumed to be of importance in development[227] (Figure 2.43). The internal pressure can even be restored if the fluid is removed and the

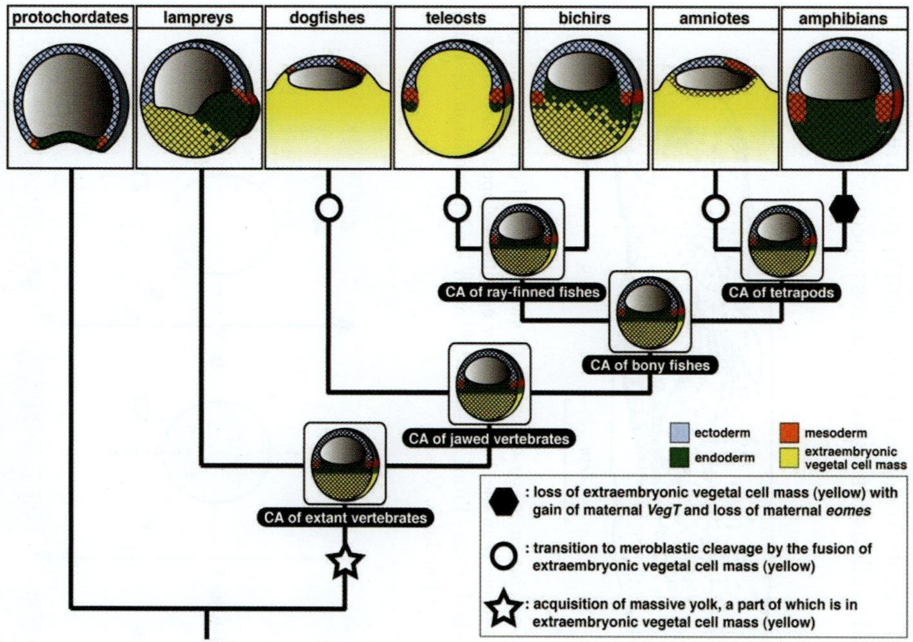

Figure 2.37. The three germ layers give rise to all of the tissues in most animals and are depicted here at the blastula/early gastrula stage for a variety of embryos. "Germ layer formation in each lineage of vertebrates and their common ancestors (CA) are illustrated schematically. In early gastrula stage embryos (dorsal at right), embryonic ectoderm, mesoderm, endoderm and extraembryonic area are indicated by sky blue, red, green and yellow, respectively. Mesh implicates the areas where cell cleavage takes place." The shaded grey area is the blastocoel. From Takeuchi et al. (2009)[228] with permission of Elsevier. Protochordates include sea squirts (tunicate, ascidians) and amphioxus. Bichirs are a primitive type of fish. Amniotes include reptiles, birds and mammals. The embryos are arranged in a low-resolution phylogenetic tree.

needle hole allowed to heal[229]. In amphibians this blastula is visually not very interesting but something critical happens. This is called the midblastula transition, which means it is at this stage that the embryo begins the transition from running pretty much automatically on stored mRNA to producing its own mRNA[230]. This is followed by the production of new proteins[231]. It is also when the synchrony of the cell divisions nearly vanishes in *Xenopus* though there is no correlation with loss of synchrony and midblastula transition across species[232,212,211].

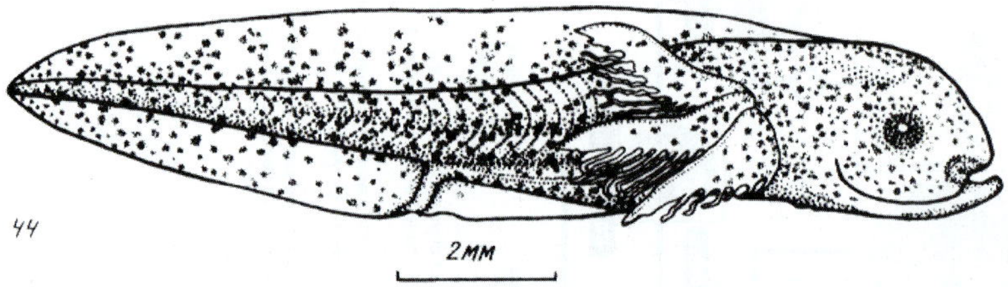

Figure 2.38. The axolotl hatches at Stage 44, ready to swim, see and eat. Legs develop later. Note that it is much larger than the 2 mm diameter egg it developed from, but is no heavier, except for absorbed water. From Sal-Site[114,203] with kind permission of Laura Muzinic, Chris Muzinic, and Randal S. Voss.

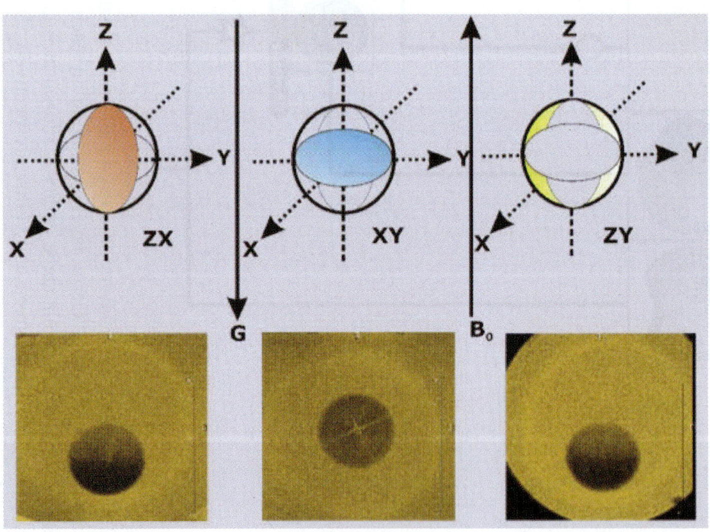

Figure 2.39. Development of the blastocoel in axolotl embryos (2 mm diameter) inside the jelly sac, as visualized with magnetic resonance imaging at 11.7 Tesla. Brightness is

proportional to local water concentration. The same live embryo at the 4-cell stage is shown as if sliced across its center in three views: the zx plane, the xy plane, and the zy plane. Note the yolk in the bottom hemisphere (zx and zy images) and the initial formation of the blastocoel (xy image). **G** is the direction of gravity and $\mathbf{B}_0$ is the direction of the magnetic field. From Gruwel *et al.* (2007)[233].

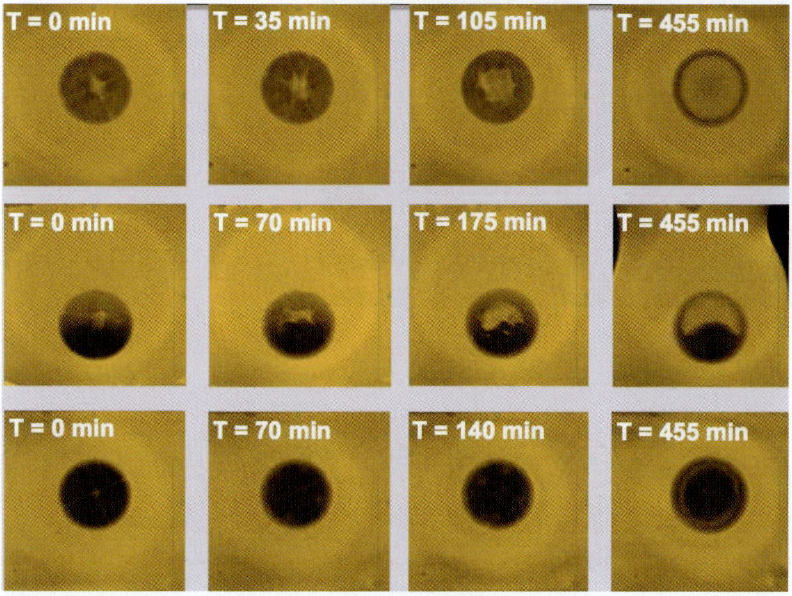

Figure 2.40. Development of a 2 mm axolotl embryo as a function of time using MRI (magnetic resonance imaging). The top row shows a horizontal slab, 0.6 mm thick near the animal pole, parallel to but above the xy plane in Figure 2.39. Development near the corresponding vegetal pole is shown in the bottom row (parallel to but below the xy plane). The middle row shows development in a plane zx (vertical slab) showing both the animal (top) and vegetal (bottom) poles of the embryo. The enlargement and shaping of the blastocoel can be seen in the first two rows. Details[233] for this figure and Figure 2.39: "Individual axolotl eggs, with jelly-coat to provide humidity and protection, as well as allowing the egg to stay oriented, were placed in a small, open-ended glass cylinder (7 mm inner diameter) and put inside a home-built MR-coil for use at 11.7 T (500 MHz for protons). Typically spin-echo experiments were carried out with TE = 12 ms, TR = 1500 ms, data matrix = 256x256, FOV [field of view] = 0.92x0.92 cm^2 and slice thickness of 0.6 mm, resulting in an in-plane resolution of 35x35 μm^2. All experiments were performed at room temperature (20±1°C). Each experiment [in Figure 2.39] collected data for two parallel slices in 6.5 min [Top]. Time-resolved experiments [shown here] were performed in a similar fashion, however, now the MR-signal was time-averaged over 35 min per acquisition to improve the signal/noise ratio. After the experiment, the embryo was monitored and data from embryos with arrested development or abnormalities was discarded."

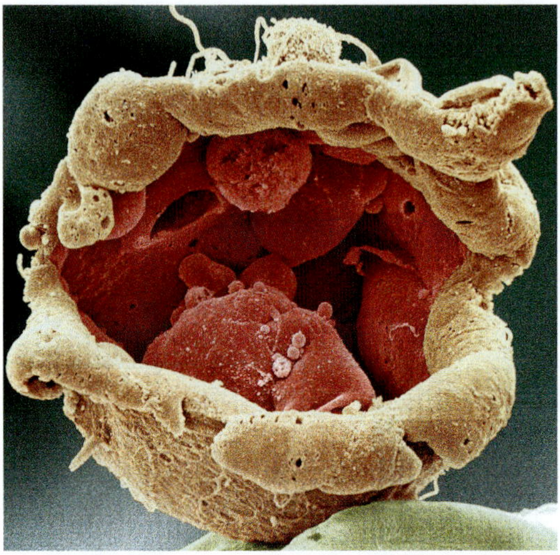

Figure 2.41. This color enhanced image of an opened day 5 human blastocyst shows the inner cell mass. The inner cell mass, which may start from a single cell that is internalized during compaction (Figure 2.44), gives rise to the embryo, whereas the outside cells become the placenta and support tissues like the amniotic sack, all of which are discarded at birth. From Eikonika[234] with kind permission of Yorgos Nikas.

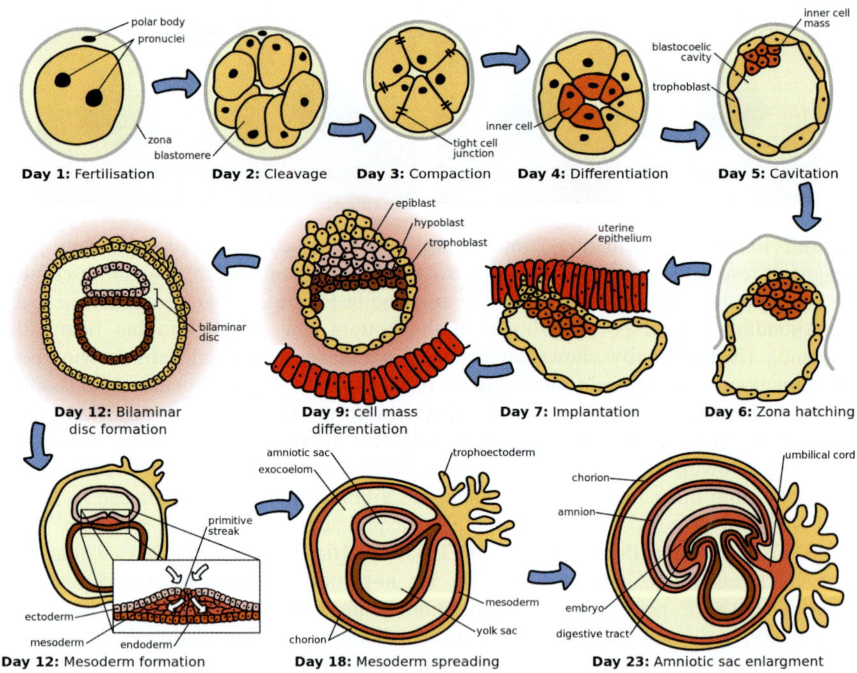

Figure 2.42. Early human embryogenesis. From Wikimedia[235] with permission under a Creative Commons Attribution-Share Alike 3.0 Unported license.

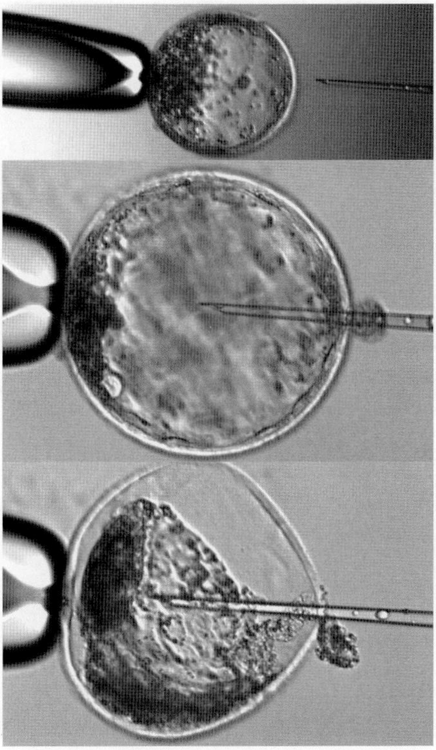

Figure 2.43. Technique for removing the blastocoel fluid from a bovine (cattle) embryo. The embryo is held with light negative pressure in the larger pipette on the left, and the microneedle entering from the right is used to extract the fluid, collapsing the blastocoel. With kind permission of Pernille Linnert Jensen. Cf.[227]. Blastula diameter[236] = 130 µm.

Birds, reptiles and fish also have a midblastula transition that happens at about 10 cycles of cleavage in the small hollow mass sitting on the yolk blob. The equivalent to the midblastula transition in mammals happens much earlier and is called compaction[237] (Figure 2.44). All mammals go through a compaction stage after a specific number of cleavages but before the blastocoel forms. Most cells end up outside at compaction and these cells then go on to form the outer support tissues. A few end up inside and form the inner cell mass (Figure 2.41). The initial number of cells in the inner cell mass of mammals can vary between a single cell and several cells depending on the species (Figure 2.45). In some mammals there is more than one round of compaction. Cells that form the inner cell mass are moved to the interior in more than one compaction event.

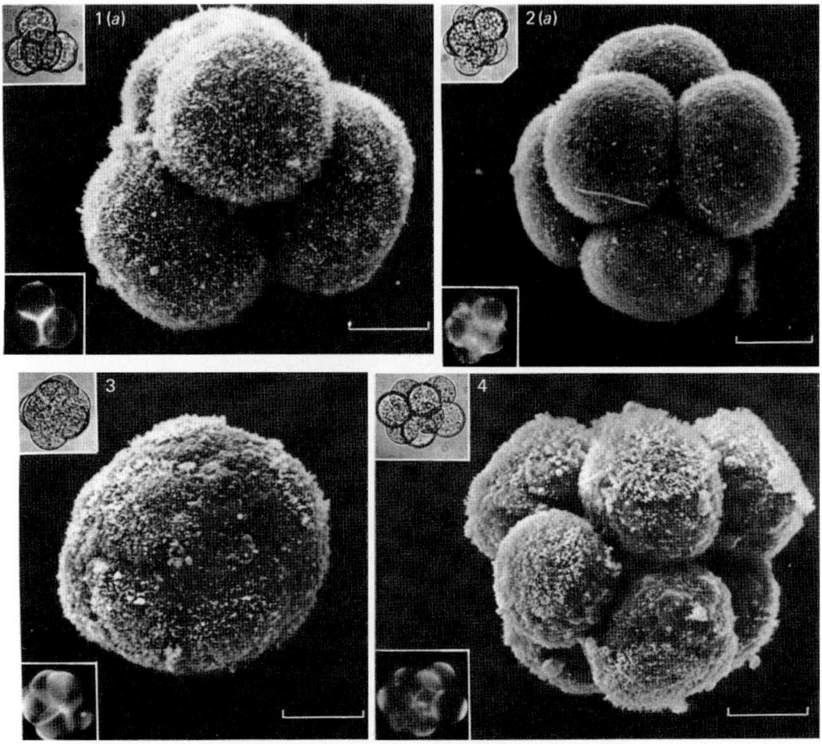

Figure 2.44. 1(a): A 4-cell mouse embryo. 2(a): An 8-cell mouse embryo. 3: An 8-cell mouse embryo that has undergone compaction, probably with one cell inside, initiating the "inner cell mass". 4: A decompacted 8-cell embryo. Bars are 10 μm. Images are scaled so that all scale bars are the same size on this page, emphasizing how strong the compaction had been. Main images are SEMs (scanning electron micrographs). Upper left insets are bright field light microscopy images. Lower left insets are fluorescence microscopy images. From Reeve & Ziomek (1981)[238] with permission of Company of Biologists.

Sea urchins, which share a far distant and more primitive common ancestor that predates the vertebrate/invertebrate line separation, also have an opening that first forms in early cleavage but that opening is not an early blastocoel. Rather, the opening allows circulation of seawater through the embryo to each cell. Cells do not make close contact except through fine filopodia. Epithelium does not form until much later, at about 200 cells. This provides clues about how the blastocoel may have originally evolved (Figures 2.46 and 2.47). In clam embryos, a proper, enclosed blastocoel forms at the four cell stage (Figure 2.48).

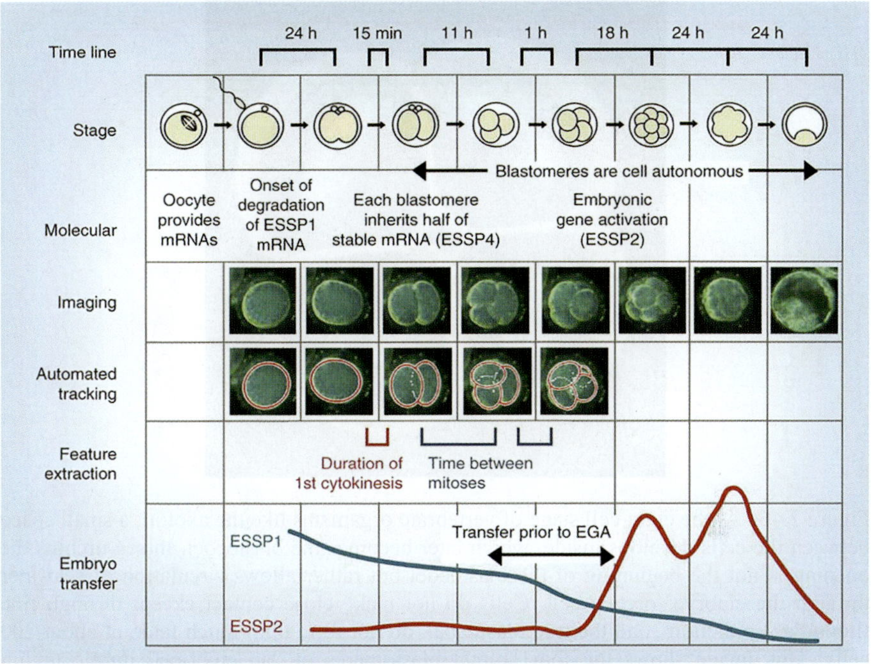

Figure 2.45. "Human embryos begin life with a set of oocyte RNAs inherited from the mother. After fertilization, a subset of maternal RNAs specific to the egg (ESSP1) must be degraded as the transition from oocyte to embryo begins. As development continues, other RNAs are partitioned equally to each blastomere (ESSP4). At EGA [Embryonic gene activation], ESSP2 genes are transcribed in a cell-autonomous manner. During the cleavage divisions, embryonic blastomeres may arrest or progress independently. 'Feature extraction' indicates the three imaging parameters for predicting successful development to the blastocyst stage: cytokinesis, the time between 1st and 2nd mitoses, and the time between 2nd and 3rd mitoses." From Wong *et al.* (2010)[239] with permission of Nature Publishing Group.

Insects such as *Drosophila* have their own variation on cleavage. Instead of having multiple cleavages, they begin by making multiple copies of the nuclei with all the DNA. Typically 6000 nuclei are made. The nuclei move to the outside of the egg spreading all over the cortex. Infolding membrane then extends around each nucleus to create multiple cells. Other species that are more closely related to insects than to us have combinations of full cleavage and the multiplication of nuclei combined with membrane infolding. A few nuclei remain with the egg's yolky material and develop into cells that digest the yolk. This method of

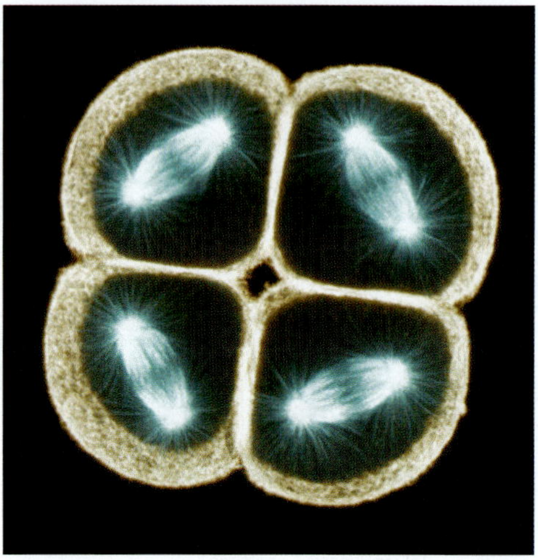

Figure 2.46. At the eight cell stage of vertebrate organisms like the axolotl, a small space between the cells develops inside, which later becomes the blastocoel. In sea urchins the opening is not the beginning of the blastocoel but rather allows circulation of seawater through the embryo to each cell. Cells do not make close contact except through fine filopodia. Epithelium, and the true blastocoel, do not form until much later, at about 200 cells. This image shows the four animal blastomeres of "an eight-cell purple urchin embryo [*Strongylocentrotus purpuratus*], fixed and stained with anti-tubulin (blue) [for microtubules] and phalloidin (orange) [for microfilaments]; projected from serial half-micron sections". From George von Dassow with his kind permission. Embryo width[240] = 90 μm.

cleavage evolved after the separation of vertebrates and invertebrates. Other invertebrates use varying combinations of full cleavage and duplication of nuclei followed by membrane infolding. In insects this membrane infolding also marks the midblastula transition (Figure 2.49).

Two rather remarkable physical signs mark the difference between Stage 8 and Stage 9 in the axolotl. At Stage 8 it is still possible to see the individual cells. If viewed from the side, the egg has a lumpy, slack look on top. At Stage 9 the individual cells are invisible to the naked eye. This is partly because the cells have divided rapidly to become even smaller. The embryo has also changed to a taut smooth top surface if viewed in profile, so that the unaided human eye can no longer distinguish the individual cell boundaries. The embryo swells up, stretching out any differences in texture for the eye to see. The change is so marked that beginning students have no trouble spotting a stage 8 embryo but will

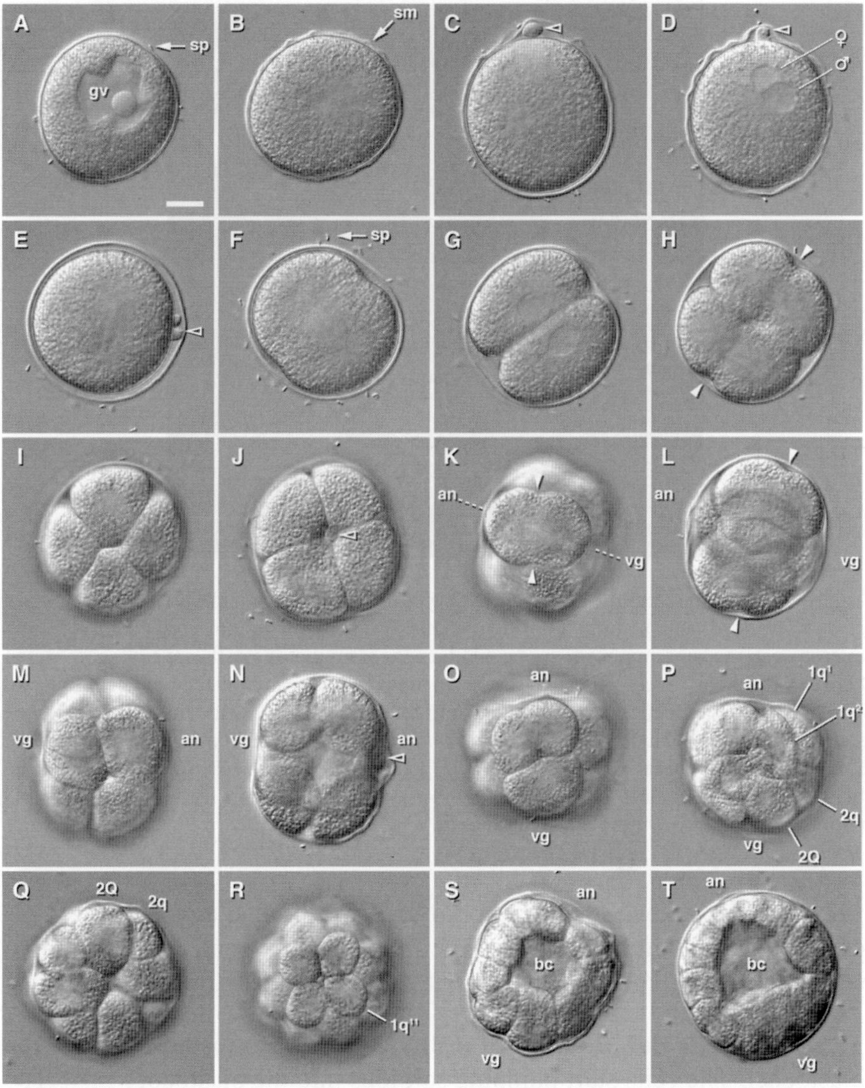

Figure 2.47. The annelid *Owenia collaris* develops a blastocoel (bc) that is initially open to seawater and then closes off by the 64-cell stage (T). gv = germinal vesicle; sm = spawning membrane; sp = sperm; an = animal pole; vg = vegetal pole. Scale bar = 20 μm. This animal has a mix of protostome and deuterostome characteristics. Reprint of Figure 1 from Smart & von Dassow (2009)[241] with kind permission of Carol Schachinger of The Biological Bulletin, Marine Biological Laboratory, Woods Hole, Massachusetts, USA and of George von Dassow.

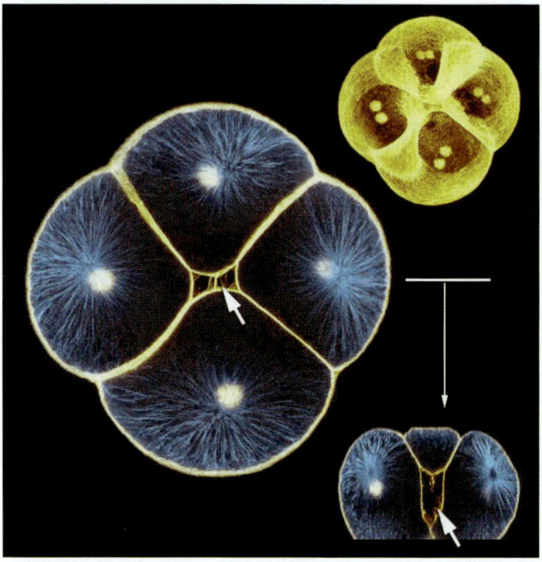

Figure 2.48. **Upper right**: Four-cell embryo of the protobranch bivalve clam *Acila castrensis* fixed and stained for microfilaments using phalloidin; projection of one-micron sections superimposed on one another, showing that the cells are fully compacted between divisions. Each cell shows a pair of centrosomes ready for the next cell division. The blastocoel is visible in the middle of the four cells. **Left**: a slice though a 3D image stained for microtubules of a four cell *Acila* embryo before centrosome duplication. **Lower right**: Vertical slice through the position indicated. Arrows show that the blastocoel in this embryo is a closed chamber. From George von Dassow with his kind permission. Embryo width[242] is about 60 μm.

frequently mistake a Stage 9 embryo for a freshly laid egg, Stage 1. This change marks Stage 9. This may be thought of as similar to inflating a soccer ball, which goes from flaccid to taut (Figure 2.50).

The inflation is due to changes in the cells that form the dome of the blastocoel. The cells start pumping salt ions into the blastocoel[243-245]. The resulting change in osmolality causes the inward migration of water molecules into the blastocoel, increasing the internal pressure[246]. The movement of sodium ions into the blastocoel is caused by a protein in the cell membranes called a sodium pump[247,248,245]. The outside cells are bound tightly together at their apical ends by "tight junctions"[249] (Figure 2.51). The total volume of the embryo increases, without much increase in surface area, which makes the surface taut.

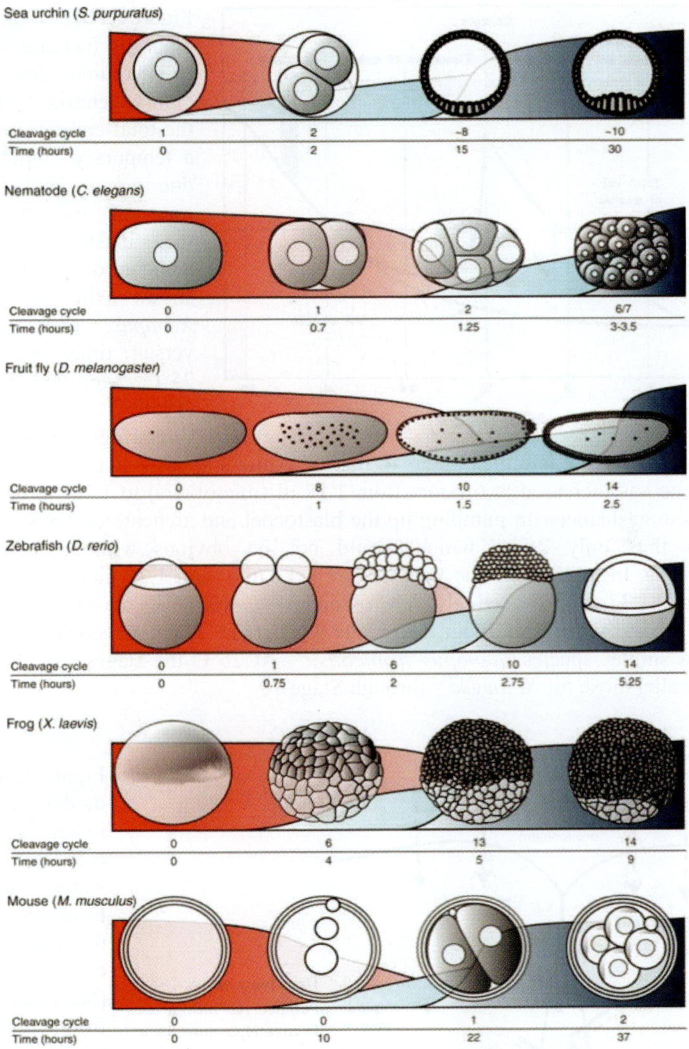

Figure 2.49. "A comparative overview of the maternal-to-zygotic transition (MZT) in several model organisms. Key embryonic stages for each model organism are depicted schematically above the corresponding cleavage cycle and time after fertilization. The red curves represent the degradation profiles of destabilized maternal transcripts in each species. The light and dark blue curves illustrate the minor and major waves, respectively, of zygotic genome activation. The last embryonic stage presented for each organism is the developmental point at which there is a major requirement for zygotic transcripts." From Tardos & Lipshitz (2009)[231] with kind permission of the Company of Biologists. The MZT (= midblastula transition) includes degradation of the maternally derived mRNAs[250-252].

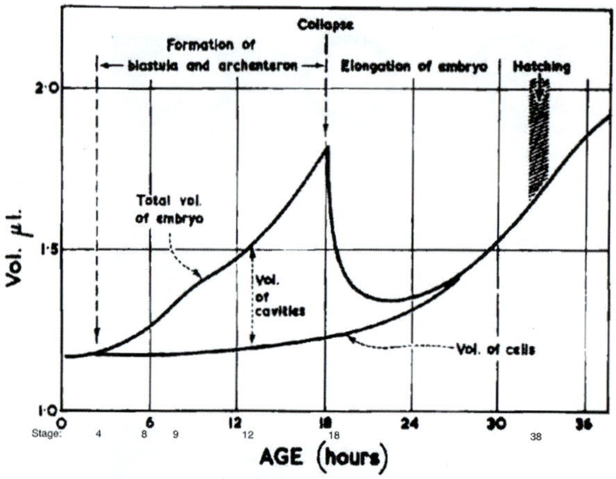

Figure 2.50. Amphibian embryos increase their total cell volume slowly during embryogenesis. However, the total embryo volume has a temporary rapid increase due to water pulled into two transient internal cavities, the blastocoel and the archenteron. This figure shows the volume of *Xenopus laevis* embryos versus time, measured at 25°C. Modified from Tuft (1958)[253] with permission of Cambridge University Press. Staging of *X. laevis*[208] is quite similar to that of axolotls at early stages. In this plot the total volume has increased at 6 hours from 1.17 μl (microliters) to 1.26 μl, or by 7.7%. The increase in diameter in pumping up the blastocoel and archenteron between Stage 8 and 9 is thus only 2.5%, which would not be obvious without these careful measurements. In millimeters the volumes are 1.17 mm^3 and 1.26 mm^3, with diameters 1.31 mm and 1.34 mm, respectively. Approximate *X. laevis* stages have been added to the diagram. The transition from Stage 8 to Stage 9 occurs at 25°C on average at ages 5 to 7 hours in a similar species, *Xenopus tropicalis*[254]. At 25°C the stages for *X. tropicalis* closely parallel those for *X. laevis*[254] through Stage 10.

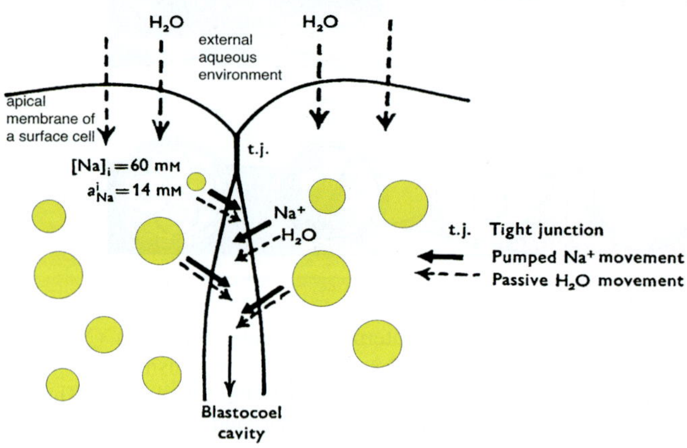

Figure 2.51. In this model the tight junctions between ectoderm cells keep water and ions from getting into the spaces between cells. Sodium ions (Na^+) are released from yolk platelets (yellow circles) and pumped into the spaces between cells by membrane proteins called sodium pumps. The increased salt concentration pulls water out of the cells into the intercellular spaces, which enlarge, forming the blastocoel. More water enters the cells from the outside environment through the apical cell membranes that are exposed to the external environment. Modified from Slack & Warner (1973)[249] with permission of John Wiley and Sons.

Even more critically, by the end of Stage 9 the cells have been determined to become three distinct cells types. The top of the ball of cells has changed into presumptive ectoderm. The middle section has changed into presumptive mesoderm, and the bottom half, the heavy yolk part, has become presumptive endoderm. Ectoderm, mesoderm and endoderm represent the three precursor cell types from which all future cell types are derived. All types of tissue in the adult organism can be traced back to one of these three tissue types. They are commonly called the "germ cell" layers[255] (Figure 2.37).

Ectoderm gives rise to the skin, brain, spinal cord and nerves. Mesoderm eventually forms all the middle parts of the embryo such as the muscles[256] and skeleton[257,258]. Endoderm forms all future interior parts such as liver[259], pancreas and intestines. The heart, like other important organs in the gut and chest, has an outer layer of mesodermal origin and an inner layer of endodermal origin[260,261].

The same determination of three essential germ cell layers also occurs in birds, fish and mammals. Mammals have that extra set of cells, the ones that form the placenta and support tissue but as those are discarded at birth their formation is not counted as a germ cell layer. In birds the embryo forms as a flattened plate on top of the undivided yolk with the three cells types. In zebrafish the three germ layers are similar with a more rounded embryo but the same basic layout. In mammals, the three layers are formed from the inner cell mass as the embryo develops. In fact, this common body plan goes all the way back to our first vertebrate ancestral forms. Invertebrates also have the same essential three layers pushing the three primary layer concept even further back along our common ancestral tree. Sea urchins have three primary layers that are remarkably similar to amphibians (Figure 2.52). Insects also have three primary layers which include mesoderm, endoderm and the ectoderm equivalent known as dorsal epidermis.

The embryo now begins two of the most marvelous and complicated stages of development imaginable. These stages are called gastrulation and neurulation. This simple three-layered embryo (Figure 2.53) changes itself to form the basic body plan of the adult organism. By the end of gastrulation the embryo will be a complex multilayered organism with a front end and a back end and a clearly defined top and bottom, with future internal organs ready to be created (Figure 2.33). By the end of neurulation a future spinal cord and brain are set in place and the whole outside is also completely covered with a layer of future skin tissue (Figures 2.54, 2.55 and 2.56).

Gastrulation in the axolotl begins with invagination[262] (also called involution, Figure 2.57), a process by which part of the outside surface tucks inside. In order to conceptualize this change in shape (and cell type, as we shall see later), let's pretend for a moment that the embryo has only two layers instead of three. We can then consider the following simplified image of gastrulation. Imagine you have a partly inflated ball. Poke your thumb into that ball, creating an indent (Figure 2.58). This is similar to how gastrulation begins, anatomically speaking. No one is poking the ball of cells with a finger but the result is similar. An indentation forms and the indentation then grows as more of the material enters the inside of the ball. With a simple ball you would end up with a half shell shape. The developing embryo is not so limited. The outside layer of the half shell shape expands as the inner layer moves inside. The inside layer tightens up at least on average until the entire lower half is tucked inside the upper half. All of the endoderm and then mesoderm ends up inside and the ectoderm is stretched to cover the outside. The closing of the ectoderm over the last visible bit of mesoderm marks the end of gastrulation.

It has long been a puzzle of just where the forces are generated that result in gastrulation. If we take a deformable hollow ball, there are at least four ways that we can get it to assume a double-layered cup shape (Figure 2.59):

1. We can push it in at one point, using a thumb. As embryos usually don't come with external thumbs, this mechanism is unlikely.
2. We can reach inside with one or more fingers and pull the opposite side in.
3. We can push in on one hemisphere.
4. A portion of the surface can squeeze in.

Except for the external thumb, one can imagine that each mechanism is plausible in embryos, and that during evolution transitions from one mechanism to another would be possible[263,264], with combinations being intermediate stages.

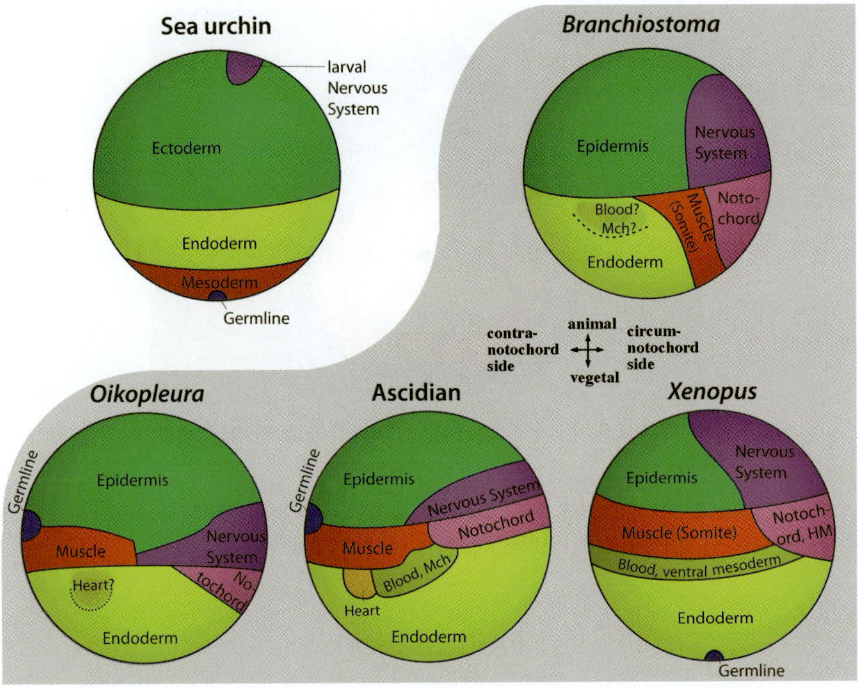

Figure 2.52. Fate maps for five organisms drawn onto their blastula stages. Ascidians are tunicates whose embryos develop into tadpoles that later become sessile adults (Figure 2.17). *Oikopleura* is a tunicate that stays in the larval, tadpole form throughout its life[265]. *Branchiostoma* is an amphioxus[266] (Figure 2.8). From Nishida & Stach (2014)[267] with kind permission of Hiroki Nishida.

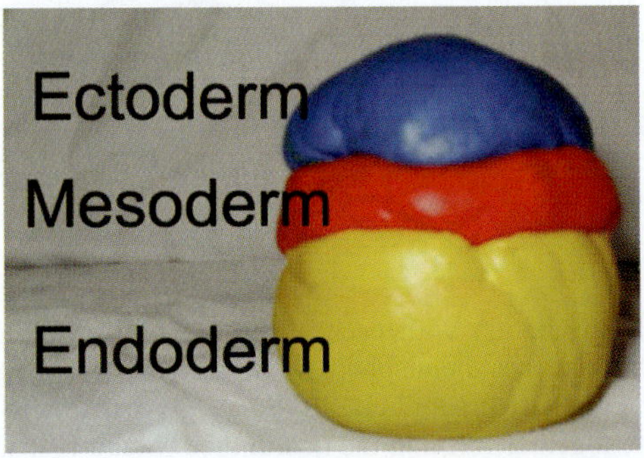

Figure 2.53. Germ cell layers begin as three distinct top to bottom layers all with surface on the outside of the embryo. Note that the epidermis + nervous system of Figure 2.52 in green and purple = ectoderm in blue here.

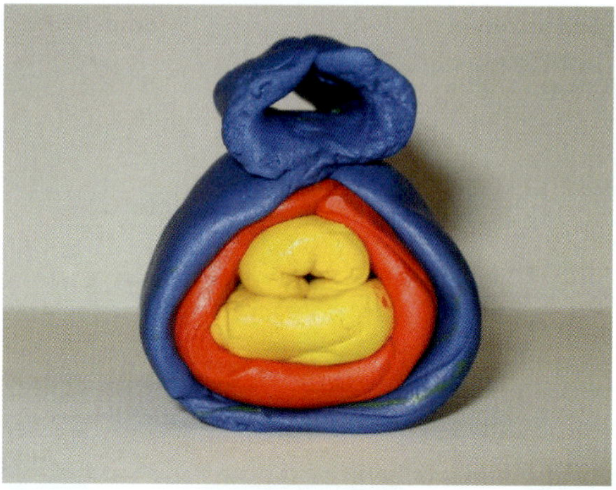

Figure 2.54. Germ cell layers end with all the yellow endoderm in the center, the ectoderm on the outside, and the mesoderm in between the ectoderm and endoderm. (The neural tube will actually be dropped inside the embryo by the time it is fully closed.)

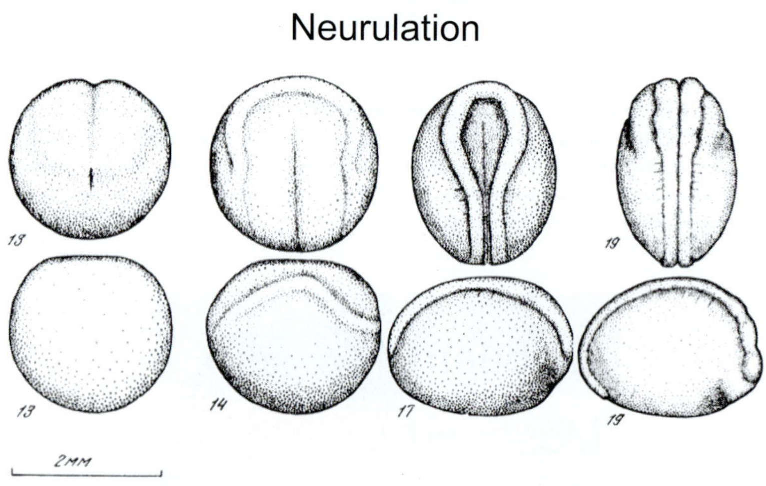

Figure 2.55. Axolotl neurulation with each stage shown from top (**top row**) and side (**bottom row**) views. The top hemisphere of the ectoderm (Stage 13) changes to a keyhole shaped neural plate (neuroepithelium) (Stage 14), narrows (Stage 17) and then its edges roll up and fuse together forming the neural tube (Stage 19). The wider region will eventually form the brain and the narrower region will form the spine. The rim of the neural plate contains the cells that will become neural crest cells. Note the beginning of the formation of the eyes in the side view on the right at Stage 19. From Sal-Site[114,203] with kind permission of Laura Muzinic, Chris Muzinic, and Randal S. Voss.

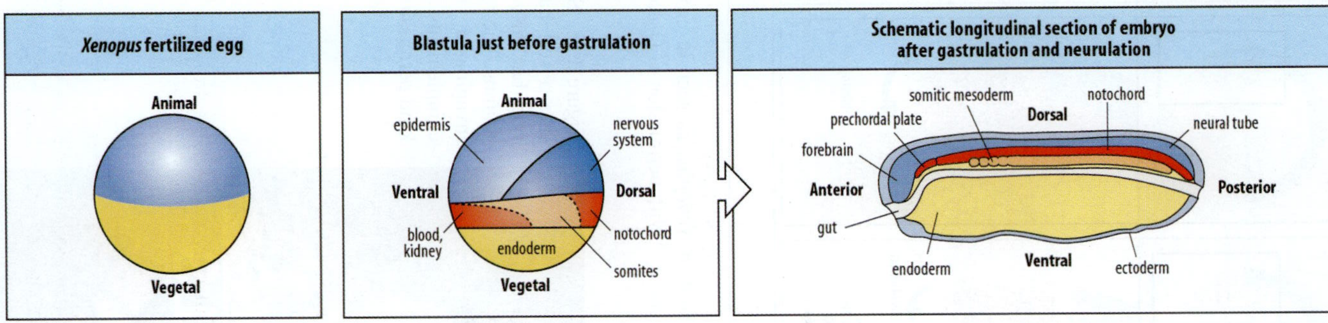

Figure 2.56. The skin (epidermis) of the frog *Xenopus* is traced here from the animal hemisphere to the ectoderm, which becomes epidermis or the nervous system (neuroepithelium) to the outer surface of the larva (called ectoderm in the third image because it is still presumptive epidermis not yet changed from ectoderm to skin). From Wolpert *et al.* (2015)[268] with permission of Oxford University Press.

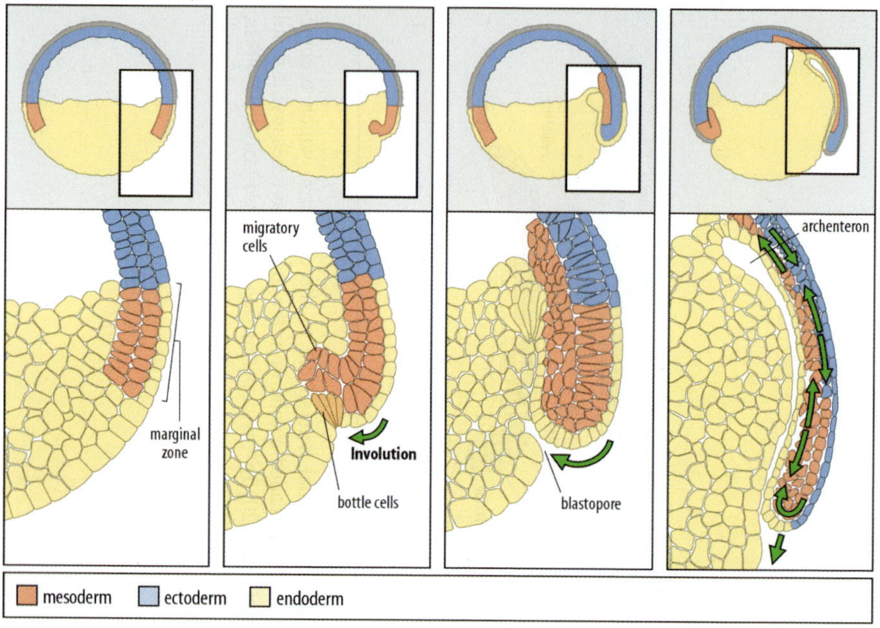

Figure 2.57. The involution movement in *Xenopus*, as seen in cross section along the midline. From *Principles of Development*[268] with permission of Oxford University Press.

Figure 2.58. Sticking your thumb into a soft ball imitates gastrulation, but early embryos don't have thumbs. The result is a cup shaped, invaginated ball. "Fruitballs"[269] complements of Suck UK, with kind permission of Xanthe.

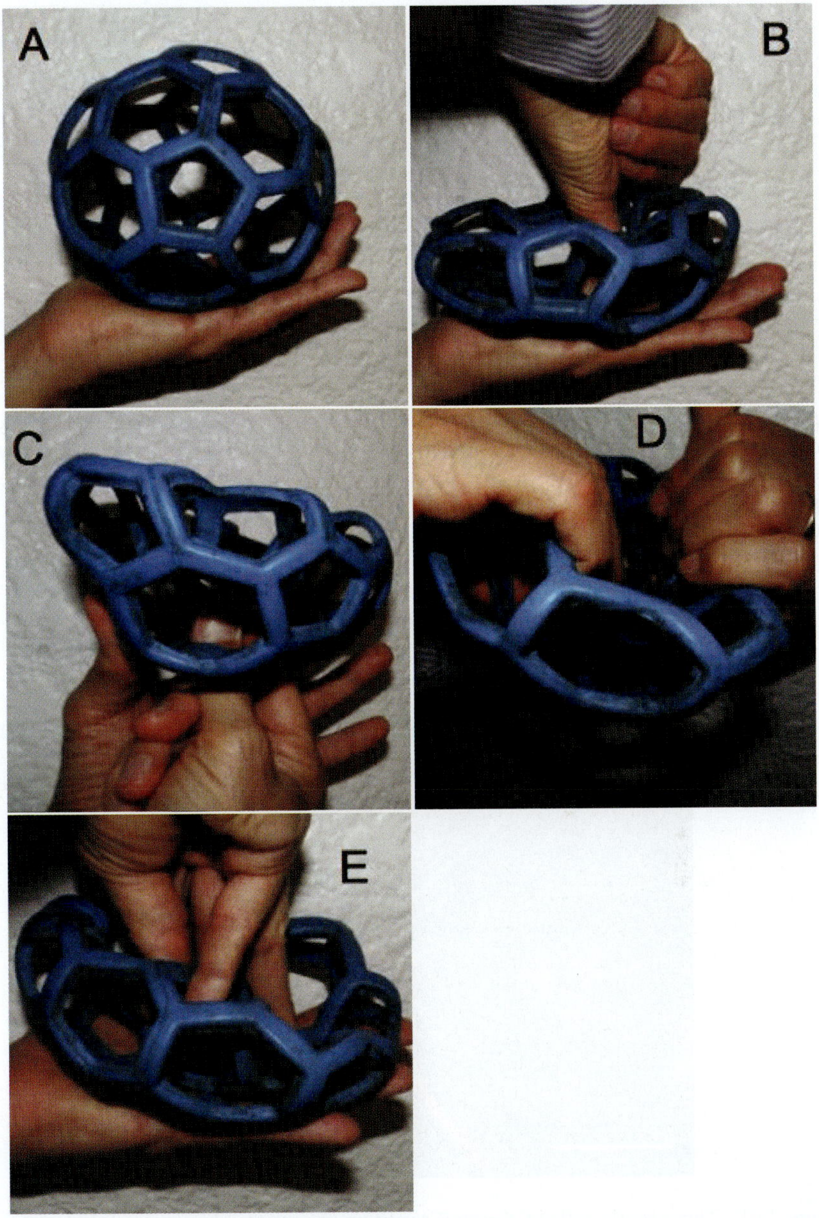

Figure 2.59. Forces can be applied in many ways to result in the deformation of a ball into a cup, emulating gastrulation. A. Undeformed ball. B. An external force (a thumb) cups the ball. C. Reaching inside to pull the opposite side in. D. Pushing in on one hemisphere. E. Pinching in a portion of the surface.

Figure 2.60. The cells in the dorsal lip of the blastopore in amphibian embryos temporarily take on a bottle shape, similar to this hand blown wine bottle. The nucleus moves down to the wider part, a process called "interkinetic migration"[270]. Interkinetic migration is not necessarily followed by cell division in salamanders[35]. From Etsy[271] with kind permission of Sunday White.

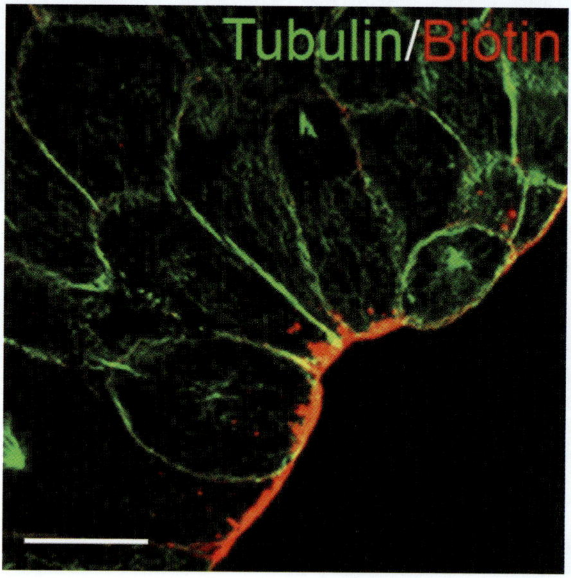

Figure 2.61. These bottle cells in *Xenopus* are lit up in green to show microtubules (made of tubulin) and red, to show where fluid from the outside has come into the apical surfaces of the cells via endocytosis. Endocytosis appears to be an important component of the mechanics of invagination. The bar is 25 μm. From Lee & Harland (2010)[272] with permission of Elsevier.

An important feature of gastrulation is the dorsal lip of the blastopore. This region is the ridge where cells are moving inside. The blastopore is the arc of cells around the "invisible poking thumb". To get over the dorsal lip and inside the cells undergo a dramatic shape change temporarily becoming what are called bottle cells. They were given this name because they resemble old-fashioned hand blown wine bottles (Figure 2.60) with the long necked shape they take on as they slip over the curve (Figure 2.61). The cells that are on the lip itself roll inside and they are replaced by the cells moving inside after them. This means that from one minute to the next which cells are present at the lip is different and always changing even though the structure they form remains present.

In spite of the fact that the cells on the dorsal lip of the blastopore are always changing, the dorsal lip is considered to have near magical properties for determining cells. It is often called "the organizer" as if this lip can somehow organize the body plan of the embryo. The organizer is said to have powers to induce differentiation. It is our opinion that the shape change from flat cell to bottle cell creates a mechanical signal to the nucleus as they roll over and that is what the organizer is. In any case, the organizer is a different set of cells from one minute to the next. It is a wave of shape change propagating through a sheet of cells, not a specific set of cells. More about that in Chapter 7. It doesn't look like a moving wave because in our frame of reference it more or less stays in the same place. Also not all cells roll inside over the dorsal lip of the blastopore. Some slide in below the dorsal lip and in doing so get pulled out long and flat instead of being changed into the bottle cell shape.

The fish, reptile, bird and mammalian equivalent of the dorsal lip of the blastopore is the primitive streak (also called Hensen's node in mammals and Kupffer's vesicle in fish). Mesoderm is formed in the primitive streak as cells round inward over the node of the primitive streak (Figure 2.62). Simpler invertebrates have a very similar process (Figure 2.18). Even the far distant and highly specialized insects have a homologue to the dorsal lip of the blastopore[273].

As the material moves inside the blastocoel becomes smaller and smaller and a new space, called the archenteron, replaces it. This new hollow space will eventually form the primitive gut which then becomes the

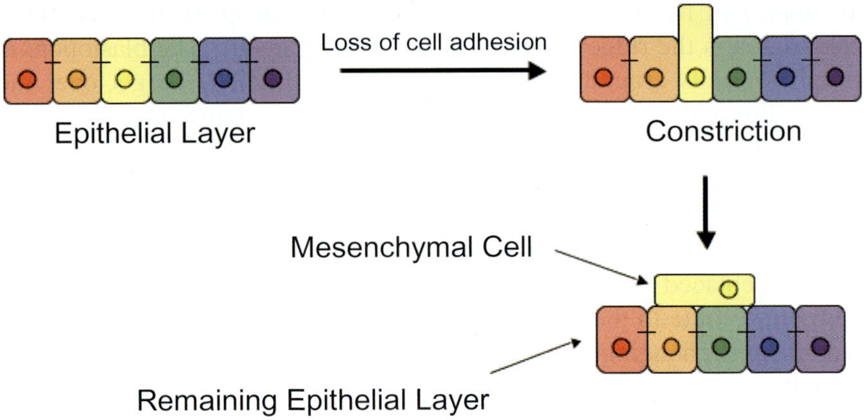

Figure 2.62. Mesoderm forms as it goes over an edge (such as the dorsal lip of the blastopore in frogs or Hensen's node in mammals) and individual cells migrate out and form the mesoderm layer in the middle. Mesenchyme is mesodermal embryonic tissue that develops into connective and skeletal tissues, including blood. From Wikipedia[274] with permission under a Creative Commons Attribution 3.0 Unported license.

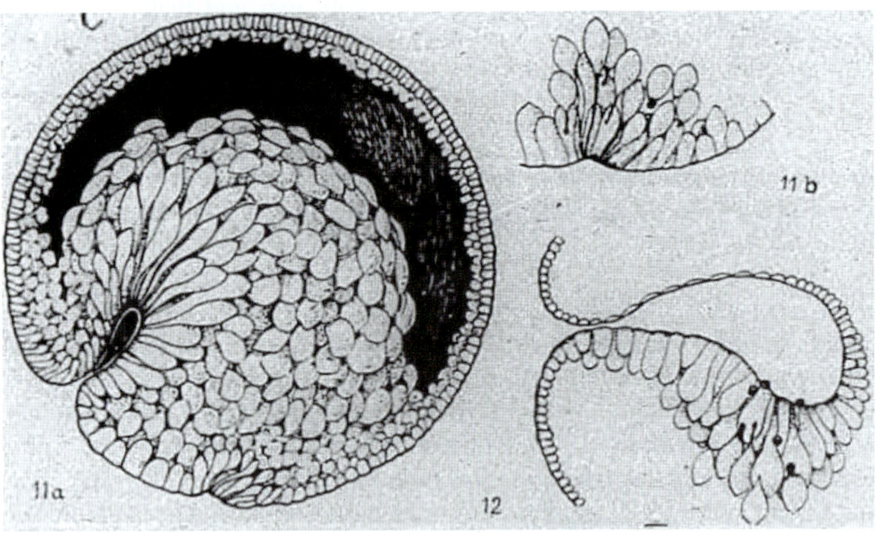

Figure 2.63. "*Amblystoma punctatum*…. Slightly schematized sections through an advanced gastrula: (11): ingression of mesoderm cells at the ventral blastoporal lip. (12): Endoderm cells lining the archenteron; some of the former blastoporal cells have given up their connection with the coat, and their pigmented neck is contracted." From Holtfreter (1943)[275] with permission of John Wiley and Sons.

inside of the stomach, intestines and colon as development continues (Figure 2.63).

Once gastrulation is finished, marked by only ectoderm left on the outside (and corresponding to axolotl Stage 13), the stage of neurulation begins (in vertebrates) (Figure 2.55). A long groove appears on top in the neural ectoderm. The neural ectoderm flattens at late Stage 13 forming a flat plate[39]. The plate widens at the head end and narrows at the tail end to form a keyhole shape. The edges of the rising neural plate meet and form a sealed tube. The tube then sinks down inside the embryo. The epithelial ectoderm continues to stretch upwards as the neural tube sinks, until the entire outer surface of the embryos is covered with epithelial ectoderm. Neurulation is completed with the full closure and sinking down of the tube of neural ectoderm on the top of the embryo, with future epithelial ectoderm (future skin tissue) stretching up to entirely cover the whole outside of the embryo by Stage 19. The neural tube and neurulation is a defining feature of all vertebrates. However, even invertebrates, such as insects, have a “neural cord” equivalent.

As this closure of the neural tube is happening, the neural crest cells are formed. Neural crest cells are individual cells that appear to be ripped out of the boundary (neural crest) between the rising epithelial ectoderm and the sinking neural ectoderm. Once these neural crest cells are free of the cell sheet they round up and differentiate into an entirely new cell type. These individual cells begin migrating throughout the body to specific parts of the embryo. Later some of these neural crest cells will differentiate into nerve cells and form connections between the brain and the rest of the body. Other neural crest cells will form pigment cells that give color to skin. It is a fair assumption that cells which form the future nerve connections to the brain (the peripheral nerves) come from the neural ectoderm ripped loose and cells that make pigment cells in the skin arise from epithelial ectoderm ripped loose but we just don’t know yet.

For the sake of completeness, we now include a word about plant embryogenesis. Plants begin with one cell that divides asymmetrically like a mosaic embryo. The smaller round cell is called the apical daughter cell. It will divide multiple times and form the apical meristem domain. This domain has a central rapidly dividing set of cells that function rather like stem cells. As the cells mature and grow out from this point, they specialize. They go on to form the different types of cells plants require in roots, leaves and shoots. The lower larger and somewhat elongated cell from original first division is called the basal daughter cell. The basal daughter cell undergoes its own asymmetric cleavage and the smaller cell becomes the apical root cap. Plants lack the three primal germ layers, ectoderm, mesoderm and endoderm but they are nonetheless layered in early embryogenesis and follow the same principal of one cell dividing and forming two distinct and different daughter cells and having regions of similar cells in clusters which grow and change during development (Figure 2.64).

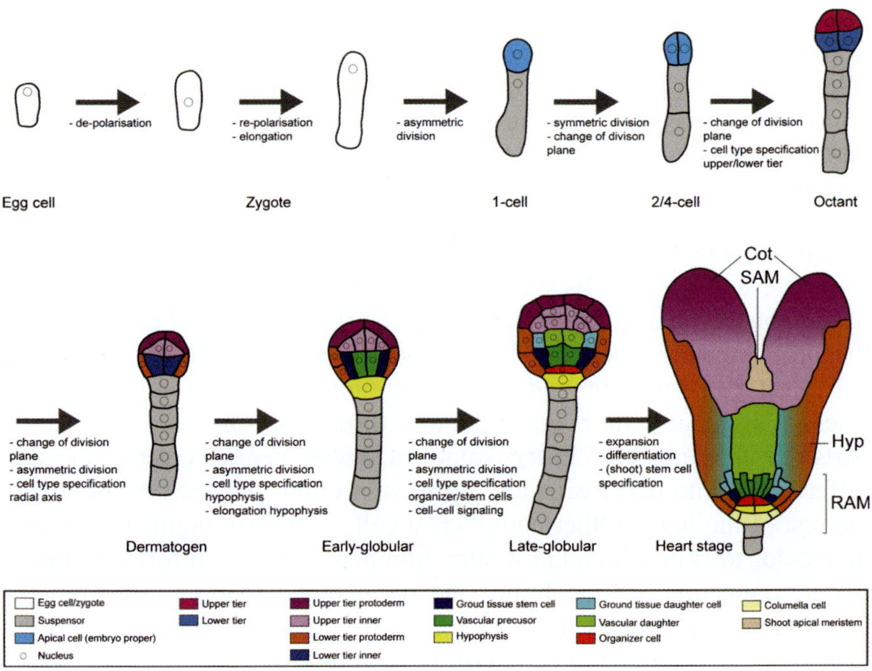

Figure 2.64. Pattern formation during embryogenesis of the plant *Arabidopsis* (Figure 2.13). "Morphogenetic processes during *Arabidopsis* embryogenesis. Schematic overview of *Arabidopsis* embryogenesis from the egg cell to the heart stage embryo, highlighting the morphogenetic processes required to progress from one stage to the next. The colors represent cells of (essentially) the same type (see color legend), based on marker gene expression and lineage analysis. Cot, cotyledon; SAM, shoot apical meristem; Hyp, hypophysis; RAM, root apical meristem." From Wendrich & Weijers (2013)[276] with permission of John Wiley and Sons.

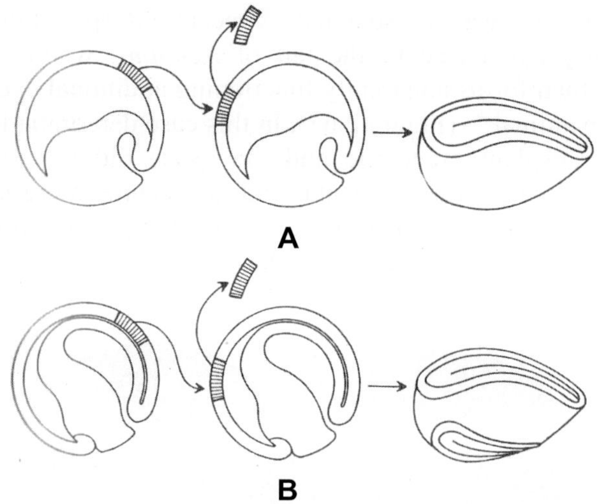

Figure 2.65. The determination experiment. "A. Piece of presumptive neural plate removed from an early newt gastrula, and grafted on to the ventral side of another of the same age; no induction occurs, but it develops in accordance with its new surroundings. B. A corresponding piece removed from a late gastrula develops in the new surroundings according to its original fate." Cf.[277]. In A the transplanted tissue is not yet determined. In B it is determined. From *Primary Embryonic Induction*[278] with permission of Elsevier under the STM (International Association of Scientific, Technical & Medical Publishers) Permissions Guidelines.

We promised we would not hit you with a lot of jargon but there are two specific terms you really need to understand both in depth and in context: determination, and differentiation. At axolotl Stage 8 (Figure 2.33) the cells are not yet "determined". This means that if cells are transplanted into another axolotl embryo at the same stage, into a different location, after healing in[279] they will develop in the new location into the cell type required there as if nothing had happened[280]. These experiments were started by Hans Spemann[281] and Otto Mangold[282-284,105,285]. Mangold showed that patches of early ectoderm and mesoderm could be switched, with normal development ensuing (Figures 2.65 and 2.66).

Thus, determination was first defined by transplanting patches of cells within or between embryos[286]. If a patch of cells would go along its merry way as if nothing had happened, and produce a bit of "wrong" tissue for the place into which it was transplanted, we say it was already determined when it was transplanted. A recent example is the transplantation of early eye tissue to the tail of *Xenopus*, where it continues growing but then forms apparently functioning additional eyes located in the tail of the animal[287] (Figure 2.67). In this case determination occurred before the transplant procedure and the tissue did not pick up cues around it and change developmental path to become more tail tissue. It had already been determined to become eye. Once determined, a tissue type cannot ordinarily be undetermined.

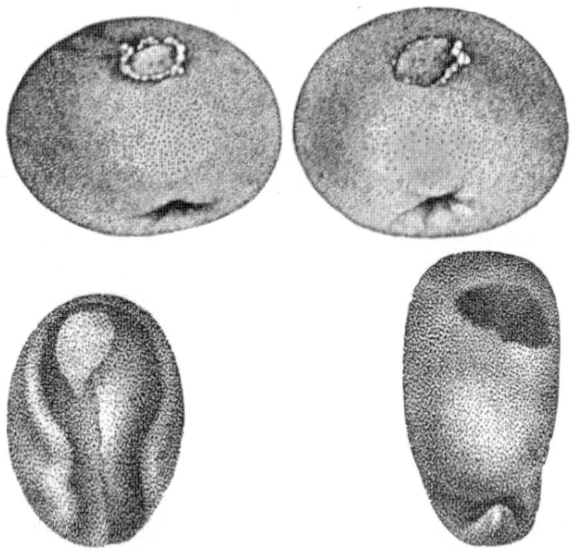

Figure 2.66. **Top**: This is the original experiment by Spemann in which he exchanged small opposite patches of ectoderm that were otherwise destined to become neural pate (left) and epidermis (right), respectively. These newt (*Triton taeniatus*) embryos are both at Stage 12 (Figure 2.33). The two views are of approximately opposite hemispheres, so the cut holes are about 45 degrees from one another, measured over the top. The blastopore in both is near the bottom of the embryo. Note the rims of probably dead cells which "quickly"[288] give way to wound healing[279]. The little disk of dark tissue on the right is then "presumptive" neural plate surrounded by "presumptive" epidermis. Their distinguishable colors reflect the variation in pigmentation over the whole embryo. **Bottom**: The two images are of the same two embryos around Stage 16 (cf. Figure 2.55). The light patch on the left embryo, which is viewed from the top, that would have become epidermis if left in place, is now part of the neural plate. The dark patch on the right embryo, viewed from the bottom, that would have become neural plate if left in place, is now epidermal tissue. Therefore these patches of tissues were not "determined" at the time of transplantation. From Spemann (1938)[288] with kind permission of Donna Anstey of Yale University Press. Cf.[281].

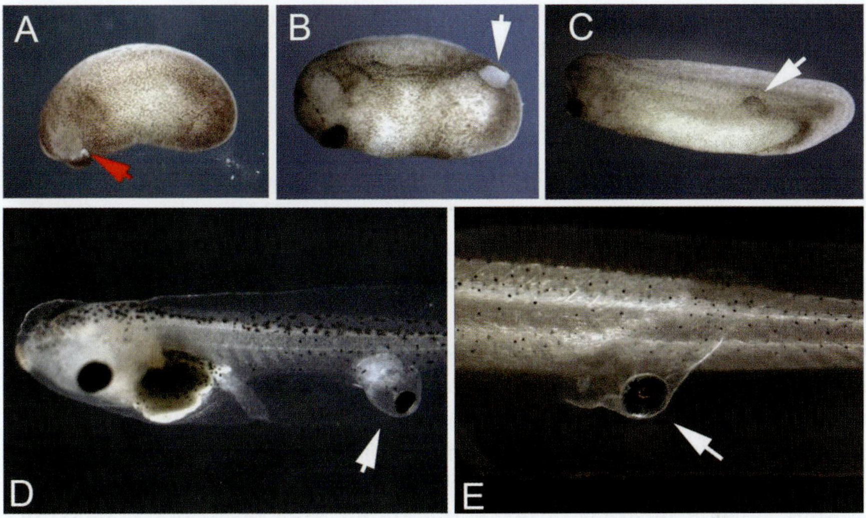

Figure 2.67. "Eye primordia transplants produce ectopic eyes at the site of graft in *Xenopus* tadpoles. Donor tissue was excised from the eye field of stage 24 embryos (**A**, red arrow), grafted along the recipient's body, and healed within 30 min (**B**, white arrow). Ectopic eyes develop at the same rate as native eyes, and wounds heal completely within 24 h post-surgery (C). Eyes transplanted to caudal regions develop within pockets of tissue (D) or tightly along the trunk of the tail (E)." From Blackiston & Levin (2013)[287] with permission of the Company of Biologists. These transplanted eyes were shown to allow responses to light, but they are not connected to the brain. Progress has been made in this regard[289,290].

Determination is followed by differentiation. Differentiation is generally defined as the process whereby different proteins and other products of gene expression appear, and often a cell changes shape or texture and/or function as it differentiates. Determination is a subtle but definite change in what a cell was into what it now becomes. The timing of determination can be very precisely estimated by observing the earliest signs of the differentiation that follow it, especially the expression of new genes, and proteins.

One way to think of this is that determination is the light switch and differentiation is the turning on of the light that is triggered by that switch. One cannot happen without the other and they are interlinked but they are also two separate things. It is possible to disconnect the switch so when the switch is thrown there is no light turned on. This is a common mechanism for certain birth defects.

What follows is another example of the importance of the process of determination and differentiation. It is important because it also shows the genes are independent of location until determination turns them on or off.

Transgenesis is the artificial process of adding genes to the genome of an organism, so that they are incorporated into the DNA and transmitted to progeny. The axolotl is the first animal to successfully be given DNA for genes from another species[291]. Axolotls have a white variant in the wild but they do not have any form of a wild albino. The white variant has dark pigment on the egg, but in lesser amounts than nonwhite eggs. The albino has no dark pigment at all. A close relative of the axolotl, the tiger salamander, does have an albino form in the wild. This albino tiger salamander is a lovely golden color and produces eggs with no pigment. An early embryologist took advantage of lack of determination of the axolotl embryo in order to insert the albino gene from one species, the tiger salamander into another, the axolotl.

Rufus Richard Humphrey crossbred an axolotl with this close cousin and created a single cleavage stage embryo with all undetermined cells. The nucleus of these undetermined cells was isolated. He also prepared eggs with the nucleus removed (enucleated eggs). He then transplanted each nucleus into the enucleated eggs. The eggs develop, though abnormally so it appeared the experiment might fail. However he took tissue from these early abnormal eggs, and transplanted this tissue back into other full axolotl embryos in cleavage stage[292]. He put the cells into the same location that he knew would be determined to differentiate into sperm and egg forming cells.

The result of this experiment was normal baby axolotls with the hybrid cells carrying the albino gene ending up in the ovaries and sperm. When the resulting axolotls were later bred the gene was carried in the sperm and eggs. Some of the new hybrid axolotls carried the albino gene of the tiger salamander. Unfortunately, the new hybrid preferred metamorphosing as other genes also came along for the ride using this crude method, but Humphrey persisted. He used repeated selective breeding to finally get a nearly pure axolotl that had just the albino gene added in. Today's transgenic methods are much more precise. Still, the entire process makes use of the fact that an undetermined cleavage stage cell (up to Stage 6) can end up differentiating normally in whatever place it finds itself. It also shows the genes are passengers along on the ride and do not know where they are without determination signals.

On a similar note, recall the black on one side and white on the other side embryo (Figure 2.21). Another and easier way to produce this pattern is to take an embryo at about 16 cells that is normally pigmented and physically squish it together with a similar white or albino embryo. If forced together, the result is a fused single embryo called a chimera that produces a single axolotl with the harlequin pattern depending on where the pigmented and unpigmented cells end up. Similar experiments have been done with black and white mice by injecting extra cells into the blastocoel where they are incorporated into the inner cell mass[293]. All this happens because of when determination happens.

This is important so let's repeat and define carefully:

Determination – cell has its fate set
Differentiation – a determined cell begins changing into the new cell type

We do know that axolotls can regenerate and that is one of their most fascinating properties. We therefore need to add another caveat. Irreversible determination means that manipulation can't change the cell's fate. In general, determination is irreversible. Sometimes determination appears to be reversible but this is generally due to the presence of stem cells, cells that were set aside and left undetermined unless and until required for regeneration and repair. There may be exceptions as our understanding of stem cells is an area of active research. In general, we consider determination is irreversible but there are exceptions that will likely be explained by the presence of stem cells.

The fact that cells are undetermined early in development but later on are determined was the basis for the development of one of the most important tools of embryology. Walther Vogt[294,295] invented the very useful concept of the fate map. In the 1920s, Vogt applied harmless indelible dyes to the exterior of the early amphibian embryo[296-298] (Figure 2.68). The stained embryo was then left to develop normally. At a later stage, the embryo was chemically fixed and cut into thin slices for viewing under the microscope. This allowed Vogt to trace where each spot of dye moved to during normal development. Vogt then mapped the

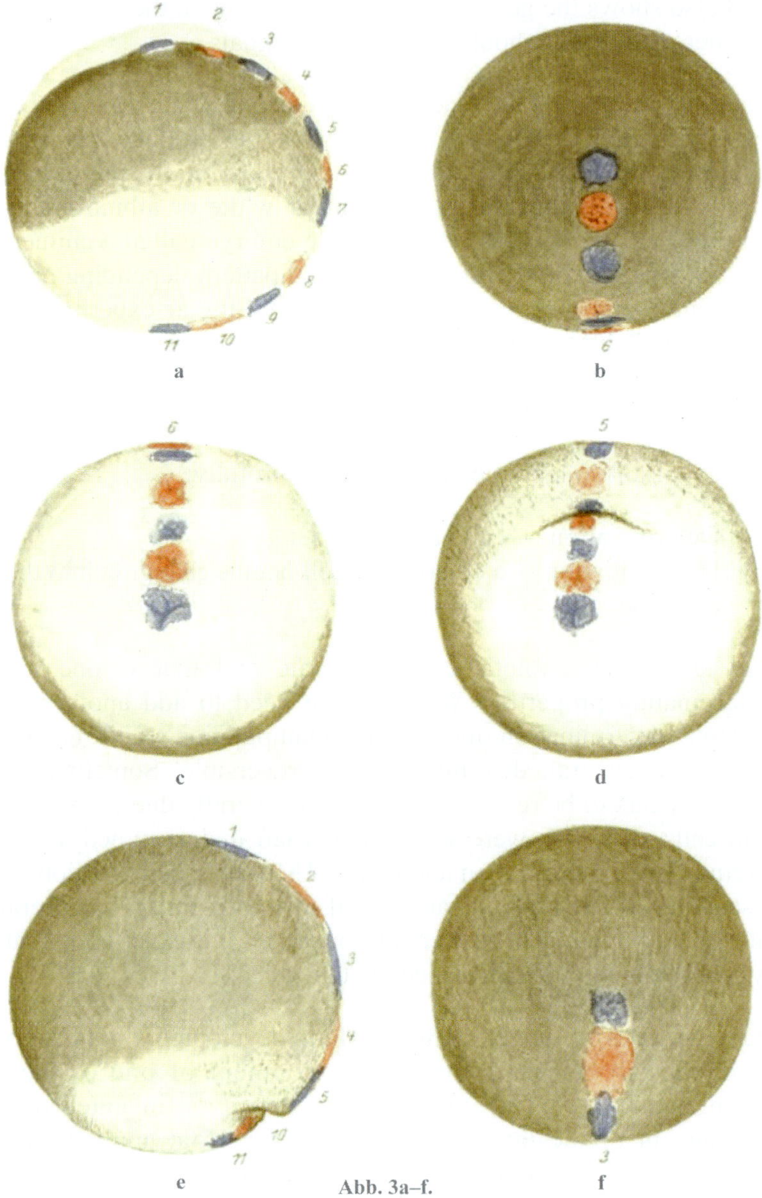

Figure 2.68. An example of Vogt's dye experiments. In this case an axolotl blastula at Stage 8 (Figure 2.33) is marked with harmless colored dyes and the spots are followed as they go inside the blastopore through Stage 10.5. The embryo would then be allowed to continue to develop until it was dissected at a much later stage. From Vogt (1929)[298].

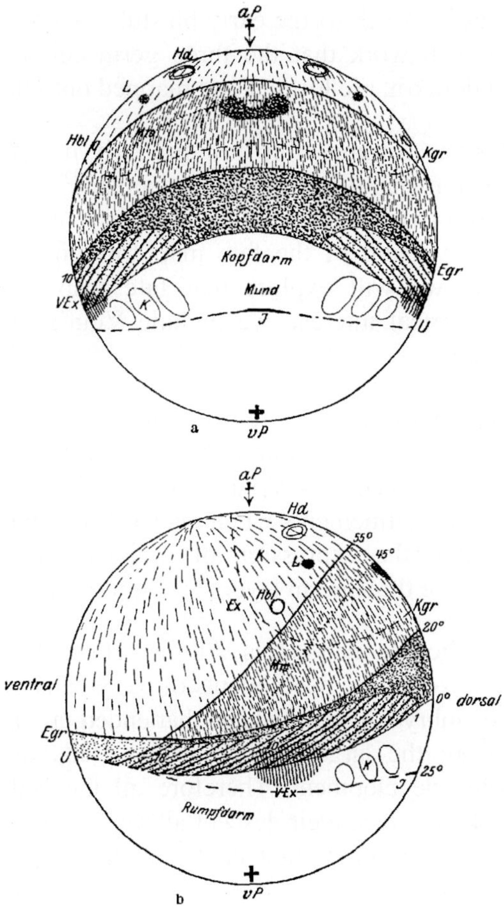

Figure 2.69. Walther Vogt's fate map. "Fig. 89. Fate map of the early blastula of *Bombinator*.... Presumptive embryo regions according to the results of the tracer experiments, **a** dorsal view, **b** view from the left. – Skin (epidermal) ectoderm bright, medullar region dashed dark. Notochord mesoderm zone ring, notochord coarse, mesoderm finely dotted; the oblique lines denote imaginary segment boundaries, the numbers 1, 10 and 18, the position corresponding to the urogenital system. Endoderm white, three ovals (*K*) denote gill endoderm. *J* and *U* presumptive invagination pit and blastopore groove. *Egr* Invagination boundary, *Kgr* limit of head and neck ectoderm. *Mw* medullary torus (border with medullary plate dotted). *A* Eye field (including chiasm). *Hd* adhesive glands. *L* lens. *Ilbl* auditory vesicle. *K* and *Ex* (in ectoderm zone) skin of the gills and limb region. *VEx* (shaded) area of the mesoderm of pronephros and front limb (below the surface, in the inner peripheral zone). *aP* animal pole. *vP* vegetative pole. – The degree designations in the profile view (89b) denote the altitude of some major limits for the dorsal meridian in degrees of latitude (0° = Equator)." From Vogt (1929)[298], page 638. Kopfdarm = foregut, Mund = mouth, Rumpfdarm = midgut.

final locations onto a sketch of the early blastula. Such a sketch is a fate map. It was via such work that the three germ cell layers, ectoderm, mesoderm and endoderm, are located and mapped onto the early embryo.

Many other scientists coming after him similarly mapped development and so we have fate maps of most model organisms[299-301]. Note that each fate map is between two different stages in embryonic development. An extremely detailed version of the fate map for urodele amphibians is available[298,302] and we will explore that fate map in great detail in Chapter 6. For historical interest, we include Vogt's original fate map here (Figure 2.69).

Fate maps for all chordates have remarkable similarities (Figures 2.52 and 2.70). The discovery of the fate map in the axolotl is still based on the dye marking experiments by Vogt, shown in Figure 2.68 with some minor corrections and refinements when his work was later replicated. In present day research the expression of various genes is also being superimposed onto the three germ layers[303] (Figure 2.71).

The importance of the fate map is this. All the cells that make the body of the newly hatched axolotl are descendants of early cells that all started on the outside of the embryo. The only cells that ended up on the inside prior to gastrulation were the large yolky nutritive cells that provided the nourishment during development. Therefore all the cells that form the body of the axolotl can trace their descent all the way back to somewhere on that outer cortical layer of the first single cell that was the fertilized egg.

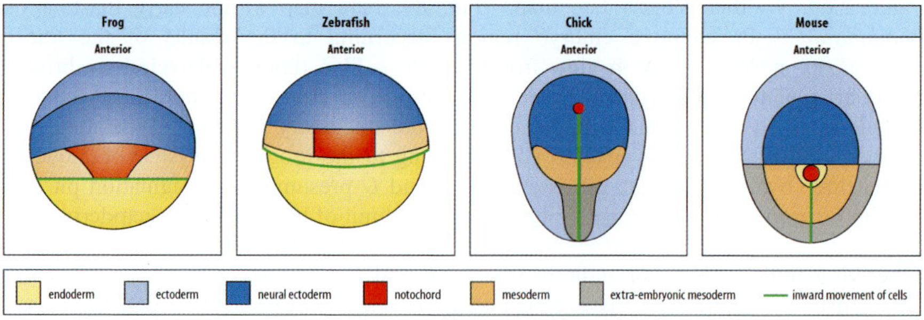

Figure 2.70. "A comparison of early vertebrate embryos reveals similar patterns of development." From *Principles of Development*[268] with permission of Oxford University Press.

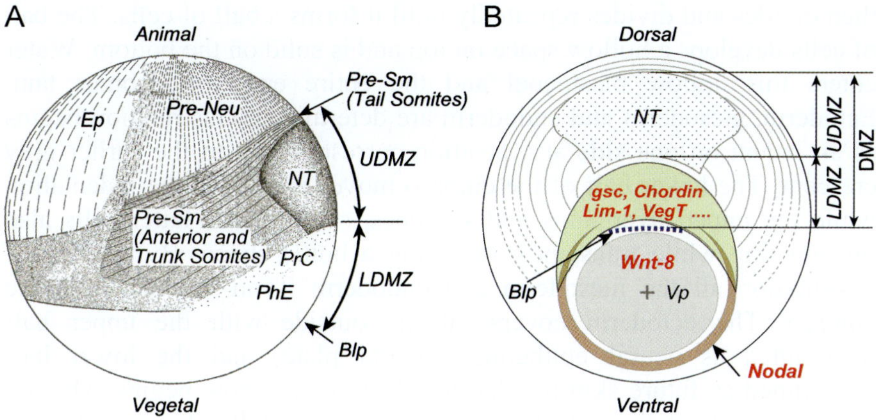

Figure 2.71. An example of current research mapping gene expression onto fate maps. "Fate map and gene expression domains of *Cynops* [the fire belly newt] early gastrula. (A) Fate map of *Cynops* late blastula/early gastrula embryo (right-lateral view, redrawn from[304]). Like other urodeles[298,305], *Cynops* has a large surface prospective mesoderm. Presumptive pharyngeal endoderm (PhE), prechordal plate (PrC), notochord (NT) and tail somites (Pre-Sm) are planarly located on the embryo's surface. Presumptive anterior and trunk somites (Pre-Sm) are located lateral to the presumptive notochord. (B) Gene expression map of early *Cynops* gastrula (dorso-vegetal view). At the early gastrula stage, expressions of Cygsc, Cychd, Cylim-1, Cynog and CyVegT are preferentially restricted to the LDMZ (light green), and their expressions do not extend below the blastopore. CyNodal[306] is expressed around the marginal zone, except for the LDMZ region. *Cynops* Wnt-8 is expressed in the ventro-vegetal part of the embryo (gray) around the vegetal pole (Vp), but Wnt-8 expression does not extend beyond the blastopore (Blp). Cybra is not expressed in any part of the embryo at this stage. The presumptive notochord (UDMZ) has epidermal self-differentiation capacity at this stage and no inducing activity. (Gene expression patterns are from Motoki *et al.*, unpublished)." From Kaneda & Motoki (2012)[307] with permission of Elsevier.

Once the mesoderm and endoderm are completely inside and the neuro-epithelium is also covered by epidermis, neurulation is finished. The embryo is still far from complete. There is a lot more differentiation and refinement required. Some of these refinements, like the budding of limbs or formation of the heart, are entire fields of embryology unto themselves. From the perspective of development, all subsequent determination and differentiation is 'mere' elaboration of the basic body plan of the embryo.

In summary, we begin with the fertilized Axolotl egg. The cortex rotates creating a window to the inside called the grey crescent. The single cell then divides and divides repeatedly until it forms a ball of cells. The ball of cells develops a hollow space on top and is solid on the bottom. Water enters this hollow blastocoel and the entire embryo becomes taut. Ectoderm, mesoderm and endoderm are determined. Gastrulation begins as an indent on one side at a position corresponding to the earlier grey crescent. The lower half of the embryo moves inside via the indentation while the upper layer thins and stretches, some of the latter cells also invaginating while transiently assuming a bottle shape. By the end of gastrulation all the mesoderm and endoderm is on the inside of the embryo. The ectoderm covers all the outside with the upper half determined as neural epithelium (neural plate) and the lower half determined as future skin (epidermis). Neurulation now begins. The top neural epithelium (neural plate) changes to a keyhole shape, sinks and forms a groove. The sides of the groove rise up to form a long tube, the neural tube. The tube seals. Neural crest cells move out from the sealed edges. The neural tube sinks down while the epidermis on the lower half stretches up until the neural tube is entirely covered. By the end of neurulation, the embryo is now a multilayered animal whose body plan is essentially laid out. There is a great deal more finishing differentiation to be done and many more refinements but the essential body plan is now there. The embryo's name also changes. The fertilized egg has moved from morula, to blastula, to gastrula to neurula and now becomes a larva.

To some extent, the embryology of all complex organisms follows this same pattern. There are shifts of when these steps occur. Some steps are more elaborated, or elongated or shortened as compared to the axolotl. However, in all organisms, morphogenesis follows the same general pattern of self-organization putting the right cell, in the right place, at that right time.

We urge you to pause at this point and carefully check the figures, and make sure you understand the fundamental anatomy that they illustrate. We will be depending on your understanding of these anatomical concepts, especially the basic movements of the three germ cell layers during gastrulation and neurulation, in all our future discussions.

1. Gordon, R. (2015). Walking the tightrope: the dilemmas of hierarchical instabilities in Turing's morphogenesis [invited]. In: *The Once and Future Turing: Computing the World*. Eds.: S.B. Cooper & A. Hodges. Cambridge, Cambridge University Press: 150-164.
2. Royal Tyrrell Museum & R. Gordon (2010). Axolotl Colony - Royal Tyrrell Museum. http://www.youtube.com/watch?v=0w-FHja-yHc
3. Caudata.org (2014). The Newt & Salamander Portal. http://www.caudata.org
4. Larson, A. (1996). Tree of Life: Ambystomatidae: Mole Salamanders. http://tolweb.org/Ambystomatidae/15448
5. Martin, C.C. & R. Gordon (1995). Differentiation trees, a junk DNA molecular clock, and the evolution of neoteny in salamanders. *J. Evol. Biol.* **8**, 339-354.
6. Huxley, J.S. (1920). Metamorphosis of axolotl caused by thyroid-feeding. *Nature* **104**(2618), 435.
7. Voss, S.R. & H.B. Shaffer (1996). What insights into the developmental traits of urodeles does the study of interspecific hybrids provide? *International Journal of Developmental Biology* **40**(4), 885-893.
8. Eycleshymer, A.C. (1895). The early development of *Amblystoma,* with observations on some other vertebrates. *Journal of Morphology* **10**, 343-418.
9. Smith, H.M. (1989). Discovery of the axolotl and its early history in biological research. In: *Developmental Biology of the Axolotl*. Eds.: J.B. Armstrong & G.M. Malacinski. New York, Oxford University Press: 3-12.
10. Wistuba, J. (2013). Der Axolotl und die Wissenschaft: eine "Entwicklungsgeschichte" [The Axolotl and the sciences – a "history of development"]. *Sekretär* **13**(1), 3-26.
11. Björklund, N.K. (1993). Small is beautiful: economical axolotl colony maintenance with natural spawnings as if axolotls mattered. In: *Handbook on Practical Methods*. Eds.: G.M. Malacinski & S.T. Duhon. Bloomington, Department of Biology, Indiana University: 38-47.
12. Björklund, N.K. & S.T. Duhon (1996). *The Mexican axolotl as a pet and a laboratory animal [unpublished, write: DickGordonCan@gmail.com].*
13. Koenig, R. (2008). Endangered species: sanctuaries aim to preserve a model organism's wild type. The axolotl, a salamander that retains unique evolutionary features and is a darling of biologists because it can regenerate limbs, faces adversity on two fronts. *Science* **322**, 1456-1457.
14. Convention on International Trade in Endangered Species of Wild Fauna and Flora (2012). *Appendices I, II and III* Geneva, CITES Secretariat.
15. Hays, J. & J. Mitchel (2011). Australia Wants to Import Axolotls For Regenerative Medicine Research. http://blog.targethealth.com/?p=8178
16. Gordon, R. (1993). Grant agencies versus the search for truth. *Accountability in Research: Policies and Quality Assurance* **2**(4), 297-301.
17. Smith, J.J., S. Putta, W. Zhu, G.M. Pao, I.M. Verma, T. Hunter, S.V. Bryant, D.M. Gardiner, T.T. Harkins & S.R. Voss (2009). Genic regions of a large salamander genome contain long introns and novel genes. *BMC Genomics* **10**, #19.
18. Lander, E.S., L.M. Linton, B. Birren, C. Nusbaum, M.C. Zody, J. Baldwin, K. Devon, K. Dewar, M. Doyle, W. FitzHugh, R. Funke, D. Gage, K. Harris, A. Heaford, J. Howland, L. Kann, J. Lehoczky, R. LeVine, P. McEwan, K. McKernan, J. Meldrim, J.P. Mesirov, C. Miranda, W. Morris, J. Naylor, C. Raymond, M. Rosetti, R. Santos, A. Sheridan, C. Sougnez, N. Stange-

Thomann, N. Stojanovic, A. Subramanian, D. Wyman, J. Rogers, J. Sulston, R. Ainscough, S. Beck, D. Bentley, J. Burton, C. Clee, N. Carter, A. Coulson, R. Deadman, P. Deloukas, A. Dunham, I. Dunham, R. Durbin, L. French, D. Grafham, S. Gregory, T. Hubbard, S. Humphray, A. Hunt, M. Jones, C. Lloyd, A. McMurray, L. Matthews, S. Mercer, S. Milne, J.C. Mullikin, A. Mungall, R. Plumb, M. Ross, R. Shownkeen, S. Sims, R.H. Waterston, R.K. Wilson, L.W. Hillier, J.D. McPherson, M.A. Marra, E.R. Mardis, L.A. Fulton, A.T. Chinwalla, K.H. Pepin, W.R. Gish, S.L. Chissoe, M.C. Wendl, K.D. Delehaunty, T.L. Miner, A. Delehaunty, J.B. Kramer, L.L. Cook, R.S. Fulton, D.L. Johnson, P.J. Minx, S.W. Clifton, T. Hawkins, E. Branscomb, P. Predki, P. Richardson, S. Wenning, T. Slezak, N. Doggett, J.F. Cheng, A. Olsen, S. Lucas, C. Elkin, E. Uberbacher, M. Frazier, R.A. Gibbs, D.M. Muzny, S.E. Scherer, J.B. Bouck, E.J. Sodergren, K.C. Worley, C.M. Rives, J.H. Gorrell, M.L. Metzker, S.L. Naylor, R.S. Kucherlapati, D.L. Nelson, G.M. Weinstock, Y. Sakaki, A. Fujiyama, M. Hattori, T. Yada, A. Toyoda, T. Itoh, C. Kawagoe, H. Watanabe, Y. Totoki, T. Taylor, J. Weissenbach, R. Heilig, W. Saurin, F. Artiguenave, P. Brottier, T. Bruls, E. Pelletier, C. Robert, P. Wincker, D.R. Smith, L. Doucette-Stamm, M. Rubenfield, K. Weinstock, H.M. Lee, J. Dubois, A. Rosenthal, M. Platzer, G. Nyakatura, S. Taudien, A. Rump, H. Yang, J. Yu, J. Wang, G. Huang, J. Gu, L. Hood, L. Rowen, A. Madan, S. Qin, R.W. Davis, N.A. Federspiel, A.P. Abola, M.J. Proctor, R.M. Myers, J. Schmutz, M. Dickson, J. Grimwood, D.R. Cox, M.V. Olson, R. Kaul, N. Shimizu, K. Kawasaki, S. Minoshima, G.A. Evans, M. Athanasiou, R. Schultz, B.A. Roe, F. Chen, H. Pan, J. Ramser, H. Lehrach, R. Reinhardt, W.R. McCombie, M. de la Bastide, N. Dedhia, H. Blocker, K. Hornischer, G. Nordsiek, R. Agarwala, L. Aravind, J.A. Bailey, A. Bateman, S. Batzoglou, E. Birney, P. Bork, D.G. Brown, C.B. Burge, L. Cerutti, H.C. Chen, D. Church, M. Clamp, R.R. Copley, T. Doerks, S.R. Eddy, E.E. Eichler, T.S. Furey, J. Galagan, J.G. Gilbert, C. Harmon, Y. Hayashizaki, D. Haussler, H. Hermjakob, K. Hokamp, W. Jang, L.S. Johnson, T.A. Jones, S. Kasif, A. Kaspryzk, S. Kennedy, W.J. Kent, P. Kitts, E.V. Koonin, I. Korf, D. Kulp, D. Lancet, T.M. Lowe, A. McLysaght, T. Mikkelsen, J.V. Moran, N. Mulder, V.J. Pollara, C.P. Ponting, G. Schuler, J. Schultz, G. Slater, A.F. Smit, E. Stupka, J. Szustakowki, D. Thierry-Mieg, L. Wagner, J. Wallis, R. Wheeler, A. Williams, Y.I. Wolf, K.H. Wolfe, S.P. Yang, R.F. Yeh, F. Collins, M.S. Guyer, J. Peterson, A. Felsenfeld, K.A. Wetterstrand, A. Patrinos & M.J. Morgan (2001). Initial sequencing and analysis of the human genome. *Nature* **409**(6822), 860-921.

19. Habermann, B., A.G. Bebin, S. Herklotz, M. Volkmer, K. Eckelt, K. Pehlke, H.H. Epperlein, H.K. Schackert, G. Wiebe & E.M. Tanaka (2004). An *Ambystoma mexicanum* EST sequencing project: analysis of 17,352 expressed sequence tags from embryonic and regenerating blastema cDNA libraries. *Genome Biology* **5**(9), R67.

20. Voss, S.R., H.H. Epperlein & E.M. Tanaka (2009). *Ambystoma mexicanum*, the axolotl: a versatile amphibian model for regeneration, development, and evolution studies. *Cold Spring Harbor Protocols* **2009**(8), pdb emo128.

21. Ruder, K., E.R. Winstead & M.S. Gibbs (2004). Quick Guide to Sequenced Genomes. http://www.genomenewsnetwork.org/resources/sequenced_genomes/genome_guide_p1.shtml

22. Wikipedia (2014). List of sequenced animal genomes. http://en.wikipedia.org/wiki/List_of_sequenced_animal_genomes
23. Šimová, I. & T. Herben (2012). Geometrical constraints in the scaling relationships between genome size, cell size and cell cycle length in herbaceous plants. *Proceedings of the Royal Society B-Biological Sciences* **279**(1730), 867-875.
24. Freeman, W.H. & B. Bracegirdle (1967). *An Atlas of Embryology*. London, Heinemann Educational Books.
25. Gerhart, J.C. (1980). Mechanisms regulating pattern formation in the amphibian egg and early embryo. In: *Biological Regulation and Development Vol. 2: Molecular Organization and Cell Function*. Ed.: R.F. Goldsberger, Plenum Press: 133-316.
26. Garcia-Lopez, R., A. Pombero & S. Martinez (2009). Fate map of the chick embryo neural tube. *Development Growth & Differentiation* **51**(3), 145-165.
27. Gordon, R. & N.K. Björklund (1996). How to observe surface contraction waves on axolotl embryos. *International Journal of Developmental Biology* **40**(4), 913-914.
28. Gasser, R.F. (1975). Atlas of Human Embryos. http://www.ehd.org/classics/gasser_pre.php
29. Gasser, R.F. (1975). The Virtual Human Embryo. http://www.ehd.org/virtual-human-embryo/
30. Nikas, G., T. Paraschos, A. Psychoyos & A.H. Handyside (1994). The zona reaction in human oocytes as seen with scanning electron microscopy. *Human Reproduction* **9**(11), 2135-2138.
31. Walker, S. (2012). Image of the Day: A light micrograph of a human egg. http://www.the-scientist.com/?articles.view/articleNo/33226/title/Image-of-the-Day/
32. Gordon, R. (1985). A review of the theories of vertebrate neurulation and their relationship to the mechanics of neural tube birth defects. *Journal of Embryology and Experimental Morphology* **89**(Suppl.), 229-255.
33. Björklund, N.K. & R. Gordon (2006). A hypothesis linking low folate intake to neural tube defects due to failure of post-translation methylations of the cytoskeleton. *International Journal of Developmental Biology* **50**(2/3), 135-141.
34. Löfberg, J. & C.O. Jacobson (1974). Effects of vinblastine sulfate, colchicine and gunaosine triphosphate on cell morphogenesis during amphibian neurulation. *Zoon* **2**, 85-98.
35. Séguin, C. & P.E. Messier (1978). Peculiar pattern of interkinetic nuclear migration in the newt *Notophtalmus viridescens*. *Anatomy and Embryology (Berlin)* **155**(1), 47-56.
36. Lundmark, C., J. Shih, P. Tibbetts & R. Keller (1984). Amphibian gastrulation as seen by scanning electron microscopy. *Scanning Electron Microscopy* **1984**(3), 1289-1300.
37. Morita, H., H. Kajiura-Kobayashi, C. Takagi, T.S. Yamamoto, S. Nonaka & N. Ueno (2012). Cell movements of the deep layer of non-neural ectoderm underlie complete neural tube closure in *Xenopus*. *Development* **139**(8), 1417-1426.

38. Burnside, M.B. & A.G. Jacobson (1968). Analysis of morphogenetic movements in the neural plate of the newt *Taricha torosa*. *Developmental Biology* **18**, 537-552.
39. Jacobson, A.G. & R. Gordon (1976). Changes in the shape of the developing vertebrate nervous system analyzed experimentally, mathematically and by computer simulation. *Journal of Experimental Zoology* **197**(2), 191-246.
40. Gordon, R. & A.G. Jacobson (1978). The shaping of tissues in embryos. *Scientific American* **238**(6), 106-113, 160.
41. Crawford-Young, S. (2007). *A Robotic Microscope for 3D Time-Lapse Imaging of Early Stage Axolotl Salamander Embryos [M.Sc. Thesis, Supervisor: R. Gordon]*. Winnipeg, Department of Electrical & Computer Engineering, University of Manitoba.
42. Mayo Foundation for Medical Education and Research (2012). Spina bifida. http://www.mayoclinic.com/health/medical/IM01543
43. Centers for Disease Control and Prevention (2012). Spina Bifida Facts. http://www.cdc.gov/ncbddd/spinabifida/facts.html
44. Lacalli, T.C. (2010). The emergence of the chordate body plan: some puzzles and problems. *Acta Zoologica* **91**(1), 4-10.
45. Stokes, M.D. & N.D. Holland (1995). Embryos and larvae of a lancelet, *Branchiostoma floridae*, from hatching through metamorphosis: Growth in the laboratory and external morphology. *Acta Zoologica* **76**(2), 105-120.
46. Jaworski, P.M. (2005). *Branchiostoma lanceolatum* anatomy scheme. https://commons.wikimedia.org/wiki/File:BranchiostomaLanceolatum_PioM.svg
47. de Bivort, B. (2007). Annotated diagram of cervical vertebra. https://commons.wikimedia.org/wiki/File:Cervical_vertebra_english.png
48. Sequeira, S.E.K. & A.R. Milner (1993). The temnospondyl amphibian *Capetus* from the Upper Carboniferous of the Czech Republic. *Palaeontology* **36**, 657-680.
49. Schoch, R.R. & A. Smith (2014). Life cycles, plasticity and palaeoecology in temnospondyl amphibians. *Palaeontology* **57**(3), 517-529.
50. Bogdanov, D. (2007). *Capetus*, a basal temnospondyl. http://en.wikipedia.org/wiki/Temnospondyls
51. Brusatte, S.L., R.J. Butler, O. Mateus & J.S. Steyer (2015). A new species of *Metoposaurus* from the Late Triassic of Portugal and comments on the systematics and biogeography of metoposaurid temnospondyls. *Journal of Vertebrate Paleontology*, e912988.
52. Viegas, J. (2015). Car-Sized Salamander Had 'Toilet Seat' Head. http://news.discovery.com/animals/car-sized-salamander-had-toilet-seat-head-150323.htm
53. Brusatte, S.L. (2015). Meet the super salamander that nearly ate your ancestors for breakfast. https://theconversation.com/meet-the-super-salamander-that-nearly-ate-your-ancestors-for-breakfast-39221
54. Wikipedia (2015). *Prionosuchus*. http://en.wikipedia.org/wiki/Prionosuchus
55. Anderson, J.S., R.R. Reisz, D. Scott, N.B. Frobisch & S.S. Sumida (2008). A stem batrachian from the Early Permian of Texas and the origin of frogs and salamanders. *Nature* **453**(7194), 515-518.
56. Wikipedia (2015). Crown group. http://en.wikipedia.org/wiki/Crown_group

57. Gao, K.Q. & N.H. Shubin (2003). Earliest known crown-group salamanders. *Nature* **422**(6930), 424-428.
58. Anderson, J.S. (2012). Fossils, molecules, divergence times, and the origin of Salamandroidea. *Proceedings of the National Academy of Sciences USA* **109**(15), 5557-5558.
59. Gao, K.-Q. & N.H. Shubin (2012). Late Jurassic salamandroid from western Liaoning, China [+Correction]. *Proceedings of the National Academy of Sciences USA* **109**(15&30), 5767-5772, 12260.
60. Roy, S. & S. Gatien (2008). Regeneration in axolotls: a model to aim for! *Experimental Gerontology* **43**(11), 968-973.
61. Kulbisky, G.P., D.W. Rickey, M.H. Reed, N.K. Björklund & R. Gordon (1999). The axolotl as an animal model for the comparison of 3-D ultrasound with plain film radiography. *Ultrasound in Medicine and Biology* **25**(6), 969-975.
62. Ghosh, S., P. Thorogood & P. Ferretti (1996). Regeneration of lower and upper jaws in urodeles is differentially affected by retinoic acid. *International Journal of Developmental Biology* **40**(6), 1161-1170.
63. Levésque, M., É. Villiard & S. Roy (2010). Skin wound healing in axolotls: a scarless process. *Journal of Experimental Zoology Part B-Molecular and Developmental Evolution* **314B**(8), 684-697.
64. Özkucur, N., H.H. Epperlein & R.H. Funk (2010). Ion imaging during axolotl tail regeneration in vivo. *Developmental Dynamics* **239**(7), 2048-2057.
65. Becker, R.O. & G. Selden (1985). *The Body Electric, Electromagnetism and the Foundation of Life*. New York, William Morrow & Co.
66. Stocum, D.L. (1995). *Wound Repair, Regeneration, and Artificial Tissues*, R.G. Landes Company.
67. Flink, I.L. (2002). Cell cycle reentry of ventricular and atrial cardiomyocytes and cells within the epicardium following amputation of the ventricular apex in the axolotl, *Amblystoma mexicanum*: confocal microscopic immunofluorescent image analysis of bromodeoxyuridine-labeled nuclei. *Anatomy and Embryology (Berlin)* **205**(3), 235-244.
68. Monaghan, J.R., J.A. Walker, R.B. Page, S. Putta, C.K. Beachy & S.R. Voss (2007). Early gene expression during natural spinal cord regeneration in the salamander *Ambystoma mexicanum*. *Journal of Neurochemistry* **101**(1), 27-40.
69. Mellinger, M.W. (1975). Reconstruction of the cerebral hemispheres in *Ambystoma mexicanum* following bilateral brain ablation and ablation/transplantation. *Anatomical Record* **181**, 424.
70. Hershkowitz, M., M. Segal & D. Samuel (1972). The acquisition of dark avoidance by transplantation of the forebrain of trained newts (*Pleurodeles waltl*). *Brain Research* **48**, 366-369.
71. Song, F., B. Li & D.L. Stocum (2010). Amphibians as research models for regenerative medicine. *Organogenesis* **6**(3), 141-150.
72. Roy, S. & M. Lévesque (2006). Limb regeneration in axolotl: is it superhealing? *ScientificWorldJournal* **6**(Suppl 1), 12-25.
73. Simon, H.-G. (2012). Salamanders and fish can regenerate lost structures - why can't we? *BMC Biology* **10**, #15.
74. Wikipedia (2014). Sawing a woman in half. http://en.wikipedia.org/wiki/Sawing_a_woman_in_half

75. Lemanski, L.F., F. Meng, S.L. Lemanski, N. Dawson, C. Zhang, D. Foster, Q. Li, M. Nakatsugawa, R.W. Zajdel, D.K. Dube & X. Huang (2001). Creation of chimeric mutant axolotls: a model to study early embryonic heart development in Mexican axolotls. *Anatomy and Embryology (Berlin)* **203**(5), 335-342.
76. Beetschen, J.C. (1996). How did urodele embryos come into prominence as a model system? *International Journal of Developmental Biology* **40**(4), 629-636.
77. Koop, D. & L.Z. Holland (2008). The basal chordate amphioxus as a simple model for elucidating developmental mechanisms in vertebrates. *Birth Defects Research* **84**(3), 175-187.
78. Holland, P.W.H. (2010). From genomes to morphology: a view from amphioxus. *Acta Zoologica* **91**(1), 81-86.
79. European Commission (2000). *Arabidopsis thaliana* - First plant genome sequenced [cobweb page]. http://ec.europa.eu/research/quality-of-life/arabidopsis.html
80. Krizek, B.A. & J.T. Anderson (2013). Control of flower size. *Journal of Experimental Botany* **64**(6), 1427-1437.
81. Kumar, K., R.A. Mella-Herrera & J.W. Golden (2010). Cyanobacterial heterocysts. *Cold Spring Harbor Perspectives in Biology* **2**(4), a000315.
82. Petricka, J.J., J.M. Van Norman & P.N. Benfey (2009). Symmetry breaking in plants: molecular mechanisms regulating asymmetric cell divisions in *Arabidopsis*. *Cold Spring Harbor Perspectives in Biology* **1**(5), a000497.
83. Kitano, H., S. Hamahashi & S. Luke (1998). The Perfect *C. elegans* Project: an initial report. *Artificial Life* **4**(2), 141-156.
84. Nelson, B.C. (2012). Establishing the Foundation for Nanogenotoxicology Investigations with *Caenorhabditis elegans*. http://www.nist.gov/mml/bbd/cell_systems/bryant-nelson.cfm - establishing
85. Sulston, J.E., E. Schierenberg, J.G. White & J.N. Thomson (1983). The embryonic cell lineage of the nematode *Caenorhabditis elegans*. *Developmental Biology* **100**(1), 64-119.
86. Sulston, J.E. (2003). *Caenorhabditis elegans*: The cell lineage and beyond (Nobel lecture). *Chembiochem* **4**(8), 688-696.
87. Alicea, B., S. McGrew, R. Gordon, S. Larson, T. Warrington & M. Watts (2014). DevoWorm: differentiation waves and computation in *C. elegans* embryogenesis. http://www.biorxiv.org/content/early/2014/10/03/009993
88. Cogburn, L.A., T.E. Porter, M.J. Duclos, J. Simon, S.C. Burgess, J.J. Zhu, H.H. Cheng, J.B. Dodgson & J. Burnside (2007). Functional genomics of the chicken - a model organism. *Poultry Science* **86**(10), 2059-2094.
89. Beckingham, K.M., J.D. Armstrong, M.J. Texada, R. Munjaal & D.A. Baker (2005). *Drosophila melanogaster* - the model organism of choice for the complex biology of multi-cellular organisms. *Gravitational and Space Biology* **18**(2), 17-29.
90. Devos, K.M. & M.D. Gale (2000). Genome relationships: the grass model in current research. *Plant Cell* **12**(5), 637-646.
91. Phoon, C.K. (2006). Imaging tools for the developmental biologist: ultrasound biomicroscopy of mouse embryonic development. *Pediatric Research* **60**(1), 14-21.
92. Vincensini, L., T. Blisnick & P. Bastin (2011). 1001 model organisms to study cilia and flagella. *Biology of the Cell* **103**(3), 109-130.

93. Grbić, M. (2003). Polyembryony in parasitic wasps: evolution of a novel mode of development. *International Journal of Developmental Biology* **47**(7-8), 633-642.
94. Corbo, J.C., A. Di Gregorio & M. Levine (2001). The ascidian as a model organism in developmental and evolutionary biology. *Cell* **106**(5), 535-538.
95. Dauga, D. (2014). Tunicate Web Portal http://tunicate-portal.org/
96. Kominami, T. & H. Takata (2004). Gastrulation in the sea urchin embryo: a model system for analyzing the morphogenesis of a monolayered epithelium. *Development, Growth & Differentiation* **46**(4), 309-326.
97. Ereskovsky, A.V., E. Renard & C. Borchiellini (2013). Cellular and molecular processes leading to embryo formation in sponges: evidences for high conservation of processes throughout animal evolution. *Development Genes and Evolution* **223**(1-2), 5-22.
98. Desnitskiy, A.G. (2009). *Volvox* (Chlorophyta, Volvocales) as a model organism in developmental biology. *Russian Journal of Developmental Biology* **40**(4), 238-241.
99. Callery, E.M. (2006). There's more than one frog in the pond: a survey of the Amphibia and their contributions to developmental biology. *Seminars in Cell & Developmental Biology* **17**(1), 80-92.
100. Fields, M.C., P. Adelfio, D. Ahmad, O. Brown, B. Cox, M. Davies, T. Esch, F. Foxley, E. Friedman, J. Hall, K. Hearing, J. Heyward, R. McElroy, M. Morin, R. Nicoletti, A. Over, A. Riley, A. Spiliotes & K. Tam (2009). *Danio rerio* in K-12 classrooms: sparking interest in the new generation of scientists. *Zebrafish* **6**(2), 145-160.
101. King, G.M., R. Gordon, K. Karmali & L.J. Biberman (1982). A new method for the immobilisation of teleost embryos for time-lapse studies of development. *Journal of Experimental Zoology* **220**(2), 147-151.
102. Gordon, R. & S.E. Levina (1983). Methionine enkephalin-induced changes in pigmentation of zebrafish (Cyprinidae, *Brachydanio rerio*) and related species and varieties, measured videodensitometrically. II. Pearl and gold danios. *General and Comparative Endocrinology* **51**(3), 378-383.
103. Levina, S. & R. Gordon (1983). Methionine enkephalin-induced changes in pigmentation of zebrafish (Cyprinidae, *Brachydanio rerio*) and related species and varieties, measured videodensitometrically. I. Zebrafish. *General and Comparative Endocrinology* **51**(3), 370-377.
104. Choudhri, S. (1984). Electron microscopy of cell-cell contacts between migrating yolk sac melanocytes of the zebrafish, *Brachydanio rerio* [Supervisor: Richard Gordon]. In: *UM Program and Proceedings, B.Sc. (Med) Presentations*. Winnipeg, Faculty of Medicine, University of Manitoba: 71-78.
105. Gordon, R. (1999). *The Hierarchical Genome and Differentiation Waves: Novel Unification of Development, Genetics and Evolution*. Singapore & London, World Scientific & Imperial College Press.
106. Tamm, S.L. (1972). Ciliary motion in *Paramecium*: Scanning electron microscope study. *Journal of Cell Biology* **55**(1), 250-255.
107. Lankester, E.R. (1901). *A Guide to the Shell and Starfish Galleries (Mollusca, Polyzoa, Brachiopoda, Tunicata, Echinodermata, and Worms), Department of Zoology, British Museum (Natural History)*, Department of Zoology, British Museum.

108. Chow, S. (2006). Ascidian (*Rhopalaea crassa*). http://commons.wikimedia.org/wiki/File:Ascidian_%28Rhopalaea_Crassa%29_%284_cm%29.png
109. Hardin, J. (2005). Echinoderms - Gastrulation: Introduction. http://worms.zoology.wisc.edu/dd2/echino/gast/intro.html
110. Höhn, S. & A. Hallmann (2011). There is more than one way to turn a spherical cellular monolayer inside out: type B embryo inversion in *Volvox globator*. *BMC Biology* **9**, #89.
111. Fox, F. (2011). *Volvox* is a genus of multicellular green algae. https://commons.wikimedia.org/wiki/File:Mikrofoto.de-volvox-4.jpg
112. Brooke, L. (2011). Salamander eggs turn off cancer. http://www.futurity.org/health-medicine/salamander-eggs-turn-off-cancer/
113. Gordon, R., C.S. Bohun, C. Breward, E. Cumberbatch, G. Djoumna, A. ElSheikh, J.F. Williams & J. Wylie (2006). Yolk dynamics in amphibian embryos. In: *Proceedings, Fields-MITACS Industrial Problem-Solving Workshop (FMIPW), August 14-18, 2006*. Ed. Toronto, Fields Institute: 33-49.
114. Bordzilovskaya, N.P., T.A. Dettlaff, S.T. Duhon & G.M. Malacinski (1989). Developmental-stage series of axolotl embryos [Erratum: Staging Table 19-1 is for 20°C, not 29°C]. In: *Developmental Biology of the Axolotl*. Eds.: J.B. Armstrong & G.M. Malacinski. New York, Oxford University Press: 201-219.
115. Tyler, M.S. (2009). *Developmental Biology: A Guide for Experimental Study*. Sunderland, Massachusetts, Sinauer Associates, 3rd.
116. Kimmel, C.B. (2015). Hatching Period (48-72 h). http://zfin.org/zf_info/zfbook/stages/hatch.html
117. Carroll Jr., E.J. & J.L. Hedrick (1974). Hatching in the toad *Xenopus laevis*: Morphological events and evidence for a hatching enzyme. *Developmental Biology* **38**(1), 1-13.
118. Nouri, C., R. Luppes, A.E.P. Veldman, J.A. Tuszynski & R. Gordon (2008). Rayleigh instability of the inverted one-cell amphibian embryo. *Physical Biology* **5**(1), 015006.
119. Gordon, R. (2009). Google Embryo for building quantitative understanding of an embryo as it builds itself: II. Progress toward an embryo surface microscope. *Biological Theory: Integrating Development, Evolution, and Cognition* **4**(4), 396-412.
120. Gordon, R. & J.E. Westfall (2009). Google Embryo for building quantitative understanding of an embryo as it builds itself: I. Lessons from Ganymede and Google Earth. *Biological Theory: Integrating Development, Evolution, and Cognition* **4**(4), 390-395 + Supplementary Appendix.
121. Sardet, C. & P. Chang (1987). The egg cortex: from maturation through fertilization. *Cell Differentiation* **21**(1), 1-19.
122. Gupta, G.S. (2005). *Proteomics of Spermatogenesis*. New York, Springer.
123. Verlhac, M.-H. & A. Villeneuve, Eds. (2010). *Oogenesis: The Universal Process*, Wiley.
124. Wall, R. (1990). *This Side Up: Spatial Determination in the Early Development of Animals*. New York, Cambridge University Press.
125. Crawford-Young, S. (2006). Effects of microgravity on cell cytoskeleton and embryogenesis. *International Journal of Developmental Biology* **50**(2/3), 183-191.

126. Dictus, W.J.A.G., E.J.J. van Zoelen, P.A.T. Tetteroo, L.G.J. Tertoolen, S.W. de Laat & J.G. Bluemink (1984). Lateral mobility of plasma membrane lipids in *Xenopus* eggs: regional differences related to animal/vegetal polarity become extreme upon fertilization. *Developmental Biology* **101**(1), 201-211.

127. Grey, R.D., D.P. Wolf & J.L. Hedrick (1974). Formation and structure of the fertilization envelope in *Xenopus laevis*. *Developmental Biology* **36**(1), 44-61.

128. Wessel, G.M., J.M. Brooks, E. Green, S. Haley, E. Voronina, J. Wong, V. Zaydfudim & S. Conner (2001). The biology of cortical granules. *International Review of Cytology - A Survey of Cell Biology* **209**, 117-206.

129. Semik, D. & W. Kilarski (1986). Surface structure of eggs and early embryos of newts. *Folia Histochemica et Cytobiologica* **24**(4), 308-309.

130. Hess, M. (2009). Chimeric axolotl. http://s3.e-monsite.com/2010/12/05/01/2009-06-24_Web_SDC139842.jpg

131. Bidinger, J. (2015). *Axolotls with left/right differences in pigmentation [personal communication]*.

132. Black, S.D., A.N. Crutchfield, M.D. Murphy & T.C. Swain (1998). Blastopore formation in the animal hemisphere: functional inversion of gastrulation by centrifugation of *Xenopus laevis* eggs. *Gravitational and Space Biology Bulletin* **11**(2), 15-21.

133. Malacinski, G.M., A.W. Neff & M. Wakahara (1987). Upside-down amphibian eggs: what can they tell us about early embryonic pattern specification. *Biologist (Charleston)* **66**(1-4), 70-87.

134. Steinke, J.H. & D.G. Benson Jr. (1970). The structure and polysaccharide cytochemistry of the jelly envelopes of the egg of the frog, *Rana pipiens*. *Journal of Morphology* **130**(1), 57-66.

135. Bonnell, B.S. & D.E. Chandler (1996). Egg jelly layers of *Xenopus laevis* are unique in ultrastructure and sugar distribution. *Molecular Reproduction and Development* **44**(2), 212-220.

136. Larabell, C.A. & D.E. Chandler (1989). Quick-freeze, deep-etch, rotary-shadow views of the extracellular matrix and cortical cytoskeleton of *Xenopus laevis* eggs. *Journal of Electron Microscopy Technique* **13**(3), 228-243.

137. Salthe, S.N. (1963). The egg capsules in the Amphibia. *Journal of Morphology* **113**(2), 161-171.

138. Humphries Jr., A.A. (1966). Observations on the deposition, structure, and cytochemistry of the jelly envelopes of the egg of the newt *Triturus viridescens*. *Developmental Biology* **13**(2), 214-&.

139. Duellman, W.E. & L. Trueb (1994). *Biology of Amphibians*. Baltimore, Johns Hopkins University Press.

140. Bonnell, B.S., D. Reinhart & D.E. Chandler (1996). *Xenopus laevis* egg jelly coats consist of small diffusible proteins bound to a complex system of structurally stable networks composed of high-molecular-weight glycoconjugates. *Developmental Biology* **174**(1), 32-42.

141. Hansson Mild, K., S. Løvtrup & T. Bergfors (1974). On the mechanical properties of the vitelline membrane of the frog egg. *Journal of Experimental Biology* **60**(3), 807-820.

142. Selman, G.G. & C.H. Waddington (1955). The mechanism of cell division in the cleavage of the newt's egg. *Journal of Experimental Biology* **32**(4), 700-733.

143. Pitari, G., S. Dupré, C. Fusco, G. Maurizi & C. Campanella (1993). Jelly plug dissolution in *Discoglossus pictus* eggs (Anura) involves peroxidase-like activity and oxidative opening of disulphide bonds. *Zygote* **1**(1), 61-69.

144. Bohun, C.S. & C.J.W. Breward (2009). Yolk dynamics in amphibian embryos. *Mathematics-in-Industry Case Studies Journal* **1**, 99-119.

145. Eisthen, H.L. (1989). Courtship and mating behavior in the axolotl. *Axolotl Newsletter*(18), 18-19.

146. Crawford-Young, S.J., A. Weatherbee, C. Fleuraru, R. Gordon, P. Thanusutiyabhorn, A. Vitkin, Y. Mao & S.S. Sherif (2016). Non-destructive imaging of live early-stage axolotl salamander embryo using Optical Coherence Tomography [Poster 255-TCSB-73]. In: *Proceedings, 18th Photonics North Conference, Québec City, May 24-26, 2016*. PhotonicsNorth: https://www.conferium.com/uploads/DocsConv_255/255-TCSB-273.doc

147. Mansour, N., F. Lahnsteiner & R.A. Patzner (2011). Collection of gametes from live axolotl, *Ambystoma mexicanum*, and standardization of *in vitro* fertilization. *Theriogenology* **75**(2), 354-361.

148. Kubo, H., M. Kotani, Y. Yamamoto & T. Hazato (2008). Involvement of sperm proteases in the binding of sperm to the vitelline envelope in *Xenopus laevis*. *Zoological Science* **25**(1), 80-87.

149. Larabell, C.A. & D.E. Chandler (1990). Stepwise transformation of the vitelline envelope of *Xenopus* eggs at activation: a quick-freeze, deep-etch analysis. *Developmental Biology* **139**(2), 263-268.

150. Larabell, C.A. & D.E. Chandler (1991). Fertilization-induced changes in the vitelline envelope of echinoderm and amphibian eggs: self-assembly of an extracellular matrix. *Journal of Electron Microscopy Technique* **17**(3), 294-318.

151. Xiang, X.Y., A. Kittelson, J. Olson, A. Bieber & D. Chandler (2005). Allurin, A 21 kD sperm chemoattractant, is rapidly released from the outermost jelly layer of the *Xenopus* egg by diffusion and medium convection. *Molecular Reproduction and Development* **70**(3), 344-360.

152. Jégo, P., J. Joly & C. Boisseau (1980). Les gangues ovulaires des Amphibiens (protéines sécrétées par l'oviducte) et leurs rôles dans la fécondation/Amphibian jelly envelopes, proteins secreted by the oviduct and surrounding the eggs: their role in fertilization. *Reproduction Nutrition Development* **20**(2), 557-567.

153. Ohta, M., H. Kubo, Y. Nakauchi & A. Watanabe (2010). Sperm motility-initiating activity in the egg jelly of the externally-fertilizing urodele amphibian, *Hynobius lichenatus*. *Zoological Science* **27**(11), 875-879.

154. Watanabe, T., H. Kubo, S. Takeshima, M. Nakagawa, M. Ohta, S. Kamimura, E. Takayama-Watanabe, A. Watanabe & K. Onitake (2010). Identification of the sperm motility-initiating substance in the newt, *Cynops pyrrhogaster*, and its possible relationship with the acrosome reaction during internal fertilization. *International Journal of Developmental Biology* **54**(4), 591-597.

155. Hiyoshi, W., T. Sasaki, E. Takayama-Watanabe, H. Takai, A. Watanabe & K. Onitake (2007). Egg jelly of the newt, *Cynops pyrrhogaster* contains a factor essential for sperm binding to the vitelline envelope. *Journal of Experimental Zoology Part A-Ecological Genetics and Physiology* **307A**(6), 301-311.

156. Elinson, R.P. (1975). Fertilization of green frog (*Rana clamitans*) eggs in their native jelly by bullfrog (*Rana catesbeiana*) sperm. *Journal of Experimental Zoology* **193**(3), 419-423.

157. Beattie, R.C. (1980). Physicochemical investigation of the jelly capsules surrounding eggs of the common frog (*Rana temporaria temporaria*). *Journal of Zoology* **190**(Jan), 1-25.

158. Born, M., E. Wolf, A.B. Bhatia, D. Gabor, A.R. Stokes, A.M. Taylor, P.A. Wayman & W.L. Wilcock (2000). *Principles of Optics: Electromagnetic Theory of Propagation, Interference and Diffraction of Light*, Cambridge University Press.

159. Seymour, R.S., F. Geiser & D.F. Bradford (1991). Gas conductance of the jelly capsule of terrestrial frog eggs correlates with embryonic stage, not metabolic demand or ambient pO_2. *Physiological Zoology* **64**(3), 673-687.

160. Seymour, R.S. (1994). Oxygen diffusion through the jelly capsules of amphibian eggs. *Israel Journal of Zoology* **40**(3-4), 493-506.

161. Pinder, A.W. & S.C. Friet (1994). Oxygen-transport in egg masses of the amphibians *Rana sylvatica* and *Ambystoma maculatum*: convection, diffusion and oxygen production by algae. *Journal of Experimental Biology* **197**(1), 17-30.

162. Marco, A. & A.R. Blaustein (2000). Symbiosis with green algae affects survival and growth of northwestern salamander embryos. *Journal of Herpetology* **34**(4), 617-621.

163. Bianchini, K., G.J. Tattersall, J. Sashaw, C.S. Porteus & P.A. Wright (2012). Acid water interferes with salamander-green algae symbiosis during early embryonic development. *Physiological and Biochemical Zoology* **85**(5), 470-480.

164. Kerney, R. (2011). Symbioses between salamander embryos and green algae. *Symbiosis* **54**(3), 107-117.

165. Smith, M.A., M. Berrill & C.M. Kapron (2002). Photolyase activity of the embryo and the ultraviolet absorbance of embryo jelly for several Ontario amphibian species. *Canadian Journal of Zoology-Revue Canadienne de Zoologie* **80**(6), 1109-1116.

166. Marco, A. & A.R. Blaustein (1998). Egg gelatinous matrix protects *Ambystoma gracile* embryos from prolonged exposure to air. *Herpetological Journal* **8**(4), 207-211.

167. Gordon, R. (1999). Pursued by the differentiation wave. In: *The Hierarchical Genome and Differentiation Waves: Novel Unification of Development, Genetics and Evolution*. Singapore & London, World Scientific & Imperial College Press: 1159-1167.

168. Gomez-Mestre, I., J.C. Touchon & K.M. Warkentin (2006). Amphibian embryo and parental defenses and a larval predator reduce egg mortality from water mold. *Ecology* **87**(10), 2570-2581.

169. Ward, D. & O.J. Sexton (1981). Anti-predator role of salamander egg membranes. *Copeia* **1981**(3), 724-726.

170. Lehman, E.M. & C.D. Campbell (2007). Developmental window of response to predator chemical cues in rough-skinned newt embryos. *Functional Ecology* **21**(5), 880-885.

171. Gunzburger, M.S. & J. Travis (2005). Critical literature review of the evidence for unpalatability of amphibian eggs and larvae. *Journal of Herpetology* **39**(4), 547-571.
172. Monello, R.J. & R.G. Wright (2001). Predation by goldfish (*Carassius auratus*) on eggs and larvae of the eastern long-toed salamander (*Ambystoma macrodactylum columbianum*). *Journal of Herpetology* **35**(2), 350-353.
173. Crump, M.L. (1983). Opportunistic cannibalism by amphibian larvae in temporary aquatic environments. *American Naturalist* **121**(2), 281-289.
174. Salthe, S.N. (1965). Increase in volume of the perivitelline chamber during development of *Rana pipiens* Schreber. *Physiological Zoology* **35**((1)), 80-98.
175. Tuszynski, J.A. & R. Gordon (2012). A mean field Ising model for cortical rotation in amphibian one-cell stage embryos. *BioSystems* **109**(3, Special Issue on Biological Morphogenesis), 381-389.
176. Larabell, C.A. (1993). A new technique for isolation and visualization of the *Xenopus* egg cortex reveals a complex cytoskeleton. *Journal of Structural Biology* **110**(2), 154-159.
177. Zhang, P., M. Zucchelli, S. Bruce, F. Hambiliki, A. Stavreus-Evers, L. Levkov, H. Skottman, E. Kerkelä, J. Kere & O. Hovatta (2009). Transcriptome profiling of human pre-implantation development. *PLoS One* **4**(11), e7844.
178. Hill, M.A. (2015). Blastocyst Development. https://embryology.med.unsw.edu.au/embryology/index.php/Blastocyst_Development
179. Klimanskaya, I., Y. Chung, S. Becker, S.-J. Lu & R. Lanza (2006). Human embryonic stem cell lines derived from single blastomeres [+ Corrigenda: 444(7118), 512; 446(7133), 342]. *Nature* **444**(7118), 481-485.
180. Flint, R.W., R. Gordon, C.C. Martin & G.W. Brodland (1989). Simulation of the inversion of amphibian eggs in a gravitational field using hollow glass spheres. In: *Microgravity as a Tool in Developmental Biology, Selected papers from a special ESA Symposium held during the 11th International Congress of the International Society of Developmental Biologists, August 20-25, 1989, Utrecht*. Eds.: G.A. Ubbels, H. Oser & T.D. Guyenne. Paris Cedex, European Space Agency: 81-83.
181. Kimball, J.W. (2012). Kimball's Biology Pages. http://users.rcn.com/jkimball.ma.ultranet/BiologyPages/
182. Vincent, J.P. & J.C. Gerhart (1987). Subcortical rotation in *Xenopus* eggs: an early step in embryonic axis specification. *Developmental Biology* **123**(2), 526-539.
183. Shkolnik, M. (2014). Mad Martian: Museum of Modern Madness. http://www.madmartian.com/
184. Morgan, T.H. & A.M. Boring (1903). The relation of the first plane of cleavage and the grey crescent to the median plane of the embryo of the frog. *Archiv für Entwicklungsmechanik der Organismen* **16**(4), 680-690 + Plate XII.
185. Le Guyader, H. (2004). *Geoffroy Saint-Hilaire: A Visionary Naturalist*. Chicago, University of Chicago Press.
186. Martín-Durán, José M., R. Janssen, S. Wennberg, Graham E. Budd & A. Hejnol (2012). Deuterostomic development in the protostome *Priapulus caudatus*. *Current Biology* **22**(22), 2161-2166.

187. Holland, L.Z. (2000). Body-plan evolution in the Bilateria: early antero-posterior patterning and the deuterostome-protostome dichotomy. *Current Opinion in Genetics & Development* **10**(4), 434-442.
188. Meinhardt, H. (2002). The radial-symmetric hydra and the evolution of the bilateral body plan: an old body became a young brain. *BioEssays* **24**(2), 185-191.
189. Geoffroy Saint-Hilaire, É. (1822). Considérations générales sur la vertebre. *Mémoires du Muséum d'histoire naturelle* **9**, 89-119, Plates V-VII.
190. De Robertis, E.M. & Y. Sasai (1996). A common plan for dorsoventral patterning in Bilateria. *Nature* **380**, 37-40.
191. Racine, V. (2013). The Embryo Project Encyclopedia - Geoffroy Saint-Hilaire, H., https://embryo.asu.edu/pages/etienne-geoffroy-saint-hilaire-1772-1844
192. Arendt, D. & K. Nübler-Jung (1994). Inversion of dorsoventral axis? *Nature* **371**(6492), 26.
193. Nübler-Jung, K. & D. Arendt (1994). Is ventral in insects dorsal in vertebrates? A history of embryological arguments favouring axis inversion in chordate ancestors. *Roux's Archives of Developmental Biology* **203**, 357-366.
194. Arendt, D. & K. NublerJung (1997). Dorsal or ventral: Similarities in fate maps and gastrulation patterns in annelids, arthropods and chordates. *Mechanisms of Development* **61**(1-2), 7-21.
195. Arendt, D. (2001). Comparison of early nerve cord development in protostomia and deuterostomia. *Journal of Morphology* **248**(3), 203-203.
196. van den Biggelaar, J.A.M., E. Edsinger-Gonzales & F.R. Schram (2002). The improbability of dorso-ventral axis inversion during animal evolution, as presumed by Geoffroy Saint Hilaire. *Contributions to Zoology* **71**(1-3), 29-36.
197. Hopwood, N. (2007). A history of normal plates, tables and stages in vertebrate embryology. *International Journal of Developmental Biology* **51**(1), 1-26.
198. Hara, K. & E.C. Boterenbrood (1977). Refinement of Harrison's normal table for the morula and blastula of the axolotl. *Development Genes and Evolution* **181**(1), 89-93.
199. Nye, H.L.D., J.A. Cameron, E.A. Chernoff & D.L. Stocum (2003). Extending the table of stages of normal development of the axolotl: limb development. *Developmental Dynamics* **226**(3), 555-560.
200. De Robertis, E.M. (2008). The molecular ancestry of segmentation mechanisms. *Proceedings of the National Academy of Sciences USA* **105**(43), 16411-16412.
201. Tetteroo, P.A.T., J.G. Bluemink, W. Dictus, E.J.J. Vanzoelen & S.W. de Laat (1984). Lateral mobility of plasma membrane lipids in dividing *Xenopus* eggs. *Developmental Biology* **104**(1), 210-218.
202. Hara, K. (1971). Cinematographic observation of "surface contraction waves" (SCW) during the early cleavage of axolotl eggs. *Wilhelm Roux' Archiv für Entwicklungsmechanik der Organismen* **167**, 183-186.
203. Muzinic, L., C. Muzinic & R.S. Voss (2012). Axolotl Embryo Staging Series. http://www.ambystoma.org/education/embryo-staging-series
204. Schreckenberg, G.M. & A.G. Jacobson (1975). Normal stages of development of the axolotl *Ambystoma mexicanum*. *Developmental Biology* **42**(2), 391-400.
205. Satoh, N. (1985). Recent advances in our understanding of the temporal control of early embryonic development in amphibians. *Journal of Embryology and Experimental Morphology* **89**(Suppl), 257-270.

206. Lefresne, J., Y. Andéol & J. Signoret (1998). Evidence for introduction of a variable G1 phase at the midblastula transition during early development in axolotl. *Development, Growth & Differentiation* **40**(5), 497-508.
207. Yang, S.C.H. & J. Bechhoefer (2008). How *Xenopus laevis* embryos replicate reliably: investigating the random-completion problem. *Physical Review E* **78**(4).
208. Nieuwkoop, P.D. & J. Faber (1956). *Normal table of Xenopus laevis Daudin. A Systematical and Chronological Survey of the Development from the Fertilized Egg Till the End of Metamorphosis*. Amsterdam, North Holland Publishing Co.
209. Nieuwkoop, P.D. & J. Faber (1994). *Normal Table of Xenopus laevis (Daudin), A Systematical and Chronological Survey of the Development from the Fertilized Egg Till the End of Metamorphosis*. New York, Garland Publishing, 4th.
210. Cebra-Thomas, J.A. (2003). *Xenopus* embryo staging. http://www.swarthmore.edu/NatSci/sgilber1/DB_lab/Frog/frog_staging.html
211. Boterenbrood, E.C. & J.M. Narraway (1986). The direction of cleavage waves and the regional variation in the duration of cleavage cycles on the dorsal side of the *Xenopus laevis* blastula. *Roux's Archives of Developmental Biology* **195**, 484-488.
212. Boterenbrood, E.C., J.M. Narraway & K. Hara (1983). Duration of cleavage cycles and asymmetry in the direction of cleavage waves prior to gastrulation in *Xenopus laevis*. *Roux's Archives of Developmental Biology* **192**(5), 216-221.
213. Crawford-Young, S.J. (2006). Effects of microgravity on cell cytoskeleton and embryogenesis. *International Journal of Developmental Biology* **50**(2-3), 183-191.
214. Yokota, H., A.W. Neff & G.M. Malacinski (1992). Altering the position of the first horizontal cleavage furrow of the amphibian (*Xenopus*) egg reduces embryonic survival. *International Journal of Developmental Biology* **36**(4), 527-535.
215. Desnitskiy, A.G. (2014). On the classification of the cleavage patterns in amphibian embryos. *Russian Journal of Developmental Biology* **45**(1), 1-10.
216. DeMill, C. & S. Ackerley (2015). Developmental Biology Online. Amphibians: Cleavage of Mesolecithal Eggs. http://www.uoguelph.ca/devobio/amphibian_1.html
217. Buchholz, D.R., S. Singamsetty, U. Karadge, S. Williamson, C.E. Langer & R.P. Elinson (2007). Nutritional endoderm in a direct developing frog: a potential parallel to the evolution of the amniote egg. *Developmental Dynamics* **236**(5), 1259-1272.
218. Elinson, R.P. (2009). Nutritional endoderm: A way to breach the holoblastic-meroblastic barrier in tetrapods. *Journal of Experimental Zoology Part B-Molecular and Developmental Evolution* **312B**(6), 526-532.
219. Elinson, R.P., M.C. Sabo, C. Fisher, T. Yamaguchi, H. Orii & K. Nath (2011). Germ plasm in *Eleutherodactylus coqui*, a direct developing frog with large eggs. *EvoDevo* **2**, #20.
220. Roberts, J.D. (1981). Terrestrial breeding in the Australian leptodactylid frog *Myobatrachus gouldii* (Gray). *Australian Wildlife Research* **8**(2), 451-462.
221. Packer, W.C. (1966). Embryonic and larval development of *Heleioporus eyrei* (Amphibia: Leptodactylidae). *Copeia* **1966**(1), 92-97.

222. del Pino, E.M. & S. Loor Vela (1990). The pattern of early cleavage of the marsupial frog *Gastrotheca riobambae*. *Development*. **110**(3), 781-789.

223. Collazo, A. & R. Keller (2010). Early development of *Ensatina eschscholtzii*: an amphibian with a large, yolky egg. *EvoDevo* **1**, #6.

224. Sawai, T. (1982). Wavelike propagation of stretching and shrinkage in the surface of the newt's egg before the first cleavage. *Journal of Experimental Zoology* **222**, 59-68.

225. Desnitskiy, A.G. & S.N. Litvinchuk (2014). Comparative and phylogenetic perspectives of the cleavage process in tailed amphibians. *Zygote* **23**(5), 722-731.

226. Gopichandran, N. & H.J. Leese (2003). Metabolic characterization of the bovine blastocyst, inner cell mass, trophectoderm and blastocoel fluid. *Reproduction* **126**(3), 299-308.

227. Jensen, P.L., M.L. Grondahl, H.C. Beck, J. Petersen, L. Stroebech, S.T. Christensen & C.Y. Andersen (2014). Proteomic analysis of bovine blastocoel fluid and blastocyst cells. *Systems Biology in Reproductive Medicine* **60**(3), 127-135.

228. Takeuchi, M., M. Takahashi, M. Okabe & S. Aizawa (2009). Germ layer patterning in bichir and lamprey; an insight into its evolution in vertebrates. *Developmental Biology* **332**(1), 90-102.

229. Stableford, L.T. (1949). The blastocoel fluid in the amphibian gastrulation. *Journal of Experimental Zoology* **112**, 529-546.

230. Collart, C., G.E. Allen, C.R. Bradshaw, J.C. Smith & P. Zegerman (2013). Titration of four replication factors is essential for the *Xenopus laevis* midblastula transition. *Science* **341**(6148), 893-896.

231. Tadros, W. & H.D. Lipshitz (2009). The maternal-to-zygotic transition: a play in two acts. *Development* **136**(18), 3033-3042, doi: 3010.1242/dev.033183.

232. Chulitskaia, E.V. (1970). Desynchronization of cell divisions in the course of egg cleavage and an attempt at experimental shift of its onset. *Journal of Embryology and Experimental Morphology* **23**(2), 359-374.

233. Gruwel, M.L.H., B. Lerner, S. Crawford-Young & R. Gordon (2007). Early development of axolotl (*Ambystoma mexicanum*) embryos using MRI micro-imaging. In: *Manitoba Institute of Child Health 3rd Annual Research Day, October 11, 2007*. Winnipeg, Manitoba Institute of Child Health: 19.

234. Nikas, Y. (2015). Inner cell mass of day 5 zonal human blastocysts #a1202. http://www.eikonika.net/v2/photo_info.php?photo_id=305

235. Wheeler, R. (2010). The first few weeks of embryogenesis in humans. Beginning at the fertilised egg, ending with the closing of the neural tube. https://commons.wikimedia.org/wiki/File:HumanEmbryogenesis.svg

236. Donnay, I. & H.J. Leese (1999). Embryo metabolism during the expansion of the bovine blastocyst. *Molecular Reproduction and Development* **53**(2), 171-178.

237. Maître, J.L., R. Niwayama, H. Turlier, F. Nédélec & T. Hiiragi (2015). Pulsatile cell-autonomous contractility drives compaction in the mouse embryo. *Nature Cell Biology* **17**, 849-855.

238. Reeve, W.J.D. & C.A. Ziomek (1981). Distribution of microvilli on dissociated blastomeres from mouse embryos: Evidence for surface polarization at compaction. *Journal of Embryology and Experimental Morphology* **62**(Apr), 339-350.

239. Wong, C.C., K.E. Loewke, N.L. Bossert, B. Behr, C.J. De Jonge, T.M. Baer & R.A. Reijo Pera (2010). Non-invasive imaging of human embryos before embryonic genome activation predicts development to the blastocyst stage. *Nature Biotechnology* **28**(10), 1115-1121.
240. Yaguchi, S., J. Yaguchi & K. Inaba (2014). Bicaudal-C is required for the formation of anterior neurogenic ectoderm in the sea urchin embryo. *Scientific Reports* **4**, #6852.
241. Smart, T.I. & G. von Dassow (2009). Unusual development of the mitraria larva in the polychaete *Owenia collaris*. *Biological Bulletin* **217**(3), 253-268.
242. Zardus, J.D. & M.P. Morse (1998). Embryogenesis, morphology and ultrastructure of the pericalymma larva of *Acila castrensis* (Bivalvia: Protobranchia: Nuculoida). *Invertebrate Biology* **117**(3), 221-244.
243. Barth, L.G. & L.J. Barth (1969). The sodium dependence of embryonic induction. *Developmental Biology* **20**, 236-262.
244. Barth, L.G. & L.J. Barth (1974). Ionic regulation of embryonic induction and cell differentiation in *Rana pipiens*. *Developmental Biology* **39**, 1-22.
245. Morrill, G.A., Kostello.Ab & J.B. Murphy (1974). Role of Na^+,K^+-ATPase in early embryonic development. *Annals of the New York Academy of Sciences* **242**(NOV29), 543-559.
246. Gordon, R., N.K. Björklund & P.D. Nieuwkoop (1994). Dialogue on embryonic induction and differentiation waves. *International Review of Cytology* **150**, 373-420.
247. Fagotto, F. & F.R. Maxfield (1994). Yolk platelets in *Xenopus* oocytes maintain an acidic internal pH which may be essential for sodium accumulation. *Journal of Cell Biology* **125**(5), 1047-1056.
248. Fagotto, F. & F.R. Maxfield (1994). Changes in yolk platelet pH during *Xenopus laevis* development correlate with yolk utilization. A quantitative confocal microscopy study. *Journal of Cell Science* **107**, 3325-3337.
249. Slack, C. & A.E. Warner (1973). Intracellular and intercellular potentials in the early amphibian embryo. *Journal of Physiology (London)* **232**(2), 313-330.
250. Giraldez, A.J. (2010). microRNAs, the cell's Nepenthe: clearing the past during the maternal-to-zygotic transition and cellular reprogramming. *Current Opinion in Genetics & Development* **20**(4), 369-375.
251. Walser, C.B. & H.D. Lipshitz (2011). Transcript clearance during the maternal-to-zygotic transition. *Current Opinion in Genetics & Development* **21**(4), 431-443.
252. Laver, J.D., X. Li, D. Ray, K.B. Cook, N.A. Hahn, S. Nabeel-Shah, M. Kekis, H. Luo, A.J. Marsolais, K.Y.Y. Fung, T.R. Hughes, J.T. Westwood, S.S. Sidhu, Q. Morris, H.D. Lipshitz & C.A. Smibert (2015). Brain tumor is a sequence-specific RNA-binding protein that directs maternal mRNA clearance during the *Drosophila* maternal-to-zygotic transition. *Genome Biology* **16**, #94.
253. Tuft, P. (1958). Changes in the osmotic activity of the blastocoel and archenteron contents during the early-development of *Xenopus laevis*. *Proceedings of the Royal Physical Society of Edinburgh* **26**(2), 42-48.
254. Khokha, M.K., C. Chung, E.L. Bustamante, L.W. Gaw, K.A. Trott, J. Yeh, N. Lim, J.C. Lin, N. Taverner, E. Amaya, N. Papalopulu, J.C. Smith, A.M. Zorn, R.M. Harland & T.C. Grammer (2002). Techniques and probes for the study of *Xenopus tropicalis* development. *Developmental Dynamics* **225**(4), 499-510.

255. Yasuo, H. & P. Lemaire (2001). Generation of the germ layers along the animal-vegetal axis in *Xenopus laevis*. *International Journal of Developmental Biology* **45**(1 Spec No), 229-235.
256. Burton, P.M. (2008). Insights from diploblasts; the evolution of mesoderm and muscle. *Journal of Experimental Zoology Part B-Molecular and Developmental Evolution* **310B**(1), 5-14.
257. Kieny, M., A. Mauger & P. Sengel (1972). Early regionalization of somitic mesoderm as studied by the development of axial skeleton of the chick embryo. *Developmental Biology* **28**(1), 142-161.
258. Swiers, G., Y.H. Chen, A.D. Johnson & M. Loose (2010). A conserved mechanism for vertebrate mesoderm specification in urodele amphibians and mammals. *Developmental Biology* **343**(1-2), 138-152.
259. Lemaigre, F. & K.S. Zaret (2004). Liver development update: new embryo models, cell lineage control, and morphogenesis. *Current Opinion in Genetics & Development* **14**(5), 582-590.
260. Yutzey, K.E. & M.L. Kirby (2002). Wherefore heart thou? Embryonic origins of cardiogenic mesoderm. *Developmental Dynamics* **223**(3), 307-320.
261. Varner, V.D. & L.A. Taber (2012). Not just inductive: a crucial mechanical role for the endoderm during heart tube assembly. *Development* **139**(9), 1680-1690.
262. Stern, C.D., Ed. (2004). *Gastrulation: From Cells to Embryo*, Cold Spring Harbor Laboratory Press.
263. Shi, D.L., M. Delarue, T. Darribère, J.F. Riou & J.C. Boucat (1987). Experimental analysis of the extension of the dorsal marginal zone in *Pleurodeles waltl* gastrulae. *Development* **100**(1), 147-161.
264. Wray, G.A. & R.A. Raff (1991). Rapid evolution of gastrulation mechanisms in a sea urchin with lecithotrophic larvae. *Evolution* **45**(8), 1741-1750.
265. Stach, T., J. Winter, J.M. Bouquet, D. Chourrout & R. Schnabel (2008). Embryology of a planktonic tunicate reveals traces of sessility [erratum **105**(29), 10268-10269]. *Proceedings of the National Academy of Sciences USA* **105**(20), 7229-7234.
266. Holland, L.Z. & N.D. Holland (2007). A revised fate map for amphioxus and the evolution of axial patterning in chordates. *Integrative and Comparative Biology* **47**(3), 360-372.
267. Nishida, H. & T. Stach (2014). Cell lineages and fate maps in tunicates: Conservation and modification. *Zoological Science* **31**(10), 645-652.
268. Wolpert, L., C. Tickle & A.M. Arias (2015). *Principles of Development.* Oxford, Oxford University Press, 5th.
269. Blisstree (2007). Fruitballs. http://www.blisstree.com/2007/09/13/food/fruitballs/
270. Spear, P.C. & C.A. Erickson (2012). Interkinetic nuclear migration: a mysterious process in search of a function. *Development, Growth & Differentiation* **54**(3), 306-316.
271. Sunday White (2012). Green Round Bottom Bottle, Vintage Chianti Bottle, Green Glass Bottle, Round Bottle, Wine Bottle. http://www.etsy.com/listing/104001050/green-round-bottom-bottle-vintage
272. Lee, J.Y. & R.M. Harland (2010). Endocytosis is required for efficient apical constriction during *Xenopus* gastrulation. *Current Biology* **20**(3), 253-258.

273. Brody, T.B. (2015). The Inter*active* Fly: *Drosophila* mesoderm migration behaviour during gastrulation. http://www.sdbonline.org/sites/fly/aimorph/gastrulation.htm - Migration
274. Wikipedia (2013). Epithelial–mesenchymal transition. https://en.wikipedia.org/wiki/Epithelial_mesenchymal_transition
275. Holtfreter, J. (1943). A study of the mechanics of gastrulation, Part I. *Journal of Experimental Zoology* **94**(3), 261-318.
276. Wendrich, J.R. & D. Weijers (2013). The *Arabidopsis* embryo as a miniature morphogenesis model. *New Phytologist* **199**(1), 14-25.
277. Gilbert, S.F. (2000). *Developmental Biology*. Sunderland, Massachusetts, Sinauer Associates.
278. Saxén, L. & S. Toivonen (1962). *Primary Embryonic Induction*. London, Logos Press, Academic Press.
279. Wyczalkowski, M.A., V.D. Varner & L.A. Taber (2013). Computational and experimental study of the mechanics of embryonic wound healing. *Journal of the Mechanical Behavior of Biomedical Materials* **28**, 125-146.
280. Needham, J. (1950). *Biochemistry and Morphogenesis*. Cambridge, Harvard University Press.
281. Spemann, H. (1918). Über die Determination der ersten Organanlagen des Amphibienembryo. I-VI/On the determination of the first organ systems of the amphibian embryo. I-VI. *Archiv für Entwicklungsmechanik der Organismen* **43**(4), 448-555.
282. Mangold, O. (1922). Transplantationsversuche zur Ermittlung der Eigenart der Keimblättern [Transplantation attempts for the investigation of the individuality of the germinal layers]. *Verhandl. deutsch. Zool. Gesell. Würzburg* **27**, 51-52.
283. Mangold, O. (1923). Transplantationsversuche zur Frage der Spezifität und der Bildung der Keimblätter/Transplantation attempts toward the question of the specificity and the formation of the germinal layers. *Archiv für mikroskopische Anatomie und Entwicklungsmechanik* **100**(1/2), 198-301.
284. Mangold, O. (1962). 50 Jahre Experimente zur Analyse der Fruhentwicklung des Molchkeims [50 years experiments on the analysis of early development of the salamander germ]. *Naturwissenschaftliche Rundschau* **15**((9/10)), 347-357.
285. Kearl, M. (2013). Otto Mangold (1891-1962). In: *The Embryo Project Encyclopedia*. http://embryo.asu.edu/pages/otto-mangold-1891-1962.
286. Oppenheimer, J.M. (1970). Hans Driesch and the theory and practice of embryonic transplantation. *Bulletin of the History of Medicine* **44**(4), 378-382.
287. Blackiston, D.J. & M. Levin (2013). Ectopic eyes outside the head in *Xenopus* tadpoles provide sensory data for light-mediated learning. *Journal of Experimental Biology* **216**(6), 1031-1040.
288. Spemann, H. (1938). *Embryonic Development and Induction*. New Haven, Yale University Press.
289. Blackiston, D.J., G.M. Anderson, N. Rahman, C. Bieck & M. Levin (2015). A novel method for inducing nerve growth via modulation of host resting potential: Gap junction-mediated and serotonergic signaling mechanisms. *Neurotherapeutics* **12**(1), 170-184.
290. Elliott, K.L., D.W. Houston & B. Fritzsch (2015). Sensory afferent segregation in three-eared frogs resemble the dominance columns observed in three-eyed frogs. *Scientific Reports* **5**, #8338.

291. Khattak, S., T. Richter & E.M. Tanaka (2009). Generation of transgenic axolotls (*Ambystoma mexicanum*). *Cold Spring Harbor Protocols* **2009**(8), doi:10.1101/pdb.prot5264.
292. Humphrey, R.R. (1967). Albino axolotls from an albino tiger salamander through hybridization. *Journal of Heredity* **58**(3), 95-101.
293. Mintz, B. & K. Illmensee (1975). Normal genetically mosaic mice produced from malignant teratocarcinoma cells. *Proceedings of the National Academy of Sciences USA* **72**(9), 3585-3589.
294. Bautzmann, H. (1941). Walther Vogt zum Gedächtnis. *Die Naturwissenschaften*(24/25), 345-347.
295. Spemann, H. (1941). Walter Vogt zum Gedächtniss/In memoriam for Walter Vogt. *Wilhelm Roux' Archiv für Entwicklungsmechanik der Organismen* **141**(1), 7-14.
296. Vogt, W. (1925). Gestaltungsanalyse am Amphibienkeim mit örtlicher Vitalfärbung. Vorwort über Wege und Ziele. I. Teil. Methodik und Wirkungsweise der örtlichen Vitalfärbung mit Agar als Farbträger/Analysis of the organization of the amphibian embryo with local vital dyes. Preface on methods and objectives. Part I. Methodology and method of operation of the local vital dye with agar as color carrier. *Wilhelm Roux' Archiv Entwicklungsmechanik Organismen* **106**, 542-610.
297. Vogt, W. (1926). Die Beziehungen zwischen Furchung, Hauptachsen des Embryo und Ausgangsstruktur im Amphibienei, nach Versuchen mit örtlicher Vitalfärbung/The relationship between cleavage, principal axes of the embryo and output structure in the amphibian egg, according to experiments with local vital staining. *Sitzungsberichte der Gesellschaft für Morphologie und Physiologie in München* **37**, 60-70.
298. Vogt, W. (1929). Gestaltungsanalyse am Amphibienkeim mit Örtlicher Vitalfärbung II. Teil. Gastrulation und Mesodermbildung bei Urodelen und Anuren/Analysis of the organization of the amphibian embryo with local vital dyes. Part II: Gastrulation and mesoderm development in urodeles and anurans. *Wilhelm Roux Archiv für Entwicklungsmechanik der Organismen* **120**(1), 384-706.
299. Smith, J.C. & G.M. Malacinski (1983). The origin of the mesoderm in an anuran, *Xenopus laevis*, and a urodele, *Ambystoma mexicanum*. *Developmental Biology* **98**(1), 250-254.
300. Cleine, J.H. & J.M.W. Slack (1985). Normal fates and states of specification of different regions in the axolotl gastrula. *Journal of Embryology and Experimental Morphology* **86**, 247-269.
301. Delarue, M., S. Sanchez, K.E. Johnson, T. Darribère & J.C. Boucaut (1992). A fate map of superficial and deep circumblastoporal cells in the early gastrula of *Pleurodeles waltl*. *Development* **114**(1), 135-146.
302. Hadorn, E., R. Wehner & A. Kühn (1977). *Algemene Zoölogie*. Utrecht, Het Spectrum, 4th.
303. Gilbert, S.F. (2007). Fate maps, gene expression maps, and the evidentiary structure of evolutionary developmental biology. In: *From Embryology to Evo-Devo: A History of Developmental Evolution*. Eds.: M.D. Laubichler & J. Maienschein. Cambridge, MIT Press: 357-374.

304. Nakamura, O. (1942). Die Entwicklung der hinteren Körperhälfte bei Urodelen [The development of the posterior half of the body of urodeles]. *Annotationes zoologicae Japonenses* **21**(4), 169-237.

305. Pasteels, J.J. (1942). New observations concerning the maps of presumptive areas of the young amphibian gastrula (*Amblystoma* and *Discoglossus*). *Journal of Experimental Zoology* **89**(2), 255-281.

306. Ito, Y., T. Oinuma, K. Takano, S. Komazaki, S. Obata & M. Asashima (2006). *CyNodal*, the Japanese newt *nodal*-related gene, is expressed in the left side of the lateral plate mesoderm and diencephalon. *Gene Expression Patterns* **6**(3), 294-298.

307. Kaneda, T. & J.-y.D. Motoki (2012). Gastrulation and pre-gastrulation morphogenesis, inductions, and gene expression: Similarities and dissimilarities between urodelean and anuran embryos. *Developmental Biology* **369**(1), 1-18.

308. Svane, I. & C.M. Young (1989). The ecology and behaviour of ascidian larvae. *Oceanography and Marine Biology Annual Review* **27**(4), 45-90.

309. Carter, A.M. & A.C. Enders (2016). Placentation in mammals: Definitive placenta, yolk sac, and paraplacenta. *Theriogenology* **86**(1), 278-287.

310. Asashima, M., G.M. Malacinski & S.C. Smith (1989). Surgical manipulation of embryos. In: *Developmental Biology of the Axolotl*. Eds.: J.B. Armstrong & G.M. Malacinski. New York, Oxford University Press: 255-263.

Chapter 3

Developmental Genetics: A Flying Tour

Since the discovery of the double helix structure of DNA (Figure 3.1)[1], our understanding of genetics has exploded like a supervolcano, blasting in multiple directions, and covering the scientific world with a broad layer of new understanding. Yet, it seems the more we learn, the more there is to learn about the way genes work. The scientific community is spending more time slowly digging out of the proverbial volcanic ash heaps than looking up to see the new landscape as a whole. A graduate student can still spend his or her entire thesis studying a single protein's effect on gene expression. And so scientists dig, often unaware of the fellow in the next block, also digging just as hard, in another little corner trying to uncover the same stretch of buried road.

This chapter will give you a flying tour of the current state of knowledge of genetics especially as it impacts embryogenesis. Some of this may well be out of date by the time this book is published because the field is changing so quickly. Other descriptions we use here will likely turn out to be gross oversimplifications if and when the whole story is completed. Nonetheless, we will attempt to get as complete a view as presently possible of the new ash covered landscape that genetics works within. We will divide our tour into three components. First, this chapter will review the basics of transcription and translation, how proteins are made and some of the simpler forms of control of gene expression. Most of this will be obvious for those who took high school biology but, hopefully, any reader will find some exciting new ideas. Second, we will examine the steps of mitosis, the process whereby a eukaryotic cell duplicates its DNA and then divides into two identical or different daughter cells. The latter is perhaps the root event in cell differentiation. According to our model of the evolution of embryogenesis, cell division is the first step and everything else in embryogenesis was built by copying, duplicating and then evolving the original system for cell replication. Therefore it is important to understand it. Third, we will look at meiosis, the process whereby germ cells or gametes (sperm and egg for humans) are created.

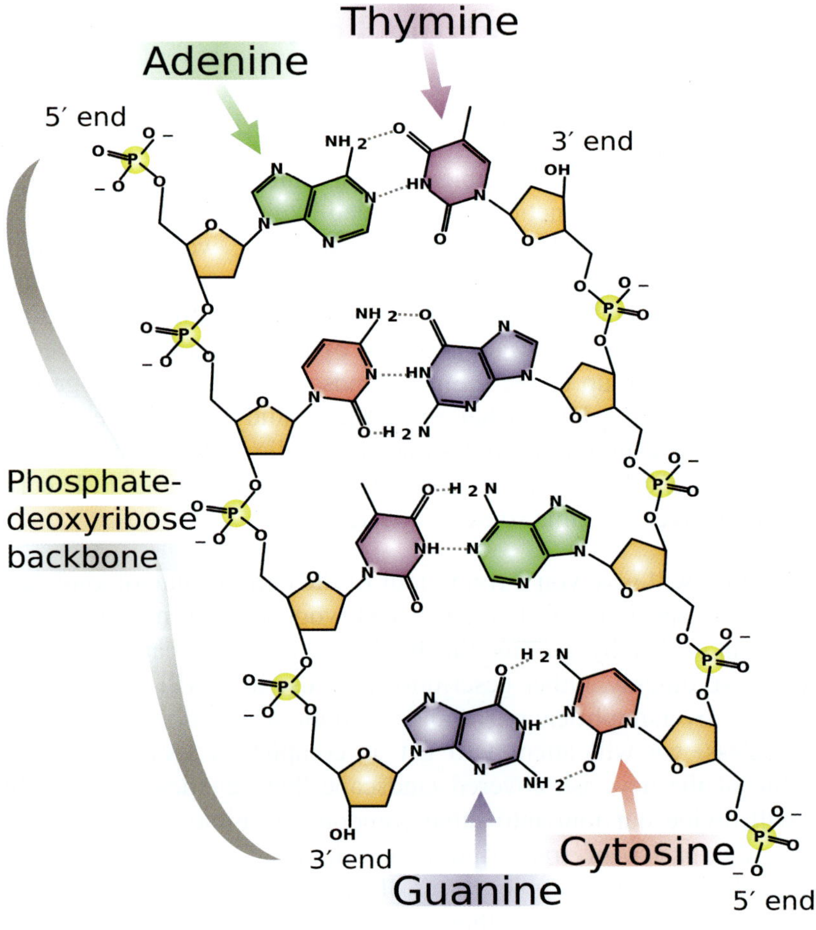

Figure 3.1. An uncoiled, flattened view of the DNA double helix showing its detailed chemical structure. Adenine (A) and thymine (T) pair up using two hydrogen bonds (dotted lines) and guanosine (G) and cytosine (C) use three hydrogen bonds, so that GC pairs are more strongly bound than AT. All atoms are shown, following the convention explained in Figure 3.15. Note that the two DNA strands are polar structures (each having different ends) with opposite orientations. From Wikipedia[2] with permission under the Creative Commons Attribution-Share Alike 3.0 Unported license. For a movie of the double helix structure of DNA see Wikipedia[3].

Normally eukaryotes all carry two copies of everything in their genome (genome = all DNA in one diploid cell). Each gamete has only one copy of the DNA instead of two. Meiosis is then the process by which the special germ cells end up having only one copy instead of two. This way, when gametes meet and join, the single cell organism (i.e., the fertilized egg) is back to having two copies. There are some exceptions to this general rule that we will touch on. More recently scientists have also discovered that information about genome functioning is embedded not just in the DNA, but also in associated peculiarities that are broadly referred to as 'epigenetic'. We will examine the breakthroughs in the new field of epigenetics in Chapter 4. The genetic supervolcano is still erupting.

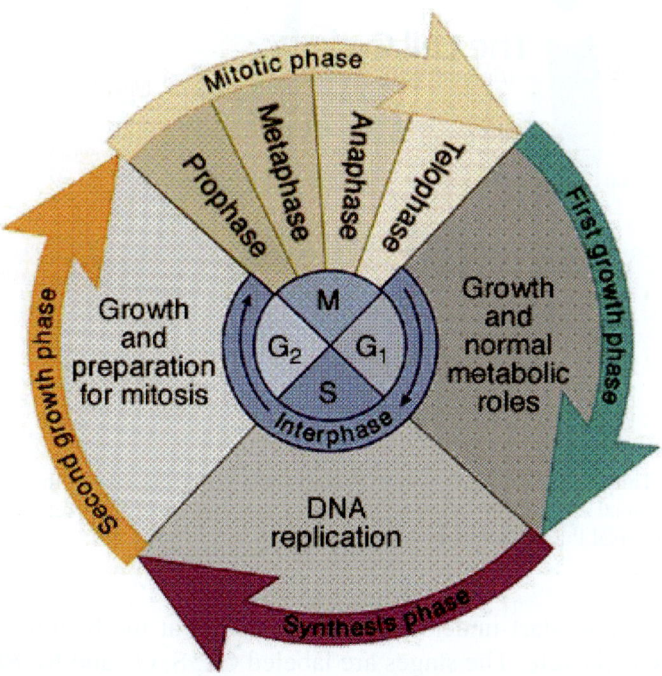

Figure 3.2. The cell cycle is shown in this diagram illustrating the ways the cycle can be labeled. This diagram is somewhat idealized. Time spent in any one phase can vary dramatically. Rapidly dividing cells of early embryos spend almost no time in interphase and go from mitosis to mitosis as quickly as possible while other cells types, such as mature cells in tissues may remain in interphase for years until some event, such as an injury, requires they duplicate to effect repairs. From St. Rosemary Educational Institution[4] with permission.

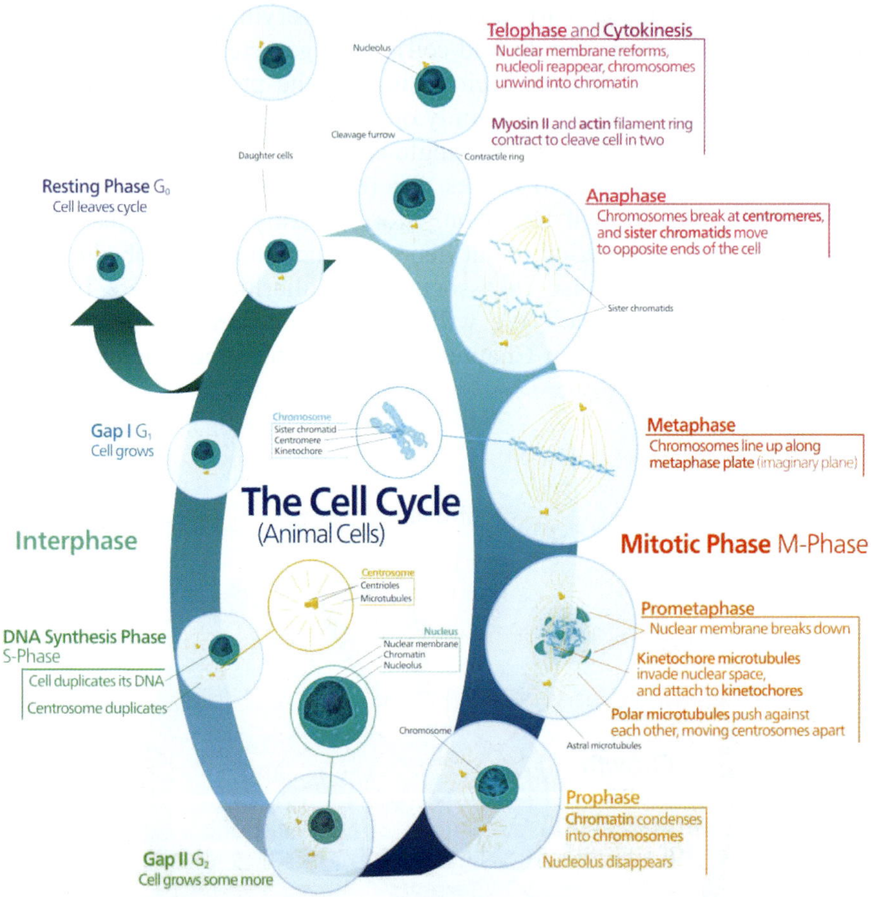

Figure 3.3. This shows details of some of the events occurring during the cell cycle in eukaryotic animal cells. From Wikipedia[8] with permission under the Creative Commons CC0 1.0 Universal Public Domain Dedication.

A good place to start understanding genetics is at the beginning of the eukaryotic cell cycle. The stages are labeled G1, S, G2 and M. Stages G1 through G2 are also referred to collectively as interphase. M is for mitosis or the short time where the chromosomes become individually visible in light microscopy. M is further divided into subsections called prophase, metaphase, anaphase and telophase (Figure 3.2). The term interphase is largely a historical one. Originally, when scientists were looking at cells using early light microscopes they couldn't see anything during G1 through G2. During the M phase, i.e., during mitosis, the chromosomes condense from being dispersed in the nucleus to a much smaller volume, so much so that they become visible even in those primitive light microscopes of the early days of genetics. Chromosomes could be seen as double strands joined at one point (centromere), which line up, and then separate at the joint, one strand of each chromosome going into one of the subsequent daughter cells (Figure 3.3). Because mitosis can be watched live, *in vivo*, this M phase was a time of fascinating activity for early scientists. Each step was carefully described and labeled. Animated[5] and real movies[6,7] of mitosis are readily available.

The visually unimpressive interphase is actually the time when the cell is doing its most interesting stuff because that is when DNA is being copied (transcribed) to create mRNA (m = messenger) and the resulting mRNA is used (translated) for guiding protein production (Figures 3.4, 3.5 and 3.6). It is also during interphase's S phase (S meaning "synthesis": Figure 3.2) that the DNA is doubled in preparation for mitosis. Newer forms of microscopy now show many of the fascinating details once hidden in the murky nucleoplasm of interphase[9]. For example, it is now known that the chromosomes swell, but are only slightly intertwined or intermingled[10], forming what are called "chromosome territories"[11] (Figure 3.7).

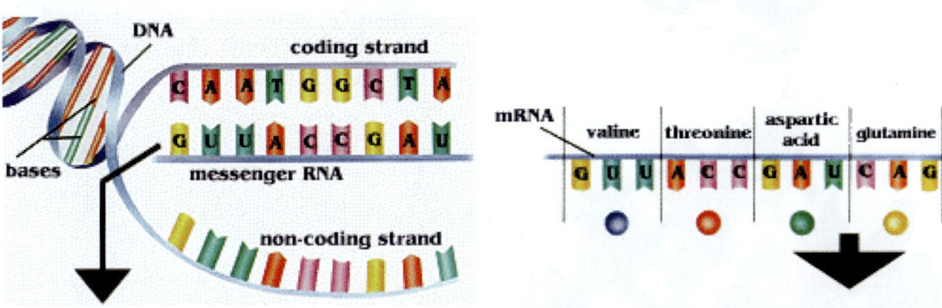

Figure 3.4. **Left**: In the nucleus a messenger RNA (mRNA) with complementary nucleotide bases to one strand of the DNA is produced, which exits the nucleus. **Right**: With the help of ribosomes in the cytoplasm, groups of 3 nucleotides (codons) are coded into a protein consisting of a specific sequence of amino acids (such as valine, threosine, aspartic acid, glutamine shown here) bound consecutively to one another in a chain. "The letters represent the bases adenine (A), thymine (T), guanine (G) and cytosine (C). In the double helix, A always pairs with T, and C with G. In the mRNA, thymine is replaced by uracil (U).... There are 20 different amino acids." From the Nobel Foundation[12] with their kind permission[13].

Figure 3.5. Proteins can have thousands of amino acids ("residues"), each made of many atoms, so depicting their structures in a comprehensible way is a fine art. Here are three depictions of the same protein, indicating atoms, secondary ribbon structure and type of amino acid, respectively, top to bottom. In the latter acidic residues are red, basic residues are blue, polar residues are green, and nonpolar residues are white. This protein, a coiled and folded string of 247 amino acids, contains 3740 atoms, not counting hydrogen atoms, and is roughly 8 nm wide[14]. From Wikipedia[15] with permission under a GNU Free Documentation License, Version 1.2. While a protein is constructed as a chain of amino acids, it is not usually depicted that way. The individual amino acids, modeled as beads that can be strung and worn, have been depicted as 3D rotating images[16]. An enzyme is a protein that catalyzes a chemical reaction, i.e., makes it go faster than it would without the protein present. Most proteins, even the cytoskeletal proteins once thought of as primarily structural, are involved in chemical reactions. Thus the distinction has been blurred.

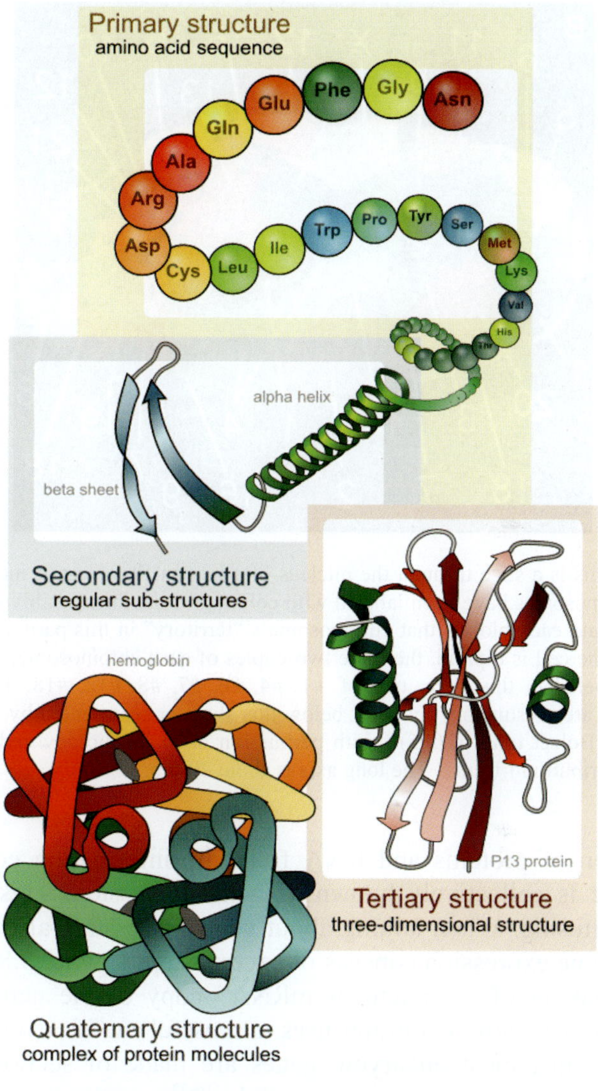

Figure 3.6. This shows how an amino acid chain (Figure 3.4) can be represented at a higher level as a ribbon that is curled or a sheet in space. These spatially organized stretches of amino acids are called "protein domains", here designated as "secondary" sub-structures. An individual protein often contains a number of domains organized in space (tertiary structure). Protein molecules of the same or different kinds may bind together, frequently into dimers or groups of four (tetramers). This level is described as quaternary structure. Sometimes that binding forms chains of proteins, as in microtubules. From Wikimedia[17] (in the public domain). Protein domains and modules are similar concepts with various definitions[333-335].

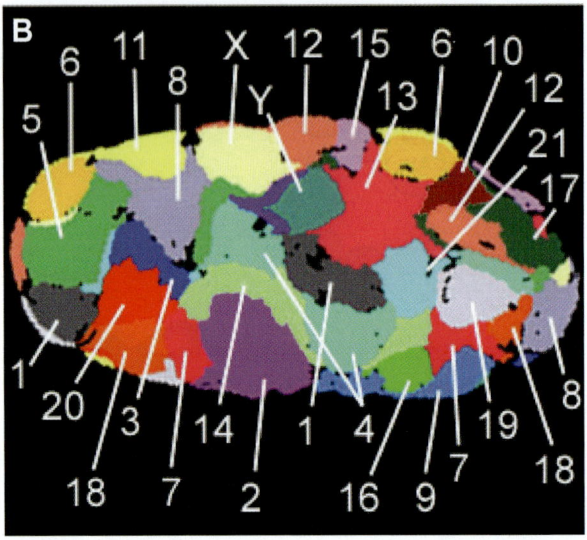

Figure 3.7. This is a slice through the nucleus of a human fibroblast at interphase stage G0. The chromosomes have been labeled with colors to make them visibly separate from one another, and each blob is that chromosome's "territory" in this particular cell. Note that because the cell is diploid, there are two copies of each chromosome, some pairs of which are labeled in this slice: two of #1, #4, #6, #7, #8, #12, #18. The "missing" chromosomes are presumably above or below this slice, or represented by the unlabeled slivers. From Bolzer *et al.* (2005)[18] with permission of PLoS Biology under a Creative Commons Attribution License. The long axis is about 20 μm.

The creation of proteins and RNA from the information contained in the genome is collectively known as gene expression. This shorthand can get confusing. When talking about gene expression, always keep in mind that gene expression consists of several steps (Figure 3.8). It begins with the making of a "precursor mRNA" copy of the gene coded in the DNA. Recall how evolution uses successful motifs over and over? Because of this, most eukaryote genes are made of sections of code usually translating to protein modules[19-22], with "empty space" (noncoding regions called "introns") in between the modules.

Bacteria cells generally have few or no introns[23] which indicates that the development of introns was an important step forward in the evolution of eukaryotes. This is the "introns late" hypothesis. The sequence of these ancient events has been hotly debated with a reverse hypothesis being that introns were always present and bacteria simply represent a more

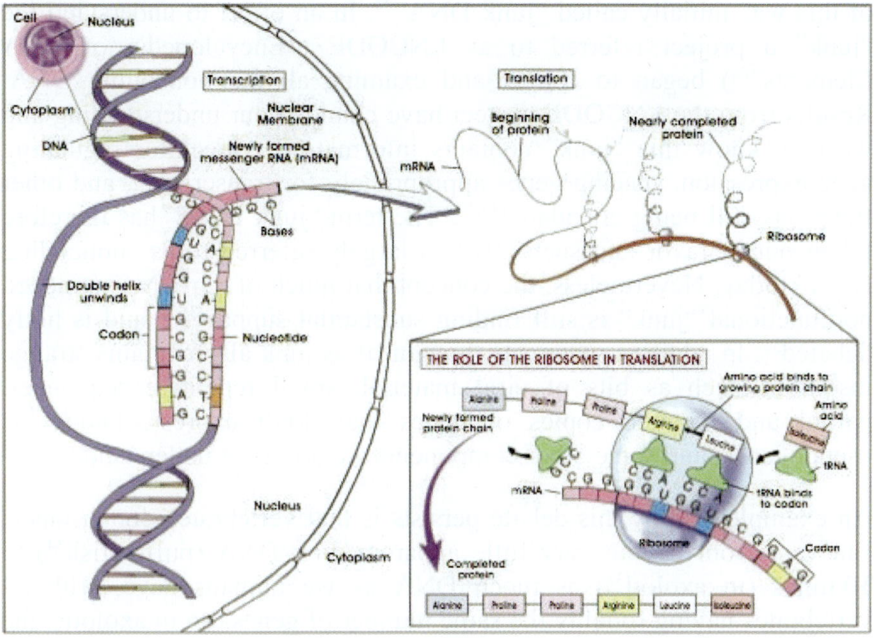

Figure 3.8. Transcription of DNA to mRNA occurs in the cell nucleus, and translation of the mRNA to protein occurs in the cytoplasm (though protein synthesis has been suggested to sometimes occur in nuclei[24]). Translation is effected inside the ribosome via use of transfer-RNAs (tRNA), one for each amino acid. A codon is a group of three consecutive nucleotides that are translated via the tRNA to a particular amino acid. From GeneticsLab[25] under a Creative Commons Attribution Share-Alike 3.0 License. The "newly formed mRNA" shown here is actually precursor or pre-mRNA, which is spliced into mRNA before exiting the nucleus (Figure 3.9).

efficient coding (the "introns early" hypothesis). The preponderance of evidence garnered from the examination of the genetic sequencing of intron signal sites seems to support the "introns late" hypothesis although some form of predecessor intron-like function was present before eukaryotes evolved. Eukaryotes certainly took the earliest intron and utilized it to create splice variation, which allows a single gene to code for multiple proteins. It also allows for the mix and matching of protein domains (Figure 3.6) to produce a far greater variety of proteins than could be accomplished with the more efficient coding used by bacteria[26].

This empty space between genes consists of not just introns but long stretches of DNA that seem to serve no coding purpose whatsoever. All of this was initially called "junk DNA"[27]. In an effort to understand the "junk" a project referred to as ENCODE ("Encyclopedia of DNA Elements"[28]) began to collect and examine all the noncoding DNA. Results from the ENCODE project have changed our understanding and we now know this "junk" contains information critical for regulating gene expression, spacing genes appropriately for transcribing, and other functions still being elucidated[29,30]. The term "junk DNA" has therefore fallen out of favor and such DNA is largely referred to as "noncoding DNA" today. Nevertheless, the concept that much of our DNA is indeed nonfunctional "junk" is still finding substantial support[31,32] and is hotly debated[33]. In addition what was thought of as junk also contains strange insertions such as bits of viral material, small repetitive sequences, partial and mutated copies of genes that don't express, known as pseudogenes, and some other components we don't yet understand.

An example of why this debate persists is that vertebrates, for instance, can vary from having very little apparent junk DNA (puffer fish[34]) to 10 times (in axolotl[35]) as much DNA as we humans have, with all vertebrates having roughly the same number of genes. So in axolotls, the proportion of junk DNA should be even higher than ours. Humans have 3.50 picograms of DNA per haploid cell, while the salamander *Necturus lewisi* has 120.60 picograms[36], a ratio of 34.5:1. Only 1.2% of the DNA in a human cell codes for protein[37], and if *Necturus* has the same number of proteins, this fraction goes down to 0.035%. This discrepancy, the widely varying ratio between the amount of coding DNA and the total amount of DNA per genome, has been called the C-value paradox[38] or C-value enigma[39], and we can say that it is still unsolved. It's just as unsolved in plants, whose genome sizes vary even more, by 2000-fold[40]. We can conclude that there likely is junk DNA in most genomes, but a lot of what we presently think of as junk probably has specific functions we haven't learned about yet. There is still much to learn about noncoding DNA.

Which genes are transcribed depends on a host of regulatory factors. These fall into the broad categories of specificity factors, repressors, general transcription factors, activators, enhancers and silencers, explored in Chapter 9. These transcriptional regulators work on a general principal of allowing or preventing correct binding of the RNA polymerase to the DNA near a gene. If the RNA polymerase is properly bound, transcription of the gene proceeds. There are specific sequences on the DNA, which guide the various regulatory factors to the DNA. These can be divided into two broad categories. One type contains long stretches of cytosine followed by guanine forming what are known as CpG "islands" along the DNA[41]. The islands tend to be in groups near genes[42]. The second broad class is the alternative promoters, which do not use CpG islands. These are much more variable and allow finer "tuning" of gene expression.

Subsequently, which introns and exons get transcribed and which get left behind by splicing during the conversion of precursor mRNA (pre-mRNA) to mRNA, is also affected by which transcription factors are present and the promoter sequences. The result can be different mRNAs from the same gene. Simply by mixing and matching exons

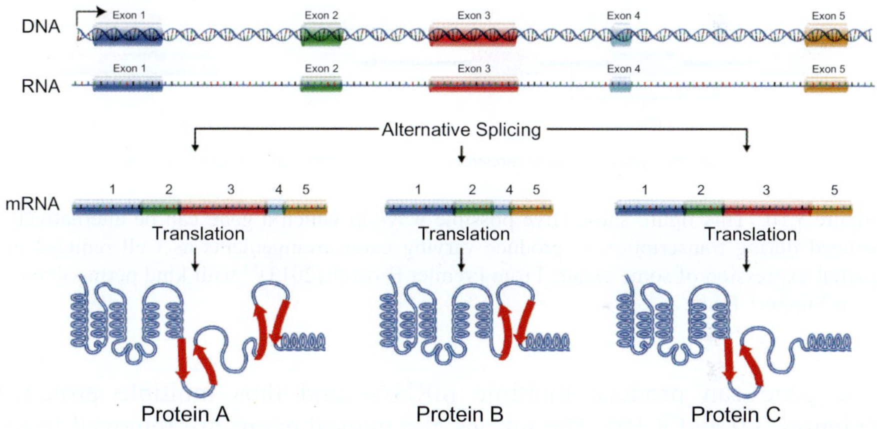

Figure 3.9. Splicing is the process of removal of parts of precursor mRNA ("RNA" in this diagram) by the spliceosome, leaving a sequence of "exons" that constitute the mRNA that exits from the nucleus to the cytoplasm. The parts removed from the RNA, and their precursors in the DNA, are called "introns". Different subsets of the exons can end up in the mRNA. This is called "alternative splicing". The result is different proteins (polypeptides) from the same DNA sequence of the genome, and sometimes nonsense[43]. In this example, three alternative mRNAs are shown, along with the three different proteins that they translate to. From Wikipedia[44], in the public domain. There is an important relationship of this to embryogenesis, since it is often the case that different alternative mRNAs are expressed in different cell types in an embryo[45-47], including in the brain[48]. Note that the use of exons to build proteins means that proteins can be quite modular in their construction, which expedites protein evolution[49]. The process of splicing can be regarded as being carried out by a "splicing code" that is analogous to the DNA code[50,51].

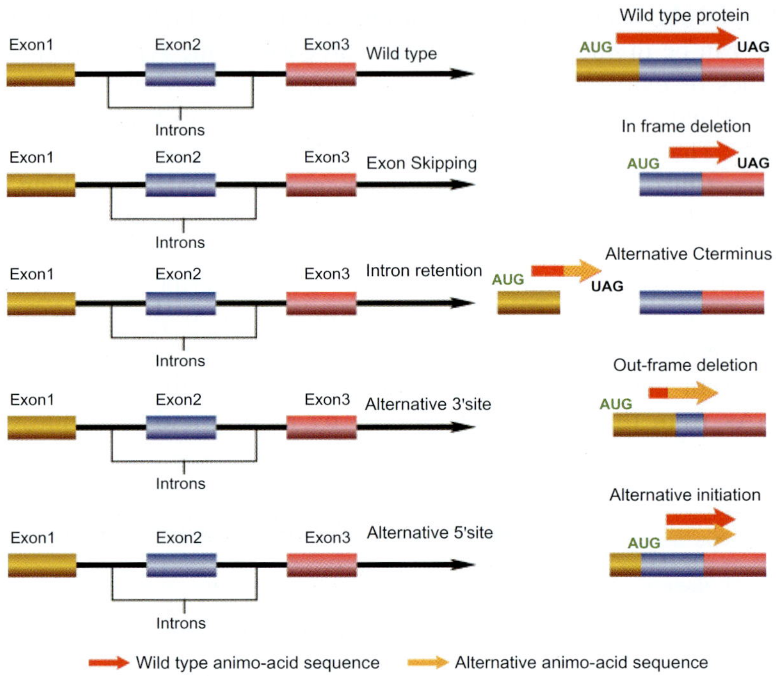

Figure 3.10. This figure shows five possible ways in which a gene can be alternatively spliced during transcription to produce varying exon arrangements as well reduced or partial expression of some exons. From Premier Biosoft (2013)[52] with kind permission of their Support Team.

one gene can produce multiple mRNAs and thus multiple proteins (Figures 3.9 and 3.10). The introns and unused exons are removed from the pre-mRNA by a dynamic molecular machine called the spliceosome, which contains about 145 proteins and RNA molecules, and whose structure and mechanism is still unsolved[53]. We now know that 95% of those human genes that have exons can also be alternatively transcribed or spliced[54]. Before the human genome was sequenced, when almost everyone thought it was one gene to one protein[55], it had been estimated based on known proteins types, that our genome likely had something in the range of 14,000 to 120,000 genes[56,57]. When the human genome project was completed everyone was stunned to find we humans actually have only about 39,000 different genes[58], later revised down to 20,000 to 25,000 genes[59,60] (Figure 3.11). Alternative splicing explains a lot of that discrepancy.

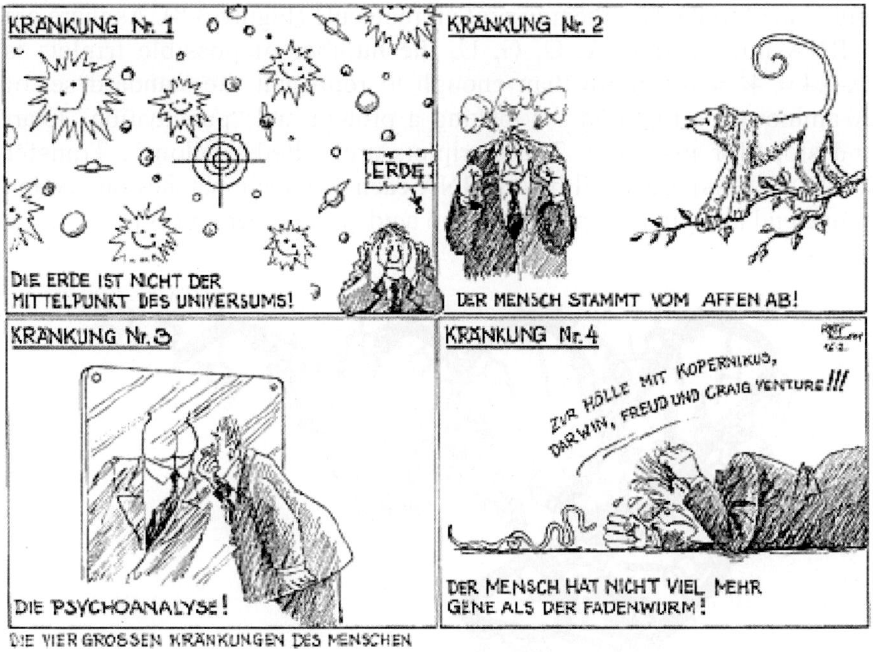

Figure 3.11. A cartoon on the decline of anthropocentrism, including the expectation that *Homo sapiens* would have the most complex genome. Translation: “Insult No. 1. The Earth is not the center of the universe! Insult No. 2. Humans descend from the ape! Insult No. 3. Psychoanalysis! Insult No. 4. Humans have not many more genes than the threadworm! To hell with Copernicus, Darwin, Freud and Craig Venter!!!” By the late Dieter Zehentmayr, from *Der Standard* with their kind permission.

The next step in gene expression is that the mRNA copy, with attached signals, is moved out of the nucleus to the ribosomes, the protein making machinery, which reside in the cytoplasm. These signals are entire sections of code in the mRNA itself, called “zipcode sequences”[61], that are not used for coding the protein but are important for signaling information. These signals can include information such as where in the cell the mRNA should be moved to before final translation occurs.

Once the mRNA is in the correct location, free or membrane-bound ribosomes[62] use the final form of the mRNA as a template to make a protein by stringing amino acids together, often with many ribosomes working in different spots along one mRNA simultaneously[63,64]. Three consecutive nucleotides in the mRNA are processed to add one specific amino acid to the lengthening protein molecule (Figure 3.8). As there are 4 RNA nucleotides: A, C, G, U, the number of possible triplets is $4x4x4 = 4^3 = 64$, more than enough to represent the standard set of 20 amino acids required for making a protein and "punctuation" (start and stop, for instance). These triplets are called "codons". Transfer RNAs (tRNAs) are small bits of RNA each of which matches one codon at one end and binds a specific amino acid at the other end (Figure 3.12).

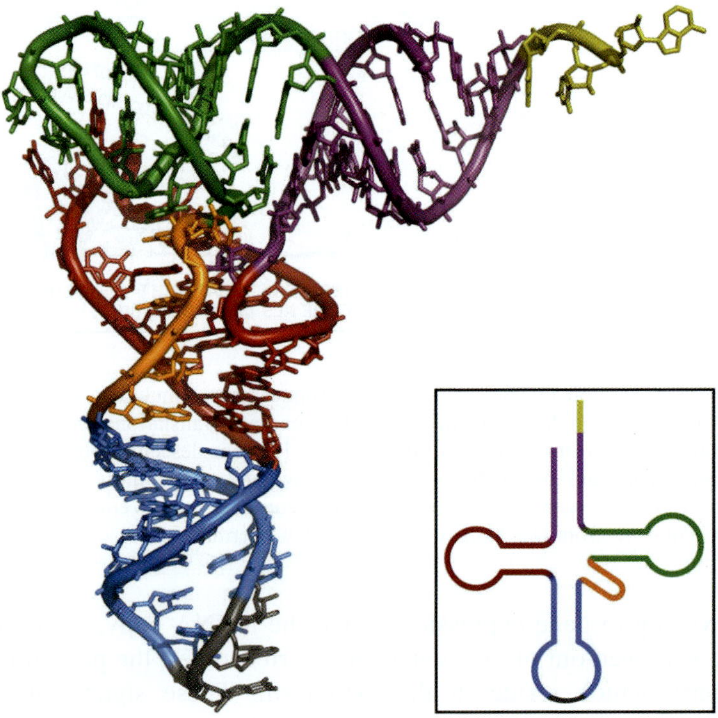

Figure 3.12. Two representations of a transfer RNA, which has an anticodon at one end (grey nucleotide base triplet at bottom) and has a site (yellow, upper right) that attaches to a specific amino acid at the other end. The tRNA enters a ribosome where it binds via the anticodon to a codon in an mRNA. The amino acid is then bound to the growing protein molecule and the remaining part of the tRNA leaves the ribosome. There are many variants of tRNAs, more than enough to cover every codon/amino acid pairing[65]. Many have other regulatory and transport functions beyond simply moving amino acids to ribosomes in translation. From Wikipedia[66] with permission under a Creative Commons Attribution-ShareAlike 3.0 Unported license. The inset is a schematic two dimensional representation showing the portions of the tRNA that have complementary nucleotide bases as parallel lines.

The ribosome works by insertion of a tRNA of the correct complement to match the mRNA codon being processed. Each codon is matched to a tRNA. The tRNA carries the specific amino acid and the amino acid is then joined to the peptide string of amino acids that make a protein. The structure of ribosomes has been worked out reasonably well (cf. our early contribution to this effort[67]), but exactly how these dynamic nanomachines function is still not completely solved[68]. The eukaryotic ribosome machine consists of 4 RNA molecules and 79 proteins[62] that are all bound together to make one functioning structure. tRNAs also turned out to be a lot more complicated than we first thought and we will explain that in Chapter 8.

Once the amino acid chain is completed, the protein that the chain makes still has to fold into its 3D structure[69,70]. Amino acids that are charged (hydrophilic) remain on the outside of the folded protein while those that are uncharged (hydrophobic) generally are found inside of the protein because the 'stuff' in the cell surrounding the protein is usually charged or has charged (electrically polarized) water molecules around. Some proteins have long stretches of uncharged amino acids so that they sit nicely inside the uncharged interior of membranes. These are the ones that are normally tagged as such by zipcode signals on the mRNA to be moved to a membrane and translated "on site" (usually the Golgi[71]) and threaded into the membrane as the amino acids are added. Mostly uncharged proteins can have a particularly difficult time folding.

If a small protein has a unique "ground state" that will be the final 3D structure, or if it is a relatively simple protein, it will normally spontaneously fold itself. The ground or "native" state is important, because it is usually only in that configuration that a protein can perform chemical reactions or specific bindings[72]. Some proteins simply do not have a single final state[73]. For others that have a ground state but are large enough to easily get trapped in higher energy (metastable) states (Figure 3.13), folding is accomplished with the assistance of other proteins such as the molecular "shaker" cages called chaperones[74,75] (Figure 3.14). These shaker cages work when an unfolded protein enters inside the barrel shaped chaperone (made of protein with a protein lid) that "shakes" the unfolded protein by hitting it with small chemical energy charges (ATP) (Figures 3.15 and 3.16) and thereby getting the protein into the correct 3D folded structure by the selective addition of force on critical portions. The chaperone then spits out a better folded protein and takes in the next unfolded protein. The longer and more complex the unfolded string is, the more likely that one or more cycles of chaperone shaking will be needed to fold the string into it's final proper form.

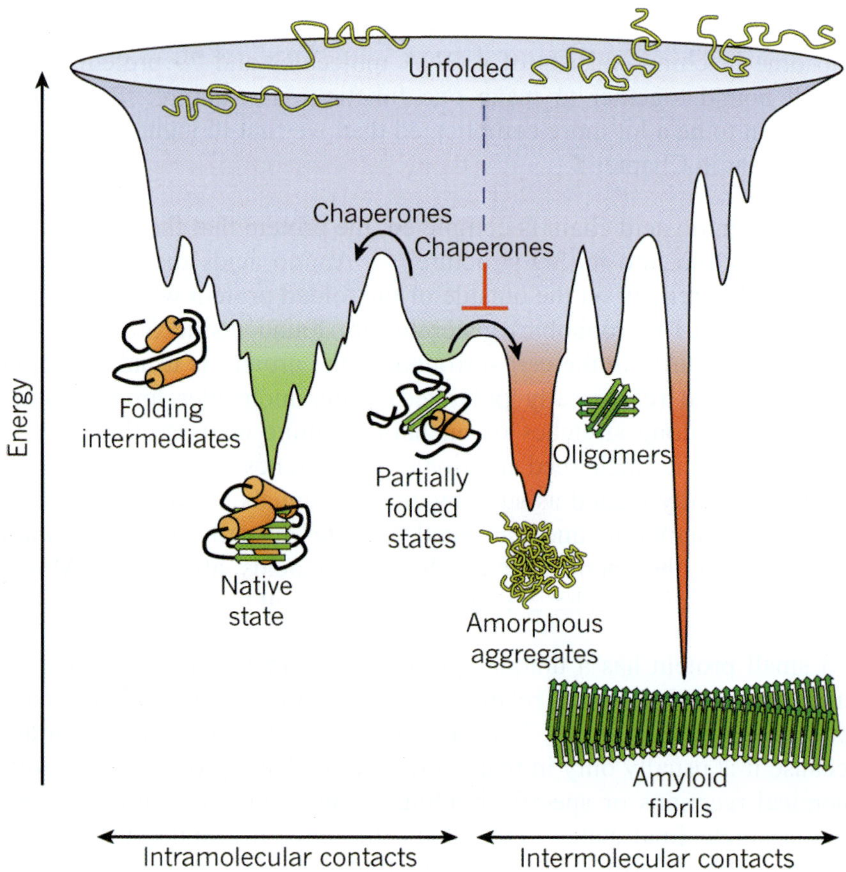

Figure 3.13. A protein's "native state" is its energy ground state, shown here as the lowest energy for intramolecular contacts, i.e., foldings where the molecule only touches itself. There are many little "metastable" valleys around the ground state where a partly folded protein can be trapped. Chaperones get proteins out of these valleys by adding enough energy (taken from the conversion of ATP to ADP) to get over the humps. The diagram also indicates energy states having to do with regular or irregular clumping of multiple protein molecules, often associated with disease[76]. Oligomers are regular structures composed of two or more proteins, which may be the same or different from one another[77]. From Hartl *et al.* (2011)[78] with permission of Nature Publishing Group.

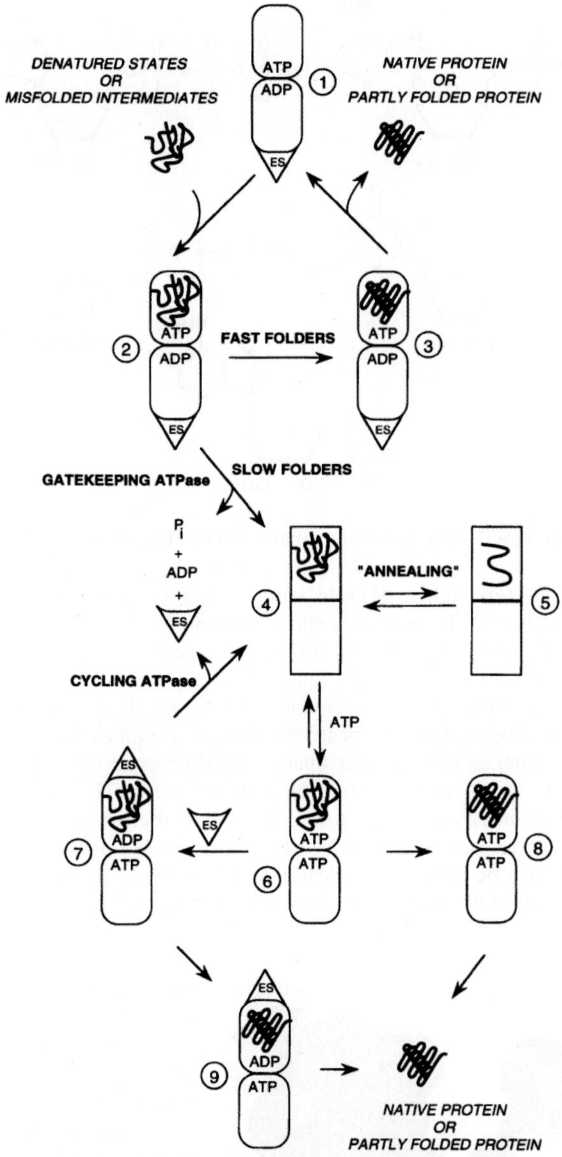

Figure 3.14. Chaperones are protein boxes that take in partly folded ("denatured") or misfolded proteins and using the energy of ATP shake them up until they settle down into a better folded or completely folded ("native") ground state. From Corrales & Fersht (1996)[79] (with permission of the National Academy of Sciences USA), where details are given. See also §1.17 *Protein Folding as a Deluding Paradigm* in *Hierarchical Genome*[74].

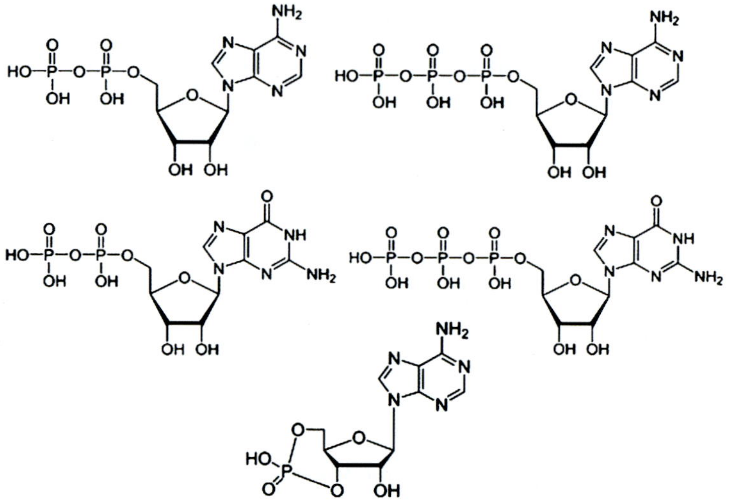

Figure 3.15. **Top row**: Adenine diphosphate (ADP, **left**) has two phosphate groups. When one phosphate group is added, we get the higher energy molecule, adenine triphosphate (ATP, **top right**). **Middle row**: guanosine diphosphate (GDP, **left**) and guanosine triphosphate (GTP, **middle right**). **Bottom**: Cyclic adenosine monophosphate (cAMP). All unlabeled intersections of line segments represent carbon (C) atoms, parallel lines (=) are double bonds and single lines are single bonds. The thicker lines merely attempt to convey some three dimensionality. N, O, P, H are nitrogen, oxygen, phosphorus and hydrogen atoms, respectively. Each C atom has 4 connections. When the sum in diagrams such as these is less than 4, the difference is made up with H atoms attached to the C atoms that are implied and not drawn. The addition of the third phosphate requires energy provided from sunlight or the breakdown of sugars during metabolism. This creates a small, easily transported packet of stored chemical energy. That energy can then be selectively released and transferred to other compounds where required when the third phosphate is removed by an enzyme.

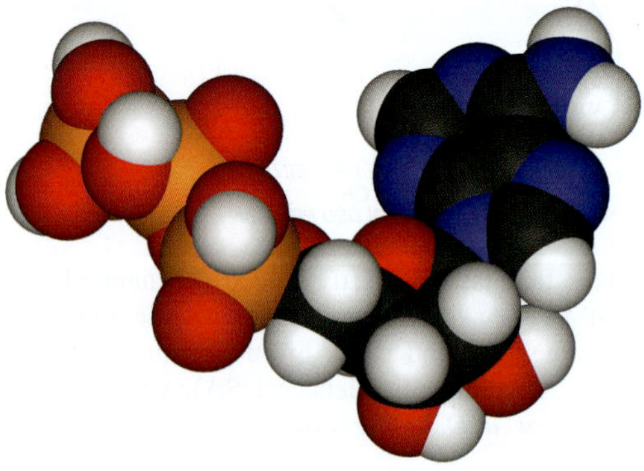

Figure 3.16. A space filling model of ATP: black = C = carbon, white = H = hydrogen, red = O = oxygen, blue = N = nitrogen, orange = P = phosphorus. From Wikipedia[80] with permission via release to public domain. These can be matched by eye with the chemical symbol version in the upper right hand corner of Figure 3.15. This molecule is used to transfer packets of energy for chemical reactions all over the organism.

Interestingly, if a protein gets unfolded in some way, such as via a heat shock (a sudden increase in temperature), it can often be refolded by the chaperone proteins. Some of these chaperone proteins are therefore referred to as heat shock proteins because they were discovered by studying the reaction of cells to heat shocks[81]. This is actually a very nice example of gene expression. The scientist shocks the cell with heat and thereby messes up a lot of the proteins. The cell responds by making many more "heat shock" proteins and their matching control RNAs and these refold the messed up proteins. Once the mess is cleaned up, the cell stops making all those extra heat shock proteins and RNAs and returns to its normal functioning state. Only a few heat shock proteins (chaperones) and RNAs remain in the cytoplasm that work on newly synthesized proteins, or they sit around waiting for the next heat shock.

It is common to observe "post-translational" chemical modifications of proteins, which can be accomplished by a large number of mechanisms[82]. Single proteins frequently bind together in pairs[83] or into working clusters[84] forming final working tertiary protein structures (Figure 3.13). For example, human hemoglobin is a tetramer is made up of two pairs of proteins[85]. Roughly 40% of all proteins "form symmetrical complexes consisting of multiple copies of themselves, called homomers"[77]. Proteins can also have attachments that are not made of amino acids, such as heme in hemoglobin[85]. We have already seen examples of larger attachments in the ribosome, which is a combination of proteins and RNA. There are combinations of proteins and long uncharged lipids that embed in membranes[86]. In addition, whether a given protein molecule can function may depend on smaller molecules called ligands attaching to it, which when attached actually alter its structure and functioning. This is called an "allosteric" interaction (cf.[87]), whose mechanism is still largely unsolved[88]. Allosteric interactions are particularly important for enzymes that process molecules in biochemical pathways. They also play a special role in regulating which protein can bind to a particular region of DNA and in that binding change gene expression.

It was once taught in biology classes that there were only three kinds of RNA: mRNA, tRNA and ribosomal RNA. That was then. Now we know there are many more kinds of RNA (Figure 3.17). Some of these new RNAs have been found by conventional techniques. More recently, "transcriptomics" and high-throughput techniques such as RNA Seq[102,103] have resulted in yet another explosion in our knowledge about gene expression and the roles of RNAs in it. Broadly conceived, the transcriptome is all RNA between transcription from DNA to translation at the ribosome. This includes RNA with a strictly regulatory function, or splice variants where the function is unknown. RNA Seq was designed to isolate and detect RNA from single-cell samples and then provide sequences for ribosomal RNA, tRNA and mRNA that map to genomic DNA sequences. RNA Seq also provides accurate counts for the total number of specific RNA molecules per cell. Imagine the general wonder

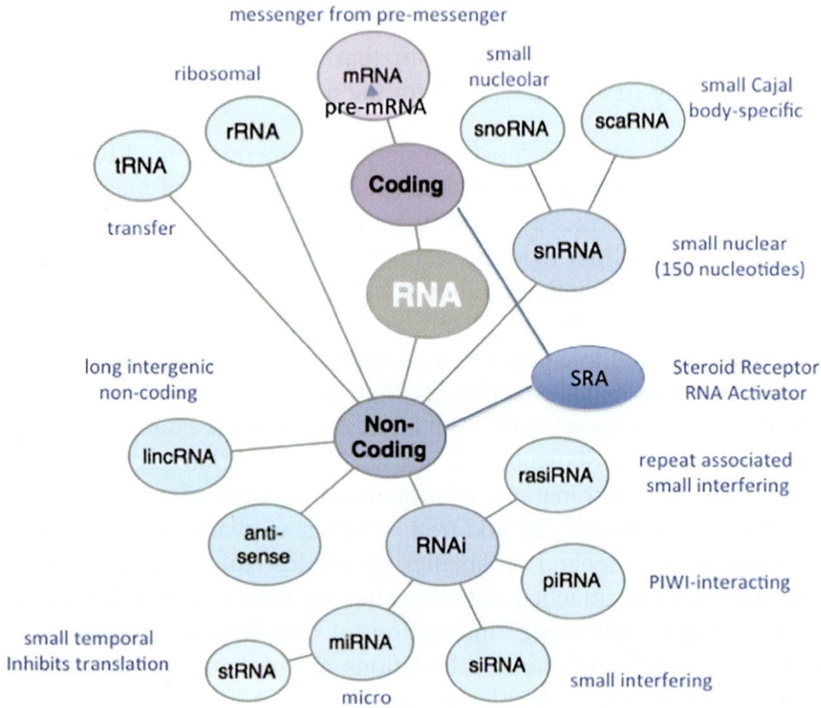

Figure 3.17. Yesterday's panoply of RNAs named after their presumed functions or locations. SRA functions both as a noncoding RNA and as a template for a protein (coding RNA)[89]. Adapted from Smith (2009)[90] with kind permission of Todd Smith, who originated this sketch, and Etienne League[91], who modified it. To this has been added: aRNA (amplified antisense RNA[92]), cRNA (complementary RNA[93]), dsRNA (double stranded RNA[94]), liRNA (long interfering RNA[95]), nanoRNA[96], spRNA (small plastid RNA[97]), tiRNA (transcription initiation RNA[98]), vRNA (viral RNA[93]), Y RNA (involved in the initiation of DNA replication[99]). With the advent of RNA nanotechnology[100], we can anticipate the introduction of artificial RNAs into cells[101] to see and manipulate what happens.

and astonishment of the scientists who were suddenly confronted with a whole lot more kinds of RNA than they ever expected to find and that this new RNA is spliced in amazing ways no one even imagined!

The functions of these multiple RNAs in cells are still being worked out, and so far, most are involved in changing some aspect of gene expression. For example, some RNA is transcribed but not translated. Instead, such an RNA molecule remains in the nucleus and binds to a section of DNA to which it is complementary. An RNA molecule undergoing such binding will change which genes are transcribed or how many copies are made. Other RNAs affect whether or not a specific exon is spliced out of a pre-mRNA. Some of these RNAs are involved in DNA repair. Some of these RNAs tag along with the mRNA out of the nucleus and slow its translation or change it. Some attach to proteins such as steroid receptors and thereby change receptor function. A form of RNA that acts as a cellular thermostat exists. Some organisms even make RNA that specifically interferes with gene functioning of invading parasites while leaving their own systems unaffected. The take home message about RNA is that each RNA strand is dynamic, can interact directly with both proteins and DNA, as well as other RNAs. All the places where RNAs are found can also be points where gene expression is regulated[104]. The central dogma of biology[105] did not account for noncoding RNA. (As an aside, it is worth noting that some chaperones also work on folding those newly discovered noncoding RNAs into their ground state configuration[106].)

When a geneticist says that "a gene is being expressed", they are referring to all intermediate steps between a gene sitting in the nucleus and the production of a functional protein. There are multiple points of regulation (alteration or "control") of that gene expression that are possible at each of those many steps that go well beyond just choosing to "turn a gene on or off". Expressions like "gene A regulates gene B" might actually mean "the snRNA made from the transcription of gene A regulates whether or not a protein is made by transcription of exons 1 and 2 in gene B instead of exons 1 and 3 in gene B".

Unfortunately expressions like "gene A regulates gene B" can also lead to sloppy thinking. Sometimes people (including scientists) will start referring to genes as if the DNA itself has mystical magical properties or that DNA is somehow able to do all kinds of neat thinking-style stuff all by itself. We are big fans of Star Trek[107] but we have frequently found ourselves chuckling at something someone in the show says that DNA can do (like a virus transforming Captain Archer into a totally different species with full alternate adult language and brain function in mere seconds, then Dr. Phlox being able to almost as easily reverse that process) and we laugh and say "Star Trek DNA". We then have to suspend our personal understanding of what DNA is and enjoy the show anyway. Magical DNA is appearing more and more often in all kinds of fictional shows. This kind of magical thinking can even creep into the minds of scientists who are not actors pretending to be scientists, or writers needing to produce a good yarn. So let's pause and just make sure we really understand what DNA is.

DNA is a string of relatively inert nucleic acids that constitutes a code for storing information on the sequence of amino acids required to make all the proteins (the proteome) required by an organism. DNA also has

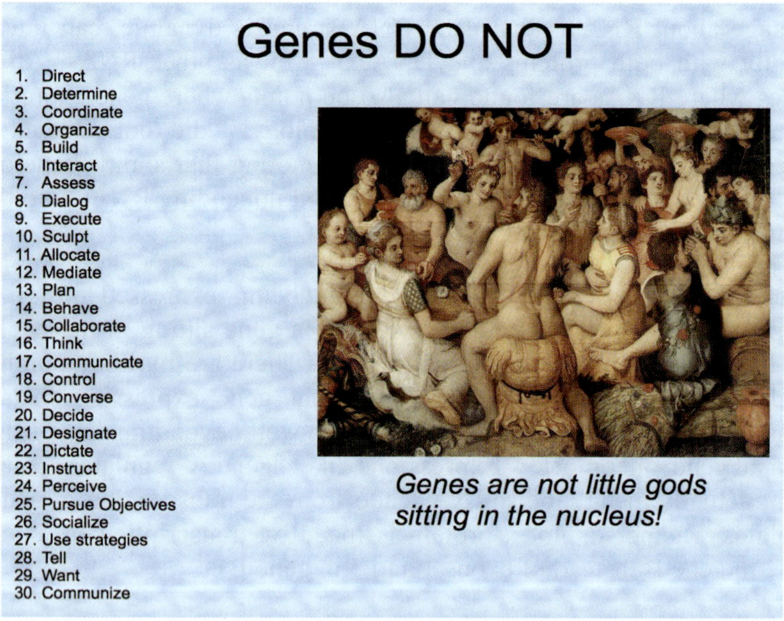

Figure 3.18. Genes are not little gods sitting in the nucleus consorting and plotting with one another. Banquet of the Gods[108] by Frans Floris about 1550, in the public domain. Genes (i.e., portions of an organism's DNA) don't do these things, all taken verbatim from the scientific literature (§1.18 *A Word on Language* in *Hierarchical Genome*[74]).

some information on when to start or stop transcribing the information, depending on outside signals, such as proteins that bind to it. These "outside" signals can actually be quite proximate, as in an RNA that is transcribed from an exon and then affects the transcription of the DNA of that same gene, or it can be as far away as a signal from outside the cell[109], perhaps from a distant part of the organism[110], and that signal is relayed all the way down to the level of the DNA. This point that DNA by itself is inert is so important that we have even made a list of the many things we have heard our fellow scientists say (and we occasionally kick ourselves for) that genes can do. These come from either sloppy language or sloppy thinking or both (Figure 3.18). Genes don't "do" anything.

The most commonly used and simplest textbook example of regulation of gene expression is the *lac* operon[111] (Figure 3.19). The *lac* operon is a set of genes that work together, found in common gut bacteria such as *Escherichia coli* (*E. coli*)[112]. Normally *E. coli* cells do not use lactose because it is a complex sugar compared to simpler, easier to digest sugars like glucose. If there is a lot of glucose and only a small amount

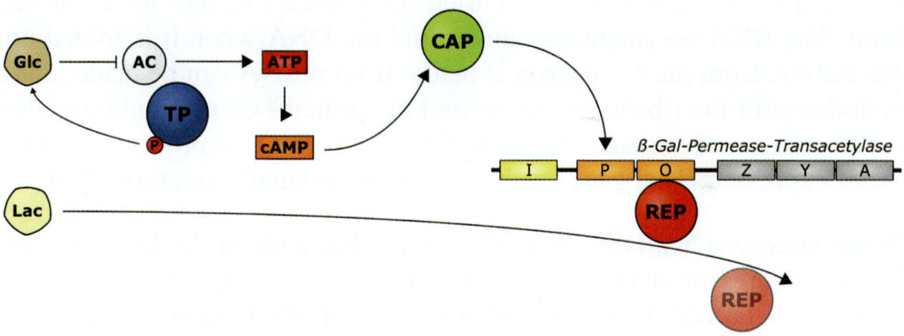

Figure 3.19. Lac = lactose, REP = lac repressor protein (Figure 3.20), Glc = glucose, AC = adenylyl cyclase, ATP (Figure 3.15), TP = glucose transport protein, P-TP = phosphorylated glucose transport protein, cAMP (Figure 3.15), CAP = cAMP receptor protein. The mRNA formed from the operon transcribes three genes: 1) Z = lacZ, which translates to a protein tetramer called β-galactosidase; 2) Y = lacY, which translates to a membrane protein that transports lactose into the *E. coli* cell from the medium; 3) A = lacA, a transacetylase. Lac binds to REP detaching it from the DNA sequence O = operator. When glucose is absent, CAP binds to promoter P and interacts with RNA polymerase facilitating its binding to the promoter[113]. I = lacI translates to REP. From Wikipedia[114] with permission under a Creative Commons Attribution/Share Alike license.

of lactose present, the *E. coli* will ignore the lactose and eat the glucose first. Think of a hungry child offered shelled versus unshelled nuts. The child eats up the ones without the shell first. If there is no glucose, the bacterium changes gene expression. It begins making copies of the proteins needed for processing lactose. How does this work?

Certain genes in *E. coli* are always accessible to polymerases[115] which attach to them. Polymerases always make a small amount of mRNA that codes for proteins corresponding to these genes. This small, fluctuating[116,117] amount is called "constitutive expression"[118]. One of the constitutively expressed genes is the *lac* repressor. The *lac* repressor protein has a specific structure called "helix turn helix". This "helix turn helix" protein is attracted to and binds two specific DNA nucleotide sequences that are located in two places in the DNA of *E. coli* (Figure 3.20). One binding site is located at the start of a set of three genes, lac X, lac Y, lac A, which are collectively called the *lac* operon. (A bacterial operon is any set of genes all adjacent on the DNA on the same side of its double strand that are transcribed as a single mRNA and are subject to the same control elements.) The other binding site is at the end of the *lac* operon. When the "helix turn helix" *lac* repressor protein is bound to DNA, the DNA is curled up, pinned together into a closed loop. The RNA polymerases cannot bind the DNA when it is folded up. No mRNA from the *lac* operon is made. If no mRNA can be made, there is nothing for the ribosomes to use and the proteins corresponding to that set of genes are not made. The gene for the enzyme that processes lactose is not expressed. That gene has been "downregulated", i.e., turned off.

When lactose is available, it binds onto the backside of the *lac* repressor. The *lac* repressor changes its shape and it can no longer bind the DNA and so it falls off. This leaves that section of DNA open. Polymerases can now move in and begin copying the DNA (which has three genes in a row) into an mRNA string. However, if the cell has both glucose and lactose, there is no need to begin processing lactose. There is a second regulatory mechanism in place for this situation. Bacterial polymerases don't just jump onto and transcribe any open bit of DNA. They are guided to bind and begin transcription by another protein called CAP. CAP refers to "cAMP receptor protein". (cAMP = cyclic AMP[119], Figure 3.15.) The chemical energy of glucose is used by the bacterium to make ATP. ATP is converted to cAMP as that chemical energy is used. When there is a lot of glucose, any cAMP is converted back to ATP as

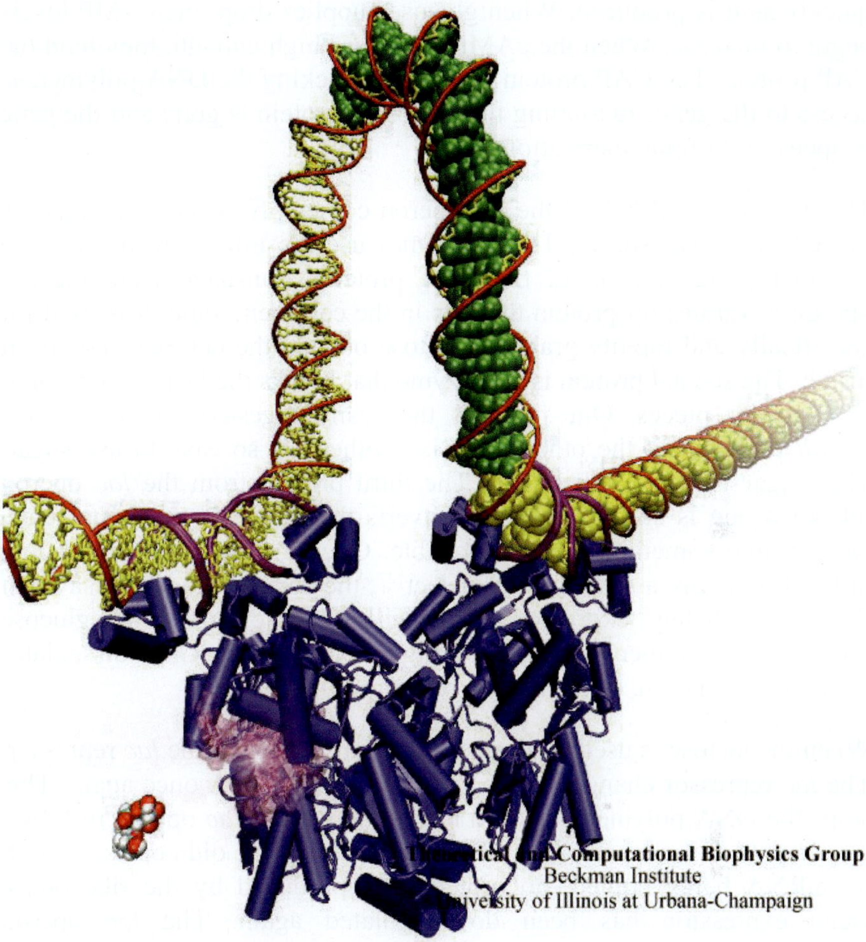

Figure 3.20. The authors of Villa *et al.* (2005)[120] describe the operation of the lac repressor rather vividly: "When *Escherichia coli* bacteria enjoy lactose and related food molecules in their environment, the cells quickly furnish proteins needed for import and metabolic digestion of the food. A set of genes, called the lac operon, is transcribed into messenger RNA that directs the synthesis of these proteins. When lactose is not available, the protein synthesis would be wasteful and, indeed, is prevented by locking the lac operon. This is achieved by a protein called lac repressor that forces the segment of the lac operon needed to initiate transcription into a loop, but that can be unlocked by a lactose molecule [red and white ligand molecule, lower left] binding to the protein as soon as the food becomes available again…. the lac repressor, a 314,000-atom protein… stretches out two 'hands' grabbing the genomic DNA and then keeps a tight grip on the DNA wrestling it into a loop". From PNAS[121] with kind permission of Klaus Schulten and David P. Brandon.

quickly as it is produced. When glucose supplies drop, the cAMP levels begin to increase. When the cAMP levels are high enough, they bind the CAP protein. The CAP protein then stops blocking the DNA polymerase access to the genes (assuming the repressor protein is gone and the gene is open) to start polymerization.

The three-gene mRNA of the *lac* operon comes off as one strand and is moved to the ribosomes. The ribosomes use the mRNA to make three different proteins. One of the three proteins translated from the *lac* operon is a transport protein that sits in the cell membrane. It is used for specifically and rapidly grabbing lactose outside the cell and moving it inside. The second protein is an enzyme that breaks the lactose sugar into two smaller pieces. One piece is the easily digested glucose, which *E. coli* prefers, but the other piece is another not so easy to use sugar. That sugar isn't wasted though. The third protein from the *lac* operon mRNA string is an enzyme that converts that difficult sugar molecule, lactose, into something more digestible. Once lactose is actually being utilized in a way a scientist can detect[122], the scientist can now say that the gene is being "expressed". This will continue until either glucose arrives from another source, lowering cAMP so that CAP dissociates from the *lac* operon, or the lactose runs out (Figure 3.19).

When the lactose is used up there is none left to bind to the *lac* repressor. The *lac* repressor changes shape and binds to the DNA once again. This stops the DNA polymerases from making any more *lac* operon mRNAs. An enzyme that chops up old mRNA removes all the old copies[123]. With no mRNA being copied, no proteins are produced by the ribosomes. Gene expression has been downregulated again. The *lac* operon represents one of the simplest forms for gene regulation in a simple organism.

In eukaryotes, such changes in gene expression regulated by the binding and unbinding of proteins and RNA to the DNA are going on all the time during the cell cycle, especially during interphase. Sometimes regulation occurs not at the gene level but at the level of mRNA. The gene is open all the time and the polymerases are actively making mRNA copies but that mRNA is destroyed before it gets to the ribosomes. There will be no proteins made until a different signal occurs that stops mRNA destruction. This type of regulation allows fast response to an external signal.

Another mechanism is that the mRNA is collected and wrapped up with proteins so it is effectively stored to be released later at need. Storage of mRNA is very important for early embryogenesis[124]. Embryos use stored mRNA at early steps of development. Since some mRNA is inserted into the oocyte as it develops into a fertilizable egg, such RNA is referred to as "maternal" although a lot of egg cells produce their own "maternal" mRNA. During cleavage stages, embryos are able to have their new cells dividing rapidly because they don't need to pause, open up the DNA, access genes and then close the DNA up over and over again.

A bacterial cell replicates by dividing into two copies of itself, each copy with identical abilities for gene expression (except for stochastic variations[125] and mutation). In simpler bacteria such as *E. coli* the genes are all available to be transcribed although some are transcriptionally active and some are not. Let's move to the more complex cyanobacterial cell *Anabaena*, which we met earlier as a model organism. *Anabaena* is a bacterial alga that forms two cell types, a regular "vegetative" cell that uses sunlight for food and the heterocyst that fixes atmospheric nitrogen N_2 into usable compounds (Figure 1.31). Recall how a third cell type, a spore, is produced under deleterious conditions (Figure 1.28). The DNA of all three cell types contains the same genes, i.e., all the genes required to produce the mRNA to make all the proteins used by the heterocysts, the vegetative cells and the spores (although heterocyst differentiation does involve gene rearrangements[126,127], which is probably reversible, since heterocysts can germinate[128,129]).

The difference between vegetative cells and heterocysts is not due to differences in their genomes. The difference lies in which part of the genome is being expressed. There are three general categories of gene expression. One set of genes is the "housekeeping genes" which express mRNA for proteins required by both the vegetative cell and the heterocyst. The second set of genes is the active cell type of genes specific to the cell type. In the vegetative cell, this would include the genes for photosynthesis (turning sunlight into food). In the heterocyst it would be the genes required for nitrogen fixation. The third set of genes is the repressed genes for each cell type. In the vegetative cell, the repressed genes are those associated with nitrogen fixation because the vegetative cell is not doing nitrogen fixation. Meanwhile, the heterocyst would have the photosynthetic genes repressed. The mature spore is presumably inactive.

In eukaryotes we can make a broad classification of all genes in a given cell type into *exposed* and *sequestered*[130], as observed by Dick's first postdoctoral mentor, the late Theodore T. Puck[131,132]. Exposed genes are available for transcription, while sequestered genes are not. Typically 50% of the DNA in a differentiated cell is exposed. Here and later when we consider the nuclear state splitter, we will elaborate on Puck's theory (Chapter 9).

Housekeeping genes exist in all organisms. (Some scientists prefer the term "reference genes" for which there is an official representative list[133].) These are genes all cells require for normal functioning such as genes for tRNA. They are expressed constitutively in all tested cell types at all stages of life and thus fall in the exposed category. Humans have 3804 known housekeeping genes[134] representing about 15-19% of all 20,000 to 25,000 genes[60]. There may be many more housekeeping proteins since both housekeeping genes and tissue specific genes often have multiple splice variants. Housekeeping genes have some fundamental differences when compared to tissue specific genes. They are highly conserved, meaning they have changed very little over evolutionary time and vary little between organisms[135]. The housekeeping genes are probably less compact[136], in the sense of containing more noncoding DNA (such as total intron length) than tissue specific genes. Housekeeping gene proteins tend to physically interact[137] with more other proteins than tissue specific proteins[138]. Most importantly they are almost all turned on and off with "CpG-dependent core promoters"[41]. The tissue specific genes use mostly alternative promoters, with some exceptions[136]. Some of these exceptions act just outside a "CpG island" at binding sites referred to as "CpG island shores"[139,140].

As an example of how the understanding of genes changes over time, the idea that housekeeping genes are "less compact" is not accepted by all scientists. The less compact idea is at odds with the notion of evolutionary selection for the efficiency of translation[141] of housekeeping genes[142]. The conundrum may be due to weak statistical correlations and a nonlinear relationship between gene length and expression level[143] or selection for independent measures of gene properties[144]. The noncoding spaces might also include critical noncoding RNA still waiting to be discovered. In any case housekeeping genes are always exposed genes.

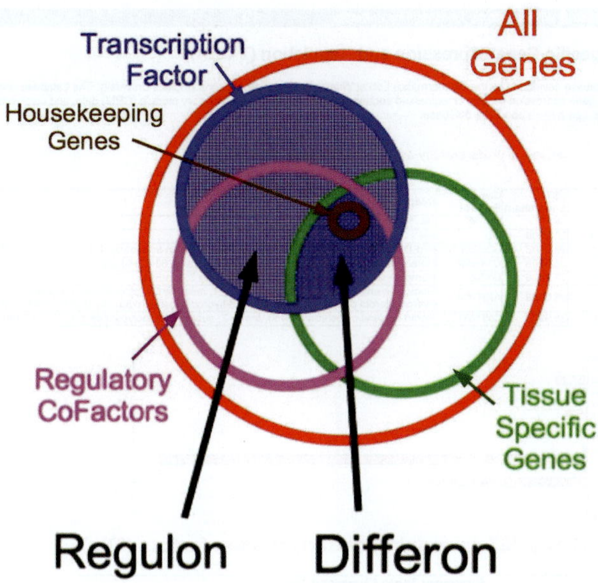

Figure 3.21. Each cell has the same set of **all genes** encoded in the DNA (**Red circle**). All cells in an organism express a set of "**Housekeeping**" or "Reference" genes. These genes represent certain elements such as ribosomes and components of the DNA polymerases as well as some of the cytoskeletal elements (**innermost brown circle**). There are also genes for specific transcription factors that are required for turning off and on gene expression (**Blue circle hatched**). **Transcription factor** gene products (i.e., proteins that are transcription factors) may be expressed in many different cell types. Many transcriptions factors require the presence of specific regulatory cofactors (**Pink circle**) such as small RNAs to perform some (or even all) of their functions. Expression of specific transcription factors and their regulatory cofactors often changes during development and thereby changes function of **tissue specific genes**. Each cell also has a set of genes that are generally specific to the tissue type that the cell belongs to but which may also be expressed in other cell types (**Green circle**). A **Regulon** is the group of a single transcription factor and all of its regulatory cofactors required to regulate a specific gene or set of genes. A regulon is defined by the genes affected by it and the **regulatory cofactors** associated with it and can be different in various cell types or when studied from different perspectives. A given regulon may also be functional in more than one cell type. A **Differon** is the set of all genes that can be expressed in a specific cell and the differon defines the cell type (**blue area** including Housekeeping genes). All the cells with the same differon are of the same type. (The differon differs from the transcriptome, the set of all mRNA expressed in a cell, because not all genes that can be expressed are always being expressed in a cell. For example, the liver may express certain genes only in response to the presence of specific toxins and so those genes will not appear in the transcriptome if the toxin is absent, but those genes are still part of the differon.)

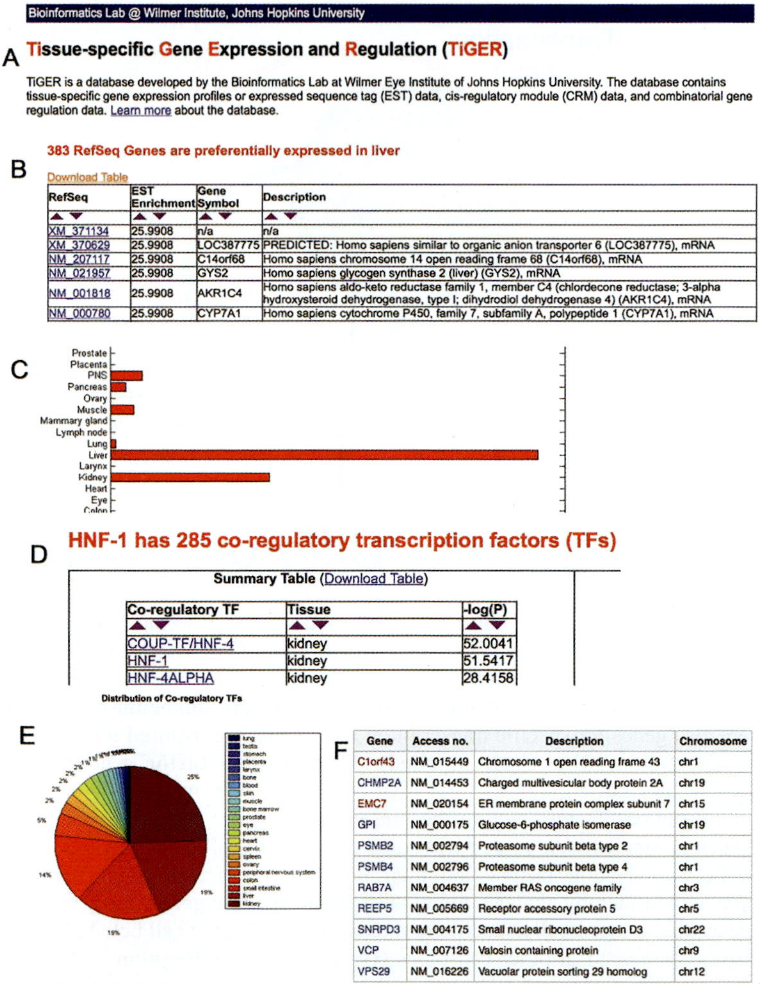

Figure 3.22. Sample search on the TiGER searchable database (Tissue Specific Gene Expression and Regulation)[146]. A. Description. B. Search on the tissue "liver" and the answer is 383 human genes known to be expressed preferentially in liver tissue. C. Although the gene is expressed preferentially in liver, it is expressed in other tissue types such as kidney. As is often the case, all the tissues in which the gene is expressed are of combined mesoderm and endoderm original except for tissues in the "peripheral nervous system". This gene is not expressed by any ectodermal derived tissues in the brain or skin and only appears in the ectodermally derived peripheral nerve tissue. D. HNF-1 is a transcription factor found in nerve tissue. There are 285 regulatory cofactors associated with HNF-1. (E-F) All cells have a set of genes found on various chromosomes that are required for functions like transcribing mRNA and forming ribosomes. These are referred to as "housekeeping" or "reference" genes for which there is a representative list[133].

The term "differon" (Figure 3.21) is not a new term[145]. We will define a *differon* as all of the exposed genes in a given cell type (Figure 3.22). The complementary set of sequestered genes we will call its *differoff*. Note that the differon includes genes that may or may not be transcribed. A cell's *transcriptome*[147,148] is the subset of DNA in a differon that is being transcribed in a given cell at a given time. The transcriptome is all the gene pre-mRNAs and spliced mRNAs, as well as all the other regulatory noncoding RNAs that may or may not be produced as part of a gene, and the untranslated RNAs such as tRNAs and ribosomal RNAs. A cell's transcriptome can be thought of as its phenotype, and thus a measure of the relationship between phenotype and genotype. However, there is no reason to assume that, for instance, every difference in cell shape or other parameters necessarily means a difference in a cell's transcriptome. As an amoeba cell crawls and changes shape it may not be changing its transcriptome. As diatom shells decrease in size and shell pattern from one generation to the next[149-151], they might not be changing their transcriptomes.

The *proteome* is the set of all proteins being expressed in a given cell at a given time[152,153]. In terms of set theory[154], then, for cell type *i*:

$$\text{proteome}_i \subseteq \text{transcriptome}_i \subseteq \text{differon}_i \subseteq \text{active\_genome} \subseteq \text{total\_DNA}$$

with:

$$\text{differon}_i \cup \text{differoff}_i = \text{active\_genome}$$

and

$$\text{differon}_i \cap \text{differoff}_i = \text{empty set}$$

where the symbol $\subseteq$ means "is a subset of", $\cup$ means "union of", and $\cap$ means "intersection of". By "active_genome" we mean the "total_DNA" except for DNA that is never used to produce anything in any kind of cell in the organism. Whether the latter is junk DNA or has structural or other functions we just haven't learned about yet remains an open question. The housekeeping genes for a given organism with *n* cell types can be expressed in set theory as:

$$housekeeping\_genes = \bigcap_{i=1}^{n} proteome_i$$

We will consider microtubules and the tubulin proteins they are made of in detail in Chapter 5. All cells require tubulin because all cells require microtubules for a variety of cell functions. So it is tempting to call all genes for tubulin in humans "housekeeping genes". This is incorrect because, if we look at humans, there are only two versions of human tubulin genes, *TUBB* and *TUBB4A*, that are present in all cells[155] and therefore qualify as housekeeping genes. Twenty four other genes for different versions of tubulin exist in addition to *TUBB* and *TUBB4*[156]. These other forms of tubulin are not expressed in all cell types (cf.[157]). They are only expressed in a subset of all possible tissue specific cells. This subset can be a significant proportion of all possible tissue types but because they do not appear in all cell types they are not true housekeeping genes. These additional *pseudohousekeeping* variants are variants of housekeeping genes which code for proteins with somewhat different properties from the standard housekeeping version[158]. Pseudohousekeeping genes are also usually regulated via alternative promoters and do not normally use the CpG-dependent core promoters of true housekeeping genes[136]. Thus in a given cell type, a specific pseudohousekeeping gene may be exposed or sequestered.

In bacteria a set of genes all controlled together is known as an operon. A similar set of genes regulated by a given transcription factor, or combination of transcription factors, triggered by a discrete signal transduction pathway, is called a *regulon.* A regulon may be thought of as a group of genes that all work together, and are controlled in concert, with a greater complexity of interactions than that found in an operon (Figure 3.21). Regulons can have significant overlaps in which genes they include. Regulons in bacteria are defined with respect to the signal that is changed and the transcription factor (usually only one) that responds to the signal. Regulon becomes a rather nonspecific term in eukaryotes with their complex gene regulation because it must be defined with respect to something. If we define a regulon with respect to transcription factors we will get certain sets of genes for each transcription factor. Since eukaryotic transcriptions factors often function differently when working in combinations with other transcription factors, eukaryotic transcription factor based regulons can have multiple, overlapping and even contradictory effects on a specific gene. If we define a eukaryotic regulon with respect to a signal like a heat shock, many transcription factors will be involved and so the set of genes could be considerably larger. Nonetheless, we would like to extend Puck's ideas to regulons by suggesting that in a given cell type, one subset of regulons is *exposed*, while the remaining regulons are *sequestered*. In this scheme, housekeeping regulons would always be exposed. In terms of set theory:

$$differon_i = \bigcup_{r=1}^{R_i} exposed\_regulon_{ir}$$

$$differoff_i = \bigcup_{r=1}^{R_i} sequestered\_regulon_{ir}$$

Determination is the process by which a cell changes from having one set of exposed/sequestered regulons to a different set of exposed/sequestered regulons. While some regulons may be available to more than one tissue type, the total set of exposed regulons is unique for each cell type. It is the subset of exposed regulons that defines the cell type.

In more complex organisms, like us, there can be hundreds to thousands of regulons being used creating hundreds to thousands of different cell types. There are plenty of potential differons: on the order of 2^N for N regulons, though generally only a small fraction of these would suffice and be actually used by an organism. In the determination process, each cell acquires a specific differon (Figure 3.21). Changing differons is what underlies and defines embryogenesis. Most organisms have cells that are all totipotent, meaning each cell has the same DNA, but use determination to select which differon to access in each cell. The main problem addressed in this book is then: how does determination occur, when considered as the selection of which cell uses which differon of genes?

Take the differentiation of the ectoderm. Begin with the genome in the ectoderm. The genes in one ectoderm cell in G1 stage can be divided into three categories as was done with the *Anabaena* cell types. However, with more complex organisms, there is a need for two extra categories. First there are housekeeping genes required in all cells. They have the matching housekeeping regulons. Second, there are some genes coding for the unique proteins required for that specific cell type such as a membrane protein of the ectodermal cell. These could be considered to be pure ectoderm genes with only ectoderm function. They can be expressing or not expressing depending on need, but they must be accessible. These would include the ectoderm regulons. The set of the housekeeping regulons plus the ectoderm regulons, and the genes they control, are the ectoderm differon. The ectoderm differon limits the transcriptome to a specific subset of RNAs and the transcriptome limits the proteome to a specific subset of mRNAs.

The differoff is all the other genes that are not being used and it is further subdivided into three subsets. The first of these differoff subsets of genes is the set of sequestered genes that were used previously in ectoderm but are no longer needed. This would include genes for the egg, the morula, for the blastocyst stage, everything up to but not including ectoderm. Since these genes are no longer needed, they need not be accessible again.

The next subset of the differoff genes is the sequestered genes that have not yet been used but will be needed later as development continues. Examples are the different membrane proteins required by migrating pigment cells that differentiate from neural ectoderm cells, part of the pigment cell differon. Given the correct determination signals, the differons containing these pigment genes will be required. The genes are turned off now because the embryo isn't ready for pigment cells, but the genes of the pigment cell differon must be available to be turned on later.

The final subset is those differoff genes and their associated regulons that have never been used by this ectoderm cell at any point in its lineage history and they will never be used by any of this cell's descendants. An example would be genes required for making liver enzymes. The ectoderm has already been determined to be ectoderm. The genes and regulons specific to liver cells will never need to be accessible.

We can now conceptually have all regulons in the genome divided into one of five categories in any given cell at any stage of development:

1. *Housekeeping Regulons*: regulons controlling genes always accessible by all cells types.
2. *Differon Regulons*: regulons controlling genes accessible in a specific cell type, including the housekeeping regulons.
3. *Differoff - Past Regulons*: regulons controlling genes that were used in the cell lineage leading up to a given cell type, but are no longer used.

4. *Differoff - Future Regulons*: regulons controlling genes that are inaccessible in a given cell type, but will become accessible when the cell or its descendants differentiate into new types of cells.
5. *Differoff - Unused Regulons*: regulons controlling genes that are inaccessible and have not been used, are not being used now, and will never be used, by that cell or any of its developmental descendants.

It is how the regulons are changed over developmental time that governs the genetic aspect of embryogenesis. Hold that thought, as we will come back to it.

Now let's return to cell division. Cells can't just abandon interphase and divide at random. Rather, there must be preparation for cell division and a great deal of it. The preparations for cell division are extensive. First among them is to duplicate all the critical parts of the cell. In particular, all of the DNA must be carefully copied exactly once, which is a good trick, because many segments are copied simultaneously[159] with different timing in different kinds of cells[160]. Since the DNA acts as the blueprint for all useful proteins, high fidelity in copying is critically important. Complex life forms have astonishing DNA repair mechanisms whose job it is to sense how faithfully the DNA is being copied and correct mistakes[161]. These repair mechanisms have evolved because the organism that can best retain faithful copies of the DNA is more likely to survive and thrive. Mutations are amazingly rare[162], and occur at lower than expected rates predicted by simple models[163]. Once the S phase is completed the cell then undergoes a pause as far as the DNA is concerned (G2) (Figure 3.2). During this working pause in DNA synthesis all the other cellular machinery is duplicated and checked[164] (by mechanisms still being worked out[165]) and preparations are made for cell division.

As the DNA is copied, information about which genes are active and which genes are not active is also recorded by a process called mitotic bookmarking[166]. Mitotic bookmarking is critical to maintaining the cell's phenotype into the next cellular generation. The daughter cells must express the same proteins from the same differon once mitosis is complete. If bookmarking fails, cells coming out of mitosis will not be the same as the parent cell. Timing of expression of the genes is also critical. For example, what if the cell suffers a heat shock during mitosis? Heat shock proteins would be needed immediately as telophase is completed. The translation of other proteins can wait if, for example, the cell had enough proteins or stored mRNA left from before mitosis.

Bookmarking is not fully understood but how tightly the DNA is wound up and compacted plays an important part of the process. Some genes that were expressing during interphase are not tightly wound and compacted and so are left accessible to the transcription/translation machinery. Other nonexpressing genes will be tightly compacted. This level of compaction is maintained through mitosis.

Cells enter mitosis (M phase) from the G2 stage of interphase with everything important doubled (Figure 3.23). Any DNA that is damaged is repaired first[161,167]. Mitosis is halted until the repairs are completed[168]. Such halts are referred to as checkpoints and there are several more of them. They occur even during the rush of early embryogenesis[169,170].

Checkpoints occur at four places in the eukaryotic cell cycle. Unrepaired DNA damage results in cells pausing at G1, S, G2 and at the start of mitosis. DNA is repaired before the cell continues. In each stage, the presence or absence of specific phosphorylation states of enzymes involved in assessing the DNA will signal whether or not DNA damage has been repaired and when/if the cell can proceed to the next stage of the cell cycle. Each checkpoint has a unique set of such enzymes specific for that checkpoint. There is also a broader class of repair enzymes for detecting and repairing damaged DNA shared at all checkpoints. In addition to the checkpoints, cells also have signal transduction pathways that are used to signal that DNA damage has occurred. The result is that the cell pauses all other functions and switches into full DNA repair mode. The mechanism of this signaling, though not entirely understood, is thought to be similar to the one used for the heat shock response which we will cover in more detail in Chapter 8.

If enough damage has accumulated that the cell can no longer function properly, the consequences are one of three possibilities. Most commonly, the cell will trigger the apoptosis pathway and commit suicide. Less often the cell will become senescent. It will remain in its location but will cease to divide, stuck at one of the checkpoints. Rarely, the cell will begin unregulated replication.

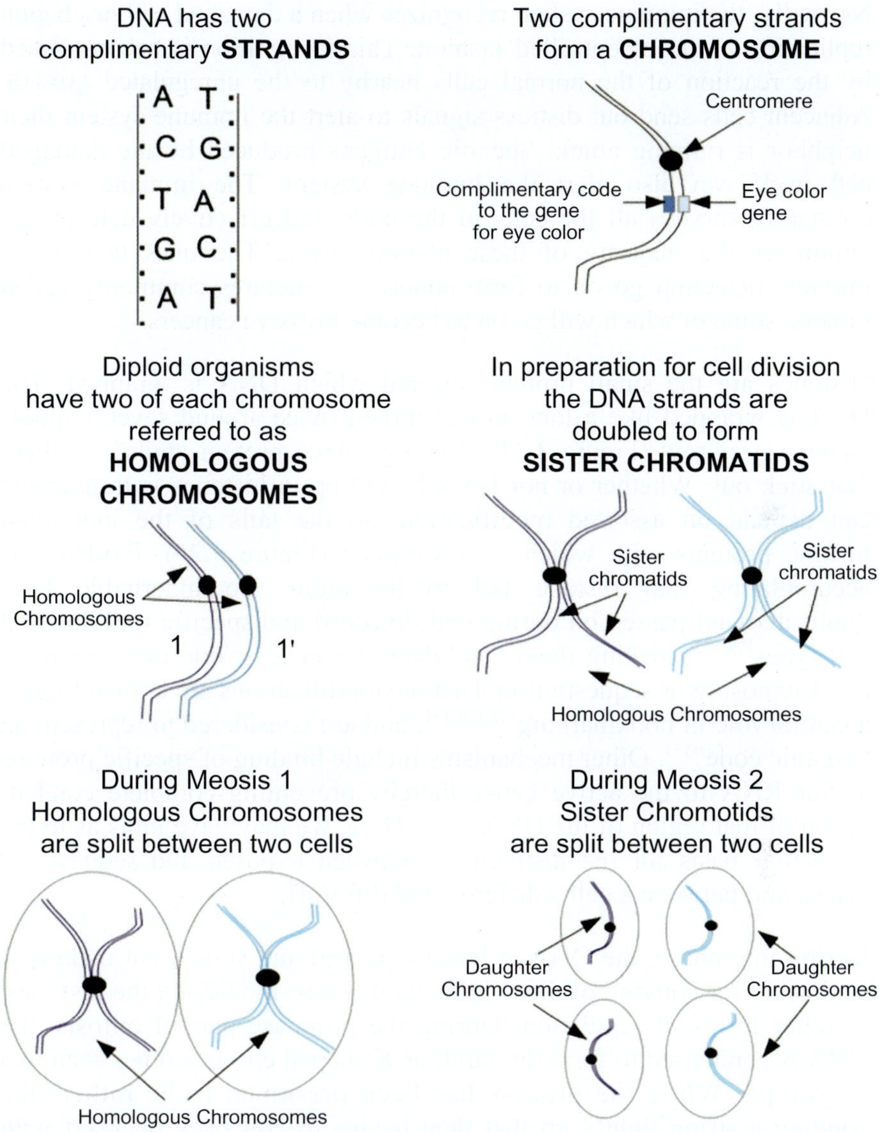

Figure 3.23. Meiosis is one of the most difficult concepts there is in genetics to teach and for students to really understand. This is almost always due to confusion between paired DNA strands, homologous chromosomes and sister chromatids. Homologous chromosomes can best be identified by following the centromeres. Where two centromeres of two matched chromosomes exist, these are homologous chromosomes whether single as in G1 or doubled as in G2. Sister chromatids share a single centromere. Homologous chromosomes have separate centromeres. This figure tries to clarify this terminology.

Normally, the immune system recognizes when a damaged cell has begun replicating in an uncontrolled fashion. This recognition is accomplished by the reaction of the normal cells nearby to the unregulated growth. Adjacent cells send out distress signals to alert the immune system their neighbor is running amok. Specific antigens produced by the damaged cell itself can also alert the immune system. The immune system constantly surveys all the cells in the body and effectively detects and eliminates the majority of these aberrant cells. The ones that evade immune detection go on to form abnormal structures commonly called tumors, some of which will go on to become invasive cancers.

Histones are the small proteins around which DNA is wrapped. The DNA is wrapped like a long thread turned twice around several spools all in a neat row (Figure 3.24). Histones have protein "tails" on them that stick out. Whether or not DNA bound up in histones is transcribed can depend on assorted modifications to the tails of the individual histone proteins with which it is wrapped (Figure 3.25). Evidence is accumulating that histone tail modifications are inheritable (i.e., duplicated and passed on during cell division) and specific to individual cell types[171,172], making them candidates for at least one mechanism of DNA exposure or sequestration. Histone modifications are known to play a central role in bookmarking[173,174,171], and are considered to represent an "organic code"[175]. Other mechanisms include binding of specific proteins and/or RNA to the active genes thereby preventing complete condensation of that region of the DNA[166,176]. Here, we may have hints as to the molecular basis for the distinctions between exposed and sequestered genes, and between a cell's differon and differoff.

During interphase, the DNA is loosely packed and spread out making it accessible for copying of those individual genes, modifying the histones, or other forms of regulation. During the prophase part of mitosis, the DNA is condensed to form the familiar X shaped chromosomes seen in a karyotype. While the process has been presumed to be rather like winding a string tightly up that then begins to coil back on itself with coils piling on coils, it may actually be more like the compression of a closing accordion[177]. Recent observations have undone the precise hierarchical coiling and compaction textbook models commonly presented[178], and some structures remain controversial while others like the 30 nm fiber have been refined[179-181], so we won't propagate the errors here. This leaves us with perhaps folding and piling up of strings of such "nucleosomes"[186], about as organized as a pile of beads on a string (Figure 3.26). Thus the DNA scaffold's structure when fully coiled up is presently uncertain, though some components that bind across adjacent strands of DNA have been found[187]. New, hierarchical models for

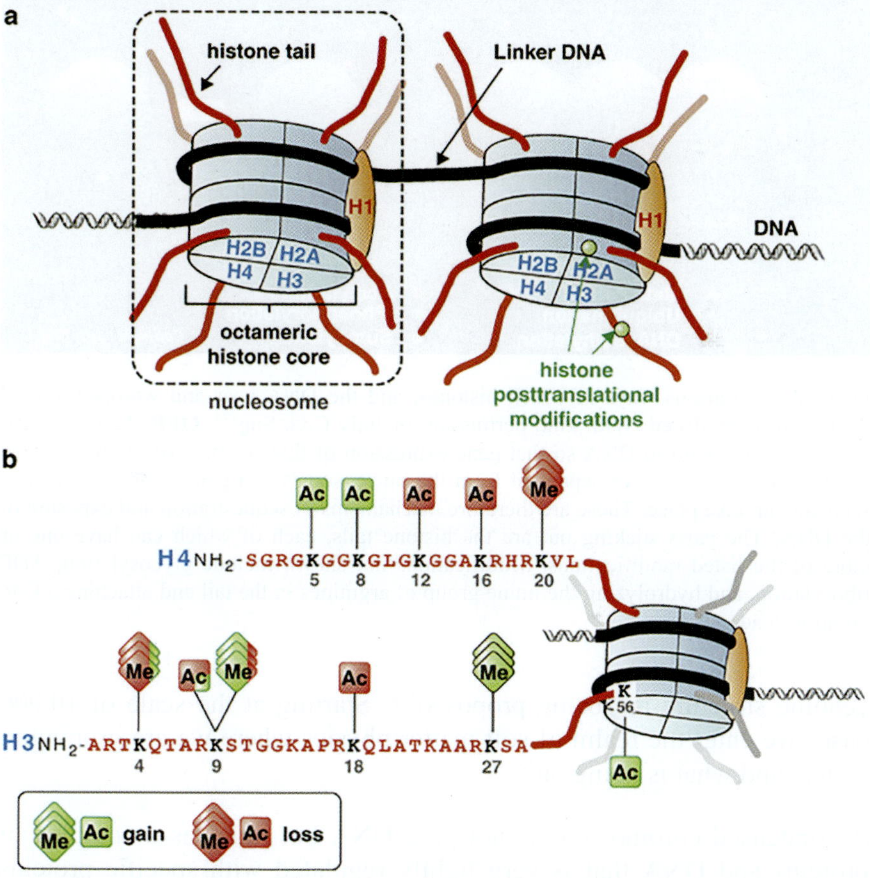

Figure 3.24. Histones are clusters of 8 proteins around which DNA in eukaryotes is wrapped. Each cluster is called a nucleosome, consisting of about 147 base pairs of DNA wrapped around two copies each of histones H2A, H2B, H3 and H4. Linker DNA varies from "roughly 10 to 100" base pairs[182]. An additional linker histone H1 appears to compact and protect the DNA[183]. The histones have a wide variety of post-translational modifications, mostly on their N-terminal tails involving acetylation (ac), methylation (me) in the example here, plus citrulination, phosphorylation and ubiquitination[172]. Each human cell, for instance, contains about 30 million nucleosomes[172]. From Füllgrabe *et al.* (2011)[184] with permission of Nature Publishing Group.

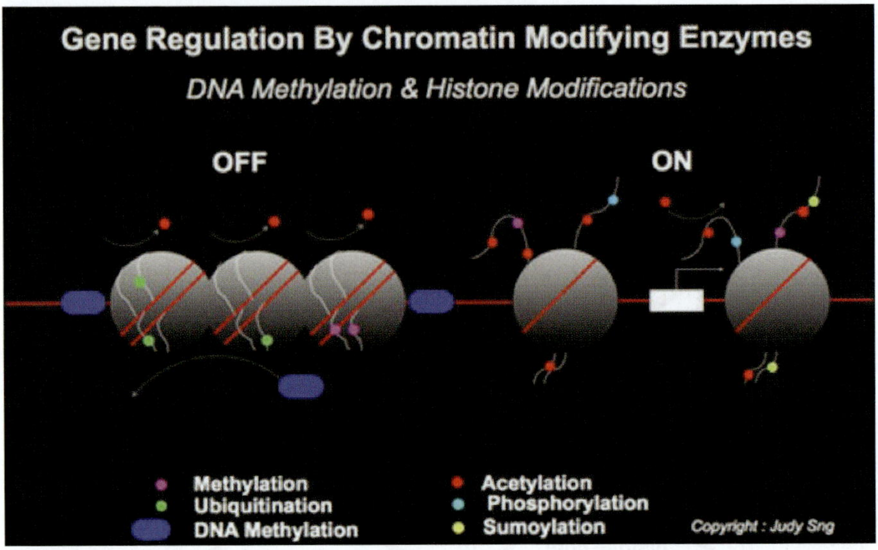

Figure 3.25. Various ways in which histones, and the DNA near and wrapped around them, can be modified. With kind permission of Judy C.G. Sng[185]. **OFF**: Three nucleosomes tightly bound to DNA so that gene expression of that segment of DNA is turned off. **ON**: DNA is loose or separated from the nucleosomes and gene expression of this segment can take place. These are therefore mechanisms of sequestration and exposure of the DNA. The parts sticking out are the histone tails, each of which can have one or more of the listed modifications. Modifications not shown include glycosylation, ADP ribosylation, and hydrolyzing the imine group of arginines in the tail and attaching a keto group instead.

genome structure are being proposed[188]. Starting at the scale of 10,000 bases we enter the realm of soft matter physics where we are groping to understand what is going on[182].

A condensed chromosome is not pure DNA but rather is a complex of proteins and DNA that is very tightly regulated with specific proteins used in precise order and place (Figure 3.27). Two examples are cohesin and condensin. Cohesin forms a circular structure around sister chromatids binding them together. Cohesins create and then stack positive supercoiling in the DNA. Cohesins work in tandem with the enzyme called type 1 topoisomerase which clips the DNA strand, winds it tighter and then rejoins the strand[192]. Proteins involved in regulating transcription are ejected from the DNA during condensation and then rebind into chromosome regions that are highly repetitive and far from any gene. This may serve to keep transcription factors stowed and moving with the DNA, ready for rapid deployment once mitosis is over and the DNA begins uncoiling again. (Maybe this also indicates yet another role for "junk" DNA.) Most histone modifications remain the same before, during and after mitosis and there are specific proteins that bind to the DNA in place of the ejected transcription factors. These maintain bookmarking signals on the DNA about which regions will be accessed after mitosis and which are not. Most transcription stops during mitosis but the hierarchy of signals which creates the regulatory milieu (or using our term, the differon) based on cell type, is retained by histone modification and bookmarking[171]. No doubt other forms of retaining this information will be found as we learn more.

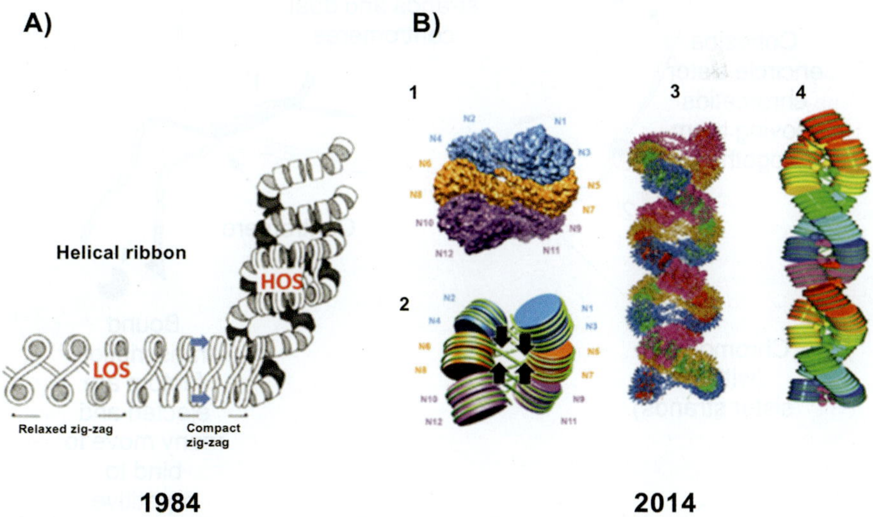

Figure 3.26. "Figure 1. A: Proposed model for the folding of the zigzag chromatin nucleofilament into a two-start helical ribbon[189]. HOS: high order structure; LOS: low order structure. The two blue arrows indicate the "two-start" nature of the helix. Adapted from[189]. B: Two-start zigzag tetranucleosomal arrangement of the chromatin fiber folding observed by 11 Å-resolution cryo-EM [low temperature electron microscopy]. (1) Overall 3D map of in vitro reconstituted nucleosome 12-mers using a 12x187 bp [base pair] DNA template. (2) Schematic representation of the same image. In both (1) and (2) an identical color coding (blue, orange, purple) has been used to highlight the tetranucleosome unit. (3 and 4) Pseudo-atomic model of nucleosome 48-mer and its schematic representation, respectively[190]. Adapted from Song *et al.* (2014)[190]." From Ausió (2015)[191] with permission of John Wiley and Sons.

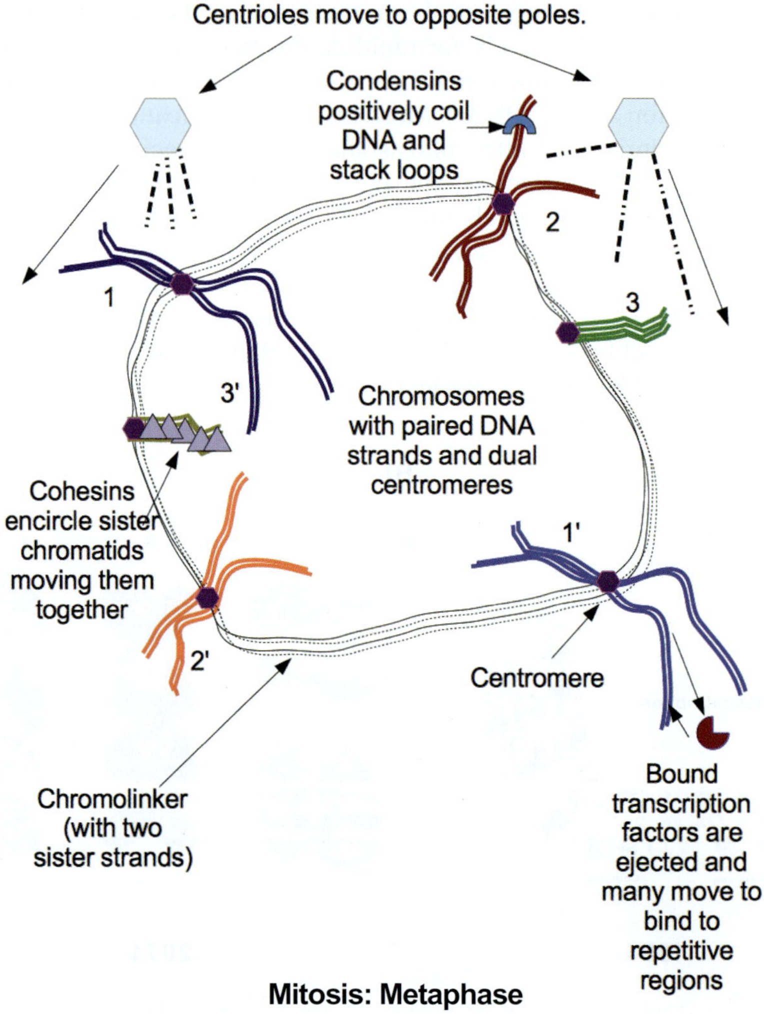

Figure 3.27. By the time prophase begins all parts of the cell have been duplicated. The division of the chromosomes requires the presence of specific proteins such as condensins and cohesions. Each serves a specific function. Multiple cohesins encircle the sister chromatids and bring them close to each other. Condensins positively coil DNA and stack loops. Transcriptions factors, which bookmark the DNA (see text) are released and moved to highly repetitive regions of the DNA. The nuclear membrane is now gone. Centrioles, previously duplicated along with all the other structures of the cell during interphase, begin moving to opposite poles of the cells. Putative chromolinkers have been included in this diagram. Lines representing the chromosomes have been thickened to represent the condensation that occurs making the chromosomes visible under a light microscope.

Once the chromosomes have been fully condensed, all of them are then lined up along a central plane of the cell during the M (mitotic) phase of the cell cycle (Figure 3.28). The arrangement of the chromosomes in a "rosette" (Figure 3.29) suggests that the chromosomes are linked together in a circle at the centromere by an unknown material, an observation that goes back to 1904[193,194]. In some cases the circles do not seem to involve all of the chromosomes[195]. Since 1923 it has been understood that the circles indicate "a constant and definite arrangement of the chromosomes in the nucleus"[196]. The more or less fixed volume occupied by each chromosome is nowadays referred to as its "chromosome territory"[197,198] (Figure 3.7).

One of the major differences between bacteria and higher organisms like us is the fact that most bacteria and bacteria-derived symbionts (chloroplasts and mitochondria) have only one single chromosome that is one long loop while we eukaryotes have many chromosomes[199,200]. That bacterial loop often has subsections that are wound up on itself forming little lumpy portions called domains (Figure 3.30). By contrast, we eukaryotes have our DNA divided into separate chromosomes. Or do we? If we use some fine microscopic level glass needles to pull out one chromosome from a cell at mitosis, intriguingly all of the chromosomes come out[201] in what appears to be a long set of beads, one "bead" per chromosome, on a chain of something. The material between the chromosomes (Figures 3.31 and 3.32), for which we coined the name "chromolinkers", seems to be made of DNA[202-204]. This actually makes real sense from an evolutionary perspective because it means that our DNA is, like the primitive bacteria, organized on one huge loop with chromosome domains. The work has only been done by one group and it remains to be followed up. If chromolinkers are indeed there, we wouldn't be able to see them at interphase because, if you calculate their probable length and the amount of coiling/compaction required to the get to the rosette of DNA, they are simply far too thin for unassisted light microscopy. Chromolinkers haven't been sequenced because they would represent only a very small portion of the total DNA and since they bind to regions that are highly repetitive, they are most likely repetitive DNA themselves. If we assume chromolinkers don't have genes, they could easily be lost as a tiny portion of DNA in the repetitive DNA that represents about half our genome[211]. A circular string of chromosomes attached via chromolinkers could be the mechanical basis for keeping the

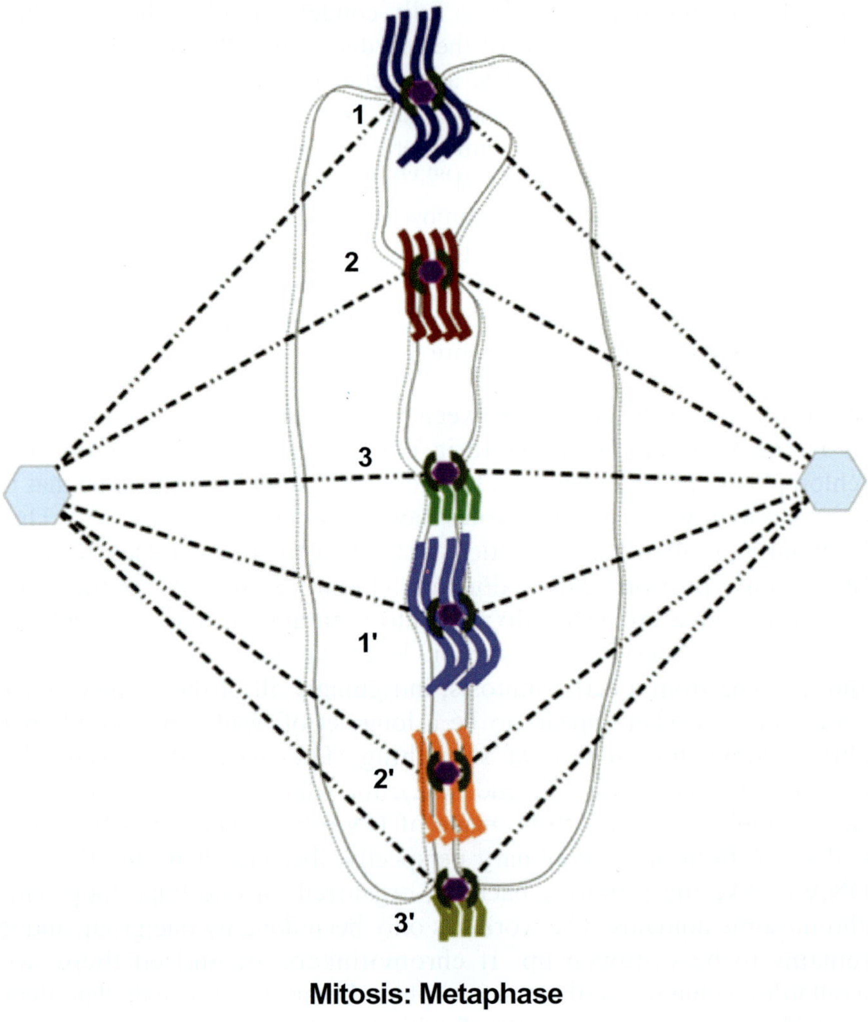

Figure 3.28. During metaphase the chromosomes are maximally condensed and they line up in the center of the cell. Each one has a microtubule attachment between the centriole and centromere created by the formation of the kinetochore represented by the dark green arcs around purple hexagons. Putative chromolinkers have been included in this diagram. The chromosomes line up along the metaphase plate. In three dimensions they actually form a circular rosette pattern but this is difficult to see in light microscopy and depict in this diagram. To visualize the rosette formation, imagine the putative chromolinkers being all of the same short length between the chromosomes, in a ring perpendicular to the page.

Figure 3.29. A model for the order of the chromosomes in a rosette at metaphase. "The diploid rosette is assembled at the first metaphase after fertilization by end-to-end fusion of the two ordered haploid sets contributed by the parents. The relative order of chromosomes within each haploid set (arrows) is identical, and haploid sets are oriented such that homologs are consistently positioned on opposite sides of the diploid rosette. This spatial arrangement of chromosomes is propagated faithfully by mitosis throughout subsequent cell generations in normal cells." From Nagele *et al.* (1998)[205] with permission of Springer-Verlag. Cf.[206-208].

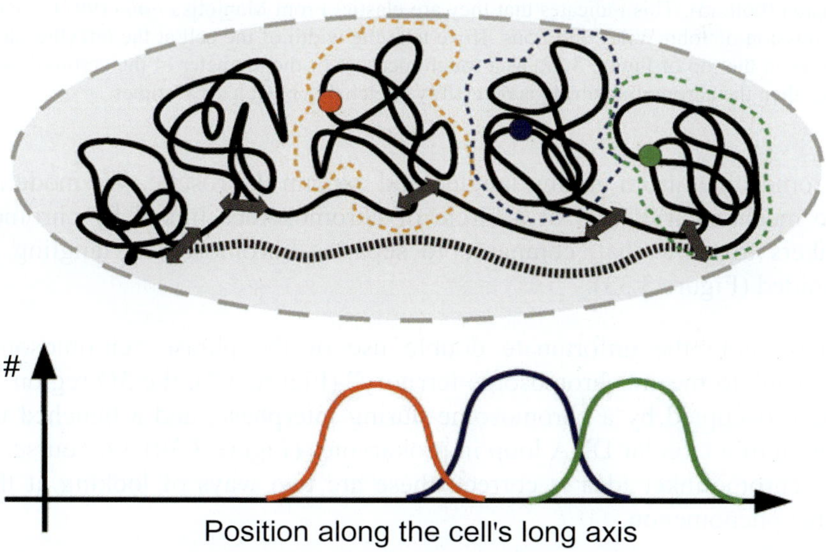

Figure 3.30. "A cartoon of the chromosomal domains being the 'building blocks' of the *E. coli* chromosome. The dark gray double arrows represent TF [transcriptional factor]–gene interactions which 'restrain' the domains. The domain 'walls' indicated by the red, blue and green dashed lines are able to diffuse and thus [are] subject to position fluctuations. The subcellular position distributions of the red, blue and green loci... exemplify the precise separate positioning of genetic sites on different domains due to their tendency of demixing. The gray dashed line represents the *ter*-proximate region[209] acting as a linker that connects the two polymer arms." The curves below represent the distribution of gene products of the three loci along the long axis of the cell. From Fritsche *et al.* (2012)[210] with permission of Oxford University Press.

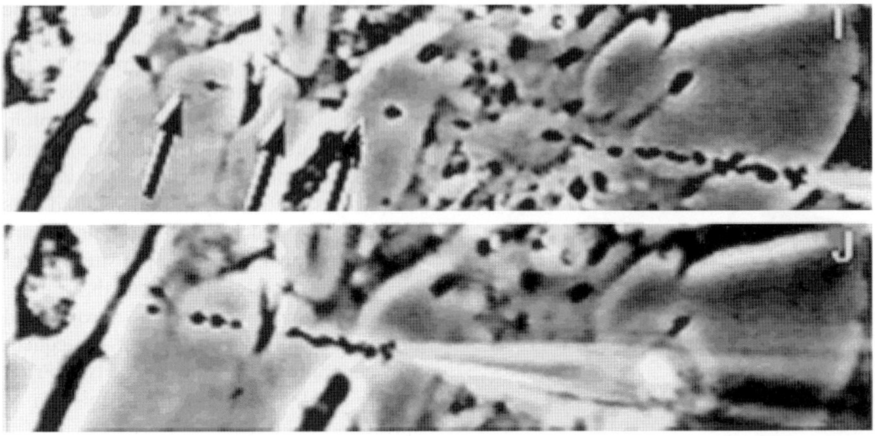

Figure 3.31. Chromolinkers may be stretched (top) and then pull back, remaining under tension (bottom). This indicates that they are elastic. From Maniotis *et al.* (1997)[203] with permission of John Wiley and Sons. If we take the width of the cell at the left (the same one as in the top of Figure 3.32) as a rough measure of the diameter of the chromolinker ring, then the chromolinker ring is reversibly stretchable by at least 15 times.

chromosomes from getting tangled and explain the rosette. We modeled the mechanical effects of a circle of chromosomes linked by chromolinkers to show that, compared to separate chromosomes, tangling is avoided (Figure 3.33).

Please note the unfortunate double use of the phrase "chromosome domain" to mean "chromosome territory" (Figure 3.7), the 3D region of space occupied by a chromosome during interphase, and a bunched up region of a circular DNA loop in prokaryotes (Figure 3.30). Of course, if the chromolinker idea is correct, these are two ways of looking at the same phenomenon.

If chromolinkers exist, how might they have evolved from the simple bacterial DNA loop? (The validity of the generalization that bacteria have loop chromosomes has itself been questioned[212] and there are exceptions[213,214].) A possible explanation lies in the domain structure of the single circle of DNA in *E. coli*. These "chromosomal loops or superhelical domains" form "four macrodomains and two less structured regions" (cf.[215]) (Figure 3.30). These loop domains result in greater efficiency by locating transcriptional factors near their targets. Protein-

protein or protein-RNA attractive interactions are hypothesized to keep each domain of the DNA clumped together[210]. In *Caulobacter crescentus* the DNA is twisted into a large number of small plectoneme loops arranged in a linear fashion like a bottle brush[219] (Figure 3.34). Side

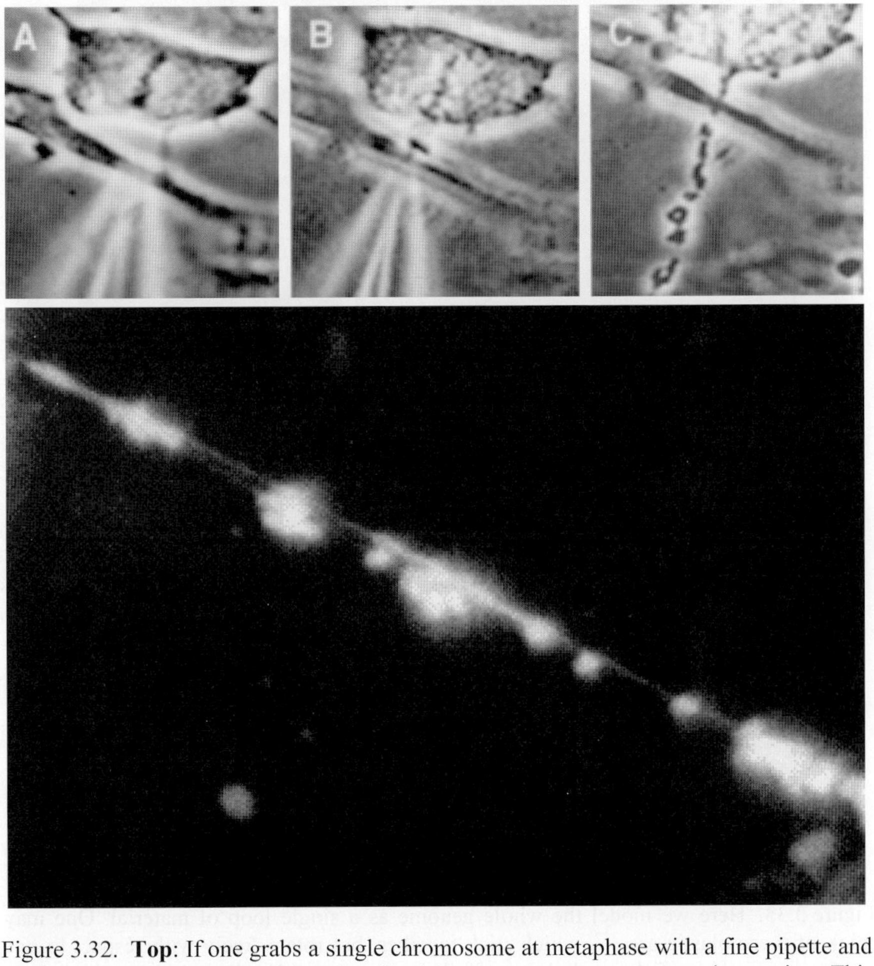

Figure 3.32. **Top**: If one grabs a single chromosome at metaphase with a fine pipette and pulls it out of its (endothelial) cell, all of the chromosomes come out in a string. This suggests that something connects each to the next. That material Dick named chromolinkers, each one binding a pair of chromosomes. **Bottom**: The two parallel strands, here stained for DNA along with their attached chromosomes, and other images[204] suggested that the chromolinkers contain DNA and to Dick that the chromolinkers at metaphase form a ring. From Maniotis *et al.* (1997)[203] with permission of John Wiley and Sons. Cf.[216].

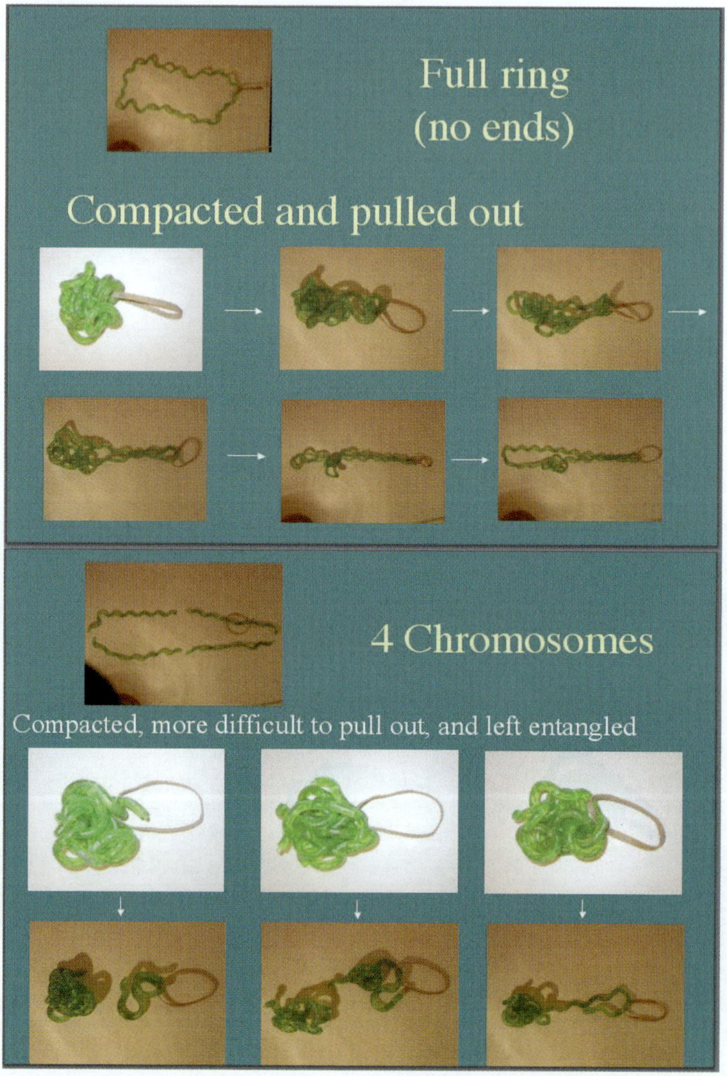

Figure 3.33. Here we model the whole genome as a single loop of material. One may think of these plastic arcs[217], which stick together, but rotate freely with respect to one another, as the ring of chromolinkers. A full ring, compacted, pulls out into a single loop (top). If we divide the ring into 4 segments (bottom), each of which would carry a chromosome, the force needed to pull out one chromosome is greater than that needed to untangle the ring, and the remaining chromosomes remain tangled. The demonstration may be done in a tactile manner by holding the compacted structures in the partially closed palm of one hand, while pulling with the other. The pulling force is localized using a rubber band.

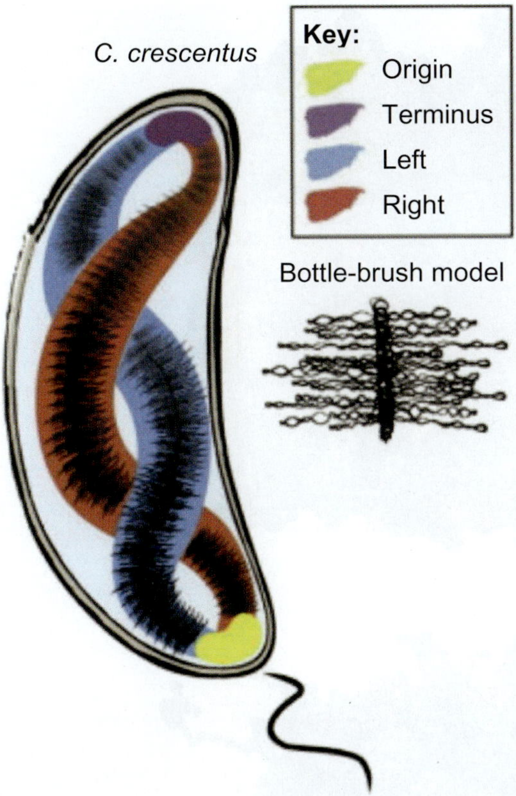

Figure 3.34. The bacterium *Caulobacter crescentus* has a single loop of double stranded DNA that is coiled so as to appear like a bottlebrush. Each side coil is called a plectoneme. While these plectonemes are too numerous to be forerunners of chromosomes, if we count the DNA along the axis of the brush as being analogous to chromolinkers, this could represent the beginning of a chromosome/chromolinker topology. The domains are color coded. From Weng & Xiao (2014)[218] with permission of Elsevier.

branching loops can clump together, again permitting a division into distinct domains connected by DNA that could be the precursor to chromolinkers. As we noted, some bacteria indeed have linear chromosomes, complete with telomere-like structures that tie up the otherwise loose ends of the DNA[213]. Thus many of the pieces for the puzzle of how circular bacteria might have evolved to circular chromolinkers decorated with chromosomes in eukaryotes are available and these observations support the chromolinker model. We therefore suggest that bacterial DNA domains are evolutionary precursors to chromosome territories in eukaryotes and the DNA segments linking those bacterial domains are the precursors to the eukaryotic chromolinkers, assuming of course, those exist.

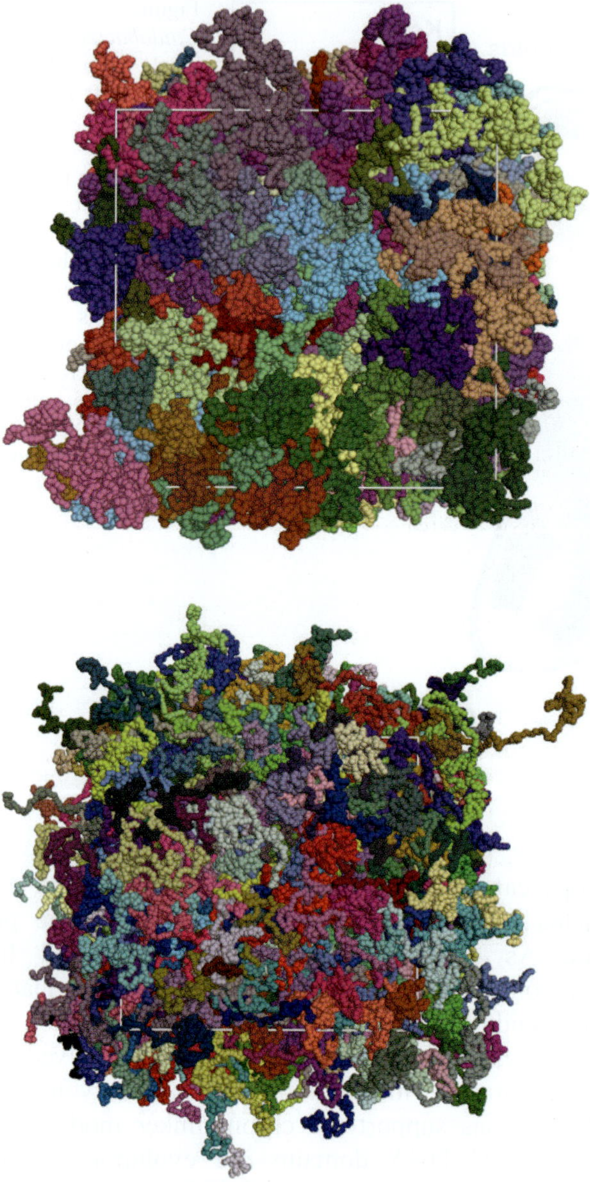

Figure 3.35. Molecular dynamics computer simulations of chromosomes represented as individually circular polymers (above) and as strings (below). The "melt of rings" simulation (top) reproduces the visual appearance of chromosome territories. From Halverson *et al.* (2014)[182] with kind permission of Alexander Y. Grosberg and ©IOP Publishing. Reproduced by permission of IOP Publishing. All rights reserved.

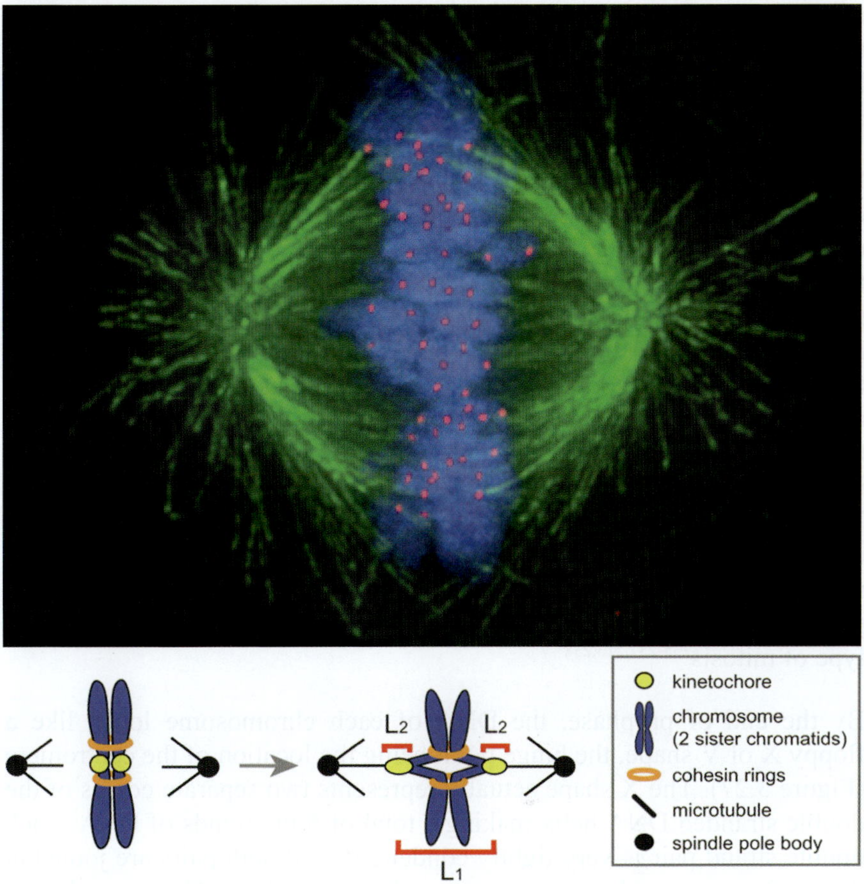

Figure 3.36. A kinetochore is a protein structure to which spindle microtubules attach that pull sister chromatids apart[220]. In the top image of the mitotic spindle in a human cell, from[221] (in the public domain), the kinetochores are marked in pink, the chromosomes in blue, and the microtubules in green. Bottom, from Nannas & Murray (2014)[222] with permission under a Creative Commons Attribution (CC BY) license: detail showing microtubules attaching to the kinetochores[223] and pulling on them. Spindle pole bodies = centrioles.

The only other model explaining the rosette observation is the recent polymer model for chromosome territories (cf.[224]) that makes an analogy to a "melt of rings"[182] (Figure 3.35). As each ring represents one chromosome, the model is not precise, since most normal eukaryotic chromosomes are linear. It has been suggested that a proxy for the modeled circularity of the chromosomes may be attachment of their ends to the inner surface of the nuclear membrane, or that cell lifetime is not long enough for linear chromosomes to "reptate"[225] or slither past one another and that is how chromosome territories are maintained[226-229].

Prophase ends and metaphase begins (Figure 3.3) with the formation of the rosette and the orientation of the fully condensed chromosomes on the metaphase plate. It is also marked by the complete disassembly of the nuclear membrane[230] in animals. Animals have "open" mitosis due to the lack of an intact nuclear membrane during cell division. The membrane is disassembled by first removing the nuclear pores and then the membrane is absorbed by the nearby endoplasmic reticulum. Plants and several other organisms have "closed" mitosis where the microtubules either form inside the nuclear membrane or pass through the intact nuclear membrane to make contact with the chromosomes. Variations between these extremes occur in fungi[231]. Some single cell eukaryotes contain two nuclei, the micronucleus undergoing closed mitosis and the macronucleus undergoing "amitotic fission"[232-234]. In prokaryotes while the process is called "fission" it has been regarded as a type of mitosis[235-237,213,238].

By the end of prophase, the DNA of each chromosome looks like a floppy X or V shape, the hinge point being the location of the centromere (Figure 3.27). The X shape actually represents two separate copies of the double stranded DNA helix making a total of four strands of DNA. Each double strand pair is very tightly condensed, and both pairs are joined at the centromere with a protein complex around it. This complex is referred to as the kinetochore[239] and it marks the spot to which the spindle apparatus of microtubules will attach (Figure 3.36).

At the other end the microtubules are attached to a microtubule organizing center (MTOC), in this context often called the spindle pole body. The microtubule organizing center[240] goes by numerous other names[74]: centrosome, basal body, nucleus associated body, parabasal body, spindle pole body, etc. Except in plants[241], it usually, but not always[242], contains a pair of "self-replicating"[243,244] centrioles[245-247]. Here we will try to stick to the three synonyms: MTOC, centrosome, pair of centrioles. The structure and replication of the microtubule organizing center are beginning to be understood[248-250] (Figure 3.37).

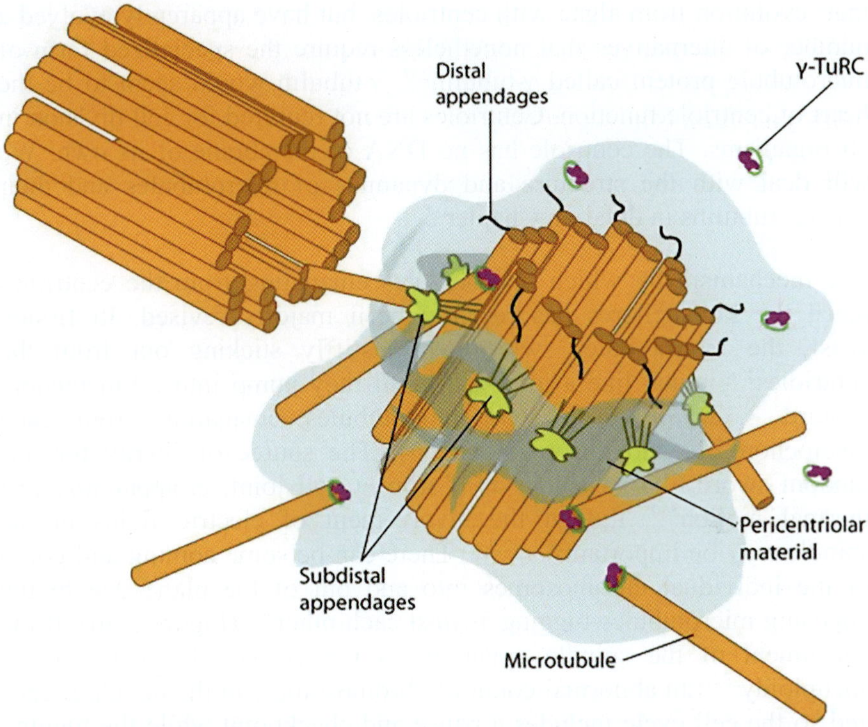

Figure 3.37. The microtubule organizing center (MTOC) consists of two mutually perpendicular centrioles and associated pericentriolar material including γ-tubulins (γ-TuRC), which nucleate the polymerization of microtubules. Image provided courtesy of the Mechanobiology Institute, National University of Singapore[251]. We previously reviewed the earlier literature arguing whether or not MTOCs, which are self-replicating, contain DNA, and listed at least 20 synonyms for them (Proposition 113 in *Hierarchical Genome*[74]). The two mutually perpendicular bundles will be referred to as centrioles in this book, and the pair of them, with the surrounding cloud of so many components, as the centrosomes. Centrosomes have been called "supercomplexes"[252]. The centrioles are described as a pair of short microtubules[253]. The centrioles may indeed be self-replicating, but based on RNA instead of DNA[254]. How the centrosome/MTOC is built and works is still baffling.

The centriole, which is critical as the starting point for the construction of flagella and cilia, but is bypassed in higher plants, fungi and the nearly normal development of animals from eggs with the centrioles experimentally removed[255], may have arisen from a symbiont[256,246] in the same way as mitochondria. Its precise origin is uncertain. Plants began their evolution from algae with centrioles, but have apparently evolved a number of alternatives that nonetheless require the specialized form of microtubule protein called γ-tubulin[257]. γ-tubulin would seem to be the heart of centriole function. Centrioles are not required for cell division in all organisms. The centriole has no DNA or membrane of its own. We will deal with the structure and dynamics of microtubules and their various tubulins in detail in Chapter 5.

The mechanism by which microtubules emanating from the centrioles reach the kinetochores has recently been majorly revised. In fission yeast, the microtubules swivel while stiffly sticking out from the centrioles[258], much like a joy stick, until they bump into a kinetochore (Figure 3.38). In *Drosophila* microtubules emanating from each kinetochore also do this pivoting[259,260]. The source of energy for this random sword-like swishing, as if from a ball joint, is apparently just thermal motion[260], though the involvement of electric fields in the spindle may be important[261,262,260]. There can be some coming and going of the individual chromosomes into and out of the plane due to the opposing microtubules tugging against each other[263] (Figure 3.36). If the attachment of the spindle apparatus is not correct this can lead to aneuploidy[264] (an abnormal count of chromosomes) in the daughter cells and so the cell cycle includes a pause and checkpoint while the tugging and pulling seems to test the whole spindle[239].

The spindle itself consists of three sets of microtubules (Figure 3.39). There are those attached directly to the kinetochore. There are microtubules that form at ninety degrees to the array and are important for determining the plane along which the cell will actually divide. These are called interpolar microtubules[265]. And there is a set of astral microtubules that are attached to the cytoskeleton underlying the cell's membrane and to the kinetochore microtubules ensuring the correct orientation of the spindle. There are proteins that can sense and respond to tension in the spindle and thereby sense when the entire spindle is correctly attached and the chromosomes properly oriented[266,267]. Mitosis normally does not proceed until all the chromosomes are in a single plane, and correctly oriented.

The central metaphase or equatorial plane predicts the future line where the cell will divide. Where that plane is located varies by both the original cell type and the eventual outcome for the two daughter cells.

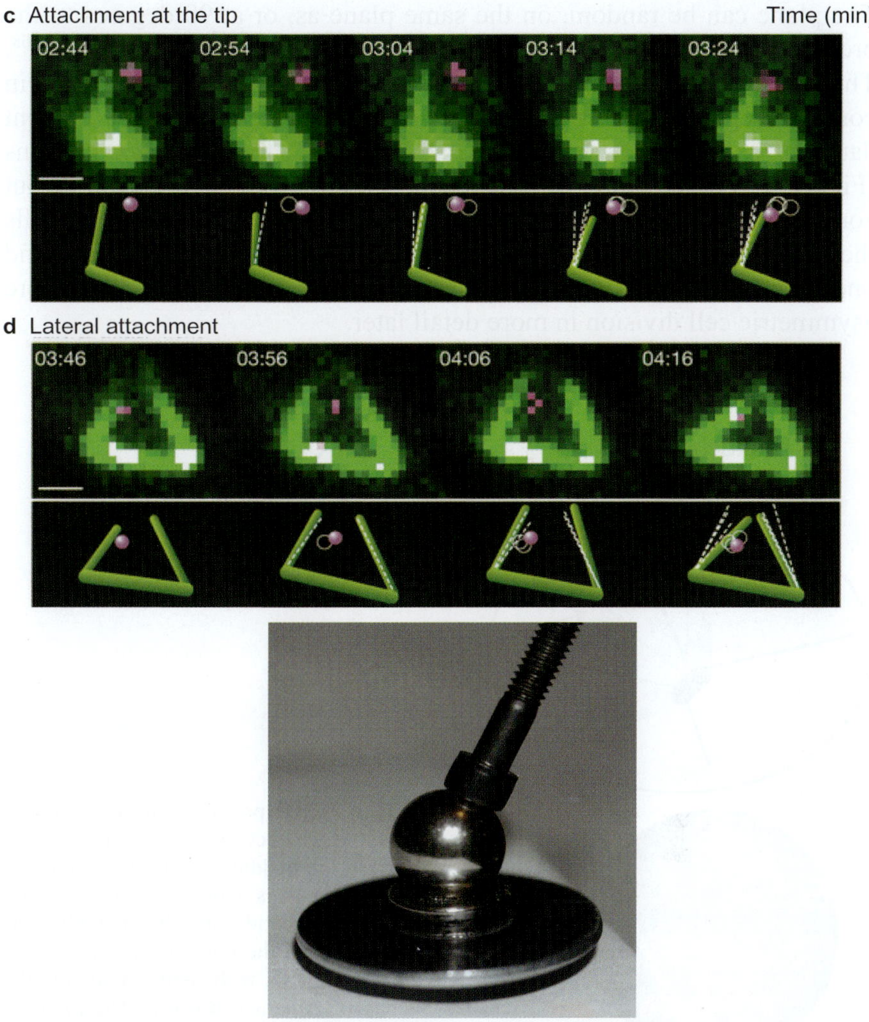

Figure 3.38. **Top**: Microtubules in the fission yeast *Schizosaccharomyces pombe* swing as if part of a joystick whose ball joint is at the centrioles (spindle pole bodies): "(c,d) Time-lapse images and the corresponding drawings of 2 examples of kinetochore capture, where the kinetochore was captured close to the microtubule tip (c) or away from the tip (d)." Microtubules are shown in green, and kinetochores are shown in magenta or white. Scale bars, 1 μm. The relative rigidity of microtubules (Chapter 5) permits the joystick analogy. From Kalinina *et al.* (2013)[258] with permission of Nature Publishing Group. **Bottom**: A homemade joystick using a magnetic ball joint. With kind permission of Edward L. French[270].

The plane can be random, on the same plane as, or at 90 degrees to the previous division depending on the type of cell or the species involved[268]. The process occurs in two steps[269]: 1) anchoring proteins end up in confined regions of the inner surface of the cell membrane of a nascent daughter cell, 2) the spindle aligns by attaching to the anchor proteins (Figure 3.39). We don't yet understand how the first step occurs[271], but cortical actin is involved[272]. Division can also be with both daughter cells the same size after division, or it can be uneven with one large cell and one tiny budlike cell, as when yeast spores are made. We will explore asymmetric cell division in more detail later.

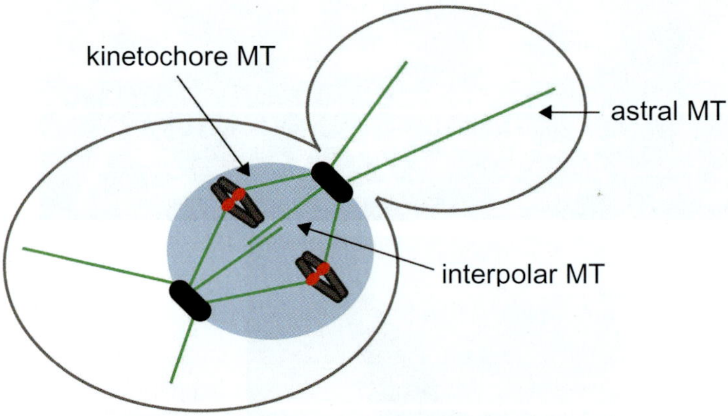

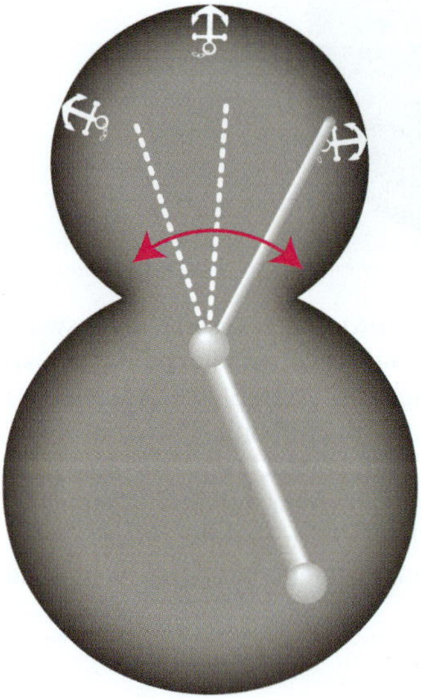

Figure 3.39. **Top**: Three positions of microtubules during mitosis, which occurs inside the nucleus (closed mitosis) in the budding yeast *Saccharomyces cerevisiae*. The nucleus is blue, the centrioles are black, two sister chromatids are shown with their kinetochores in red, and the microtubules are green. From Biggins (2013)[239] with permission of the Genetics Society of America. **Bottom**: The pivoting astral microtubules attach to the cortical anchor protein Num1 that is attached to the inner membrane of the budding cell (the smaller one here). Cf.[271,269]. From Baumgärtner & Tolić (2014)[273] with permission under a Creative Commons Attribution License.

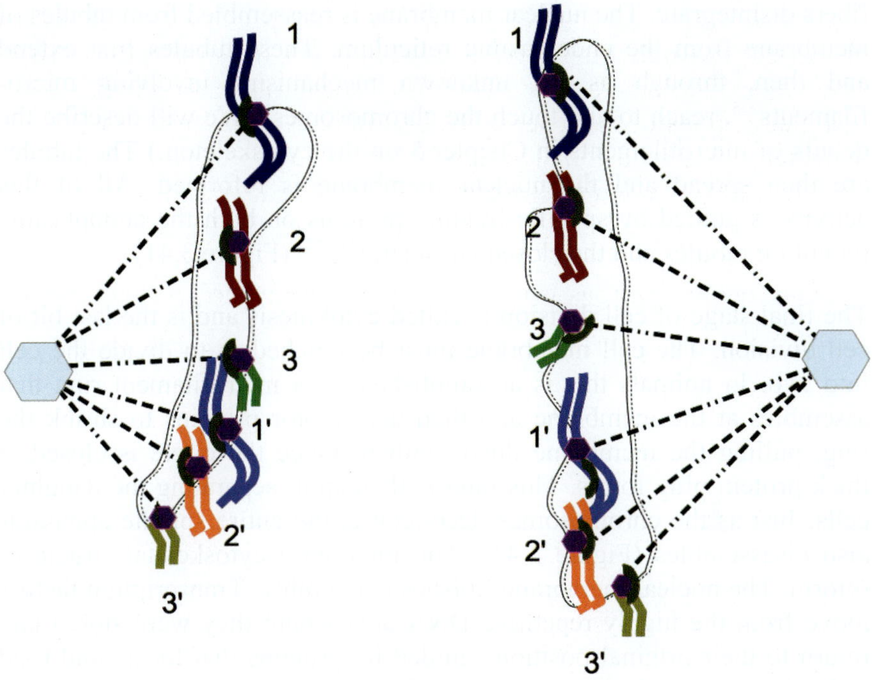

Figure 3.40. During telophase the chromosomes are divided at the centromeres and sister chromatids are moved to opposite ends of the cell. In the final stage, anaphase (not shown) the rest of the cytoplasm is divided as the cell membrane descends and divides the cell. (Note complementary DNA strands are shown here.)

Recall that the preanaphase chromosome is a double chromosome made of paired sister chromatids. During anaphase, the sister chromatids are parted and move to opposite poles of the cell (Figure 3.36). They are then sometimes referred to as daughter chromosomes. Sister chromatids become daughter chromosomes (Figure 3.40).

During telophase cell duplication is almost completed. The cell membrane begins to pinch down dividing the cell into two. All the duplicated chromatin proteins are also segregated to each of the two sides of the cell, so that each daughter cell has a single full complement of cellular machinery. In cases of asymmetric division[274] the components are unevenly divided so one daughter cell gets a different set of components from the other daughter cell but both still have the same chromosome content. The decondensing of the DNA begins. Spindle fibers disintegrate. The nuclear membrane is reassembled from tubules of membrane from the endoplasmic reticulum. These tubules first extend and then, through as yet unknown mechanisms involving micro-filaments[275], reach to and touch the chromosomes. (We will describe the details of microfilaments in Chapter 5 on the cytoskeleton.) The tubules are then spread and the nuclear membrane is reformed. All of this activity is guided by specific binding proteins on both the endoplasmic reticulum tubules and the closed chromatin[276,277] (Figure 3.41).

The final stage of cell division is called cytokinesis and is the last bit of cell division. The cell membrane must be pinched in to divide the cell into two. In animals this is accomplished by a microfilament ring that assembles at the membrane and then uses motor proteins to shrink the ring, pulling the membrane down with it. Once the circle is closed, a thick protein plug forms. This plug is then split, separating the daughter cells. Just as the chromosomes decondense, the entire spindle apparatus also disassembles (Figure 3.41). The interphase cytoskeletal structures reform. The nuclear membrane finishes reassembly. Transcription factors move from the highly repetitive DNA area where they were stored and return to their original positions guided by proteins that locate and bind to bookmark proteins[171] or specific DNA sequences. Gene expression is returned to normal. The two new cells each enter G1 interphase (Figure 3.42) and then duplication of the genome starts again in the next S phase (Figures 3.43 and 3.44).

It is important to note that during the time the DNA is tightly wound up, the genes (with a few exceptions) are not available for transcription. The cell has to make all the required proteins and mRNA messages it might need for cell division before condensing the DNA into visible chromosomes. At the end of mitosis, the DNA is unwound, or decondensed, and the visible chromosomes vanish, at least for the light microscopist, but remain localized as chromosome territories[11,10] (Figure 3.7), albeit with considerable variation between individual cells of the same type[278,228,279]. It is possible that chromosome territory structure varies in time, depending on which functions are being performed in a cell, a phenomenon that has been described as analogous to crowdsourcing[280]. There is some correlation between extended stretches along a chromosome that have particular properties, also called chromosome domains, and their three dimensional chromosome territories[281]. Generally, the portions of the chromosome that have genes

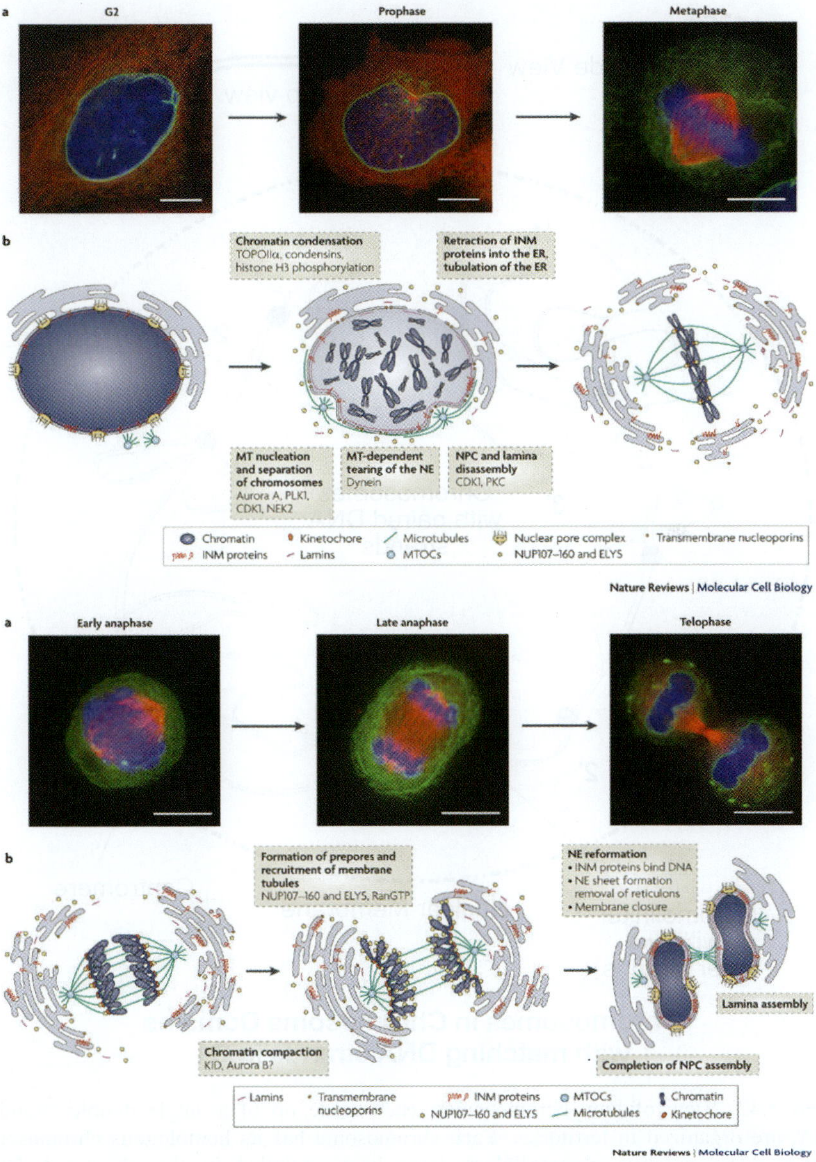

Figure 3.41. In open mitosis the nuclear membranes break down and then reassemble. The micrographs of HeLa cells[282] show the sequence of events. The inner nuclear membrane is green, DNA is shown in blue and microtubules are red. INM = inner nuclear membrane, ER = endoplasmic reticulum, MT = microtubule, NE = nuclear envelope, NPC = nuclear pore complex, MTOC = microtubule organizing center. From Güttinger *et al.* (2009)[276] with permission of Nature Publishing Group.

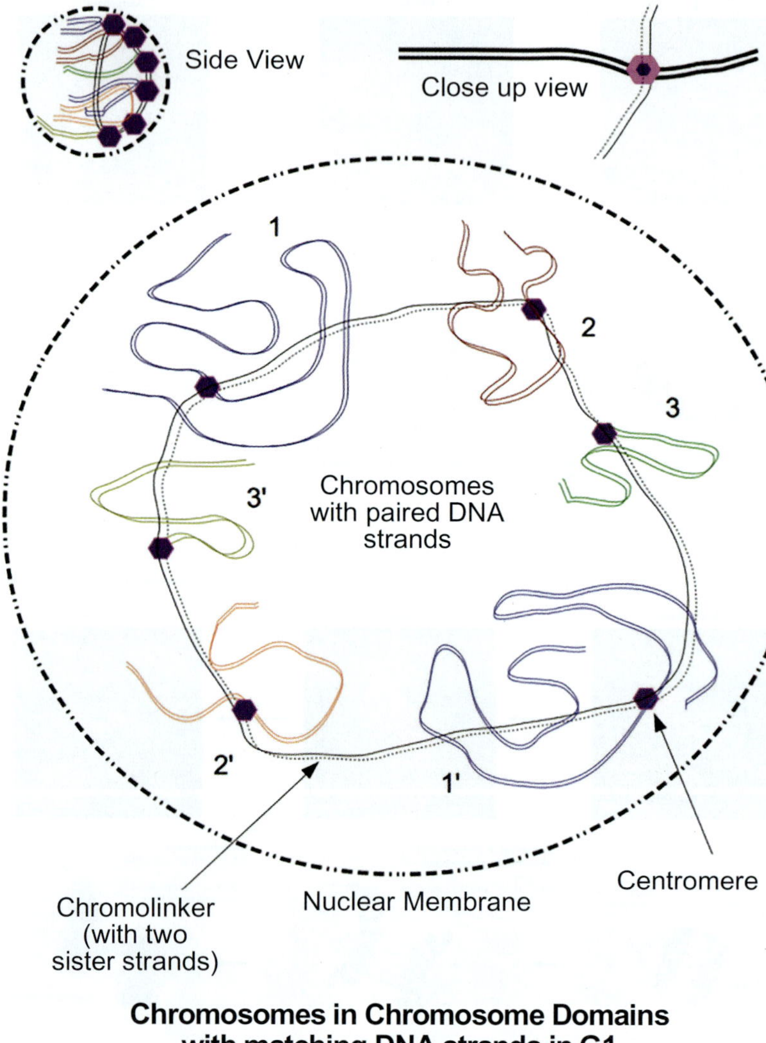

Chromosomes in Chromosome Domains with matching DNA strands in G1

Figure 3.42. View of how chromosomes, each made up of a single double stranded DNA, are organized in territories. Each chromosome has its homologous chromosome partner. (The putative chromolinkers have been included in this diagram.) Each centromere (purple hexagons) is not obvious at this stage but it is a region of DNA with a specific sequence that attracts proteins that will eventually form the kinetochore, the protein structure where microtubules attach. The chromolinker is assumed to remain attached to the chromosomes at the centromere site throughout the cell cycle.

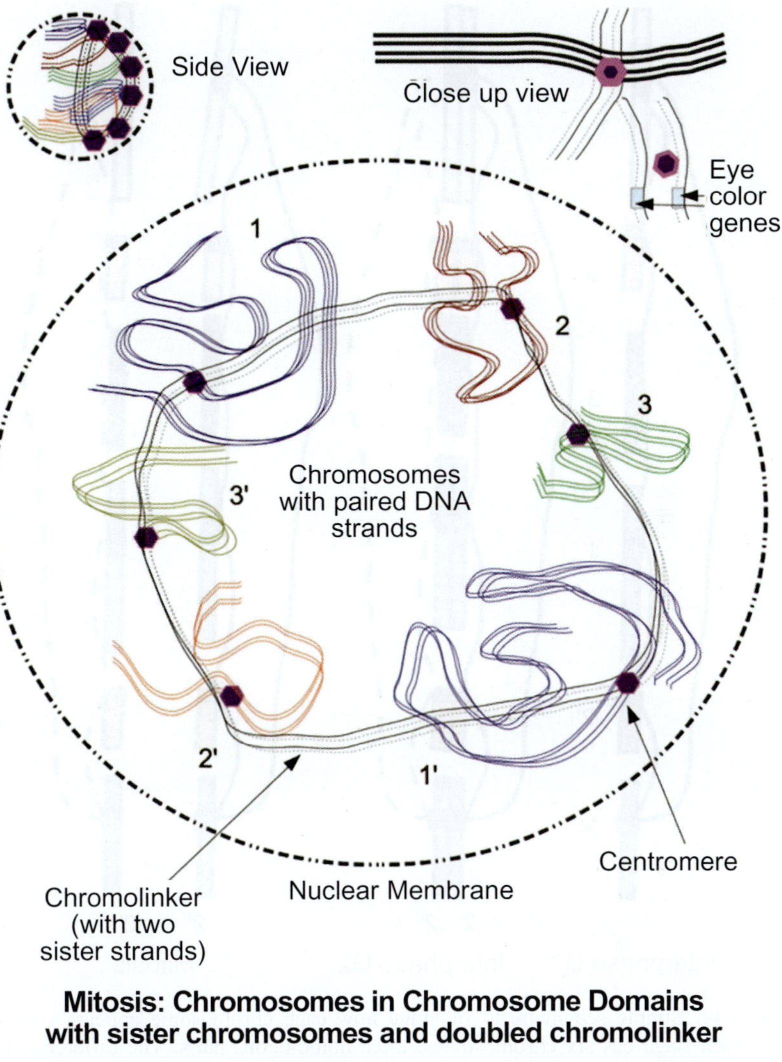

Figure 3.43. Before the cell enters mitosis all the chromosomes are doubled but still joined by a single now visible centromere, the region where the two sister chromatid strands are closest and where the kinetochore will form. (Doubling of the putative chromolinkers is presumed to also occur if they are present.)

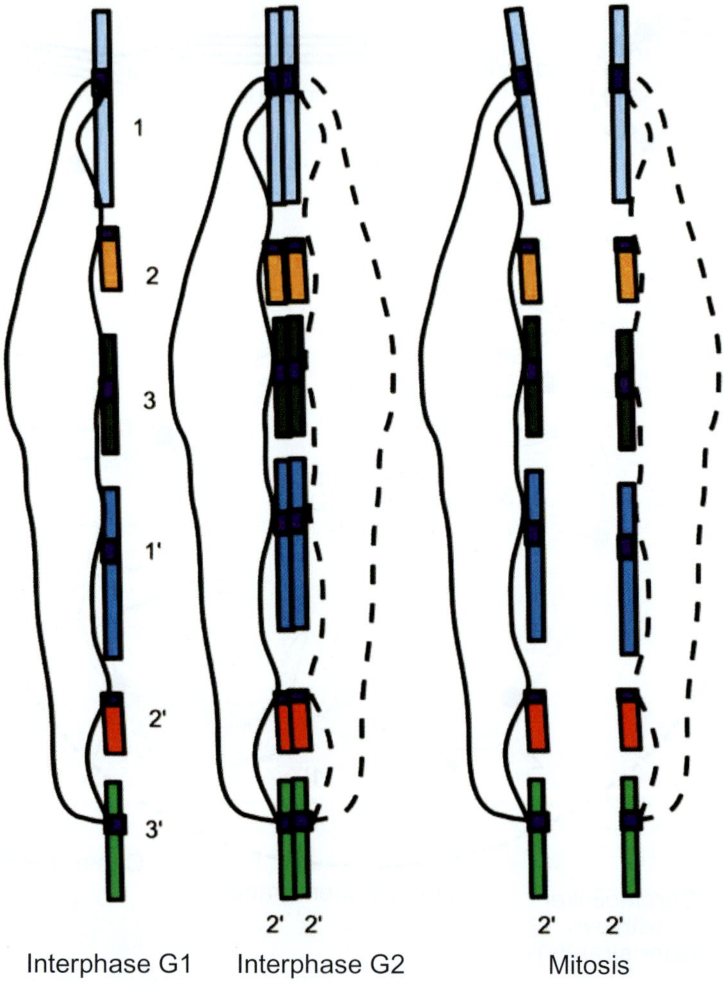

Figure 3.44. Simple schematic to illustrate how each chromosome appears versus cell stage. Each stage has six chromosomes, three homologous pairs. The difference lies in the presence of identical sister chromatids. (The putative chromolinkers are also included in this illustration.) (Note complementary DNA strands are not shown here.)

that are being actively expressed tend to be found in the center of the nucleus[283]. Genes that are not being used tend to be found more on the outer edges of the nucleus at or near the nuclear membrane. At this writing, models of the "higher order structure of the genome" are in tremendous flux[284].

In humans we have 46 chromosomes, 22 pairs, called autosomes, that may differ in their precise coding of specific genes but are identical in terms of what genes they code for and the organization of those genes. In addition to the 22 pairs, there are also two sex chromosomes, with human females having two X chromosomes and males have one X and one Y chromosome. From the perspective of gene regulation, however, the higher order division of the genome into chromosomes seems to be irrelevant. The best proof of this is the muntjac deer. These are small deer of many species common in Asia and all of them look somewhat alike. The Indian muntjac (*Muntiacus muntjak vaginalis*) has the lowest recorded chromosome number of any mammal. It has 7 chromosomes in males and 6 in females. In contrast Reeves's muntjac (*Muntiacus reevesi*), has 46 chromosomes. Yet both species of the deer are closely related and look much alike (Figure 3.45). Chromosomes provide some kind of higher organization important for mitosis and yet they are seemingly not responsible for creating subsets of gene expression (differons). Considering that the chromosomes are loosely unwound and each can extend from the outer edges of the nucleus to down into the center part it is easier to see why chromosomes themselves are not important for creating the functional subsets of genes.

This brings us to consider a cell type that is created if the cell goes into G0 instead of G1 after mitosis (Figure 3.3). A cell that goes from mitosis into G0 is usually a terminally differentiated cell. This cell will never again undergo mitosis. It will spend the rest of its life in interphase. It will not reproduce and divide again. When it finally accumulates enough small bits of damage that it cannot repair, it ceases to function and dies. Sometimes it is subjected to some external force it can't recover from, and it dies. Some brain cells are examples of terminally differentiated cells. If they are damaged, they cannot replace themselves. What makes certain cells go into G0 instead of G1 is the determination signal they receive and the differentiation they undergo in response to that signal. There is also programmed cell death and that comes in two types. One is where a cell accumulates a certain amount or type of damage and a regulon is activated that results in death of the cell, a suicide regulon. Programmed cell death is also very important during embryogenesis for things like creating space between fingers and toes in mammals. In that case a differon will initiate the suicide regulon.

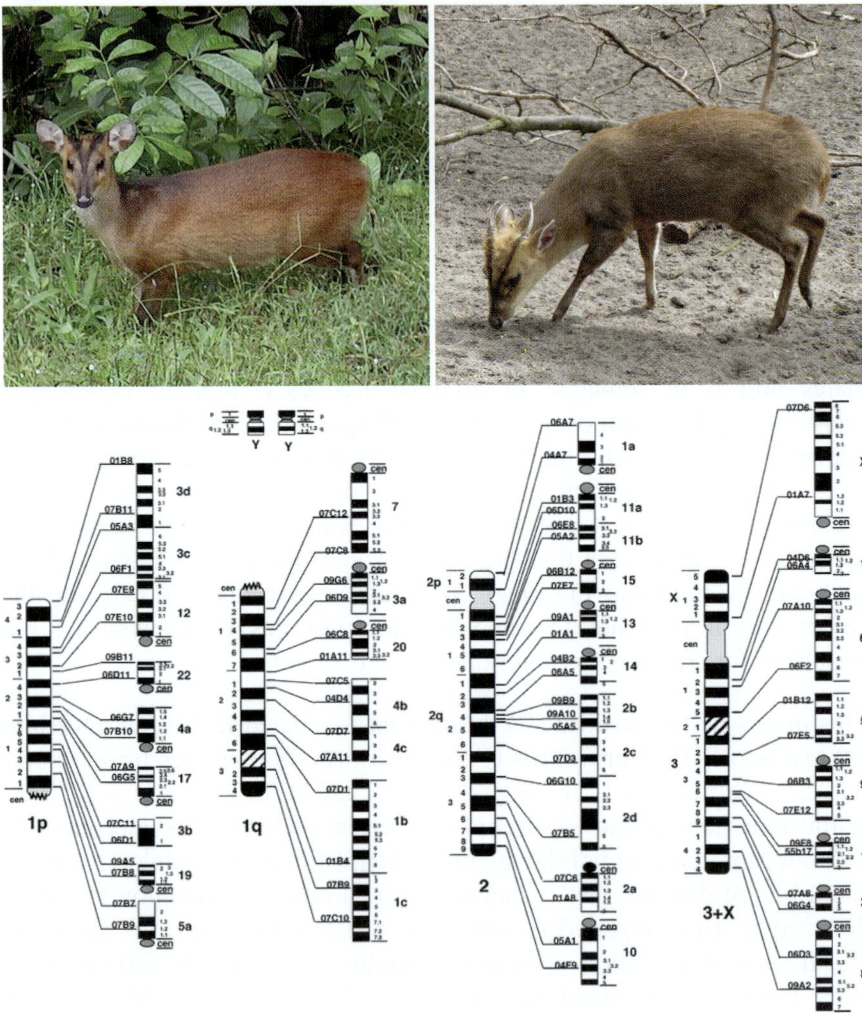

Figure 3.45. In each column the haploid set of chromosomes is shown from the Indian muntjac deer (*Muntiacus muntjak vaginalis*) on the left (2N = 6 in the female and 7 in the male), and its proposed origin from fused smaller chromosomes in an ancestor of the Chinese muntjac (*Muntiacus reevesi*), on the right (2N = 70). (The two ends of chromosome 1 of the Indian muntjac are separated as 1p and 1q.) Note that "All tandem fusions… are of the centromere-telomere (head-tail) type", i.e., all of "the same orientation, [i.e.], … oriented toward the centromere of the [Indian muntjac] chromosomes". From Chi *et al.* (2005)[285] with permission of Springer. Images of Indian[286] and Chinese (Reeve's)[287] muntjacs with permission under Creative Commons Attribution-Share Alike 3.0 Unported licenses. A deer is a deer is a deer, despite huge differences in numbers of chromosomes.

There is an exception to the G0 rule though, that being stem cells[288]. Stem cells will sit quietly in one place and basically wait until something stimulates them. These stem cells will then leave G0 enter G1, cycling through to M and then undergo an unequal division. This leaves one daughter cell, which is a new stem cell that goes back to G0 to sit and wait again, and one new different cell that undergoes further differentiation. Stem cells are found in many tissues such as skin and liver and evidence for stem cells continues to grow in places like heart and brain where we previously assumed there were none. It is as if a subset of cells simply step out of the dance of embryogenesis and sit on the sidelines and do nothing but wait until someone invites them back in. More on this in Chapter 11.

The main purpose of this book is to put forth a new theory of how, when and where cells become different from one another during normal embryo development even though they have the same genome. Most cells have a nucleus that is generally assumed to be able to replace and function in an enucleated egg cell like the nucleus in an untampered egg. Such a nucleus will go on to develop into a complete, fertile adult organism. This property is called totipotency. Certain adult cell types have indeed been shown to have this property[289,290,291]. Stem cells can be totipotent or more often can only produce a subset of cell types (pleuripotency), all related to when in development they stepped out of the dance. So stem cells from skin cannot be used to repair and regenerate liver because skin stem cells are ectodermal when set aside and not endodermal. It is as if the endodermal pathway is closed to the skin stem cells. Most cells for most organisms retain the whole genome whether or not the individual cell is totipotent. There are a few exceptions worth mentioning.

Theodor and Marcella Boveri, in the late 1890s, worked on a nematode worm that parasitizes horses[292], *Ascaris megalocephala* (now called *Parascaris aequorum*), in which all but one of the cells at the 16 cell stage has gone through fragmentation of the two pairs of chromosomes into smaller chromosomes, with successive loss of the smaller chromosomes[293]. This process is called “chromosome diminution”. The one cell that avoids chromosome diminution becomes a germ cell, carrying intact chromosomes to the next generation. An analogous phenomenon occurs in single cell animals called ciliates. Each cell has two nuclei, a “micronucleus” in which the chromosomes remain intact, and a "macronucleus" in which major chromosome fragmentation and other phenomena occur[294]. The surface of ciliates is quite complex, and the subject of much morphogenetic investigation[295-297]. Perhaps differentiation of the ciliate's surface is related to chromosome changes. If, as has been proposed, multicellular animals are descended from ciliate-like single celled animals (§5.05 *The Ciliate Origin of Multicellular Organisms* in *Hierarchical Genome*[74]), then the nematodes that exhibit chromosome diminution in their somatic cells may be hangovers of an intermediate step in the evolution of animals. But if chromosome diminution is the original form of differentiation in embryos, why would a shift to totipotency have occurred? One reason could be that this shift allowed the organism to be populated with stem cells, i.e., cells that retain the ability to differentiate into many other kinds of cells in the adult. Whatever the evolutionary reason, chromosome diminution is the exception and not the rule during embryogenesis.

We have covered many aspects of gene and chromosome functioning and we have touched on the passing of information from one generation to the next. We will now have a look at meiosis, which is a special type of mitosis unique to creating germ cells. The sperm and egg are each formed with only one copy of the total genome instead of two. That way, when two gametes combine, the result is a fertilized zygote (one-cell embryo) with two copies of each gene. Once fertilization occurs, embryogenesis can begin again. The preparation of egg and sperm is what meiosis accomplishes. Meiosis is very similar to mitosis but there are two steps of separation of DNA material instead of one. The two steps are referred to as meiosis 1 and meiosis 2. Just as in mitosis, the visible chromosome movements of meiosis are divided into named steps. Meiosis 1 has a prophase 1, metaphase 1, anaphase 1 and telophase 1. In addition, prophase 1 is further subdivided into stages. These prophase 1 stages are leptotene 1, zygotene 1, pachytene 1, diplotene 1, and diakinesis. Once meiosis 1 is complete, the resulting two cells are haploid. The single copy of each chromosome was doubled before meiosis began, and the single chromosome is made up of two identical sister chromatids. Meiosis 2 is essentially the same as mitosis. The end result is that the sister chromatids are split. The final result is four haploid cells. Normally, in complex organisms, all cells are diploid except for the haploid gametes.

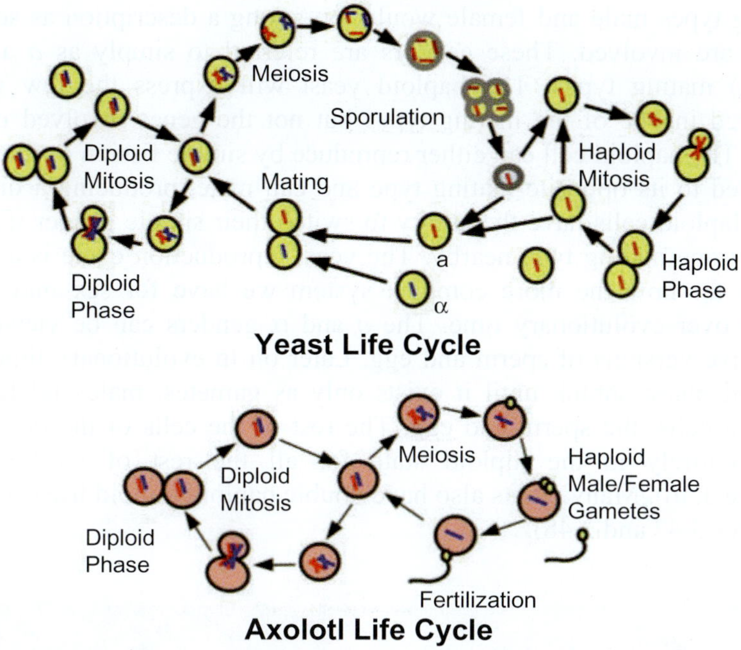

Figure 3.46. The yeast has a diploid life cycle and a haploid life cycle. In both, the yeast itself looks and functions essentially the same and can continue through multiple generations. In diploid organisms like axolotls and ourselves, the haploid life cycle has been reduced to the production of sperm and egg by meiosis and ends with fertilization, i.e. the haploid stage is therefore only found in the gametes.

Yeast are simple unicellular organisms from which we can learn a lot about how this system may have evolved[298]. Yeast also have both a diploid and haploid state (Figure 3.46). However, yeast are functional as both diploid and haploid. Both the diploid yeast and the haploid yeast can reproduce by a simple mitosis type of cell division called fission, which results in a bud. The bud on a parent cell breaks free and grows. The diploid cell has the advantage of being better able to survive by the masking of any deleterious alleles. If the environment were to become hostile, it can undergo a cell division into four separate haploid spores that can wait until conditions improve. Once conditions improve the haploid spores will begin to grow. The haploid yeast cannot make spores. While the haploid cell has one full complement of genes it also expresses most but not all of all possible genes. The haploids have two mating types. There are a few genes that determine mating type. To call these mating types male and female would too strong a description as so few genes are involved. These genders are referred to simply as *a* and α (alpha) mating types. The haploid yeast will express the few genes involved in one of the mating types but not the genes involved in the other. The haploid cell can either reproduce by simple fission or it can be attracted to its opposite mating type and can mate, producing a diploid cell. Haploid cells have the ability to switch their simple gender if there is only one mating type nearby. The yeast reproduction cycle is a good model for how the more complex system we have for sex may have arisen over evolutionary time. The *a* and α genders can be viewed as primitive versions of sperm and egg. Later on in evolutionary time, the haploid phase shrank until it exists only as gametes, male and female haploid cells, the sperm and egg. The rest of the cells of the organism exist entirely in the diploid state for all the rest of its lifecycle (Figure 3.46). Many plants also have double haploid/diploid life cycles[299] (Figures 3.47 and 3.48).

Figure 3.47. Gametophyte (haploid) and sporophyte (diploid) life stages of many plants undergo very different kinds of morphogenesis. Leaflike gametophyte (left) 2 cm wide and filamentous sporophyte (right, filaments about 5 μm wide) of the red alga *Pyropia yezoensis*[300], from Friedman (2013)[299] with kind permission of Koji Mikami (cf.[301]).

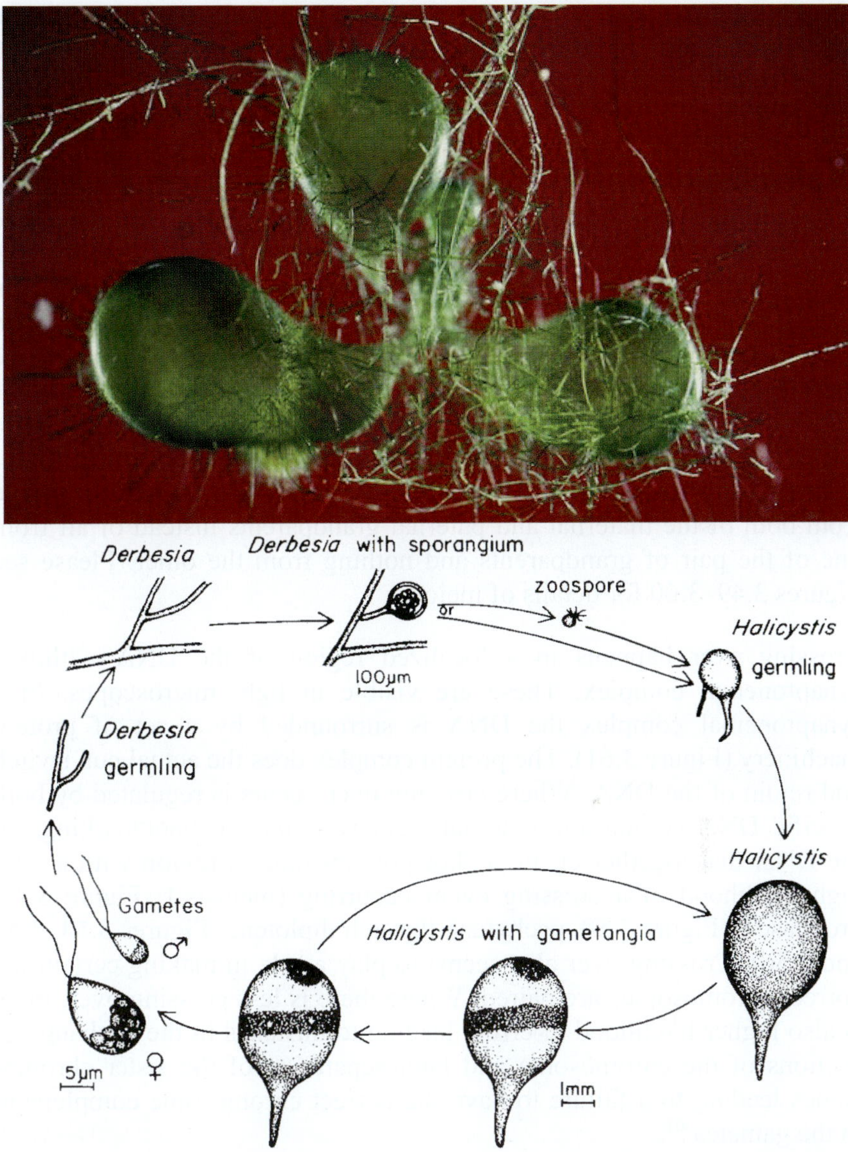

Figure 3.48. **Top**: The green alga *Derbesia* has bulbous, haploid gametophytes about 1 cm wide, while its diploid sporophyte stage is filamentous. Top photo from Friedman (2013)[299] with kind permission of Kjersti Sjøtun. **Bottom**: Life cycle from Graves & Gutknecht (1976)[302] ©1976 James S. Graves and John Gutknecht with kind permission of Suzanne O'Donnell, Rockefeller University Press. The two morphologies are so different that they were initially identified as separate species!

Meiosis begins for all sexually reproducing organisms with the genome duplicated, just as in mitosis. Four copies of each gene are then present, two identical pairs, each identical pair on the two homologous chromosomes, joined at a centromere (Figure 3.49). In mitosis, the chromosomes line up single file around a rosette at a plane called the anaphase plate. In meiosis 1, they line up too, but in pairs of homologous chromosomes. The two centromeres of each homologous chromosome move together and DNA from all four strands condenses. At this stage, entire sections of the DNA are exchanged between two strands on opposite homologous chromosomes in a process called crossing over. Interestingly, crossing over between the sister chromatids on the same chromosome is strongly suppressed, while crossing over on the strands from the chromosomes of the homologous pair always happens at least once. Organisms inherit a blended mixture of each parent's chromosomes due to the crossing over event. This means there is inheritance from both of the maternal and paternal grandparents instead of all from one of the pair of grandparents and nothing from the other. Please see Figures 3.49–3.60 for details of meiosis.

Crossing over happens in a localized region of the DNA within a synaptonemal complex. These are visible in light microscopes. In a synaptonemal complex the DNA is surrounded by a set of protein machinery (Figure 3.61). The protein complex does the actual cut, switch and rejoin of the DNA. Where crossing over occurs is regulated by both specific DNA coding and external signals, similar to bookmarking, on the DNA that together create a "hotspot" meaning a region with a very high likelihood of a crossing event occurring (meiosis 1: Figure 3.49, prophase 1: Figure 3.50, pachytene through diplotene: Figures 3.54, 3.55 and 3.56). Crossing over also seems to play a role in making certain the correct chromosomes are paired. Where there is less crossing over, there is also higher likelihood of errors leading to mistakes in the exchange of portions of the chromosome and later separation of the sister chromosomes leading to a failure to have the correct chromosome complement in the gametes[303].

The first cell division separates the paired sister chromosomes, not the sister chromatids (Figure 3.58). The result is two new haploid cells with each having half the number of chromosomes, although each chromosome still has two full copies of the DNA. Because of crossing over, this single chromosome is a blend of the original two,

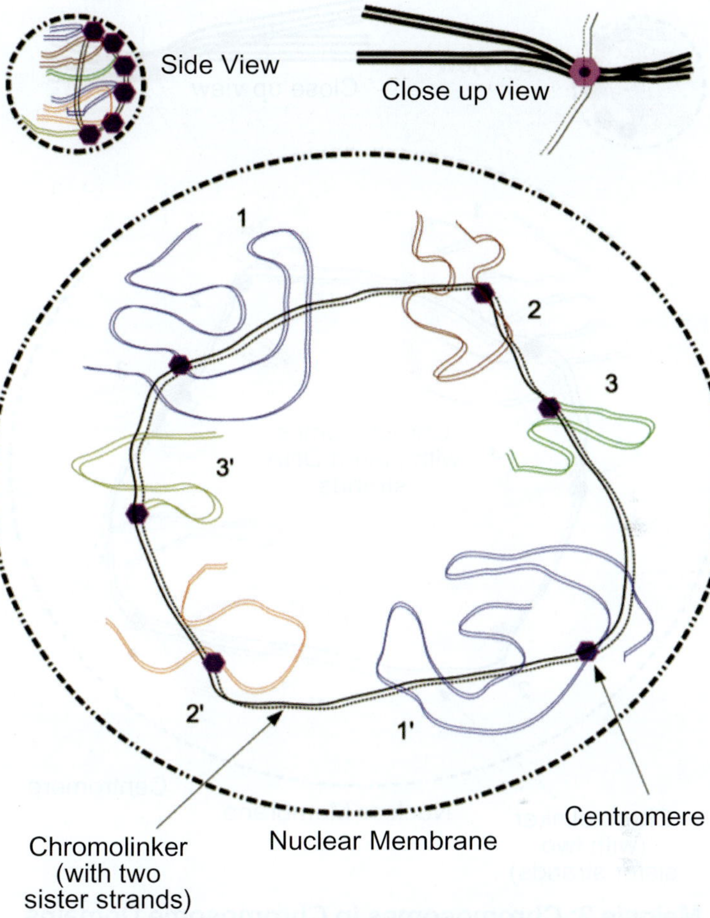

Meiosis 1: Chromosomes During Interphase in Chromosome Domains with paired DNA strands

Figure 3.49. Interphase. In the following set of figures we illustrate the steps of meiosis based on comprehensive reviews[304-309]. We include our own model of how the proposed (but unproven) chromolinkers could fit into the meiosis cycle. This first figure illustrates interphase of a pre-sperm or pre-egg cell. To keep the figures from being too busy, we use a 2N = 6 genome. The diploid chromosomes are each located in their individual chromosome territories within the nucleus. (We ignore sex chromosomes for the purpose of this illustration.) Centromeres are represented in the **Side View** as oriented so that the centromeres are found on one side of the nucleus with the telomeres on the other.

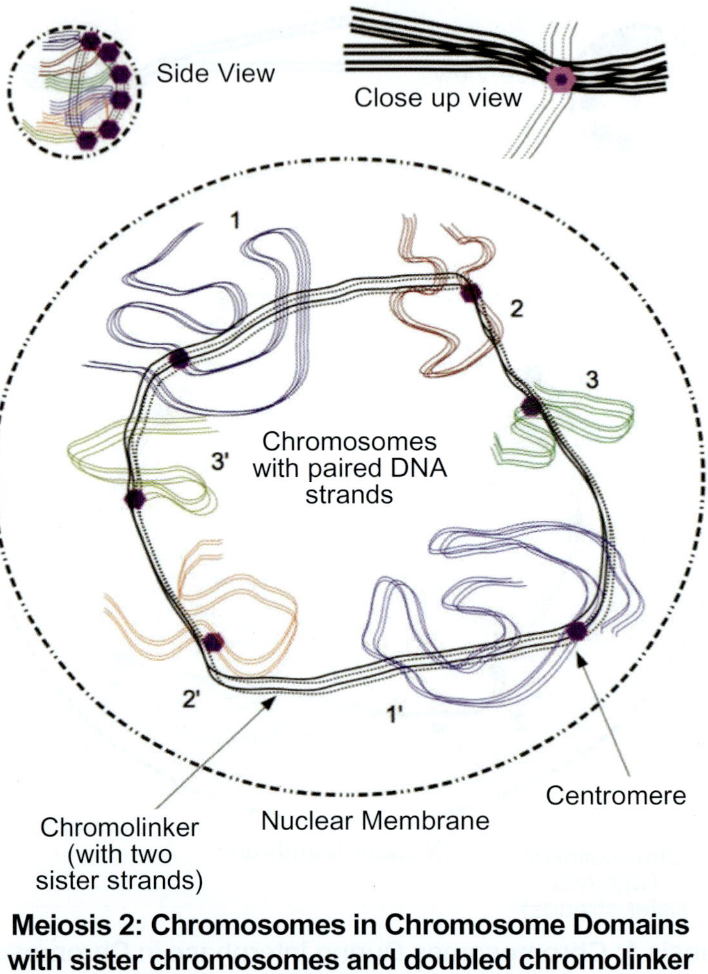

Figure 3.50. Prophase 1: chromosome doubling. At the beginning of prophase 1 each individual chromosome is doubled but still joined at the centromere. The two copies are called sister chromatids as long as they remain joined. Sister chromatids are counted as a single chromosome until they are separated. Once separation occurs (during anaphase, Figure 3.58) these sister chromatids are each called chromosomes. This is important because at the beginning of prophase 1 there are actually four copies of each chromosome. Two copies of the double stranded DNA are on one chromosome because of the joined sister chromatids, and these are identical. The other two are on the other homologous chromosome, also on the joined sister chromatids, and also identical to each other. While the paired sister chromatids are always identical, the sequence of nucleic acids on the homologous pairs may or may not be identical. Putative chromolinkers are represented between chromosomes joining them at the centromere. The chromolinkers are presumed to also be doubled at this stage.

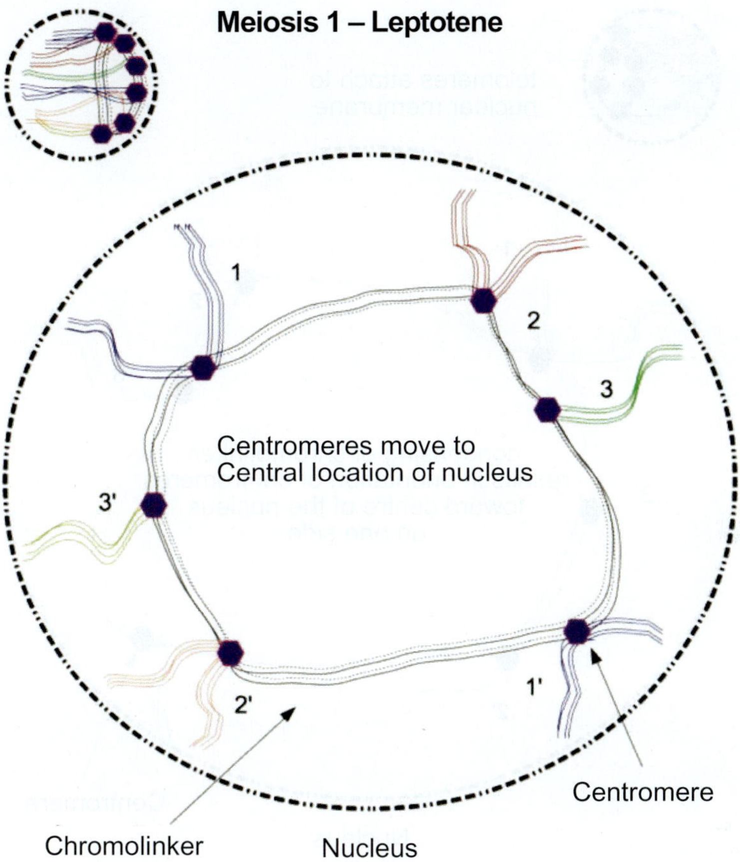

Figure 3.51. Prophase 1: Leptotene. The chromosomes consisting of paired sister chromatids begin to condense and organize their centromeres together on one region of the nucleus. Note how chromosomes 1, 1' are metacentric (the centromere is in the middle of the chromosome). Chromosomes 2 and 2' are submetracentric (the arms are not of equal length). Chromosomes 3 and 3' are acrocentric (one arm is so short as to be invisible). During evolutionary periods whole arms of the chromosomes can be mixed and matched. For example, the domestic horse has one metacentric chromosome that has the same content as two specific acrocentric horse chromosomes in a related wild horse species[310]. Chromosomes themselves are less structurally important than the clusters of genes on their arms. (See Figure 3.45.)

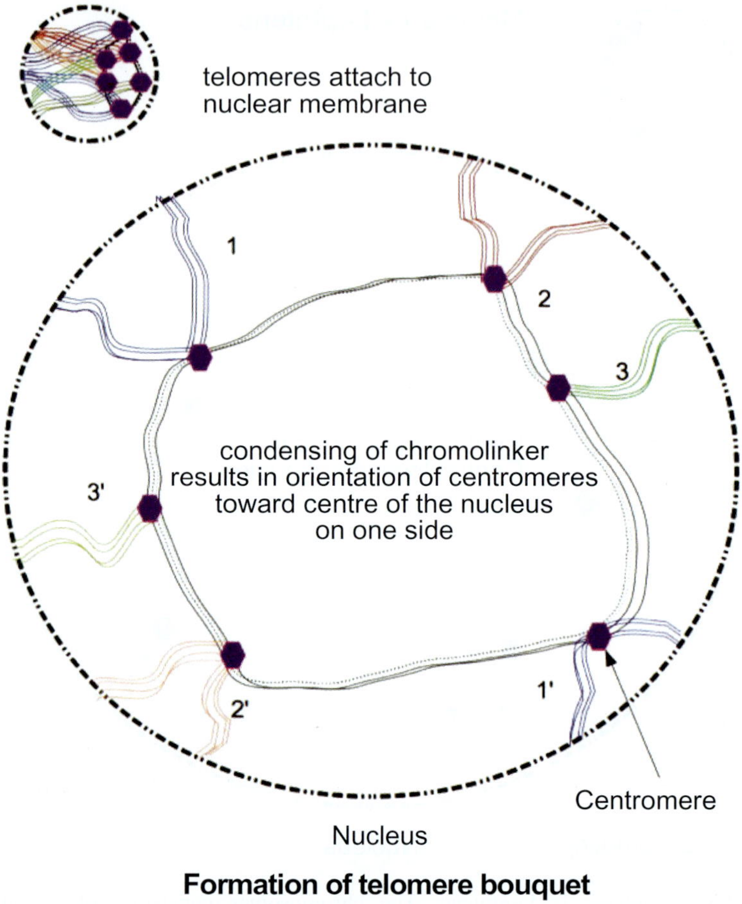

Figure 3.52. Prophase 1: Zygotene. During the leptotene to zygotene phases of meiosis, chromosomes undergo condensation and shortening. One fascinating aspect of meiosis is the formation of the telomere bouquet where all the telomeres attach to the nuclear membrane and cluster together in an organized fashion on one side of the nucleus[311]. The clustering is nonrandom with telomeres of sister chromosomes being closer to each other than to other telomeres. This clustering is likely related to properly lining up homologous pairs because during mitosis there is no telomere bouquet formation. Also not all organisms have telomere bouquets. Rather, organisms with large genomes are more likely to have larger bouquet areas while organisms with smaller genomes will have smaller bouquets or use a similar attachment of telomere to the membrane but the attachment points may be widely spaced for nonhomologous telomere pairs. The general rule is that homologous pairs attach their telomeres close together and sister chromatids do so even more closely. Most, but not all, organisms also cluster their all telomeres together more or less tightly on one side.

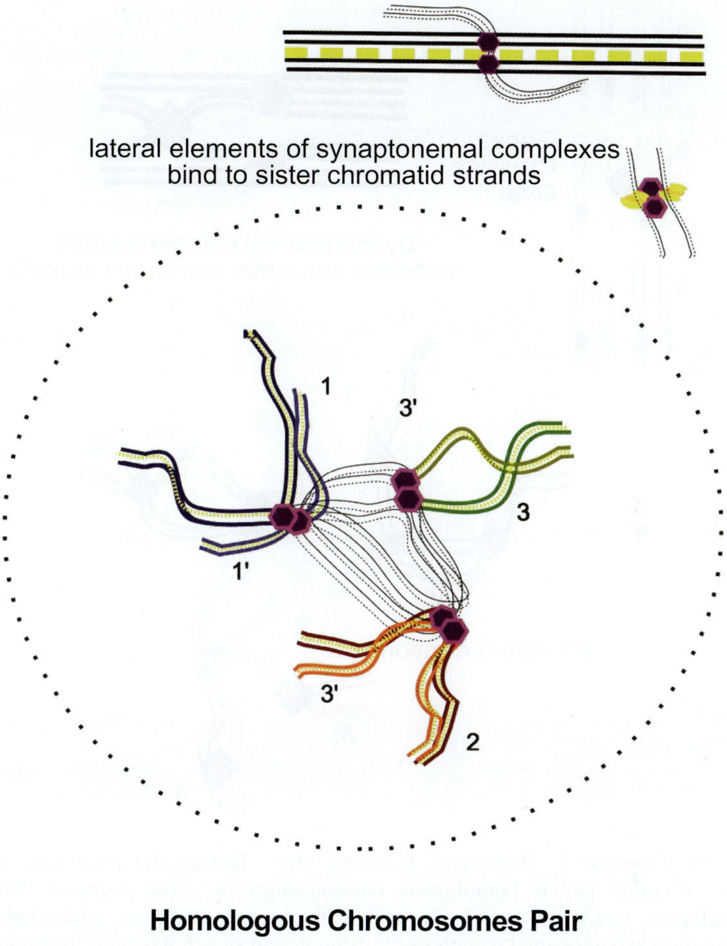

Figure 3.53. Prophase 1: Condensation. Sister chromatids are closely paired by the formation of synaptonemal complexes which are proteins between the sister chromatids (Figure 3.61). Further condensation of the chromosomes and chromolinkers occurs. The most visually interesting feature is that the nuclear membrane has begun to be disassembled. It will be completely disassembled by the end of prophase 1.

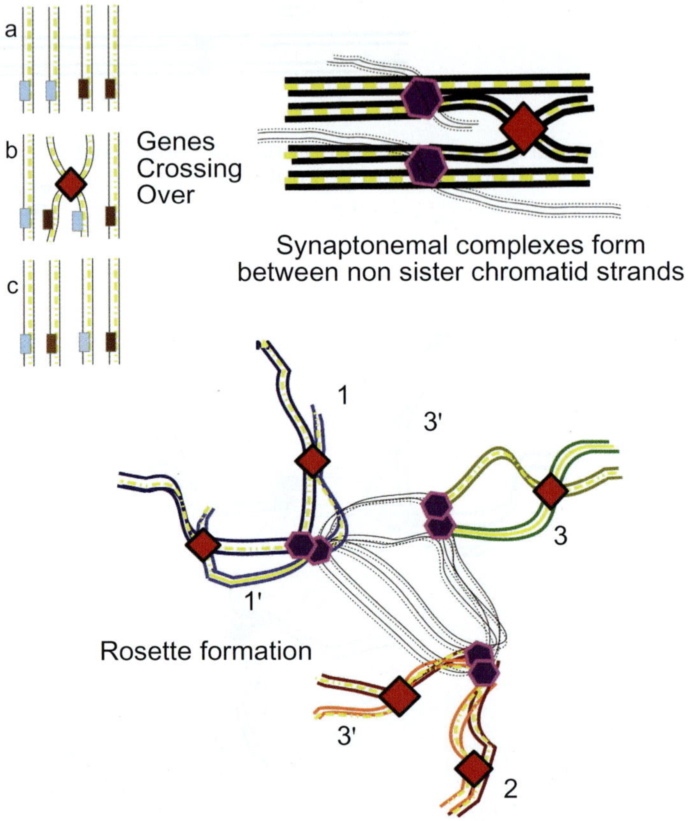

Figure 3.54. Prophase 1: Pachytene, Crossing Over. During the pachytene stage of meiosis individual paired homologous chromosomes lie close together. Non-sister chromatid pairs touch and cross over and at the cross over point, additional protein structures form synaptonemal complexes[312]. Strands can switch which of the homologous chromosome pairs they are on, leading to even more genetic mixing. In this image we are no longer showing the double stranded DNA because it is now tightly compacted and bound up in the proteins forming the synaptonemal complex. However, the double stranded DNA is still there. To illustrate that the double stranded DNA is still present we have also included a figure following a pair of genes through a crossing over. Synaptonemal complexes seem to be important in chromosome organization because at least one will form between each arm of homologous pairs. If synaptonemal complexes do not form because of, for example, a mutation in the gene that produces binding proteins required for their formation, chromosomes will fail to separate properly during later stages at a much higher frequency. If synaptonemal complexes form between nonsister chromatids crossing over can lead to chromosome abnormalities resulting in the loss of entire sections of genes in subsequent cell divisions. These types of errors are known as translocations. Translocations usually result in nonviable embryos or cause severe abnormalities in the offspring. Pachytene ends with diplotene (Figure 3.56).

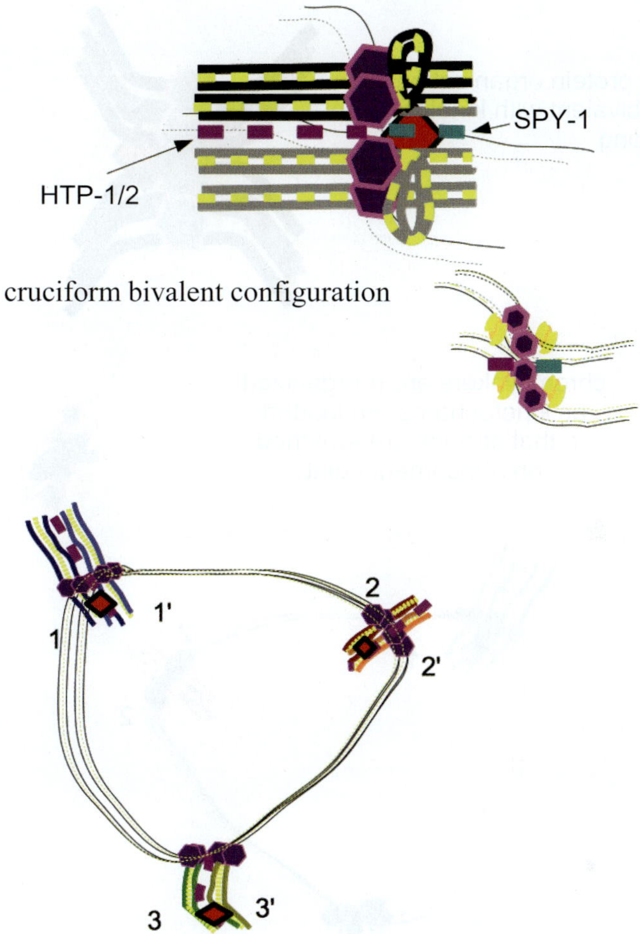

Figure 3.55. Prophase 1: During this last portion of prophase 1, the paired homologous chromosomes become even more condensed. The synaptonemal complexes move towards the centromeres resulting in the cruciform shape. The chromosomes have become so tightly condensed that the cruciform is visible in a high powered light microscope. The proteins binding the sister chromatids together are now joined by more proteins that bind homologous pairs even more tightly. HPT 1/2 and SPY 1 are examples of these proteins. At this later stage a final doubling occurs, this time of the centromeres. We assume, at some point during this process, the chromolinkers will similarly be bound with proteins that keep them organized and paired.

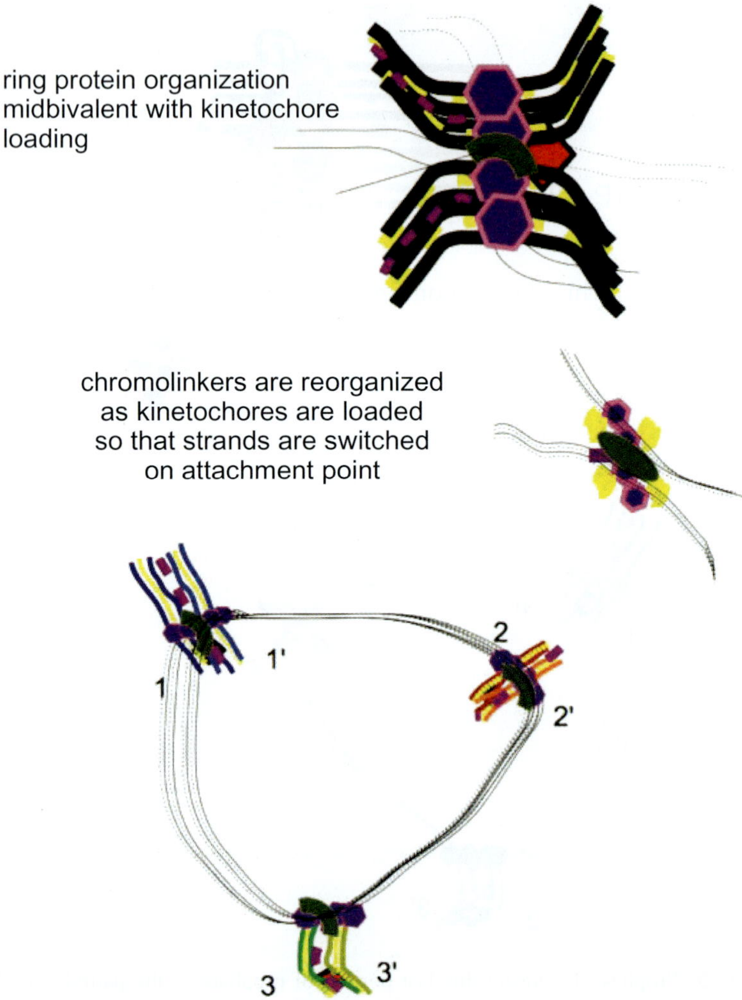

Figure 3.56. Prophase 1: Kinetochore Formation/Diplotene. The last step of prophase is the final bit of packing and reorganization of the paired homologous chromosomes before they are separated. Chromosomes remain attached at the synaptonemal complexes but otherwise begin to repel one another in this phase. A ring of proteins attaches to the homologous pairs of chromosomes. This ring forms the kinetochores, the attachment points for the microtubules that will pull the chromosomes apart. We propose that the chromolinkers, if they exist, are crossed over and rearranged as well during diplotene so that they now form into two matched rings lying side by side.

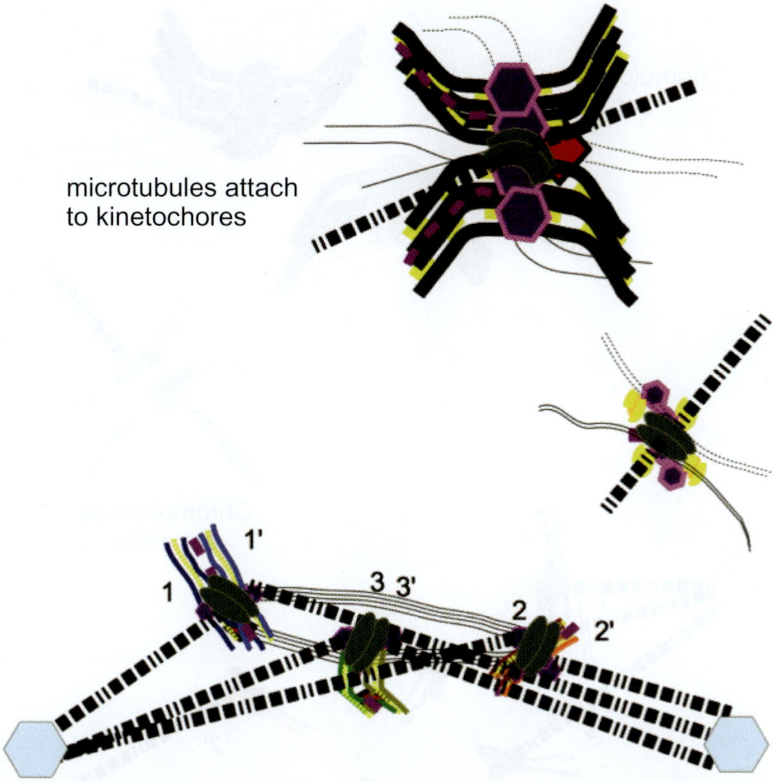

Figure 3.57. Metaphase 1. During metaphase 1 of meiosis the chromosomes are lined up in pairs and microtubules attach. The nuclear membrane has completely vanished. The centrioles in nonplant eukaryotes have moved to opposite sides of the cell and the microtubules connecting the kinetochores and centrioles have formed a spindle apparatus. Plants have anastral meiosis metaphase 1 and mitotic metaphase spindles and use microtubule organizing centers (MTOCs) which are associated with γ-tubulin left at specific points in the cell membrane at each cell division where the cell closed down and divided into two[256]. In plants each spindle organizes at 90 degrees to the previous one.

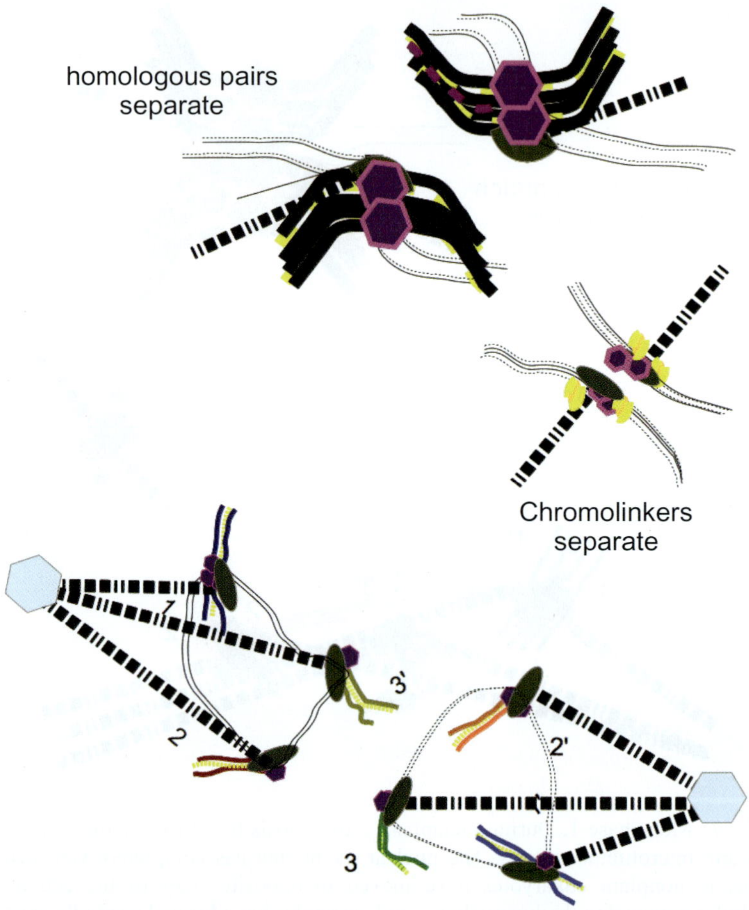

Figure 3.58. Telophase 1. Telophase 1 is when the separation of homologous chromosomes occurs. One of each pair of homologous pairs is pulled by the spindle apparatus towards opposite poles. Which one of the pair moves to which pole appears to be random. Recall that each of the homologous pairs is still doubled with two paired sister chromatids. These sister chromatid pairs still count as only being a single chromosome because they are still joined together. The important part of meiosis, the reduction of chromosome numbers to half (commonly referred to by the shorthand 2N to 1N with N referring to particular species chromosome number), is accomplished in meiosis 1. The final stage of meiosis is anaphase 1 (not shown) where the cell is divided into two.

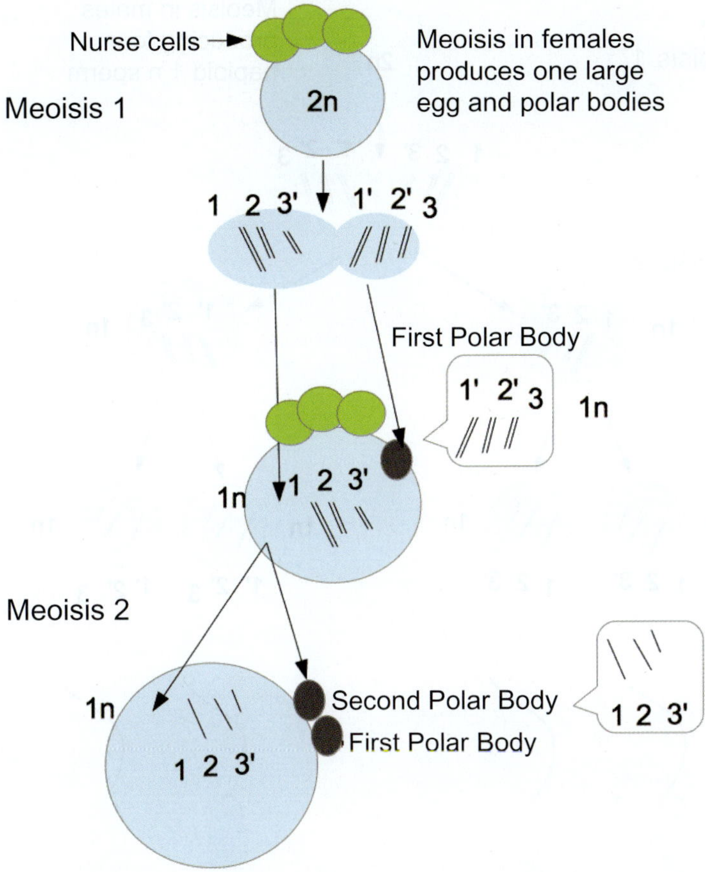

Figure 3.59. In females, anaphase 1 and telophase 1 of meoisis 1 involve division of the 2n germ cell (top, in blue) during which the homologous chromosomes are separated. The labels 1, 1', etc., are used to indicate the randomness of which daughter cell gets which chromosome. The result of the meiosis 1 asymmetric cell division, which includes the stages referred to as prophase 1, metaphase 1, telophase 1 and anaphase 1, is a large 1n cell, the egg (blue), and a small 1n cell called the first polar body (black). At fertilization, the egg then undergoes another asymmetric division called meoisis 2, via prophase 2, metaphase 2, telophase 2 and anaphase 2, producing the second polar body. The sister chromatids are separated in meiosis 2. Developing egg cells can pause during meiosis and wait for a long time. In human females, the cells that will form the eggs will be set aside and begin meiosis while the female is still in utero as a fetus. The eggs pause at prophase 1. It can be decades, before these cells eventually complete meiosis 1. When the egg begins to prepare for fertilization it will extrude the first polar body and then finish meiosis 1 and start meiosis 2 pausing again but this time at metaphase 2, waiting for the arrival of a sperm before dividing and extruding the second polar body.

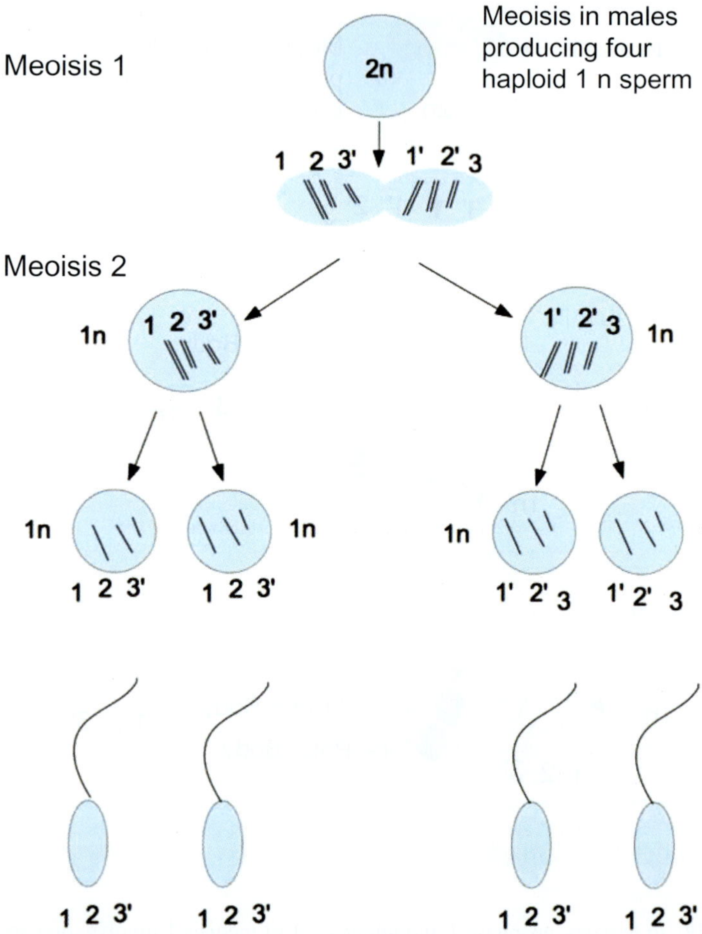

Figure 3.60. In the male germ cell the cells divide evenly during anaphase 1 and anaphase 2, eventually producing four haploid (1N) cells. These cells then undergo further specialization to form four haploid sperm.

not a perfect copy of either. The cells now do a standard mitosis style division. The chromosomes line up, this time single file around a rosette on the anaphase plate and the two sister chromatids of DNA separate, coming apart at the centromere. Each cell that started meiosis will produce 4 haploid cells, each with a unique one quarter subset of DNA due to the crossing over event. If the four subsets are used to make sperm, the result is four sperm (Figure 3.60).

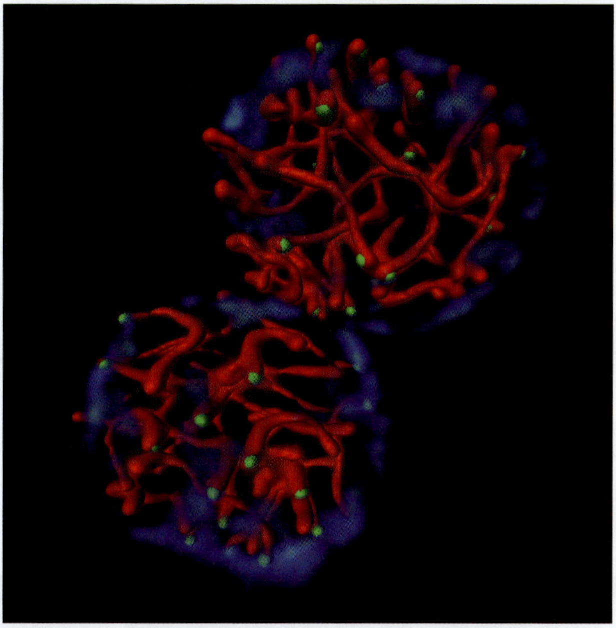

Figure 3.61. This fluorescent light micrograph shows the synaptonemal complexes (in red, surface rendered to smooth the image) of two mouse spermatocytes taken from a 16-18 day old mouse. The centromeres, i.e., the centers of the chromosomes, are shown, smoothed, in green. The rest of each chromosome is unstained and therefore invisible here. The nuclear pores are shown in blue, unsmoothed. The proteins fluorescently stained are SYCP3, CREST and Mab414, respectively. The nuclear pores outline each nucleus, showing us that each synaptonemal complex extends through the whole volume inside its nucleus. Note that some of the centromeres are adjacent to a synaptonemal complex while others are not, so that crossing over, at least at the moment of fixation of these nuclei, is apparently not occurring in all chromosomes. The diameter of these nuclei is 8 μm and each synaptonemal complex is a strand tethered at both ends (where the chromosomes' telomeres are) to the inside of the nuclear membrane at pachytene[313]. Kindly provided by Mark Lessard[314].

Division for eggs is asymmetric. All of the cell material remains with one subset of DNA and the rest of the DNA is discarded in tiny cells resembling yeast buds (Figure 3.62), called polar bodies. This is the most extreme form of asymmetric cell division[315]. Polar bodies cannot reproduce, though their chromosomes can be transplanted to an egg whose nucleus has been removed, which upon fertilization can produce normal offspring[316]. Unmanipulated, they degenerate[317]. In most organisms, in addition to getting most of the cytoplasmic material in the cell division (perhaps 80% of the mitochondria, for instance, by an unsolved mechanism[318]), the single egg gamete is further enlarged by surrounding nurse cells that pass the egg even more cytoplasm[319,320]. This creates one super sized egg cell with one haploid set of DNA. The egg cell is packed full of extra RNAs[321], especially mRNA, as well as additional proteins and nutrients (Figure 3.59). The extra mRNA is either synthesized by the egg itself or is provided by the nurse cells during oogenesis. When the sperm and egg meet, the two single copies of the DNA are combined. The newly created diploid cell now has a full double complement of DNA. The sperm provides one copy of the DNA and some paternal RNA, such as sncRNAs (Figure 3.17) and RNAs that bind to nuclear matrix[322] (Figure 3.63). Some of the sperm DNA is histone-bound, perhaps readied for use early in embryogenesis[322]. The mother provides the other copy of the DNA, and all the other material required for early growth and development.

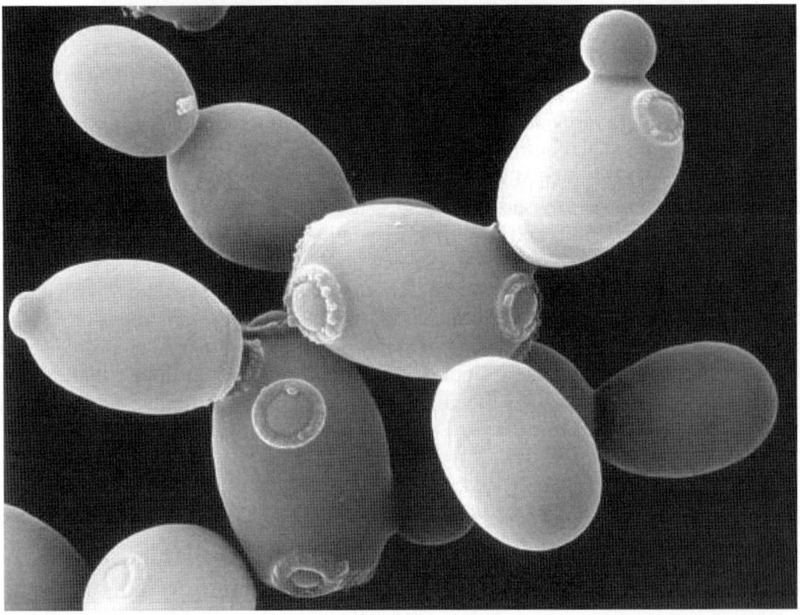

Figure 3.62. The budding yeast *Saccharomyces cerevisiae* is an ovoid 5 to 10 μm long that is used to brew ale[323]. Each budded new cell leaves a scar. Once the surface is covered with scars, it can bud no more. We now know where the advertising slogan "Where there's life… there's Bud"[324] comes from. From Hoffmann (1996)[325], attributed to Hoffmann-La Roche.

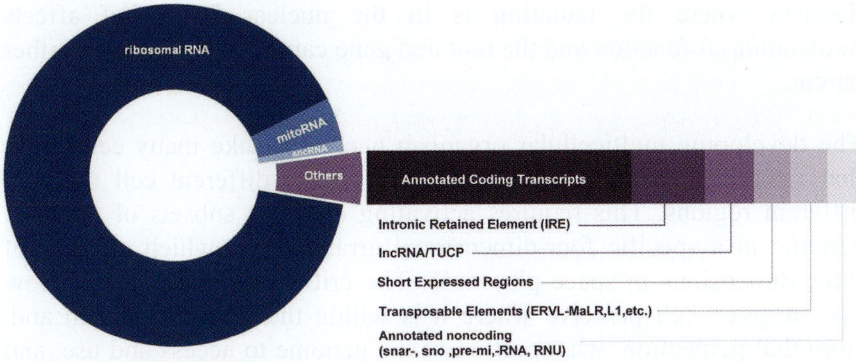

Figure 3.63. In addition to its DNA, a sperm contributes many kinds of RNA when it fertilizes an egg. Their roles during spermatogenesis and early embryogenesis are being worked out. However, note that none are messenger RNA (mRNA). From Jodar *et al.* (2013)[326] with permission of Oxford University Press.

Of particular note is that the mitochondria at fertilization are almost exclusively from the egg. A few mitochondria may come into the egg from the sperm in some species but they are so few in number they usually end up not contributing and vanish instead[327]. In some plants, such as geraniums, both chloroplasts and mitochondria can come from both parents, albeit in different proportions[328]. There is a lively history to the discovery of these "exceptions" to strict maternal inheritance of these organelles[329]. These are exceptions, however. Maternal inheritance is the rule.

Mitochondria and chloroplasts do not have a full prokaryotic genome. Many of the proteins they require to function are made in the eukaryotic nucleus and transported to them. In some cases, such as the intracellular parasitic microsporidia fungi, all of the mitochondrial genes have been transferred to the nucleus[330]. Many ideas for why some genes remain on an organelle's single chromosome have been proposed[331,332]. Large, complex membrane proteins remain on the prokaryotic genome. An analogy may be made with robots, which can be controlled by an onboard computer or by an external computer linked to the robot, or both, sharing separate functions. There are human genetic diseases of the mitochondria that can only be inherited from the mother and are invariably passed on to the offspring. This happens because the mutated gene is on the mitochondrial DNA. Alternatively there are mitochondrial diseases where the mutation is in the nuclear DNA but affects mitochondrial function and the mutated gene can be inherited from either parent.

The developing multicellular organism needs to make many cells from that first fertilized cell. It also needs to create different cell types in different regions. This requires activating different subsets of the total genome in a specific four-dimensional arrangement, which consists of three dimensions in space plus time. The critical question is this. How does a given cell perceive where it is within the whole organism and, from that perception, what portion of the genome to access and use, and what portion of the genome to ignore or suppress? Is it perception, or something else? We are back to our original question. How does an egg become a complex of cells producing the right cell, in the right place, at the right time?

1. Watson, J.D. & F.H. Crick (1953). Molecular structure of nucleic acids; a structure for deoxyribose nucleic acid. *Nature* **171**(4356), 737-738.
2. Ball, M.P. (2013). DNA chemical structure. http://en.wikipedia.org/wiki/DNA - mediaviewer/File:DNA_chemical_structure.svg
3. Wikipedia (2014). DNA. http://en.wikipedia.org/wiki/DNA
4. St. Rosemary Educational Institution. (2013). Cell Cycles: Interphase, Mitosis, Cytokinesis. http://schoolworkhelper.net/cell-cycles-interphase-mitosis-cytokinesis/
5. McClean, P., N. Olson, K. Reindl & A. White (2011). Mitosis: The Movie. http://vcell.ndsu.edu/animations/mitosis/movie-flash.htm
6. Skeen, V. & E.D. Salmon (1993). Mitosis and Cytokinesis in a Newt Lung Cell: DIC microscopy of cell division in a newt lung cell. http://labs.bio.unc.edu/Salmon/moviesmitosis.html
7. Skibbens, R.V., V.P. Skeen & E.D. Salmon (1993). Directional instability of kinetochore motility during chromosome congression and segregation in mitotic newt lung cells: a push-pull mechanism. *Journal of Cell Biology* **122**(4), 859-875.
8. Wikipedia (2013). Mitosis. http://en.wikipedia.org/wiki/Mitosis
9. Righolt, C.H., A. Guffei, H. Knecht, I.T. Young, S. Stallinga, L.J. van Vliet & S. Mai (2014). Differences in nuclear DNA organization between lymphocytes, Hodgkin and Reed-Sternberg cells revealed by structured illumination microscopy. *Journal of Cellular Biochemistry* **115**(8), 1441-1448.
10. Olivares-Chauvet, P., D. Fennessy, D.A. Jackson & A. Maya-Mendoza (2011). Innate structure of DNA foci restricts the mixing of DNA from different chromosome territories. *PLoS One* **6**(12), e27527.

11. Cremer, T. & M. Cremer (2010). Chromosome territories. *Cold Spring Harbor Perspectives in Biology* **2**(3), a003889.
12. Nobelprize.org (2012). DNA makes RNA makes protein. http://www.nobelprize.org/nobel_prizes/medicine/laureates/1993/illpres/dna-rna.html
13. Nobelprize.org (2012). DNA-RNA-Protein: Introduction. http://www.nobelprize.org/educational/medicine/dna/index.html
14. Banner, D.W., A. Bloomer, G.A. Petsko, D.C. Phillips & I.A. Wilson (1976). Atomic coordinates for triose phosphate isomerase from chicken muscle. *Biochemical and Biophysical Research Communications* **72**(1), 146-155.
15. Opabinia regalis (2006). Protein views. http://en.wikipedia.org/wiki/File:Proteinviews-1tim.png
16. Johnson, G. & J. Huntoon (2008). Amino Acid Beads. http://moltoys.blogspot.ca/p/amino-acid-beads.html
17. Villarreal, M.R. (2008). Main protein structures levels. https://commons.wikimedia.org/wiki/File:Main_protein_structure_levels_en.svg
18. Bolzer, A., G. Kreth, I. Solovei, D. Koehler, K. Saracoglu, C. Fauth, S. Muller, R. Eils, C. Cremer, M.R. Speicher & T. Cremer (2005). Three-dimensional maps of all chromosomes in human male fibroblast nuclei and prometaphase rosettes. *PLoS Biology* **3**(5), 826-842.
19. Bornberg-Bauer, E. & M. Mar Alba (2013). Dynamics and adaptive benefits of modular protein evolution. *Current Opinion in Structural Biology* **23**(3), 459-466.
20. Javadi, Y. & L.S. Itzhaki (2013). Tandem-repeat proteins: regularity plus modularity equals design-ability. *Current Opinion in Structural Biology* **23**(4), 622-631.
21. Main, E.R.G., J.J. Phillips & C. Millership (2013). Repeat protein engineering: creating functional nanostructures/biomaterials from modular building blocks. *Biochemical Society Transactions* **41**, 1152-1158.
22. Schultz, J.E. & J. Natarajan (2013). Regulated unfolding: a basic principle of intraprotein signaling in modular proteins. *Trends in Biochemical Sciences* **38**(11), 538-545.
23. Koonin, E.V. (2006). The origin of introns and their role in eukaryogenesis: a compromise solution to the introns-early versus introns-late debate? *Biology Direct* **1**, #22.
24. Iborra, F.J., D.A. Jackson & P.R. Cook (2004). The case for nuclear translation. *Journal of Cell Science* **117**(24), 5713-5720.
25. GeneticsLab (2013). Transcription and Translation. http://geneticslab.wikispaces.com/Transcription+and+Translation
26. Rogozin, I.B., L. Carmel, M. Csuros & E.V. Koonin (2012). Origin and evolution of spliceosomal introns. *Biology Direct* **7**, #11.
27. Ohno, S. (1972). So much "junk" DNA in our genome. *Brookhaven Symposia in Biology* **23**, 366-370.
28. Wikipedia (2014). ENCODE. http://en.wikipedia.org/wiki/ENCODE
29. Ecker, J.R., W.A. Bickmore, I. Barroso, J.K. Pritchard, Y. Gilad & E. Segal (2012). ENCODE explained. *Nature* **489**, 52-55.
30. Maher, B. (2012). ENCODE: the human encyclopaedia. First they sequenced it. Now they have surveyed its hinterlands. But no one knows how much more

information the human genome holds, or when to stop looking for it. *Nature* **489**, 46-48.

31. Doolittle, W.F. (2013). Is junk DNA bunk? A critique of ENCODE. *Proceedings of the National Academy of Sciences USA* **110**(14), 5294-5300.
32. Niu, D.-K. & L. Jiang (2013). Can ENCODE tell us how much junk DNA we carry in our genome? *Biochemical and Biophysical Research Communications* **430**(4), 1340-1343.
33. Eddy, S.R. (2012). The C-value paradox, junk DNA and ENCODE. *Current Biology* **22**(21), R898-R899.
34. Brenner, S., G. Elgar, R. Sandford, A. Macrae, B. Venkatesh & S. Aparicio (1993). Characterization of the pufferfish (*Fugu*) genome as a compact model vertebrate genome. *Nature* **366**(6452), 265-268.
35. Martin, C.C. & R. Gordon (1995). Differentiation trees, a junk DNA molecular clock, and the evolution of neoteny in salamanders. *J. Evol. Biol.* **8**, 339-354.
36. Gregory, T.R. (2005). Animal Genome Size Database. http://www.genomesize.com
37. Hall, S.S. (2012). Hidden Treasures in Junk DNA. What was once known as junk DNA turns out to hold hidden treasures, says computational biologist Ewan Birney. http://www.scientificamerican.com/article.cfm?id=hidden-treasures-in-junk-dna
38. Thomas Jr., C.A. (1971). The genetic organization of chromosomes. *Annual Review of Genetics* **5**(1), 237-256.
39. Wikipedia (2013). C-value enigma. http://en.wikipedia.org/wiki/C-value_enigma
40. Greilhuber, J., T. Borsch, K. Müller, A. Worberg, S. Porembski & W. Barthlott (2006). Smallest angiosperm genomes found in Lentibulariaceae, with chromosomes of bacterial size. *Plant Biology* **8**(6), 770-777.
41. Deaton, A.M. & A. Bird (2011). CpG islands and the regulation of transcription. *Genes & Development* **25**(10), 1010-1022.
42. Gushchanskaya, E.S., A.V. Artemov, S.V. Ulyanov, M.D. Logacheva, A.A. Penin, E.S. Kotova, S.B. Akopov, L.G. Nikolaev, O.V. Iarovaia, E.D. Sverdlov, A.A. Gavrilov & S.V. Razin (2014). The clustering of CpG islands may constitute an important determinant of the 3D organization of interphase chromosomes. *Epigenetics* **9**(7), 951-963.
43. Filichkin, S.A. & T.C. Mockler (2012). Unproductive alternative splicing and nonsense mRNAs: A widespread phenomenon among plant circadian clock genes. *Biology Direct* **7**, #20.
44. Wikipedia (2014). DNA alternative splicing. http://en.wikipedia.org/wiki/File:DNA_alternative_splicing.gif
45. Biamonti, G., S. Bonomi, S. Gallo & C. Ghigna (2012). Making alternative splicing decisions during epithelial-to-mesenchymal transition (EMT). *Cellular and Molecular Life Sciences* **69**(15), 2515-2526.
46. Ellis, J.D., M. Barrios-Rodiles, R. Colak, M. Irimia, T. Kim, J.A. Calarco, X. Wang, Q. Pan, D. O'Hanlon, P.M. Kim, J.L. Wrana & B.J. Blencowe (2012). Tissue-specific alternative splicing remodels protein-protein interaction networks. *Molecular Cell* **46**(6), 884-892.
47. Liu, H., L. He & L. Tang (2012). Alternative splicing regulation and cell lineage differentiation. *Current Stem Cell Research & Therapy* **7**(6), 400-406.

48. Norris, A.D. & J.A. Calarco (2012). Emerging roles of alternative pre-mRNA splicing regulation in neuronal development and function. *Frontiers in Neuroscience* **6**, #122.
49. Moore, A.D., S. Grath, A. Schüler, A.K. Huylmans & E. Bornberg-Bauer (2013). Quantification and functional analysis of modular protein evolution in a dense phylogenetic tree. *Biochimica et Biophysica Acta (BBA) - Proteins and Proteomics* **1834**(5), 898-907.
50. Barbieri, M. (2003). *The Organic Codes. An Introduction to Semantic Biology*. Cambridge, Cambridge University Press.
51. Barbieri, M. (2009). Three types of semiosis. *Biosemiotics* **2**(1), 19-30.
52. Premier Biosoft (2013). Gene Splicing Overview & Techniques. http://www.premierbiosoft.com/tech_notes/gene-splicing.html
53. van der Feltz, C., K. Anthony, A. Brilot & D.A.P. Krummel (2012). Architecture of the spliceosome. *Biochemistry* **51**(16), 3321-3333.
54. Pan, Q., O. Shai, L.J. Lee, J. Frey & B.J. Blencowe (2008). Deep surveying of alternative splicing complexity in the human transcriptome by high-throughput sequencing. *Nature Genetics* **40**(12), 1413-1415.
55. Wikipedia (2014). One gene-one enzyme hypothesis. http://en.wikipedia.org/wiki/One_gene-one_enzyme_hypothesis
56. Fields, C., M.D. Adams, O. White & J.C. Venter (1994). How many genes in the human genome? *Nat. Genet.* **7**(3), 345-346.
57. Dunham, I. (2000). The gene guessing game. *Yeast* **17**(3), 218-224.
58. Venter, J.C., M.D. Adams, E.W. Myers, P.W. Li, R.J. Mural, G.G. Sutton, H.O. Smith, M. Yandell, C.A. Evans, R.A. Holt, J.D. Gocayne, P. Amanatides, R.M. Ballew, D.H. Huson, J.R. Wortman, Q. Zhang, C.D. Kodira, X.H. Zheng, L. Chen, M. Skupski, G. Subramanian, P.D. Thomas, J. Zhang, G.L. Gabor Miklos, C. Nelson, S. Broder, A.G. Clark, J. Nadeau, V.A. McKusick, N. Zinder, A.J. Levine, R.J. Roberts, M. Simon, C. Slayman, M. Hunkapiller, R. Bolanos, A. Delcher, I. Dew, D. Fasulo, M. Flanigan, L. Florea, A. Halpern, S. Hannenhalli, S. Kravitz, S. Levy, C. Mobarry, K. Reinert, K. Remington, J. Abu-Threideh, E. Beasley, K. Biddick, V. Bonazzi, R. Brandon, M. Cargill, I. Chandramouliswaran, R. Charlab, K. Chaturvedi, Z. Deng, V. Di Francesco, P. Dunn, K. Eilbeck, C. Evangelista, A.E. Gabrielian, W. Gan, W. Ge, F. Gong, Z. Gu, P. Guan, T.J. Heiman, M.E. Higgins, R.R. Ji, Z. Ke, K.A. Ketchum, Z. Lai, Y. Lei, Z. Li, J. Li, Y. Liang, X. Lin, F. Lu, G.V. Merkulov, N. Milshina, H.M. Moore, A.K. Naik, V.A. Narayan, B. Neelam, D. Nusskern, D.B. Rusch, S. Salzberg, W. Shao, B. Shue, J. Sun, Z. Wang, A. Wang, X. Wang, J. Wang, M. Wei, R. Wides, C. Xiao, C. Yan, A. Yao, J. Ye, M. Zhan, W. Zhang, H. Zhang, Q. Zhao, L. Zheng, F. Zhong, W. Zhong, S. Zhu, S. Zhao, D. Gilbert, S. Baumhueter, G. Spier, C. Carter, A. Cravchik, T. Woodage, F. Ali, H. An, A. Awe, D. Baldwin, H. Baden, M. Barnstead, I. Barrow, K. Beeson, D. Busam, A. Carver, A. Center, M.L. Cheng, L. Curry, S. Danaher, L. Davenport, R. Desilets, S. Dietz, K. Dodson, L. Doup, S. Ferriera, N. Garg, A. Gluecksmann, B. Hart, J. Haynes, C. Haynes, C. Heiner, S. Hladun, D. Hostin, J. Houck, T. Howland, C. Ibegwam, J. Johnson, F. Kalush, L. Kline, S. Koduru, A. Love, F. Mann, D. May, S. McCawley, T. McIntosh, I. McMullen, M. Moy, L. Moy, B. Murphy, K. Nelson, C. Pfannkoch, E. Pratts, V. Puri, H. Qureshi, M. Reardon, R. Rodriguez, Y.H. Rogers, D. Romblad, B. Ruhfel, R. Scott, C. Sitter, M. Smallwood, E. Stewart, R. Strong, E. Suh, R. Thomas, N.N. Tint, S. Tse, C.

Vech, G. Wang, J. Wetter, S. Williams, M. Williams, S. Windsor, E. Winn-Deen, K. Wolfe, J. Zaveri, K. Zaveri, J.F. Abril, R. Guigo, M.J. Campbell, K.V. Sjolander, B. Karlak, A. Kejariwal, H. Mi, B. Lazareva, T. Hatton, A. Narechania, K. Diemer, A. Muruganujan, N. Guo, S. Sato, V. Bafna, S. Istrail, R. Lippert, R. Schwartz, B. Walenz, S. Yooseph, D. Allen, A. Basu, J. Baxendale, L. Blick, M. Caminha, J. Carnes-Stine, P. Caulk, Y.H. Chiang, M. Coyne, C. Dahlke, A. Mays, M. Dombroski, M. Donnelly, D. Ely, S. Esparham, C. Fosler, H. Gire, S. Glanowski, K. Glasser, A. Glodek, M. Gorokhov, K. Graham, B. Gropman, M. Harris, J. Heil, S. Henderson, J. Hoover, D. Jennings, C. Jordan, J. Jordan, J. Kasha, L. Kagan, C. Kraft, A. Levitsky, M. Lewis, X. Liu, J. Lopez, D. Ma, W. Majoros, J. McDaniel, S. Murphy, M. Newman, T. Nguyen, N. Nguyen, M. Nodell, S. Pan, J. Peck, M. Peterson, W. Rowe, R. Sanders, J. Scott, M. Simpson, T. Smith, A. Sprague, T. Stockwell, R. Turner, E. Venter, M. Wang, M. Wen, D. Wu, M. Wu, A. Xia, A. Zandieh & X. Zhu (2001). The sequence of the human genome. *Science* **291**(5507), 1304-1351.

59. International Human Genome Sequencing Consortium (2004). Finishing the euchromatic sequence of the human genome. *Nature* **431**(7011), 931-945.
60. Wikipedia (2014). Human genome. http://en.wikipedia.org/wiki/Human_genome
61. Xing, L. & G.J. Bassell (2013). mRNA localization: an orchestration of assembly, traffic and synthesis. *Traffic* **14**(1), 2-14.
62. Wikipedia (2015). Ribosome. http://en.wikipedia.org/wiki/Ribosome
63. Gordon, R. (1969). Polyribosome dynamics at steady state. *Journal of Theoretical Biology* **22**(3), 515-532.
64. Chu, D. & T. von der Haar (2012). The architecture of eukaryotic translation. *Nucleic Acids Research* **40**(20), 10098-10106.
65. Goodenbour, J.M. & T. Pan (2006). Diversity of tRNA genes in eukaryotes. *Nucleic Acids Res* **34**(21), 6137-6146.
66. Yikrazuul (2010). Tertiary structure of tRNA. https://en.wikipedia.org/wiki/Transfer_RNA - /media/File:TRNA-Phe_yeast_1ehz.png
67. Bender, R., S.H. Bellman & R. Gordon (1970). ART and the ribosome: a preliminary report on the three-dimensional structure of individual ribosomes determined by an Algebraic Reconstruction Technique. *Journal of Theoretical Biology* **29**, 483-488.
68. Agirrezabala, X. & J. Frank (2010). From DNA to proteins via the ribosome: structural insights into the workings of the translation machinery. *Human Genomics* **4**(4), 226-237.
69. Mishra, S. (2011). "Chargaff's Rules" for Protein Folding: Stoichiometric Leitmotif Made Visible. *Journal of Biomolecular Structure & Dynamics* **28**(4), 649-652.
70. Wikipedia (2014). Protein folding. http://en.wikipedia.org/wiki/Protein_folding
71. Wikipedia (2014). Golgi apparatus. http://en.wikipedia.org/wiki/Golgi_apparatus
72. Tiroli-Cepeda, A.O. & C.H. Ramos (2011). An overview of the role of molecular chaperones in protein homeostasis. *Protein and Peptide Letters* **18**(2), 101-109.

73. Uversky, V.N. (2013). Unusual biophysics of intrinsically disordered proteins. *Biochimica et Biophysica Acta-Proteins and Proteomics* **1834**(5), 932-951.
74. Gordon, R. (1999). *The Hierarchical Genome and Differentiation Waves: Novel Unification of Development, Genetics and Evolution*. Singapore & London, World Scientific & Imperial College Press.
75. Jewett, A.I. & J.E. Shea (2010). Reconciling theories of chaperonin accelerated folding with experimental evidence. *Cellular and Molecular Life Sciences* **67**(2), 255-276.
76. Wikipedia (2014). Amyloid. http://en.wikipedia.org/wiki/Amyloid
77. Levy, E.D. & S. Teichmann (2013). Structural, evolutionary, and assembly principles of protein oligomerization. In: *Oligomerization in Health and Disease*. Eds.: J. Giraldo & F. Ciruela. **117**: 25-51.
78. Hartl, F.U., A. Bracher & M. Hayer-Hartl (2011). Molecular chaperones in protein folding and proteostasis. *Nature* **475**(7356), 324-332.
79. Corrales, F.J. & A.R. Fersht (1996). Toward a mechanism for GroEL-GroES chaperone activity: an ATPase-gated and -pulsed folding and annealing cage. *Proceedings of the National Academy of Sciences USA* **93**(9), 4509-4512.
80. Wikipedia (2006). File:ATP-3D-vdW.png. http://en.wikipedia.org/wiki/File:ATP-3D-vdW.png
81. De Maio, A. (1999). Heat shock proteins: Facts, thoughts, and dreams. *Shock* **11**(1), 1-12.
82. Bond, A.E., P.E. Row & E. Dudley (2011). Post-translation modification of proteins; methodologies and applications in plant sciences. *Phytochemistry* **72**(10), 975-996.
83. Klingström, T. & D. Plewczynski (2011). Protein-protein interaction and pathway databases, a graphical review. *Briefings in Bioinformatics* **12**(6), 702-713.
84. Braun, P. & A.-C. Gingras (2012). History of protein-protein interactions: From egg-white to complex networks. *Proteomics* **12**(10), 1478-1498.
85. Wikipedia (2014). Hemoglobin. http://en.wikipedia.org/wiki/Hemoglobin
86. Nakayama, H., K. Kurokawa & B.L. Lee (2012). Lipoproteins in bacteria: structures and biosynthetic pathways. *FEBS Journal* **279**(23), 4247-4268.
87. Gordon, R. (1968). Steady-state properties of Ising lattice membranes. *Journal of Chemical Physics* **49**, 570-580.
88. Gunasekaran, K., B.Y. Ma & R. Nussinov (2004). Is allostery an intrinsic property of *all* dynamic proteins? *Proteins-Structure Function and Bioinformatics* **57**(3), 433-443.
89. Leygue, E. (2007). Steroid receptor RNA activator (SRA1): unusual bifaceted gene products with suspected relevance to breast cancer. *Nuclear Receptor Signaling* **5**, e006.
90. Smith, T. (2009). Small RNAs Get Smaller. http://finchtalk.blogspot.com/2009/05/small-rnas-get-smaller.html
91. Leygue, E. (2013). "Bi-functional" Steroid Receptor RNA Activator (SRA) … an update [slides]. In: *Symposium on RNA in Physiology and Disease, Winnipeg, June 18, 2013*. Ed.: J. Xie. Winnipeg, University of Manitoba.
92. McClain, K.L., Y.H. Cai, J. Hicks, L.E. Peterson, X.T. Yan, S. Che & S.D. Ginsberg (2005). Expression profiling using human tissues in combination with RNA amplification and microarray analysis: assessment of Langerhans cell histiocytosis. *Amino Acids* **28**(3), 279-290.

93. Wang, R., Y. Xiao & J.K. Taubenberger (2014). Rapid sequencing of influenza A virus vRNA, cRNA and mRNA non-coding regions. *Journal of Virological Methods* **195**, 26-33.
94. Nellimarla, S. & K.L. Mossman (2014). Extracellular dsRNA: its function and mechanism of cellular uptake. *Journal of Interferon and Cytokine Research* **34**(6), 419-426.
95. Sajeesh, S., J.Y. Choe, T.Y. Lee & D.-K. Lee (2015). Guanidine modified polyethyleneimine-g-polyethylene glycol nanocarriers for long interfering RNA (liRNA) based advanced anticancer therapy. *Journal of Materials Chemistry B* **3**(2), 207-216.
96. Nickels, B.E. & S.L. Dove (2011). NanoRNAs: A Class of Small RNAs That Can Prime Transcription Initiation in Bacteria. *Journal of Molecular Biology* **412**(5), 772-781.
97. Sugita, M., Z. Svab, P. Maliga & M. Sugiura (1997). Targeted deletion of *sprA* from the tobacco plastid genome indicates that the encoded small RNA is not essential for pre-16S rRNA maturation in plastids. *Molecular and General Genetics* **257**(1), 23-27.
98. Taft, R.J., P.G. Hawkins, J.S. Mattick & K.V. Morris (2011). The relationship between transcription initiation RNAs and CCCTC-binding factor (CTCF) localization. *Epigenetics & Chromatin* **4**, #13.
99. Collart, C., C.P. Christov, J.C. Smith & T. Krude (2011). The midblastula transition defines the onset of Y RNA-dependent DNA replication in *Xenopus laevis*. *Molecular and Cellular Biology* **31**(18), 3857-3870.
100. Grabow, W.W. & L. Jaeger (2014). RNA self-assembly and RNA nanotechnology. *Accounts of Chemical Research* **47**(6), 1871-1880.
101. Ishikawa, J., H. Furuta & Y. Ikawa (2013). RNA Tectonics (tectoRNA) for RNA nanostructure design and its application in synthetic biology. *Wiley Interdisciplinary Reviews-RNA* **4**(6), 651-664.
102. Wang, Z., M. Gerstein & M. Snyder (2009). RNA-Seq: a revolutionary tool for transcriptomics. *Nature Reviews Genetics* **10**(1), 57-63.
103. Marguerat, S. & J. Bähler (2010). RNA-seq: from technology to biology. *Cellular and Molecular Life Sciences* **67**(4), 569-579.
104. Wikipedia (2015). Non-coding RNA. https://en.wikipedia.org/wiki/Non-coding_RNA
105. Wikipedia (2015). Central dogma of molecular biology. https://en.wikipedia.org/wiki/Central_dogma_of_molecular_biology
106. Woodson, S.A. (2010). Taming free energy landscapes with RNA chaperones. *RNA Biology* **7**(6), 677-686.
107. Drum, R.W. & R. Gordon (2003). Star Trek replicators and diatom nanotechnology. *Trends in Biotechnology* **21**(8), 325-328.
108. Wikipedia (2013). Feast of the Gods (art): Banquet of the Gods - Frans Floris. http://en.wikipedia.org/wiki/Feast_of_the_Gods_%28art%29
109. Dinger, M.E., T.R. Mercer & J.S. Mattick (2008). RNAs as extracellular signaling molecules. *Journal of Molecular Endocrinology* **40**(4), 151-159.
110. Spiegelman, Z., G. Golan & S. Wolf (2013). Don't kill the messenger: Long-distance trafficking of mRNA molecules. *Plant Science* **213**, 1-8.
111. Santillán, M. & M.C. Mackey (2008). Quantitative approaches to the study of bistability in the lac operon of *Escherichia coli*. *Journal of the Royal Society Interface* **5**, S29-S39.

112. Wikipedia (2014). *Escherichia coli.* http://en.wikipedia.org/wiki/Escherichia_coli
113. Wikipedia (2014). *lac* operon. http://en.wikipedia.org/wiki/Lac_operon
114. Wikipedia (2014). Cyclisches Adenosinmonophosphat. http://de.wikipedia.org/wiki/Cyclisches_Adenosinmonophosphat
115. Wikipedia (2013). RNA polymerase RNA polymerase
116. Fraser, D. & M. Kærn (2009). A chance at survival: gene expression noise and phenotypic diversification strategies. *Molecular Microbiology* **71**(6), 1333-1340.
117. Viñuelas, J., G. Kaneko, A. Coulon, G. Beslon & O. Gandrillon (2012). Towards experimental manipulation of stochasticity in gene expression. *Progress in Biophysics and Molecular Biology* **110**(1), 44-53.
118. Liang, L.W., R. Hussein, D.H.S. Block & H.N. Lim (2013). Minimal effect of gene clustering on expression in *Escherichia coli. Genetics* **193**(2), 453-465.
119. Wikipedia (2014). Cyclic adenosine monophosphate. http://en.wikipedia.org/wiki/Cyclic_AMP
120. Villa, E., A. Balaeff & K. Schulten (2005). Structural dynamics of the *lac* repressor-DNA complex revealed by a multiscale simulation. *Proceedings of the National Academy of Sciences USA* **102**(19), 6783-6788.
121. Villa, E., A. Balaeff & K. Schulten (2007). Gene Lock. http://www.ks.uiuc.edu/Highlights/?section=2005&highlight=2005-05
122. Friedman, N. (2013). Studying gene expression at the level of the single cell. *Briefings in Functional Genomics* **12**(2), 73-74.
123. Wikipedia (2013). Ribonuclease. http://en.wikipedia.org/wiki/RNase
124. Kelso-Winemiller, L.C. & M.M. Winkler (1991). 'Unmasking' of stored maternal mRNAs and the activation of protein synthesis at fertilization in sea urchins. *Development* **111**(2), 623-633.
125. Cai, L., N. Friedman & X.S. Xie (2006). Stochastic protein expression in individual cells at the single molecule level. *Nature* **440**(7082), 358-362.
126. Haselkorn, R. (1992). Developmentally regulated gene rearrangements in prokaryotes. *Annual Review of Genetics* **26**, 113-130.
127. Kumar, K., R.A. Mella-Herrera & J.W. Golden (2010). Cyanobacterial heterocysts. *Cold Spring Harbor Perspectives in Biology* **2**(4), a000315.
128. Wolk, C.P. (1965). Heterocyst germination under defined conditions. *Nature* **205**(4967), 201-202.
129. Wilcox, M., Mitchiso.Gj & R.J. Smith (1973). Pattern formation in the blue-green-alga *Anabaena*. II. Controlled proheterocyst regression. *Journal of Cell Science* **13**(3), 637-649.
130. Puck, T.T., P. Webb & R. Johnson (1998). Genome exposure and regulation in mammalian cells. *Somatic Cell and Molecular Genetics* **24**(5), 291-301.
131. Patterson, D. (2009). *Theodore Thomas Puck 1916-2005, A Biographical Memoir*. Washington, DC, USA, National Academy of Sciences.
132. Gordon, R. (2011). Cosmic Embryo #1: My Erdös Number Is 2*i*. http://www.science20.com/cosmic_embryo/cosmic_embryo_1_my_erd%C3%B6s_number_2i
133. Wikipedia (2015). Reference genes. https://en.wikipedia.org/wiki/Reference_genes
134. Eisenberg, E. & E.Y. Levanon (2013). Human housekeeping genes, revisited [Corrigendum: **30**(3), 119-120 (2014)]. *Trends in Genetics* **29**(10), 569-574.

135. Schaefer, M.H., J.S. Yang, L. Serrano & C. Kiel (2014). Protein conservation and variation suggest mechanisms of cell type-specific modulation of signaling pathways. *PLoS Computational Biology* **10**(6), e1003659.
136. Zhu, J., F. He, S. Hu & J. Yu (2008). On the nature of human housekeeping genes. *Trends in Genetics* **24**(10), 481-484.
137. Wikipedia (2014). Interactome. http://en.wikipedia.org/wiki/Interactome
138. Bossi, A. & B. Lehner (2009). Tissue specificity and the human protein interaction network. *Molecular Systems Biology* **5**, #260.
139. Illingworth, R.S. & A.P. Bird (2009). CpG islands - 'A rough guide'. *FEBS Letters* **583**(11), 1713-1720.
140. Irizarry, R.A., C. Ladd-Acosta, B. Wen, Z. Wu, C. Montano, P. Onyango, H. Cui, K. Gabo, M. Rongione, M. Webster, H. Ji, J.B. Potash, S. Sabunciyan & A.P. Feinberg (2009). The human colon cancer methylome shows similar hypo- and hypermethylation at conserved tissue-specific CpG island shores. *Nature Genetics* **41**(2), 178-186.
141. Akashi, H. & A. Eyre-Walker (1998). Translational selection and molecular evolution. *Current Opinion in Genetics & Development* **8**(6), 688-693.
142. Ma, L.N., P. Cui, J. Zhu, Z.H. Zhang & Z. Zhang (2014). Translational selection in human: more pronounced in housekeeping genes. *Biology Direct* **9**, #17.
143. Rao, Y.S., Z.F. Wang, X.W. Chai, G.Z. Wu, M. Zhou, Q.H. Nie & X.Q. Zhang (2010). Selection for the compactness of highly expressed genes in *Gallus gallus*. *Biology Direct* **5**, #35.
144. Choi, S.S. & S. Hannenhalli (2013). Three independent determinants of protein evolutionary rate. *Journal of Molecular Evolution* **76**(3), 98-111.
145. Nigon, V. & J. Godet (1977). Genetic-analysis of cell differentiation: Possible extension of the hemoglobin differon model to *Drosophila* morphogenesis and immunoglobulin determination. *Journal of Theoretical Biology* **64**(1), 97-111.
146. Liu, X., X. Yu, D.J. Zack, H. Zhu & J. Qian (2008). TiGER: A database for tissue-specific gene expression and regulation. *BMC Bioinformatics* **9**, #271.
147. Tang, F., K. Lao & M.A. Surani (2011). Development and applications of single-cell transcriptome analysis. *Nature Methods* **8**(4), S6-S11.
148. Gerstein, M.B., J. Rozowsky, K.-K. Yan, D. Wang, C. Cheng, J.B. Brown, C.A. Davis, L. Hillier, C. Sisu, J.J. Li, B. Pei, A.O. Harmanci, M.O. Duff, S. Djebali, R.P. Alexander, B.H. Alver, R. Auerbach, K. Bell, P.J. Bickel, M.E. Boeck, N.P. Boley, B.W. Booth, L. Cherbas, P. Cherbas, C. Di, A. Dobin, J. Drenkow, B. Ewing, G. Fang, M. Fastuca, E.A. Feingold, A. Frankish, G. Gao, P.J. Good, R. Guigo, A. Hammonds, J. Harrow, R.A. Hoskins, C. Howald, L. Hu, H. Huang, T.J.P. Hubbard, C. Huynh, S. Jha, D. Kasper, M. Kato, T.C. Kaufman, R.R. Kitchen, E. Ladewig, J. Lagarde, E. Lai, J. Leng, Z. Lu, M. MacCoss, G. May, R. McWhirter, G. Merrihew, D.M. Miller, A. Mortazavi, R. Murad, B. Oliver, S. Olson, P.J. Park, M.J. Pazin, N. Perrimon, D. Pervouchine, V. Reinke, A. Reymond, G. Robinson, A. Samsonova, G.I. Saunders, F. Schlesinger, A. Sethi, F.J. Slack, W.C. Spencer, M.H. Stoiber, P. Strasbourger, A. Tanzer, O.A. Thompson, K.H. Wan, G. Wang, H. Wang, K.L. Watkins, J. Wen, K. Wen, C. Xue, L. Yang, K. Yip, C. Zaleski, Y. Zhang, H. Zheng, S.E. Brenner, B.R. Graveley, S.E. Celniker, T.R. Gingeras & R. Waterston (2014). Comparative analysis of the transcriptome across distant species. *Nature* **512**(7515), 445-448.

149. Round, F.E. (1972). The problem of reduction of cell size during diatom cell division. *Nova Hedwigia* **23**, 291-303.
150. Jewson, D.H. (1992). Size reduction, reproductive strategy and the life cycle of a centric diatom. *Philosophical Transactions of the Royal Society B-Biological Sciences* **336**, 191-213.
151. Gordon, R. (2010). Diatoms and nanotechnology: early history and imagined future as seen through patents. In: *The Diatoms: Applications for the Environmental and Earth Sciences*. Eds.: J.P. Smol & E.F. Stoermer. Cambridge, Cambridge University Press. **2nd**: 585-602.
152. Dominguez, F., A. Pellicer & C. Simón (2009). The human embryo proteome. *Reproductive Sciences* **16**(2), 188-190.
153. Arnold, G.J. & T. Frohlich (2011). Dynamic proteome signatures in gametes, embryos and their maternal environment. *Reproduction Fertility and Development* **23**(1), 81-93.
154. Wikipedia (2014). Set theory. http://en.wikipedia.org/wiki/Set_theory
155. Wikipedia (2014). Housekeeping gene. http://en.wikipedia.org/wiki/Housekeeping_gene - Common_housekeeping_genes_in_human
156. HGNC (2014). Tubulins. http://www.genenames.org/genefamilies/TUB
157. Nielsen, M.G., S.R. Gadagkar & L. Gutzwiller (2010). Tubulin evolution in insects: gene duplication and subfunctionalization provide specialized isoforms in a functionally constrained gene family. *BMC Evolutionary Biology* **10**, #113.
158. Raynaud-Messina, B., A. Debec, Y. Tollon, M. Gares & M. Wright (2001). Differential properties of the two *Drosophila* γ-tubulin isotypes. *European Journal of Cell Biology* **80**(10), 643-649.
159. Yekezare, M., B. Gómez-González & J.F.X. Diffley (2013). Controlling DNA replication origins in response to DNA damage - inhibit globally, activate locally. *Journal of Cell Science* **126**(6), 1297-1306.
160. Kolesnikova, T.D. (2013). Regulation of DNA replication timing. *Molecular Biology* **47**(1), 12-33.
161. Spies, M. (2013). There and back again: new single-molecule insights in the motion of DNA repair proteins. *Current Opinion in Structural Biology* **23**(1), 154-160.
162. Drake, J.W., B. Charlesworth, D. Charlesworth & J.F. Crow (1998). Rates of spontaneous mutation. *Genetics* **148**(4), 1667-1686.
163. Clune, J., D. Misevic, C. Ofria, R.E. Lenski, S.F. Elena & R. Sanjuan (2008). Natural selection fails to optimize mutation rates for long-term adaptation on rugged fitness landscapes. *PLoS Computational Biology* **4**(9), e1000187.
164. Tyson, J.J., B. Novák, K. Chen & J. Val (1995). Checkpoints in the cell cycle from a modeler's perspective. *Progress in Cell Cycle Research* **1**, 1-8.
165. Yasutis, K.M. & K.G. Kozminski (2013). Cell cycle checkpoint regulators reach a zillion. *Cell Cycle* **12**(10), 1501-1509.
166. Arampatzi, P., M. Gialitakis, T. Makatounakis & J. Papamatheakis (2013). Gene-specific factors determine mitotic expression and bookmarking via alternate regulatory elements. *Nucleic Acids Research* **41**(4), 2202-2215.
167. Yi, C. & C. He (2013). DNA repair by reversal of DNA damage. *Cold Spring Harbor Perspectives in Biology* **5**(1), a012575.
168. Peng, A. (2013). Working hard for recovery: mitotic kinases in the DNA damage checkpoint. *Cell and Bioscience* **3**, #20.

169. Crest, J., N. Oxnard, J.Y. Ji & G. Schubiger (2007). Onset of the DNA replication checkpoint in the early *Drosophila* embryo. *Genetics* **175**(2), 567-584.

170. Blythe, S.A. & E.F. Wieschaus (2015). Zygotic genome activation triggers the DNA replication checkpoint at the midblastula transition. *Cell* **160**(6), 1169-1181.

171. Kadauke, S. & G.A. Blobel (2013). Mitotic bookmarking by transcription factors. *Epigenetics & Chromatin* **6**, #6.

172. Kimura, H. (2013). Histone modifications for human epigenome analysis. *Journal of Human Genetics* **58**(7), 439-445.

173. Delcuve, G.P., M. Rastegar & J.R. Davie (2009). Epigenetic control. *Journal of Cellular Physiology* **219**(2), 243-250.

174. Zaidi, S.K., D.W. Young, M.A. Montecino, J.B. Lian, A.J. van Wijnen, J.L. Stein & G.S. Stein (2010). Mitotic bookmarking of genes: a novel dimension to epigenetic control. *Nature Reviews Genetics* **11**(8), 583-589.

175. Kühn, S. (2014). *Organic Codes and their Identication: Is the Histone Code a True Organic Code? [Master's Thesis]*. Matieland, South Africa, Department of Biochemistry, University of Stellenbosch.

176. Devaiah, B.N. & D.S. Singer (2013). Two faces of brd4: mitotic bookmark and transcriptional lynchpin. *Transcription* **4**(1), 13-17.

177. Naumova, N., M. Imakaev, G. Fudenberg, Y. Zhan, B.R. Lajoie, L.A. Mirny & J. Dekker (2013). Organization of the mitotic chromosome. *Science* **342**(6161), 948-953.

178. Uhlmann, F. (2013). Open questions: Chromosome condensation - Why does a chromosome look like a chromosome? *BMC Biology* **11**, #9.

179. Hansen, J.C. (2012). Human mitotic chromosome structure: what happened to the 30-nm fibre? *EMBO Journal* **31**(7), 1621-1623.

180. Joti, Y., T. Hikima, Y. Nishino, F. Kamada, S. Hihara, H. Takata, T. Ishikawa & K. Maeshima (2012). Chromosomes without a 30-nm chromatin fiber. *Nucleus* **3**(5), 404-410.

181. Nishino, Y., M. Eltsov, Y. Joti, K. Ito, H. Takata, Y. Takahashi, S. Hihara, A.S. Frangakis, N. Imamoto, T. Ishikawa & K. Maeshima (2012). Human mitotic chromosomes consist predominantly of irregularly folded nucleosome fibres without a 30-nm chromatin structure. *EMBO Journal* **31**(7), 1644-1653.

182. Halverson, J.D., J. Smrek, K. Kremer & A.Y. Grosberg (2014). From a melt of rings to chromosome territories: the role of topological constraints in genome folding. *Reports on Progress in Physics* **77**(2), #022601.

183. Xiao, B.T., B.S. Freedman, K.E. Miller, R. Heald & J.F. Marko (2012). Histone H1 compacts DNA under force and during chromatin assembly. *Molecular Biology of the Cell* **23**(24), 4864-4871.

184. Füllgrabe, J., E. Kavanagh & B. Joseph (2011). Histone onco-modifications. *Oncogene* **30**(31), 3391-3403.

185. Sng, J. (2013). Welcome to Neuroepigenetics Laboratory: Gene Regulation By Chromatin Modifying Enzymes. http://www.neuroepigenetics.com/snglab/Our_work.html [cobweb page]

186. Fussner, E., M. Strauss, U. Djuric, R. Li, K. Ahmed, M. Hart, J. Ellis & D.P. Bazett-Jones (2012). Open and closed domains in the mouse genome are configured as 10-nm chromatin fibres. *EMBO Reports* **13**(11), 992-996.

187. Vogel, B., A. Löschberger, M. Sauer & R. Hock (2011). Cross-linking of DNA through HMGA1 suggests a DNA scaffold. *Nucleic Acids Research* **39**(16), 7124-7133.
188. Phillips-Cremins, J.E. (2014). Unraveling architecture of the pluripotent genome. *Current Opinion in Cell Biology* **28**, 96-104.
189. Woodcock, C.L.F., L.L.Y. Frado & J.B. Rattner (1984). The higher-order structure of chromatin: Evidence for a helical ribbon arrangement. *Journal of Cell Biology* **99**(1), 42-52.
190. Song, F., P. Chen, D.P. Sun, M.Z. Wang, L.P. Dong, D. Liang, R.M. Xu, P. Zhu & G.H. Li (2014). Cryo-EM study of the chromatin fiber reveals a double helix twisted by tetranucleosomal units. *Science* **344**(6182), 376-380.
191. Ausió, J. (2015). The shades of gray of the chromatin fiber: Recent literature provides new insights into the structure of chromatin. *BioEssays* **37**(1), 46-51.
192. Hagstrom, K.A. & B.J. Meyer (2003). Condensin and cohesin: More than chromosome compactor and glue. *Nature Reviews Genetics* **4**(7), 520-534.
193. Montgomery Jr., T.H. (1904). The main facts in regard to the cellular basis of heredity. *Proceedings of the American Philosophical Society* **43**(175), 5-14.
194. Cleland, R.E. (1926). Meiosis in the pollen mother cells of *Oenothera biennis* and *Oenothera biennis sulfurea*. *Genetics* **11**(2), 127-162.
195. Cleland, R.E. (1922). The reduction divisions in the pollen mother cells of *Oenothera franciscana*. *American Journal of Botany* **9**(7), 391-413.
196. Cleland, R.E. (1923). Chromosome arrangements during meiosis in certain *Oenothera*s. *American Naturalist* **57**, 562-566.
197. Cremer, T. & C. Cremer (2006). Rise, fall and resurrection of chromosome territories: a historical perspective. Part I. The rise of chromosome territories. *European Journal of Histochemistry* **50**(3), 161-176.
198. Cremer, T. & C. Cremer (2006). Rise, fall and resurrection of chromosome territories: a historical perspective Part II. Fall and resurrection of chromosome territories during the 1950S to 1980s. Part III. Chromosome territories and the functional nuclear architecture: experiments and models from the 1990s to the present. *European Journal of Histochemistry* **50**(4), 223-272.
199. Thanbichler, M., S.C. Wang & L. Shapiro (2005). The bacterial nucleoid: A highly organized and dynamic structure. *Journal of Cellular Biochemistry* **96**(3), 506-521.
200. Benza, V.G., B. Bassetti, K.D. Dorfman, V.F. Scolari, K. Bromek, P. Cicuta & M.C. Lagomarsino (2012). Physical descriptions of the bacterial nucleoid at large scales, and their biological implications. *Reports on Progress in Physics* **75**(7), #076602.
201. McClendon, J.F. (1907). Experiments on the eggs of *Chaetopterus* and *Asterias* in which the chromatin was removed. *Wood's Hole Massachusetts Marine Biology Laboratory Bulletin* **12**, 141-145.
202. Myhra, S. & A. Brøgger (1975). Interchromosomal DNA-containing fibres in human cells. *Humangenetik* **29**(3), 183-189.
203. Maniotis, A.J., K. Bojanowski & D.E. Ingber (1997). Mechanical continuity and reversible chromosome disassembly within intact genomes removed from living cells. *Journal of Cellular Biochemistry* **65**(1), 114-130.
204. Maniotis, A. (2011). Packaging DNA. In: *Chimeras and Consciousness: Evolution of the Sensory Self.* Eds.: L. Margulis, C.A. Asikainen & W.E. Krumbein, MIT Press: 167-172.

205. Nagele, R.G., T. Freeman, J. Fazekas, K.M. Lee, Z. Thomson & H.Y. Lee (1998). Chromosome spatial order in human cells: evidence for early origin and faithful propagation. *Chromosoma* **107**(5), 330-338.
206. Nagele, R., T. Freeman, L. McMorrow & H.Y. Lee (1995). Precise spatial positioning of chromosomes during prometaphase: evidence for chromosomal order. *Science* **270**(5243), 1831-1835.
207. Nagele, R.G., T. Freeman, L. McMorrow, Z. Thomson, K. Kitson-Wind & H.Y. Lee (1999). Chromosomes exhibit preferential positioning in nuclei of quiescent human cells. *Journal of Cell Science* **112**(Pt 4), 525-535.
208. Zalenskaya, I.A. & A.O. Zalensky (2004). Non-random positioning of chromosomes in human sperm nuclei. *Chromosome Research* **12**(2), 163-173.
209. Meile, J.-C., R. Mercier, M. Stouf, C. Pages, J.-Y. Bouet & F. Cornet (2011). The terminal region of the *E. coli* chromosome localises at the periphery of the nucleoid. *BMC Microbiology* **11**, #28.
210. Fritsche, M., S. Li, D.W. Heermann & P.A. Wiggins (2012). A model for *Escherichia coli* chromosome packaging supports transcription factor-induced DNA domain formation. *Nucleic Acids Research* **40**(3), 972-980.
211. Treangen, T.J. & S.L. Salzberg (2012). Repetitive DNA and next-generation sequencing: computational challenges and solutions. *Nature Reviews Genetics* **13**(1), 36-46.
212. Bendich, A.J. (2007). The size and form of chromosomes are constant in the nucleus, but highly variable in bacteria, mitochondria and chloroplasts. *BioEssays* **29**(5), 474-483.
213. Bendich, A.J. & K. Drlica (2000). Prokaryotic and eukaryotic chromosomes: what's the difference? *BioEssays* **22**(5), 481-486.
214. Rand, D.M. (2009). 'Why genomes in pieces?' revisited: Sucking lice do their own thing in mtDNA circle game. *Genome Research* **19**(5), 700-702.
215. Junier, I., F. Boccard & O. Espéli (2014). Polymer modeling of the *E. coli* genome reveals the involvement of locus positioning and macrodomain structuring for the control of chromosome conformation and segregation. *Nucleic Acids Research* **42**(3), 1461-1473.
216. Glanz, J. (1997). Force-carrying web pervades living cell. *Science* **276**(5313), 678-679.
217. Tangle Creations (2014). Promotional Tangles. http://www.tanglecreations.com/collections/promotional-tangles
218. Weng, X.L. & J. Xiao (2014). Spatial organization of transcription in bacterial cells. *Trends in Genetics* **30**(7), 287-297.
219. Le, T.B.K., M.V. Imakaev, L.A. Mirny & M.T. Laub (2013). High-resolution mapping of the spatial organization of a bacterial chromosome. *Science* **342**(6159), 731-734.
220. Wikipedia (2014). Kinetochore. http://en.wikipedia.org/wiki/Kinetochore
221. Afunguy (2006). File:Kinetochore.jpg. http://en.wikipedia.org/wiki/File:Kinetochore.jpg
222. Nannas, N.J. & A.W. Murray (2014). Tethering sister centromeres to each other suggests the spindle checkpoint detects stretch within the kinetochore. *PLoS Genetics* **10**(8), e1004492.
223. McIntosh, J.R., E. O'Toole, K. Zhudenkov, M. Morphew, C. Schwartz, F.I. Ataullakhanov & E.L. Grishchuk (2013). Conserved and divergent features of

kinetochores and spindle microtubule ends from five species. *Journal of Cell Biology* **200**(4), 459-474.

224. Nicodemi, M. & A. Pombo (2014). Models of chromosome structure. *Current Opinion in Cell Biology* **28**, 90-95.
225. Wikipedia (2013). Reptation. https://en.wikipedia.org/wiki/Reptation
226. Socher, S.H. & D. Davidson (1970). Heterogeneity in G_2 duration during lateral root development. *Chromosoma* **31**(4), 478-484.
227. Halverson, J.D., W.B. Lee, G.S. Grest, A.Y. Grosberg & K. Kremer (2011). Molecular dynamics simulation study of nonconcatenated ring polymers in a melt. I. Statics. *Journal of Chemical Physics* **134**(20), #204904.
228. Nagano, T., Y. Lubling, T.J. Stevens, S. Schoenfelder, E. Yaffe, W. Dean, E.D. Laue, A. Tanay & P. Fraser (2013). Single-cell Hi-C reveals cell-to-cell variability in chromosome structure. *Nature* **502**(7469), 59-64.
229. Aranda-Anzaldo, A., M.A.R. Dent & A. Martinez-Gómez (2014). The higher-order structure in the cells nucleus as the structural basis of the post-mitotic state. *Progress in Biophysics & Molecular Biology* **114**(3), 137-145.
230. Cotter, L., T.D. Allen, E. Kiseleva & M.W. Goldberg (2007). Nuclear membrane disassembly and rupture. *Journal of Molecular Biology* **369**(3), 683-695.
231. De Souza, C.P. & S.A. Osmani (2007). Mitosis, not just open or closed. *Eukaryotic Cell* **6**(9), 1521-1527.
232. Tucker, J.B., J. Beisson, D.L. Roche & J. Cohen (1980). Microtubules and control of macronuclear 'amitosis' in *Paramecium*. *Journal of Cell Science* **44**, 135-151.
233. Raikov, I.B. (1994). The diversity of forms of mitosis in protozoa - a comparative review. *European Journal of Protistology* **30**(3), 253-269.
234. Cole, E. & T. Sugai (2012). Developmental progression of *Tetrahymena* through the cell cycle and conjugation. *Methods in Cell Biology* **109**, 177-236.
235. Sagan, L. (1967). On the origin of mitosing cells. *Journal of Theoretical Biology* **14**(3), 255-274.
236. Heath, I.B. (1980). Variant mitoses in lower eukaryotes - indicators of the evolution of mitosis. *International Review of Cytology-A Survey of Cell Biology* **64**, 1-80.
237. Wise, D. (1988). The diversity of mitosis: the value of evolutionary experiments. *Biochemistry and Cell Biology-Biochimie et Biologie Cellulaire* **66**(6), 515-529.
238. Dolan, M.F., H. Melnitsky, L. Margulis & R. Kolnicki (2002). Motility proteins and the origin of the nucleus. *Anatomical Record* **268**(3), 290-301.
239. Biggins, S. (2013). The composition, functions, and regulation of the budding yeast kinetochore. *Genetics* **194**(4), 817-846.
240. Mignot, J.P. (1996). The centrosomal big bang: From a unique central organelle towards a constellation of MTOCs. *Biology of the Cell* **86**(2-3), 81-91.
241. Brown, R.C. & B.E. Lemmon (2007). The pleiomorphic plant MTOC: An evolutionary perspective. *Journal of Integrative Plant Biology* **49**(8), 1142-1153.
242. Sluder, G., F.J. Miller & C.L. Rieder (1989). Reproductive capacity of sea urchin centrosomes without centrioles. *Cell Motility and the Cytoskeleton* **13**(4), 264-273.

243. Loncarek, J. & A. Khodjakov (2009). Ab ovo or de novo? Mechanisms of centriole duplication. *Molecules and Cells* **27**(2), 135-142.
244. Guichard, P., D. Chretien, S. Marco & A.-M. Tassin (2010). Procentriole assembly revealed by cryo-electron tomography. *EMBO Journal* **29**(9), 1565-1572.
245. Hodges, M.E., N. Scheumann, B. Wickstead, J.A. Langdale & K. Gull (2010). Reconstructing the evolutionary history of the centriole from protein components. *Journal of Cell Science* **123**(Pt 9), 1407-1413.
246. Carvalho-Santos, Z., J. Azimzadeh, J.B. Pereira-Leal & M. Bettencourt-Dias (2011). Evolution: Tracing the origins of centrioles, cilia, and flagella. *Journal of Cell Biology* **194**(2), 165-175.
247. Tsou, M.F.B., C. Fong, W.J. Wang, M. Kim & R.K. Soni (2011). Dissecting the molecular difference between MTOC-competent and non-competent centrioles. *Molecular Biology of the Cell* **22**(Abstracts), #124.
248. Pihan, G.A. (2013). Centrosome dysfunction contributes to chromosome instability, chromoanagenesis, and genome reprograming in cancer. *Frontiers in Oncology* **3**, #277.
249. Firat-Karalar, E.N. & T. Stearns (2014). The centriole duplication cycle. *Philosophical Transactions of the Royal Society B-Biological Sciences* **369**(1650), #20130460.
250. Mennella, V., D.A. Agard, B. Huang & L. Pelletier (2014). Amorphous no more: subdiffraction view of the pericentriolar material architecture. *Trends in Cell Biology* **24**(3), 188-197.
251. MBInfo & P.T.P. Thao (2012). Structure of the centrosome. http://www.mechanobio.info/figure/1384243918921.jpg.html
252. Zheng, Y.Y., J.J. Guo, X. Li, Y.B. Xie, M.M. Hou, X.Y. Fu, S. Dai, Y.Y. Miao & J. Ren (2014). An integrated overview of spatiotemporal organization and regulation in mitosis in terms of the proteins in the functional supercomplexes. *Frontiers in Microbiology* **5**, #573.
253. Winey, M. & E. Noole (2014). Centriole structure. *Philosophical Transactions of the Royal Society B-Biological Sciences* **369**(1650).
254. Alliegro, M.C., M.A. Alliegro & R.E. Palazzo (2006). Centrosome-associated RNA in surf clam oocytes. *Proceedings of the National Academy of Sciences USA* **103**(24), 9034-9038.
255. Stearns, T. (2015). Centrioles, in absentia. *Science* **348**(6239), 1091-1092.
256. Brown, R.C. & B.E. Lemmon (2011). Dividing without centrioles: innovative plant microtubule organizing centres organize mitotic spindles in bryophytes, the earliest extant lineages of land plants. *AoB Plants* **2011**, plr028.
257. Shimamura, M., R.C. Brown, B.E. Lemmon, T. Akashi, K. Mizuno, N. Nishihara, K.I. Tomizawa, K. Yoshimoto, H. Deguchi, H. Hosoya, T. Horio & Y. Mineyuki (2004). γ-tubulin in basal land plants: characterization, localization, and implication in the evolution of acentriolar microtubule organizing centers. *Plant Cell* **16**(1), 45-59.
258. Kalinina, I., A. Nandi, P. Delivani, M.R. Chacón, A.H. Klemm, D. Ramunno-Johnson, A. Krull, B. Lindner, N. Pavin & I.M. Tolić-Nørrelykke (2013). Pivoting of microtubules around the spindle pole accelerates kinetochore capture. *Nature Cell Biology* **15**(1), 82-87.

259. Maiato, H., C.L. Rieder & A. Khodjakov (2004). Kinetochore-driven formation of kinetochore fibers contributes to spindle assembly during animal mitosis. *The Journal of Cell Biology* **167**(5), 831-840.

260. Pavin, N. & I.M. Tolić-Nørrelykke (2014). Swinging a sword: how microtubules search for their targets. *Systems and Synthetic Biology* **8**(3), 179-186.

261. Zhao, Y. & Q. Zhan (2012). Electric fields generated by synchronized oscillations of microtubules, centrosomes and chromosomes regulate the dynamics of mitosis and meiosis. *Theoretical Biology & Medical Modelling* **9**(1), #26.

262. Havelka, D., O. Kučera, M.A. Deriu & M. Cifra (2014). Electro-acoustic behavior of the mitotic spindle: A semi-classical coarse-grained model. *PLoS One* **9**(1), e86501.

263. Oliveira, R.A. & K. Nasmyth (2010). Getting through anaphase: splitting the sisters and beyond. *Biochemical Society Transactions* **38**(6), 1639-1644.

264. Gregan, J., S. Polakova, L. Zhang, I.M. Tolić-Nørrelykke & D. Cimini (2011). Merotelic kinetochore attachment: causes and effects. *Trends in Cell Biology* **21**(6), 374-381.

265. Fridman, V., A. Gerson-Gurwitz, N. Movshovich, M. Kupiec & L. Gheber (2009). Midzone organization restricts interpolar microtubule plus-end dynamics during spindle elongation. *EMBO Reports* **10**(4), 387-393.

266. Stephens, A.D., R.A. Haggerty, P.A. Vasquez, L. Vicci, C.E. Snider, F. Shi, C. Quammen, C. Mullins, J. Haase, R.M. Taylor, J.S. Verdaasdonk, M.R. Falvo, Y. Jin, M.G. Forest & K. Bloom (2013). Pericentric chromatin loops function as a nonlinear spring in mitotic force balance. *Journal of Cell Biology* **200**(6), 757-772.

267. Almonacid, M., M.-E. Terret & M.-H. Verlhac (2014). Actin-based spindle positioning: new insights from female gametes. *Journal of Cell Science* **127**(3), 477-483.

268. Morin, X. & Y. Bellaiche (2011). Mitotic spindle orientation in asymmetric and symmetric cell divisions during animal development. *Developmental Cell* **21**(1), 102-119.

269. Lu, M.S. & C.A. Johnston (2013). Molecular pathways regulating mitotic spindle orientation in animal cells. *Development* **140**(9), 1843-1856.

270. French, E.L. (2012). What are the axes on a ball joint? , http://bbs.homeshopmachinist.net/threads/53106-What-are-the-axes-on-a-ball-joint?s=f05ebe24f8af412b75e5bb211b244404

271. Peyre, E. & X. Morin (2012). An oblique view on the role of spindle orientation in vertebrate neurogenesis. *Development Growth & Differentiation* **54**(3), 287-305.

272. Kojo, K.H., T. Higaki, N. Kutsuna, Y. Yoshida, H. Yasuhara & S. Hasezawa (2013). Roles of cortical actin microfilament patterning in division plane orientation in plants. *Plant and Cell Physiology* **54**(9), 1491-1503.

273. Baumgärtner, S. & I.M. Tolić (2014). Astral microtubule pivoting promotes their search for cortical anchor sites during mitosis in budding yeast. *PLoS One* **9**(4), e93781.

274. Bisgrove, S.R., D.C. Henderson & D.L. Kropf (2003). Asymmetric division in fucoid zygotes is positioned by telophase nuclei. *Plant Cell* **15**(4), 854-862.

275. Higaki, T., N. Kutsuna, T. Sano & S. Hasezawa (2008). Quantitative analysis of changes in actin microfilament contribution to cell plate development in plant cytokinesis. *BMC Plant Biology* **8**, #80.
276. Güttinger, S., E. Laurell & U. Kutay (2009). Orchestrating nuclear envelope disassembly and reassembly during mitosis. *Nature Reviews Molecular Cell Biology* **10**(3), 178-191.
277. Zhang, D. & S. Oliferenko (2013). Remodeling the nuclear membrane during closed mitosis. *Current Opinion in Cell Biology* **25**(1), 142-148.
278. Dekker, J. & L. Mirny (2013). Biological techniques: Chromosomes captured one by one. *Nature* **502**(7469), 45-46.
279. Junker, J.P. & A. van Oudenaarden (2014). Every cell is special: Genome-wide studies add a new dimension to single-cell biology. *Cell* **157**(1), 8-11.
280. Wood, A.M., A.G. Garza-Gongora & S.T. Kosak (2014). A crowdsourced nucleus: Understanding nuclear organization in terms of dynamically networked protein function. *Biochimica et Biophysica Acta-Gene Regulatory Mechanisms* **1839**(3), 178-190.
281. Bickmore, W.A. & B. van Steensel (2013). Genome architecture: domain organization of interphase chromosomes. *Cell* **152**(6), 1270-1284.
282. Skloot, R. (2010). *The Immortal Life of Henrietta Lacks*, Crown Publishers New York.
283. Ganai, N., S. Sengupta & G.I. Menon (2014). Chromosome positioning from activity-based segregation. *Nucleic Acids Research* **42**(7), 4145-4159.
284. Belmont, A.S. (2014). Large-scale chromatin organization: the good, the surprising, and the still perplexing. *Current Opinion in Cell Biology* **26**, 69-78.
285. Chi, J.X., L. Huang, W. Nie, J. Wang, B. Su & F. Yang (2005). Defining the orientation of the tandem fusions that occurred during the evolution of Indian muntjac chromosomes by BAC mapping. *Chromosoma* **114**(3), 167-172.
286. Nyanatusita (2006). Indian Muntjac, female, in Udawattakele, Sri Lanka. http://en.wikipedia.org/wiki/File:Indian_Muntjac,_Barking_Deer_2.jpg
287. Margoz (2008). *Muntiacus reevesi* in Zoo-Botanical Garden in Toruń. http://en.wikipedia.org/wiki/File:Muntiacus_reevesi_Toru%C5%84_.jpg
288. Wikipedia (2014). Stem cell. http://en.wikipedia.org/wiki/Stem_cell
289. Gurdon, J.B. (2013). The egg and the nucleus: A battle for supremacy (Nobel lecture). *Angewandte Chemie-International Edition* **52**(52), 13890-13899.
290. Wilmut, I. (2013). A tribute to Keith Campbell: The birth of the first clone of an adult vertebrate, "Dolly" the sheep. *Cellular Reprogramming* **15**(5), 339-343.
291. Condic, M.L. (2014). Totipotency: What it is and what it is not. *Stem Cells and Development* **23**(8), 796-812.
292. Boveri, T. (1889). Fertilization and segmentation in *Ascaris megalocephala*. *Journal of the Royal Microscopical Society*, 220-223.
293. Hall, B.K. & W.M. Olson, Eds. (2003). *Keywords and Concepts in Evolutionary Developmental Biology*. Cambridge, Harvard University Press.
294. Riley, J.L. & L.A. Katz (2001). Widespread distribution of extensive chromosomal fragmentation in ciliates. *Molecular Biology and Evolution* **18**(7), 1372-1377.
295. Tartar, V. (1961). *The Biology of Stentor*. Oxford, Pergamon Press.
296. Frankel, J. (1989). *Pattern Formation, Ciliate Studies and Models*. New York, Oxford University Press.

297. Grimes, G.W. & K.J. Aufderheide (1991). *Cellular Aspects of Pattern Formation: The Problem of Assembly*. Basel, Karger.

298. Knop, M. (2011). Yeast cell morphology and sexual reproduction - A short overview and some considerations. *Comptes Rendus Biologies* **334**(8-9), 599-606.

299. Friedman, W.E. (2013). One genome, two ontogenies. *Science* **339**(6123), 1045-1046.

300. Nakamura, Y., N. Sasaki, M. Kobayashi, N. Ojima, M. Yasuike, Y. Shigenobu, M. Satomi, Y. Fukuma, K. Shiwaku, A. Tsujimoto, T. Kobayashi, I. Nakayama, F. Ito, K. Nakajima, M. Sano, T. Wada, S. Kuhara, K. Inouye, T. Gojobori & K. Ikeo (2013). The first symbiont-free genome sequence of marine red alga, susabi-nori (*Pyropia yezoensis*). *PLoS One* **8**(3), e57122.

301. Mikami, K. (2014). Structural divergence and loss of phosphoinositide-specific phospholipase C signaling components during the evolution of the green plant lineage: implications from structural characteristics of algal components. *Frontiers in Plant Science* **5**, doi: 10.3389/fpls.2014.00380.

302. Graves, J.S. & J. Gutknecht (1976). Ion transport studies and determination of the cell wall elastic modulus in the marine alga *Halicystis parvula*. *The Journal of General Physiology* **67**(5), 579-597.

303. Handel, M.A. & J.C. Schimenti (2010). Genetics of mammalian meiosis: regulation, dynamics and impact on fertility. *Nature Reviews Genetics* **11**(2), 124-136.

304. Klein, F., P. Mahr, M. Galova, S.B. Buonomo, C. Michaelis, K. Nairz & K. Nasmyth (1999). A central role for cohesins in sister chromatid cohesion, formation of axial elements, and recombination during yeast meiosis. *Cell* **98**(1), 91-103.

305. Schvarzstein, M., S.M. Wignall & A.M. Villeneuve (2010). Coordinating cohesion, co-orientation, and congression during meiosis: lessons from holocentric chromosomes. *Genes & Development* **24**(3), 219-228.

306. Brunet, S. & M.H. Verlhac (2011). Positioning to get out of meiosis: the asymmetry of division. *Human Reproduction Update* **17**(1), 68-75.

307. Lee, K.-Y., T. Davies & M. Mishima (2012). Cytokinesis microtubule organisers at a glance. *Journal of Cell Science* **125**(15), 3495-3500.

308. Sheehan, M.J. & W.P. Pawlowski (2012). Imaging chromosome dynamics in meiosis in plants. *Methods in Enzymology* **505**, 125-143.

309. Tiang, C.-L., Y. He & W.P. Pawlowski (2012). Chromosome organization and dynamics during interphase, mitosis, and meiosis in plants. *Plant Physiology* **158**(1), 26-34.

310. Myka, J.L., T.L. Lear, M.L. Houck, O.A. Ryder & E. Bailey (2003). FISH analysis comparing genome organization in the domestic horse (*Equus caballus*) to that of the Mongolian wild horse (*E. przewalskii*). *Cytogenetic and Genome Research* **102**(1-4), 222-225.

311. Richards, D.M., E. Greer, A.C. Martin, G. Moore, P.J. Shaw & M. Howard (2012). Quantitative dynamics of telomere bouquet formation. *PLoS Computational Biology* **8**(12), e1002812.

312. Hirose, Y., R. Suzuki, T. Ohba, Y. Hinohara, H. Matsuhara, M. Yoshida, Y. Itabashi, H. Murakami & A. Yamamoto (2011). Chiasmata promote monopolar attachment of sister chromatids and their co-segregation toward the proper pole during meiosis I. *PLoS Genetics* **7**(3), e1001329.

313. Fritsche, M., L.G. Reinholdt, M. Lessard, M.A. Handel, J. Bewersdorf & D.W. Heermann (2012). The impact of entropy on the spatial organization of synaptonemal complexes within the cell nucleus. *PLoS One* **7**(5), e36282.
314. Lessard, M. (2014). *Synaptonemal complex [personal communication]*. Bar Harbor, Maine, USA, The Jackson Laboratory.
315. Maro, B. & M.H. Verlhac (2002). Polar body formation: new rules for asymmetric divisions. *Nature Cell Biology* **4**(12), E281-E283.
316. Wakayama, T. & R. Yanagimachi (1998). The first polar body can be used for the production of normal offspring in mice. *Biology of Reproduction* **59**(1), 100-104.
317. Miao, Y.L., S.F. Ma, X.Y. Liu, D.Q. Miao, Z.L. Chang, M.J. Luo & J.H. Tan (2004). Fate of the first polar bodies in mouse oocytes. *Molecular Reproduction and Development* **69**(1), 66-76.
318. Dalton, C.M. & J. Carroll (2013). Biased inheritance of mitochondria during asymmetric cell division in the mouse oocyte. *Journal of Cell Science* **126**(13), 2955-2964.
319. Jaglarz, M.K., M. Zelazowska & S.M. Bilinski (2000). Alternative ways to guarantee transfer of cytoplasm from nurse cells into the oocyte. *Acta Biologica Cracoviensia Series Botanica* **42**(Suppl. 1), 44-44.
320. Alexandrova, O., M. Schade, A. Böttger & C.N. David (2005). Oogenesis in *Hydra*: Nurse cells transfer cytoplasm directly to the growing oocyte. *Developmental Biology* **281**(1), 91-101.
321. Mische, S., M. Li, M. Serr & T.S. Hays (2007). Direct observation of regulated ribonucleoprotein transport across the nurse cell/oocyte boundary. *Molecular Biology of the Cell* **18**(6), 2254-2263.
322. Johnson, G.D., C. Lalancette, A.K. Linnemann, F. Leduc, G. Boissonneault & S.A. Krawetz (2011). The sperm nucleus: chromatin, RNA, and the nuclear matrix. *Reproduction* **141**(1), 21-36.
323. Wikipedia (2014). Yeast. http://en.wikipedia.org/wiki/Yeast
324. Quitney, J. (2013). Budweiser Beer Commercial "Where There's Life, There's Bud" circa 1956 Anheuser Busch. https://www.youtube.com/watch?v=NgicCqbIvCU
325. Hoffmann, N. (1996). Bäckerhefe - ein lebendes Reagens für die organisch-chemische Synthese. *Chemie in unserer Zeit* **30**(4), 201-213.
326. Jodar, M., S. Selvaraju, E. Sendler, M.P. Diamond & S.A. Krawetz (2013). The presence, role and clinical use of spermatozoal RNAs. *Human Reproduction Update* **19**(6), 604-624.
327. Luo, S.-M., H. Schatten & Q.-Y. Sun (2013). Sperm mitochondria in reproduction: Good or bad and where do they go? *Journal of Genetics and Genomics* **40**(11), 549-556.
328. Nagata, N., Sodmergen, C. Saito, A. Sakai, H. Kuroiwa & T. Kuroiwa (1997). Preferential degradation of plastid DNA with preservation of mitochondrial DNA in the sperm cells of *Pelargonium zonale* during pollen development. *Protoplasma* **197**(3-4), 217-229.
329. Lane, N. (2005). *Power, Sex, Suicide: Mitochondria and the Meaning of Lfe*, Oxford University Press, UK.
330. Bullerwell, C.E. & B.F. Lang (2005). Fungal evolution: the case of the vanishing mitochondrion. *Current Opinion in Microbiology* **8**(4), 362-369.

331. Allen, J.F. (2003). Why chloroplasts and mitochondria contain genomes. *Comparative and Functional Genomics* **4**(1), 31-36.

332. Kleine, T., U.G. Maier & D. Leister (2009). DNA transfer from organelles to the nucleus: the idiosyncratic genetics of endosymbiosis. *Annual Review of Plant Biology* **60**, 115-138.

333. Zehfus, M.H. (1997). Identification of compact, hydrophobically stabilized domains and modules containing multiple peptide chains. *Protein Science* **6**(6), 1210-1219.

334. Koide, S. (2009). Generation of new protein functions by nonhomologous combinations and rearrangements of domains and modules. *Current Opinion in Biotechnology* **20**(4), 398-404.

335. Porter, L.L. & G.D. Rose (2012). A thermodynamic definition of protein domains. *Proceedings of the National Academy of Sciences of the United States of America* **109**(24), 9420-9425.

Chapter 4

Epigenetics: Higher Order Gene Control

We know that there is a hierarchy of genes. We know that some genes are able to regulate and control many other genes. In genetics such genes are referred to as "master genes". If a specific master gene is turned on, it makes some of the genes below it within that hierarchy exposed (available for transcription) in a given cell. At the same time it will make other genes below it in the hierarchy sequestered (not available for transcription) in that same cell. If we imagine the genome is organized into multiple subsets, each subset controlled by a master gene, or key master genes, organized into regulons and differons, then a possible explanation of higher order organization becomes clearer. When the hierarchy is disrupted normal development can be affected.

There can be many errors that occur during mitosis and meiosis. These errors can cause sections of the genome to be rearranged (Figure 4.1). For example, during mitosis a chromosome can loop back over on itself and have one section copied backwards to the usual arrangement (called an inversion[1]). Sometimes sections of DNA can be dropped entirely (deletions[2]) or moved to another chromosome (translocations[3,4]). Whole sections of chromosomes may be duplicated[5]. "Transposable elements" are segments of DNA with complementary ends that can move from one place to another (Figure 4.2), the so-called "jumping genes" which can inactivate other genes by jumping into them (or regions involved in regulating them) and thereby produce patterns such as the multicolored pattern of Indian corn[6,7].

During meiosis 1, a chromosome can fail to separate from its partner and instead both chromosomes of the pair move to the same side. When two gametes meet, and one is carrying an extra chromosome, the resulting embryo has an extra chromosome. Conversely it is also possible to have embryos with missing chromosomes. Two non-identical chromosomes

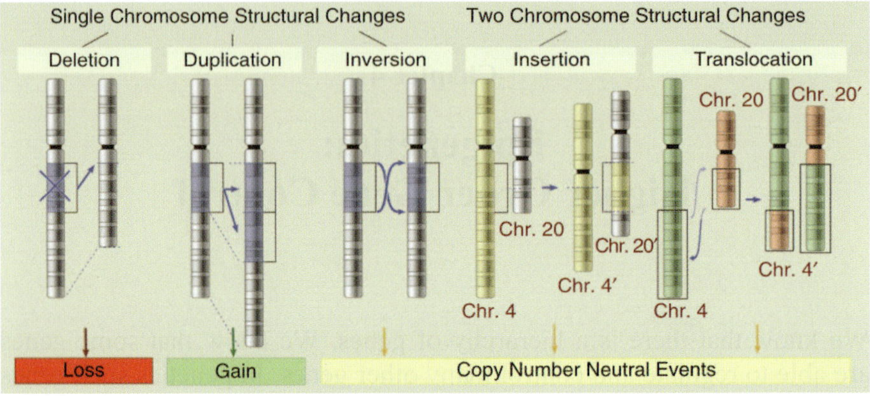

Figure 4.1. Examples of rearrangements that can occur within and between chromosomes. The "copy number" is the number of copies of an individual gene or stretch of DNA. This can be a loss, a gain, or no change ("neutral event"). In the case of a gain (duplication) the two copies can subsequently evolve independently. This is important in the evolution of differentiation trees. The length of DNA involved can range from tiny to huge[2]. From *Recent Advances in Autism Spectrum Disorders*[8] with permission under a Creative Commons Attribution 3.0 Unported license.

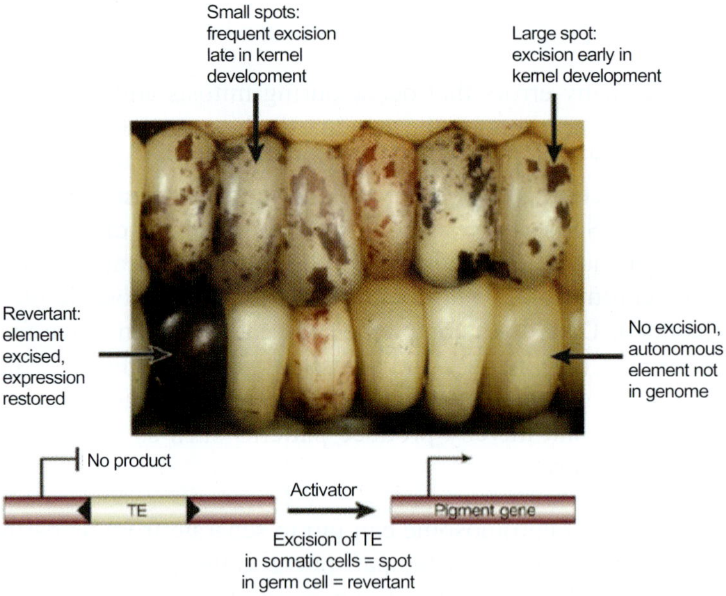

Figure 4.2. Variegated corn (maize) has transposable elements (TE) that, when they are excised from the middle of a pigment gene, restore its function. From Feschotte, Jiang & Wessler (2002)[9] with permission of Nature Publishing Group.

can also inappropriately cross over resulting in a gamete with one chromosome made of a blend of two others. If such a gamete combines with another during fertilization the result is an embryo with one normal chromosome and one blended one. It is from studying the results of such mistakes in genome functioning that many mechanisms for higher order genome control can be better understood.

Down Syndrome is probably the most common and universally recognized aneuploidy (aneuploidy = incorrect chromosome number) in humans. Down Syndrome is considered a complex syndrome because there are effects on many different biological systems and these effects can occur at any point in the life of an individual (Figure 4.3). Because there are so many possible effects, there is also high variability in how severe these effects are between one individual and the next. For example, some individuals with Down Syndrome suffer profound intellectual impairment while others have near to normal, or even normal intellectual functioning[10].

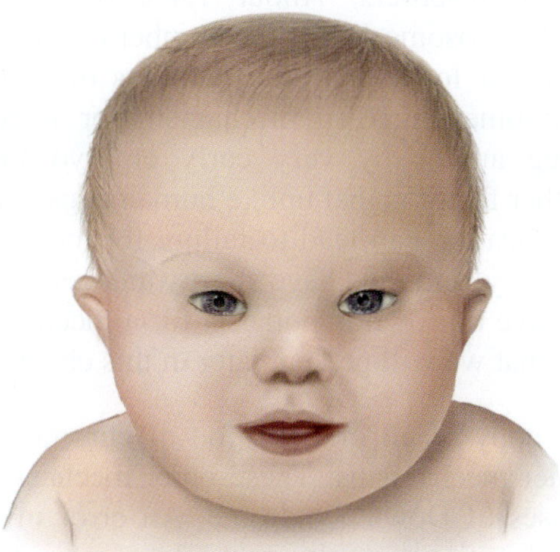

Figure 4.3. Symptoms of the Down Syndrome phenotype include varying degrees of mental impairment, stunted growth, umbilical hernia, increased skin back of neck, low muscle tone, abnormal teeth, slanted eyes, shortened hands, short neck, obstructive sleep apnea, flat head, flexible ligaments, proportionally large and protruding tongue, separation of first and second toes, bent fifth finger tip, Brushfield spots in the iris, strabismus, congenital heart disease, undescended testicles. From Centers for Disease Control and Prevention[11], in Public Domain.

The Down Syndrome phenotype appears with high variability in individuals with three copies of all or part of the 21st chromosome, hence the other name, Trisomy 21. Most commonly, Down Syndrome occurs due to an error in meiosis in the egg where one of the homologous chromosome 21 pairs fails to move with the chromosomes that form the polar body at telophase 1. This results in an egg that ends up with two copies of chromosome 21 instead of the normal one copy. When fertilized, the sperm will add its copy of chromosome 21 and so the fertilized egg will have three copies. Maternal origin errors resulting in Down Syndrome occur randomly, and are maternal age related. Women have all their eggs set aside and paused at meiosis 1 when they are still fetuses themselves. The eggs sit and wait in this form until ovulation. Older eggs are statistically more prone to this particular error for reasons we don't understand. Age related Down Syndrome does not increase risk for Down Syndrome in normal children of parents with a Downs Syndrome child.

Sperm have their own potential for problems. There are several paternal age related genetic disorders. About 1.4% of all sperm have an abnormality[12] of chromosome 1. Yet the number of babies born with that kind of problem is far lower than the numbers of faulty sperm would suggest. There is some unknown mechanism for screening out faulty sperm by the egg and/or by very early embryo loss shortly after fertilization. Another factor is that unlike human eggs, whose long period of waiting for fertilization can lead to failure for many reasons[13], sperm are made continuously, and thus have short waits between formation and fertilization. We have our own hypothesis for a mechanism for ensuring early embryo loss that we will present later in this chapter.

Inheriting a third copy of all or a portion of chromosome 21 can also occur through other mechanisms. These other mechanisms explain only about 2.5% of cases of Down Syndrome. An error in meiosis called a Robertsonian translocation[14] results in the long arm of chromosome 21 being removed from its usual place and attached to the long arm of chromosome 14 (or more rarely chromosome 15). The individual with such a translocation has the normal complement of genes and so is normal, a fascinating example of how irrelevant chromosomes are as long as gene dosage is balanced. Unfortunately, depending on when and how the changed long arm is transferred to the next generation, such a normal individual can produce children with Down Syndrome. There is also a tiny group of families in whom errors in meiosis occur much more commonly and the women produce children with Down Syndrome and other errors of meiosis at a higher frequency than expected, at all ages. The reasons such families are afflicted is unknown but may be due to an error in the meiotic process such as faulty enzymes of DNA repair or reduced rates of crossing over due to faulty enzymes in the synaptonemal complexes with faulty cohesion currently topping the list of suspects[13].

In the most common age related form of Down Syndrome there are a lot of puzzles. Where is the extra chromosome 21 carried on the hypothesized chromolinker ring? It is known that about 1/3 of embryos with an extra chromosome 21 spontaneously abort early in pregnancy. Another roughly 1/3 die before birth. Only about 1/3 of Down Syndrome affected fetuses actually make it to birth. With prenatal screenings even the 1/3 that could survive often don't, since most parents choose termination. It would be a fascinating study to determine whether or not the placement of the extra chromosome in a specific location on the chromolinker ring, beside or distant from the other two chromosome 21, might be related to this survival rate. The difference in survival may be due instead to other effects that cause the high degree of variability found within the Down Syndrome spectrum linked to mechanisms that compensate for the extra chromosome. There is a lot we don't know. So what exactly do we know is going on?

Chromosome 21 is a very small chromosome that contains only 352 known genes. With splice variants and assorted RNAs, the actual number of possible proteins and RNAs is certainly much higher. Scientists now have the ability to measure exactly how many mRNAs are being made by a specific gene. Since there are three sets of 352 genes instead of two, we could reasonably expect 1.5 times the number of mRNA transcripts for those 352 genes. This concept is called "gene dosage affect".

People with Down Syndrome are at very high risk for developing Alzheimer's. A gene with the name APP (for amyloid beta (A4) precursor protein) is found on chromosome 21. Mutations of APP in otherwise normal individuals are associated with early onset of Alzheimer's. It was originally assumed that having three copies of the APP gene results in 50% more transcripts and that explained the higher risk for Alzheimer's in people with Down Syndrome. More recent evidence assessing the actual number of transcripts indicates the entire gene dosage issue is not that simple. APP expression levels were actually very close to normal in people with Down Syndrome. However, some 40 other RNA transcripts of related genes were present in abnormal levels including eight genes that were significantly down regulated instead of being overexpressed. The gene dosage effect means that the amount of protein produced can be much higher or more much lower than the simple 1.5X expression we would expect if the number of genes strictly governed the number of RNAs and proteins produced. This system is highly nonlinear.

To further muddy the waters, a careful study of the actual number of proteins from these down regulated RNA transcripts showed only a single protein of the eight was found to be present at a concentration lower than normal[15]. Conversely, there are about three dozen other genes (depending on the study) that are overexpressed, some of them much more than the 1.5 times expected by having three copies instead of two. These genes are connected to a variety of cell functions including cytoskeletal regulation, lipid metabolism, intracellular protein transport, transmembrane receptor protein tyrosine kinase signal transduction pathways, and regulation of immune functioning. Even more startling, only some of these overabundant transcripts came from chromosome 21. A large number of genes from other chromosomes were also found to be dysregulated.

All this indicates that there are compensatory mechanisms that adjust the number of transcripts produced by the three gene copies or even change protein production rates to correct some imbalances. These compensatory mechanisms keep many protein levels normal even if the gene dosage is wrong. This also indicates that there are proteins or RNAs present from genes on chromosome 21 that interact with genes, or their RNA transcripts, on other chromosomes. And so the gene dosage effect plays an important role in the phenotype of individuals with Downs Syndrome by affecting the gene dosage over the entire genome.

Some gene dosage effects can be compensated for while others cannot[16,10,17]. This is why the number of chromosomes is irrelevant to gene expression patterns as long as the gene dosage is correct. The regulation of genes is spread over the entire genome and the organization is based on units of expression that interact and not on a chromosome to chromosome basis. And so we know that the genome functions as a whole unit with one part on one chromosome affecting other parts on other chromosomes. There is a wholeness here of the kind we discuss in Chapter 12, in which the whole genome can, so to speak, affect its parts and be in some sense causative in itself. The eukaryotic genome is *trans* acting across chromosomes instead of just *cis* acting as in the bacterial *lac* operon (Figure 3.19).

In what follows, we will try to be consistent by using the terminology of Theodore Puck[18] that we introduced in Chapter 3: any segment of DNA can be regarded as *exposed* (available for transcription) or *sequestered* (not available for transcription) at a given time in a given cell. There are many synonyms for these states of DNA in the literature: sequestered = inaccessible, closed, silent, coiled up, compressed, inactive, shut down, heterochromatic; exposed = accessible, open, uncoiled, uncompressed, euchromatic. Those are all adjectives. Each one has verb forms, which would indicate a process of change from sequestered to exposed, or *vice versa*. But we can't be completely consistent, because some terms, like "X inactivation", are in such widespread use that replacement by "X sequestration" would just not be accepted. The word "coiled" implies a particular geometry to DNA, which has recently been challenged, so we'll avoid that too unless we are talking about well understood geometry, such as the coiling of DNA around histones in a nucleosome or the coiling of strands of DNA to bring enhancers and their genes together physically with a DNA polymerase. We will describe that kind of coiling at DNA polymerase in detail in Chapter 9.

Although gene dosage is more important than chromosome numbers, there are also nevertheless high levels of control for gene expression of large sections of chromosomes and even whole chromosomes. We began to understand these via a phenomenon called X inactivation which affects all mammalian females but is most clearly seen in the calico or tortoiseshell cat (Figure 4.4). The classic calico cat has lovely bright orange and black patches with white paws, white chest, a white strip down her middle and a pretty white chin. The attractive patches of red and black on such a cat are considered highly desirable. There are many superstitions about calico cats bringing good luck and protecting your home from natural disasters. Breeders have tried for generations to get the calico pattern to breed true but they have always failed. The pattern occurs only in female cats. Very occasionally, a male cat would appear with the pattern and the breeders would get excited and hopeful but such

Figure 4.4. Tortoiseshell (left) and calico (right) cats. From *The Book of the Cat*[19] with online permission of Chest of Books http://chestofbooks.com/. Painting by English artist William Luker (1851-1889).

a male would invariably be sterile[20]. The explanation for this coat color, and its stubborn refusal to breed true, was found by Mary Lyon in 1961[21] (cf.[22]) and it has to do with both gene methylation and higher order regulation of gene groups.

Methylation is the changing of some of the cytosine or adenine[23-27] nucleotides of DNA (Figure 4.5) by replacing a single hydrogen atom with a carbon bound with three hydrogens (a methyl group)[28]. Both the methylation of the DNA by a methyltransferase and interaction of methylated DNA with other proteins may involve a single cytosine nucleotide swinging out from the DNA helix[29] (Figure 4.6). The larger methyl group sticks out on the surface of its groove in DNA (Figure 4.7), changing how that region of DNA interacts with the proteins that regulate access for transcription and also with functional RNAs. We can anticipate finding other forms of tagging of DNA that work in ways similar to methylation as research progresses.

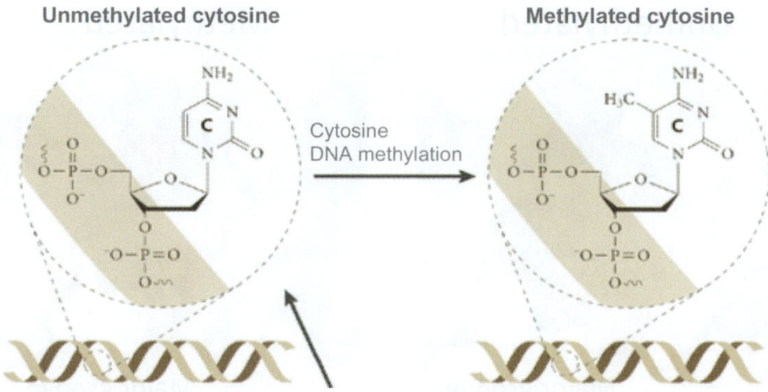

Figure 4.5. DNA methyltransferase can take a methyl group from S-adenosyl-methionine and add it to cytosine (labeled C) while it is part of a strand of DNA. Most DNA methylation is of some of the cytosines. From Law & Jacobsen (2010)[30] with permission of Nature Publishing Group.

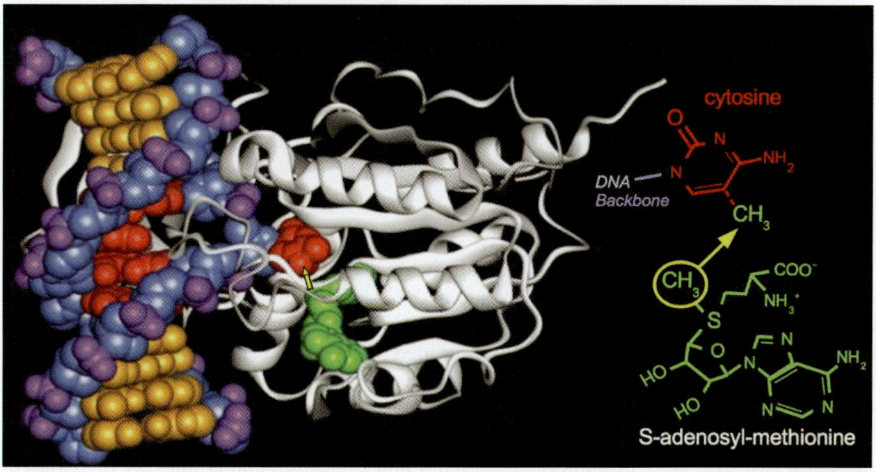

Figure 4.6. The methylation of DNA occurs when a cytosine nucleotide (red) is flipped out of the DNA double helix by a methyltransferase enzyme (protein) molecule (white). The methyl group is transferred to the DNA from a molecule of S-adenosyl methionine (AdoMet) pictured in green. Figure modified from the original[31,32] with kind permission and assistance of Xiaodong Cheng and Richard J. Roberts. Cf.[33]. The arrows indicate the transfer of the methyl (-CH_4) group, before the transfer actually occurs. Most hydrogen (H) atoms attached to carbon atoms (C) are not shown, per convention. Most C atoms in chains and rings are just indicated by the points where line segments, representing chemical bonds between atoms, meet. Parallel lines indicate double bonds. N = nitrogen, O = oxygen, S = sulfur. Blue and purple represent the DNA backbone and gold represents other nucleotides.

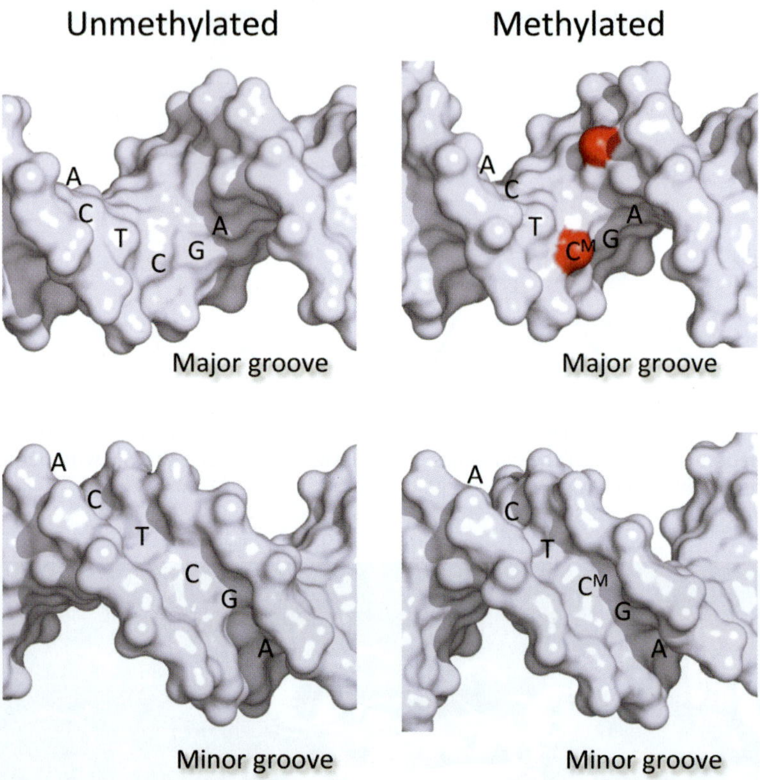

Figure 4.7. The additional methyl group in methylated cytosine (C^M) changes the chemistry of the base in the major groove of DNA and alters the local DNA shape with changes in groove width. These changes alter which proteins can and cannot bind to the changed region of the DNA. Figure is provided with kind permission of Remo Rohs[34] and Ana Carolina Dantas Machado[35], and with permission of Oxford University Press. Cf.[36]. Cytosines in mammals only get methylated (with rare exceptions) when they occur in CpG dinucleotides (i.e., one C/G base pair followed by a G/C base pair). Consequently there are two cytosines in adjacent base pairs and opposite strands that get methylated. These two methyl groups are the two red balls in the figure. One methylated cytosine is shown in its position in the minor groove, but its methyl group is not visible from this side. DNA diameter is about 2 nm.

During mitosis, each double helix contains one strand that is methylated and a new strand that does not have a methylation pattern. A protein called DNA methyltransferase copies the pattern of methylation to the unmethylated strand after the strand is made[37]. As we mentioned earlier, methylation can also be directly applied to the histone tails and this version of methylation/demethylation is also faithfully copied. Methylation of the DNA and/or the histones is used in loosening or tightening the DNA and thereby also affecting which genes are expressed. Methylation also affects the accessibility of CpG promoter sequences so that a promoter that is methylated is generally no longer able to act as a promoter, although there are exceptions[37].

X inactivation represents a specialized form of methylation of DNA that essentially sequesters most of an X chromosome for transcription. The methylation signal is on the chromosome itself in the form of a gene called Xist (X-inactive specific transcript). When Xist is transcribed it makes an RNA that binds to and coats the rest of the X chromosome. When the Xist RNA is present, other proteins and RNAs required for silencing the X chromosome are recruited to the area. The genes are then silenced by being heavily methylated. The last step in the silencing process is that changes to the histones make the inactivation permanent (Figure 4.8). The DNA of this almost totally inactivated X chromosome is methylated and is tightly coiled up and then shoved off to one side of the nucleus right under the nuclear membrane. (Recall that expressed genes are more often found deep in the nucleus whereas sequestered genes are located at the edges of the nucleus so this movement of the inactivated X fits the overall pattern.) It sits there, most of it sequestered[38], for the rest of female's life (Figure 4.9). If Xist is deleted on one X chromosome, X inactivation does not happen on that chromosome and the other X is always inactivated instead. Xist can be considered the master gene, or even a regulon, for X inactivation.

Female mammals have two X chromosomes. (There are a few individual and species exceptions, as always[39].) Two X chromosomes is one too many and so one has to be shut down in order (presumably) to keep gene dosage at the same level as that in the male[40]. Mechanisms of dosage compensation vary between species but it is X inactivation that accounts for the pattern on female calico cats[41]. Males have but one X chromosome because they also have a wee little Y chromosome with only a few genes, only 3% of those present when the Y chromosome first appeared in mammals between 180 to 300 million years ago[42-44]. All the other genes once present on the Y have migrated to other locations in the same way many of the genes of mitochondria have migrated to the cell nucleus. Those few remaining Y chromosome genes are both essential to making a male and to maintain the gene dosage for some related

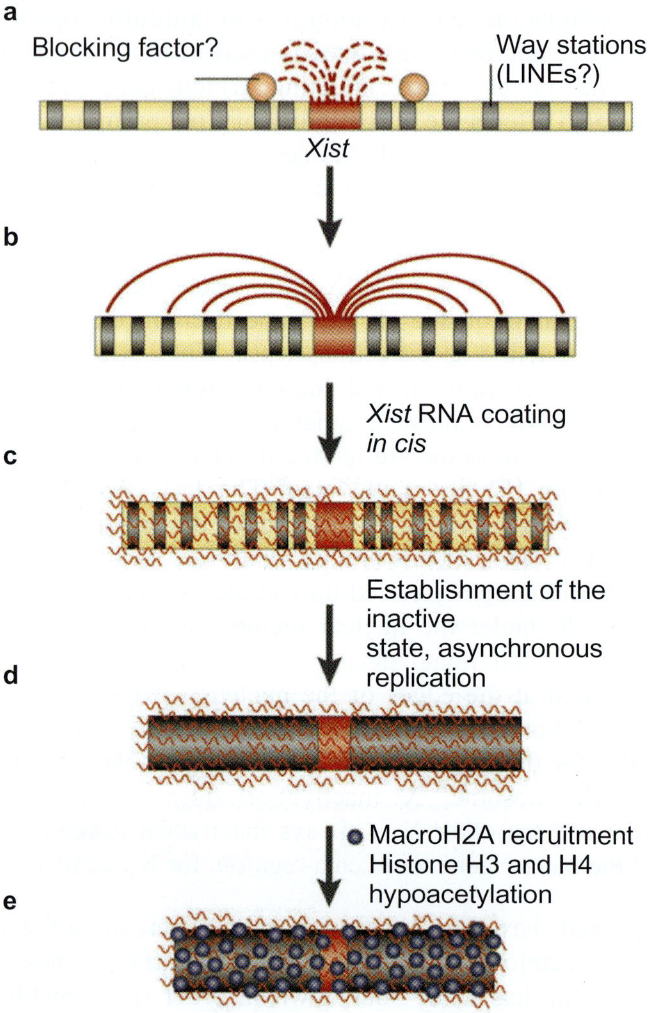

Figure 4.8. Details of the process of X inactivation (X sequestration). "**a)** Before inactivation, Xist RNA is expressed in an unstable form (dotted red lines) and the postulated blocking factor(s) (red) prevents Xist upregulation and/or its association with the chromosome in cis. **b)** Xist RNA becomes upregulated[45]. **c)** Stabilized Xist RNA coats the X chromosome. **d)** Transcriptional silencing of genes on the X chromosome occurs as a result of Xist RNA and is rapidly followed by a shift to asynchronous replication timing of the X chromosome. **e)** Chromatin modifications are presumed to transform the Xist RNA-coated chromosome into a stably inactive and condensed chromatin state". From Avner & Heard (2001)[45] with permission of Nature Publishing Group. In humans 15% of genes on the X chromosome escape X inactivation[46], in mice 3%[38].

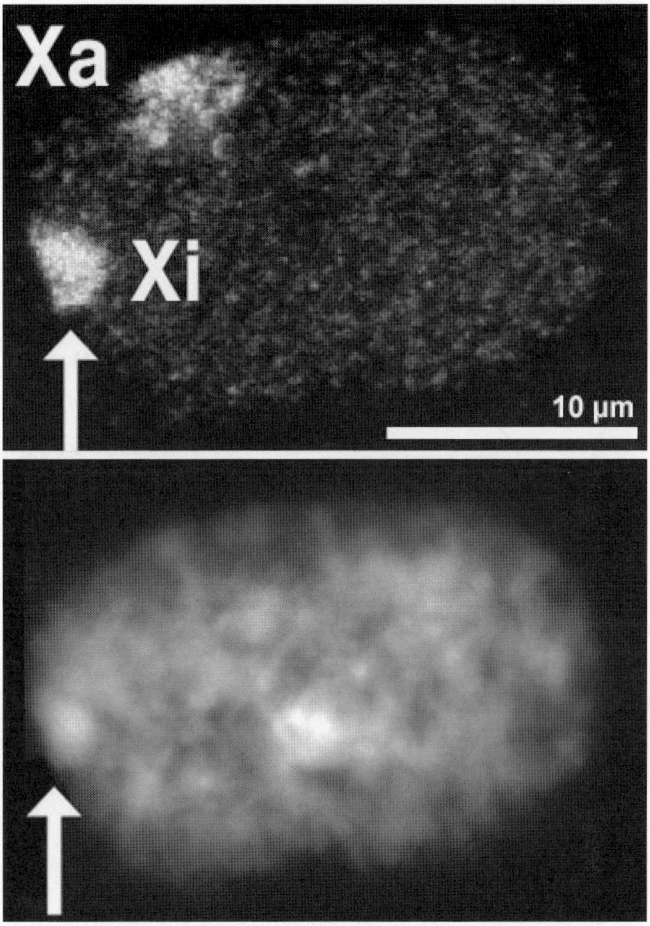

Figure 4.9. The Barr body is a condensed, mostly inactive X chromosome, located at the inner surface of the nuclear membrane. **Top**: Confocal microscopy image showing both X chromosomes in a human female cell (Xa = active X chromosome). **Bottom**: Only the inactive X chromosome (Xi) = the Barr body, shows up in the same nucleus when stained with DAPI. From Wikipedia[47] with permission under a Creative Commons Attribution-Share Alike 3.0 Unported license.

regulatory proteins[42,43]. Very early in embryogenesis, when a female mammalian embryo is a small ball of cells, one X chromosome is inactivated (sequestered) in each cell (Figure 4.10). The timing varies between species so when female cats have X inactivation is different in development from when female humans have it[48].

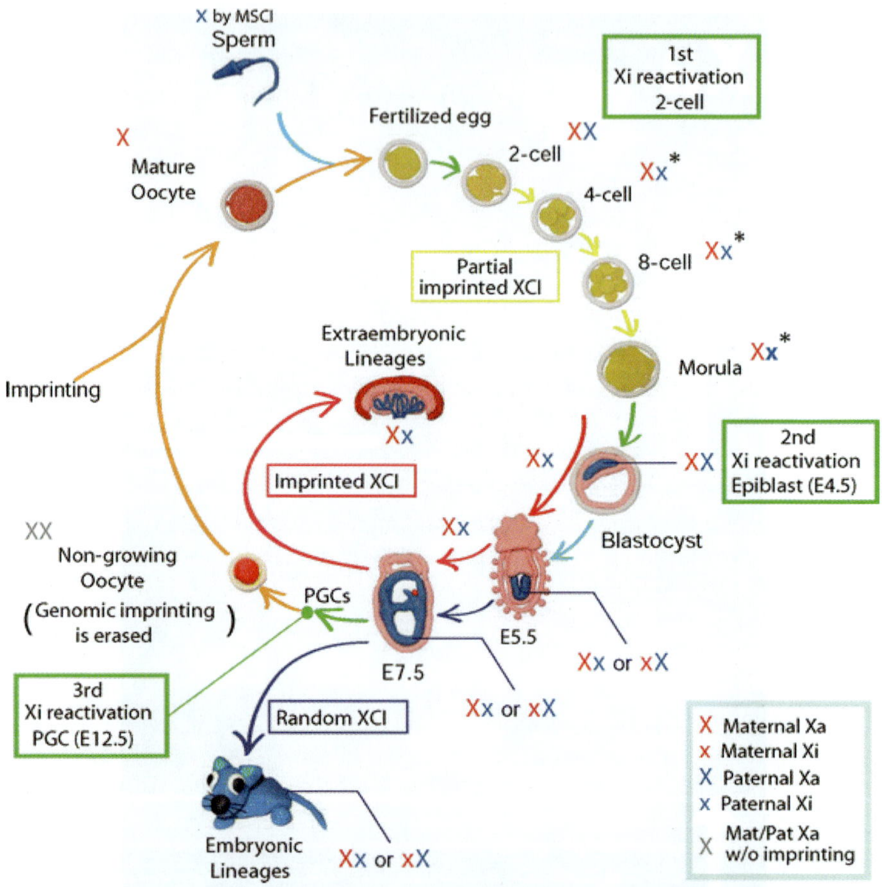

Figure 4.10. X chromosome inactivation (sequestration) and reactivation (exposure) in the life cycle of the female mouse. From Ohhata & Wutz (2013)[49] with permission of Springer under a Creative Commons Attribution License. E = embryonic day[50]; MSCI = meiotic sex chromosome inactivation; PGCs = primordial germ cells; Xa = active X chromosome; XCI = X chromosome inactivation; Xi = inactive X chromosome.

Which of the two X chromosomes is silenced in a particular embryonic cell is normally random[51]. Once silenced, that inactivated X and its inactivation signals are copied. Thus this X and its daughter chromosomes are silenced (sequestered) during all future mitoses. Here is how all of this affects pigmentation. There is a gene for coat pigmentation in cats that is located on the X chromosome. In the female cat, one gene for coat color is shut off when the X is inactivated. But genes come in pairs called "alleles", and in a given cell, after X inactivation, one allele is sequestered and the other is exposed. If a female cat happens to get an orange allele for the gene from one parent and a black allele for the gene from the other, then during early embryogenesis, the cat will have some cells with only the working orange pigment gene and other cells will have only the black gene working. Both genes, one from the mother and one from the father, are present. Only one is used in each cell because the other is coiled up and sequestered on the inactive X.

The gene for coat pigmentation doesn't actually "work" until much later in development. Recall the neural crest cells that break free from the boundary zone of the neural ectoderm and the epithelial ectoderm during neural tube closure (Chapter 2). These neural crest cells migrate out and later on in development some of them form the pigment cells that give us the color we see on the cat. The pigment cells migrate to their place and then do several rounds of replication to produce patches that are all clones of the first cell to migrate into an area. These patches of replicating cells expand until they meet cells from another clone. When it is time for the genes to be turned on about half of the calico cat's pigment cells have the orange gene turned on and half have the black gene turned on. The cat continues developing and the result is the orange and black blotch pattern we see on her coat (Figure 4.11).

A mutation on a gene from another chromosome causes the white patches on the calico cat. When this altered gene is active, pigment cell migration is slowed and the cells don't migrate all the way out from the midline of the back to all edges of the cat. They also mix up less than they otherwise would for reasons not really understood. Hence, the white pattern, not associated with the genes on the X chromosome, is usually pretty symmetrical on both sides causing the attractive "dipped in milk" look of the classic calico cat. The patches of black and orange are random but larger than on a cat without the white. If the cat has no white mutation on this gene, the pigment cell migration is not slowed and she develops into a plainer tortoiseshell cat usually (though not always) with smaller and more blended patches of color (Figure 4.4).

The calico cat has one X chromosome inactivated in each cell and the inactivation of that one X is for her entire life. The inactivated state of this X chromosome is transmitted to all[21,49] copies of that chromosome during subsequent cell divisions and their mitoses. All mechanisms, including DNA methylation, histone modifications and coating by various RNAs such as Xist (Figure 4.8), that are involved in the transmission of the state of a gene from a cell to its daughter cell at mitosis, are transmitted by bookmarking[52]. Bookmarking is transmission of a cell's "memory" to its progeny[53,54].

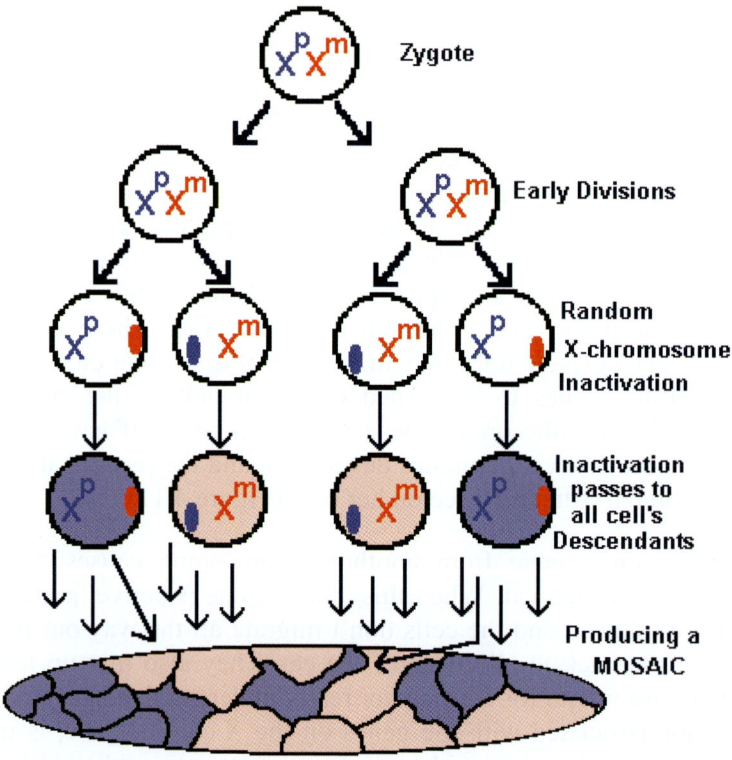

Figure 4.11. "Perhaps this diagram will help you to understand the Lyon hypothesis and how it causes female mammals, including women, to be mosaic. Early in development the X chromosomes are randomly inactivated in each cell. In any particular cell it could be the paternal (X^p) or the maternal (X^m) that remains active. The inactivated X chromosome becomes a tiny blob of dense, useless chromatin called the Barr body – shown here as a dark oval. These cells pass their inactivation patterns to all their descendants. These descendants contribute to the body plan of the organism, often in a complicated way that produces a mottled pattern. The complete organism is a mosaic of cells expressing alleles from either the X^p or the X^m chromosome, depending on which one has escaped inactivation." From Love (2010)[55] with kind permission of Jamie Love, who retains the copyright.

Yet, when a calico cat has kittens she can pass on either the orange gene or the black gene to each of her babies (Figure 4.10) and her female kittens will not have the same pattern of inactivation. The male kittens she has will each get one X chromosome from her and one Y chromosome from their father (Figure 4.12). Because the pigment gene is only on the X chromosome any male kitten will be either orange or black and he will inherit only one of his mother's coat color genes. What color her female kittens are depends on the both parents contribution. If the father is black, he will always contribute one X chromosome with the black gene. The mother can contribute either a black or orange gene. The female kittens will be either black and orange

Figure 4.12. The calico cat is a female cat ♀ with white underparts (middle of top row). The gene for color is located on the X chromosome (long line with oval). The female cat has two X chromosomes and therefore two color pigment genes (two long lines with orange and black ovals). At about 64 cells X inactivation of one X in each cell occurs[56]. The result is patches of yellow and patches of black in the cat with two different color genes. Male cats ♂ have only one X chromosome and one Y chromosome (grey short line) and so are either all black or all orange. The gene that creates white patches is inherited independently of the color genes and is not located on the X chromosome. When a calico cat has kittens she can have an all black or an all yellow daughter with probability ¼, but not both. Which solid color daughter she has depends on the color gene on the X chromosome contributed by the father (left: black father, right: orange father).

like her or all black, depending on which X they got from their mother. If the father is orange, the reverse happens and the female kittens will be orange or orange and black. The all orange and all black female kittens will have their own X inactivation in early development and be just as much mosaics as their prettier calico sisters but the X inactivation pattern will be invisible. X inactivation is therefore stripped off and then reset with each generation.

Another interesting effect of X inactivation and gene dosage regulation is the changes in color vision in humans. The evolution of color vision in mammals is not a simple progression. Each vertebrate has up to five functional visual photopigment opsin genes that clump into two gene

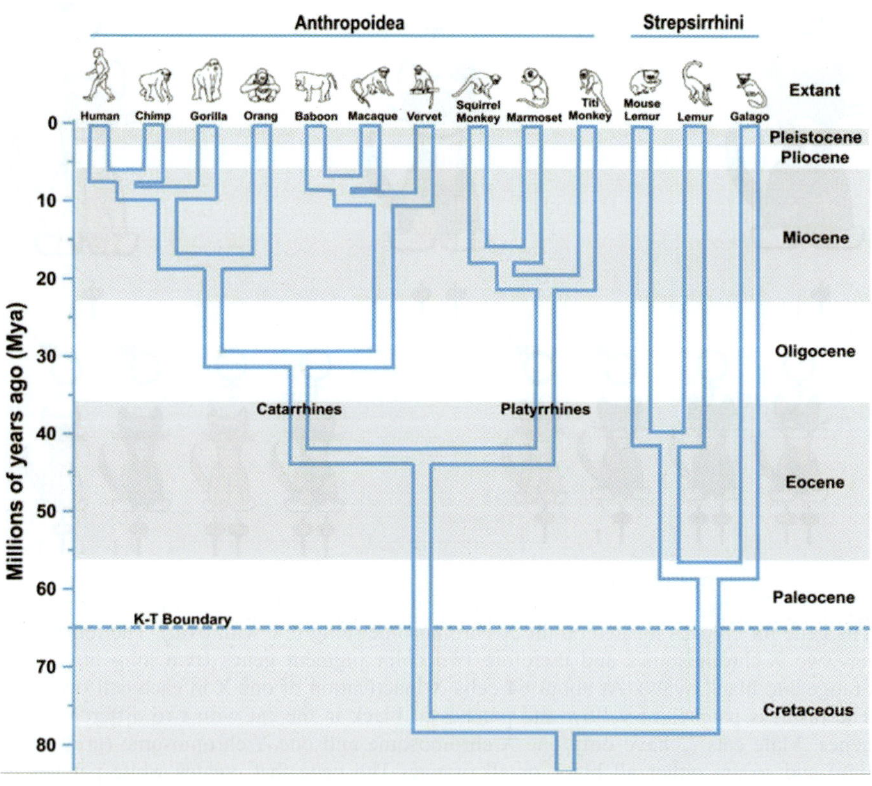

Figure 4.13. The two color opsin genes are side by side on the X chromosome in all catarrhine primates. From Jacobs (2008)[58] with permission of Cambridge University Press.

families that can be traced back up to 540 million years ago. Four of these genes are expressed in the cone cells in the retina[57]. The other is expressed in the rod cells, which are generally associated with black and white vision but in low light can also have a role in color vision[59]. For the past 35 million years the catarrhine primates, which include humans (Figure 4.13), have had two cone opsin genes side by side on the X chromosome, probably as a result of a gene duplication followed by evolutionary divergence. Another opsin gene[58] is located on chromosome 7.

One of the opsin genes on the X chromosome is called *OPN1LW* and it codes for protein pigment used by cone cells that respond to the yellow/orange part of the visible spectrum (*LW* = long-wavelength red light). The other gene on the X chromosome is called *OPN1MW* and it codes for a protein pigment used to detect light at *MW* = middle wavelengths (yellow/green light). The gene on chromosome 7 codes for the third opsin gene and is called *OPN1SW* and it codes for a protein pigment that is used to see the blue/violet part of the visible spectrum (*SW* = short-wavelength light). Thus there are three different kinds of cone cells, heavily concentrated in the central, foveal region of the retina, yet each individual cell is expressing only one of the three possible opsins[60]. Somehow there is regulation that assures one opsin gene, and only one opsin gene expressed per cone cell. These cells are arranged in a patchy manner, including some large patches of cone cells that all express the same pigment gene (Figure 4.14).

With the combination of the three cone cell types we humans have trichromatic vision. A few people have tetrachromatic vision meaning they have four different types of cones and that gives them even more ability to distinguish color. Males can have tetrachromatic vision only if they have two different variations (heterozygous alleles) of the gene on chromosome 7. They can express one of the two genes on their single X and the two different alleles they have on chromosome 7. Females can have tetrachromatic vision in two ways. They can have heterozygous alleles on chromosome 7, like males, or they can also have tetrachromatic vision if they have a mosaic pattern in their retina due to X inactivation[61,62,63]. In theory, a rare women could even have hexachromatic vision but there is no evidence of that having ever occurred because we simply don't test beyond tetrachromacy. It would seem that the X inactivation clones are usually too big for tetrachromacy, except

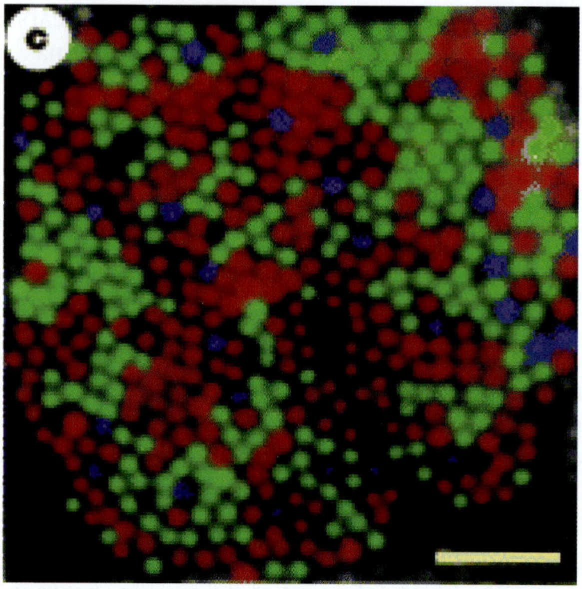

Figure 4.14. A region of male human retina showing the patchy distribution of the three kinds of cone cells. Blue = S cones, green = M cones, red = L cones, where S, M, L = short, medium and long wavelength opsins, respectively. Bar = 5 arcmin of visual angle. From Roorda & Williams (1999)[66] with permission of Nature Publishing Group. Cone cells range from 0.5 to 4 μm in diameter from the fovea outwards[60].

perhaps one woman found in an investigated group of 24 women[64,65]. Even so, it is well established that on average, women have much better color vision than men because of X inactivation.

Mice ordinarily have only one opsin per X chromosome. Genetically engineered mice have been created in which a second opsin was introduced that has a different wavelength response. When they were on the same X chromosome, little changed. Some gene regulatory mechanism in mice meant that only one opsin gene was accessed on the X chromosome and the other was always inactivated even though two were present and available. Some form of dosage compensation mechanism exists in mice for opsin genes that does not exist in humans. However, when heterozygous female mice (having one gene per X chromosome but different opsin genes on each X) were made, some of the mice were able to distinguish colors. This is because X inactivation left a mosaic of the two opsins. The fact that only a few of these engineered female mice acquired color vision stumped the investigators[58]. However, we can see that the explanation may lie in the size of the patches of clones of X inactivated cells. Color vision requires small overlapping patches, and X inactivation generates large patches[67], often bigger than whole eyes. Only the few female mice with overlapping mosaic cone pigment cells could see color. The problem of patchiness due to clones of different X inactivation[67] is usually, but not always[68], overcome by mixing of cone cells during their hypothesized migration to the foveal region[69]. Humans apparently have more mixing than mice which makes sense since they normally have two opsin genes and mixing them means better vision[69].

Humans are fascinated by the differences between men and women, trying to discern which are important and which are not. There has thus been a long interest in quantifying and explaining differences in visual perception. These have been hypothesized to have arisen during the Pleistocene (11,700 to 2.5 million years ago[70]) in hunter/gatherer societies, correlating with men being hunters and women gatherers. When spatial abilities were tested, men scored higher on spatial abilities and women higher in their ability to locate objects. The tests were done without considering color perception[71,72]. However, other work has focused on color vision[73-75]. Humans have superb color vision, with color blindness being at the lowest frequency in hunter/gatherer societies[76]. Red/green color blindness, the most common form found in men, is present in all cultures and appears in many other mammals[77]. This may be a product of natural selection because partial color blindness leads to improved ability to detect subtle differences in contrast and the ability to assess distance, and therefore an ability to not be fooled by color camouflage[78,79]. Being color blind is a relative advantage to a group of hunters (opposite to earlier suggestions[80,81]). It gives the hunter an edge for spotting prey and accurately aiming at it with a rock, spear or gun no matter what camouflage the prey has. In contrast, being able to see color is a big advantage in a gatherer[82]. The ability to pick out bright red color on a green background is an important trick for someone who is out picking berries[83] (though its importance to overall survival is still being debated[77]). It will be interesting to see what happens as new technology for correcting partial color blindness[84-87] comes into widespread use. The invention of these new eyeglasses stems from the hypothesis that it is not red berries, but the color of skin, which reflects emotions via blushing, that trichromatic vision evolved for in primates[88]. Color vision may improve nonverbal communication. In our technological society where communication is more important than hunting we can use our technology to compensate for poor color vision.

Mammals started 200 million years ago with one opsin gene on the X chromosome[89,90,57], which itself originated from our tetrapod ancestors about 300 million years ago[91,92]. A tandem duplication about 30 to 40 million years ago resulted in two opsin genes on the X chromosome of later humans[93]. With two X chromosomes, human females have two pairs of genes for color vision on them. Males have only one pair. Color blindness is therefore much more common among human males than among human females. The carrier of color blindness genes, one good gene and one bad, will have excellent color vision. She will most likely produce female offspring who have color vision. All male offspring will have a 50:50 probability of being red/green color blind. If the offspring are color blind, it is more to their benefit than to their detriment as hunters.

Now suppose selection for the presence of red/green color blindness were to continue until the frequency of the gene was so common that woman began having the color blindness allele appear on both their X chromosomes? If a few women in the group are color blind, it isn't that much of a problem because if the women work together and share their food, the color blind woman could do guard duty watching out for women eating tigers while the others pick. But if the frequency of the color blindness allele reached the point where everyone in the group was color blind, then the females would begin to have trouble gathering and that group would be selected against by starvation because they won't see a lot of berries, even if they can avoid tigers. From a whole group, or population point of view, it is advantageous to have some color blindness genes, but not too many.

Having two genes with different alleles, like red/green colorblindness and normal color vision on two chromosomes, is called heterozygosity (from heterozygous alleles). When heterozygosity for an allele is being selected both for and against in a population at the same time, that allele can end up being present at a low but constant frequency over many generations. This is called "heterozygote advantage"[94] or "heterosis"[95]. It has been speculated that many recessive genes for deleterious traits in humans are actually genes that provide a heterozygote advantage to the whole population[96]. Any heterozygous alleles affected by X inactivation can, as a consequence, cause a trait that is always much more common in males than in females and is therefore also called "sex linked". X inactivation is therefore also an excellent example of how large scale regulation of gene expression at the level of whole chromosomes can cause evolutionary selection of whole populations and the frequencies with which certain types of gene differences are maintained in that whole population[79].

This then leads to the question of why exactly did the gene for color blindness and the gene for coat pigmentation both end up on the X chromosome? We know that the genome as a whole can have entire segments of genes moved around by errors in meiosis and mitosis and yet still function normally. The movement of those genes to the X chromosome may have to do with active selection by evolution. In the past somewhere, some translocation of a cluster of genes resulted in the genes for color, and therefore for color blindness, being moved to the X chromosome. Because of X inactivation, those families of animals with that arrangement of genetic material did better than those who did not and so the positioning persisted because they reproduced more until it became a fixed genetic trait of primates.

X inactivation is unique to mammals. All mammals do it in some fashion. Other organisms have simpler versions of X inactivation but none to the degree that mammals do. The other thing all mammals have is a highly advanced and adaptable brain compared to other organisms. Mammals have greater intelligence as a group. By intelligence we mean the general ability to perceive, to determine patterns, the capacity for problem solving, learning, memory, communication and in higher mammals planning, creativity, abstract thought, and even self-awareness. What does that have to do with X inactivation? Genes for color blindness and pigmentation are on the X chromosome and the tissues (central nervous system and skin) in which they are activated are ectodermal in their embryonic origin (Figure 2.54). Why is this important?

Disproportionately, many of the genes associated with intelligence in humans also reside on the X chromosome[97-100]. Most of these genes, when mutated, act in a recessive fashion in females due to mosaicism, meaning cells with the normal gene on the active X will compensate for the effects of the faulty gene. (There are some exceptions to this[101,102].) The result is that if a woman has a gene for intelligence that is defective on one X and normal on the other, she will be normal or near to normal, depending on the personal X mosaic pattern in her brain. If she has a gene for some aspect of intellectual functioning that is superior, she will have normal to high functioning, depending on her unique pattern of X inactivation. The boy who has inherited the X chromosome with his mother's superior gene will have a higher advantage than his mother since he only has the one gene to express. In contrast, if the mother has defective genes for some aspect of intellectual ability, she (and her daughters) will most likely be normal because of their other gene located on the other X chromosome, but half her sons will show the effects and have some obvious intellectual disability.

Take a further step back and imagine mammals as a group. They have different and complex brains and they have intelligence. Intelligence represents quite an evolutionary experiment and it is one that has been highly successful. Intelligence leads to adaption to rapidly changing environmental conditions and the ability to live successfully in a wider range of environments. While forms of intelligence have appeared in other clades, the octopus being an obvious example[103], intelligence is nonetheless the common feature of the mammalian clade. Many species of mammals are intelligent. The same cannot be said of most species of mollusk (except octopus). Because of X inactivation any evolutionary experiment on intelligence genes on the X chromosome (due to a mutation in the gene expression patterns associated with intelligence) will be clearly shown in male offspring in the very first generation.

Maybe in the distant past some lucky genetic accident moved a whole chunk of ectoderm genes onto the X chromosome or there was ongoing selection for such transfers[104]. Being ectodermal and therefore some affecting brain functioning, a lot of those genes were involved in intelligence. X inactivation may have allowed novel intelligence to be quickly revealed in males and rapidly selected for by females who picked the best and brightest males for mates[98,100,105]. Intelligence then took off and thrived in mammals in ways it just couldn't in other organisms. Of course, that is just a little wild speculation[106,107] for the sheer fun of it, albeit with consequences. Geneticists are prone to their own kind of "just so" stories, and there is inconclusive evidence that the expected greater variability in male intelligence at the high end may not be correct[108]. Will this empirical genetic result generalize to human intelligence? Time will tell.

Meanwhile we left those poor kittens without explaining how their X chromosomes got reset. So why didn't the X inactivation bookmarking get passed to the next generation? Methylation varies in the course of the life cycle (Figures 4.15 and 4.10). At fertilization all the Xist methylation on the inactivated X chromosome is removed[109]. A whole lot of methylation all over the genome is also removed during both sperm and egg formation and at fertilization. This occurs all over the genome. This is analogous to restarting a computer and clearing the cache[110]. The developing embryo makes its own epigenomic memories during development. Not all the methylation is removed, however. Some methylation remains so that the chromosomes of paternal origin can be distinguished from those of maternal origin by the functioning genome machinery of the embryo.

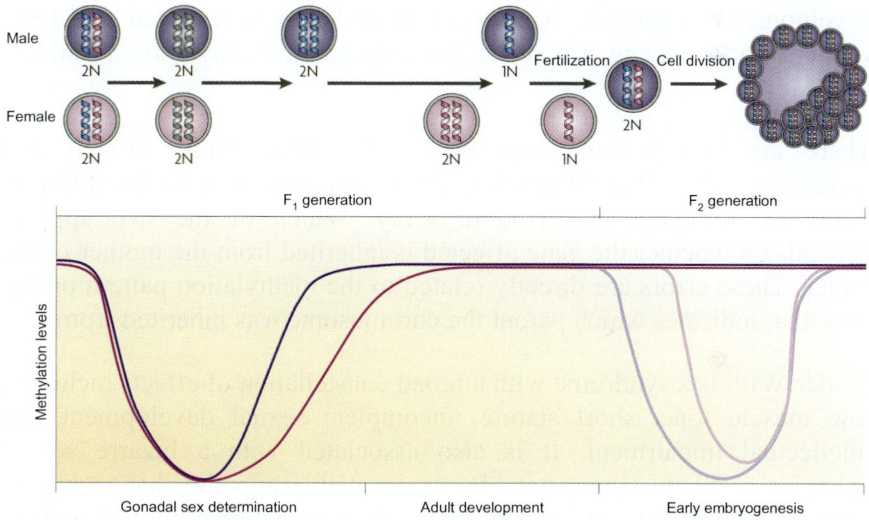

Figure 4.15. "Alterations in methylation status during development. During embryonic development and gonadal sex determination, primordial germ cells undergo genome-wide demethylation, which erases previous parental-specific methylation marks that regulate imprinted gene expression. In the male (colored purple) germ line, paternal methylation marks in imprinted genes are laid down very early during sperm formation. The female (colored pink) germ line establishes maternal methylation marks in imprinted genes at a later stage. After fertilization, the parental genomes are demethylated. Genome-wide remethylation occurs on both parental genomes before implantation. However, certain imprinted genes maintain their methylation marks throughout this reprogramming, allowing for parental-specificity of certain genes." From Jirtle & Skinner (2007)[111] with kind permission of Randy L. Jirtle and Nature Publishing Group. For a more elaborate DNA life cycle diagram see Strachan & Read (1999)[112].

Keeping track of whether a chromosome is paternal or maternal is known as genetic imprinting and it is only one type of cell memory that is caused by methylation.

Methylation is not just critical to X inactivation, imprinting and bookmarking old and new strands. Methylation is also a form of higher DNA sequestration control that is used for smaller portions of the genes on other chromosomes. It also controls entire functional groupings of genes throughout the genome. Methylation helps to up regulate or down regulate gene expression at appropriate times. This can either be locus-dependent (often leading to splice variants), or in several places on large sections of DNA. Methylation of DNA sequestering it over large segments, combined with methylation and/or acetylation of histones, is also used for permanent sequestration of entire sections of genes as development progresses. Almost all methylation is removed and reset when sperm and egg are made but imprinting methylation is added. Why?

There are two human syndromes called Prader-Willi (PWS) and Angelman (AS), where imprinting errors can cause two wildly different errors of embryogenesis (Figure 4.16). Which of the two appears depends on whether the gene affected is inherited from the mother or the father. These errors are directly related to the methylation pattern on the gene that indicates which parent the chromosome was inherited from.

Prader-Willi is a syndrome with a broad constellation of effects including low muscle tone, short stature, incomplete sexual development and intellectual impairment. It is also associated with a bizarre set of behavior problems, most notably, an insatiable appetite that results in over eating and extreme obesity unless there is external control such as the family quite literally keeping the refrigerator door locked.

Angelman Syndrome is also associated with intellectual impairment, usually more profound than in Prader-Willi syndrome. There are characteristic hand flapping motions, jerky movements, sleep disturbance, seizures, and a happy smiling and laughing disposition. Unlike Prader-Willi, those with Angelman tend to be small and thin with poor appetites and prominent tongues that often roll about and stick out in uncontrolled movements accompanied by a lot of drooling.

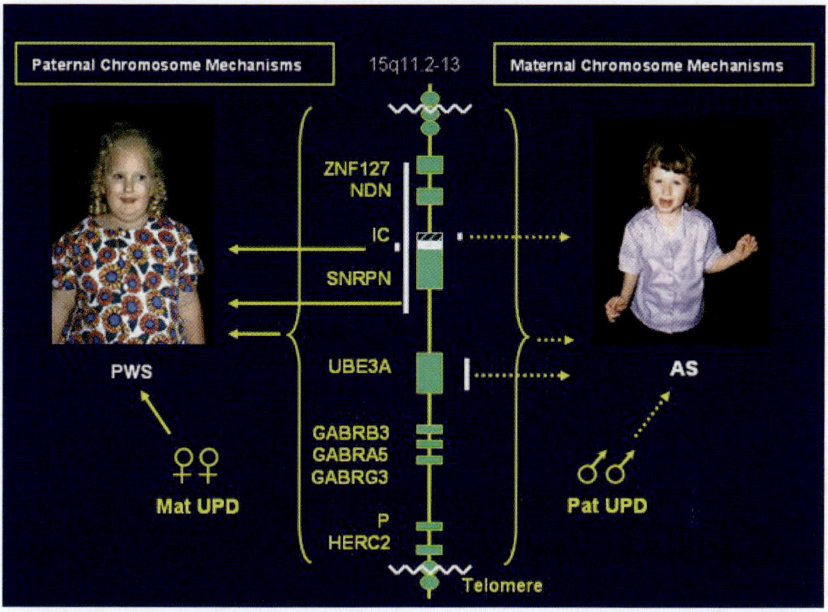

Figure 4.16. Normally there are two identical expressing genes, one from each homologous chromosome. If one fails, the other can compensate. However, in the case of imprinted genes, there are differences in the expression pattern between the genes on the homologous chromosomes because of different methylation patterns. Prader-Willi (PWS) and Angelman (AS) are two very different genetic syndromes that are caused when the 15q11.2-13 region of Chromosome 15 is changed. If the chromosome inherited from the father is affected by some mutation (such as a microdeletion within region 15q11.2-13) then the result is Prader-Willi. If the chromosome affected is inherited from the mother, the result will be Angelman. If one parent's contribution is lost and replaced with the contribution of the other parent through homologous recombination, the result is uniparental disomy. For many genes, such uniparental disomy does not cause a problem but in imprinted regions, like 15q11.2-13, the result of uniparental disomy can be a genetic syndrome. Which syndrome will depend on which parental contribution, and the matching methylation pattern, was lost during the homologous recombination repair process. From Williams (2015)[113] with kind permission of Charles A. Williams. UPD = uniparental disomy.

Yet in both these very different syndromes, the same region of a chromosome is affected. The region is found on chromosome 15 within a set of genes in the region of the longer arm of the chromosome called q11-13. Stretches of DNA in this region contains the SNURF–SNRPN set of genes. SNRPN codes for the protein known as "small nuclear ribonucleoprotein-associated protein N". This region of chromosome 15 also codes for a cluster of 82 (and counting) RNAs known as small nucleolar RNAs (snoRNAs). The snoRNAs affect other functional RNAs by methylating specific ribonucleic acids but only appear in those nerve cells where SNURF–SNRPN is expressed. This family of functional RNAs and N protein match specific RNA sequences on other mRNAs from different genes. They are critical to creating differing splice variants on other genes and thereby making novel proteins. SNURF–SNRPN is only expressed in certain nerve cells.

Another gene in the same Prader-Willi/Angelman region is UBE3A. The UBE3A gene codes for a specific protein called "Ubiquitin-protein ligase E3A". All the cells of the body, except the brain, produce this small protein enzyme. This enzyme puts an informational "tag" on other proteins that directs those proteins to begin degradation. In other words, the enzyme marks specific proteins for destruction. And so the region of chromosome 15q11-13, referred to as the PWS/AS region, is a region that contains a set of genes with *trans* effects on a lot of others genes and the proteins they make by changing gene splicing, sending certain proteins off for rapid degradation, and affecting the function of a host of other RNAs. The region can therefore be considered a regulon containing master genes. But why is the expression only different in certain nerve cells? Why do only these nerve cells stop expressing UBE3A and start expressing SNURF–SNRPN instead?

Recall that the preexisting pattern of methylation of the X chromosome is removed from gametes during meiosis. After fertilization, the gamete specific methylation is almost entirely removed but information as to which chromosome was paternal in origin and which was maternal is nonetheless retained (Figure 4.17). This is used to guide new gender specific patterns of methylation that are added to the gametes after meiosis, one methylation pattern for the sperm derived genome and a different one for the egg genome. During embryogenesis, many methylation events occur as some portions of the DNA are exposed and some are sequestered. The location is largely DNA sequence dependent, i.e., certain sequences "invite" methylation. This pattern continues for each differon. After embryogenesis is complete, methylation patterns are relatively fixed in each tissue type. Methylation represents one critically important type of higher control of multiple subsets of the genes (differons) that change as the organism changes during development (Figure 4.15).

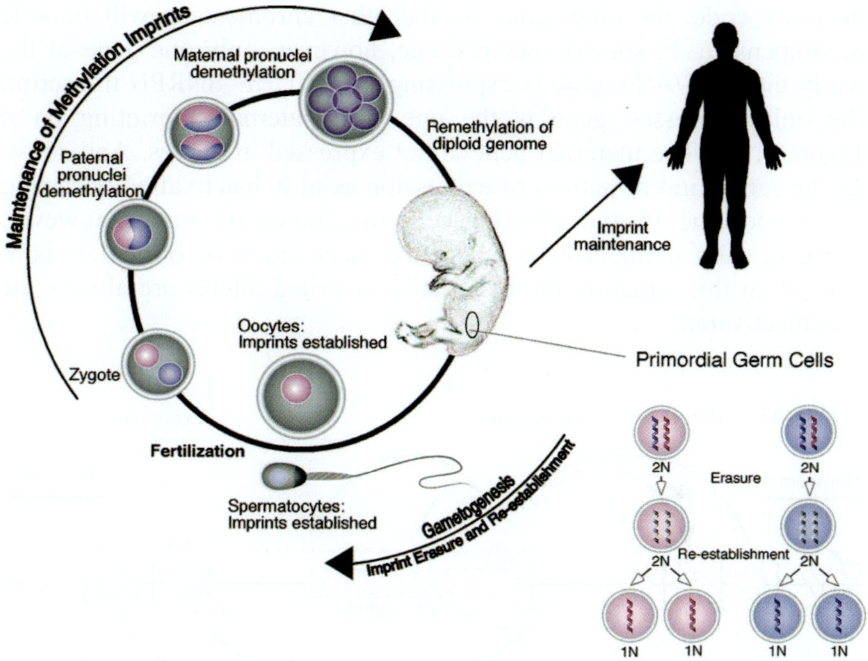

Figure 4.17. "Imprinting throughout development. Methylation marks associated with imprinted genes are established at two distinct phases of development. During gametogenesis, the imprint marks present on the maternal (*pink*) and paternal (*blue*) chromosomes are erased (*grey*), followed by establishment of primary imprints which then reflect only the sex of the individual in which they reside. Just after fertilization, a global demethylation event occurs in the zygote, first in the paternal pronucleus (*blue*), followed by the maternal pronucleus (*pink*). Imprint marks that were established in the gametes must resist this demethylation process. Remethylation of the diploid genome occurs post-implantation, and includes setting of the secondary imprints, which are maintained throughout the lifespan of the individual." From Murphy & Jirtle (2003)[114] and Dolinoy *et al.* (2007)[115] with permission of Wiley Periodicals and Nature Publishing Group.

In the case of the PWS/AS region, there is heavy parental gender specific imprinting. The SNURF–SNRPN transcript is only expressed from the paternally imprinted gene and only when that gene is being expressed in specific nerve cells outside of the hippocampus and cerebellum regions of the brain. In the rest of body, imprinting does not affect UBE3A. There is no SNURF–SNRPN transcript being made in the rest of the body. In all the cells not involved in nerves, if one of the UBE3A genes is nonfunctional from something like a microdeletion or a mutation of the DNA code, the other gene on the other chromosome will be able to compensate. In specific nerve tissue, however, only one gene of the two in the PWS/AS region is expressing the SNURF–SNRPN transcript. The only expressed gene is the one with paternal imprinting on it (Figure 4.18). The maternal gene is not expressed in nerves. And so we see the same kind of pattern of expression as in X inactivation occurring on chromosome 15 and affecting cells of ectodermal origin. However, unlike in cat's pigment cells where the inactivation of one allele is at random, in this situation the maternally inherited alleles are always the ones inactivated.

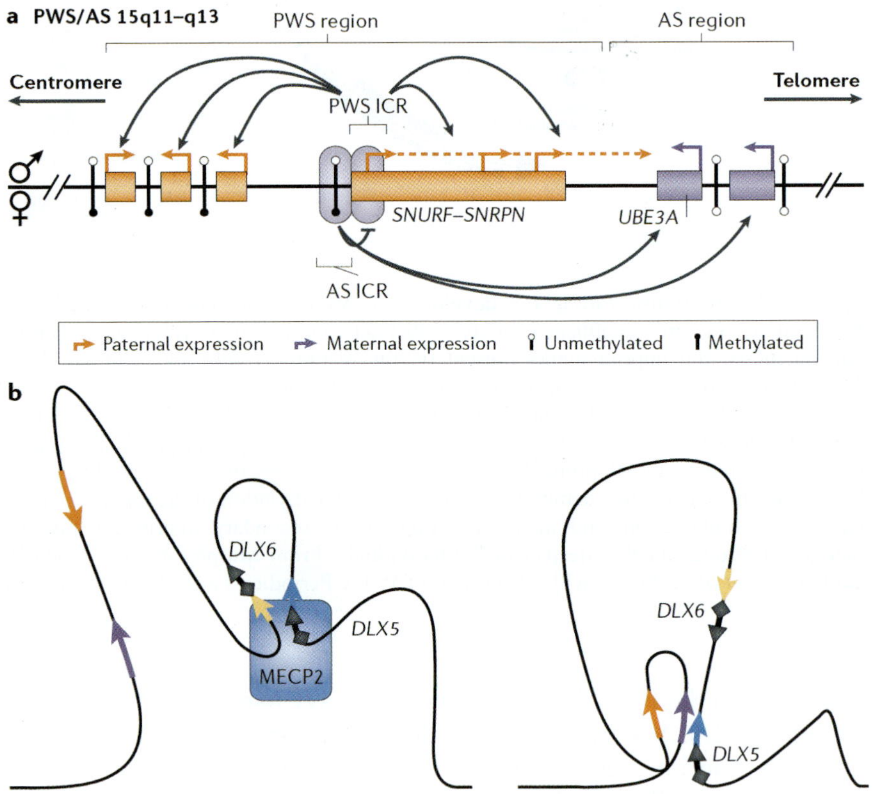

Figure 4.18. There are two imprinted control regions (ICRs) in the PWS–AS locus. If maternally imprinted, the control region allows for the expression of the UBE3A gene by the activity of transcripts from the AS-ICR. The paternally imprinted pattern can lead to expression SNURF-SNRPN. SNRPN produces a long and complex transcript that leads

to the expression of not only SNURF–SNRPN genes, but also several small nucleolar RNAs (snoRNAs). This transcript is also thought to inhibit the expression of UBE3A from the paternal allele because a bit of the RNA binds to the UBE3A gene and prevents it from being transcribed. Expression of SNURF–SNRPN also results in changes in chromatin structure, with increased acetylation and methylation of H3K4 (histone 3 lysine 4), and reduced methylation of H2K9 at the PWS–AS imprinting center[116]. Whether or not the expression of SNURF–SNRPN occurs on paternally methylated DNA depends on a different DNA binding protein called MeCP2 (methyl-CpG-binding protein). MeCP2, if present, binds to sequences with specific paternal methylation patterns, and in doing so, changes the conformation of the DNA forming a sequestered chromatin loop that includes DLX5 and DLX6 from the SNRPN upstream reading frame. MeCP2 prevents the paternal expression pattern. If MeCP2 is absent, as it is in specific types of nerve cells, then and only then is SNURF–SNRPN expressed. And thus control of the SNURF–SNRPN depends on both a preexisting paternal methylation pattern and the absence of the MeCP2 protein[117]. From Bienvenu & Chelly (2006)[118] with permission of Nature Publishing Group.

The seven genes (and a multitude of snoRNAs) in the PWS/AS region are associated with Prader-Willi only when the paternal imprinting is missing or the paternally inherited gene is not expressed because of a mutation (Figure 4.18). And so in the brain, the PWS/AS region acts as a regulon with master genes and master RNAs. It also acts like a single unpaired gene due to imprinting only when it is being expressed in brain tissue. Otherwise it functions like any nonimprinted gene. Thus an even higher order effect acts on the PWS/AS region and results in it functioning in a different way in brain tissue compared to how it functions in the rest of the body.

The nature of that higher order function is due to the DNA binding protein MeCP2 (methyl-CpG-binding protein). MeCP2, if present, binds to DNA sequences with specific paternal methylation patterns, and in doing so, changes the conformation of the DNA strand to form a sequestered chromatin loop that includes DLX5 and DLX6 from the SNRPN upstream reading frame (Figure 4.18). If MeCP2 is absent, as it is in specific types of nerve cells, then, and only then, is SNURF–SNRPN expressed and UBE3A silenced. And thus control of the SNURF–SNRPN depends on both a preexisting paternal methylation pattern combined with the absence of the MeCP2 protein. MeCP2 is only turned off in nerve cells. It is always expressed in other cells[118]. MECP2 is a higher order regulon which controls the SNURF–SNRPN regulon which controls a host of other genes in nerve cells.

Angelman syndrome occurs when the maternally derived UBE3A is mutated or deleted. Since the UBE3A paternal gene is silenced in the nerve cells of the hippocampus and cerebellum, they cannot produce UBE3A and the products of the maternal UBE3A are missing altogether in that part of the brain. While at least four proteins that are bound by the protein coded by the UBE3A gene are known, how exactly not having UBE3A in the hippocampus and cerebellum leads to the syndrome is still unknown. The whole puzzling Prader-Willi/Angelman clinical picture has revealed that imprinting is very important for human brain functioning.

Methylation is also critically important to the constant and ongoing process of gene repair. Life causes DNA breaks. In any given cell on any given day there will be about a million lesions in the genome of each and every cell[119]. These errors and breaks introduced into our genome occur for a variety of reasons. Reactive oxygen species produced as a byproduct of normal metabolism are always present and always have the potential for creating a spontaneous mutation. A huge class of external sources of DNA damage exists ranging from the sun's UV and Earth's background radiation to specific toxins in the plants we eat. The typical result of such toxins is that the correct nucleic acid is present but some of the individual nucleic acids are damaged. They are missing side groups and/or they are inappropriately crosslinked to adjacent nucleic acids.

The processes of mitosis and meiosis also lead to replication errors. The insertion of an incorrect nucleic acid during DNA replication occurs once for every one billion base pairs. The enzymatic machinery doing DNA replication will occasionally slip, particularly in highly repetitive regions, and either skip replicating a section thereby shortening it, or cover the same section too many times, lengthening it. This lengthening or shortening of highly repetitive regions can result in faulty gene regulation. An example of this is found in Fragile X Syndrome. Instability in the region is introduced by having a few extra CGG repeats. Sometimes the extra CGGs are reduced in number in meiosis and sometimes more are added. When the CGG section is lengthened Fragile X Syndrome is the result. Individuals with Fragile X syndrome have a long face, large ears, high palette, large testicles (most are males due to the gene being on the X chromosome and mosaic females being less affected), flat feet, soft skin and low muscle tone combined with varying degrees of intellectual impairment and learning disabilities.

Fragile X was given this name because in the early days of trying to understand what was going on, it was found that the X chromosome would have what looked like a broken end on one arm in light microscopy. There is a gene in this region known as FMR1. This gene codes for fragile X mental retardation protein, or FMRP. This protein is critical for normal intellectual development and has other functions as well. Mutations and changes in this gene are associated with a host of intellectual impairments including autism, tremor and ataxia (lack of voluntary coordination of muscle movements) disorders, as well as some forms of Parkinson's Disease and premature ovarian failure. And so it is a very important protein. The mutations prevent or alter the way the FMRP binds to ribosomes which then down regulates the translation of certain proteins. And so FMRP is a regulon that works at the translation level, not the transcription level. In Fragile X syndrome, an extra section of CGG repeats inserted during meiosis results in the gene producing an mRNA that signals the entire gene should be inappropriately methylated and completely shut down. The precise mechanism of that shut down is uncertain but it is known to be epigenetic in nature. The normal epigenetic regulation of the gene is changed by adding repeat units[120]. There are several other forms of these kinds of repeat number dependent epigenetic disorders arising from errors in DNA replication.

Direct repair of DNA is used for nucleic acids that are missing side groups or inappropriately crosslinked to adjacent nucleic acids. Photolyase is one of the oldest such repair enzymes[121]. When DNA is subjected to ultraviolet light (UVB: 290–320 nm wavelength range[122]) two thymines in sequence on a strand can lose their respective methyl groups. The two then form an inappropriate crosslink consisting of a uracil-uracil cyclobutane dimer. Photolyase binds such dimers with high affinity and uses light energy and cofactors containing the methyl group to directly break the covalent bond between the two uracils and replace the missing methyl groups, converting them back to thymines. Humans and other placental mammals have lost the photolyase enzymes but these enzymes are widespread in other phyla, especially in plants.

Humans and other placental mammals use the second most common form of DNA repair of the uracil-uracil cyclobutane dimer. This form is called nucleotide excision repair. DNA repair enzymes are part of complexes of several proteins that bind directly to the DNA. Mismatched nucleic acids or those with missing side groups or having inappropriate cross bridges produce bulges and twists in the normally smooth DNA helix. The mismatch repair complex detects misshapen DNA and binds to it. The repair complex then removes the faulty nucleotides and replaces them with the correct ones using the opposite strand as the template. A related DNA repair mechanism removes and replaces short (~12 base) sections of DNA instead of just a single base. Again the opposing strand is used as a template. Methylation markers on the original DNA strand mark the older (and more likely to be correct) DNA strand. The newer strand is the one that usually has the nucleic acid(s) replaced[123]. Methylation patterns are added to the new strand only after it has been checked.

The process of crossing over during meiosis involves creating double strand breaks and repairing them. Double strand breaks can occur at other times in the cell's cycle. Double stranded breaks are both particularly dangerous since they can lead to large rearrangements of the genome, and particularly hard to repair because there is no joined opposite strand to use as a template for repair. There are three ways to repair double stranded DNA breaks. First, if the two breaks of the DNA strand are not opposite each other, the broken DNA will have two single stranded tails. This kind of break is repaired by the nonhomologous double stranded DNA repair mechanism. Second, if the strands differ, additional nucleotides are added or replaced until the ends match and then they are rejoined.

If there is a double stranded break but there are no single stranded tails to guide the rejoining then there is no way to be certain the entire code is present. If the two ends are simply joined there is a high probability that some chunk of the code will be lost between the blunt ends. To avoid this, for double stranded DNA with blunt ends, one of two forms of homologous double stranded break repair is used[124] (Figure 4.19). In a mechanism very similar to crossing over at meiosis, the homologous chromosome is pulled close, cut on one strand, joined to the broken DNA and then the homologous strand is used to guide the construction of a new copy of the DNA over the broken section. The third type of DNA repair is specific to double stranded breaks caused by DNA helicases accidentally cutting two strands while doubling the DNA in preparation for cell division. These are repaired by a mechanism that is still poorly understood but is related to the one using homologous chromosomes. Again methylation patterns are used to determine old and new strands. And so we see that methylation is also critical for DNA repair.

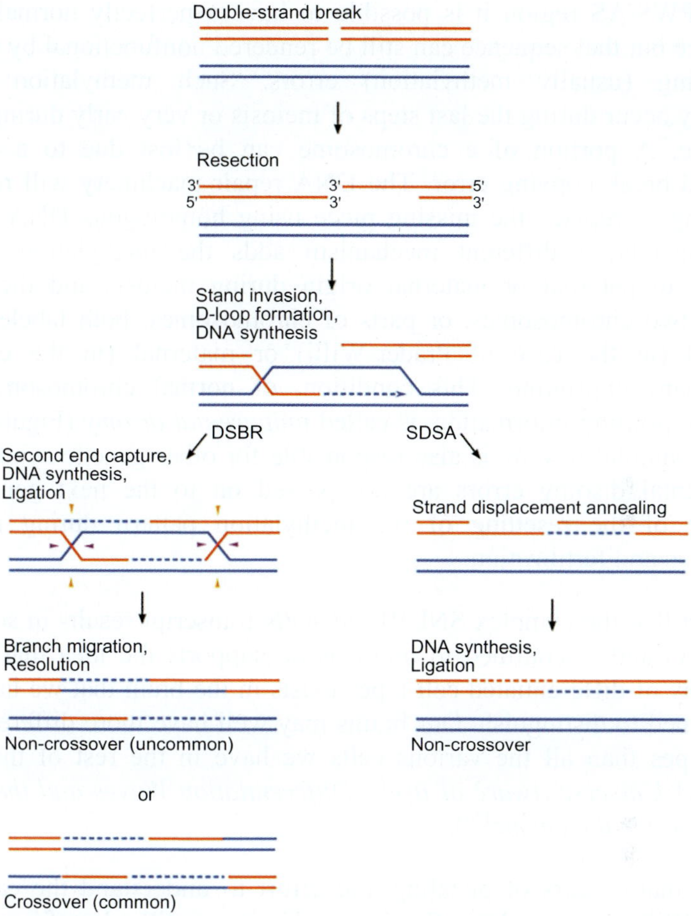

Figure 4.19. Repair of DNA double-strand breaks. Double-strand breaks can be repaired by homologous recombination, including double-strand break repair (DSBR) and synthesis-dependent strand annealing (SDSA). In both pathways, repair is initiated by resection of the break site to provide a single-stranded DNA overhang. "Strand invasion" is when the single strand ends overhang onto a homologous sequence in the matching homologous chromosome. This "invasion" is followed by DNA synthesis at the invading end. During the synthesis a "D loop" forms. After strand invasion and synthesis, the second double stranded end can be captured to form an intermediate at two places. After gap-repair, DNA synthesis and ligation, the structure is resolved. This may or may not result in crossing over. Alternatively, the reaction occurs by strand displacement, annealing of the extended single-strand end to the single stranded DNA on the other break end, followed by gap-filling DNA synthesis and ligation. The alternative form will not mean crossing over. From Sung & Klein (2006)[125] with permission of Nature Publishing Group.

In the PWS/AS region it is possible to have a perfectly normal DNA sequence but that sequence can still be rendered nonfunctional by having imprinting (usually methylation) errors. Such methylation errors typically occur during the last steps of meiosis or very early during rapid cleavage. A portion of a chromosome can be lost due to a double stranded break copying error. The DNA repair machinery will respond by trying to replace the missing piece using homologous DNA repair. Unfortunately, a different mechanism adds the methylation pattern specific to paternal or maternal origin during meiosis and the result can be two chromosomes, or parts of chromosomes, both labeled with paternal (in the case of Prader-Willi) or maternal (in the case of Angelman) imprinting. This condition, of normal chromosomes but faulty imprinting information, is called *uniparental disomy* (Figure 4.16) and uniparental disomy is also responsible for other genetic syndromes. Uniparental disomy errors are not passed on to the next generation because of the resetting of the methylation pattern during gamete formation and fertilization.

The fact that the complex SNURF–SNRPN transcript results in so many snoRNAs and is confined to brain tissue supports the idea that a vast hierarchy of differentiated cell types exists in the brain that we have not yet learned to distinguish. Our brains may well have more differentiated cells types than all the various cells we have in the rest of the body (§6.06 *A Universe Aware of Itself: Differentiation Waves and the Brain* in *Hierarchical Genome*[126]).

It took many years of puzzling and effort to understand the cause of Prader-Willi and Angelman Syndrome. There are still a lot of issues that need explaining. One puzzle is why does imprinting even exist? In the mammal, gender of origin specific silencing of genes (imprinting) has evolved to a high degree. For example, in mammals, nearly all of the genes required to create and run the trophoblast (Figure 4.20), the cells that eventually form the placenta and all the support tissues for the embryo, are run almost entirely from proteins created by genes of paternal origin (Figure 4.10). Conversely, the inner cell mass, which eventually forms the embryo proper (Figure 2.41), runs almost entirely on maternal genes in the earliest stages. There are some exceptions: imprinted genes are known in trophoblast where a maternal gene is the expressing gene in mouse[128]. Interestingly, in mice there is also X inactivation in the trophoblast but it is selective and not random. The

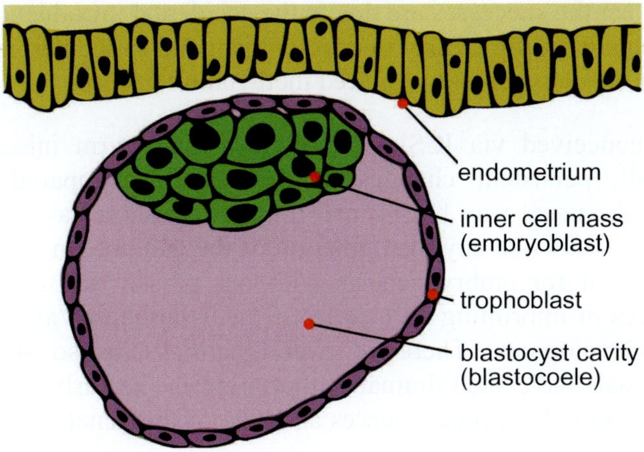

Figure 4.20. Most mammalian embryos implant at an early stage in the wall of the uterus (endometrium) at a stage in which the fertilized egg has divided and produced two tissues, the inner cell mass, which will later form the embryo, and the trophoblast, which will later become the placenta. From Wikipedia[127] with permission under a GNU Free Documentation License.

paternal X is always the one silenced (sequestered). The maternal X is always the active X.

Mammals cannot reproduce parthenogenetically, probably because of imprinting. Imprinting arrests development of artificially activated, unfertilized eggs at different stages, depending on the species[129-132]. An exception is when parthenogenetic cells are mixed in with normal cells in an embryo[130]. The normal cells sometimes somehow compensate for and "control" the abnormal ones well enough to get them through nearly normal embryogenesis. This can also be done with some cancer cells[129]. Imprinting is also behind why cloning of adult mammals, by transferring cell nuclei to an egg with the nucleus removed[133,134], is so far of very limited success[135,136]. If you take a nucleus from a later cell and transplant it into an egg, it carries with it incorrect methylation patterns that often shorten the life of the animal. Dolly[137], the first cloned mammal[138-144] lived only half the normal lifespan of a typical sheep and she had a host of health problems attributed to methylation errors[141], though still poorly understood[140]. The success rate remains low, with frequent developmental anomalies occurring in clones[139], but some animals conceived this way do appear to be healthy and normal[142]. We speculate that the successes are due to the accidental introduction of stem cells which are set aside early in development and so would have little in the way of developmentally acquired methylation patterns.

Children conceived via ICSI (intra-cytoplasmic sperm injection) also have small, persistent changes in methylation compared to those conceived the usual way but no real long term effects have been found yet[144]. So it looks like any manipulation of the genome, in attempting to "reprogram" it for embryogenesis with our present technology, often leaves traces of imprinting such as DNA methylation that are deleterious to normal development. There are several theories ('just so' stories) as to why mammals have such dramatic imprinting in the early embryo. It is our opinion that this process serves as a test of the genetic health of the embryo.

A female axolotl releases 300-4000 fertilized eggs per spawning (Chapter 2). About 10% to 70% will die before hatching[145] presumably because of lack of fertilization or genetic errors. Of those that do survive in the wild, most will be eaten. Only a very small number of young will survive to breed. This means the axolotl has little invested in any one fertilized egg. With mammals, the opposite it true. A typical human female will mature only about 400 eggs in her entire life. Only a few of these will be fertilized and even fewer of them will be successfully birthed and raised to become reproducing adults. During the time a woman is pregnant, she gives up a lot of her own nutrition and, as pregnancy progresses, carrying the unborn makes her bulkier and potentially an easier target for predators, and birthing her offspring can be fatal if things go wrong. She has much more invested in each egg from an evolutionary perspective. The advantage of being a diploid organism is that a deleterious gene is masked by the presence of a second potentially normal gene in the gene pairs. The disadvantage is that faulty genes are indeed masked by heterozygosity and there is little selection against them except for the rare appearance of an offspring with two deleterious alleles.

During the very early phase of mammalian development, imprinting creates two effectively haploid organisms for a short space of time. One such haploid "organism" is the trophoblast (Figure 4.20) run on paternal genes. The other is the inner cell mass run on maternal genes. If the father's genes are faulty, the embryo will not implant and there will be no pregnancy. If the mother's genes are faulty then no embryo will form to develop inside the trophoblast. The pregnancy will continue for a few days or weeks and then fail. Human embryos at the fertilized egg stage have a high proportion of chromosome abnormalities[146]. Imprinting thereby becomes an important quality control measure for assessing the developing embryo. The female mammal that uses imprinting will be much more likely to produce healthy offspring after such a heavy investment of resources into that baby. She will have more successful babies than one who does not use imprinting. This likely explains why so many chromosomal abnormalities are not passed by men to their children even if they carry them. Because of imprinting, the early embryo is functionally a haploid organism with critical genes exposed. Imprinting therefore exposes any abnormal embryos so early in development that embryos are lost often before a woman even knows she was pregnant.

Imprinted genes in mice also affect both maternal and suckling behavior which are critical during the period when the mother is nursing[147]. This also supports our 'screening hypothesis' about embryos. An embryo that has faulty genes will not nurse properly and these effects are also genetically emphasized where the mother has poor mothering behaviors[148]. Both the mouse with poor mothering skills, and the defective offspring in a population, are selected against by the effects of imprinting. Partial support for our hypothesis can be found in the fact that imprinting of placental genes, which would lead to more early placental failures, is more complete in those mammals that have evolved to carry few offspring. In mammals with multiple offspring in each pregnancy, like mice, imprinting effects on behavior at birth are more common. So with a litter of many successful embryos but with one defective embryo, the pregnancy will go ahead but the mother will not nurse the single defective offspring[149]. Imprinting also varies in other ways between different mammalian species[150]. Elaborate imprinting is a relatively new adaptation for mammals and is undergoing rapid evolutionary change[151].

Mammals are not the only ones using imprinting although they use it the most. A form of imprinting also exists in the higher plant class known as angiosperms, or seed bearing plants. This plant imprinting is both similar and different from mammals. In the ovule there are two cells existing together, one haploid and one diploid. Each is fertilized by sperm from pollen. This results in a diploid embryonic plant from the haploid egg and a triploid cell beside it (Figure 4.21). This triploid cell produces the

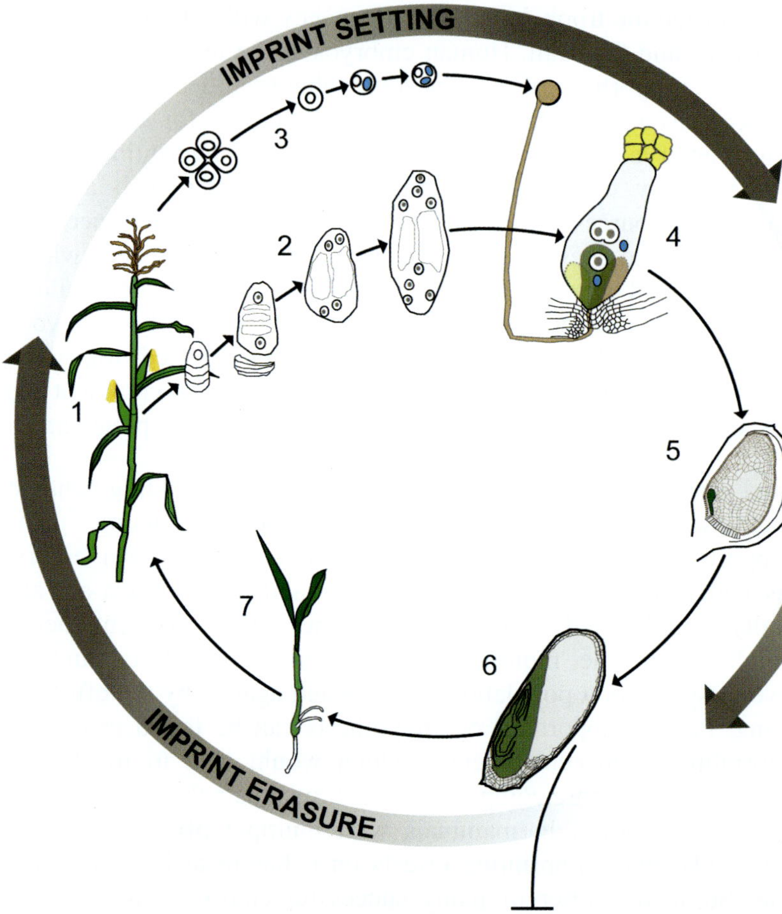

Figure 4.21. "Imprinting, double fertilization, and the plant life history. A plant at flowering (1) develops female and male reproductive organs from somatic tissue in which spore mother cells undergo meiosis and undergo a haploid phase of mega- (2) and micro- (3) gametophyte development. The megagametophyte, or embryo sac, forms two female gametes over the course of three mitotic divisions—the central cell and the egg cell—in addition to two synergid cells that assist in the fertilization process. The micro-gametophyte, or pollen grain, undergoes two mitotic divisions to form a vegetative cell and two male gametes, the sperm cells (blue). At fertilization (4) the pollen tube delivers the two sperm cells to the receptive synergid cell. One sperm cell fertilizes the egg cell (green) and the other sperm cell nucleus combines with the two polar nuclei of the large central cell of the embryo sac. Within the ovule (5) the fertilized central cell subsequently develops into the endosperm (grey), and the fertilized egg cell forms the embryo (green). In the seed at maturity (6), the endosperm is terminally differentiated

and does not contribute genetically to the next generation, whereas the embryo germinates (7) and develops into the mature plant. Genomic imprinting in the embryo requires erasure and resetting of imprinting marks according to the sex of the gametes. Clearly, parental imprints need to be set during gametophyte development and to regulate allelic gene expression during early embryo and endosperm development. Currently available data indirectly indicate that the erasure of imprints occurs during late embryo development or germination. *Color key:* generative cell, sperm cell, and sperm cell nuclei: blue; central and egg cell nuclei: white with grey nucleolus [cf.[154,155]]; pollen grain, pollen tube, and receptive synergid: brown; egg cell, embryo, seedling, and plant: green; central cell and endosperm: grey." From Kaverne (2015)[156] with permission under a Creative Commons Attribution License.

plant equivalent of the trophoblast known as endosperm, the protective and nutritive part of the seed[152]. In that seed, many male genes are silenced. In the endosperm, many female genes are silenced. This imprinting in plants arose from the same basic mechanism in a common ancestor with mammals but it is different from that of mammals. It may serve some other function important to plants, such as something relating to nutrition. However, in general, the role of polyploid cells in embryogenesis, i.e., those cells in specific tissues with more than two copies of the genome per cell, is still being researched[153].

Insects also have their own form of imprinting. A variant imprinting exists in many insects for sex determination. These insects have no X or Y-chromosomes. Rather in the female both maternally and paternally inherited genes remain active. In males, the paternally inherited genes are inactivated. This system is thought to have arisen because females can reproduce parthenogenetically. If a male is present the female will mate and produce both male and female offspring. If no male is present to mate with, a male will arise from an unfertilized egg[157]. Some insects have also evolved to utilize imprinting for other purposes. Certain insects live by using the larvae of other insects as a food source. Interestingly, these same parasitizing insects use a form of imprinting of specific genes to avoid being a second parasite on a larva their close male relatives are already eating. Instead, they will move along and find a new larva to eat. Why paternal relatives and not maternal is not known[158].

In birds and butterflies the males have two larger sex chromosomes and the females have but one. At least for body size, greater variability shows up in the females. For other insects and mammals with sex chromosomes like ours (including *Drosophila*), size variability is greater in males[105]. If we go further back in evolution of vertebrates to the reptiles, the system is based on temperature during embryogenesis with warmer temperatures making females and cooler making males and not differences in the sex chromosomes or imprinting.

Epigenetics then, is the collection of all the various mechanisms, including methylation patterns, which can cause some portions of the genome to be expressed and other portions to not be expressed during the life of an organism in all cells or in specific cell types. This information is added without changing the actual sequence of the DNA itself. Some of this epigenetic information for regulation of gene expression can even be inherited beyond simple cell division and can be passed from one generation to the next. For example we now know that if either your grandmother or your mother smoked during pregnancy, nicotine induced deleterious epigenetic changes can cause you to have asthma[159]. In 2008, the definition of epigenetic trait was defined as a "stably heritable phenotype resulting from changes in a chromosome without alterations in the DNA sequence"[160]. This definition is not universally accepted but it is the definition used in this book. Epigenetics, like our understanding of RNA function, is new and quickly changing with fascinating new discoveries leading the way.

Imagine then, a system where the DNA is loose and exposed if it is being used, and somewhat loose but sequestered if it is not being used now, but might be required soon. The DNA of genes that are not needed now, but may be required later, is sequestered but the sequestration is reversible with the correct signal. Those genes that never need to be accessed can be sequestered permanently in much the same manner as one X chromosome is inactivated in mammalian females.

We can now return to our conceptual model of how all regulons of genes can be divided into one of five categories in any given cell:

1. *Housekeeping Regulons*: regulons always exposed in all cells types.
2. *Differon Regulons*: regulons exposed in a specific cell type, (and by that definition would include the housekeeping regulons).
3. *Differoff - Past Regulons*: regulons that were used in the cell lineage leading up to a given cell type, but have become sequestered in that cell type and cannot be accessed again.

4. *Differoff - Future Regulons*: regulons that are sequestered in a given cell type, but will become exposed when the cell or its descendants differentiate into new types of cells.
5. *Differoff - Unused Regulons*: regulons that are sequestered in a given cell type and have not been used, are not being used now, and will never be used, by that cell or any of its developmental descendants.

Further, imagine a system where there are proteins and RNA that use the control elements that exist in the DNA sequence of genes in order to up regulate and down regulate them. There are also specific signal sequences in the DNA to sequester or expose whole segments of the DNA. All of this is organized into a hierarchy. The hierarchy was created over evolutionary time as we moved from simple to more complex organisms. During development we can imagine a cell with the whole genome at the beginning of the process. Most of the DNA is reversibly sequestered and not in use. There is a small exposed section that is regulated by proteins and RNA from the maternal (and to a lesser extent possibly paternal) preloading of the egg. At each signal to differentiate, the regulating proteins and RNAs respond by exposing one section of DNA and resequestering another section and then proteins use epigenetic control such as methylation to seal the sequestration once a section is no longer in use. Since most of the genes will never be used in any one tissue, most of the DNA that is sequestered is never exposed again.

This entire hierarchical system has the epigenetic sequestration removed and reset with each round of meiosis except for a few used for keeping track of paternal versus maternal origin or for DNA repair. These are required for functions such as testing the fitness of the mammalian embryo. We can now imagine how the genome is used during embryogenesis. The only missing piece is what is the actual signal that tells each cell what part of the genome to sequester and what part to leave exposed and why is that signal confined to a particular subset of cells and why is the signal received only at certain times during embryogenesis? Thanks to genetics and epigenetics we know a lot about how we get the right cell. But genetics and epigenetics can only tell us that one part. This is why embryogenesis is still a largely unsolved problem in spite of the huge increase in our understanding of genetics and epigenetics. The missing part in our understanding is why do we get the right cell in the right place, and only at the right time? We think we have found the missing signal. It is differentiation waves.

1. Kirkpatrick, M. & N. Barton (2006). Chromosome inversions, local adaptation and speciation. *Genetics* **173**(1), 419-434.
2. Shapira, S.K. (1998). An update on chromosome deletion and microdeletion syndromes. *Current Opinion in Pediatrics* **10**(6), 622-627.
3. Roukos, V. & T. Misteli (2014). The biogenesis of chromosome translocations. *Nature Cell Biology* **16**(4), 293-300.
4. Lönnig, W.E. & H. Saedler (2002). Chromosome rearrangements and transposable elements. *Annual Review of Genetics* **36**, 389-410.
5. Lupski, J.R., J.R. Roth & G.M. Weinstock (1996). Chromosomal duplications in bacteria, fruit flies, and humans. *American Journal of Human Genetics* **58**(1), 21-27.
6. Keller, E.F. (1983). *A Feeling for the Organism, The Life and Work of Barbara McClintock*. New York, W.H. Freeman and Co.
7. Rebollo, R., B. Horard, B. Hubert & C. Vieira (2010). Jumping genes and epigenetics: Towards new species. *Gene* **454**(1-2), 1-7.
8. Alqallaf, A.K., F.M. Alkoot & M.e.S. Aldabbous (2013). Discovering the genetics of autism. In: *Recent Advances in Autism Spectrum Disorders*. Ed.: M. Fitzgerald, InTech. **I**: http://www.intechopen.com/books/recent-advances-in-autism-spectrum-disorders-volume-i/discovering-the-genetics-of-autism.
9. Feschotte, C., N. Jiang & S.R. Wessler (2002). Plant transposable elements: where genetics meets genomics. *Nature Reviews Genetics* **3**(5), 329-341.
10. Yahya-Graison, E.A., J. Aubert, L. Dauphinot, I. Rivals, M. Prieur, G. Golfier, J. Rossier, L. Personnaz, N. Créau, H. Bléhaut, S. Robin, J.M. Delabar & M.C. Potier (2007). Classification of human chromosome 21 gene-expression variations in down syndrome: Impact on disease phenotypes. *American Journal of Human Genetics* **81**(3), 475-491.
11. Centers for Disease Control and Prevention (2012). A drawing of the facial features of Down syndrome. https://commons.wikimedia.org/wiki/File:Down_syndrome_lg.jpg
12. Baumgartner, A., P. Van Hummelen, X.R. Lowe, I.D. Adler & A.J. Wyrobek (1999). Numerical and structural chromosomal abnormalities detected in human sperm with a combination of multicolor FISH assays. *Environmental and Molecular Mutagenesis* **33**(1), 49-58.
13. Nagaoka, S.I., T.J. Hassold & P.A. Hunt (2012). Human aneuploidy: mechanisms and new insights into an age-old problem. *Nature Reviews Genetics* **13**(7), 493-504.
14. Wikipedia (2015). Robertsonian translocation. https://en.wikipedia.org/wiki/Robertsonian_translocation
15. Lockstone, H.E., L.W. Harris, J.E. Swatton, M.T. Wayland, A.J. Holland & S. Bahn (2007). Gene expression profiling in the adult Down syndrome brain. *Genomics* **90**(6), 647-660.

16. FitzPatrick, D.R., J. Ramsay, N.I. McGill, M. Shade, A.D. Carothers & N.D. Hastie (2002). Transcriptome analysis of human autosomal trisomy. *Human Molecular Genetics* **11**(26), 3249-3256.
17. Briggs, J.A., E.A. Mason, D.A. Ovchinnikov, C.A. Wells & E.J. Wolvetang (2013). Concise review: New paradigms for Down syndrome research using induced pluripotent stem cells: Tackling complex human genetic disease. *Stem Cells Translational Medicine* **2**(3), 175-184.
18. Puck, T.T., P. Webb & R. Johnson (1998). Genome exposure and regulation in mammalian cells. *Somatic Cell and Molecular Genetics* **24**(5), 291-301.
19. Simpson, F. (1903). *The Book of the Cat*. New York, Cassell and Company, Limited.
20. Centerwall, W.R. & K. Benirschke (1973). Male tortoiseshell and calico (T-C) cats: animal models of sex chromosome mosaics, aneuploids, polyploids, and chimerics. *Journal of Heredity* **64**, 272-278.
21. Lyon, M.F. (1961). Gene action in the X-chromosome of the mouse (*Mus musculus* L.). *Nature* **190**, 372-373.
22. Ohno, S., W.D. Kaplan & R. Kinosita (1959). Formation of the sex chromatin by a single X-chromosome in liver cells of *Rattus norvegicus*. *Experimental Cell Research* **18**(2), 415-418.
23. Fu, Y., G.Z. Luo, K. Chen, X. Deng, M. Yu, D. Han, Z. Hao, J. Liu, X. Lu, L.C. Dore, X. Weng, Q. Ji, L. Mets & C. He (2015). N^6-methyldeoxyadenosine marks active transcription start sites in chlamydomonas. *Cell* **161**(4), 879-892.
24. Greer, E.L., M.A. Blanco, L. Gu, E. Sendinc, J. Liu, D. Aristizábal-Corrales, C.H. Hsu, L. Aravind, C. He & Y. Shi (2015). DNA methylation on N^6-adenine in *C. elegans*. *Cell* **161**(4), 868-878.
25. Heyn, H. & M. Esteller (2015). An adenine code for DNA: A second life for N6-methyladenine. *Cell* **161**(4), 710-713.
26. Rood, J. (2015). New Epigenetic Mark Found on Metazoan DNA: Three studies identify roles for N6-methyladenosine in algae, worms, and flies. http://www.the-scientist.com//?articles.view/articleNo/42858/title/New-Epigenetic-Mark-Found-on-Metazoan-DNA/
27. Zhang, G., H. Huang, D. Liu, Y. Cheng, X. Liu, W. Zhang, R. Yin, D. Zhang, P. Zhang, J. Liu, C. Li, B. Liu, Y. Luo, Y. Zhu, N. Zhang, S. He, C. He, H. Wang & D. Chen (2015). N^6-methyladenine DNA modification in *Drosophila*. *Cell* **161**(4), 893-906.
28. Plongthongkum, N., D.H. Diep & K. Zhang (2014). Advances in the profiling of DNA modifications: cytosine methylation and beyond. *Nature Reviews Genetics* **15**(10), 647-661.
29. Roberts, R.J. (1994). An amazing distortion in DNA induced by a methyltransferase (Nobel Lecture). *Angewandte Chemie International Edition in English* **33**(12), 1222-1228.
30. Law, J.A. & S.E. Jacobsen (2010). Establishing, maintaining and modifying DNA methylation patterns in plants and animals. *Nature Reviews Genetics* **11**(3), 204-220.
31. Cheng, X., S. Kumar, J. Posfai, J.W. Pflugrath & R.J. Roberts (1993). Crystal structure of the Hhal DNA methyltransferase complexed with S-adenosyl-l-methionine. *Cell* **74**(2), 299-307.

32. Roberts, R.J. (2014). A Modern Day Gene Genie Sir Richard Roberts on Rebase. https://www.neb.com/tools-and-resources/feature-articles/a-modern-day-gene-genie-sir-richard-roberts-on-rebase
33. Roberts, R.J. & X. Cheng (1998). Base flipping. *Annual Review of Biochemistry* **67**, 181-198.
34. Rohs, R. (2014). The Rohs Lab. http://rohslab.cmb.usc.edu/
35. Dantas Machado, A.C., T. Zhou, S. Rao, P. Goel, C. Rastogi, A. Lazarovici, H.J. Bussemaker & R. Rohs (2014). Evolving insights on how cytosine methylation affects protein–DNA binding. *Briefings in Functional Genomics*, doi: 10.1093/bfgp/elu1040.
36. Lazarovici, A., T. Zhou, A. Shafer, A.C. Dantas Machado, T.R. Riley, R. Sandstrom, P.J. Sabo, Y. Lu, R. Rohs, J.A. Stamatoyannopoulos & H.J. Bussemaker (2013). Probing DNA shape and methylation state on a genomic scale with DNase I. *Proceedings of the National Academy of Sciences USA* **110**(16), 6376-6381.
37. Spruijt, C.G. & M. Vermeulen (2014). DNA methylation: old dog, new tricks? *Nature Structural & Molecular Biology* **21**(11), 949-954.
38. Berletch, J.B., F. Yang, J. Xu, L. Carrel & C.M. Disteche (2011). Genes that escape from X inactivation. *Human Genetics* **130**(2), 237-245.
39. Fredga, K. & M.G. Bulmer (1988). Aberrant chromosomal sex-determining mechanisms in mammals, with special reference to species with XY females. *Philosophical Transactions of the Royal Society of London Series B-Biological Sciences* **322**(1208), 83-94.
40. Schulz, E.G. & E. Heard (2013). Role and control of X chromosome dosage in mammalian development. *Current Opinion in Genetics & Development* **23**(2), 109-115.
41. Disteche, C.M. (2012). Dosage compensation of the sex chromosomes. *Annual Review of Genetics* **46**, 537-560.
42. Bellott, D.W., J.F. Hughes, H. Skaletsky, L.G. Brown, T. Pyntikova, T.-J. Cho, N. Koutseva, S. Zaghlul, T. Graves, S. Rock, C. Kremitzki, R.S. Fulton, S. Dugan, Y. Ding, D. Morton, Z. Khan, L. Lewis, C. Buhay, Q. Wang, J. Watt, M. Holder, S. Lee, L. Nazareth, J. Alföldi, S. Rozen, D.M. Muzny, W.C. Warren, R.A. Gibbs, R.K. Wilson & D.C. Page (2014). Mammalian Y chromosomes retain widely expressed dosage-sensitive regulators. *Nature* **508**(7497), 494-499.
43. Cortez, D., R. Marin, D. Toledo-Flores, L. Froidevaux, A. Liechti, P.D. Waters, F. Grützner & H. Kaessmann (2014). Origins and functional evolution of Y chromosomes across mammals. *Nature* **508**(7497), 488-493.
44. Pessia, E., J. Engelstädter & G.A.B. Marais (2014). The evolution of X chromosome inactivation in mammals: the demise of Ohno's hypothesis? *Cellular and Molecular Life Sciences* **71**(8), 1383-1394.
45. Avner, P. & E. Heard (2001). X-chromosome inactivation: Counting, choice and initiation. *Nature Reviews Genetics* **2**(1), 59-67.
46. Brown, C. & J. Greally (2003). A stain upon the silence: genes escaping X inactivation. *Trends in Genetics* **19**(8), 432-438.
47. Dietzel, S. (2008). Barr body: File:Sd4hi-unten-crop.jpg. https://en.wikipedia.org/wiki/Barr_body
48. Dupont, C. & J. Gribnau (2013). Different flavors of X-chromosome inactivation in mammals. *Current Opinion in Cell Biology* **25**(3), 314-321.

49. Ohhata, T. & A. Wutz (2013). Reactivation of the inactive X chromosome in development and reprogramming. *Cellular and Molecular Life Sciences* **70**(14), 2443-2461.
50. Hill, M.A. (2014). Embryology: Mouse Development. https://php.med.unsw.edu.au/embryology/index.php?title=Mouse_Development
51. Gordon, R. (2011). Epilogue: the diseased breast lobe in the context of X-chromosome inactivation and differentiation waves. In: *Breast Cancer: A Lobar Disease*. Ed.: T. Tot. London, Springer: 205-210.
52. Zaidi, S.K., D.W. Young, M.A. Montecino, J.B. Lian, A.J. van Wijnen, J.L. Stein & G.S. Stein (2010). Mitotic bookmarking of genes: a novel dimension to epigenetic control. *Nature Reviews Genetics* **11**(8), 583-589.
53. Sarge, K.D. & O.-K. Park-Sarge (2009). Mitotic bookmarking of formerly active genes Keeping epigenetic memories from fading. *Cell Cycle* **8**(6), 818-823.
54. Arampatzi, P., M. Gialitakis, T. Makatounakis & J. Papamatheakis (2013). Gene-specific factors determine mitotic expression and bookmarking via alternate regulatory elements. *Nucleic Acids Research* **41**(4), 2202-2215.
55. Love, J. (2010). Sex Determination, Chromosomes and Mosaicism. http://www.synapses.co.uk/genetics/sexone.html
56. Nowack, R. (1993). Curious X-inactivation facts about calico cats. *Journal of NIH Research* **5**, 60-65.
57. Jacobs, G.H. (2009). Evolution of colour vision in mammals. *Philosophical Transactions of the Royal Society B-Biological Sciences* **364**(1531), 2957-2967.
58. Jacobs, G.H. (2008). Primate color vision: A comparative perspective. *Visual Neuroscience* **25**(5-6), 619-633.
59. Sugita, Y. & K. Tasaki (1988). Rods also participate in human color vision. *Tohoku Journal of Experimental Medicine* **154**(1), 57-62.
60. Wikipedia (2014). Cone cell. http://en.wikipedia.org/wiki/Cone_cell
61. Conway, B.R. (2009). Color vision, cones, and color-coding in the cortex. *Neuroscientist* **15**(3), 274-290.
62. Rister, J. & C. Desplan (2011). The retinal mosaics of opsin expression in invertebrates and vertebrates. *Developmental Neurobiology* **71**(12), 1212-1226.
63. McCrone, J. (2002). Tetrachromats. *Lancet Neurology* **1**(2), 136.
64. Jordan, G., S.S. Deeb, J.M. Bosten & J.D. Mollon (2010). The dimensionality of color vision in carriers of anomalous trichromacy. *Journal of Vision* **10**(8), #12.
65. Jakab, Z. & K. Wenzel (2004). Detecting tetrachromacy in human subjects. *Perception* **33**(ECVP Abstract Supplement), 64.
66. Roorda, A. & D.R. Williams (1999). The arrangement of the three cone classes in the living human eye. *Nature* **397**(6719), 520-522.
67. Falconer, D.S. & P.J. Avery (1978). Variability of chimaeras and mosaics. *Journal of Embryology and Experimental Morphology* **43**, 195-219.
68. Hofer, H.J. & D.R. Williams (2014). Color vision and the retinal mosaic. In: *New Visual Neurosciences*. Eds.: J.S. Werner & L.M. Chalupa. Cambridge, MIT Press: 469-483.
69. Hofer, H., J. Carroll, J. Neitz, M. Neitz & D.R. Williams (2005). Organization of the human trichromatic cone mosaic [+ erratum: **26**(2), 722]. *Journal of Neuroscience* **25**(42), 9669-9679.
70. Wikipedia (2014). Pleistocene. https://en.wikipedia.org/wiki/Pleistocene

71. Silverman, I. & M. Eals (1992). Sex differences in spatial abilities: Evolutionary theory and data. In: *The Adapted Mind: Evolutionary Psychology and the Generation of Culture*. Eds.: J.H. Barkow, L. Cosmides & J. Tooby. New York, NY, US, Oxford University Press: 533-549.

72. Silverman, I., J. Choi & M. Peters (2007). The hunter-gatherer theory of sex differences in spatial abilities: Data from 40 countries. *Archives of Sexual Behavior* **36**(2), 261-268.

73. Pickford, R.W. (1965). The genetics of colour blindness. In: *CIBA Foundation Symposium on Colour Vision, Physiology and Experimental Psychology*. Eds.: A.V.S. DeRelick & J. Knight. Boston, Litte, Brown & Co.: 228-248.

74. Ling, Y., A. Hurlbert & L. Robinson (2006). Sex differences in colour preference. In: *Progress in Colour Studies: Psychological Aspects*. Eds.: C.P. Biggam, C. Kay & N. Pitchford. Amsterdam, John Benjamins Publishing Co.: 173-188.

75. Abramov, I., J. Gordon, O. Feldman & A. Chavarga (2012). Sex & vision I: Spatio-temporal resolution. *Biology of Sex Differences* **3**, #20.

76. Pickford, R.W. (1963). Natural selection and colour blindness. *The Eugenics Review* **55**(2), 97-101.

77. Yokoyama, S. & N. Takenaka (2005). Statistical and molecular analyses of evolutionary significance of red-green color vision and color blindness in vertebrates. *Molecular Biology and Evolution* **22**(4), 968-975.

78. Judd, D.B. (1943). Colorblindness and the detection of camouflage. *Science* **97**(2529), 544-546.

79. Morgan, M.J., A. Adam & J.D. Mollon (1992). Dichromats detect colour-camouflaged objects that are not detected by trichromats. *Proceedings of the Royal Society of London B* **248**(1323), 291-295.

80. Neel, J.V. & R.H. Post (1963). Transitory “positive” selection for colorblindness? *Eugenics Quarterly* **10**(1), 33-35.

81. Adam, A. (1969). A further query on color blindness and natural selection. *Biodemography and Social Biology* **16**(3), 197-202.

82. Mollon, J.D. (1989). "Tho' she kneel'd in that place where they grew...": The uses and origins of primate color-vision. *Journal of Experimental Biology* **146**(1), 21-38.

83. Osorio, D. & M. Vorobyev (1996). Colour vision as an adaptation to frugivory in primates. *Proceedings of the Royal Society of London. Series B, Biological Sciences* **263**(1370), 593-599.

84. Wsiaki, A. (2000). *Design of a Computer Graphics Viewer for Colour Deficiency Correction [Undergraduate Thesis, Supervisor: Richard Gordon]*. Winnipeg, Department of Electrical & Computer Engineering, University of Manitoba.

85. Dishman, L. (2013). The amazing story of the $300 glasses that correct colorblindness: Mark Changizi and Tim Barber turned research on human vision and blood flow into colorblindness-correcting glasses you can buy on Amazon. Here’s how they did it. http://www.popsci.com/science/article/2013-02/amazing-story-300-glasses-can-cure-colorblindness

86. Gordon, N.K. (2015). Genetics of Color Vision. http://embryogenesisexplained.org/2015/09/25/genetics-of-color-vision/

87. Scott, E.Z. (2015). Colorblind Man Sees Purple for the First Time. https://www.youtube.com/watch?v=WCcxwieuDH0

88. Changizi, M.A., Q. Zhang & S. Shimojo (2006). Bare skin, blood and the evolution of primate colour vision. *Biology Letters* **2**(2), 217-221.
89. Tovee, M.J. (1994). The molecular genetics and evolution of primate colour vision. *Trends in Neurosciences* **17**(1), 30-37.
90. Dulai, K.S., M. von Dornum, J.D. Mollon & D.M. Hunt (1999). The evolution of trichromatic color vision by opsin gene duplication in New World and Old World primates. *Genome Research* **9**(7), 629-638.
91. Bachtrog, D. (2006). A dynamic view of sex chromosome evolution. *Current Opinion in Genetics & Development* **16**(6), 578-585.
92. Mácha, J., R. Teichmanová, A.K. Sater, D.E. Wells, T. Tlapáková, L.B. Zimmerman & V. Krylov (2012). Deep ancestry of mammalian X chromosome revealed by comparison with the basal tetrapod *Xenopus tropicalis*. *BMC Genomics* **13**, #315.
93. Nathans, J., D. Thomas & D.S. Hogness (1986). Molecular genetics of human color vision: the genes encoding blue, green, and red pigments. *Science* **232**(4747), 193-202.
94. Wikipedia (2014). Heterozygote advantage. http://en.wikipedia.org/wiki/Heterozygote_advantage
95. Brues, A.M. (1969). Genetic load and its varieties. *Science* **164**(3884), 1130-1136.
96. Wills, C. (2011). Rapid recent human evolution and the accumulation of balanced genetic polymorphisms. *High Altitude Medicine & Biology* **12**(2), 149-155.
97. Lehrke, R. (1972). A theory of X-linkage of major intellectual traits. *American Journal of Mental Deficiency* **76**(6), 611-619.
98. Lubs, H.A. (1999). The other side of the coin: A hypothesis concerning the importance of genes for high intelligence and evolution of the X chromosome. *American Journal of Medical Genetics* **85**(3), 206-208.
99. Gécz, J., C. Shoubridge & M. Corbett (2009). The genetic landscape of intellectual disability arising from chromosome X. *Trends in Genetics* **25**(7), 308-316.
100. Johnson, W., A. Carothers & I.J. Deary (2009). A role for the *X* chromosome in sex differences in variability in general intelligence? *Perspectives on Psychological Science* **4**(6), 598-611.
101. Hoenigsberg, H.F., M. Ordoñez, M.M.E. Depolanco & A. Espinel (1982). Population genetics in the American tropics .XVI. Data on partial dominance of recessives in *Drosophila willistoni*. *Theoretical and Applied Genetics* **61**(2), 183-191.
102. Crow, J.F. (1998). Overdominance - A half-century later. *Evolutionary Biology* **30**, 1-13.
103. Montgomery, S. (2015). *The Soul of an Octopus: A Surprising Exploration into the Wonder of Consciousness*, Atria Books.
104. Craig, I.W., C.M.A. Haworth & R. Plomin (2009). Commentary on "A role for the X chromosome in sex differences in variability in general intelligence?" (Johnson et al., 2009). *Perspectives on Psychological Science* **4**(6), 615-621.
105. Reinhold, K. & L. Engqvist (2013). The variability is in the sex chromosomes. *Evolution* **67**(12), 3662-3668.

106. Johnson, W., A. Carothers & I.J. Deary (2009). Speculation to inform and speculation to explore: Response to Craig et al. (2009) and Turkheimer & Halpern (2009). *Perspectives on Psychological Science* **4**(6), 622-623.
107. Turkheimer, E. & D.F. Halpern (2009). Sex differences in variability for cognitive measures: Do the ends justify the genes? (Commentary on Johnson et al., 2009). *Perspectives on Psychological Science* **4**(6), 612-614.
108. Giummo, J. & W. Johnson (2012). Testing for evidence of an X-linked genetic basis for a greater proportion of males with high cognitive ability. *Behavior Genetics* **42**(5), 808-819.
109. Iqbal, K., S.G. Jin, G.P. Pfeifer & P.E. Szabó (2011). Reprogramming of the paternal genome upon fertilization involves genome-wide oxidation of 5-methylcytosine. *Proceedings of the National Academy of Sciences USA* **108**(9), 3642-3647.
110. Wikipedia (2015). Cache (computing). https://en.wikipedia.org/wiki/Cache_(computing)
111. Jirtle, R.L. & M.K. Skinner (2007). Environmental epigenomics and disease susceptibility. *Nature Reviews Genetics* **8**(4), 253-262.
112. Strachan, T. & A.P. Read (1999). *Figure 8.19 Changes in DNA methylation during mammalian development in: Human Molecular Genetics 2*. New York, Wiley-Liss, 2nd.
113. Williams, C.A. (2015). University of Florida Department of Pediatrics: Division of Genetics and Metabolism: Teaching Resources: Syndrome Gene Maps: Angelman and Prader-Willi Gene Region. https://www.peds.ufl.edu/divisions/genetics/teaching/syndrome_gene_maps.htm
114. Murphy, S.K. & R.L. Jirtle (2003). Imprinting evolution and the price of silence. *Bioessays* **25**(6), 577-588.
115. Dolinoy, D.C., R. Das, J.R. Weidman & R.L. Jirtle (2007). Metastable epialleles, imprinting, and the fetal origins of adult diseases. *Pediatric Research* **61**(5 Part 2), 30R-37R.
116. Makedonski, K., L. Abuhatzira, Y. Kaufman, A. Razin & R. Shemer (2005). MeCP2 deficiency in Rett syndrome causes epigenetic aberrations at the PWS/AS imprinting center that affects UBE3A expression. *Human Molecular Genetics* **14**, 1049-1058.
117. Horike, S., S. Cai, M. Miyano, J.F. Cheng & T. Kohwi-Shigematsu (2005). Loss of silent-chromatin looping and impaired imprinting of DLX5 in Rett syndrome. *Nature Genetics* **37**(1), 31-40.
118. Bienvenu, T. & J. Chelly (2006). Molecular genetics of Rett syndrome: when DNA methylation goes unrecognized [+ Corrigendum]. *Nature Reviews Genetics* **7**(6), 415-426, 583.
119. Wikipedia (2015). DNA repair. https://en.wikipedia.org/wiki/DNA_repair
120. Wikipedia (2015). Fragile X syndrome. https://en.wikipedia.org/wiki/Fragile_X_syndrome
121. Wikipedia (2015). Photolyase. http://en.wikipedia.org/wiki/Photolyase
122. Wikipedia (2015). Ultraviolet. https://en.wikipedia.org/wiki/Ultraviolet
123. Wikipedia (2015). Nucleotide excision repair. http://en.wikipedia.org/wiki/Nucleotide_excision_repair
124. Wikipedia (2015). Homologous recombination. http://en.wikipedia.org/wiki/Homologous_recombination

125. Sung, P. & H. Klein (2006). Mechanism of homologous recombination: mediators and helicases take on regulatory functions. *Nature Reviews Molecular Cell Biology* **7**(10), 739-750.
126. Gordon, R. (1999). *The Hierarchical Genome and Differentiation Waves: Novel Unification of Development, Genetics and Evolution*. Singapore & London, World Scientific & Imperial College Press.
127. Seans Potato Business (2007). Blastocyst. http://commons.wikimedia.org/wiki/File:Blastocyst_English.svg
128. Mancini-Dinardo, D., S.J. Steele, J.M. Levorse, R.S. Ingram & S.M. Tilghman (2006). Elongation of the *Kcnq1ot1* transcript is required for genomic imprinting of neighboring genes. *Genes & Development* **20**(10), 1268-1282.
129. Mintz, B. & K. Illmensee (1975). Normal genetically mosaic mice produced from malignant teratocarcinoma cells. *Proceedings of the National Academy of Sciences USA* **72**(9), 3585-3589.
130. Strain, L., J.P. Warner, T. Johnston & D.T. Bonthron (1995). A human parthenogenetic chimera. *Nature Genetics* **11**(2), 164-169.
131. Rougier, N. & Z. Werb (2001). Minireview: parthenogenesis in mammals. *Molecular Reproduction and Development* **59**(4), 468-474.
132. Pennarossa, G., A. Paffoni, G. Ragni, F. Gandolfi & T.A.L. Brevini (2011). Parthenogenesis in mammals: pros and cons in pluripotent cell derivation. *Central European Journal of Biology* **6**(5), 770-775.
133. Galli, C., I. Lagutina & G. Lazzari (2003). Introduction to cloning by nuclear transplantation. *Cloning and Stem Cells* **5**(4), 223-232.
134. Stocum, D.L. (2006). Somatic cell nuclear transfer (SCNT). http://www.britannica.com/EBchecked/topic/1382860/somatic-cell-nuclear-transfer-SCNT
135. Meng, L., R.-X. Jia, Y.-Y. Sun, Z.-Y. Wang, Y.-J. Wan, Y.-L. Zhang, B.-S. Zhong & F. Wang (2014). Growth regulation, imprinting, and epigenetic transcription-related gene expression differs in lung of deceased transgenic cloned and normal goats. *Theriogenology* **81**(3), 459-466.
136. Zhang, X., D. Wang, Y. Han, F. Duan, Q. Lv & Z. Li (2014). Altered imprinted gene expression and methylation patterns in mid-gestation aborted cloned porcine fetuses and placentas. *Journal of Assisted Reproduction and Genetics* **31**(11), 1511-1517.
137. Wilmut, I. (2001). *The Second Creation: Dolly and the Age of Biological Control*. cambridge, Harvard University Press.
138. Wilmut, I., A.E. Schnieke, J. McWhir, A.J. Kind & K.H.S. Campbell (1997). Viable offspring derived from fetal and adult mammalian cells. *Nature* **385**, 810-813.
139. Campbell, K.H.S., R. Alberio, I. Choi, P. Fisher, R.D.W. Kelly, J.H. Lee & W. Maalouf (2005). Cloning: Eight years after Dolly. *Reproduction in Domestic Animals* **40**(4), 256-268.
140. Vajta, G. & M. Gjerris (2006). Science and technology of farm animal cloning: State of the art. *Animal Reproduction Science* **92**(3-4), 211-230.
141. Wilmut, I., G. Sullivan & J. Taylor (2009). A decade of progress since the birth of Dolly. *Reproduction Fertility and Development* **21**(1), 95-100.
142. Akagi, S., M. Geshi & T. Nagai (2013). Recent progress in bovine somatic cell nuclear transfer. *Animal Science Journal* **84**(3), 191-199.

143. Wilmut, I. (2013). A tribute to Keith Campbell: The birth of the first clone of an adult vertebrate, "Dolly" the sheep. *Cellular Reprogramming* **15**(5), 339-343.
144. Whitelaw, N., S. Bhattacharya, G. Hoad, G.W. Horgan, M. Hamilton & P. Haggarty (2014). Epigenetic status in the offspring of spontaneous and assisted conception. *Human Reproduction* **29**(7), 1452-1458.
145. Armstrong, J.B. & S.T. Duhon (1989). Induced spawnings, artificial insemination, and other genetic manipulations. In: *Developmental Biology of the Axolotl*. Eds.: J.B. Armstrong & G.M. Malacinski. New York, Oxford University Press: 228-235.
146. Koot, Y.E.M., G. Teklenburg, M.S. Salker, J.J. Brosens & N.S. Macklon (2012). Molecular aspects of implantation failure. *Biochimica et Biophysica Acta-Molecular Basis of Disease* **1822**(12), 1943-1950.
147. Curley, J.P. (2011). Is there a genomically imprinted social brain? *BioEssays* **33**(9), 662-668.
148. Isles, A.R. & A.J. Holland (2005). Imprinted genes and mother - offspring interactions. *Early Human Development* **81**(1), 73-77.
149. Wood, A.J. & R.J. Oakey (2006). Genomic imprinting in mammals: emerging themes and established theories. *PLoS Genetics* **2**(11), 1677-1685.
150. Bischoff, S.R., S. Tsai, N. Hardison, A.A. Motsinger-Reif, B.A. Freking, D. Nonneman, G. Rohrer & J.A. Piedrahita (2009). Characterization of conserved and nonconserved imprinted genes in swine. *Biology of Reproduction* **81**(5), 906-920.
151. Keverne, E.B. (2014). Significance of epigenetics for understanding brain development, brain evolution and behaviour. *Neuroscience* **264**, 207-217.
152. Feil, R. & F. Berger (2007). Convergent evolution of genomic imprinting in plants and mammals. *Trends in Genetics* **23**(4), 192-199.
153. Fox, D.T. & R.J. Duronio (2013). Endoreplication and polyploidy: insights into development and disease. *Development* **140**(1), 3-12.
154. Brangwynne, C.P., T.J. Mitchison & A.A. Hyman (2011). Active liquid-like behavior of nucleoli determines their size and shape in *Xenopus laevis* oocytes. *Proceedings of the National Academy of Sciences USA* **108**(11), 4334-4339.
155. Marko, J.F. (2012). The liquid drop nature of nucleoli. *Nucleus-Austin* **3**(2), 115-117.
156. Dickinson, H. & S. Scholten (2013). And baby makes three: Genomic imprinting in plant embryos. *PLoS Genetics* **9**(12), e1003981.
157. Nur, U. (1990). Heterochromatization and euchromatization of whole genomes in scale insects (Coccoidea: Homoptera). *Development*(Suppl.), 29-34.
158. Lizé, A., A.M. Cortesero, A. Atlan & D. Poinsot (2007). Kin recognition in *Aleochara bilineata* could support the kinship theory of genomic imprinting. *Genetics* **175**(4), 1735-1740.
159. Leslie, F.M. (2013). Multigenerational epigenetic effects of nicotine on lung function. *BMC Medicine* **11**, #27.
160. Berger, S.L., T. Kouzarides, R. Shiekhattar & A. Shilatifard (2009). An operational definition of epigenetics. *Genes & Development* **23**(7), 781-783.

Chapter 5

The Cytoskeleton

If one consults any of the standard developmental biology textbooks, the cytoskeleton will be briefly presented as the support structure of the cell and about as interesting to most biologists as the floor struts in a ballet school would be to an artistic director seeking the next prima ballerina. However, to think of cytoskeleton only as the structure on which cells do their remarkable things, leaves out a good part of the story of how cells work.

We prefer to think of the cytoskeleton as a troupe of acrobats. They run about in the cell, come together, stack to form some remarkable structure and then as rapidly as it appears, the structure falls apart and/or moves and the acrobats appear elsewhere on the stage. These acrobats are the individual proteins that make up the cytoskeleton, and the many proteins that bind to cytoskeleton, transiently hold it together, or move along it. The surface area available on the cytoskeleton filaments for molecules to electrostatically and chemically bind is huge, far exceeding that of all cell membranes in an organism[1]. Ignoring the cytoskeleton while doing molecular biology would be like trying to choreograph a classical ballet while sharing the same stage with a troupe from Cirque de Soleil[2,3].

This unfortunate attitude towards the cytoskeleton is beginning to change. For example, several of the pharmaceutical scientific supply companies have begun manufacturing kits designed specifically to detect the phenomenon of "cytoskeletal rearrangements" because all kinds of interesting other stuff seems to happen precisely when these rearrangements occur and therefore the rearrangements can be used to time such events[4,5]. We find it ironic that only a few people seem to actually think this is more than a convenient coincidence.

The cytoskeleton is, as the name implies, the skeleton of the cell giving it shape and form. The concept goes back to Nikolai K. Koltsov in 1903[6]. Koltsov formulated the idea that the deviation of the shape of a cell from the simple ball, that one would expect if it were a liquid drop, is caused by a stiff but elastic cytoskeleton within the cell[7]. Cytoskeleton is nowadays more precisely defined as the various structures that are filamentous polymers of a single class of protein. These filamentous polymers have long-range order within the cell[8]. They should have been called metapolymers, because the proteins they are made of are themselves polymers of amino acids. These cytoskeletal polymers are commonly classified as supramolecular structures. They are polymers of polymers, roughly topologically linear in their structure.

There are three types of cytoskeletal filaments, arranged in order of decreasing diameter as seen by transmission electron microscopy: microtubules (MTs, 25 nm), intermediate filaments (IFs, 11 nm), and microfilaments (MFs, 7 nm) (nm = nanometer = 10^{-9} meter = 10^{-3} μm). For things this small, we find it convenient to think in terms of the size of the hydrogen atom, whose diameter is 0.1 nm = 1 Å (Angstrom), so a microtubule, for instance, is 250 hydrogen atoms wide. For something so thin, microtubules can be incredibly long, with reported lengths[9] up to 68 μm (micrometers). In *Drosophila* (fruit fly) species (Preface Figure 0.1) the length of the sperm can be[10] 5.8 cm (58,000 μm), implying equally long

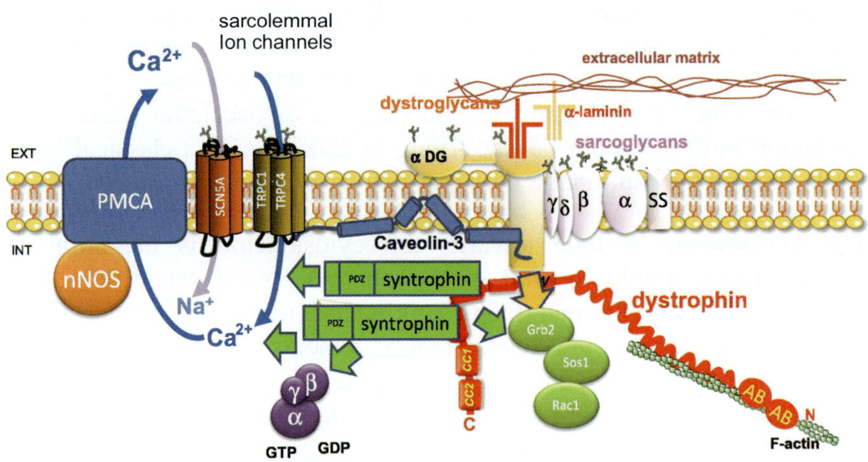

Figure 5.1. Different interactions between the dystrophin complex and signaling molecules organize a signalplex anchored to the subsarcolemmal dystrophin. The importance of calcium ions in the structure is illustrated. The F actin (microfilament) is linked into the complex allowing actin to respond to signals from the membrane complex. From Constantin (2014)[11] with permission of Elsevier.

microtubules. Such microtubules are therefore over 2 million times longer than their diameter. If we imagine a microtubule as a piece of cooked spaghetti with a width of 3 mm then the length of the spaghetti would be as long as 7 kilometers! The three cytoskeletal filaments are found throughout the cell. They are connected to each other at many points, as well as to the inner surface of the cell membrane (Figure 5.1), to the outer nuclear membrane and to the inner surface of the nucleus via specialized attachment proteins (Figures 5.2, 5.3 and 5.4).

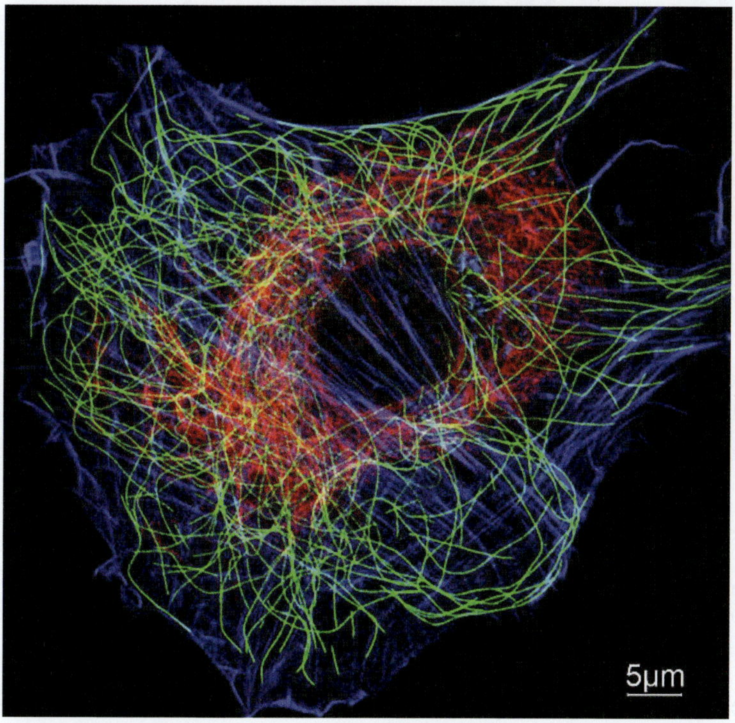

Figure 5.2. A Rat-2 fibroblast cell[12], fluorescently labeled with green = microtubules (anti-tubulin antibody), red = intermediate filaments (anti-vimentin), blue = microfilaments (F-actin, phalloidin). With kind permission of Rosmarie Sütterlin and Cora-Ann Schoenenberger, University of Basel, Switzerland[13]. Many proteins exist that attach the three cytoskeletal fibers (MT = microtubule, IF = intermediate filament, MF = microfilament) to one another and to the cell membrane (CM) and the outer nuclear membrane (ONM): MT-MT[14,15], MT-IF[16], MT-MF[17-20], IF-IF[17], IF-MF[18], MF-MF[21,22], MT-CM[17,23], MF-CM[17,23], MT-ONM[24,25], IF-ONM[26,24,25], MF-ONM[26,24,25]. MT-MT bundling can also occur via electrostatic or steric interactions that align the filaments[27].

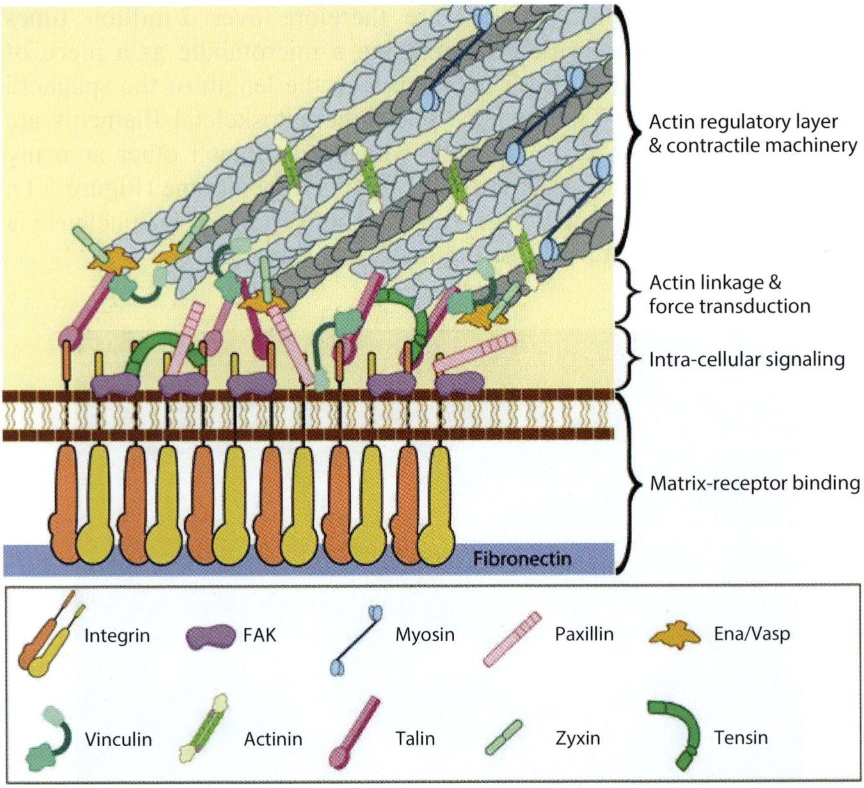

Figure 5.3. This is the structure of a focal adhesion, an example of protein molecules connecting cytoskeletal filaments (microfilaments tilted at 45° made of actin, in two shades of grey) to the cell membrane (pair of horizontal lines) and outwards to the extracellular material (fibronectin, here). From MBInfo[28] with permission under a Creative Commons Attribution-Noncommercial 4.0 International License.

There is a growing awareness of other proteins that may well also be cytoskeletal in nature and deserving of addition to this triumvirate[29-31]. The cytoskeleton, as a concept, can also go beyond just filaments to include structural sheets, patches and meshes[32]. For example, septins, which are found in animals, are an example of these other cytoskeletal-like proteins. Septins use a protein called anillin to attach to micro-filament rings[33] and thereby form rings themselves. These rings line the inner surface of the cell membrane, increasing the membrane's rigidity[34,35] (Figure 5.5). Another example is CTP synthase, which forms filaments in bacteria, yeast and animals[30,36] (Figure 5.6). We will never-

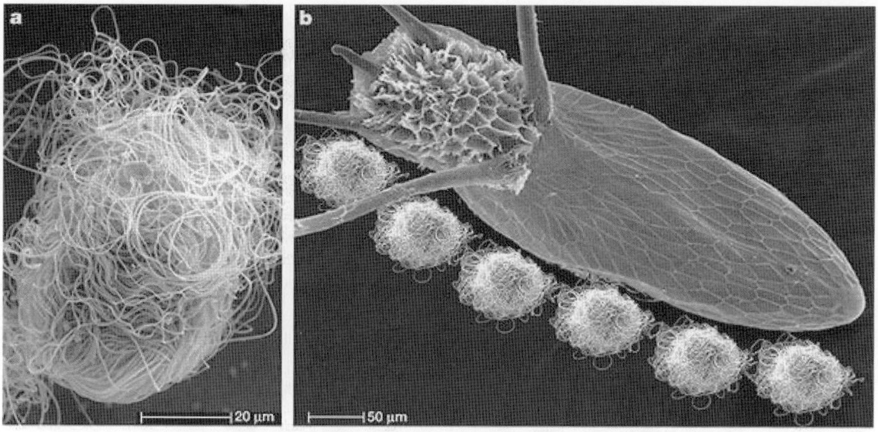

Figure 5.4. The fly *Drosophila bifurca* has giant sperm that are 20 times longer than the male. Most of this length is a microtubule bundle, which implies that, except for the small sperm head, individual microtubules are nearly 6 cm in length here. Growth of the sperm tail involves giant mitochondria[37]. On the right is an egg of this species, with the image of a sperm duplicated 6 times for comparison. From Bjork & Pitnick (2006)[38] with permission of Nature Publishing Group.

theless concentrate on the three best known, best characterized and universally accepted cytoskeletal components. However, the reader should keep in mind that there are likely to be more types of cytoskeletal-like structures that need to be considered in embryogenesis. We will also focus on the mechanical properties of cytoskeleton. There are poorly understood electrical[39-41] and proposed memory related/ epigenetic properties[42] of cytoskeleton. These properties may also become important in explaining aspects of embryogenesis.

The cytoskeleton is an extremely dynamic structure that can depolymerize, move and repolymerize within the cytoplasm in a matter of minutes. The three components are often interlinked in a meshwork that is required for cell movements, maintaining or altering cell shape, and for organizing and powering mitosis and meiosis. The cytoskeleton is required for transporting and sequestering proteins to specific regions of the cell. Cytoskeletal associated proteins act as levers and motors pulling cargo such as the vesicles involved in nerve transmission[43,44] (Figures 5.7 and 5.8), chloroplasts[45] and whole nuclei[46-48].

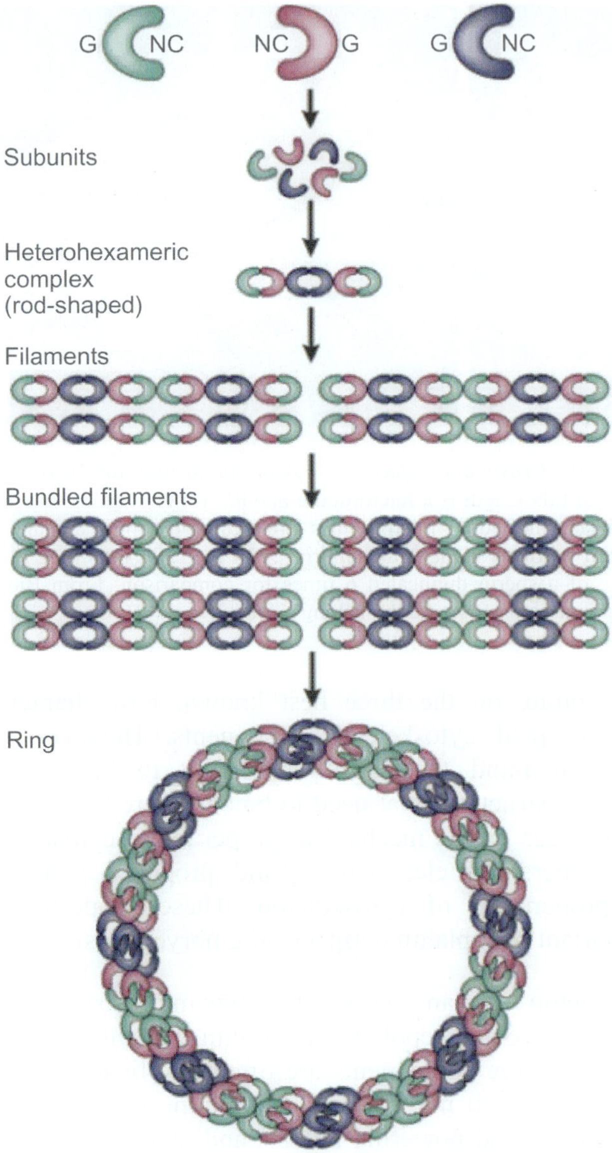

Figure 5.5. Septin proteins of different kinds bind to one another in the middle (G interface = GTP-binding domain) and at their ends (NC) to form rods, which can then aggregate into filaments, bundles, and rings. Such rings are bound to microfilament rings via the protein anillin and the contraction of the microfilament rings slows down if septin is absent[49]. From Mostowy & Cossart (2012)[35] with permission of Nature Publishing Group.

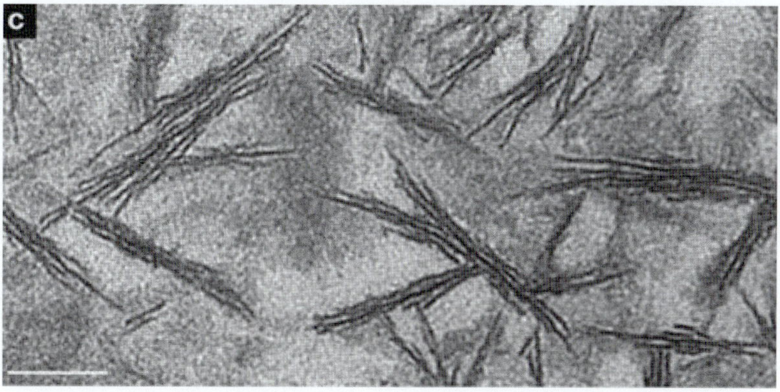

Figure 5.6. Bundled filaments of CTP synthase, which might be regarded as another cytoskeletal component. Scale bar = 100 nm. From Ingerson-Mahar *et al.* (2010)[50] with permission of Nature Publishing Group.

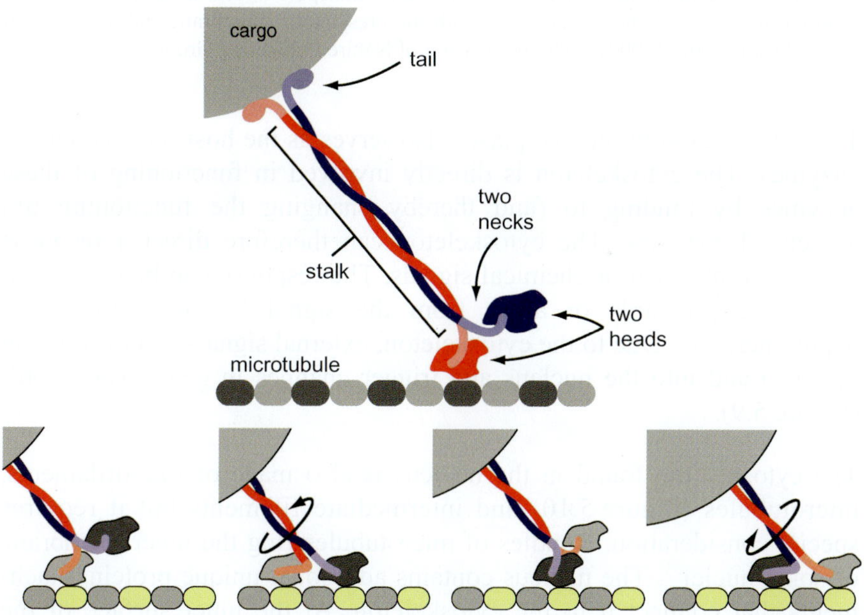

Figure 5.7. Kinesin walks along microtubules and can drag an enormous load such as synaptic vesicles (Figure 5.8). **Top**: detailed structure of this motor molecule. **Bottom**: The hand-over-hand mechanism by which kinesin walks. Note that unlike some cartoon movies of kinesin set to music with a regular beat, the motion would be stochastic (jittery). From a web site by Steven M. Block[51], where an actual light microscope movie can be viewed, with his kind permission. Cf.[52].

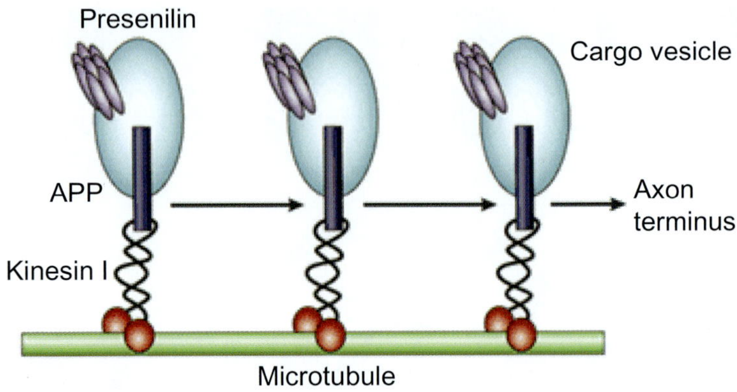

Figure 5.8. Microtubules in nerve axons can deliver large loads, such as synaptic vesicles, from the cell body to synapses even a meter away. Here "…the cargo-receptor model proposes that APP tethers vesicles containing presenilin and other synaptic proteins to microtubules through an interaction between the APP intracellular domain and kinesin I. Microtubule-dependent fast anterograde transport directs these vesicles to the axon terminus, where the vesicles fuse with the presynaptic membrane and deliver their cargo" From Fortini (2002)[53] with permission of Nature Publishing Group.

The cytoskeleton in the cytoplasm also serves as the host for a variety of enzymes. The cytoskeleton is directly involved in functioning of these enzymes by binding to (and thereby changing the functioning of) associated enzymes. The cytoskeleton can therefore direct a response to external physical or chemical signals. The response can be movement of the cell towards or away from the signal by polymerizing or depolymerizing. Due to the cytoskeleton, external signals can even travel down to and into the nucleus and trigger changes in gene expression[54] (Figure 5.9).

The cytoskeleton found in the nucleus is also made of microfilaments, microtubules (Figure 5.10) and intermediate filaments but it requires special consideration. Bundles of microtubules ring the inner membrane of some nuclei[55]. The nucleus contains additional unique protein attachments that connect it to the cytoskeleton. At the inner surface of the double nuclear membrane is a sheet of intermediate filaments called lamins (Figure 5.9), whose functions may include maintenance or change of nuclear shape[56,57] (but see Chapter 10), protection of the nucleus from mechanical shocks[58], and intranuclear rotation (rotation of the contents of the nucleus)[67-69]. Intranuclear rotation includes both sides of the double layer nuclear membranes[70], but without causing the cytoplasm to rotate. Therefore there may be a layer just outside the outer nuclear membrane that has low viscosity, like the cell cortex (Figure 2.29). If this is the case, the motive force likely requires the cytoskeleton and is occurring

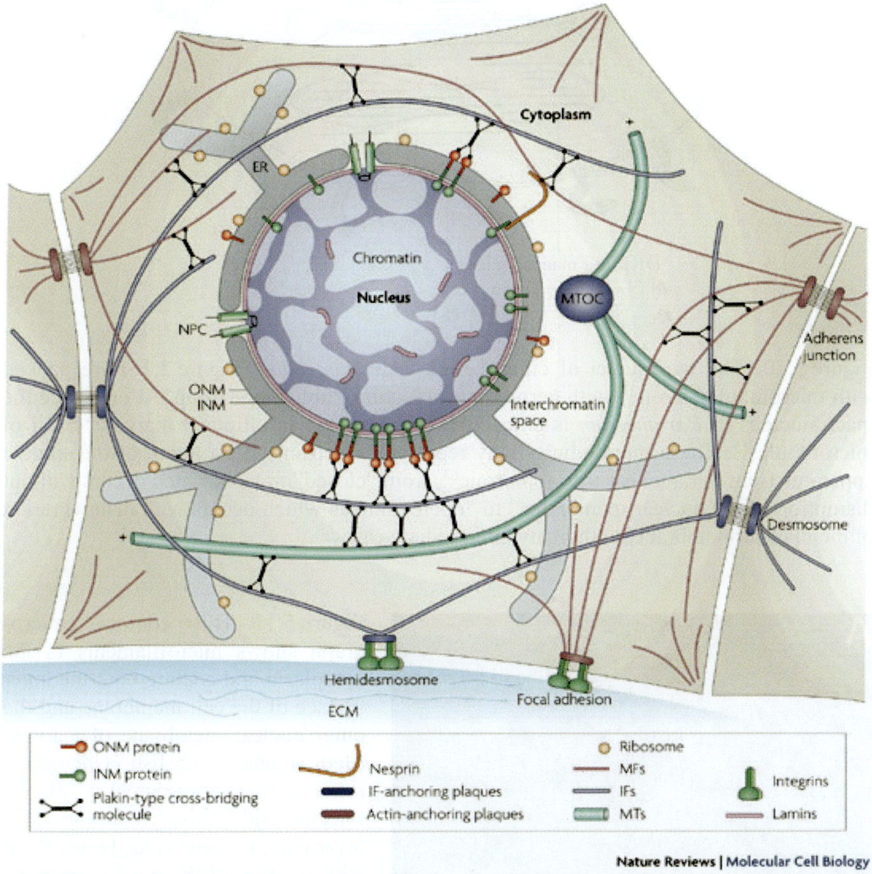

Figure 5.9. An epithelial cell illustrates the three components of the cytoskeleton, microfilaments (MFs), microtubules (MTs) and intermediate filaments (IFs), and some of their many connections with one another and other cellular structures. For example, microtubules are connected to desmosomes, which both act as rivets binding cells to one another, and may be involved in mechanical signaling between cells[59,60], such as we shall discuss later with differentiation waves. Abbreviations: inner nuclear membrane (INM), outer nuclear membrane (ONM), nuclear pore complex (NPC), endoplasmic reticulum (ER), microtubule organizing center (MTOC), extracellular matrix (ECM). From Herrmann *et al.* (2007)[61] with permission of Nature Publishing Group.

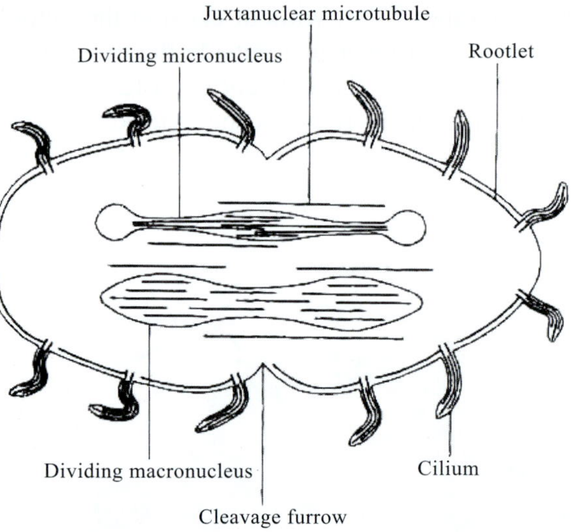

Figure 5.10. The two nuclei of ciliates such as *Paramecium* (Figure 2.16) are packed with microtubules. From[62] with permission of Nature Publishing Group. At one stage the macronucleus in *Paramecium* is ellipsoidal and contains an elliptical marginal band of microtubules[55]. These microtubules may represent a step in the evolution of the spindle apparatus from intranuclear to cytoplasmic[63], from "closed" mitosis which occurs without disruption of the nuclear membrane[64] to "open" mitosis which occurs with disintegration of the nuclear membrane, with many intermediate cases[65].

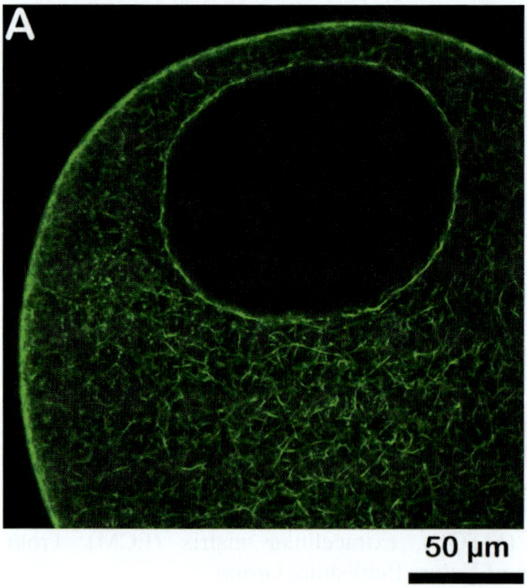

Figure 5.11. This confocal micrograph shows microfilaments in the cytoplasm and coating both the inner surface of the cell membrane and the outer nuclear membrane of a living Mediterannean starfish *Astropecten aranciacus* transparent oocyte injected with Alexa-phalloidin. This layer may, in some cells, be involved with rotation of the outer membrane of the nucleus relative to the inner nuclear membrane. From Santella *et al.* (2015)[66] with permission of Elsevier. Original kindly provided by Luigia Santella.

on the outside surface of the outer nuclear membrane (Figure 5.11). In some cells nuclear rotation and cell membrane rotation are coupled (Figure 5.12). These rotations may be required for changes in the conformation of the DNA and could form the basis of exposing and sequestering of genes but the question has not been well explored. Resolving this puzzle will require the skills of molecular biologists, engineers and physicists.

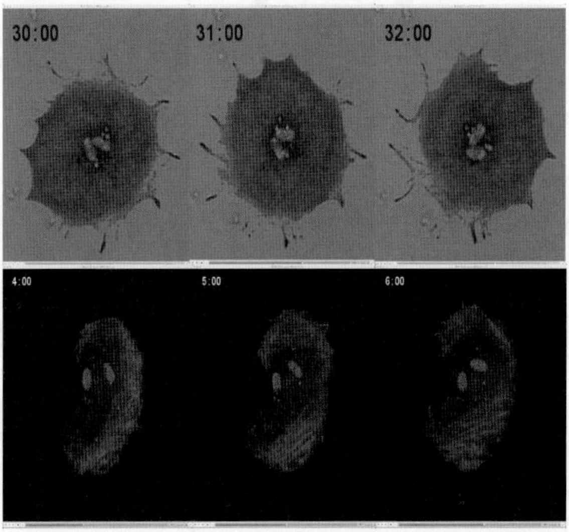

Figure 5.12. **Top**: Zebrafish (*Danio rerio*) pigment cells (melanophores) exhibit counter-clockwise rotation of the nuclei (purple) and the apical cell membrane (attached tiny green fluorescent beads). **Bottom**: The cytoplasmic microfilaments rotate in the opposite direction. Cell diameter is about 70 μm. Micrographs are 1 hour apart. These cells often have two nuclei. A possible function of the melanophore rotation is to straighten the stripes on the zebrafish[71]. From Yamanaka & Kondo (2015)[72] with permission of John Wiley and Sons.

The cytoskeletal elements in the nucleus combine with a variety of nuclear proteins forming a "nuclear matrix" which acts like a scaffold on which the DNA is attached[75] (Figure 5.13). For this reason the whole lot is referred to as the "nucleoskeleton". The nucleoskeleton itself is then connected through a putative low viscosity exterior nuclear membrane region to the cytoskeleton of the cytoplasm via proteins that are embedded in the two nuclear membranes (Figure 5.14).

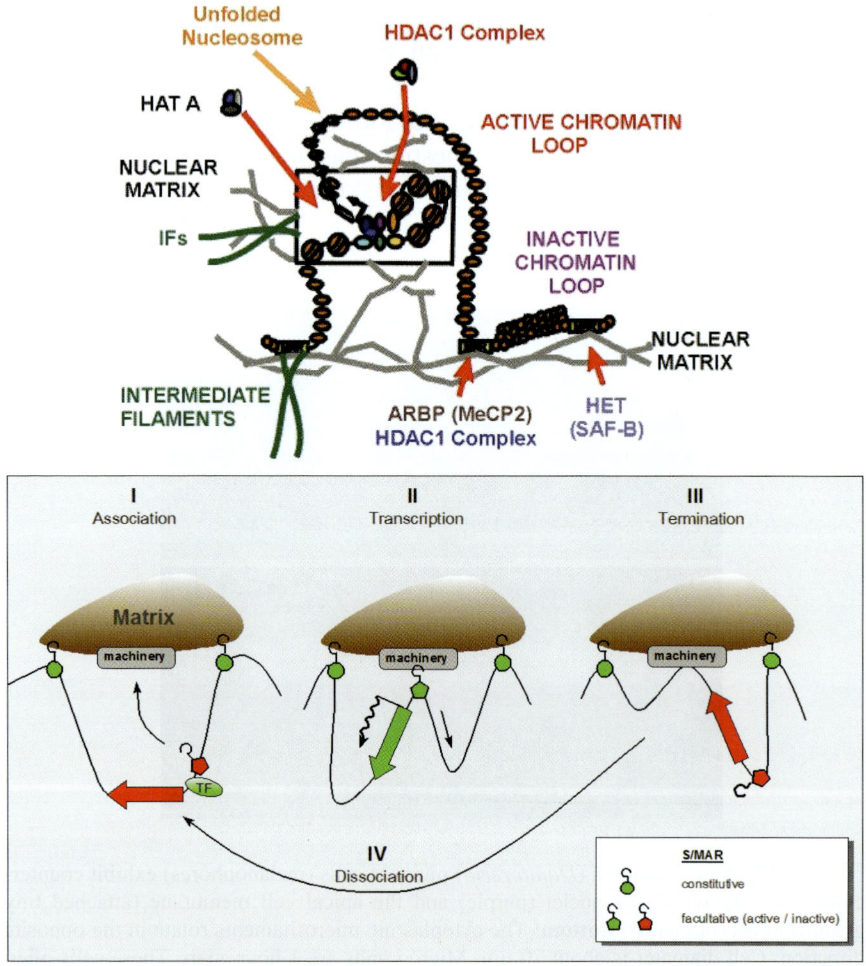

Figure 5.13. **Top**: DNA is organized in chromatin loops. Depending on the degree of binding, which is determined by the state of acetylation/methylation of the histone tails, the conformation of the DNA will either allow or not allow transcription factors to enter and transcription to occur. Active chromatin loops have regions of DNA both wrapped around histones and region where no histones are present. Inactive chromatin loops are tightly packed. From James R. Davie with his kind permission. Cf.[73]. **Bottom**: Within the active chromatin loop, DNA is further organized into smaller secondary loops based on the presence of specific binding sites such as S/Mars (see text). For transcription to occur a gene must be both on an active chromatin loop and on an open facultative loop. This is why S/Mar binding sites are only functionally important in specific cell types. S/Mar binding sites within inactive chromatin loops are inaccessible to any machinery. From Wikipedia[74] with permission under a Creative Commons Attribution-ShareAlike 3.0 License.

These three filaments, microtubules, intermediate filaments and microfilaments, were once thought to be unique to eukaryotes. Many old textbooks included a table of differences between eukaryotes and prokaryotes and the presence of cytoskeletal elements figured prominently as a defining characteristic of the eukaryotic cell. This has proven to be incorrect, and as we showed in Chapter 1, many characteristics of eukaryotes are found in, if not inherited from, prokaryotes. Homologs of all three cytoskeletal proteins have been found in bacteria, albeit usually having different functions[8]. There is a form of tubulin found in bacteria called FtsZ that is also required for cell division in most prokaryotes (but not all[76,77]). FtsZ may represent the evolutionary precursor of all of the tubulins[78,79] (Figure 5.15).

The first cells with a cytoskeleton[80] thus predate our last universal common ancestor (LUCA) and were likely autotrophs consuming environmental carbon dioxide, sulfuric acid, water, ammonia and trace elements[81]. They had some form of cell division using the earliest cytoskeleton to perform the division. Since they made their own energy from the breakdown of chemical compounds like sulfides into elemental sulfur they would have to deal with waste. Elemental sulfur is insoluble and would require excretion from the cell. The first new form of cytoskeleton not used for cell division was likely copied from the one used for cell division but then modified to be used for excretion. The first movement (Figure 1.13) may well have come about simply because cells used these cytoskeletal proteins to excrete waste out of one end and in the process were moved in the other direction[82].

If the cell encountered some obstacle to this movement, it would have to flow around the obstacle. Initially, this was likely material the early cell simply encountered accidentally as it moved while excreting waste. The ability to actively engulf prey would later evolve from this simple movement (Figure 1.14). Where useful, the items would be engulfed and digested instead of avoided. The evolution of the cytoskeleton thus allowed cells to move towards food sources or away from danger and to engulf prey. When cells began the first simplest forms of differentiation they again used the machinery of asymmetric cell division (Figure 1.19), modified it, and as a result changes in gene expression were tied to cell division. That primitive early system has evolved into a complex multifaceted one used by all cells on Earth today in some fashion.

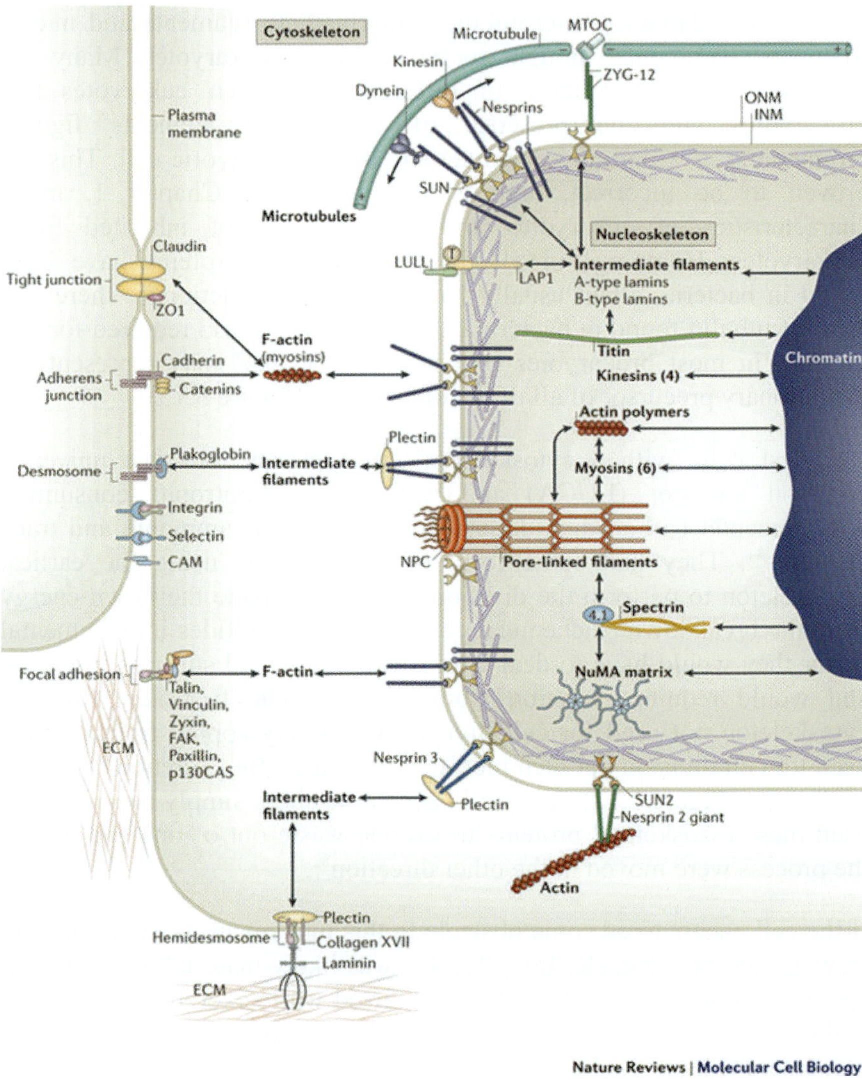

Figure 5.14. This sketch takes us from the cytoplasm, across the double nuclear membrane, and into the nucleus, where we find intermediate filaments and microfilaments (actin polymers). The DNA in the nucleus is just pictured here as a lump of chromatin, and microtubules (Figure 5.10) have been left out, although the microtubule motor molecule, kinesin (Figure 5.7), is indicated. MTOC = microtubule organizing center, ONM = outer nuclear membrane, INM = inner nuclear membrane, NPC = nuclear pore complex, Plasma membrane = cell membrane, ECM = extracellular matrix, F-actin and actin polymers = microfilaments. The ONM and INM taken together are called the nuclear envelope. From Simon & Wilson (2011)[25] with permission of Nature Publishing

Group. The SUN proteins are part of the **LI**nkers of **N**ucleoskeleton and **C**ytoskeleton complexes (LINCs)[83] that bridge the gap (perinucleus[84]) between the two nuclear membranes[24,84] and mechanically couple the nucleus to the cytoplasm[83], which is mechanically linked to the cell membrane[26].

Microtubules, the largest of the three proteins, are formed by the polymerization of dimers of tubulin. The most common dimer pair is one alpha and one beta tubulin bound together[85] as an αβ-heterodimer (Figure 5.16). It takes thirteen αβ-heterodimers to form one full helical turn of the most common form of microtubule. However, the heterodimers are not added one by one up the helix in the manner in which DNA is formed. Instead, the individual heterodimers first form firmer bonds to the heterodimers that will eventually be above and below them along the length of the whole microtubule. Microtubules are thus lengthened by forming strings of heterodimer pairs called protofilaments.

The protofilaments are initially curved away from the microtubule while the heterodimers bind. The protofilaments close up into the tube a short distance below the tip as the tip grows. The closing is accomplished by slight changes in the structure of tubulins on binding of a high energy phosphate group[86]. The change is enough to alter the curvature of the protofilaments[87] from flaring to tube-like with a sequential snapping shut (Figures 5.17 and 5.18). This means there is a force against the binding of protofilaments to one another and it is overcome by the addition of energy. When the microtubules depolymerize, the process is reversed and the spiral bonds are lost first. The protofilaments form frayed ends before the entire microtubule falls apart. The result is that the microtubules appear to have a flared and fuzzy frayed end in an electron micrograph (Figure 5.19).

Microtubules are hollow tubes and they therefore have an inside and outside. Most molecules that bind to or change the surface of micro-tubules do so on the outside surface. However, there is one remarkable enzyme that diffuses into the ends and acts inside the microtubule (cf.[88]) by acetylating α-tubulin's "lumen" (Figure 5.20). It moves slowly to the interior, and so this acetylation occurs mostly in long lived stable microtubules[99]. Somehow, yet to be worked out[100], this acetylation alters the binding and movement of molecules on the outside[101], perhaps by an allosteric mechanism. There is a branch of research about microtubules

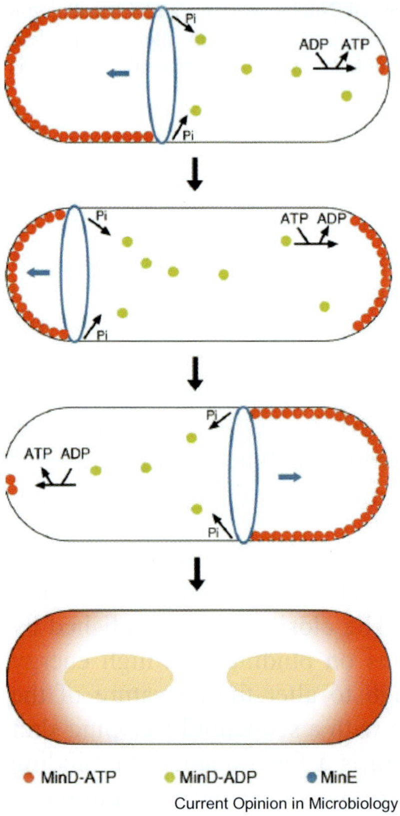

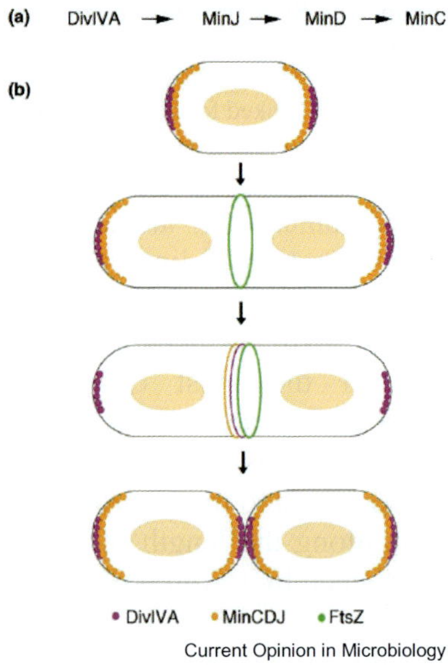

Figure 5.15. **Left**: "Division site selection in *E. coli* is governed by a dynamic Min system. The MinCDE system in *E. coli* is oscillating between the cell poles. MinDATP (red spheres) binds to the cell membrane and recruits MinC (not shown). MinE (blue ellipse) binds to MinD and activates ATP hydrolysis and inorganic phosphate (Pi) is released. This reaction leads to effective removal of MinD-ADP (green spheres) and, as passive passengers, MinC and MinE from the membrane. Nucleotide exchange in the cytoplasm reactivates the membrane affinity of MinD and subsequently, MinD-ATP assembles at membrane areas that are far distant from the highest MinE concentration (usually the opposite pole). This simple oscillation generates over time a MinCD gradient which is lowest at midcell. The last cell in the panel illustrates this gradient (red shading). The replicated and segregated nucleoids are illustrated in this cell as light orange ellipsoids. The nucleoid occlusion factor (SlmA in *E. coli*) is bound to the DNA and leads to the inhibition of efficient Z-ring formation in this area. Together, the dynamic Min system and the nucleoid occlusion allow assembly of a functional cytokinetic ring only at midcell." **Right**: "Revised model for division site selection in *Bacillus subtilis*. A summary of the interactions between the division site selection proteins in *B. subtilis* is given in (a). MinJ interacts with DivIVA and MinD, while MinC only interacts with MinD and DivIVA only interacts with MinJ. The order of proteins also shows the dependency for their localization in vivo. The revised model for division site selection in *B. subtilis* takes into account that MinC (and presumably MinD and MinJ) is dynamically relocalized to the maturing divisome (b). In newborn cells DivIVA (purple spheres) recruits the MinCDJ complex (orange spheres) to the poles. Note that only MinJ binds

directly to DivIVA. Upon polymerization of FtsZ (green spheres) into a Z-ring (green ellipse) at midcell, MinCDJ is recruited away from the cell poles to the site of active division. A reason for the recruitment of the inhibitory MinCDJ complex to the divisome is likely to prevent a new round of division. After completion of septation, MinCDJ is distributed evenly to both poles. The orange ellipses represent the segregated nucleoids, which impose the nucleoid occlusion effect (here executed by the Noc protein) and together with the Min system restrict division to midcell, only." From Bramkamp & van Baarle (2009)[89] with permission of Elsevier.

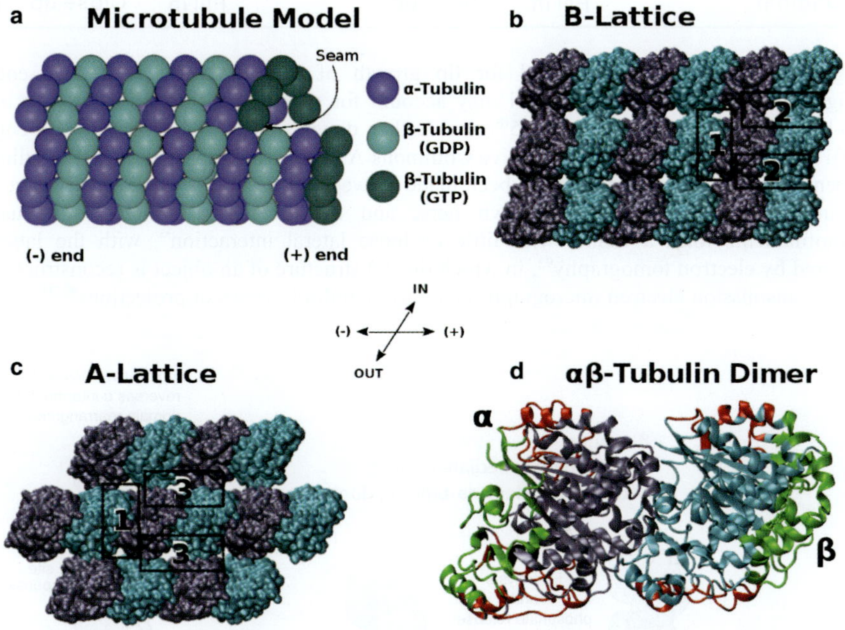

Figure 5.16. A small segment of a microtubule is shown (a). The αβ-heterodimers are attached head to tail along each protofilament (horizontally oriented) leaving all α-tubulins at the left, defining what is called the minus (−) end, and leaving all β-tubulins at the right or positive (+) end. This nomenclature is unrelated to electrical charge. Each α-tubulin always has a GTP bound to it, which is therefore generally not indicated in diagrams. Each β-tubulin starts out with GTP bound, but it can lose a phosphate so that GDP is bound instead. (b,c): The protofilaments have two ways of binding sideways to one another, so that either they match (B-lattice) or alternate (A-lattice) as to which tubulin touches which laterally. In the Microtubule Model with the usual 13 protofilaments most of the lateral contacts are as in a B-lattice, but here is a seam at which the configuration is as in an A-lattice. While the model (a) represents the most common configuration for microtubules (B-lattice), some microtubules form predominantly as an A-lattice. The ribbon depiction of an αβ-heterodimer (d) is labeled red for parts that make contacts between protofilaments and green for parts that bind the αβ-heterodimer into protofilaments. From Ayoub *et al.* (2014)[90] with permission of Elsevier.

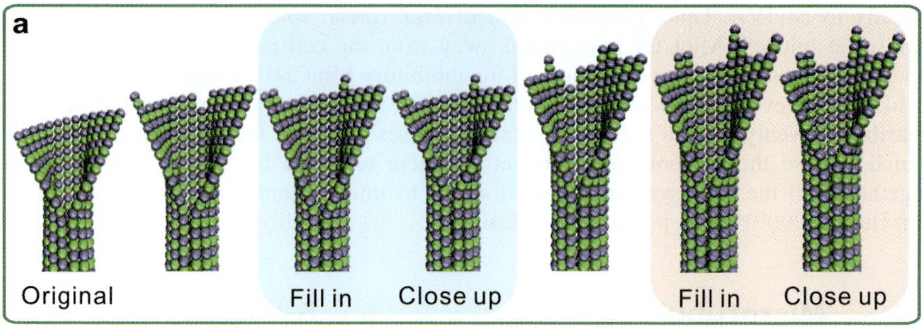

Figure 5.17. A mechanical model for tip growth of a microtubule at its plus-end (Figure 5.16). Note that this model may account for seam formation. When a row of αβ-heterodimers is complete, the 5th row further down zips closed. From Ji & Feng (2011)[91] with permission via a Creative Commons Attribution License. Blue = β-tubulin, green = α-tubulin. A distinction has been made between a "single curved sheet structure" giving the flared appearance shown here, and a structure in which "individual protofilaments form a funnel with little or loose lateral interaction", with the latter favored by electron tomography[92], in which the 3D structure of an object is reconstructed from transmission electron micrographs taken from multiple views or projections[93,94].

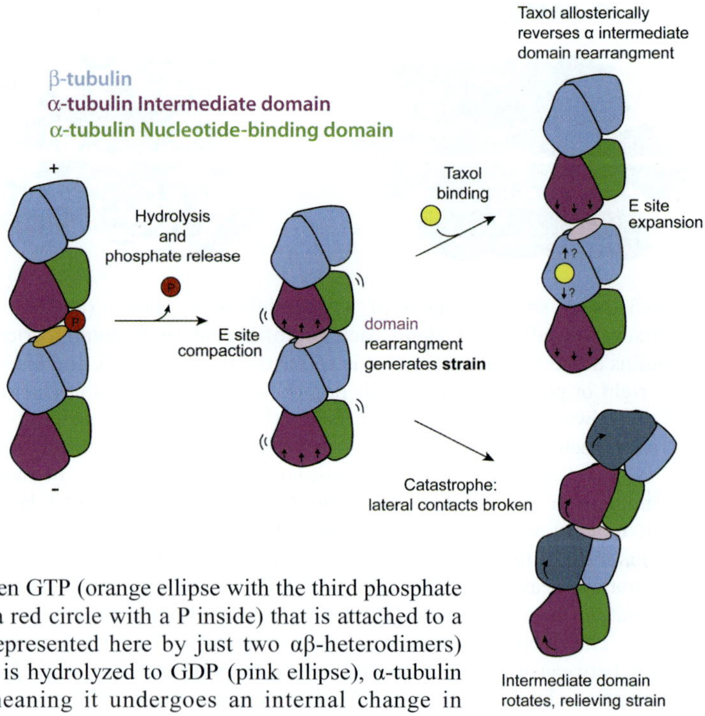

Figure 5.18. When GTP (orange ellipse with the third phosphate group shown as a red circle with a P inside) that is attached to a protofilament (represented here by just two αβ-heterodimers) in a microtubule is hydrolyzed to GDP (pink ellipse), α-tubulin "rearranges", meaning it undergoes an internal change in

structure, i.e., an allosteric change. This bends the protofilament, allowing it to break free from adjacent protofilaments (not shown here), leading to "catastrophe", which means depolymerization. If the drug taxol[95] is bound to the β-tubulin (dark blue), it can push the other way, preventing the protofilaments from curving; it thereby prevents depolymerization. This makes the microtubule stable, and it no longer undergoes dynamic instability. The E site is the place on the β-tubulin where GTP binds. From Alushin *et al.* (2014)[86] with permission of Elsevier.

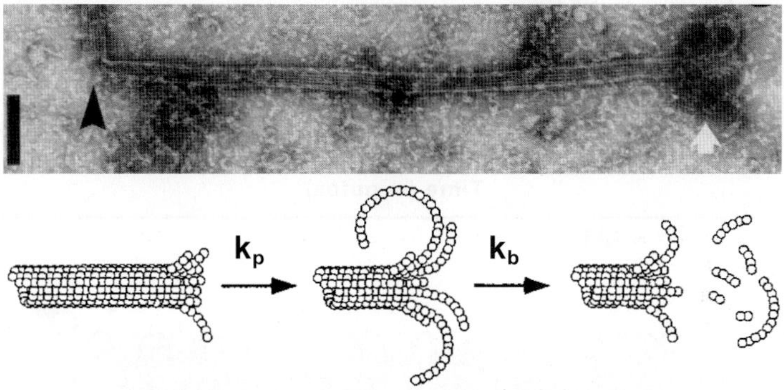

Figure 5.19. When microtubules depolymerize at their plus-end (Figure 5.16) their protofilaments curve away from the microtubule and break apart. Scale bar in the transmission electron micrograph is 100 nm. From Tran *et al.* (1997)[96] with permission of Elsevier. As suggested by the way protofilaments peel away from one another, the strength of microtubules is along, rather than across the microtubules[97]. The curvature may be a cooperative effect[98], i.e., a form of phase transition.

that studies the possibility that microtubules take advantage of quantum mechanics to transport messages nearly instantly across cells. This internal acetylation may alter the quantum mechanical effect of coherent energy transfer inside microtubules[102]. In any case, microtubules are polyelectrolytes, accumulating a counterion distribution both inside and out, which allows them to act somewhat like electric wires[103,42,104], as do microfilaments[105,104] and DNA[106] (cf. Figure 1.43).

Most microtubules have a remarkable property: they are in a state of continuous cycles of growth by polymerization alternating with depolymerization, gaining or losing αβ-heterodimers at one or both ends. This means that their lengths are continuously fluctuating on the time scale of minutes (Figure 5.21). This is called dynamic instability. To understand this "dynamic instability" of microtubules, we need an aside about the chemistry involved.

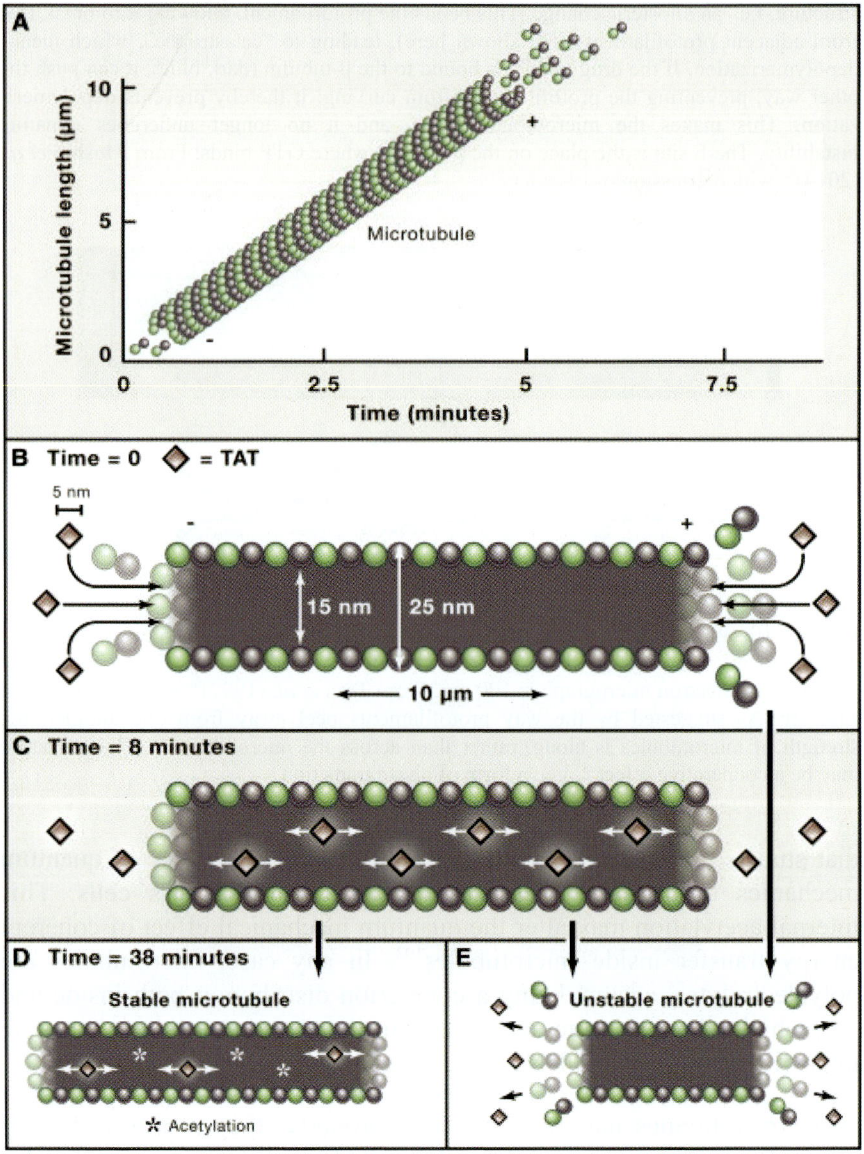

Figure 5.20. The enzyme tubulin acetyl transferase (TAT) diffuses inside the lumen of microtubules where it acetylates tubulin from the inside. This stabilizes the microtubule against depolymerization. From Kull & Sloboda (2014)[100] with permission of Elsevier. It has been hypothesized that oxygen gas (O_2) is also present in high concentration in the microtubule lumen[107].

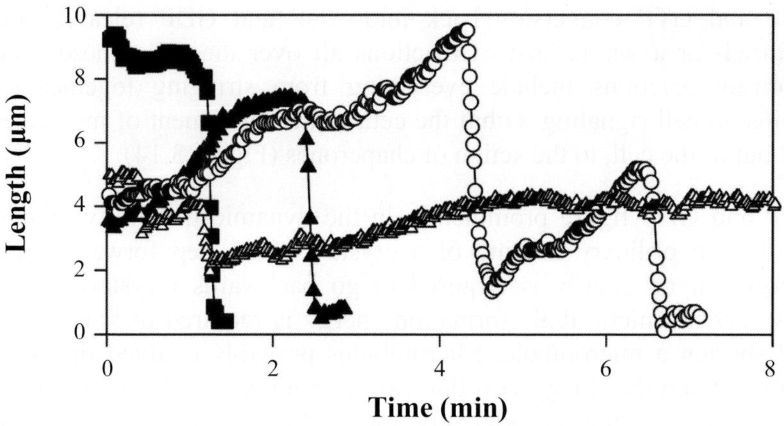

Figure 5.21. Here the lengths of 4 microtubules were recorded every 5 seconds. They generally fit the dynamic instability model of relatively slow growth followed by (a "catastrophe" of) rapid shortening, then "rescue" to growth again. From Panda *et al.* (2005)[108] with permission of the National Academy of Sciences, USA, copyright 2005.

It is one of those "laws that rule the universe", and our lives, that things tend to disorder unless energy is applied to keep them ordered or to create more order. Creating order by converting tubulin dimers into microtubules takes the addition of energy. We already showed in Chapter 3 how ATP levels control metabolism in the *lac* operon and how the conversion of ATP to ADP releases energy used by chaperone shaker cages to fold proteins (Figure 3.14). Microtubules use a guanosine based high energy phosphate carrier (guanosine diphosphate or GDP) instead of an adenine based one. When energy is applied and the third phosphate is added, GDP becomes guanosine triphosphate (GTP) (Figures 3.15 and 3.16). The third phosphate is the high energy carrier.

ATP can be made in two ways. The first is directly from light energy by plants or by the breakdown of chemical bonds such as those in sulfides found in hot springs used by certain prokaryotes, especially of the Archaea phylum. ATP and GTP can both be produced by adding energy to ADP or GDP along with a free phosphate, using the same energy released from the breakdown of carbohydrates, fats and proteins into carbon dioxide and water (Figure 5.22). GTP can also be used to make ATP from ADP and phosphate. And so the GTP and ATP perform very similar functions in the cell and are almost interchangeable.

ATP and GTP conversion back into ADP and GDP releases energy required for a whole host of reactions all over the cell. These energy-requiring reactions include everything from stringing together DNA chains, to cell signaling within the cell, to the movement of molecules in and out of the cell, to the action of chaperones (Figure 3.14).

GTP and GDP figure prominently in the dynamic instability of microtubules. In ordinary melting of a crystal, every step forward requires energy but no energy is required to go backwards (crystallization)[109]. However, in microtubule formation energy is required to both lengthen and shorten a microtubule. Microtubules probably evolved this way to form nonequilibrium systems that can respond, via feedback, to maintain their state in spite of relatively minor fluctuations[110,111]. This gives greater control over when and how a microtubule is constructed or disassembled and provides stability and reliability of function. It works as follows.

Only GTP-αβ-heterodimers can be added to the end of a protofilament. GDP-αβ-heterodimers fall off. But GTP hydrolyzes to GDP over time. So if there's lots of GTP-αβ-heterodimer at the tip, the microtubule can keep growing. If there's not enough, then the end GDP-αβ-heterodimers fall off faster than the GTP-αβ-heterodimers can be added. Fluctuations in the amount of GTP-αβ-heterodimer arriving at the tip can then lead to growth if GTP levels are high and shrinkage if low. That's the basic idea, summarized as a fluctuating "GTP-cap", which has been developed mathematically[112-115] starting with Dick's PhD mentor, the late Terrell L. Hill[116,117] (Figure 5.23). Note that in the viscid cytoplasm, the diffusion of tubulin towards a growing microtubule tip may be rate limiting[118]. This is called diffusion-limited precipitation[119,120], which can have effects on the pattern of arrangement of microtubules in a cell, and may account for the so-called "exploratory" behavior of microtubules. Microtubules appear to reach out, withdraw and reach out again over and over until they hit something and bind[121,122]. Tubulin may also adsorb to the surface of a microtubule and undergo a possibly biased 1D random walk along the microtubule, speeding up its delivery to the tip compared to 3D diffusion[126,127]. This is an example of dimensionality reduction[128], a mechanism used widely by many organisms for a wide variety of ways to scavenge and transport scarce molecules[129].

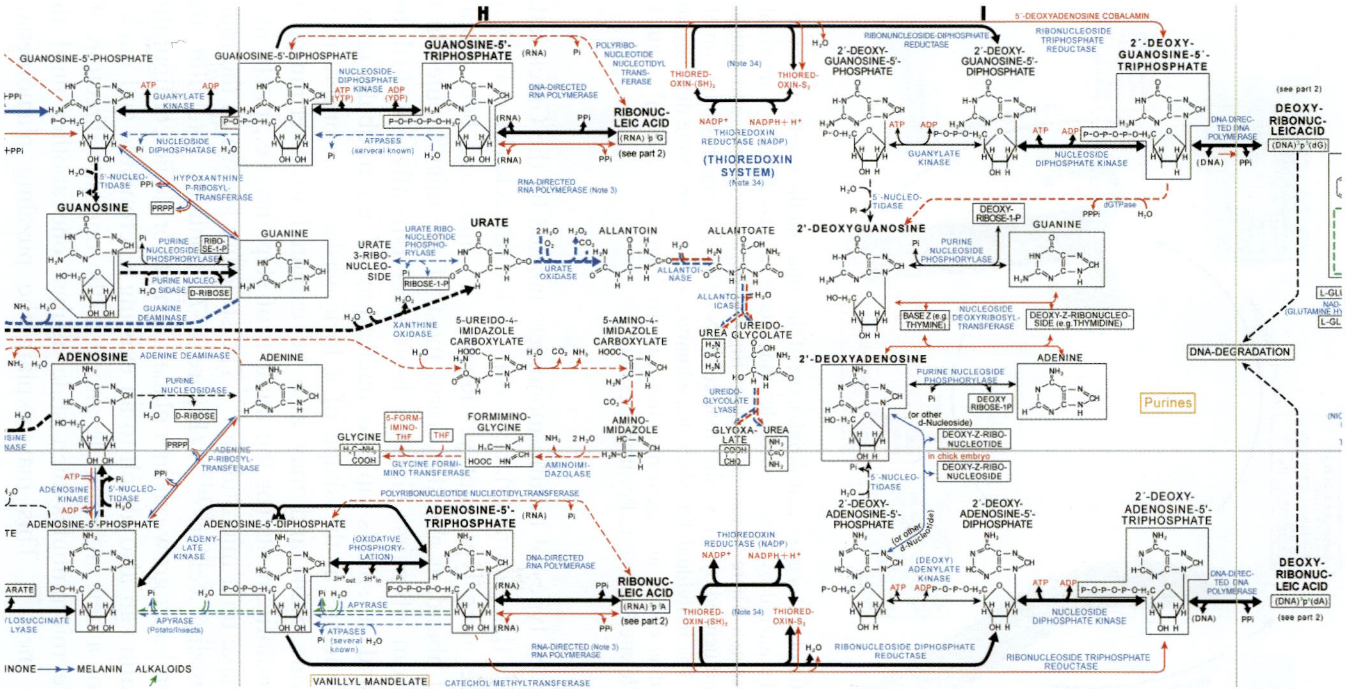

Figure 5.22. A small portion of an online map of the metabolic pathways, showing synthesis of ATP (adenosine-5'-triphosphate) and GTP (guanosine-5'-triphosphate) and their relationships to RNA and DNA synthesis. From Roche[123], detailed in[124], with permission of John Wiley and Sons.

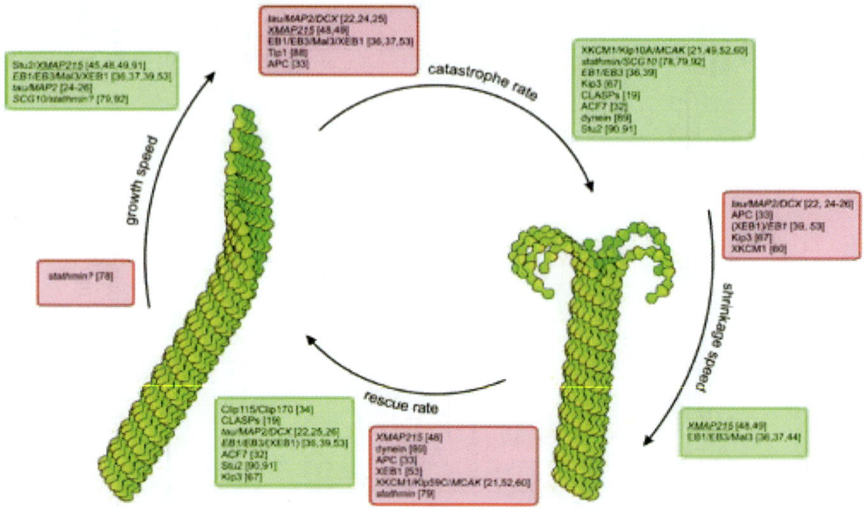

Figure 5.23. Microtubules grow and shrink most of the time, but the speeds of growth and shrinkage can be altered by many molecules that attach to them (MAPs = microtubule attachment proteins). There is also a random flipping between growth and shrinkage, which itself is influenced by many other molecules. Note the peeling or winding up of the protofilaments. The individual dumbbell shaped units are αβ-dimers. Note that they are in a helical, and thus chiral, arrangement. From van der Vaart *et al.* (2009)[125] with kind permission of Helene Pettersson, Portland Press Ltd.

There are seven types of tubulin monomers labeled by the Greek letters alpha (α), beta (β), gamma (γ), delta (δ), epsilon (ε), zeta (ζ), eta (η), and there may be more types not yet labeled[130]. Alpha and beta tubulins are the most abundant and they form most of the microtubules in the cytoplasm from αβ-pairs. Gamma tubulin is typically attached to a membrane and it speeds up the nucleation of microtubule growth, forming a structure in the shape of a lock washer, i.e., one turn of a helix[131] (Figure 5.24) and may also be essential to determining the number of protofilaments (or the number of steps up, i.e., pitch, in the helix) that make up a microtubule[132]. The structure known as the microtubule organizing center[133-135] (Figure 3.37) contains γ-tubulin from which microtubules grow out in a planar, conical or radial configuration. Gamma tubulin can also attach to an existing microtubule and nucleate the growth of a new microtubule aimed in the opposite direction. The two opposite microtubules then slide against one another until their minus ends meet[137]. This may explain the puzzling phenomena of nucleation of free microtubules[138-140]. Delta and epsilon are also involved in mitosis and meiosis and are found only in the microtubule organizing center. Zeta is a tubulin monomer peculiar to one set of protists and required by them for cell division.

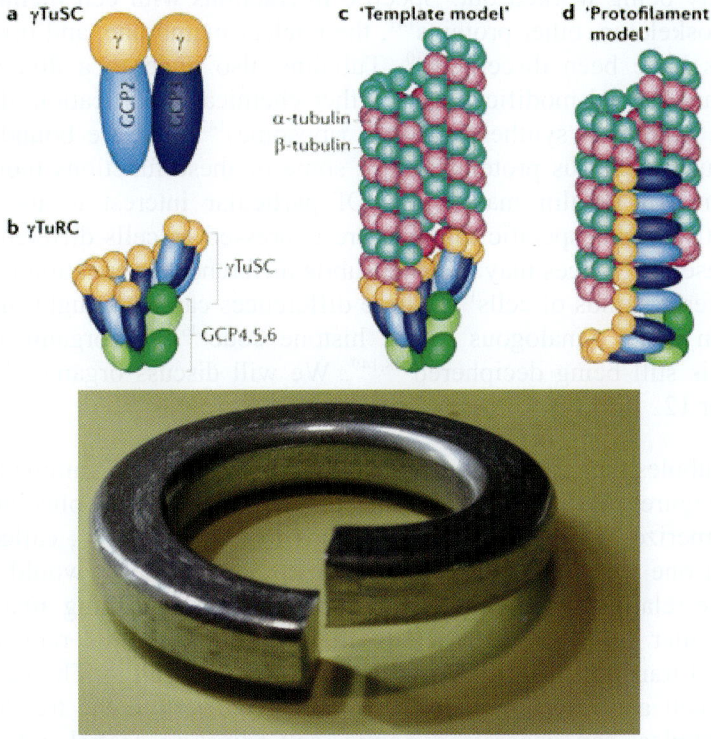

Figure 5.24. The gamma ring (b) is probably an open loop, shaped like a lock washer (below), of γ-tubulins (γTu) attached to other proteins (a and b) in a small complex (SC) which is part of the ring complex (RC). αβ-heterodimers (pairs consisting of an α-tubulin attached to a β-tubulin) stick to these at their minus-end (Figure 5.16), resulting in rapid growth of a microtubule (c). Nucleation in general is the process by which molecules or atoms that can form crystals, amorphous solids or liquid droplets get started in the process more quickly by adhering to another substance[136]. An alternative model for how microtubule nucleation by γ-tubulins might occur is given in (d). Microtubule structures from Kollman *et al.* (2011)[131] with permission of Nature Publishing Group. In (c) note the seam where there is an imperfect match between the frontmost vertical columns. Each vertical column of αβ-heterodimers is called a protofilament. The end of a microtubule that can attach to a γ-tubulin ring is called the minus (–) end, and the other end is the plus (+) end. These designations have nothing to do with electric charge.

The seven types of tubulin come in additional subtypes, with varying amino acid sequences, called isoforms, each with matching individual genes. We are confronted with both a high degree of evolutionary conservation of tubulin sequences and quite a variety of tubulins. The special functions of these variations of the most common tubulins are presently being worked out. Specific interactions with cell membranes, the cytoskeleton, other proteins[141], the nuclear membrane, and inside the nucleus have been discerned[130]. Tubulins also undergo a diversity of post-translational modifications (further chemical modification after the tubulin protein is synthesized on a ribosome)[142], and are bound by or even cut by various proteins[143]. For some of these functions more than one kind of tubulin may work. Of particular interest to us is that alternative tissue-specific tubulins are expressed as cells differentiate[144] and these differences may indicate subtle differences in function required for different kinds of cells[145]. These differences can be thought of as the "tubulin code" (analogous to the histone code[146]), an organic code[147] which is still being deciphered[148,149]. We will discuss organic codes in Chapter 12.

Microtubules that are not attached at their minus-end to a gamma tubulin ring (Figures 5.16 and 5.24) can polymerize at their plus-end and depolymerize at both ends[140]. This leads to treadmilling[150], called such since if one were to follow a particular αβ-heterodimer it would appear to move relative to the ends of the microtubule[151]. Anything attached to a particular αβ-heterodimer could move with it, so this represents a potential transport mechanism in the confines of a cell[150]. The geometry of the cell and its membranes, in creating barriers to the treadmilling microtubules, can generate dynamic, self-adjusting spatial patterns of arrangement of the microtubules[152-154].

There are many important enzymes that cannot do their catalytic reaction unless they first attach to a microtubule. In this way, signals for changes in gene expression can be localized to certain areas of the cell where microtubules are present. There are other proteins called kinesins and dyneins, which attach to microtubules and move along them carrying cellular components. By using these "motor proteins", microtubules can act as two-way highways[155,465] for cellular traffic complete with traffic jams[156] (cf.[157]). There are still other proteins that cap ends of the microtubule[158] or in other ways[125,159] stabilize it against polymerization or depolymerization[159]. In addition to providing direct force due to polymerization[160], microtubules can act as fulcrums for other proteins to resist. The tail of the sperm is able to lash because of bundles of microtubules sliding relative to each other via attached proteins called dyneins[161]. A collection of motor and binding molecules attach microtubules to one another with many options for sliding or sticking together[162] (Figure 5.25).

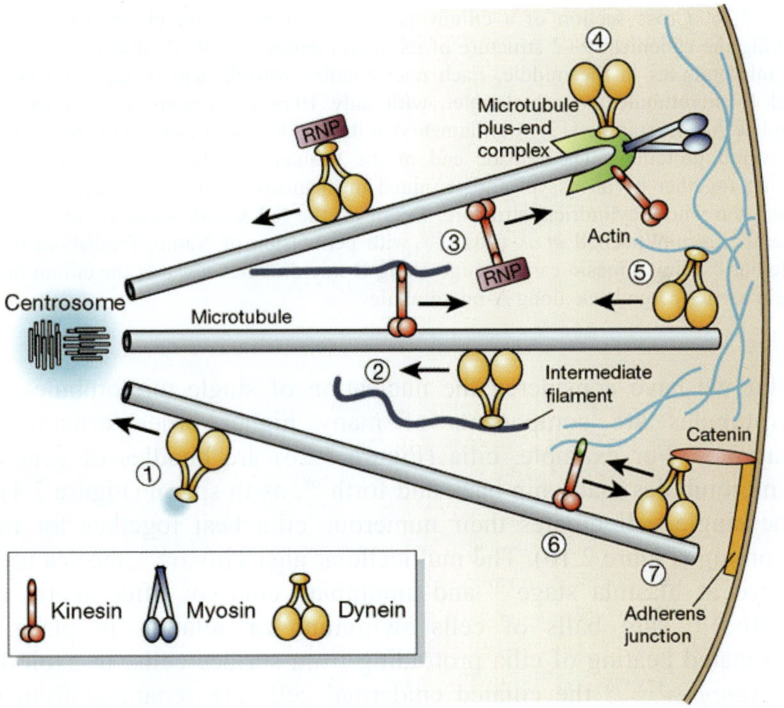

Figure 5.25. "Role of cytoskeletal motors beyond membrane transport. 1, Retrograde transport of centrosomal components. 2, Anterograde and retrograde transport of intermediate filaments. 3, Anterograde and retrograde transport of ribonucleoprotein (RNP) complexes. 4, Myosin, kinesin and dynein motors interact with components of the microtubule plus-end complex. 5, Anchorage of dynein at the actin-rich cell cortex. 6, Interaction of a kinesin-like protein with actin. 7, Catenin-mediated anchorage of dynein at adherens junctions." From Schliwa & Woehlke (2003)[163] with permission of Nature Publishing Group. New techniques permit the construction of protein motors that can be turned on and off by light, which should increase our understanding of how they work and what they do[164].

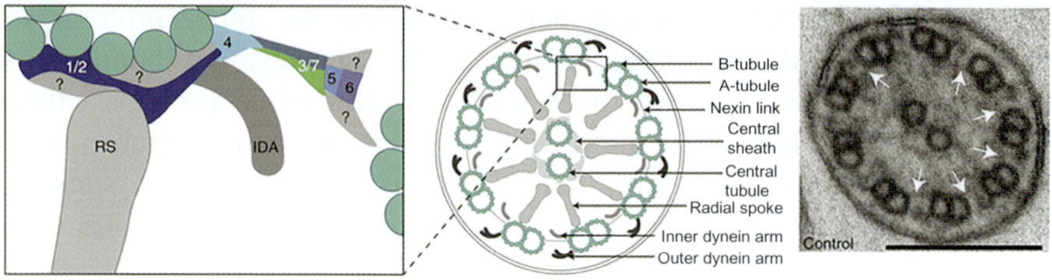

Figure 5.26. Cross section of a cilium as seen by transmission electron microscopy, showing the canonical 9+2 structure of its microtubules, i.e., 9 "doublets" around, plus two microtubules in the middle. Each microtubule protofilament is seen end on. The arced B-microtubule in each doublet, with only 10 protofilaments, is attached to its complete A-microtubule (13 protofilaments) at the inside end of its C-shape by a recently discovered protein[165]. The outside end of the C-shape involves three protofilaments sticking together by an as yet undetermined mechanism[166]. There is a cell membrane around the whole cylindrical structure. Detail on the left shows locations of additional proteins. From Wirschell *et al.* (2013)[167] with permission of Nature Publishing Group. Scale bar 0.2 μm. Kinesin carries cargo along B-microtubules towards the cilium tip and dynein carries cargo back along A-microtubules[465].

So far we have considered the nucleation of single microtubules. But microtubules are components of many higher order structures in eukaryotes. For example, cilia (Figure 5.26) are bundles of generally 20 microtubules that whip back and forth[168], as in sperm (Figure 2.4). In some single cell ciliates their numerous cilia beat together for rapid swimming (Figure 2.16). The multicellular alga *Volvox*[169], the sea urchin embryo at blastula stage[170] and amphibian embryos after neural tube closure[171-175] are balls of cells swimming or rotating in place via coordinated beating of cilia protruding from surface cells. In axolotls[173] and *Xenopus*[175,176] the ciliated epidermal cells are separated from one another by nonciliated cells, making their coordinated beating a bit dumbfounding. In us, clearance of our lungs[177] and movement of eggs through the Fallopian tubes[178] occur via these "metachronic" waves of cilia beating in these epithelia (Proposition 118 in[179]). Another microtubule structure is found in the parallel, helical or complex arrays lining the inner surface (cortex) of many plant[180] and ciliate[181] cells. Normal growth of cells or organisms has its basis in these microtubule arrays, which can be disrupted via tubulin mutations[182,183] (Figure 5.27).

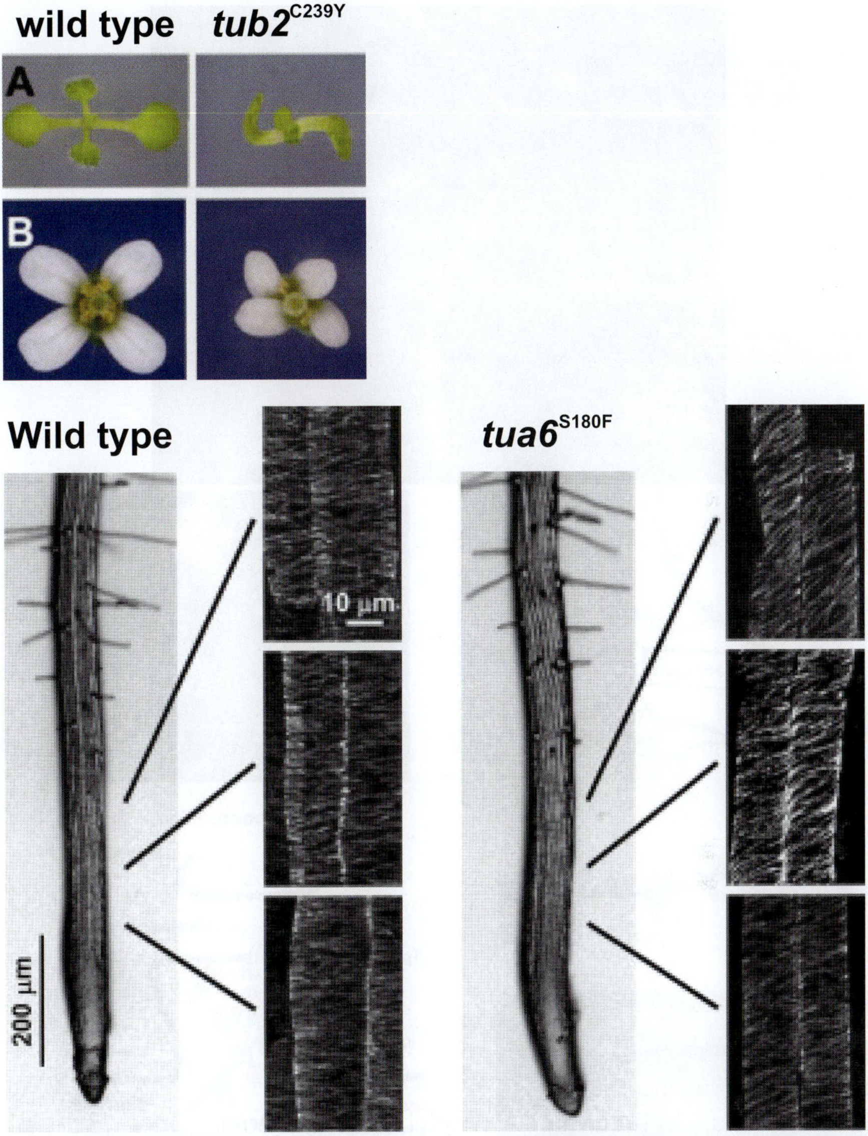

Figure 5.27. **Top**: *Arabidopsis* is a small plant related to mustard that has become a widely used model organism (Figure 2.13). Here change of a single amino acid in one of the nine β-tubulins[186] results in altered A) cotyledons, first leaves and B) flowers. **Bottom**: The cortical microtubule arrays in the roots show both twisting and disruption in the mutant. From Hashimoto (2013)[183] with permission of John Wiley and Sons.

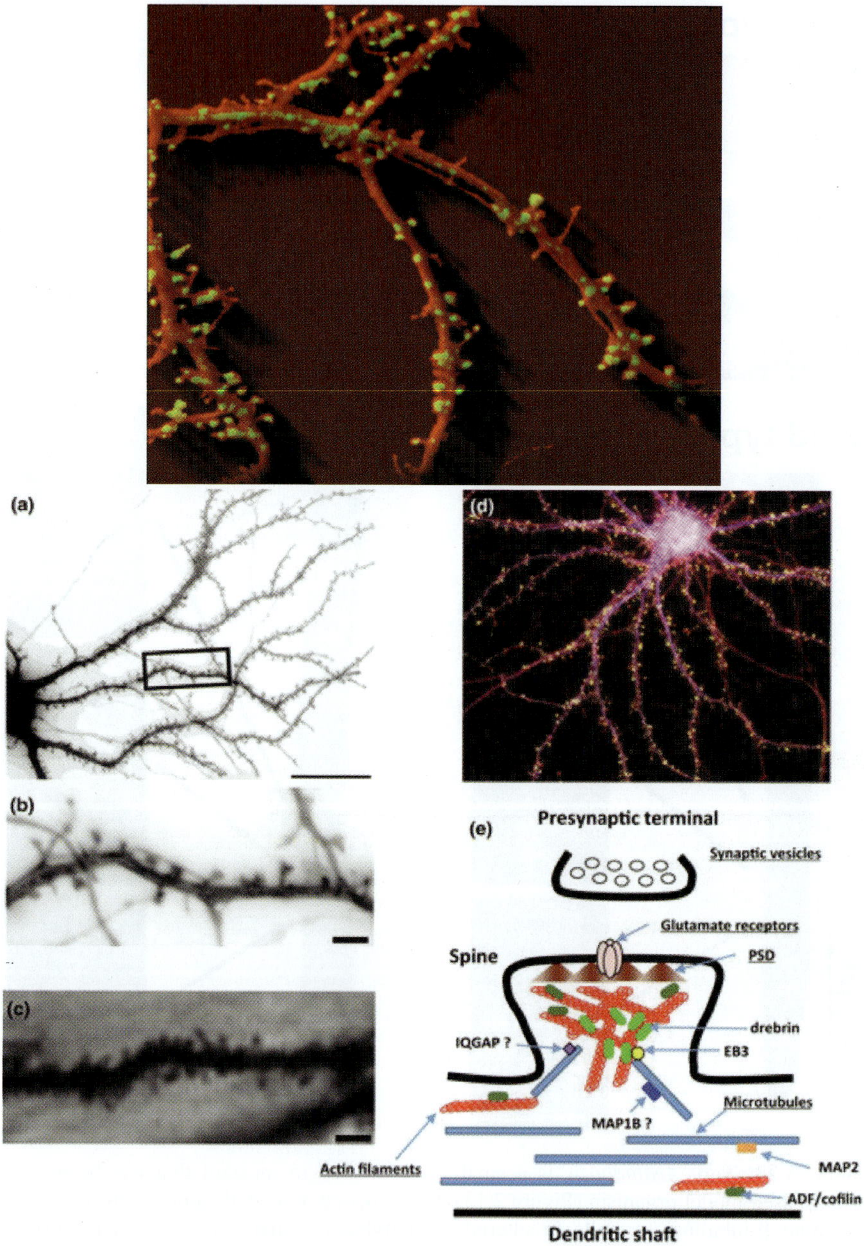

Figure 5.28. The dendrites of nerve cells in the brain are packed with microtubules in both orientations and studded with small projections called dendritic spines, which are packed with microfilaments (actin filaments). Some of the dendritic spines make contact

with other nerve cells through synapses (presynaptic terminal here). Top image (brown background) is from Korkotian (2009)[187] with kind permission of Eduard Korkotian at the Department of Neurobiology, Weizmann Institute, Rehovot, Israel. Bottom images (a-e) from Shirao & González-Billault (2013)[188] with permission of John Wiley and Sons. Scale bars are 20 μm in (a) and 2 μm in (b), which is a cultured hippocampal neuron; (c) is from an adult rat brain.

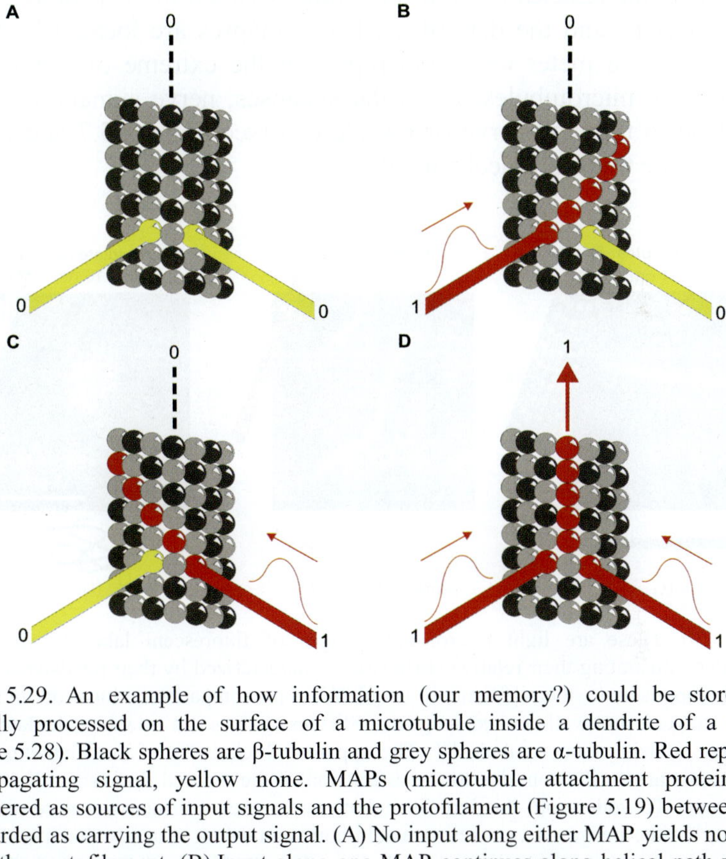

Figure 5.29. An example of how information (our memory?) could be stored and logically processed on the surface of a microtubule inside a dendrite of a neuron (Figure 5.28). Black spheres are β-tubulin and grey spheres are α-tubulin. Red represents a propagating signal, yellow none. MAPs (microtubule attachment proteins) are considered as sources of input signals and the protofilament (Figure 5.19) between them is regarded as carrying the output signal. (A) No input along either MAP yields no output along the protofilament. (B) Input along one MAP continues along helical path with no output along the protofilament. (C) Input along one MAP continues along an anti-helical path with no output along the protofilament. (D) Inputs along both MAPs arrive simultaneously canceling their lateral contributions and sending a signal along the protofilament. This corresponds to the logical AND operation[189], known as an AND gate in computer electronics. An XOR (exclusive OR) logical operation[190] can likewise be conceived, which together with AND could form the basis for any type of logical computation[184]. The propagating signal could be phosphorylation of the proteins. From Craddock *et al.* (2012)[184] with permission under a Creative Commons Attribution (CC BY) license.

The branching dendrites of brain cells contain masses of microtubules (Figure 5.28), which may be the recording medium for our mental memories[42,184] (Figure 5.29) and essential for consciousness[185]. Smaller structures called dendritic spines branch from the dendrites[202] and contain microfilaments[203]. The axons of neurons both inside and outside the brain[204] have bundles of microtubules. These microtubules deliver molecules and vesicles in both directions between the cell body, where the nucleus is, and the dendrites where synapses are located[205]. Axons can be up to a meter long, and represent the extreme of intracellular transport by microtubules. From the synapses, nerve signals are transmitted on to the next nerve or muscle cell (see Figures 5.7 and 5.8 for details of the transport mechanism).

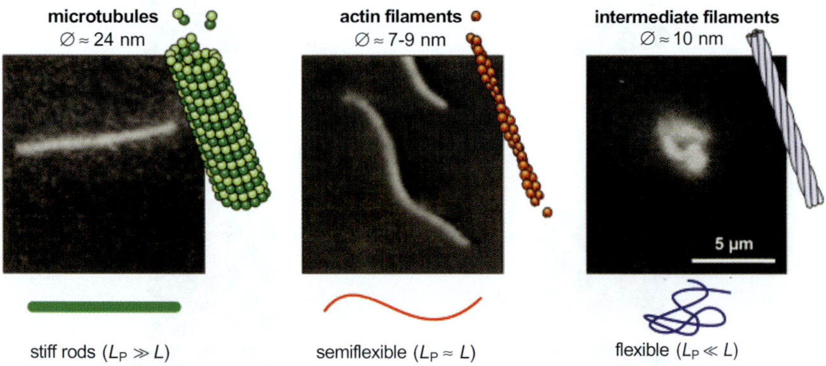

Figure 5.30. These are light microscope images of fluorescent labeled cytoskeletal molecules, illustrating their relative stiffnesses, "characterized by their persistence length L_P in comparison to their [filament] length L". For comparison, think of uncooked, slightly cooked, and well-cooked spaghetti. Stiffness values can now be calculated from computer models of microtubules[191]. The apparent thicknesses of the filaments in the micrographs are an illusion, due to the diffraction of the emitted fluorescent light in the microscope. To understand this, consider that one can see a distant wet wire reflecting sunlight, but the image on your retina would correspond to a structure much wider than the actual wire, due to diffraction of the light in the eye. The models are magnified 125 times compared to the micrographs, i.e., actual widths of the filaments in the micrographs would be 125 times narrower than the models. Note that microfilaments are also called "actin filaments". From a thesis by Steve Pawlizak[192] with his kind permission and from Käs *et al.* (1996)[193] with permission of Elsevier. Attached proteins can raise or lower the persistence length, as has been reviewed for microfilaments[194]. In absolute terms, estimates of persistence lengths are: microfilaments[193]: 1.8 μm; intermediate filaments[195]: 2 μm; microtubules[196]: 1000 to 8000 μm, and for comparison: double stranded DNA[197,198]: 0.05 μm; single stranded DNA[199]: 0.002 μm; interphase chromatin[467]: ~0.2 μm; FtsZ[200]: 0.25 μm; bacterial pili[201] (Figure 1.43): 0.25 μm.

Microtubules are the straight actors in the cytoplasm (with the greatest persistence length: Figure 5.30), the nice guys who (eventually) get the girl, while microfilaments and especially intermediate filaments are the squiggles, much like in the book "*The Dot and the Line: A Romance in Lower Mathematics*"[206], which was made into a cartoon movie[207]. Nevertheless, it is now time to give these less refined squiggly characters their due.

Microfilaments are polymers in the same way that microtubules are. The basic unit of the microfilament (filamentous F-actin) is the protein subunit called actin (or "globular" G-actin)[208]. In the scientific literature it is common to find the terms actin and microfilament used interchangeably but it is important to remember that actin is the single subunit that forms the microfilament even if we often read of "actin filaments" or just plain actin.

Microfilaments are generally short filaments (about 0.5 μm in length[209]) and they are usually bound together and to other cellular structures (Figure 5.31) by assorted proteins[210]. Microfilaments also appear in the cell nucleus[211]. They can form different shapes including branching structures, nets, rings, bundles[212] and cables, with high turnover of

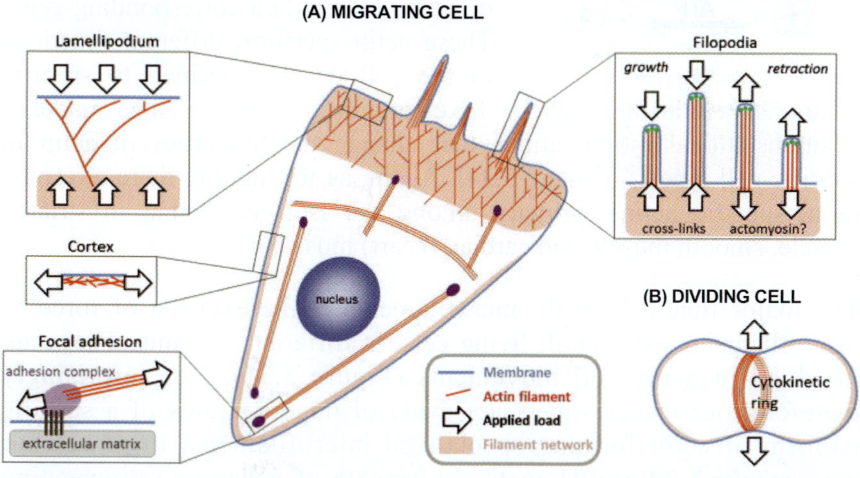

Figure 5.31. Some of the configurations taken on by microfilaments (actin filaments) in cells. Note that cell migration can involve pressure effects, not just changes in the cytoskeleton[214,215]. From Romet-Lemonne & Jégou (2013)[194] with permission of Elsevier.

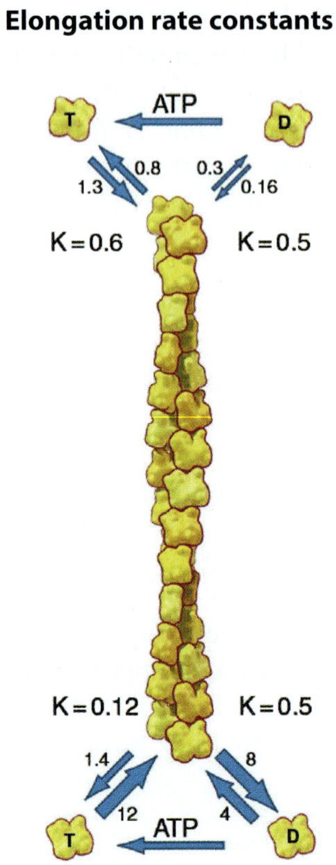

Figure 5.32. Treadmilling is the norm for microfilaments, because the rates of attachment and detachment to and from the two ends are not identical. This keeps microfilaments relatively short, fluctuating in length, and in continuous treadmilling. D = ADP bound to actin, T = ATP bound to actin. From Pollard & Borisy (2003)[213] with permission of Elsevier.

individual filaments due to the same phenomenon of treadmilling as occurs in microtubules (though with different rate constants)[213,209] (Figure 5.32). The ability of microfilaments to provide a pushing force via their polymerization[213,216,194] was again first worked out by Dick's PhD mentor Terrell Hill[160]. Actins can be so numerous as to contribute to what is called the crowded cytoplasm[217-223] (Figure 5.33), which requires special hydrodynamic methods for its computer simulation[224]. There are several isoforms of actin, each with a corresponding gene. These actins perform different functions in the cell or are confined to specific locations within the cell[225]. Like tubulin, actin can also undergo posttranslational modifications that change the properties of actin in specific cell types[226] (Figure 5.34). Again, as for tubulin, there are some tissue-specific actins, primary among the isoforms found in striated muscle, smooth muscle and cardiac (heart) muscle[227,228].

The major function of all microfilaments is the exertion of force[229]. Microfilaments exist in all living cells in differing amounts. They are involved in many cell movements (Figure 5.31). Not surprisingly therefore, our muscle fibers are made of huge amounts of a specific isoform of actin forming specialized microfilaments to power our movements[230]. Microfilaments are capable of generating propagating waves[231] at various scales down to nanometers[232] that can be transmitted cell to cell. The mechanism for this propagation is unknown but could possibly be accomplished via cell junctions joined to microfilaments[233,234].

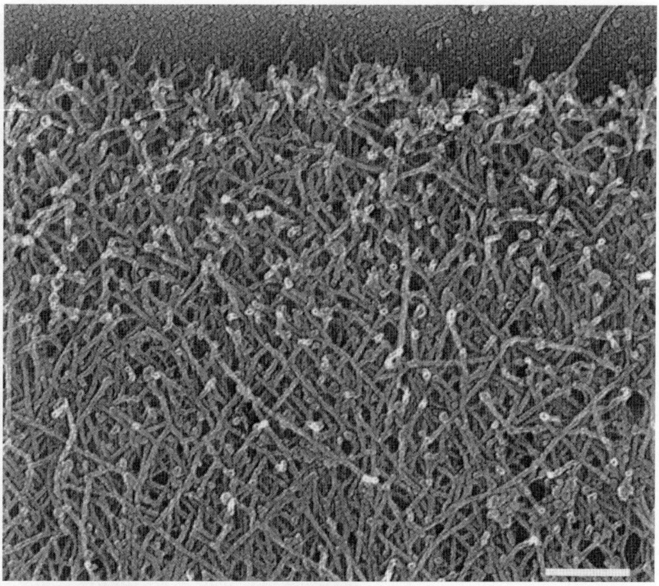

Figure 5.33. The crowding of proteins in the cytoplasm is well illustrated by this electron micrograph of microfilaments at the leading edge (lamellipodium: Figure 5.31) of a crawling keratocyte. By Tatyana Svitkina. Scale bar = 0.2 μm. From Small *et al.* (1999)[235] with permission of John Wiley and Sons.

Like microtubules, actin and its actin binding proteins are also involved in how the cell responds to signals such as mechanical stresses and chemical signals from other cells. This signaling can be from one part of the cell to another or from the exterior of the cell to the interior.

No discussion of microfilaments would be complete without mentioning myosin. Myosin is not considered to be a cytoskeletal protein but microfilaments are normally found paired with myosin and myosin drives much of what the microfilament does (Figure 5.25). Myosin was originally characterized in muscle fibers (hence the myo part of the name) where it has the structure of a movable head on the end of a long stick. The head cocks back like the trigger of a gun and attaches to the microfilament when ATP is provided (Figure 5.35). The ATP is converted to ADP. The head then loses the ADP and folds back against the stick and, in the process, pulls on the microfilament. Hence we can generalize that microtubules are powered by GTP and microfilaments are powered by ATP.

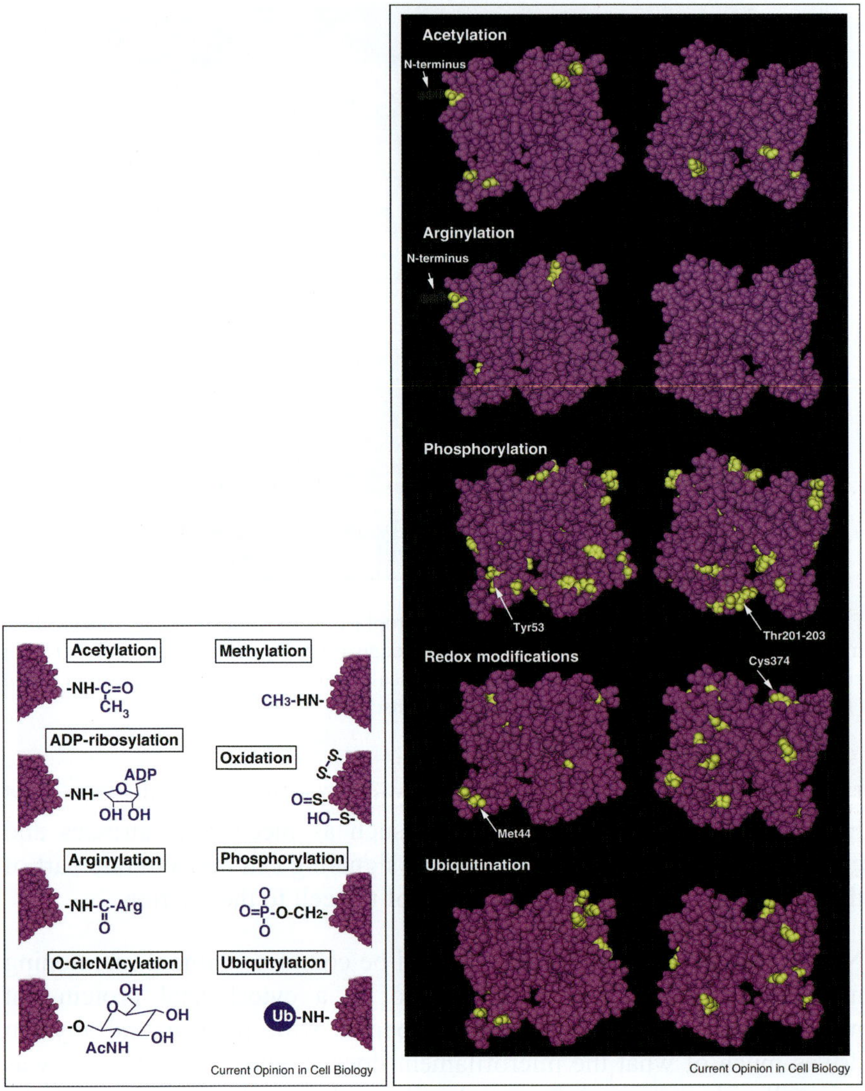

Figure 5.34. On the left are the chemical forms of the major posttranslational modifications of actin. On the right are "front" and "back" views of actin showing where these posttranslational modifications are placed. From Terman & Kashina (2013)[226] with permission of Elsevier. For the structure and general functions of ubiquitin, a small protein of 76 amino acids, see Vijay-Kumar *et al.* (1987) and Goodsell (2004)[236,237].

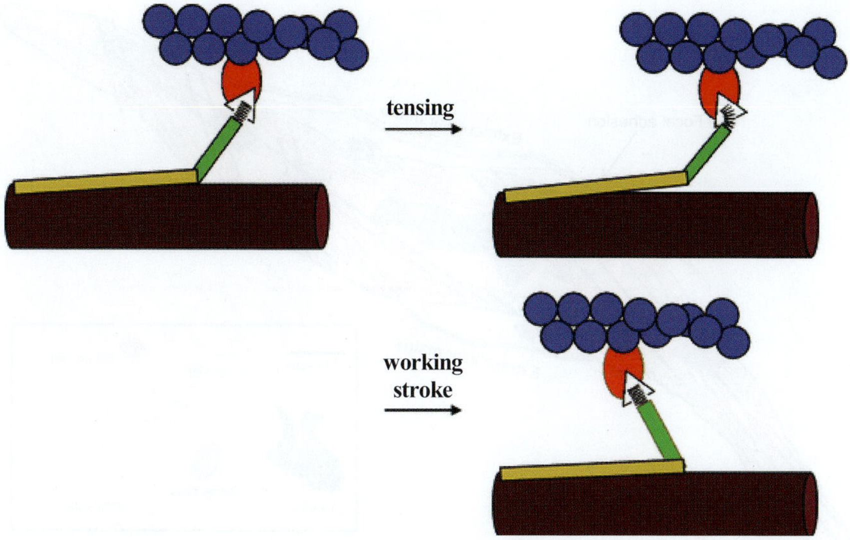

Figure 5.35. The myosin motor protein changes shape pushing a microfilament one step at a time in a given direction by swinging its "neck" domain like a lever arm[241]. (Red oval, white triangle and green rectangle = "head", yellow rectangle = "stick" in text.) The action involves a tension generating step with no motion followed by release of the tension in the working stroke, which moves the microfilament (shown in blue) by one step from right to left in this diagram. From Offer & Ranatunga (2010)[242] with permission of Springer. The exact role of ATP in providing the energy for this mechanism is still being worked out[243].

In muscle, many of these myosin sticks are bundled together in the middle, oriented in both directions with the heads poking out all over at both ends of the bundle. The bundle forms the thick filament of muscle sarcomeres. These thick filaments are surrounded by the thinner microfilaments. The cocking and release of myosin against the microfilaments creates the force generated by our muscles. Smooth muscle cells have much less organized bundles of microfilaments and myosin but these also exert force and that force can be considerable as women who have given birth can attest[238] (Figure 5.36).

As is so often the case in biology, the myosin first discovered in muscle turned out to be a highly specialized version of a protein used within living cells for a wide variety of functions and encoded by many genes throughout the genomes in essentially all organisms. Nature loves to reuse that which works well and myosin is a stunning example of this.

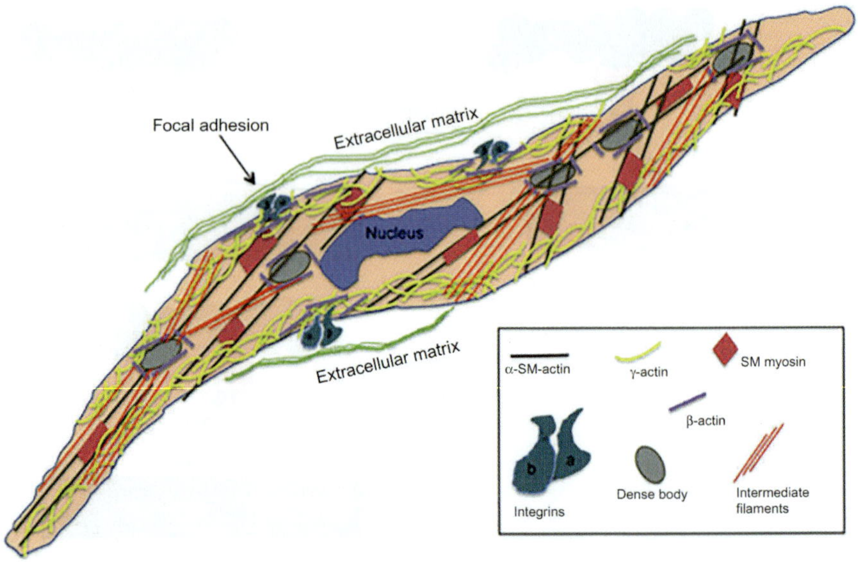

Figure 5.36. The cytoskeleton of a smooth muscle cell as found in the walls of contractile blood vessels. From Yamin & Morgan (2012)[238] with permission of John Wiley and Sons. SM = smooth muscle. Note that the microfilaments are of three kinds, each made of a different variant of actin.

The basic structure of a myosin head that binds microfilaments and performs the power stroke is so highly conserved that myosin of a rabbit will bind to microfilaments of amoeba[239,240]. The stick or tail end of myosin is the most highly variable part and different sequences in the tail end (which is not always stick shaped) give myosins the ability to bind to a host of other proteins, depending on the required function. There are some 40 recognized myosin genes in humans. Myosin[244] is also one of those gene types that is translated into multiple splice variants[245]. Therefore, the number of genes for myosin translates to many more kinds of myosin proteins present in our cells because of all the possible splice variants.

There are over a dozen classes of myosins in cells[244]. Some of these classes are based on function and some are based on the phyla they appear in. Other classes are blends of both. These many myosins are required for functions as diverse as moving plant chloroplasts nearer to the sunlit side of a leaf to the functioning of the tiny hairs inside human ears that allow us to hear sound. Wherever actin is present and functioning, there is a myosin somewhere nearby nodding its well-conserved head away from its tail, binding to a microfilament and then falling back, and thereby using ATP to power whatever it is the microfilament is doing.

Between microtubules and microfilaments in diameter are the intermediate filaments (Figure 5.30). Intermediate filaments generally provide mechanical stability and shape[246] to cells and resist external forces that would alter cell shape[247]. Like rubber bands, they are elastic and can stretch three times their resting length before breaking, while microtubules and microfilaments snap if even slightly stretched[248]. Hair and wool owe their properties to keratin, an intermediate filament[249]. Intermediate filaments range from the stiffening structures of long nerve cell axons to the lamin structures lining the nucleus, through to the proteins that form reptilian scales. An excellent example of intermediate filaments is vimentin[255,256], which is found in cells of mesodermal origin, in connective tissue and in nerve cells, and is involved in cell-cell adhesion, migration and cell signaling[257]. Intermediate filaments also stiffen microtubules against bending and buckling by forming cross-linking structures that act like the struts of an old fashioned train bridge[252,253,258,259]. Their mechanical properties are now being modeled quantitatively[260] (Figure 5.37). Unlike both microtubules and microfilaments, intermediate filaments lack a polarity and so cannot be used for directed transport, and they are not active enzymatically[267]. They are also unable to exert any force by themselves. They assemble in small packages that stick together (Figures 5.38 and 5.39) and then squeeze or compact into a narrower filament (Figure 5.40). Intermediate filaments interact with microfilaments and microtubules by binding to various cell components, including organelles[268,256]. Motor molecules such as kinesins and the myosins that are attached to or travel along the microfilaments and microtubules also attach to intermediate filaments, causing the latter to move (Figure 5.41). Apical intermediate filaments in epithelial cells are attached directly or indirectly to oriented paraxial microtubules (Figure 5.42).

There are about 70 different genes for intermediate filaments, many of them tissue specific[267,268] (Figure 5.43). Even so, intermediate filaments all have the same general structure. They have four helical regions connected by three linker regions and highly variable ends. The central

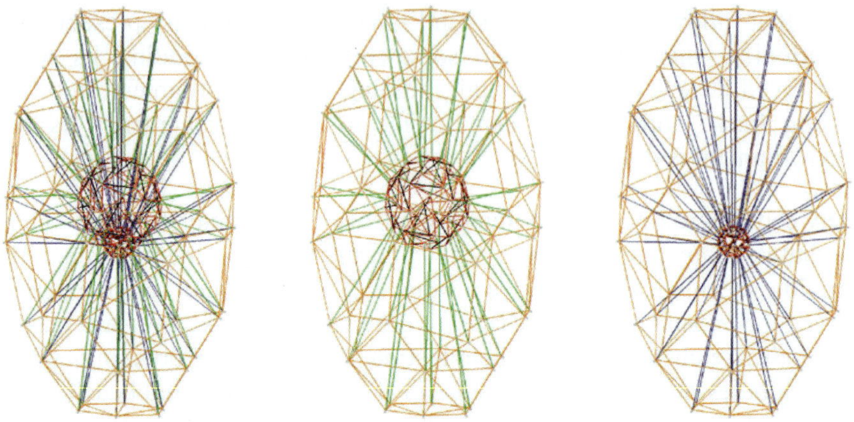

Figure 5.37. A tensegrity model is one in which there are stiff and contractile parts that balance one another to produce a flexible structure under tension that tends towards and retains a stable 3D shape without collapsing, i.e., has "integrity"[250]. This is a computer simulation of such a 3D model for a whole cell, in this case a bone-making osteocyte, seen here in projection. The ellipsoid representing the cell is 10 μm tall, 5 μm wide, and 2.5 μm thick. In this case, the cell membrane is presumed attached to hard bone matrix surrounding the cell, and does not change shape. However, the nucleus (large sphere) increases in size relative to its initial size due to simulated prestress of the intermediate filaments. The computer drawing on the left shows the microfilaments in orange, intermediate filaments in green, and microtubules in blue. The nucleus is the larger sphere above the centrosome, which is a microtubule organizing center. The middle drawing is the same with the centrosome and microtubules not shown, and the right is the same as the left with the nucleus and intermediate filaments not shown. From Kardas *et al.* (2013)[251] with permission of Springer. Note that this simulation does not allow for changes in the connections or resting lengths of the filaments, let alone the bending of microtubules and its amelioration by attached intermediate filaments[252,253]. Tensegrity simulations have now been done that allow for length changes of components of tensegrity robots[254].

linker regions are elastic and can give and pull back making the filaments deformable but elastic, a property known as viscoelasticity[275]. Most of us have encountered viscoelasticity in Silly Putty[276]. A ball of Silly Putty will bounce, but if left to the gentle pull of gravity, it flows slowly like a thick liquid. The highly variable ends of intermediate filaments allow them to selectively bind to and associate with all sorts of structures. Phosphorylation (the addition of a phosphate group to specific amino acids) in intermediate filaments changes their binding properties. These changes are required for the changes in the whole cytoskeleton used for movement of the chromosomes during mitosis. Other phosphorylations are required for other functions of intermediate filaments, especially for their interactions with enzymes.

Intermediate filaments have a neat property known as the alpha/beta transition, which occurs spectacularly in hagfish when they are disturbed and explosively secrete slime[279-281]. The proteins involved can be in a short, tight coil form (α) or they can be in a zigzag straight string (β). They can flip between these two configurations. Some external force has

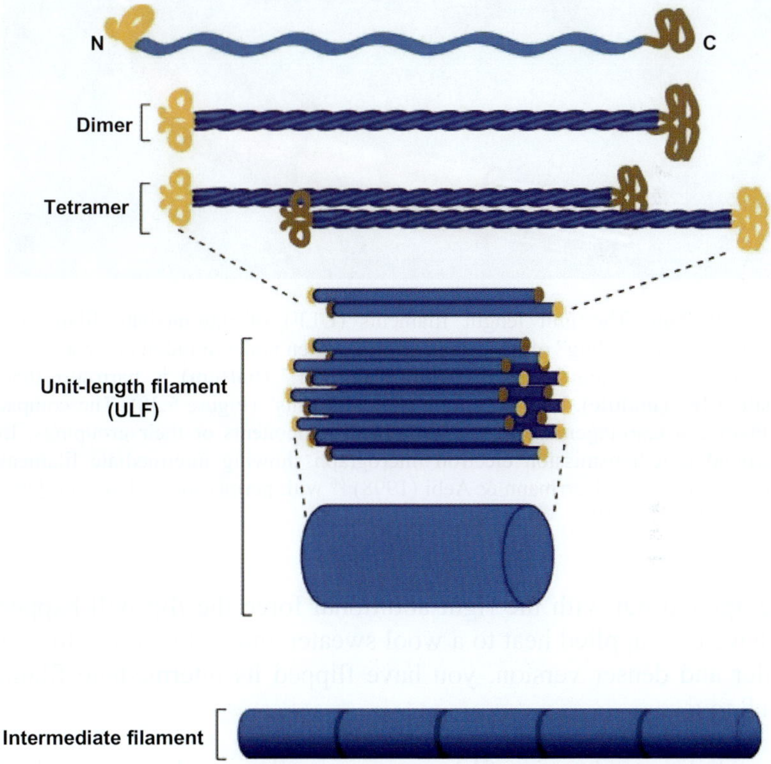

Figure 5.38. Intermediate filaments (IFs) such as vimentin assemble in a modular fashion[261]. "Intermediate filaments are built from monomers that associate with each other form[ing] dimers. Pairs of dimers then associate in an anti-parallel fashion to form staggered tetramers. Lateral associations between eight tetramers form unit-length filaments, which are able to anneal to each other, end-to-end, to form intermediate filaments"[262]. Image provided courtesy of the Mechanobiology Institute, National University of Singapore. Cf.[256]. N = Nitrogen terminal end[263] of the IF protein, C = Carboxyl terminal end[264].

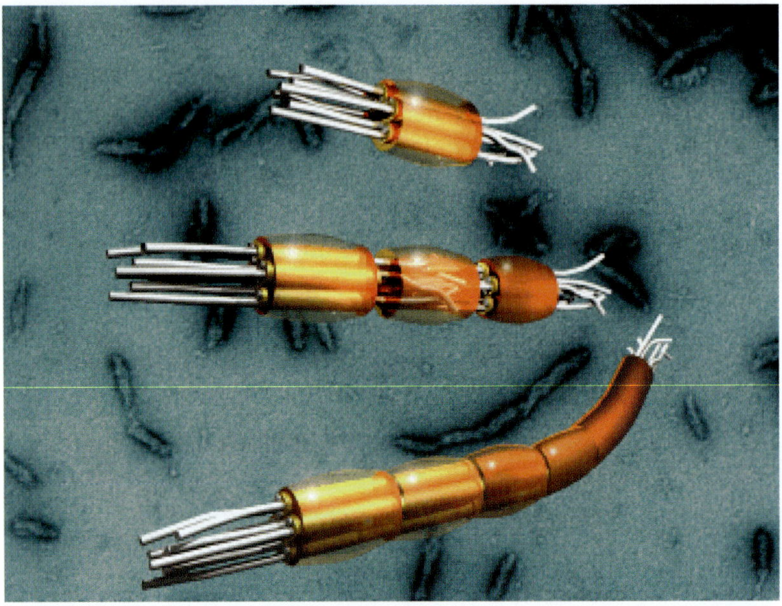

Figure 5.39. **Top**: The unit length filaments (ULF) of intermediate filaments (IF) polymerize via "annealing" of their loose ends and then undergo radial compaction over a period of about an hour[265], so that the "mature IF" (**bottom**) is narrower than the "immature IF" (**middle**). ULF = "single protofilaments" (Figure 5.38). The compaction may involve a rearrangement of the individual components or their groupings. In the background is a transmission electron micrograph showing intermediate filaments of various lengths. From Herrmann & Aebi (1998)[266] with permission of Elsevier, based on models by Daniel Stoffler.

to be applied but with the right additional force the flip will happen. If you have ever applied heat to a wool sweater and had it shrink to a much smaller and denser version, you have flipped its intermediate filaments from β to α.

Microtubules and microfilaments normally work in mechanical opposition to each other. It is assumed that histones, microtubules and microfilaments all evolved from a single precursor molecule[282]. This molecule was duplicated and then the three forms diverged over evolutionary time. Fundamental to this divergence is that microtubules use GTP and microfilaments use ATP. Their function requires mechanical opposition that is regulated at least in part, by the levels of GTP and ATP changing in opposition.

Figure 5.40. A model of a cross section of the intermediate filament vimentin as it undergoes compaction, leading to the mature filament. The initial structure contains dimers (Figure 5.38) that may be oriented towards or away from the axis (+/–), and octamers afterwards, as shown by their tails (Figure 5.41). The compaction process may propagate irregularly (stochastically) along the filament[267]. From Strelkov *et al.* (2003)[269] with permission of John Wiley and Sons. (This molecular level compaction should not be confused with mammalian embryo compaction discussed in Chapter 7.)

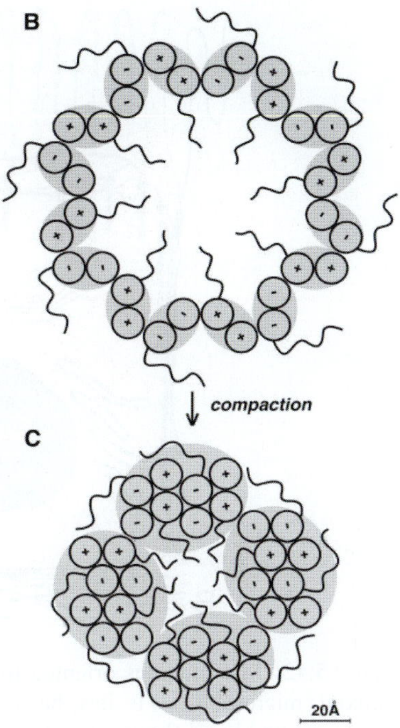

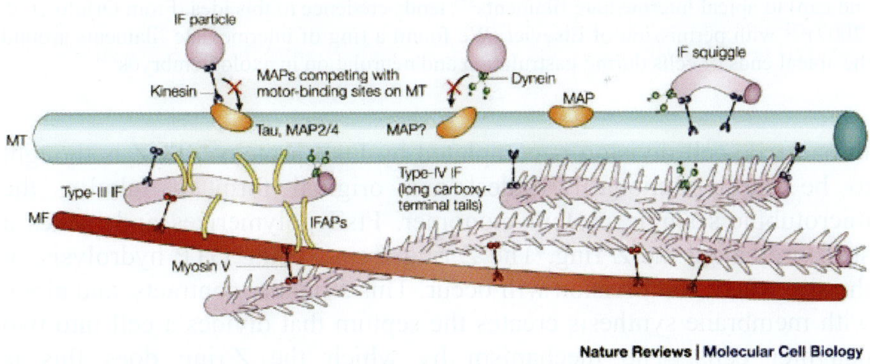

Figure 5.41. While intermediate filaments (IF) are not motile in themselves, nor have been shown to exert forces via polymerization or depolymerization, they do move along microtubules (MT) and microfilaments (MF), carried by myosin, kinesin or dynein. Microtubule-associated proteins (MAPs) may compete with motor proteins for binding sites on microtubules, altering the traffic patterns. Some intermediate filaments have tails or crossbridges[270] off of their individual proteins that give the appearance of bristles and are involved in attachments between IFs[271]. From Chang & Goldman (2004)[267] with permission of Nature Publishing Group.

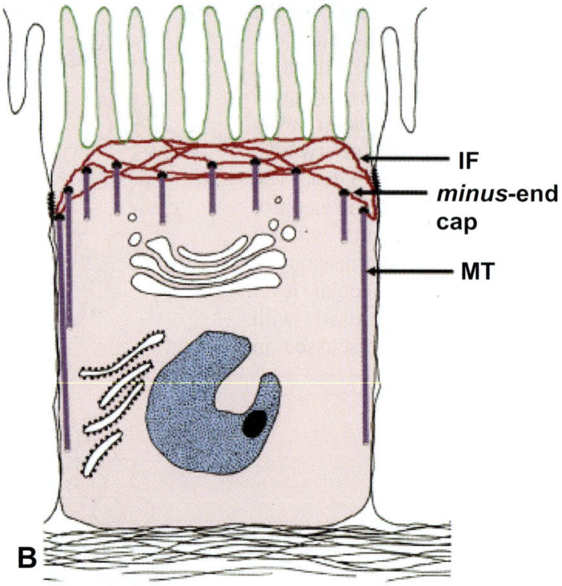

Figure 5.42. Microtubules oriented in the vertical direction in epithelial cells are called paraxial microtubules. It has been hypothesized that they push against the apical (top, here) and basal (bottom) ends of such cells[272]. The observation that one end of each paraxial microtubule may be anchored directly or indirectly via its γ-tubulin ring (minus-end cap) to apical intermediate filaments[273] lends credence to this idea. From Oriolo *et al.* (2007)[273] with permission of Elsevier. We found a ring of intermediate filaments around the apical ends of cells during gastrulation and neurulation in axolotl embryos[274].

In bacteria, cell division is completed by FtsZ protein[78]. FtsZ is thought to be the evolutionarily oldest and original form of tubulin, the microtubule's single protein monomer. FtsZ polymerizes and makes a structure called the Z ring. The Z ring forms, using GTP hydrolysis, at the site where cell division will occur. This structure contracts, and along with membrane synthesis creates the septum that divides a cell into two daughter cells. The mechanism by which the Z ring does this is unknown. What is known is that the formation of the Z ring always precedes cell division and its position is always where the cell divides. Two additional proteins seem to be involved, depending on the species of bacterium: FtsA and ZipA[283]. ZipA is a primitive form of an absolutely enormous class of microtubule binding proteins[284,285], and FtsA is a primitive actin homologue that forms actin-like filaments[286,283]. Both function to anchor the FtsZ ring firmly to the cell membrane.

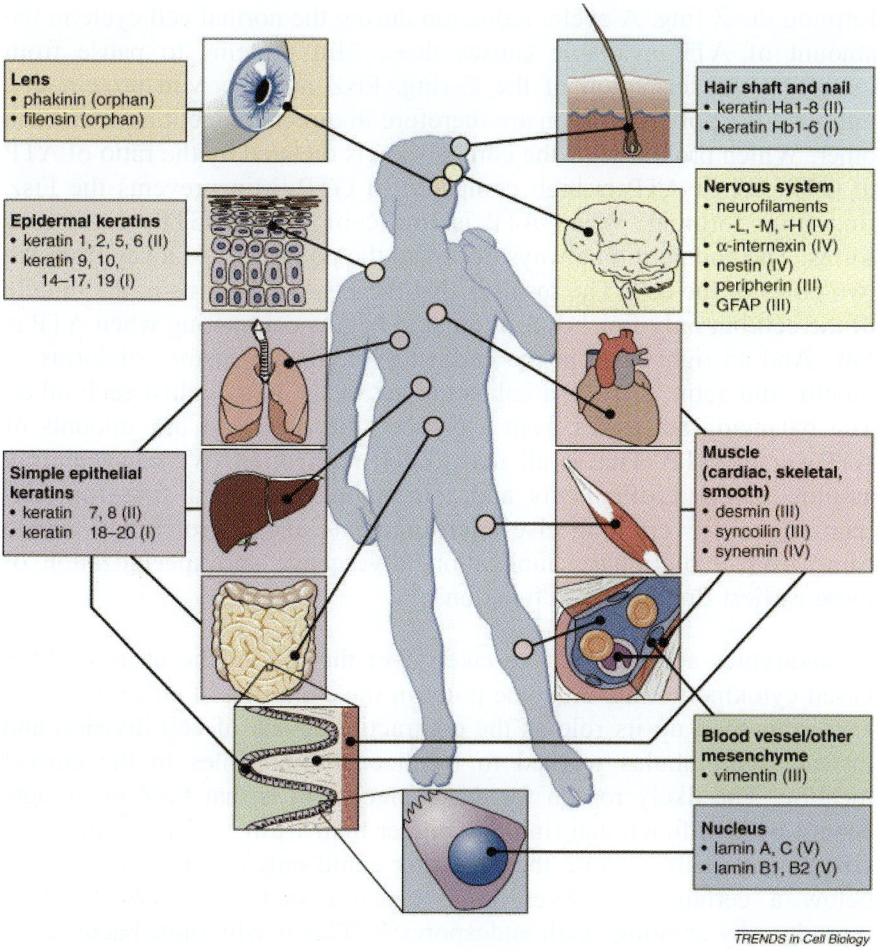

Figure 5.43. A body map showing where different kinds of tissue-specific intermediate filaments are expressed. From Toivola *et al.* (2005)[268] with permission of Elsevier. For a body map of the whole human proteome see[277], where it is shown that about 50% of proteins are from housekeeping genes and 50% are tissue-specific to varying extents. The reader could construct similar maps for other cytoskeletal proteins or protein classes from the accompanying interactive database[278].

Where the Z ring forms depends on the activity of another class of proteins collectively known as Min in some bacteria (Figure 5.15). These Min proteins, which use ATP hydrolysis, nucleate at opposite ends of the cell and polymerize towards the middle. Min proteins inhibit FtsZ from forming the Z ring. A cyclic reduction during the normal cell cycle in the amount of ATP available causes these Min proteins to pause from inhibiting the formation of the Z ring. FtsZ and the Min protein that inhibits FtsZ polymerization are therefore in direct competition with each other. Which protein wins the competition is dictated by the ratio of ATP to GTP. When ATP is high compared to GTP, Min prevents the FtsZ ring from forming. When ATP is low compared to GTP, the Z ring forms. The competition always occurs at the point farthest away from the two poles of the cell. The result is that the FtsZ ring forms in the middle of the cell but only finishes forming and begins contracting when ATP is low. And so right back in the earliest evolutionary history of forms of tubulin and actin, there is a balancing act of the two against each other. The balancing act arises from a competition based on the amounts of ATP versus GTP. This is all tied to cell replication. We can therefore assume that microfilaments and microtubules evolved first for self-reproduction. Everything else microfilaments and microtubules can do came from evolutionary duplication, divergence and specialization of these earliest first form and function.

In eukaryotes actin rings have taken over the role of the bacterial FtsZ based cytokinetic ring. At some point in the evolution of eukaryotes, the FtsZ ring gave up its role as the contractile device of cell division and instead microtubules worked to localize chromosomes to the correct location. The likely reason for this replacement is that FtsZ monomers cannot make a functional ring any bigger than 1 μm[287]. The result was a size limit to cells because the FtsZ ring could only contract across cells below a certain size. Even huge bacteria such as *Epulopiscium*[288] reproduce by creating small endospores[287]. This is why most bacteria are small relative to eukaryotic cells. With replacement of the FtsZ ring by a microfilament ring, the upper boundary for cell size was dramatically increased. This switch from FtsZ to actin for the cytokinetic ring of cell division[289] may have been one of the major steps in allowing eukaryotes to take off evolutionarily by removing the size restriction for cell division.

The spindle apparatus[290-294], which separates the chromosomes during mitosis or meiosis, is perhaps the most spectacular structure inside cells that is made of both microtubules and microfilaments[295]. In animals, the entire netlike “spindle apparatus” that moves the chromosomes is organized by microtubule organizing centers which act as nucleation sites for microtubule formation[296]. The precise place where cell division occurs is also dictated by the orientation of the microtubule spindle in eukaryotes[297-299]. Cell division always occurs along the same plane in the cell that the metaphase plate forms on. There are forces involved in aligning and separating the chromosomes so that a full set goes to each daughter cell. These are gradually being worked out in yeast, where a single microtubule is attached to each chromosome's kinetochore of the budding yeast *Saccharomyces cerevisiae* compared to 16 to 20 microtubules per kinetochore in mammalian spindles and so yeast are easier to conduct such studies on[300]. Sometimes it's best to pick the simplest model organism[301].

Yeast is a model organism that has been intensely studied and much of its structure and function are well understood. We've had a long relationship with yeast given their historically ancient uses in beer, wine and bread making[302]. It is therefore useful to look at yeast to gain insights into the dynamic nature of cytoskeletal functioning and the linking of

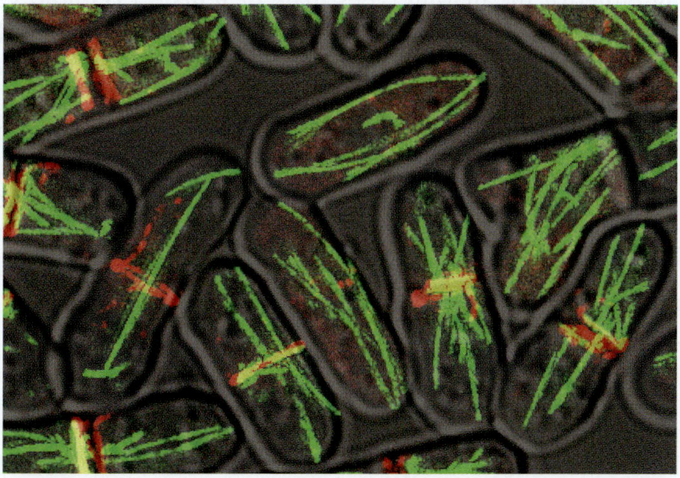

Figure 5.44. "Merge micrograph of fission yeast cells [*Schizosaccharomyces pombe*] observed by phase contrast [cell walls in grey], by Rlc1-RFP fluorescence of the myosin II regulatory light chain showing the cytokinetic contractile ring [red] and by Atb2-GFP fluorescence of the alpha-tubulin showing the microtubule cytoskeleton [green]. The GFP [green fluorescent protein] and RFP [red fluorescent protein] images are the maximal projection of 15 Z-series [stacked serial sections] imaged at 0.2 μm intervals." From Ribas & Cortes (2014)[305] with kind permission of Juan Carlos Ribas. Cell diameters are 3 to 4 μm[304].

structure to function. Recall the yeast life cycle from (Figure 3.46). Yeast are fungi that do not normally produce filamentous threads (although there is a form that can make filaments called dimorphic yeast[303]). Nondimorphic yeast remain in the single cell state throughout their life. There are several types of yeast but they can be generally divided into fission and budding yeast. Fission yeast produce two equal sized daughter cells[304] (Figure 5.44). Budding yeast first form a small bud that then grows to equal the original cell in size before splitting off[306] (Figure 3.62). The cytoskeletal functioning of budding yeast represents an intermediate between the simplest cytoskeletal function, namely cell division in bacteria, and the more complex cytoskeletal functioning in eukaryotic multicellular development.

All yeast cells can reproduce and function through an entire G1 to mitosis cycle (Figure 3.2) in either the diploid form or the haploid form. Alternatively, the diploid yeast cell can undergo meiosis and produce four haploid spore cells. The spores can develop into functioning haploid yeast with either an "alpha" (α) or "*a*" mating type. (We have no basis for calling either of the mating types male or female.) Haploid yeast can mate with an opposite mating type and create a new diploid cell. With that reviewed, it is time to look at the cytoskeleton during this process.

In all yeast, the condensation of chromosomes never reaches the point that the chromosomes are clearly visible in light microscopy. Nonetheless, all yeast do have a proper metaphase (Figures 3.39 and 5.45). The conclusion that yeast have a true metaphase is based on the cell cycle proteins that are present. Another way that yeast are simpler than us is that there are only four tubulin genes in fission yeast. There are one beta (β), one gamma (γ) and two alpha (α) tubulins.

During interphase, microtubules are found only in the cytoplasm and they form long structures running the axial length of the cell from microtubule organizing centers located at the ends of the cell. There is an additional microtubule organizing center, called a spindle pole body (centriole) that is specific to forming the spindle apparatus and is required for mitosis. The spindle pole body appears at the end of G2 and then duplicates at the beginning of mitosis. (This duplication is a process that is still a mystery[307,308].) The microtubules that run the length of the cell depolymerize and vanish at this time and the subunits (tubulin) from these axial microtubules are then used to make the spindle apparatus, or

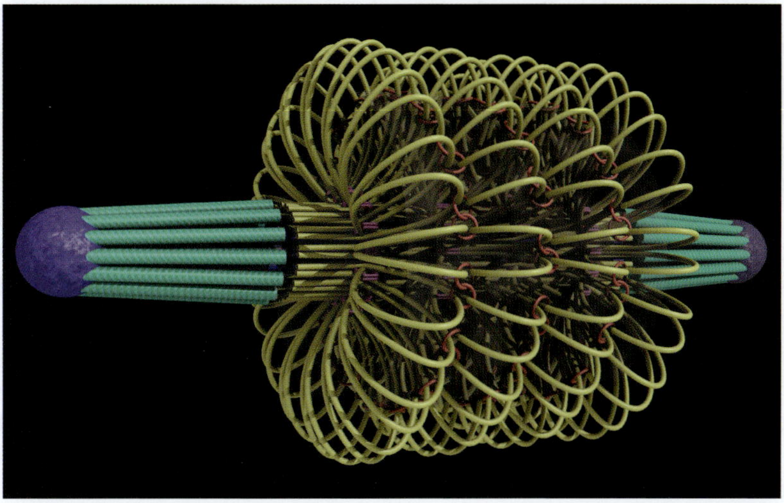

Figure 5.45. "The yeast is *Saccharomyces cerevisiae* as used for baking and brewing.... Proposed structure of yeast mitotic spindle in metaphase.... This model of the yeast mitotic spindle shows the spindle-pole bodies as blue spheres, the kinetochore microtubules as green cylinders, the DNA as yellow tubes, cohesin as linked red rings, and condensin linking molecules in purple. The translucent gray shell around the spindle shows the center of the region that contains cohesin as seen in fluorescence microscopy images taken of the spindle." From Taylor II *et al.* (2014)[310] with kind permission of Russell M. Taylor II and Kerry Bloom. For a movie see YouTube[311].

alternatively move as whole clusters of microtubules to the spindle[309]. The formation of other microtubules is also suppressed at the same time preventing any accidental interactions that might mess up the orientation of the mitotic spindle.

The old spindle pole body is connected to the new spindle pole body by a thick set of bundles of microtubules called core microtubules. As soon as the core microtubule link is formed, astral microtubules begin to radiate out from the spindle pole bodies. These astral kinetochore microtubules attach the kinetochores of the chromosomes to the spindle pole body (Figure 3.36).

At metaphase, the centromeres of sister chromosomes are split by an as yet unknown mechanism and actually separate for about ten minutes before being rejoined, presumably by some protein binding between them. (Not all yeast have this ten minute split.) Once this split/rejoining has occurred, two bundles of microtubules extend from the spindle pole bodies to the cell wall. The link between the older spindle pole bodies forms first and connects to the top of the new cell bud via special proteins that have moved there during interphase. After the old spindle pole body is connected to the top of the bud, another of these astral microtubules connects the new spindle pole body to the neck of the bud. The older spindle pole body always goes into the new bud because it is attached to the top of the bud while the new spindle pole body always goes into the mother cell because it is attached to the old cell. This specifically timed connection of microtubules cleanly confines the spindle pole bodies to their correct respective sides. If the proteins at the top of the bud are experimentally removed, the spindle pole body attachment is random. If the sequence and timing of microtubule formation is changed, mitosis fails and the bud does not get any chromosomal material[312].

At anaphase, the kinetochore microtubules (attached to the chromosomes) are shortened while those in the link microtubules (between spindle pole bodies) are elongated. The spindle pole bodies move to the far ends of the cell, retaining their microtubule connections to the cell cortex at their respective places. The cortex microtubule connection to the old spindle pole body does not move. The cortex connection in the mother cell moves from the bud neck region to the far end of the mother cell. The chromosomes are now permanently separated and pulled behind their respective spindle pole bodies. The chromosomal material is thus divided equally into two portions and each portion has one spindle pole body. The cell exits from anaphase into telophase and cytokinesis occurs by disassembling the microtubules and assembling a cytokinetic actin ring. The final division of the cell occurs during telophase by the activity of this actin ring.

There are three major microfilament structures in all yeast. These are the cortical actin patches, polarized actin cables, and cytokinetic actin rings, all attached to the cell cortex, the layer of cytoplasm just inside the cell membrane (Figure 5.46). Cortical actin patches are clusters of branching microfilaments assembled into pouch-like forms that hold material from outside of the cell and form delivery vesicles (by a process called endocytosis[313]). These patches then move the vesicles to the long polarized[314] microfilament cables. Polarized microfilament cables (often called polarized actin cables) are multiple small overlapping sections of microfilaments that are crosslinked (Figure 5.47). A delivery vesicle travels along the cables until it is delivered to internal membrane compartments in the cell. Once the vesicle reaches the internal membrane, it binds with the internal membrane and dumps its contents. The actin in the patch disassembles during this dumping and returns to the surface of the cell. The actin reassembles again as a new patch and then cycles again. The actin patches and the cable delivery system are required by the yeast to "eat" all the material it needs to survive.

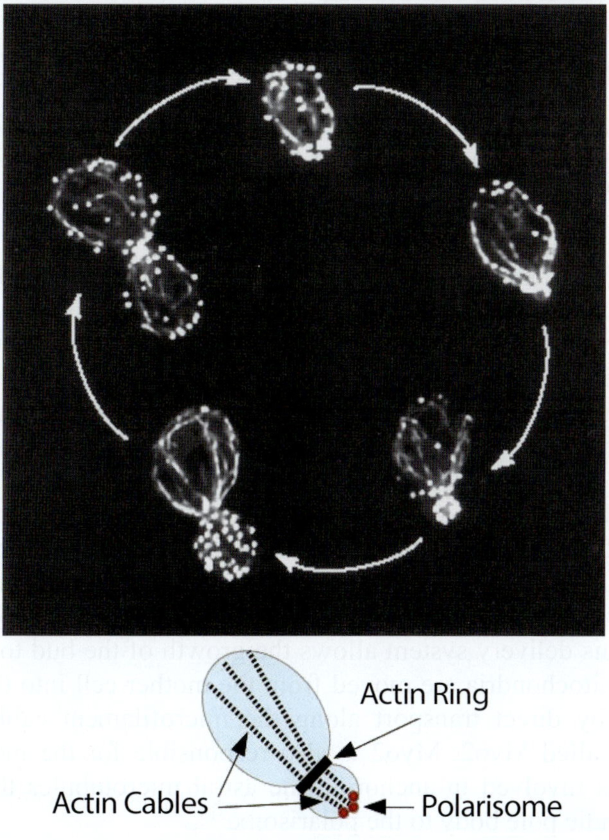

Figure 5.46. **Top**: Fluorescence images of the actin/microfilaments in the budding yeast *Saccharomyces cerevisiae* through its cell cycle. Note that most of the microfilaments, including patches, actin cables and microfilament rings, are located at or attached to the cortex of the cell. From Amberg (1998)[316] with permission of the American Society for Cell Biology. **Bottom**: Sketch of the polarisome corresponding to the micrograph at 5 o'clock.

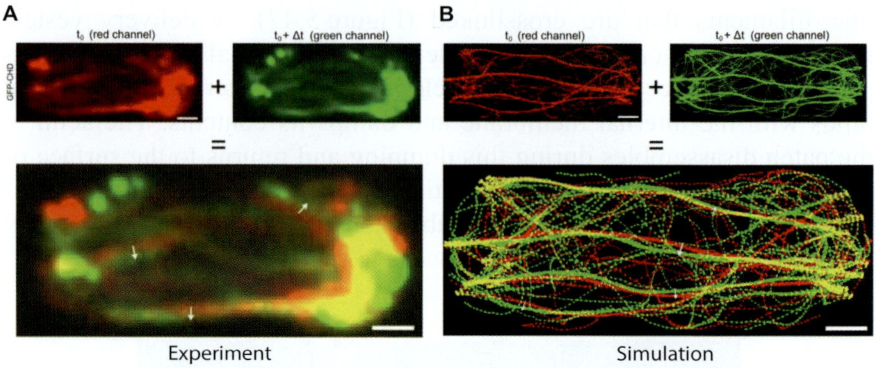

Figure 5.47. Movement of actin cables in the fission yeast *Schizosaccharomyces pombe* (Figure 5.44) from one moment (red) to the next (green), $\Delta t = 8$ seconds later. Yellow means no movement. A) experimental, B) computer simulation. Scale bar: 1 μm. From Tang *et al.* (2014)[317] with permission of the American Society for Cell Biology and CCC Republication.

During bud formation, protein structures localize to the tip of the newly forming bud and create a polarity cap structure at the far end of the bud. This multiprotein structure is called a polarisome[315]. The polarisome acts as a marker placed in the yeast cell's inner membrane surface which indicates where to build the cytoskeletal structure required for budding. Another protein structure forms at the circular mouth end of the bud where it is joined to the mother cell. Long microfilament cables form, similar to the ones used to transport vesicles. The cables run the length of the bud to the circular mouth and from the bud mouth along the length of the mother cell (Figure 5.46). These specialized bud actin cables form a delivery system for various cellular components from the mother cell to the bud. This delivery system allows the growth of the bud to occur. For example, mitochondria are moved from the mother cell into the growing yeast bud by direct transport along the microfilament cables using a myosin[318] called Myo2. Myo2 is also responsible for the movement of the proteins involved in anchoring the astral microtubules that connect the old spindle pole body to the polarisome[319].

Of greater interest to embryogenesis is the microfilament ring, formed in yeast from microfilament cables at what will eventually become the neck of the new bud[320]. In yeast, the location of the microfilament ring is selected during interphase. There are over 100 proteins involved in the process, most of which are well characterized[321]. Specific proteins accumulate and bind to the future ring location[322]. These proteins then recruit additional proteins. A specific enzyme called Rho is critical. Rho is a GTPase, meaning it requires GTP to be in an active form and only where Rho is activated by GTP can it do its specific reaction (Figure 5.48). The yeast form of this Rho GTPase[323] is called Rho1. The activity of Rho1 and various other proteins recruit yet another protein, formin Cdc12p[324], to the neck of the bud. After anaphase, formin Cdc12p binds to the future ring location and thereby creates nucleating nodes. These nodes capture free actin. Short overlapping microfilaments form the ring. These are similar to the long microfilament cables, but unlike the cables that are always bound in only one direction, these filaments can be bound together in two directions (cf.[325]). Myosins and other proteins crosslink the filaments and stabilize the ring as it forms.

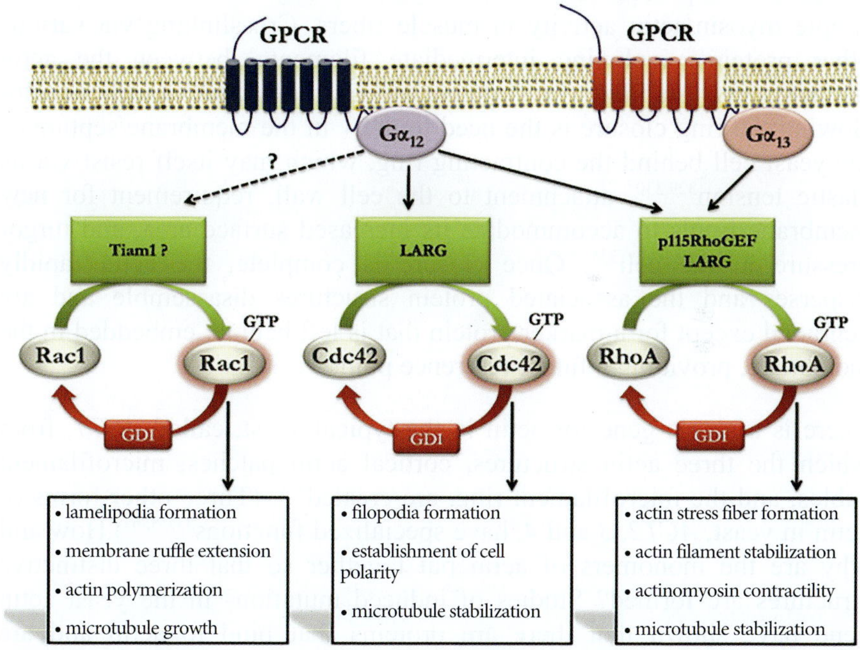

Figure 5.48. GTPases are critical for the activation of Rac1, Cdc42 and RhoA, which then participate in cytoskeletal remodeling. These molecules cycle between an inactive (GDP-bound) form and an active (GTP-bound) form. GPCR is a G protein coupled receptor with seven transmembrane domains. We will describes these in detail in Chapter 8. From Schappi *et al.* (2014)[326] with permission from Elsevier.

Once this ring has formed, it begins contracting and in contracting creates the neck of the bud. Meanwhile the bud continues growing as material is carried into it along the microfilament cables to the polarisome. By controlling what is carried over into the bud and what is left in the mother cell, different concentrations of certain proteins in the mother and in the bud are created. When the bud has separated and begins accessing its own genes, it contains a different concentration of proteins. The result is the bud can have a different gene expression pattern (a new differon) from the mother cell and so the budding process is also a form of very simple differentiation[327,328] that is always firmly linked to mitosis. Budding yeast therefore provide a model of how an asymmetric division (Chapter 6) with differentiation can occur.

The thickness of the yeast microfilament ring remains constant as the ring contracts indicating that the actin is depolymerizing as the ring contracts. The process is fueled by myosins and yet is much slower than simple myosin/actin activity in muscle fibers. Crosslinking via various other proteins, including intermediate filaments, between the actin filaments may stabilize the structure as it closes down. Another factor slowing the ring closure is the need to draw in the membrane/septum of the yeast cell behind the contracting ring, which may itself resist via its elastic tension[329,330], attachment to the cell wall, requirement for new membrane lipids to accommodate its increased surface area, and turgor pressure in the cell[331]. Once closure is complete, the actin rapidly disperses and the associated protein structures disassemble and are degraded except for a marker protein that is left behind, embedded in the membrane, providing a future reference point[332].

There is a single gene for actin in the typical yeast, called *ACT1*, from which the three actin structures, cortical actin patches, microfilament cables, and the microfilament ring, are created[321]. (Three other forms of actin in yeast, *ACT2*, *3* and *4*, have specialized functions[333-336].) How and why are the monomers of actin put together so that three distinctive structures are formed? Studies of induced mutations in the yeast actin gene have shown that there are proteins that bind to actin and are essential to its filamentous structure. For example, if the region of actin that is normally used to bind myosin II is altered by a mutation, the cables fail to form[337]. This indicates that actin polymerization can be organized around specific nucleating sites, analogous to microtubule organizing centers. The type of protein acting as a nucleating site for the polymerization of microfilaments determines which actin structure is formed.

There is a common cytoskeletal theme occurring here. In bacteria the presumed predecessor protein to the cytoskeleton is involved in cell division but not in any sort of differentiation. In the simple eukaryotic organism, the yeast, the cytoskeleton is used to do cell division and that cell division is also used to create differential protein movement resulting in a simple form of differentiation linked to cell division. The bud of the yeast can be different and have a different set of active genes, a different differon, from the mother cell. The cytoskeletal structure of the yeast during cell divisions is therefore essential to the differentiation of the yeast bud. The bud forming apparatus of budding yeast is therefore analogous to, and we expect to be evolutionarily related to, the structure we refer to as the cell state splitter, which we will cover in greater detail in Chapter 6.

Cytoskeletal evolution led from cell division to movement to cell differentiation. Not surprisingly then, microfilaments also interact with microtubules via motor and stationary attachment proteins that tether them together for cell movement[340,341]. An excellent example of microtubule/microfilament interaction occurs during cell motility in motile endothelial cells (Figure 5.49). The front end of such a cell extends long membrane ruffles supported by microfilaments[342]. When the ruffling membrane contacts a surface microfilaments guide the placement of membrane proteins, which then bind to the surface. Their inner, cytoplasmic portions act as attachment sites ("tip attachment complexes" = TACs[343]) for the plus ends of elongating microtubules. The center of the cell is stabilized by a microtubule and microfilament meshwork. The back end of the moving cell is mainly microtubules, many attached to the nucleus. These microtubules must be disassembled and the microtubule organizing center proteins must be detached from the surface before the membrane can be pulled up behind like a trailing skirt. Therefore, the moving amoeboid type cell has a polarity. This polarity includes rotation of the nucleus so that either its long or short axis is aligned with the direction of motion, apparently via attached microtubules that provide the torque for rotation[344] (Figure 5.50).

There is also polarity in the microtubules themselves. The plus ends of the microtubules are located in the outer parts of the cell. The minus ends

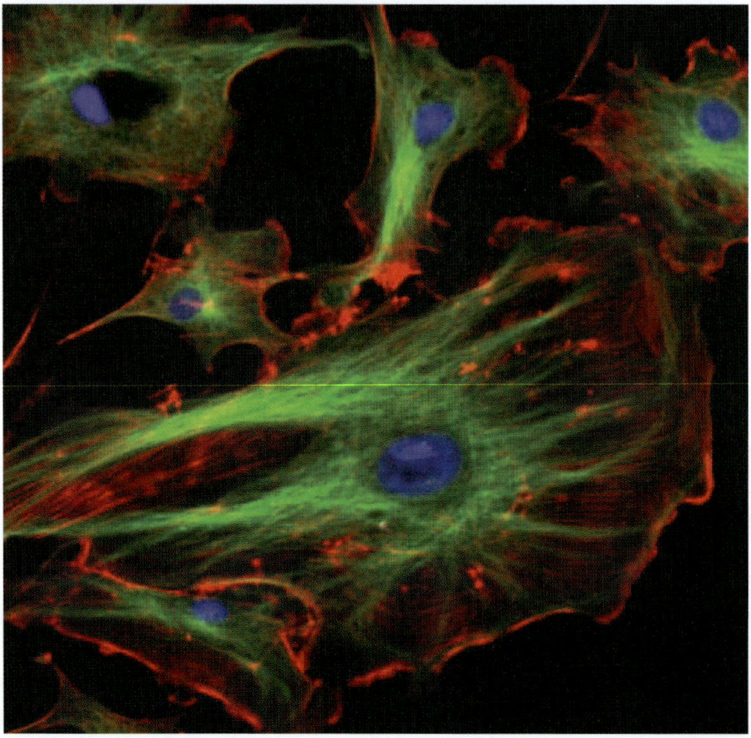

Figure 5.49. In this fluorescence image of migrating endothelial cells[338], the leading edge has a great deal of microfilaments (red) while microtubules trail (green). Nuclei are blue. From Wikipedia[339] with permission via Wikimedia Commons, under public domain. Image width is about 50 μm. Cf.[466].

connect to the nucleus (Figures 5.31 and 5.49). If we compare migrating cells to epithelial cells, we see that the basal end of an epithelial cell is attached to the plus end of the paraxial microtubules. When cells leave an epithelium they do so at their basal ends, so the polarization of the microtubules of an epithelial cell and that of a migrating cell may have a common evolutionary basis (Figure 5.51).

The cytoskeleton also functions in the freshly fertilized egg of all complex organisms. The cytoskeleton is critical in establishing mechanical biases that in later development are used by the embryo to determine front and back and top and bottom (also known abstractly as symmetry breaking[345-347]). When the axolotl egg, properly known as the

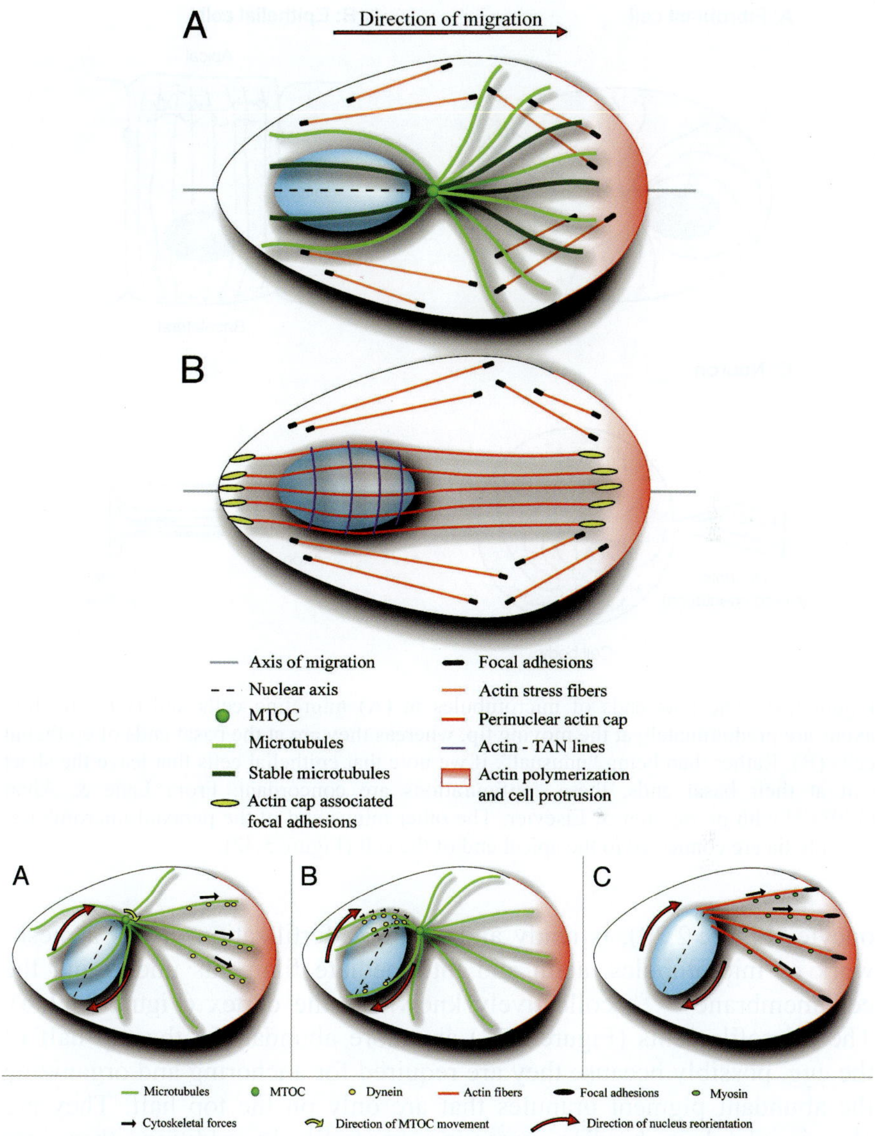

Figure 5.50. **Top**: In a migrating cell the nucleus shown here is elongated in the direction of motion. Thus the orientation of the nucleus is part of what constitutes the polarity of such a cell. **Bottom**: Hypothetical mechanisms that could provide the mechanical torque that aligns the nucleus. From Maninová *et al.* (2014)[344] with kind permission of Mary Ann Muller, US Journals Division, Taylor & Francis Group. The pink zone at the leading edge is full of actin (Figure 5.31: A, Figure 5.33). Compare these diagrams with the real thing in Figure 5.49.

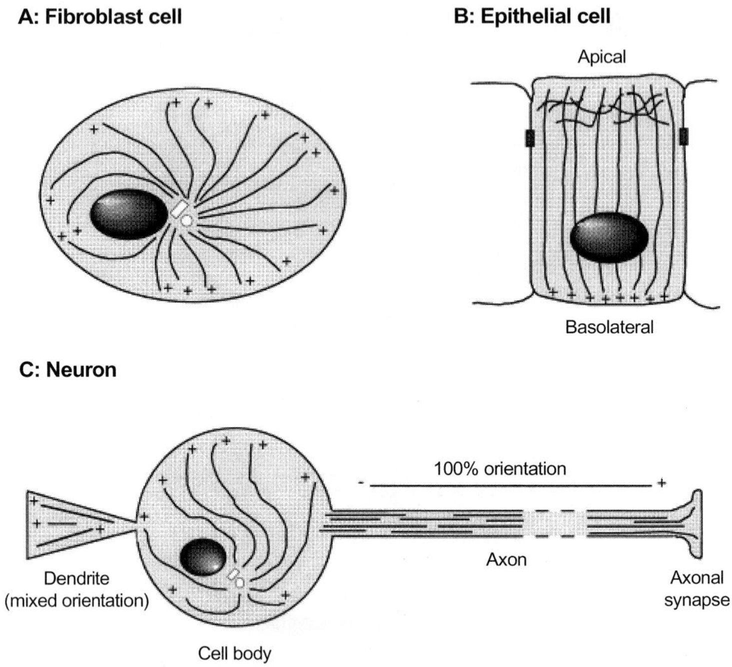

Figure 5.51. The plus ends of microtubules in (A) migrating cells and (C) extending axons are predominately at the moving tip, whereas they are at the basal ends of epithelial cells (B). Rather than being "unusual", if we note that epithelial cells that leave the sheet exit at their basal ends, these configurations are concordant. From Lane & Allan (1998)[350] with permission of Elsevier. The other minus end of the paraxial microtubules in epithelia are connected to the apical end of the cell (Figure 5.42).

oocyte (Figure 2.20), is ready and awaiting fertilization, it has a meshwork of microtubules, actin and intermediate filaments underlying the cell membrane[348,349], collectively known as the cortex (Figure 5.52: a). The microfilaments (Figure 5.53) are more abundant in the top half of the egg, possibly because they are required for anchoring and organizing the abundant pigment granules that are only on the top half. They are also involved in binding incoming sperm[351]. In addition, there are vesicles filled with sequestered calcium underlying the entire membrane of the oocyte. These vesicles also require cytoskeletal anchoring in the cortex. Other vesicles, with other functions are also present[352]. The outer surface of the egg is studded with receptors ready to bind with the sperm. The nucleus of the axolotl oocyte is paused in meiosis, as are all mature eggs cells awaiting the sperm, at metaphase of meiosis 2 (Figure 3.59).

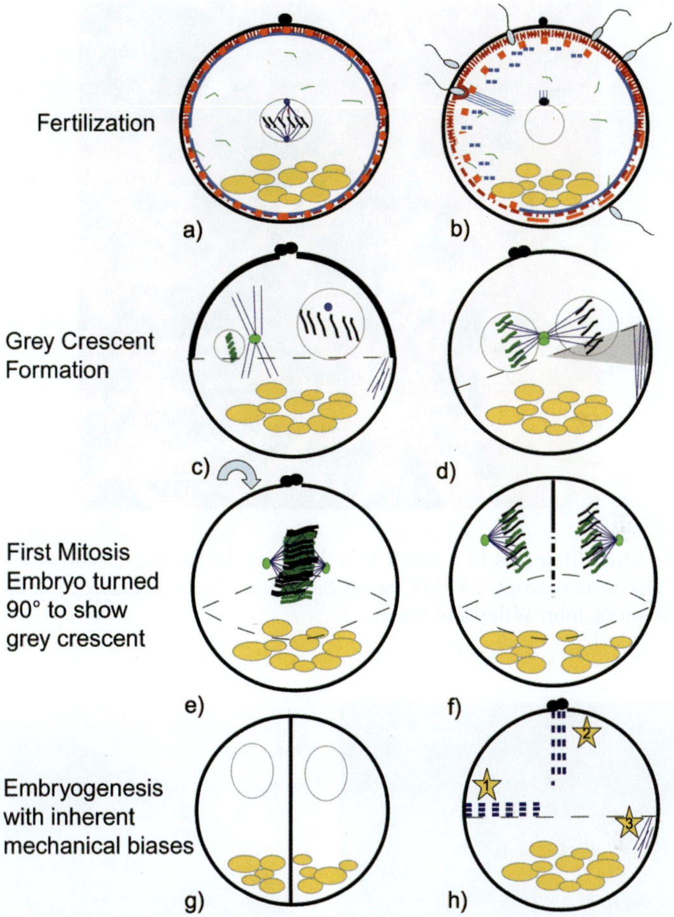

Figure 5.52. Axolotl egg, fully mature and ready to be fertilized with important components illustrated (see details in text). Grey crescent formation and polar body extrusion (small black circles) are described in detail in Chapter 2. At each step mechanical biases are created or reinforced. These biases act to guide the location of subsequent important developmental events such as the location of the dorsal lip of the blastopore.

The axolotl egg is already somewhat polarized. Heavy yolk vesicles are localized towards the bottom of the egg inside and there is little to no pigment on the bottom half of the cortex. There is more than twice as much tubulin in the top (animal) hemisphere as in the bottom (vegetal) hemisphere[353]. The nucleus is large (Figure 5.54), 100,000 times the volume of a later "normal" (somatic) nucleus, stabilized by internal

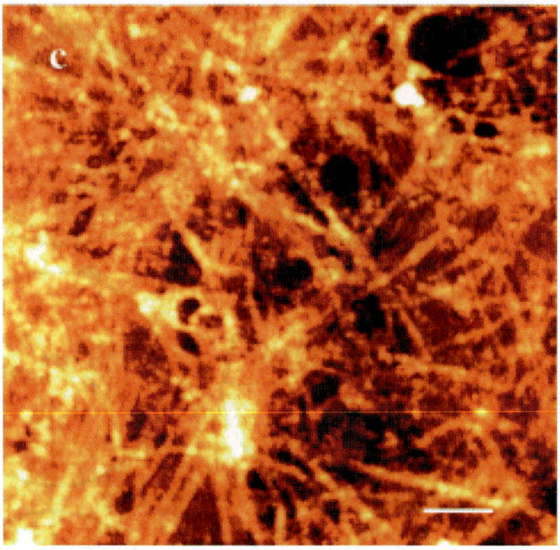

Figure 5.53. Microfilaments in a patch of a *Xenopus laevis* oocyte cortex as visualized by atomic force microscopy (AFM). Scale bar 1 μm. From Santacroce *et al.* (2006)[356] with permission of John Wiley and Sons.

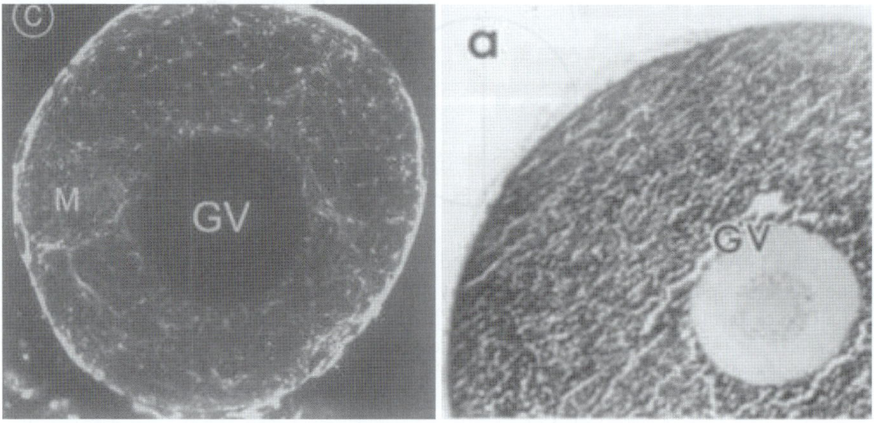

Figure 5.54. **Left**: Confocal microscope section of a *Xenopus laevis* oocyte nucleus (also known as the "germinal vesicle"). The oocyte is about 1000 μm (1 mm) in diameter[355], with the nucleus about 450 μm in diameter in the mature oocyte. From Gard (2002)[357] with permission of Elsevier Science (USA). **Right**: A 400 μm diameter germinal vesicle in a vertical slice of a fixed oocyte of the fire belly newt *Cynops* of egg diameter 1700 μm (1.7 mm). From Imoh & Miyazaki (1984)[355] with permission of John Wiley and Sons.

nuclear actin[354]. The nucleus is suspended above the mass of yolk platelets near the animal pole[355]. The oocyte, ready for fertilization, packs enough nuclear proteins for 15,000 cells[354], or about 14 generations ($2^{14} = 16{,}384$), though it has only one copy of its DNA. By neural plate stage in amphibians there are about 100,000 cells so by back calculating rates of cell division we guesstimate that those 14 generations mean that the egg is packed with enough nuclear proteins to get to about stage 8 without having to make any more[358,359,360] (Figure 2.33).

As we discussed in Chapter 2, a great deal of RNA from the mother has also been imported into the egg via nurse cells, or prepared by the egg itself during oogenesis. Much of this is messenger RNA but other forms of the newly discovered small functional RNAs are also present, postulated to affect the spatial distribution of mRNA[361,362]. All this RNA is organized in top to bottom or bottom to top gradients[363-365]. Some specific mRNAs are known to be attached to the (ventral) bottom of the embryo's cortex (Figure 5.55). Some maternal RNAs are localized to the mitochondria and it is important to reiterate that some RNA that is considered "maternal" was actually produced by the egg itself as part of the process of its "ripening" during the maturation process of oogenesis[366]. These differences between the top and bottom half of the egg result in a diffuse and imprecise, but nonetheless extremely important top/bottom polarity. This polarity is not firmly fixed.

The pigmented top half of the egg is called the animal hemisphere or animal cap and the bottom half the vegetal hemisphere (Figure 5.52: a). The egg can be experimentally rotated so the contents slosh around inside. If the pigmented top ends up at 90° to horizontal then the egg will still develop normally (Figure 2.21). The exception is a 180° inversion, which can lead to a symmetric flow rather than sloshing[367,368] (Figures 5.56 and 5.57) and frequent cessation of development. Embryos whose eggs are inverted and survive have reversed pigmentation: light on the dorsal side and dark on their bellies[369]. This clearly shows that the mRNA gradients are simply not the essential element of embryo body plan organization.

The much smaller sperm is similarly loaded with specific small RNAs in addition to its haploid genetic complement (Figure 3.63). Depending on the species, the sperm may also carry its own centrosome. The centrosome is analogous to the yeast spindle pole body. A centrosome is two

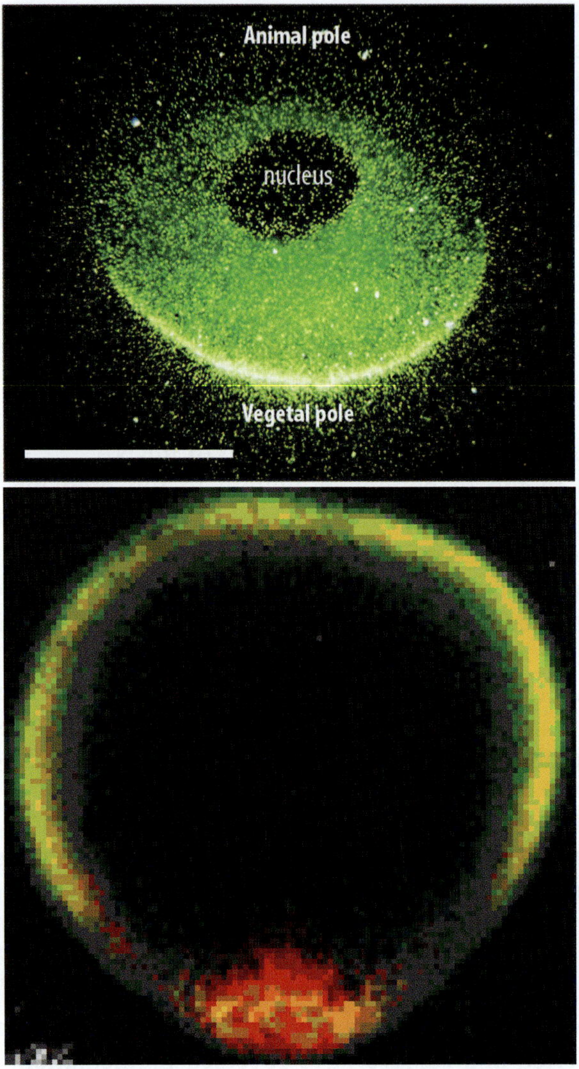

Figure 5.55. **Top**: Distribution of the mRNA Vg-1 in a mature *Xenopus laevis* oocyte. Scale bar 0.5 mm. From Staveley (2015)[372]. The fluorescence outside the oocyte may be background and/or smearing when this slice was cut using a microtome. While such mRNA, at least that attached to the cortex, as shown in the bright arc at the vegetal pole, rotates with the cortical rotation[373], direct measurement mRNA of sliced embryos before and after rotation shows that the mRNA inside, in the cytoplasm, does not rotate[363]. This is consistent with the "Martian eye" model for cortical rotation (Figure 2.29). **Bottom**: Another mRNA, *Xlsirt* (red), is also localized at the vegetal pole in *Xenopus*. From Kloc *et al.* (1996)[374] with kind permission of Malgorzata Kloc.

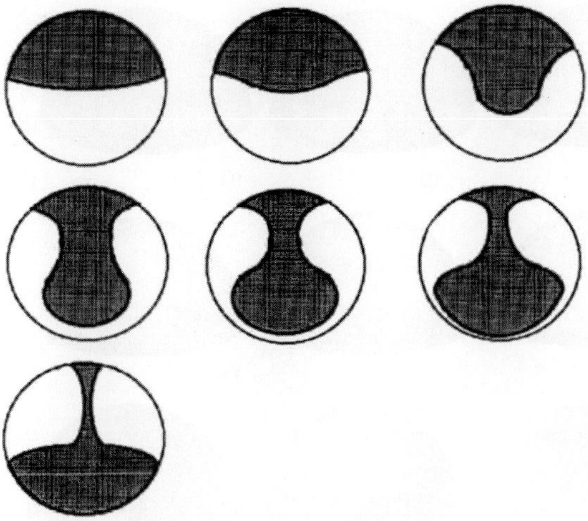

Figure 5.56. Sequence of outlines of the flow of glycerol into gear oil after 180° inversion of a spherical glass flask, made from photographs. From Flint *et al.* (1989)[367]. This may explain the observations that "major yolk zones"[368,369] shift asymmetrically in *Xenopus* eggs that have been rotated by angles less than 175°, and symmetrically in eggs that have been inverted 175° to 180°.

centrioles joined together which separate during mitosis and then each one duplicates before the next mitosis (Figure 3.37). In humans, as in most mammals[370], the centrosome is supplied by the father via the sperm[371]. If the centrosome is defective in some way, fertilization fails and this may account for the very high failure rate of artificial injection of sperm into the egg in assisted reproduction for humans[371]. In mice, the centrosome is maternally derived and the paternal centrosome degenerates during spermatogenesis[370]. The origin of the centrosome for urodeles, such as the axolotl, is the sperm[375].

The axolotl egg is fertilized internally inside the female. The female does a mating dance with the male[376,377]. He lays a packet of sperm (Figure 2.26) on a rock or other suitable device. He then guides her in the mating dance until her cloaca is over the sperm packet whereupon she picks it up. As was evident in our albino axolotls, both sexes flush red in the process, a characteristic of orgasm[378] that we may have inherited from our common ancestor with them although for us it is generally more confined.

Preventing polyspermy is critically important for all species. Extra sperm will result in gametes with the wrong complement of genes, a condition known as polyploidy. Plants are better able to tolerate accidental polyploidy[379] and often use it to advantage in their evolution.[380] There are a few examples of polyploidy in animals including the axolotl which can produce normal looking though slow growing individuals with up to

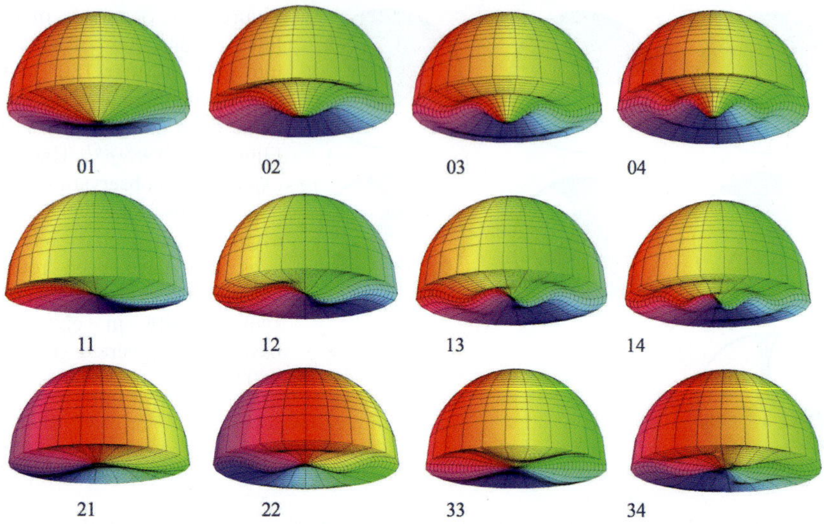

Figure 5.57. Bessell functions are the modes of oscillation of the surface of a circular drum. Here we place that circle on an inverted hemisphere to represent the initial state of a bottom heavy egg that has been inverted. To make the image visually clear, the top hemisphere (which is at the bottom when the egg is inverted) is not shown. The upside down Bessell functions then show how a heavy fluid on top could start to drip down. Some of these dripping modes are axially symmetric, and others are not. The symmetric ones could lead to large scale dripping patterns such as that in Figure 5.56. The others could lead to large scale sloshing down one side. The pairs of numbers index the different Bessel function modes. Direct 3D time-lapse observations of sloshing inside inverted amphibian eggs or during cortical rotation have yet to be made. From Nouri *et al.* (2008)[368] with permission of IOP Publishing. The speeds at which these different modes drip down, calculated[368] using ComFlo computational fluid dynamics software[383], led to the prediction that about 62% of inverted embryos would slosh down the side and thus develop normally, whereas 38% would drip symmetrically and thus not develop further[368]. The internal viscosity of the cytoplasm varies from one spawning to another, yielding survival rates for embryos whose eggs were inverted of 25% to 60%, albeit often with reversed pigmentation. Those that didn't survive exhibited abnormal morphogenesis[369]. So the theory and simulations seem to roughly fit the facts.

seven sets (7N) instead of the usual two sets of chromosomes (2N). Polyploid axolotls generally do not reproduce, though some can[381]. While these axolotls are about normal in size (contrary to the situation in polyploidy plants where plant size increases with ploidy)[382], they have larger cells than diploids (Figure 5.58), including larger eggs[384]. Apparently, because polyploid urodeles have fewer (albeit larger) brain cells, they are not as smart as normal diploids[385]. Polyploidy can occur in

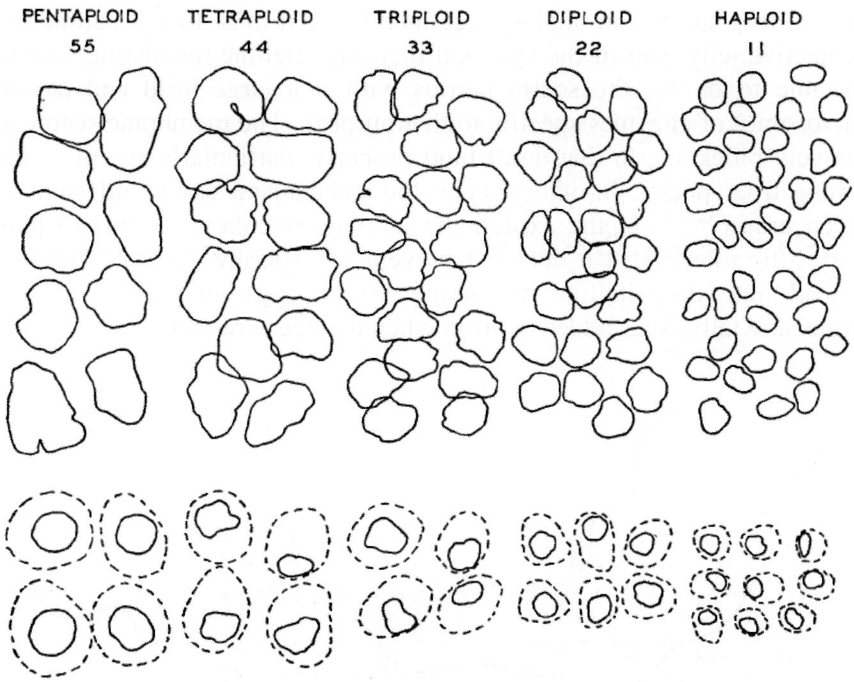

Figure 5.58. Cell size (above, tailfin epidermal cells) and nuclear size (below, Leydig gland cells) increase right to left as the ploidy (number of chromosome sets) increases in larvae of the newt *Triturus viridescens*. From Fankhauser (1945)[388] with kind permission of Perry Cartwright of the University of Chicago Press. In monolayer epithelia such as this the cell thickness is nearly independent of the polyploidy, so the size increase is in the other two dimensions. Pronephric tubules have a diameter independent of polyploidy, so that not only cell size, but also cell shape (curvature) change with polyploidy[382].

human embryos as often as 7%[386] to 15%[387], mostly due to polyspermy[386]. In humans a triploid embryo is usually miscarried early in pregnancy and even if the baby reaches term, it will live at most a few months past delivery[387].

The vitelline membrane (called the *zona pellucida* in mammals) forms around the egg during maturation of the oocyte. It consists of a fibrous layer of glycoproteins and in some species it is studded with receptors designed to bind sperm of only the correct species and thereby prevent any others from entering. In *Xenopus*, which has external fertilization, the vitelline membrane is complex, and also studded with receptors.

Before frog sperm can bind the egg membrane it must first penetrate the protective jelly coat of the egg, and then the vitelline membrane, and to be able to do so, the sperm carries with it a large head end packet (acrosome) of enzymes specific to that purpose. The membrane seems to actively block sperm that don't bind properly, particularly sperm of the wrong frog species. In other species the mechanisms can be different or is uncertain[389-391]. In the axolotl the vitelline membrane seems to play very little role in that it does not prevent polyspermy. Axolotl sperm is placed onto the vitelline membrane of the eggs internally with the protective jelly coats added overtop when the eggs are laid.

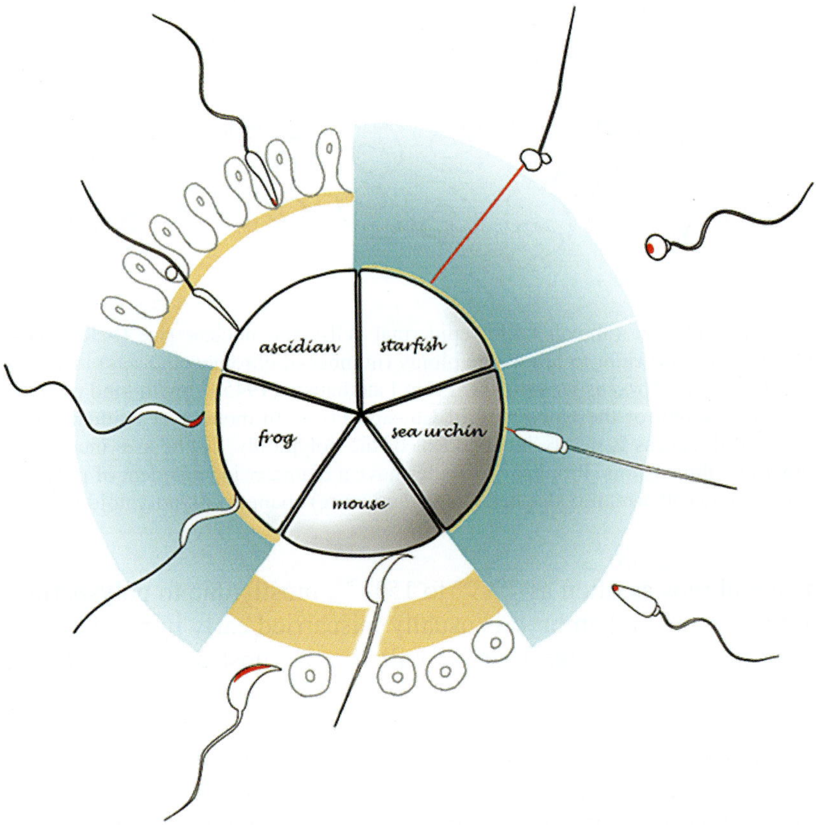

Figure 5.59. Various ways sperm get through the obstacles outside an egg. Note that in starfish "filamentous actin protrudes at the tip of sperm head". Cyan = jelly coat, orange = protein, such as the vitelline membrane in frogs and other amphibians. From Hirohashi *et al.* (2008)[394] with permission of John Wiley and Sons.

Fertilization begins with the sperm head binding to the cell membrane of the egg (Figure 5.59). As with the vitelline membrane, this binding process is highly individualized by species[392,393], which prevents sperm from the wrong species from being able to fertilize the egg. The sperm membrane then fuses with the egg membrane[395,396]. This fusion requires binding to actin in the egg cortex[397,391]. In many animals ordinarily only one sperm enters the egg. In axolotls, there is no such blocking which is why there can be more than one sperm penetrating the egg. Even so, only one of the sperm nuclei that gets into the egg will bind with the female nucleus. The others will remain just below the cortex and eventually disintegrate.

Once the successful sperm has contacted the egg several things happen (Figure 5.52: b). At sperm binding, the calcium containing vesicles ("cortical granules") under the surface of the axolotl egg's membrane dump their contents into the cytoplasm just under the membrane. The dumping action begins at the point of sperm entry and travels outward in a widening circular wave until the entire surface of the egg has been covered. This calcium flux wave is known as the "activation wave". The activation wave, important as it is, can be started by an experimental pinprick in some species, like *Xenopus*. The activation wave is fast, taking 3 to 14 minutes to start, traverse the egg, and settle down, and consists of a propagating light zone followed by a propagating dark zone[398] accompanied by a calcium wave[399] (Figure 5.60). A given point on the surface undergoes an expansion, followed by a contraction, then another expansion[398].

The activation wave varies by species. It was shown initially by Ernest Just[400] in sea urchins. Activation is typically followed by a wave of separation of the "fertilization" or vitelline membrane[401]. The vitelline membrane in some species actually rises up from the egg, the rising coinciding with the activity going on within the egg cell membrane at fertilization. In mammals, the same vesicles which are dumping calcium ions under the egg cell membrane also release proteases that are transported across the cell membrane and break the connections between the *zona pellucida* and the egg and release the vitelline membrane receptors and any other bound sperm remaining. The activation wave and the separation of the vitelline membrane are now known as fast and slow blocks to polyspermy respectively[402,403]. The blocking is accompanied by a change in membrane electric conductance and depolarization[395].

For embryos such as the axolotl, which allow many sperm to enter the eggs, usually only one sperm survives. This elimination process has been called a "very slow" or "cytoplasmic" block to polyspermy[404]. The eggs of various animals exhibit quite a variety of calcium waves after fertilization[405] (Figure 5.61, see also Chapter 10). One curious variant is that in invertebrates the calcium ion concentration in the whole cortex rises all at once, producing a "cortical flash" over the whole egg[406] (Figure 5.62). We don't understand the mechanisms or purposes for a lot of this activity and the variability between species only adds to our bewilderment.

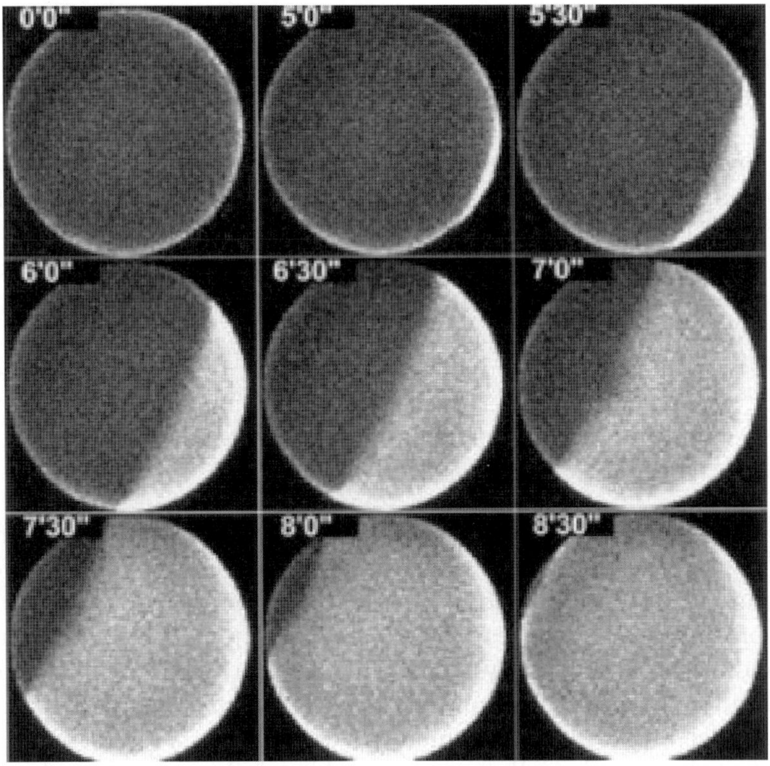

Figure 5.60. Calcium wave component of the activation wave, triggered by a sperm entering at lower right, traversing a *Xenopus* oocyte. Note the delay of 5 minutes, followed by the fluorescently visible wave travelling in the cortex at 8.9 μm/second = 534 μm/minute. The egg diameter is reported as 1200 μm = 1.2 mm. From Fontanilla & Nuccitelli (1998)[407] with permission of Elsevier. A similar activation wave traverses urodele amphibians[375], but does not block polyspermy.

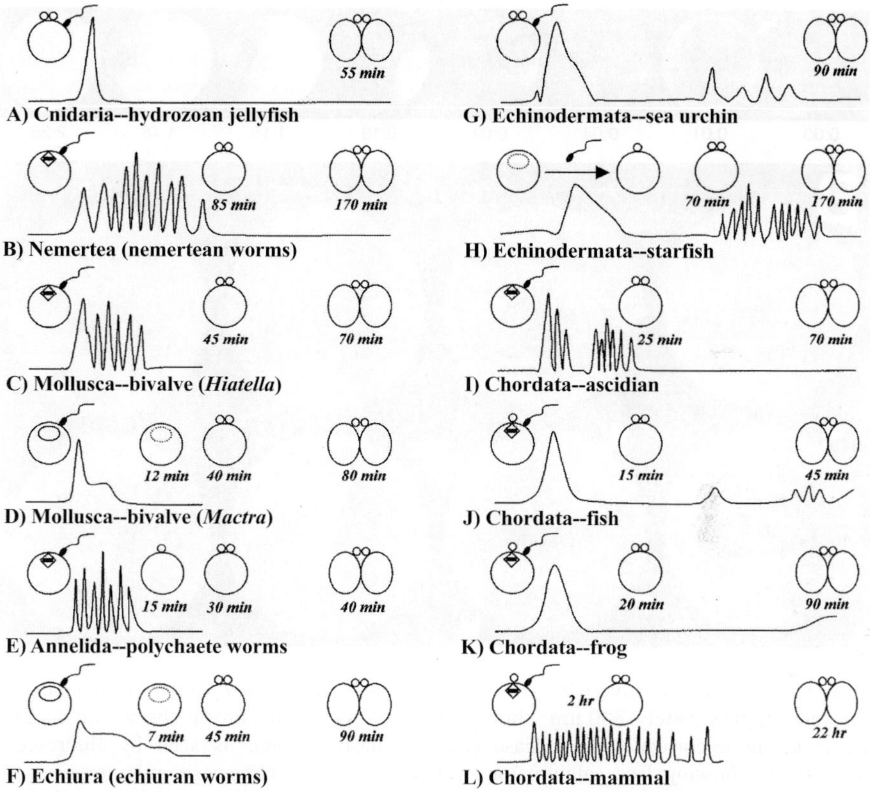

Figure 5.61. Calcium waves triggered by fertilization vary from simple spikes to oscillations and combinations of both. This shows the tremendous variety in the temporal course of calcium waves, but no embryos lack them. The calcium changes are not drawn to scale. Timings of first and second polar body extrusion and first cleavage are given. From Stricker (1999)[408] with permission of Elsevier. The spatial trajectory of these waves generally follows that depicted in Figure 5.60. Note that there is no obvious correlation between whether one-cell embryo waves are single or multiple and the position of the animal in the tree of life. In mammals the Golgi apparatus is involved in multiple waves[409].

We do know that the calcium flux(es) in the axolotl egg triggers three different cytoskeletal assembly cascades resulting in three large internal cytoskeletal rearrangements. Each of these large internal cytoskeletal rearrangements has visible consequences that can be seen from the outside.

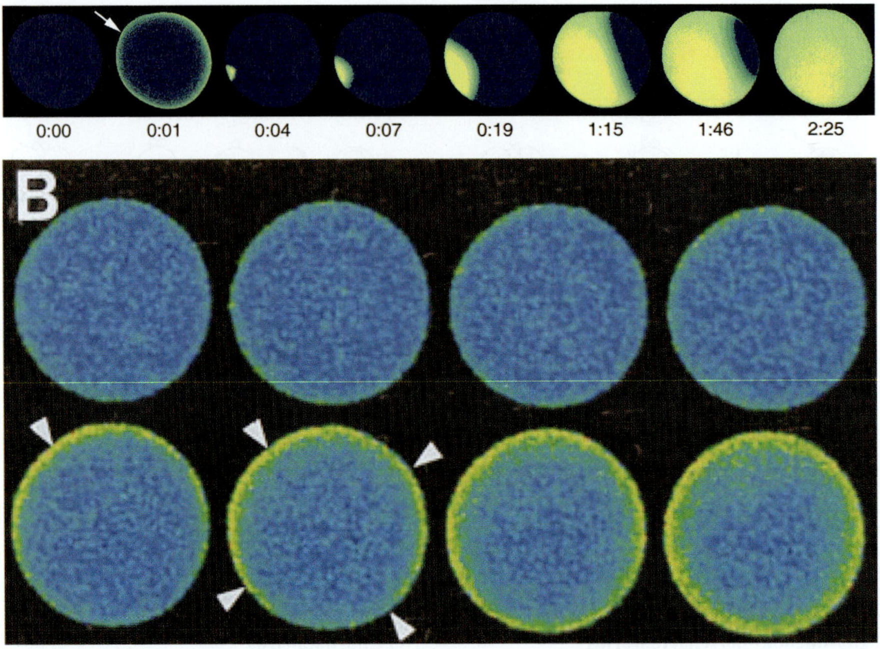

Figure 5.62. **Top**: In invertebrates such as this starfish *Astropecten aranciacus* of diameter approximately 300 μm, shortly after the sperm attachment the whole surface lights up in a cortical Ca^{2+} release (cortical flash, arrow), as seen by fluorescent microscopy showing free calcium ions versus time[406,397]. Time is in minutes:seconds. Thus in this zygote the fast cortical flash, which decays in 2 to 8 seconds[66], precedes the global Ca^{2+} wave from the point of sperm/egg interaction to the opposite pole. In starfish perturbation of the cortical actin cytoskeleton affects many aspects of intracellular Ca^{2+} signaling and leads to multiple sperm/egg interactions, most of which nevertheless do not result in sperm entry[410]. From Puppo *et al.* (2008)[397] with permission under a Creative Commons Attribution License (CCAL). with details kindly provided by Luigia Santella, who won a prize in the 5th Annual Nikon Small World in Motion for an accompanying movie. Cf.[410] **Bottom**: The cortical flash is caught propagating over the cortex in consecutive frames taken 0.8 seconds apart in a fertilized egg of the nemertean *Cerebratulus lacteus*. Egg diameter is 200 μm. From Stricker (1996)[411] with permission of Elsevier. Its speed is thus about 100 μm/second = 6000 μm/minute. "Proper physiological conditions (Just's legacy[400]) are necessary to produce a cortical flash as a result of the unintentional side effect of experimental stimulation of the cortical actin cytoskeleton which may play an important role for the embryonic development more than the global Ca^{2+} wave" (Luigia Santella, personal communication, 2015).

The first of these cytoskeletal changes begins with the activation wave and causes the tail of the sperm to fall off, releasing the head into the interior of the egg. Calcium release also causes the microtubules underlying the cortex to depolymerize beginning where the first successful sperm enters the cell (Figure 5.52: b). The fact that the depolarization occurs from that one point on one side and then travels over the entire surface of the egg to the other side means that for a short period of time the egg has a lot of free tubulin at the sperm entry point and a meshwork of polymerized microtubules on the opposite side. The tubulin released by the activation wave reforms microtubules around the sperm head. These microtubules assist in directing the sperm nucleus, with its accompanying centrosome (where present) and any sperm RNA[412], towards the center of the egg. When a sperm centrosome is present it comes with a small amount of its own gamma tubulin to initiate the first microtubule rearrangement and it also grabs and uses any maternal gamma tubulin in the immediate area. If more than one sperm has made it inside, this snarfing up of available tubulin and gamma tubulin[404] by one sperm prevents any other sperm from being moved inward because there is nothing left to make a microtubule structure out of in the vicinity of the losing sperm. This is why in spite of the polyspermy of axolotls, only one sperm nucleus moves inside. Any other sperm are left, paused, under the cortex where they degenerate. While the incorporation of sperm into the egg requires actin[413,414,397,391], the migration of the male pronucleus towards the female pronucleus and the movement of both to the center of the animal cap occur on the recently constructed microtubules[413-415] (cf.[416]). This first structure of microtubules is called the sperm aster.

Even as the sperm head is being drawn in, the second cytoskeletal rearrangement starts. The activation wave's released calcium binds to a protein called calmodulin and then the calcium/calmodulin complex binds to and activates the enzyme CaMKinase II. CaMKinase II then causes the destruction of a cyclin protein generally known as p34, whose presence blocks the progression of meiosis in the egg. With the p34 destroyed, meiosis, which can be paused for years in long lived females like us, can finally finish. In a stunning display of the universality of this process to life on our planet, all eukaryotes use some form of a p34 protein for regulation of meiosis[417]. We also see in this second cytoskeletal arrangement, a rapid signal transduction, a message carried from the outside of the cell to the inside of the nucleus. That signal is that a sperm has successfully bound and the egg nucleus responds by finally completing meiosis.

The spindle apparatus of the egg nucleus becomes active when the p34 is destroyed. The second assembly cascade with cytoskeletal rearrangements takes place and the result is that microtubules and microfilament cables form and connect the maternal nucleus to the exterior of the cell. The maternal nucleus, stuck in meiosis 2 for so long now, finally finishes meiosis 2 and this creates a second polar body. The second polar body is carried upward by myosin motors along this second set of cytoskeletal elements (microfilaments: Figure 5.52: b,c). In the lab the sighting of the second polar body (usually but not always next to the first polar body) is a welcome event indicating planned experimental work for the batch of eggs can now proceed because fertilization has been sucessful. The second polar body is extruded from the egg as the sperm head reaches the interior. As the polar body is expelled pigment granules are shoved aside. In the axolotl the topmost area of the egg membrane, where the second polar body is extruded, remains clear with very little pigment. This clear zone is variable but is about 3 μm in diameter and persists as development continues.

The maternal spindle apparatus vanishes once maternal meiosis 2 is completed and a maternal pronucleus forms. The pronucleus is essentially a nucleus but contains only the haploid maternal genetic material. A second pronucleus, complete with a nuclear membrane forms around the sperm head as the sperm chromosomes decondense (Figure 5.52: c). The first two cytoskeletal rearrangements, sperm ingress with formation of the sperm aster, and completion of maternal meiosis 2 with polar body extrusion, use many of the tubulin heterodimers freed by the calcium wave. After these two structures disassemble, the heterodimers are freed again. This affects the third important cytoskeletal rearrangement, the one that results in cortical rotation and the formation of the grey crescent.

Cortical rotation is a major morphogenetic event in amphibian embryogenesis (Figure 2.28). As mentioned in Chapter 2, the whole cortex rotates 15° to 30° relative to the cytoplasm underneath[418]. Before cortical rotation begins, individual microtubules form from the ventral surface and extend upward at random angles. However, because of the previous cytoskeletal rearrangements (ingress of the sperm nuclei, completion of maternal meiosis 2 and the formation the sperm aster) there is a relative shortage of free αβ-tubulin heterodimers on the sperm entry point side and a relative abundance on the opposite side. This results in the microtubules on the opposite hemisphere growing more quickly and they make contact between the yolk cortex region and the pigmented region first. As these first microtubules make contact and begin the rotation of the cortex, the movement of the cortex also aligns other microtubules nearby.

Cortical rotation has major consequences for the cytoskeleton. The microtubules in the cortex start out short and randomly oriented, and end up aligned at the end of the process, with 89% having the same polarity[419]. They carry an assortment of molecules along with them during the rotation, perhaps using microtubule motor proteins[420]. Despite the huge literature on cortical rotation (over 400 papers), we still do not know exactly what is going on. There are a few alternative models, all based on the idea that something about microtubules is essential to the mechanism. These models are not mutually exclusive, so much work remains to sort them out. They exemplify how much we have yet to learn about how cytoskeleton actually works in embryos:

1. Flow induced alignment: The microtubules start out embedded in the egg cortex and randomly oriented. As they grow each one pushes the cortex in its direction. However, the outer cortex is reasonably rigid, so that it has a net rotatory motion which is the net vector sum of the tangential torques. In other words, the whole cortex starts rotating, much like the Martian Eye toy[421,422] (Figure 2.29). This has been phrased as a "positive feedback" mechanism (Figure 5.63).
2. Alignment by bundling: due to dipole-dipole interactions between nearby microtubules, which have large surface concentrations of electric charge, a phase transition occurs resulting in their parallel alignment[422].
3. Rotation of adjacent whole microtubules along their individual axes can lead to their bundling[423,424] (cf.[425]). Motor molecules can either move along one protofilament or follow the helix of a microtubule[426], and thus could provide the torque for microtubule rotation if it were anchored to something else or subject to viscous drag.

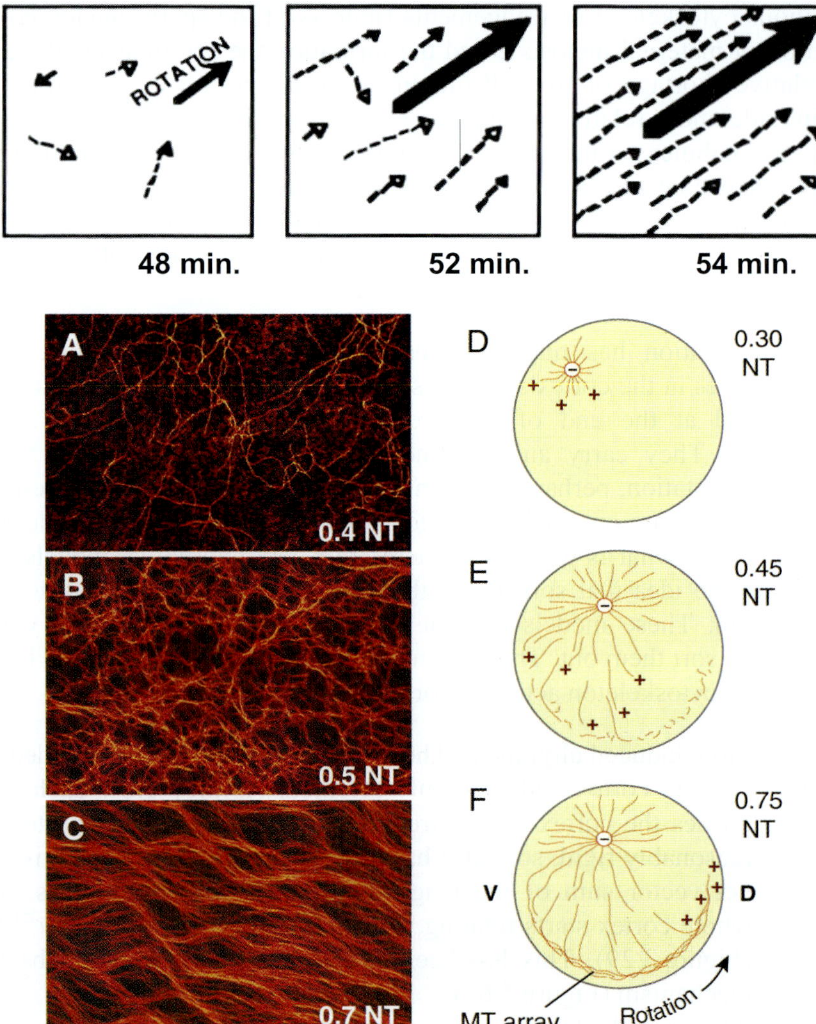

Figure 5.63. **Top**: As cortical rotation begins, each microtubule anchored in the cortex contributes a small push against the underlying cytoplasm, resulting in a tiny torque ("Rotation"). This is a torque because the force would be equal and opposite on the opposite side of the spherical one-cell embryo. The torque vector is shown growing here as cortical rotation proceeds. It would also initially show some randomness in direction. From Rowning & Gerhart (1990)[430] with kind permission of John Gerhart and the late Eric H. Davidson. This is a positive feedback[431] mechanism[430,421]. **Bottom**: "Formation of the microtubule (MT) array. (A-C) Vegetal views of microtubule array formation during the first cell cycle. Times shown are normalized times (NT), with 0.0 NT representing fertilization and 1.0 NT representing the first cleavage division. (A) At 0.4 NT, short

disorganized MT polymers have started to appear in the vegetal shear zone. (B) By 0.5 NT, more MTs are present, but are not yet aligned. (C) By 0.7 NT, during peak cortical rotation, the vegetal shear zone is populated by a parallel array of MT bundles that are aligned along the axis of rotation. (D-F) MTs of the vegetal array arise from several sources. (D) The sperm centriole introduces polarity by acting as a minus-end MT-organizing center (–). The resulting radial array of MTs is called the sperm aster. (E) MTs from the sperm aster grow toward the periphery of the egg, as do additional MTs from unknown sources in the core cytoplasm. In addition, short disorganized MT polymers arise in the vegetal shear zone. (F) During rotation, MTs from deep in the cytoplasm bend into the vegetal shear zone and align with peripheral MTs to form the parallel array, with the plus-ends (+) of the growing MTs pointing towards the future dorsal (D) side of the embryo. V, ventral. (A-C) Reproduced, with permission of Elsevier, from Cha & Gard (1999)[432]. (D-F) Adapted, with permission, from Houliston & Elinson (1991)[433]." From Weaver & Kimelman (2004)[429] with kind permission of the Company of Biologists and from Cha & Gard (1999)[432] with permission of Elsevier.

4. Motor molecules moving along the microtubules could provide the force for cortical rotation. This may occur by alternating layers of aligned microtubules and attached motor molecules, which would then sum their velocities[427], much like cells in the colonial diatom *Bacillaria paradoxa* sum their velocities[428] (Figure 1.42). This "layered movement" model would explain why the zone of shear between the cortex and the underlying cytoplasm is not sharp[427]. During cortical rotation material bound at the vegetal pole, including some maternal mRNA[373], also rotates[429] about 30°, although some material there goes two to three times faster[434], probably because of layered movement. Dynein and kinesin-14 are motor molecules capable of moving one microtubule along another when they both have the same orientation[435]. (A similar layered movement of microfilaments may also explain why cytoplasmic streaming in algal cells can be ten times faster than the fastest myosin motor molecules[436,437].)
5. The sloshing back of the yolk as the rotation lifts it against gravity could also contribute to microtubule flow induced alignment[421].
6. The microtubules elongate, which itself could provide the force of rotation[160]. (Elongating microfilaments also exert forces[438,439].)
7. An interaction with the cytoplasm (perhaps with the micro-filaments) has been postulated that causes short microtubules that are misaligned to depolymerize[430].

Alignment of the microtubules can be biased in a given direction by manually rotating the cortex before the cortical microtubules appear[430]. Without external interference, the sperm entry point creates a subtle mechanical bias towards a front and back (posterior/anterior) axis. We know that this mechanical bias exists because the midline of *Xenopus* embryos normally forms within 8° of the sperm entry point (see[440] and Proposition 52 in[179]) (Figure 5.52: f), not precisely through it (which would be 0°). There is recent work on alignment of microfilaments in nematode egg cortex via a proposed rolling mechanism[346]. There is similar work in cells in tissue culture growing on patterned surfaces[441]. These two results, like the manual rotation of the cortex, indicate that if one aligns the microfilaments first, this would bias the direction of alignment of the subsequently appearing microtubules. Intermediate filaments also align during cortical rotation[442,422], and are probably involved as well.

Axolotls routinely produce hundreds to thousands of embryos in a spawning. If one watches all the embryos in these normal spawnings, peculiar exceptions to the axes predicted by the mechanical biases often appear. For example, we once had a hatchling embryo that was all white on the left side and darkly pigmented on the right side as if it had completed all early development while the egg was lying on its side (as in Figure 2.21). (This can happen when eggs get crowded in an improper jelly coat lacking the normal perfect round chamber for each egg.) The two polar bodies normally extrude out of the same spot at the top of the egg. However, on rare occasion one polar body will appear to be at 90 degrees from the other one on the egg, yet the embryo that forms from such an egg develops normally. These exceptions to the rule indicate that while there are strong mechanical biases created during oogenesis and fertilization, they are only biases. These axes are not fixed until later in development. Confusion between a mere bias and the final determination of the axis is a source of considerable confusion in the history of our understanding of embryogenesis.

So why does cortical rotation happen? Michael Danilchik made the remarkable observation that disruption of cortical microfilaments during cortical rotation driven by microtubules results in substantial reversal of which side the heart is on in *Xenopus*[443]. Dick proposed that torque later in embryogenesis, acting on any chiral filaments such as microtubules or microfilaments aligned by cortical rotation, could lead to a left/right difference in supercoiling of those filaments.

To make this concrete, think in terms of microtubules to which the motor molecule kinesin or dynein binds. In the relaxed state, if we follow the αβ-tubulins around a microtubule, they conform to a helix, which has a particular handedness, and is thus chiral (Figures 5.23 and 5.24). This means that the aligned microtubules on the left and right sides of an egg cell or an organism are not mirror images of one another (Figure 5.64). In other words, as soon as an organism aligns its microtubules, it has ceased to be bilaterally symmetric (Figure 5.65).

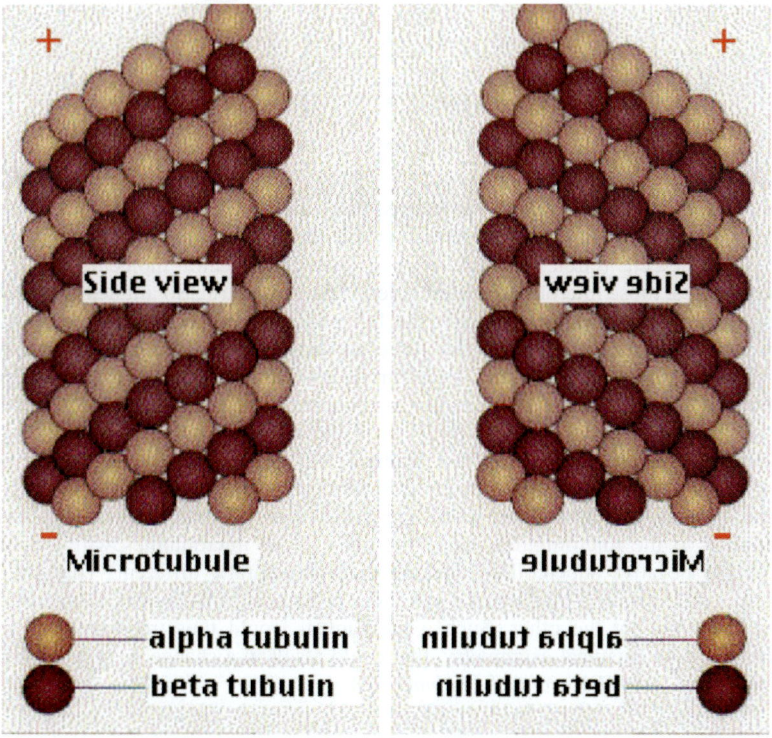

Figure 5.64. The mirror image of a microtubule cannot be rotated in 3D space to superimpose with itself. Keep in mind that the + and − ends at the top and bottom are different. Each vertical alpha/beta top/bottom pair retains this +/− intrinsic polarity. This is what makes a microtubule chiral. The mirror image microtubule on the right does not exist in organisms. Adapted from Kimball (2013)[444] with permission under a Creative Commons Attribution 3.0 Unported (CC BY 3.0) license.

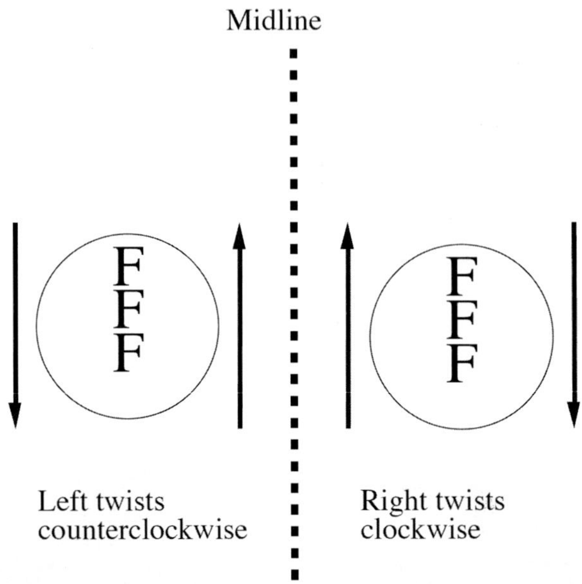

Figure 5.65. This schematically represents an embryo or egg that is bilaterally symmetric, i.e., there is a midline about which the embryo is bilaterally symmetric. However, if the cells (circles) contain aligned, filamentous chiral molecules (FFF) such as microtubules, those microtubules are not left/right symmetric. So aligned microtubules create a geometric difference between the left and right halves. Torque is a twisting action, which can be represented as two forces in opposite directions applied on opposite sides of a material object. For example, during involution of the ectoderm, due to the streaming of that tissue over the smaller blastopore, shears will develop in the tissue. There will be a gradient of shear away from the midline, and that gradient will be bilaterally symmetric. But due to the gradient, any particular cell will experience a torque. Cells located at mirror image positions will experience mirror image, i.e., opposite, torques, as shown here, where the circles represent single cells. If this torque gets applied to the microtubules inside the cells, those microtubules will experience, positive and negative supercoiling, making them structurally different. That structural difference could result in binding of a particular protein, such as dynein, to microtubules on one side, but not the other. Such a mechanism would thus translate a geometric asymmetry into a chemical asymmetry, that could result in a change in gene expression. Copy of Figure 54 in *Hierarchical Genome*[179].

How is this geometric left/right asymmetry, based on the chirality of the microtubules, turned into a left/right asymmetry in gene expression? Let us represent the coiling of a microtubule by C. Suppose that a torque is applied to each aligned microtubule, twisting it a little bit, and that the torque field (Figure 5.66), i.e., the distribution of torque over the cell or embryo, is left/right mirror symmetric (i.e., bilaterally symmetric). Then on the right side, each microtubule would be twisted into a supercoiled state that we can represent as $C+S$, and on the left side each microtubule would be twisted into a supercoiled state that we can represent as $C-S$. The structure of the microtubules is now different left and right, and so we hypothesize that a motor molecule such as dynein or kinesin binds, say, to microtubules supercoiled $C+S$ but not to microtubules supercoiled $C-S$. Thus one side of the embryo has bound dynein or

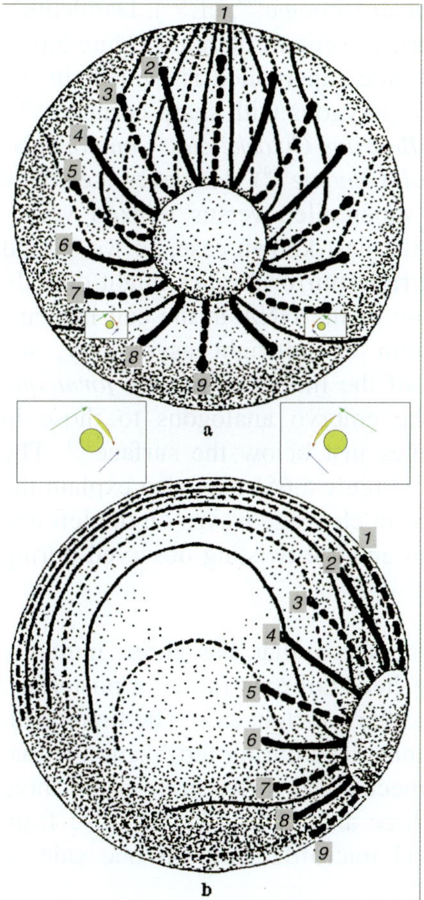

Figure 5.66. Involution movements during gastrulation in a urodele embryo, sketched on and inside a "time averaged" gastrulation stage embryo[455], which does not depict the changing arc and size of the blastopore lip. The outside movements are shown by thicker solid and dashed lines as surface ectoderm cells move towards and enter the blastopore. The corresponding trajectories inside are shown by thinner solid and dashed lines. This 1929 work bears repeating by modern methods[456]. **a**) Bottom view, with the midline between #1 and #9. Note that the external movement is not uniform in velocity around the blastopore, resulting in a shear gradient, indicated by exaggerated larger (green) and smaller (blue) vectors on either side of the two cells added to the original diagram on the outer surface and shown to scale in mirror symmetric positions, and enlarged in insets. The difference between these vectors generates a torque (curved red arrow) on the cell as it moves towards the blastopore. The torques on these two cells are mirror symmetric. However, they could lead to positive and negative supercoiling of chiral filamentous molecules in the cells, and thus to left/right differences in gene expression. Simplifications of this sketch have been published that don't connect the outer to the inner lines correctly[457,458]. **b**) Left side view, implying even greater shearing gradients after the cells have turned inside.

kinesin and the other side has kinesin or dynein floating free in the cytoplasm. This of course requires that the embryo consist of two or more cells, so that the dynein or kinesin, being intracellular, cannot freely diffuse across the embryo. Thus the left/right difference could set in at any time after first cleavage. The binding or lack therefore, would then be involved in signaling to the nuclei that are on the left or right side of the embryo. Genes unique for left or right sidedness could then be turned off or on as development continues.

While Dick proposed that this happens with "asymmetric fibrils, such as microtubules" (Figure 54 in *Hierarchical Genome*[179], cf.[445]), Danilchik's evidence points to microfilaments during cortical rotation (Figure 5.67). The fact that microfilaments can become supercoiled is evident in the observations of rotating microfilament polygons with sharp corners[446,447] (§3.08 *Force Generating and Load Bearing Cytoskeletal Components: Microfilaments (MF)* in *Hierarchical Genome*[179]). However, selective binding of proteins to positively[448] supercoiled versus relaxed versus negatively[449-451] supercoiled DNA is the best known case, so we should not ignore the possibility that the left/right mirror torques might directly affect the supercoiling of some portion of the DNA in the cell nucleus. (For alternative models of the origin of left/right asymmetry, see Vandenberg *et al.*[452,453].) In embryos of the limpet *Credipula fornicate*, spiraling folds on the surface of the embryo analogous to those in Figure 5.67 correlate with microtubules just below the surface[454]. The left/right asymmetry created by shear (Figure 5.65) may also explain the observation of the later different speed in cleavage waves on the left and right side of the embryo (Figure 2.36) and the twisting observed during first cleavage (Figure 5.67).

Normal development results in an embryo that has three mutually perpendicular axes. The axes are the head/tail axis (anterior/posterior), the back/belly axis (dorsal/ventral) and the left/right axis. The final result of the three major cytoskeletal reorganizations during fertilization is that the egg is left with three biases or mechanical differences in the three directions which later produce the three axis. These biases arise from differences between microtubules and microfilaments on one side of each axis versus the other side.

To review, the first of these biases is the sperm entry point and wider grey crescent region which create a bias in the midsection of the blastula

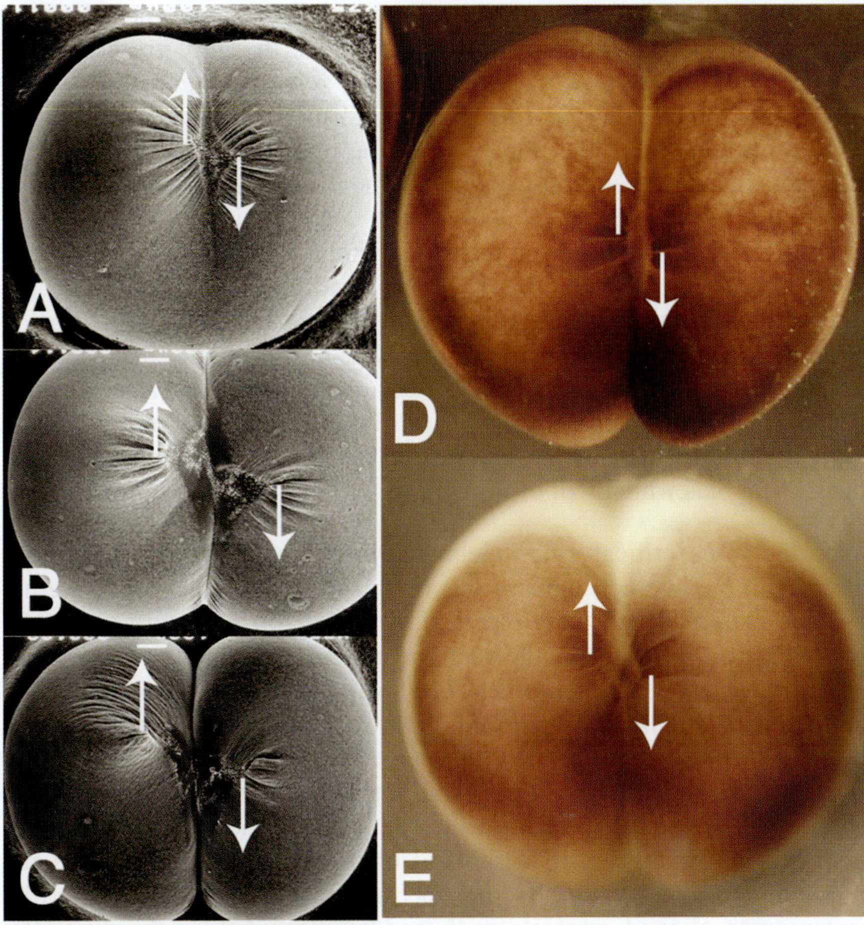

Figure 5.67. **Top**: The first cleavage of a *Xenopus* one-cell embryo is chiral, i.e., not left/right symmetric, and exhibits a twist attributable to microfilaments[443]. A, B, C: scanning electron micrographs, D: light micrograph of dejellied embryo, E: embryo with undisturbed jelly sac. From Xenbase[459] with permission under a Creative Commons Attribution License. Diameter 1 mm. **Bottom**: This photo illustrates two concepts. First, it shows a torque being applied to a plastic bottle, by applying a pair of forces in opposite directions. Second, it shows how a cell division might involve twisting. From WikiHow[460] with kind permission from Chris Hadley under Creative Commons CC BY-NC-SA 3.0 License http://creativecommons.org/licenses/by-nc-sa/3.0/.

in the portion of the embryo that will eventually become mesoderm so that one side of the future mesoderm is wider (the grey crescent side) and the other is narrower (the sperm entry region) (Figure 2.69). This predisposes the orientation of the eventual formation of the dorsal/ventral axis by biasing where the dorsal lip of the blastopore eventually forms (Chapter 7). The second bias region is where the polar body is extruded (Figure 3.59). This region is relatively devoid of actin and pigmentation appears clear compared to the rest of the embryo[461]. This is a visual cue which persists to gastrulation. The top is mechanically weak from the perspective of actin deposition because the actin has been shoved aside by the extrusion of the polar body. This weakness will later bias mechanical movements resulting in the firm establishment of anterior/posterior axis as blastulation proceeds. The third bias is the alignment of the cortical microtubules by grey crescent rotation which creates a third mechanical bias which will eventually result in left/right asymmetry.

Once the three critical cytoskeletal rearrangements of fertilization are completed, the one-cell embryo has two haploid pronuclei, one paternal and one maternal, and three mechanical biases for the future axes. The pronuclei fuse. Meanwhile, the centrosome replicates and the two centrosomes move opposite each other. The sperm aster then disassembles. If an imaginary line is drawn between the two centrosomes once they reach their final position, the line will be perpendicular to the path the sperm moved from entry into the egg to the interior within a mere 4 degrees. The exact mechanism for moving the centrosomes into position is unknown but it appears to be a kind of mechanical tug-of-war of the microtubules and dynein of the sperm aster that resolves with the centrosomes aligned opposite each and in place before the pronuclei membranes of sperm and egg disintegrate and the genetic material comes together to form the first diploid nucleus of the new organism. While the nucleus doubles the genetic material and prepares for the first mitosis (including many changes in methylation of histones and DNA), the centrosomes double again but remain linked. Therefore the centrosome orientation guides the cleavage planes but centrosome orientation came from the sperm entry bias (Figure 5.68).

The first cleavage of the egg follows, beginning from the top in the clear zone of polar body extrusion. The furrow exhibits a slight twisting (Figure 5.67) as it travels downward, so that the first cleavage is precisely aligned to the centrosomes and also goes through the approximate center of the grey crescent dividing it between the two daughter cells (plus or minus eight degrees) and in line with the sperm entry point (plus or minus four degrees)[463,464]. With the completion of that first cleavage, embryogenesis is well underway (Figure 5.52: e,f,g).

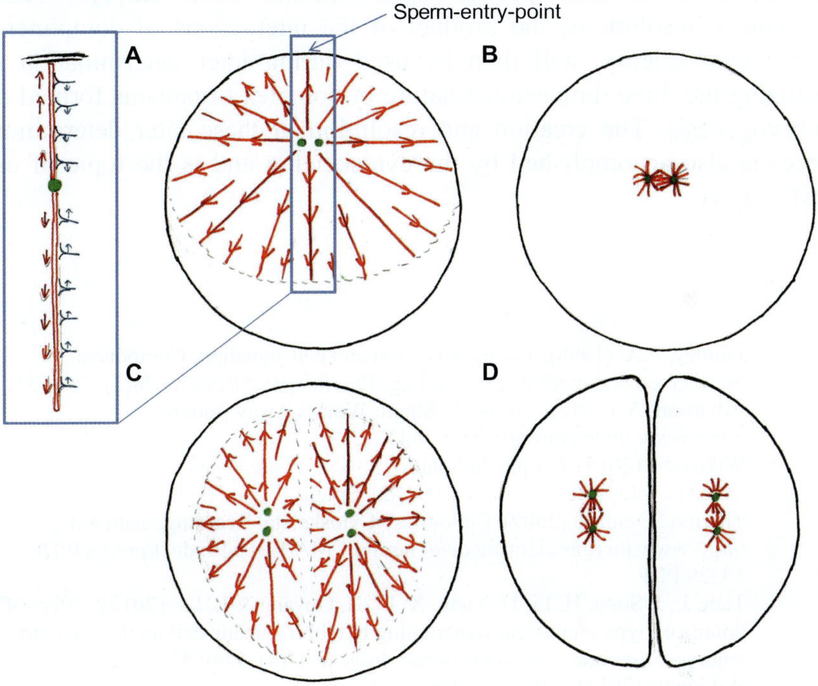

Figure 5.68. "Model for Cleavage Plane Determination in Large Cells: Astral microtubules (red) pull with dynein (blue) on cytoplasm to determine center and longest axis for cell division. [The figure corresponds a top down view of E,F in Figure 5.52.] (A) Sperm enters at periphery. Cellular boundary causes asymmetry in sperm aster. Numbers of dynein bound is proportional to microtubule length, resulting in net force on centrosome toward cell's center, but strongest stress on duplicated centrosomes (green) is perpendicular to movement. (B) Sperm aster breaks down; small first mitotic spindle forms. (C) At onset of anaphase, asters expand but do not grow into each other. The free interaction zone between the microtubules generates the asymmetry in the aster, leading to a net force on the pair of centrosomes toward the future centers of the daughter cells. The forces on the individual centrosomes cause the largest stress perpendicular to this movement, resulting in the alignment of the linked centrosomes with the aster's longest axis. (D) The cytokinetic furrow divided the cell into two, where the telophase asters overlapped, cutting through the sperm entry point. Asters break down; small mitotic spindles form at center and along longest axis of daughter cells." From Wühr *et al.* (2010)[462] with permission of Elsevier.

These three mechanical biases do not, in and of themselves, determine the axes of development. The embryo remains remarkably plastic so that interference with these biases does not always end normal embryogenesis. Rather, these mechanical biases will result in future mechanical instabilities created by mechanical forces being biased to resolution in specific directions and specific regions of the whole embryo. These mechanical resolutions, the product of the interactions of components of the cytoskeleton, will then be used during later determination in finalizing the three dimensional nature of complex organisms formed by embryogenesis. The creation and resolution of these later determining forces is also accomplished by the cytoskeleton and is the topic of our next chapter.

1. Janmey, P.A. (1998). The cytoskeleton and cell signaling: Component localization and mechanical coupling. *Physiological Reviews* **78**(3), 763-781.
2. Adamson, A. (2012). Cirque du Soleil: Worlds Away [movie]. http://www.imdb.com/title/tt1792647/
3. Wikipedia (2015). Cirque du Soleil. http://en.wikipedia.org/wiki/Cirque_du_Soleil
4. Thermo Scientific (2009). Cellomics® Cytoskeletal Rearrangement Kit. http://www.thermoscientific.de/eThermo/CMA/PDFs/Product/productPDF_59329.PDF
5. Tian, J., J. Song, H. Li, D. Yang, X. Li, H. Ouyang & L. Lai (2012). Effect of donor cell type on nuclear remodelling in rabbit somatic cell nuclear transfer embryos. *Reproduction in Domestic Animals* **47**(4), 544-552.
6. Wikipedia (2013). Nikolai Koltsov. http://en.wikipedia.org/wiki/Nikolai_Koltsov
7. Babkov, V.V. (2002). The theoretical-biological concept of Nikolai K. Kol'tsov. *Russian Journal of Developmental Biology* **33**(4), 255-262.
8. Shih, Y.L. & L. Rothfield (2006). The bacterial cytoskeleton. *Microbiology and Molecular Biology Reviews* **70**(3), 729-754.
9. Mickey, B. & J. Howard (1995). Rigidity of microtubules is increased by stabilizing agents. *Journal of Cell Biology* **130**(4), 909-917.
10. Pitnick, S., T.A. Markow & G.S. Spicer (1995). Delayed male maturity is a cost of producing large sperm in *Drosophila. Proceedings of the National Academy of Sciences USA* **92**(23), 10614-10618.
11. Constantin, B. (2014). Dystrophin complex functions as a scaffold for signalling proteins. *Biochimica et Biophysica Acta* **1838**(2), 635-642.
12. Plodinec, M., M. Loparic, R. Suetterlin, H. Herrmann, U. Aebi & C.-A. Schoenenberger (2011). The nanomechanical properties of rat fibroblasts are modulated by interfering with the vimentin intermediate filament system. *Journal of Structural Biology* **174**(3), 476-484.

13. Sütterlin, R. & C.-A. Schoenenberger (2014). Rat-2 fibroblast. http://www.biozentrum.unibas.ch/typo3temp/pics/schoenenberger_overview_6666954fa0.jpg
14. MacRae, T.H. (1992). Microtubule organization by cross-linking and bundling proteins. *Biochimica et Biophysica Acta* **1160**(2), 145-155.
15. Ribbeck, K., A.C. Groen, R. Santarella, M.T. Bohnsack, T. Raemaekers, T. Kocher, M. Gentzel, D. Gorlich, M. Wilm, G. Carmeliet, T.J. Mitchison, J. Ellenberg, A. Hoenger & I.W. Mattaj (2006). NuSAP, a mitotic RanGTP target that stabilizes and cross-links microtubules. *Molecular Biology of the Cell* **17**(6), 2646-2660.
16. Bloom, G.S., F.C. Luca & R.B. Vallee (1985). Cross-linking of intermediate filaments to microtubules by microtubule-associated protein 2. *Annals of the New York Academy of Sciences* **455**, 18-31.
17. Fuchs, E. & I. Karakesisoglou (2001). Bridging cytoskeletal intersections. *Genes & Development* **15**(1), 1-14.
18. Röper, K., S.L. Gregory & N.H. Brown (2002). The 'spectraplakins': cytoskeletal giants with characteristics of both spectrin and plakin families. *Journal of Cell Science* **115**(22), 4215-4225.
19. Sonnenberg, A. & R.K.H. Liem (2007). Plakins in development and disease. *Experimental Cell Research* **313**(10), 2189-2203.
20. Wang, J., X. Xue & H. Ren (2012). New insights into the role of plant formins: regulating the organization of the actin and microtubule cytoskeleton. *Protoplasma* **249**, 101-107.
21. Tilney, L.G. & D.J. DeRosier (2005). How to make a curved *Drosophila* bristle using straight actin bundles. *Proceedings of the National Academy of Sciences USA* **102**(52), 18785-18792.
22. Khurana, S. & S.P. George (2011). The role of actin bundling proteins in the assembly of filopodia in epithelial cells. *Cell Adhesion & Migration* **5**(5), 409-420.
23. Jefferson, J.J., C.L. Leung & R.K.H. Liem (2004). Plakins: Goliaths that link cell junctions and the cytoskeleton. *Nature Reviews Molecular Cell Biology* **5**(7), 542-553.
24. Mellad, J.A., D.T. Warren & C.M. Shanahan (2011). Nesprins LINC the nucleus and cytoskeleton. *Current Opinion in Cell Biology* **23**(1), 47-54.
25. Simon, D.N. & K.L. Wilson (2011). The nucleoskeleton as a genome-associated dynamic 'network of networks'. *Nature Reviews, Molecular Cell Biology* **12**(11), 695-708.
26. Lombardi, M.L. & J. Lammerding (2011). Keeping the LINC: the importance of nucleocytoskeletal coupling in intracellular force transmission and cellular function. *Biochemical Society Transactions* **39**, 1729-1734.
27. Liu, Y., Y. Guo, J.M. Valles Jr & J.X. Tang (2006). Microtubule bundling and nested buckling drive stripe formation in polymerizing tubulin solutions. *Proceedings of the National Academy of Sciences USA* **103**(28), 10654-10659.
28. MBInfo (2014). iPALM in Mechanobiology: Focal adhesion proteins are stratified along the axial dimension. http://www.mechanobio.info/topics/methods/super-resolution-microscopy-intro/7_super-resolution-microscopy

29. Kwok, R. (2011). The new cell anatomy. A menagerie of intriguing cell structures, some long-neglected and others newly discovered, is keeping biologists glued to their microscopes. *Nature* **480**(7375), 26-28.
30. Ingerson-Mahar, M. & Z. Gitai (2012). A growing family: the expanding universe of the bacterial cytoskeleton. *FEMS Microbiology Reviews* **36**(1), 256-266.
31. Moseley, J.B. (2013). An expanded view of the eukaryotic cytoskeleton. *Molecular Biology of the Cell* **24**(11), 1615-1618.
32. Pilhofer, M. & G.J. Jensen (2013). The bacterial cytoskeleton: more than twisted filaments. *Current Opinion in Cell Biology* **25**(1), 125-133.
33. Oegema, K., M.S. Savoian, T.J. Mitchison & C.M. Field (2000). Functional analysis of a human homologue of the *Drosophila* actin binding protein anillin suggests a role in cytokinesis. *Journal of Cell Biology* **150**(3), 539-552.
34. Estey, M.P., M.S. Kim & W.S. Trimble (2011). Septins. *Current Biology* **21**(10), R384-R387.
35. Mostowy, S. & P. Cossart (2012). Septins: the fourth component of the cytoskeleton. *Nature Reviews Molecular Cell Biology* **13**(3), 183-194.
36. Azzam, G. & J.L. Liu (2013). Only one isoform of *Drosophila melanogaster* CTP synthase forms the cytoophidium. *PLoS Genetics* **9**(2), e1003256.
37. Noguchi, T., M. Koizumi & S. Hayashi (2011). Sustained elongation of sperm tail promoted by local remodeling of giant mitochondria in *Drosophila. Current Biology* **21**(10), 805-814.
38. Bjork, A. & S. Pitnick (2006). Intensity of sexual selection along the anisogamy-isogamy continuum. *Nature* **441**(7094), 742-745.
39. Satarić, M.V., N. Bednar, B.M. Satarić & G. Stojanović (2009). Actin filaments as nonlinear RLC transmission lines. *International Journal of Modern Physics B* **23**(22), 4697-4711.
40. Satarić, M.V., D.I. Ilić, N. Ralević & J.A. Tuszynski (2009). A nonlinear model of ionic wave propagation along microtubules. *European Biophysics Journal with Biophysics Letters* **38**(5), 637-647.
41. Sahu, S., S. Ghosh, K. Hirata, D. Fujita & A. Bandyopadhyay (2013). Multi-level memory-switching properties of a single brain microtubule. *Applied Physics Letters* **102**(12), #123701.
42. Priel, A., J.A. Tuszynski & N.J. Woolf (2010). Neural cytoskeleton capabilities for learning and memory. *Journal of Biological Physics* **36**(1), 3-21.
43. Vale, R.D., B.J. Schnapp, T.S. Reese & M.P. Sheetz (1985). Organelle, bead, and microtubule translocations promoted by soluble factors from the squid giant axon. *Cell* **40**(3), 559-569.
44. Schlager, M.A. & C.C. Hoogenraad (2009). Basic mechanisms for recognition and transport of synaptic cargos. *Molecular Brain* **2**, #25.
45. Chen, S.T. & C.W. Li (1991). Relationships between the movements of chloroplasts and cytoskeletons in diatoms. *Botanica Marina* **34**(6), 505-511.
46. Spear, P.C. & C.A. Erickson (2012). Interkinetic nuclear migration: a mysterious process in search of a function. *Development, Growth & Differentiation* **54**(3), 306-316.
47. Spear, P.C. & C.A. Erickson (2012). Apical movement during interkinetic nuclear migration is a two-step process. *Developmental Biology* **370**(1), 33-41.
48. Szikora, S., I. Gaspar & J. Szabad (2013). 'Poking' microtubules bring about nuclear wriggling to position nuclei. *Journal of Cell Science* **126**(1), 254-262.

49. Guillot, C. & T. Lecuit (2013). Adhesion disengagement uncouples intrinsic and extrinsic forces to drive cytokinesis in epithelial tissues. *Developmental Cell* **24**(3), 227-241.
50. Ingerson-Mahar, M., A. Briegel, J.N. Werner, G.J. Jensen & Z. Gitai (2010). The metabolic enzyme CTP synthase forms cytoskeletal filaments. *Nature Cell Biology* **12**(8), 739-746.
51. Block, S.M. (2003). Public information for: Kinesin Moves by an Asymmetric Hand-Over-Hand Mechanism. http://web.stanford.edu/group/blocklab/ScienceLimping/
52. Asbury, C.L., A.N. Fehr & S.M. Block (2003). Kinesin moves by an asymmetric hand-over-hand mechanism. *Science* **302**(5653), 2130-2134.
53. Fortini, M.E. (2002). γ-secretase-mediated proteolysis in cell-surface-receptor signalling. *Nature Reviews Molecular Cell Biology* **3**(9), 673-684.
54. Maniotis, A.J., C.S. Chen & D.E. Ingber (1997). Demonstration of mechanical connections between integrins, cytoskeletal filaments, and nucleoplasm that stabilize nuclear structure. *Proceedings of the National Academy of Sciences USA* **94**(3), 849-854.
55. Tucker, J.B., J. Beisson, D.L. Roche & J. Cohen (1980). Microtubules and control of macronuclear 'amitosis' in *Paramecium*. *Journal of Cell Science* **44**, 135-151.
56. Wren, N.S., Z.X. Zhong, R.S. Schwartz & K.N. Dahl (2012). Modeling nuclear blebs in a nucleoskeleton of independent filament networks. *Cellular and Molecular Bioengineering* **5**(1), 73-81.
57. Funkhouser, C.M., R. Sknepnek, T. Shimi, A.E. Goldman, R.D. Goldman & M.O. de la Cruz (2013). Mechanical model of blebbing in nuclear lamin meshworks. *Proceedings of the National Academy of Sciences USA* **110**(9), 3248-3253.
58. Swift, J. & D.E. Discher (2014). The nuclear lamina is mechano-responsive to ECM elasticity in mature tissue. *Journal of Cell Science* **127**(Pt 14), 3005-3015.
59. Harmon, R.M. & K.J. Green (2013). Structural and Functional Diversity of Desmosomes. *Cell Communication and Adhesion* **20**(6), 171-187.
60. Kowalczyk, A.P. & K.J. Green (2013). Structure, Function, and Regulation of Desmosomes. In: *Molecular Biology of Cadherins*. Ed.: F. VanRoy. **116**: 95-118.
61. Herrmann, H., H. Bär, L. Kreplak, S.V. Strelkov & U. Aebi (2007). Intermediate filaments: from cell architecture to nanomechanics. *Nature Reviews Molecular Cell Biology* **8**(7), 562-573.
62. Eichenlaub-Ritter, U. & J.B. Tucker (1984). Microtubules with more than 13 protofilaments in the dividing nuclei of ciliates. *Nature* **307**(5946), 60-62.
63. Pickett-Heaps, J.D. (1975). Aspects of spindle evolution. *Annals of the New York Academy of Sciences* **253**, 352-361.
64. Heath, I.B. (1980). Variant mitoses in lower eukaryotes: indicators of the evolution of mitosis? *International Review of Cytology* **64**, 1-80.
65. De Souza, C.P. & S.A. Osmani (2007). Mitosis, not just open or closed. *Eukaryotic Cell* **6**(9), 1521-1527.
66. Santella, L., N. Limatola & J.T. Chun (2015). Calcium and actin in the saga of awakening oocytes. *Biochemical and Biophysical Research Communications* **460**, 104-113.

67. Onyshko, D. (1989). A possible relationship between nuclear rotation and gene activation [B.Sc. (Med) Thesis, supervisor: Richard Gordon]. In: *Program and Proceedings, B.Sc.(Med) Presentations*. Winnipeg, Faculty of Medicine, University of Manitoba: 142-148.
68. Hay, M. & U. De Boni (1991). Chromatin motion in neuronal interphase nuclei: changes induced by disruption of intermediate filaments. *Cell Motility and the Cytoskeleton* **18**(1), 63-75.
69. Strickfaden, H., A. Zunhammer, S. van Koningsbruggen, D. Köhler & T. Cremer (2010). 4D chromatin dynamics in cycling cells: Theodor Boveri's hypotheses revisited. *Nucleus-Austin* **1**(3), 284-297.
70. Ji, J.Y., R.T. Lee, L. Vergnes, L.G. Fong, C.L. Stewart, K. Reue, S.G. Young, Q. Zhang, C.M. Shanahan & J. Lammerding (2007). Cell nuclei spin in the absence of lamin B1. *Journal of Biological Chemistry* **282**(27), 20015-20026.
71. Chen, T.H., J.J. Hsu, X. Zhao, C.Y. Guo, M.N. Wong, Y. Huang, Z.W. Li, A. Garfinkel, C.M. Ho, Y. Tintut & L.L. Demer (2012). Left-right symmetry breaking in tissue morphogenesis via cytoskeletal mechanics. *Circulation Research* **110**(4), 551-559.
72. Yamanaka, H. & S. Kondo (2015). Rotating pigment cells exhibit an intrinsic chirality. *Genes to Cells* **20**(1), 29-35.
73. He, S., K.L. Dunn, P.S. Espino, B. Drobic, L. Li, J. Yu, J.-M. Sun, H.Y. Chen, S. Pritchard & J.R. Davie (2008). Chromatin organization and nuclear microenvironments in cancer cells. *Journal of Cellular Biochemistry* **104**(6), 2004-2015.
74. Bode, J. (2007). Scaffold/matrix attachment region: File:SMARs facultative.png. http://en.wikipedia.org/wiki/File:SMARs_facultative.png
75. Razin, S.V., O.V. Iarovaia & Y.S. Vassetzky (2014). A requiem to the nuclear matrix: from a controversial concept to 3D organization of the nucleus. *Chromosoma* **123**(3), 217-224.
76. Bernander, R. & T.J.G. Ettema (2010). FtsZ-less cell division in archaea and bacteria. *Current Opinion in Microbiology* **13**(6), 747-752.
77. Miyagishima, S., M. Nakamura, A. Uzuka & A. Era (2014). FtsZ-less prokaryotic cell division as well as FtsZ- and dynamin-less chloroplast and non-photosynthetic plastid division. *Frontiers in Plant Science* **5**, #459.
78. Dyer, N. (2009). Tubulin and its prokaryotic homologue FtsZ: a structural and functional comparison. *Science Progress* **92**(2), 113-137.
79. McIntosh, J.R., V. Volkov, F.I. Ataullakhanov & E.L. Grishchuk (2010). Tubulin depolymerization may be an ancient biological motor. *Journal of Cell Science* **123**(20), 3425-3434.
80. Erickson, H.P. (2007). Evolution of the cytoskeleton. *BioEssays* **29**(7), 668-677.
81. Smith, E., H.J. Morowitz & V. Spinivasan (2006). Experimental search for minimal organisms and the last universal common ancestor. *Complexity* **12**(1), 11-12.
82. Searcy, D.G. & W.G. Hixon (1991). Cytoskeletal origins in sulfur-metabolizing archaebacteria. *BioSystems* **25**(1-2), 1-11.
83. Sosa, B.A., U. Kutay & T.U. Schwartz (2013). Structural insights into LINC complexes. *Current Opinion in Structural Biology* **23**(2), 285-291.
84. Shaiken, T.E. & A.R. Opekun (2014). Dissecting the cell to nucleus, perinucleus and cytosol. *Scientific Reports* **4**, #4923.

85. Nogales, E., S.G. Wolf & K.H. Downing (1998). Structure of the αβ tubulin dimer by electron crystallography. *Nature* **391**(6663), 199-203, erratum: 393(6681), 6191.

86. Alushin, G.M., G.C. Lander, E.H. Kellogg, R. Zhang, D. Baker & E. Nogales (2014). High-resolution microtubule structures reveal the structural transitions in αβ-tubulin upon GTP hydrolysis. *Cell* **157**(5), 1117-1129.

87. Amos, L.A. & J. Löwe (2014). The subtle allostery of microtubule dynamics. *Nature Structural & Molecular Biology* **21**(6), 505-506.

88. Gordon, R. (1970). On Monte Carlo algebra. *Journal of Applied Probability* **7**, 373-387.

89. Bramkamp, M. & S. van Baarle (2009). Division site selection in rod-shaped bacteria. *Current Opinion in Microbiology* **12**(6), 683-688.

90. Ayoub, Ahmed T., Travis J.A. Craddock, M. Klobukowski & J. Tuszynski (2014). Analysis of the strength of interfacial hydrogen bonds between tubulin dimers using quantum theory of atoms in molecules. *Biophysical Journal* **107**(3), 740-750.

91. Ji, X.-Y. & X.-Q. Feng (2011). Mechanochemical modeling of dynamic microtubule growth involving sheet-to-tube transition. *PLoS One* **6**(12), e29049.

92. Höög, J.L., S.M. Huisman, Z. Sebö-Lemke, L. Sandblad, J.R. McIntosh, C. Antony & D. Brunner (2011). Electron tomography reveals a flared morphology on growing microtubule ends. *Journal of Cell Science* **124**(5), 693-698.

93. Gordon, R., R. Bender & G.T. Herman (1970). Algebraic Reconstruction Techniques (ART) for three-dimensional electron microscopy and x-ray photography. *Journal of Theoretical Biology* **29**(3), 471-481.

94. Gordon, R. (2011). Stop breast cancer now! Imagining imaging pathways towards search, destroy, cure and watchful waiting of premetastasis breast cancer. In: *Breast Cancer - A Lobar Disease*. Ed.: T. Tot. London, Springer: 167-203.

95. Wikipedia (2014). Paclitaxel. http://en.wikipedia.org/wiki/Paclitaxel

96. Tran, P.T., P. Joshi & E.D. Salmon (1997). How tubulin subunits are lost from the shortening ends of microtubules. *Journal of Structural Biology* **118**(2), 107-118.

97. Heussinger, C., F. Schüller & E. Frey (2010). Statics and dynamics of the wormlike bundle model. *Physical Review E* **81**(2), #021904.

98. Mohrbach, H., A. Johner & I.M. Kulić (2012). Cooperative lattice dynamics and anomalous fluctuations of microtubules. *European Biophysics Journal with Biophysics Letters* **41**(2), 217-239.

99. Szyk, A., A.M. Deaconescu, J. Spector, B. Goodman, M.L. Valenstein, N.E. Ziolkowska, V. Kormendi, N. Grigorieff & A. Roll-Mecak (2014). Molecular basis for age-dependent microtubule acetylation by tubulin acetyltransferase. *Cell* **157**(6), 1405-1415.

100. Kull, F.J. & R.D. Sloboda (2014). A slow dance for microtubule acetylation. *Cell* **157**(6), 1255-1256.

101. Reed, N.A., D. Cai, T.L. Blasius, G.T. Jih, E. Meyhofer, J. Gaertig & K.J. Verhey (2006). Microtubule acetylation promotes kinesin-1 binding and transport. *Current Biology* **16**(21), 2166-2172.

102. Craddock, T.J.A., D. Friesen, J. Mane, S. Hameroff & J.A. Tuszynski (2014). The feasibility of coherent energy transfer in microtubules. http://arxiv.org/abs/1405.3170
103. Priel, A., A.J. Ramos, J.A. Tuszynski & H.F. Cantiello (2006). A biopolymer transistor: electrical amplification by microtubules. *Biophysical Journal* **90**(12), 4639-4643.
104. Friesen, D., T. Craddock, A. Priel & J.A. Tuszynski (2015). Cytoskeletal electrostatic and ionic conduction effects in the cell. In: *Fields of the Cell*. Eds.: D. Fels, M. Cifra & F. Scholkmann. Kerala, India, Research Signpost: 247-270.
105. Lin, E.C. & H.F. Cantiello (1993). A novel method to study the electrodynamic behavior of actin filaments. Evidence for cable-like properties of actin. *Biophysical Journal* **65**(4), 1371-1378.
106. Chakraborty, T., Ed. (2007). *Charge Migration in DNA: Perspectives from Physics, Chemistry and Biology*. Berlin, Springer.
107. Denis, P.A. (2014). Heuristic consequences of a load of oxygen in microtubules. *BioSystems* **118**, 17-30.
108. Panda, D., K. Rathinasamy, M.K. Santra & L. Wilson (2005). Kinetic suppression of microtubule dynamic instability by griseofulvin: implications for its possible use in the treatment of cancer. *Proceedings of the National Academy of Sciences USA* **102**(28), 9878-9883.
109. Wikipedia (2015). Crystallization. https://en.wikipedia.org/wiki/Crystallization - Thermodynamic_view
110. Shen, T.Y. & P.G. Wolynes (2005). Nonequilibrium statistical mechanical models for cytoskeletal assembly: Towards understanding tensegrity in cells. *Physical Review E* **72**(4), #041927.
111. Gatenby, R.A. & B.R. Frieden (2013). The critical roles of information and nonequilibrium thermodynamics in evolution of living systems. *Bulletin of Mathematical Biology* **75**(4), 589-601.
112. Bolterauer, H., H.J. Limbach & J.A. Tuszynski (1999). Models of assembly and disassembly of individual microtubules: stochastic and averaged equations. *Journal of Biological Physics* **25**(1), 1-22.
113. Rezania, V. & J.A. Tuszynski (2008). A first principle (3+1)-dimensional model for microtubule polymerization. *Physics Letters A* **372**(47), 7051-7056.
114. Hinow, P., V. Rezania & J.A. Tuszynski (2009). Continuous model for microtubule dynamics with catastrophe, rescue, and nucleation processes. *Physical Review E* **80**(3), #031904.
115. Buxton, G.A., S.L. Siedlak, G. Perry & M.A. Smith (2010). Mathematical modeling of microtubule dynamics: Insights into physiology and disease. *Progress in Neurobiology* **92**(4), 478-483.
116. Chen, Y.D. & T.L. Hill (1987). Theoretical studies on oscillations in microtubule polymerization. *Proceedings of the National Academy of Sciences USA* **84**(23), 8419-8423.
117. Chamberlin, R.V. & W.A. Eaton (2015). Terrell L. Hill, 1917-2014, A Biographical Memoir. Washington, DC, USA, National Academy of Sciences.
118. Odde, D.J. (1997). Estimation of the diffusion-limited rate of microtubule assembly. *Biophysical Journal* **73**(1), 88-96.
119. Gordon, R. & R.W. Drum (1994). The chemical basis of diatom morphogenesis. *International Review of Cytology* **150**, 243-372, 421-422.

120. Parkinson, J., Y. Brechet & R. Gordon (1999). Centric diatom morphogenesis: a model based on a DLA algorithm investigating the potential role of microtubules. *Biochimica et Biophysica Acta - Molecular Cell Research* **1452**(1), 89-102.
121. Kirschner, M.W. & J.C. Gerhart (2005). *The Plausibility of Life: Resolving Darwin's Dilemma*. New Haven, Yale University Press.
122. Gerhart, J. & M. Kirschner (2007). The theory of facilitated variation. *Proceedings of the National Academy of Sciences USA* **104**(Suppl. 1), 8582-8589.
123. Michal, G. (2014). Biochemical Pathways. http://www.roche.com/responsibility/supporting_communities/science_education/pathways.htm; http://www.roche.com/sustainability/for_communities_and_environment/philanthropy/science_education/pathways.htm
124. Wünschiers, R., M. Jahn, D. Jahn, I. Schomburg, S. Peifer, E. Heinzle, H. Burtscher, J. Garbe, A. Steen, M. Schobert, D. Oesterhelt, W. Josef & A. Chang (2012). Metabolism. In: *Biochemical Pathways: An Atlas of Biochemistry and Molecular Biology*. Eds.: G. Michal & D. Schomburg, Wiley: 37-209. 2nd.
125. van der Vaart, B., A. Akhmanova & A. Straube (2009). Regulation of microtubule dynamic instability. *Biochemical Society Transactions* **37**(Pt 5), 1007-1013.
126. Mechulam, A., K.G. Chernov, E. Mucher, L. Hamon, P.A. Curmi & D. Pastre (2009). Polyamine sharing between tubulin dimers favours microtubule nucleation and elongation via facilitated diffusion. *PLoS Computational Biology* **5**(1), e1000255.
127. Yadav, V. & S. Mukherji (2015). Polymerization of microtubules in the presence of diffusing tubulins: a boundary layer analysis. *Journal of Statistical Mechanics-Theory and Experiment*(February), P02009.
128. Adam, G. & M. Delbrück (1968). Reduction of dimensionality in biological diffusion processes. In: *Structural Chemistry and Molecular Biology*. Eds.: A. Rich & N. Davidson. New York, W.H. Freeman: 198-215.
129. Peters, R. (2005). Translocation through the nuclear pore complex: selectivity and speed by reduction-of-dimensionality. *Traffic* **6**(5), 421-427.
130. Ludueña, R.F. (2013). A hypothesis on the origin and evolution of tubulin. *International Review of Cell and Molecular Biology* **302**, 41-185.
131. Kollman, J.M., A. Merdes, L. Mourey & D.A. Agard (2011). Microtubule nucleation by γ-tubulin complexes. *Nature Reviews Molecular Cell Biology* **12**(11), 709-721.
132. Cheng, S.F. & M.J. Stevens (2014). Self-assembly of chiral tubules. *Soft Matter* **10**(3), 510-518.
133. Brinkley, B.R. (1985). Microtubule organizing centers. *Annual Review of Cell Biology* **1**, 145-172.
134. Luders, J. & T. Stearns (2007). Opinion - Microtubule-organizing centres: a re-evaluation. *Nature Reviews Molecular Cell Biology* **8**(2), 161-167.
135. Murata, T., T. Tanahashi, T. Nishiyama, K. Yamaguchi & M. Hasebe (2007). How do plants organize microtubules without a centrosome? *Journal of Integrative Plant Biology* **49**(8), 1154-1163.
136. Wikipedia (2014). Nucleation. http://en.wikipedia.org/wiki/Nucleation

137. Becker, B.E. & L. Cassimeris (2005). Cytoskeleton: Microtubules born on the run. *Current Biology* **15**(14), R551-R554.
138. Vorobjev, I.A., T.M. Svitkina & G.G. Borisy (1997). Cytoplasmic assembly of microtubules in cultured cells. *Journal of Cell Science* **110**(21), 2635-2645.
139. Yvon, A.M.C. & P. Wadsworth (1997). Non-centrosomal microtubule formation and measurement of minus end microtubule dynamics in A498 cells. *Journal of Cell Science* **110**(19), 2391-2401.
140. Dammermann, A., A. Desai & K. Oegema (2003). The minus end in sight. *Current Biology* **13**(15), R614-R624.
141. Sirajuddin, M., L.M. Rice & R.D. Vale (2014). Regulation of microtubule motors by tubulin isotypes and post-translational modifications. *Nature Cell Biology* **16**(4), 335-344.
142. Garnham, C.P. & A. Roll-Mecak (2012). The chemical complexity of cellular microtubules: Tubulin post-translational modification enzymes and their roles in tuning microtubule functions. *Cytoskeleton* **69**(7), 442-463.
143. Katsetos, C.D., M.J. Reginato, P.W. Baas, L. D'Agostino, A. Legido, J.A. Tuszyński, E. Dráberová & P. Dráber (2015). Emerging microtubule targets in glioma therapy. *Seminars in Pediatric Neurology* **22**(1), 49-72.
144. Guo, J., C. Walss-Bass & R.F. Ludueña (2010). The β isotypes of tubulin in neuronal differentiation. *Cytoskeleton* **67**(7), 431-441.
145. Gardiner, J., R. Overall & J. Marc (2012). Plant microtubule cytoskeleton complexity: microtubule arrays as fractals. *Journal of Experimental Botany* **63**(2), 635-642.
146. Kühn, S. (2014). *Organic Codes and their Identication: Is the Histone Code a True Organic Code? [Master's Thesis]*. Matieland, South Africa, Department of Biochemistry, University of Stellenbosch.
147. Barbieri, M. (2003). *The Organic Codes. An Introduction to Semantic Biology*. Cambridge, Cambridge University Press.
148. Verhey, K.J. & J. Gaertig (2007). The tubulin code. *Cell Cycle* **6**(17), 2152-2160.
149. Janke, C. & J.C. Bulinski (2011). Post-translational regulation of the microtubule cytoskeleton: mechanisms and functions [+ erratum (2012) **13**(4) 276]. *Nature Reviews Molecular Cell Biology* **12**(12), 773-786.
150. Hill, T.L. & M.W. Kirschner (1982). Subunit treadmilling of microtubules or actin in the presence of cellular barriers: possible conversion of chemical free energy into mechanical work. *Proceedings of the National Academy of Sciences USA* **79**, 490-494.
151. Grego, S., V. Cantillana & E.D. Salmon (2001). Microtubule treadmilling in vitro investigated by fluorescence speckle and confocal microscopy. *Biophysical Journal* **81**(1), 66-78.
152. Maly, I.V. & G.G. Borisy (2002). Self-organization of treadmilling microtubules into a polar array. *Trends in Cell Biology* **12**(10), 462-465.
153. Shaw, S.L., R. Kamyar & D.W. Ehrhardt (2003). Sustained microtubule treadmilling in *Arabidopsis* cortical arrays. *Science* **300**(5626), 1715-1718.
154. Lloyd, C. (2011). Dynamic microtubules and the texture of plant cell walls. *International Review of Cell and Molecular Biology* **287**, 287-329.
155. Ebbinghaus, M., C. Appert-Rolland & L. Santen (2010). Bidirectional traffic on microtubules. *Lecture Notes in Computer Science* **6350**, 542-551.

156. Conway, L., D. Wood, K.E. Daly, A.M. O' Neill, E. Tuzel & J. Ross (2011). Microtubule motor traffic jams. *Molecular Biology of the Cell* **22**(Meeting Abstracts), #1480.

157. Gordon, R. (1969). Polyribosome dynamics at steady state. *Journal of Theoretical Biology* **22**(3), 515-532.

158. Hendricks, A.G., J.E. Lazarus, E. Perlson, M.K. Gardner, D.J. Odde, Y.E. Goldman & E.L.F. Holzbaur (2012). Dynein tethers and stabilizes dynamic microtubule plus ends. *Current Biology* **22**(7), 632-637.

159. Hawkins, T.L., D. Sept, B. Mogessie, A. Straube & J.L. Ross (2013). Mechanical properties of doubly stabilized microtubule filaments. *Biophysical Journal* **104**(7), 1517-1528.

160. Hill, T.L. (1981). Microfilament or microtubule assembly or disassembly against a force. *Proceedings of the National Academy of Sciences USA* **78**, 5613-5617.

161. Hayashi, S. & C. Shingyoji (2008). Mechanism of flagellar oscillation - bending-induced switching of dynein activity in elastase-treated axonemes of sea urchin sperm. *Journal of Cell Science* **121**(17), 2833-2843.

162. Subramanian, R. & T.M. Kapoor (2012). Building complexity: Insights into self-organized assembly of microtubule-based architectures. *Developmental Cell* **23**(5), 874-885.

163. Schliwa, M. & G. Woehlke (2003). Molecular motors. *Nature* **422**(6933), 759-765.

164. Nakamura, M., L. Chen, S.C. Howes, T.D. Schindler, E. Nogales & Z. Bryant (2014). Remote control of myosin and kinesin motors using light-activated gearshifting. *Nature Nanotechnology* **9**(9), 693-697.

165. Yanagisawa, H.-a., G. Mathis, T. Oda, M. Hirono, E.A. Richey, H. Ishikawa, W.F. Marshall, M. Kikkawa & H. Qin (2014). FAP20 is an inner junction protein of doublet microtubules essential for both the planar asymmetrical waveform and stability of flagella in *Chlamydomonas*. *Molecular Biology of the Cell* **25**(9), 1472-1483.

166. Nicastro, D., X. Fu, T. Heuser, A. Tso, M.E. Porter & R.W. Linck (2011). Cryo-electron tomography reveals conserved features of doublet microtubules in flagella. *Proceedings of the National Academy of Sciences USA* **108**(42), E845-E853.

167. Wirschell, M., H. Olbrich, C. Werner, D. Tritschler, R. Bower, W.S. Sale, N.T. Loges, P. Pennekamp, S. Lindberg, U. Stenram, B. Carlén, E. Horak, G. Köhler, P. Nürnberg, G. Nürnberg, M.E. Porter & H. Omran (2013). The nexin-dynein regulatory complex subunit DRC1 is essential for motile cilia function in algae and humans. *Nature Genetics* **45**(3), 262-268.

168. Pigino, G., A. Maheshwari, K.H. Bui, C. Shingyoji, S. Kamimura & T. Ishikawa (2012). Comparative structural analysis of eukaryotic flagella and cilia from *Chlamydomonas*, *Tetrahymena*, and sea urchins. *Journal of Structural Biology* **178**(2), 199-206.

169. Ueki, N., S. Matsunaga, I. Inouye & A. Hallmann (2010). How 5000 independent rowers coordinate their strokes in order to row into the sunlight: phototaxis in the multicellular green alga *Volvox*. *BMC Biology* **8**, #103.

170. Katow, H. & S. Ooka (2010). Development of swimming regulation systems in sea urchin: From blastulae to larvae. *Developmental Biology* **344**(1), 450-451.

171. Steinman, R.M. (1968). An electron microscope study of ciliogenesis in developing epidermis and trachea in the embryo of *Xenopus laevis*. *American Journal of Anatomy* **122**(1), 19-56.

172. Billett, F.S. & T.H. Courtenay (1973). A stereoscan study of the origin of ciliated cells in the embryonic epidermis of *Ambystoma mexicanum*. *Journal of Embryology and Experimental Morphology* **29**(3), 549-558.

173. Landström, U. (1977). On the differentiation of prospective ectoderm to a ciliated pattern in embryos of *Ambystoma mexicanum*. *Journal of Embryology and Experimental Morphology* **41**(OCT), 23-32.

174. Fox, H. (1986). Early development of caecilian skin with special reference to the epidermis. *Journal of Herpetology* **20**(2), 154-167.

175. König, G. & P. Hausen (1993). Planar polarity in the ciliated epidermis of *Xenopus* embryos. *Developmental Biology* **160**(2), 355-368.

176. Park, T.J., S.L. Haigo & J.B. Wallingford (2006). Ciliogenesis defects in embryos lacking inturned or fuzzy function are associated with failure of planar cell polarity and Hedgehog signaling. *Nature Genetics* **38**(3), 303-311.

177. Tashkin, D.P. (2013). Effects of marijuana smoking on the lung. *Annals of the American Thoracic Society* **10**(3), 239-247.

178. Chung, M.I., T. Kwon, F. Tu, E.R. Brooks, R. Gupta, M. Meyer, J.C. Baker, E.M. Marcotte & J.B. Wallingford (2014). Coordinated genomic control of ciliogenesis and cell movement by RFX2. *Elife* **3**, #e01439.

179. Gordon, R. (1999). *The Hierarchical Genome and Differentiation Waves: Novel Unification of Development, Genetics and Evolution*. Singapore & London, World Scientific & Imperial College Press: [http://www.worldscientific.com/worldscibooks/10.1142/2755].

180. Prodhan, A., R. Funada, J. Ohtani, H. Abe & K. Fukazawa (1995). Orientation of microfibrils and microtubules in developing tension-wood fibers of Japanese ash (*Fraxinus mandshurica* var. *japonica*). *Planta* **196**(3), 577-585.

181. Adoutte, A. & A. Fleury (1996). Cytoskeleton of ciliates. In: *Ciliates: Cells as Organisms*. Eds.: K. Hausmann & P.C. Bradbury. Stuttgart, Gustav Fischer: 41-50.

182. Thitamadee, S., K. Tuchihara & T. Hashimoto (2002). Microtubule basis for left-handed helical growth in *Arabidopsis*. *Nature* **417**(6885), 193-196.

183. Hashimoto, T. (2013). Dissecting the cellular functions of plant microtubules using mutant tubulins. *Cytoskeleton* **70**(4), 191-200.

184. Craddock, T.J., J.A. Tuszynski & S. Hameroff (2012). Cytoskeletal signaling: is memory encoded in microtubule lattices by CaMKII phosphorylation? *PLoS Computational Biology* **8**(3), e1002421.

185. Craddock, T.J.A., S.R. Hameroff, A.T. Ayoub, M. Klobukowski & J.A. Tuszynski (2015). Anesthetics act in quantum channels in brain microtubules to prevent consciousness. *Current Topics in Medicinal Chemistry* **15**(6), 523-533.

186. Snustad, D.P., N.A. Haas, S.D. Kopczak & C.D. Silflow (1992). The small genome of *Arabidopsis* contains at least nine expressed β-tubulin genes. *Plant Cell* **4**(5), 549-556.

187. Korkotian, E. (2009). Neuron - real and "thinking". http://www.iscb.org/cms_addon/conferences/ismbeccb2009/artscience.php

188. Shirao, T. & C. González-Billault (2013). Actin filaments and microtubules in dendritic spines. *Journal of Neurochemistry* **126**(2), 155-164.

189. Wikipedia (2015). Logical conjunction. https://en.wikipedia.org/wiki/Logical_conjunction
190. Wikipedia (2015). Exclusive or. https://en.wikipedia.org/wiki/Exclusive_or
191. Zhang, J. & C. Wang (2014). Molecular structural mechanics model for the mechanical properties of microtubules. *Biomechanics and Modeling in Mechanobiology*, doi: 10.1007/s10237-10014-10564-x.
192. Pawlizak, S. (2009). *Mechanosensitive Behavior of Neuronal Growth Cones [Diploma Thesis]*. Leipzig, Germany, Faculty of Physics and Earth Sciences, University of Leipzig.
193. Käs, J., H. Strey, J.X. Tang, D. Finger, R. Ezzell, E. Sackmann & P.A. Janmey (1996). F-actin, a model polymer for semiflexible chains in dilute, semidilute, and liquid crystalline solutions. *Biophysical Journal* **70**(2), 609-625.
194. Romet-Lemonne, G. & A. Jégou (2013). Mechanotransduction down to individual actin filaments. *European Journal of Cell Biology* **92**(10-11), 333-338.
195. Nöding, B. & S. Köster (2012). Intermediate filaments in small configuration spaces. *Physical Review Letters* **108**(8), #088101.
196. Portran, D., M. Zoccoler, J. Gaillard, V. Stoppin-Mellet, E. Neumann, I. Arnal, J.L. Martiel & M. Vantard (2013). MAP65/Ase1 promote microtubule flexibility. *Molecular Biology of the Cell* **24**(12), 1964-1973.
197. Rittman, M., E. Gilroy, H. Koohy, A. Rodger & A. Richards (2009). Is DNA a worm-like chain in Couette flow? In search of persistence length, a critical review. *Science Progress* **92**(2), 163-204.
198. Wikipedia (2014). Persistence length. http://en.wikipedia.org/wiki/Persistence_length
199. Chi, Q., G. Wang & J. Jiang (2013). The persistence length and length per base of single-stranded DNA obtained from fluorescence correlation spectroscopy measurements using mean field theory. *Physica A-Statistical Mechanics and Its Applications* **392**(5), 1072-1079.
200. Turner, D.J., I. Portman, T.R. Dafforn, A. Rodger, D.I. Roper, C.J. Smith & M.S. Turner (2012). The mechanics of FtsZ fibers. *Biophysical Journal* **102**(4), 731-738.
201. Rheinlaender, J., A. Gräbner, L. Ott, A. Burkovski & T.E. Schäffer (2012). Contour and persistence length of *Corynebacterium diphtheriae* pili by atomic force microscopy. *European Biophysics Journal with Biophysics Letters* **41**(6), 561-570.
202. Kasai, H., M. Fukuda, S. Watanabe, A. Hayashi-Takagi & J. Noguchi (2010). Structural dynamics of dendritic spines in memory and cognition. *Trends in Neurosciences* **33**(3), 121-129.
203. Urbanska, M., L. Swiech & J. Jaworski (2012). Developmental plasticity of the dendritic compartment: focus on the cytoskeleton. *Advances in Experimental Medicine and Biology* **970**, 265-284.
204. Yuasa-Kawada, J., R. Suzuki, F. Kano, T. Ohkawara, M. Murata & M. Noda (2003). Axonal morphogenesis controlled by antagonistic roles of two CRMP subtypes in microtubule organization. *European Journal of Neuroscience* **17**(11), 2329-2343.
205. Kapitein, L.C. & C.C. Hoogenraad (2011). Which way to go? Cytoskeletal organization and polarized transport in neurons. *Molecular and Cellular Neuroscience* **46**(1), 9-20.

206. Juster, N. (1963). *The Dot and the Line: A Romance in Lower Mathematics*. New York, Random House.
207. Jones, C., M. Noble, L. Goldman & N. Juster (1965). The Dot and the Line: A Romance in Lower Mathematics [Movie]. http://www.bcdb.com/bcdb/detailed.cgi?film=2958; https://www.youtube.com/watch?v=4ZEuXA8Wr2g
208. Wikipedia (2015). Actin. https://en.wikipedia.org/wiki/Actin
209. Moseley, J.B. & B.L. Goode (2006). The yeast actin cytoskeleton: From cellular function to biochemical mechanism. *Microbiology and Molecular Biology Reviews* **70**(3), 605-645.
210. Michelot, A. & D.G. Drubin (2011). Building distinct actin filament networks review in a common cytoplasm. *Current Biology* **21**(14), R560-R569.
211. Zheng, B., M. Han, M. Bernier & J.-K. Wen (2009). Nuclear actin and actin-binding proteins in the regulation of transcription and gene expression. *FEBS Journal* **276**(10), 2669-2685.
212. Nambiar, R., R.E. McConnell & M.J. Tyska (2010). Myosin motor function: the ins and outs of actin-based membrane protrusions. *Cellular and Molecular Life Sciences* **67**(8), 1239-1254.
213. Pollard, T.D. & G.G. Borisy (2003). Cellular motility driven by assembly and disassembly of actin filaments. *Cell* **112**(4), 453-465.
214. DeSimone, D.W. & A.R. Horwitz (2014). Many modes of motility. *Science* **345**(6200), 1002-1003.
215. Petrie, R.J., H. Koo & K.M. Yamada (2014). Generation of compartmentalized pressure by a nuclear piston governs cell motility in a 3D matrix. *Science* **345**(6200), 1062-1065.
216. Dickinson, R.B. (2009). Models for actin polymerization motors. *Journal of Mathematical Biology* **58**(1-2), 81-103.
217. Fulton, A.B. (1982). How crowded is the cytoplasm? *Cell* **30**, 345-347.
218. Cuneo, P., E. Magri, A. Verzola & E. Grazi (1992). 'Macromolecular crowding' is a primary factor in the organization of the cytoskeleton. *Biochemical Journal* **281**(2), 507-512.
219. Jefferys, B.R., L.A. Kelley & M.J. Sternberg (2010). Protein folding requires crowd control in a simulated cell. *Journal of Molecular Biology* **397**(5), 1329-1338.
220. Ernst, D., M. Hellmann, J. Koehler & M. Weiss (2012). Fractional Brownian motion in crowded fluids. *Soft Matter* **8**(18), 4886-4889.
221. Leduc, C., K. Padberg-Gehle, V. Varga, D. Helbing, S. Diez & J. Howard (2012). Molecular crowding creates traffic jams of kinesin motors on microtubules. *Proceedings of the National Academy of Sciences USA* **109**(16), 6100-6105.
222. Trevors, J.T. (2012). A perspective on the mobilization, localization and delivery of molecules in the crowded bacterial cytoplasm. *Current Issues in Molecular Biology* **14**(2), 39-45.
223. Foffi, G., A. Pastore, F. Piazza & P.A. Temussi (2013). Macromolecular crowding: chemistry and physics meet biology (Ascona, Switzerland, 10-14 June 2012). *Physical Biology* **10**(4), #040301.
224. Ando, T., E. Chow & J. Skolnick (2013). Dynamic simulation of concentrated macromolecular solutions with screened long-range hydrodynamic interactions: Algorithm and limitations. *Journal of Chemical Physics* **139**(12), #121922.

225. Perrin, B.J. & J.M. Ervasti (2010). The actin gene family: Function follows isoform. *Cytoskeleton* **67**(10), 630-634.
226. Terman, J.R. & A. Kashina (2013). Post-translational modification and regulation of actin. *Current Opinion in Cell Biology* **25**(1), 30-38.
227. Storti, R.V., D.M. Coen & A. Rich (1976). Tissue-specific forms of actin in the developing chick. *Cell* **8**(4), 521-527.
228. Khaitlina, S.Y. (2001). Functional specificity of actin isoforms. *International Review of Cytology - A Survey of Cell Biology* **202**, 35-98.
229. Graumann, P.L. & H.J.D. Soufo (2004). An intracellular actin motor in bacteria? *BioEssays* **26**(11), 1209-1216.
230. Koubassova, N.A. & A.K. Tsaturyan (2011). Molecular mechanism of actin-myosin motor in muscle. *Biochemistry-Moscow* **76**(13), 1484-1506.
231. Shlomovitz, R. & N.S. Gov (2007). Membrane waves driven by actin and myosin. *Physical Review Letters* **98**(16), 168103.
232. Zouani, O.F., V. Gocheva & M.-C. Durrieu (2014). Membrane nanowaves in single and collective cell migration. *PLoS One* **9**(5), e97855.
233. Ivanov, A.I. (2008). Actin motors that drive formation and disassembly of epithelial apical junctions. *Frontiers in Bioscience* **13**, 6662-6681.
234. Ivanov, A.I. & N.G. Naydenov (2013). Dynamics and regulation of epithelial adherns junctions: recent discoveries and controversies. In: *International Review of Cell and Molecular Biology*. Ed.: K.W. Jeon. **303**: 27-99.
235. Small, J., K. Rottner, P. Hahne & K.I. Anderson (1999). Visualising the actin cytoskeleton. *Microscopy Research and Technique* **47**(1), 3-17.
236. Vijay-Kumar, S., C.E. Bugg & W.J. Cook (1987). Structure of ubiquitin refined at 1.8 Å resolution. *Journal of Molecular Biology* **194**(3), 531-544.
237. Goodsell, D. (2004). Ubiquitin. http://www.rcsb.org/pdb/101/motm.do?momID=60
238. Yamin, R. & K.G. Morgan (2012). Deciphering actin cytoskeletal function in the contractile vascular smooth muscle cell. *Journal of Physiology-London* **590**(17), 4145-4154.
239. Taylor, E.W. (1972). Chemistry of muscle contraction. *Annual Review of Biochemistry* **41**, 577-616.
240. McMahon, T.A. (1984). *Muscles, Reflexes, and Locomotion*. Princeton, New Jersey USA, Princeton University Press.
241. Uyeda, T.Q.P., P.D. Abramson & J.A. Spudich (1996). The neck region of the myosin motor domain acts as a lever arm to generate movement. *Proceedings of the National Academy of Sciences USA* **93**(9), 4459-4464.
242. Offer, G. & K.W. Ranatunga (2010). Crossbridge and filament compliance in muscle: implications for tension generation and lever arm swing. *Journal of Muscle Research and Cell Motility* **31**(4), 245-265.
243. Debold, E.P., S. Walcott, M. Woodward & M.A. Turner (2013). Direct observation of phosphate inhibiting the force-generating capacity of a miniensemble of myosin molecules. *Biophysical Journal* **105**(10), 2374-2384.
244. Wikipedia (2014). Myosin. http://en.wikipedia.org/wiki/Myosin
245. Schevzov, G., N.M. Curthoys, P.W. Gunning & T. Faths (2012). Functional diversity of actin cytoskeleton in neurons and its regulation by tropomyosin. *International Review of Cell and Molecular Biology* **298**, 33-94.
246. Lazarides, E. (1980). Intermediate filaments as mechanical integrators of cellular space. *Nature* **283**, 249-256.

247. Iwatsuki, H. & M. Suda (2010). Seven kinds of intermediate filament networks in the cytoplasm of polarized cells: structure and function. *Acta Histochemica et Cytochemica* **43**(2), 19-31.
248. Wagner, O.I., S. Rammensee, N. Korde, Q. Wen, J.-F. Leterrier & P.A. Janmey (2007). Softness, strength and self-repair in intermediate filament networks. *Experimental Cell Research* **313**(10), 2228-2235.
249. Oshima, R.G. (2007). Intermediate filaments: a historical perspective. *Experimental Cell Research* **313**(10), 1981-1994.
250. Levin, S.M. (2010). Biotensegrity: a new way of modeling biologic forms. http://biotensegrity.com
251. Kardas, D., U. Nackenhorst & D. Balzani (2013). Computational model for the cell-mechanical response of the osteocyte cytoskeleton based on self-stabilizing tensegrity structures. *Biomechanics and Modeling in Mechanobiology* **12**(1), 167-183.
252. Brodland, G.W. & R. Gordon (1990). Intermediate filaments may prevent buckling of compressively-loaded microtubules. *Journal of Biomechanical Engineering* **112**(3), 319-321.
253. Mehrbod, M. & M.R.K. Mofrad (2011). On the significance of microtubule flexural behavior in cytoskeletal mechanics. *PLoS One* **6**(10), e25627.
254. Iscen, A., A. Agogino, V. SunSpiral & K. Tumer (2014). Flop and roll: Learning robust goal-directed locomotion for a Tensegrity Robot. In: *2014 IEEE/RSJ International Conference on Intelligent Robots and Systems (IROS 2014)*. IEEE: 2236-2243.
255. Norlén, L., S. Masich, K.N. Goldie & A. Hoenger (2007). Structural analysis of vimentin and keratin intermediate filaments by cryo-electron tomography. *Experimental Cell Research* **313**(10), 2217-2227.
256. Minin, A.A. & M.V. Moldaver (2008). Intermediate vimentin filaments and their role in intracellular organelle distribution. *Biochemistry-Moscow* **73**(13), 1453-1466.
257. Ivaska, J., H.-M. Pallari, J. Nevo & J.E. Eriksson (2007). Novel functions of vimentin in cell adhesion, migration, and signaling. *Experimental Cell Research* **313**(10), 2050-2062.
258. Jin, M.Z. & C.Q. Ru (2013). Localized buckling of a microtubule surrounded by randomly distributed cross linkers. *Physical Review E* **88**(1), #012701.
259. Jin, M.Z. & C.Q. Ru (2014). Localized vibration of a microtubule surrounded by randomly distributed cross linkers. *Journal of Biomechanical Engineering-Transactions of the ASME* **136**(7), #071002.
260. Qin, Z. & M.J. Buehler (2012). Computational and theoretical modeling of intermediate filament networks: Structure, mechanics and disease. *Acta Mechanica Sinica* **28**(4), 941-950.
261. Kushner, A.M. & Z.B. Guan (2011). Modular design in natural and biomimetic soft materials. *Angewandte Chemie-International Edition* **50**(39), 9026-9057.
262. MBInfo & P.T.P. Thao (2012). Intermediate Filament Assembly. http://www.mechanobio.info/figure/1384241639768.jpg.html
263. Wikipedia (2014). N-terminus. http://en.wikipedia.org/wiki/N-terminus
264. Wikipedia (2014). C-terminus. http://en.wikipedia.org/wiki/C-terminus
265. Georgakopoulou, S., D. Möller, N. Sachs, H. Herrmann & U. Aebi (2009). Near-UV circular dichroism reveals structural transitions of vimentin subunits

during intermediate filament assembly. *Journal of Molecular Biology* **386**(2), 544-553.

266. Herrmann, H. & U. Aebi (1998). Intermediate filament assembly: fibrillogenesis is driven by decisive dimer-dimer interactions. *Current Opinion in Structural Biology* **8**(2), 177-185.
267. Chang, L. & R.D. Goldman (2004). Intermediate filaments mediate cytoskeletal crosstalk. *Nature Reviews Molecular Cell Biology* **5**(8), 601-613.
268. Toivola, D.M., G.Z. Tao, A. Habtezion, J. Liao & M.B. Omary (2005). Cellular integrity plus: organelle-related and protein-targeting functions of intermediate filaments. *Trends in Cell Biology* **15**(11), 608-617.
269. Strelkov, S.V., H. Herrmann & U. Aebi (2003). Molecular architecture of intermediate filaments. *BioEssays* **25**(3), 243-251.
270. Gotow, T. (2011). Neurofilament cross-bridge - a structure associated specifically with the neurofilament among the intermediate filament family. In: *Cytoskeleton of the Nervous System*. Eds.: R.A. Nixon & A. Yuan. **3**: 225-247.
271. Pawelzyk, P., N. Mücke, H. Herrmann & N. Willenbacher (2014). Attractive interactions among intermediate filaments determine network mechanics in vitro. *PLoS One* **9**(4), e93194.
272. Burnside, M.B. (1971). Microtubules and microfilaments in newt neurulation. *Developmental Biology* **26**, 416-441.
273. Oriolo, A.S., F.A. Wald, V.P. Ramsauer & P.J.I. Salas (2007). Intermediate filaments: A role in epithelial polarity. *Experimental Cell Research* **313**(10), 2255-2264.
274. Martin, C.C. & R. Gordon (1997). Ultrastructural analysis of the cell state splitter in ectoderm cells differentiating to neural plate and epidermis during gastrulation in embryos of the axolotl *Ambystoma mexicanum*. *Russian Journal of Developmental Biology* **28**(2), 71-80.
275. Wikipedia (2014). Viscoelasticity. http://en.wikipedia.org/wiki/Viscoelasticity
276. Wikipedia (2014). Silly Putty. http://en.wikipedia.org/wiki/Silly_Putty
277. Uhlén, M., L. Fagerberg, B.M. Hallström, C. Lindskog, P. Oksvold, A. Mardinoglu, A. Sivertsson, C. Kampf, E. Sjöstedt, A. Asplund, I. Olsson, K. Edlund, E. Lundberg, S. Navani, C.A. Szigyarto, J. Odeberg, D. Djureinovic, J.O. Takanen, S. Hober, T. Alm, P.H. Edqvist, H. Berling, H. Tegel, J. Mulder, J. Rockberg, P. Nilsson, J.M. Schwenk, M. Hamsten, K. von Feilitzen, M. Forsberg, L. Persson, F. Johansson, M. Zwahlen, G. von Heijne, J. Nielsen & F. Pontén (2015). Tissue-based map of the human proteome. *Science* **347**(6220), 394, #1260419.
278. Uhlén, M., F. Pontén, C. Lindskog, H. Tegel, P. Nilsson, K. von Feilitzen, E. Lundberg, A. Sivertsson & T. Alm (2015). The Human Protein Atlas. http://www.proteinatlas.org/humanproteome
279. Fudge, D.S., K.H. Gardner, V.T. Forsyth, C. Riekel & J.M. Gosline (2003). The mechanical properties of hydrated intermediate filaments: Insights from hagfish slime threads. *Biophysical Journal* **85**(3), 2015-2027.
280. Winegard, T., J. Herr, C. Mena, B. Lee, I. Dinov, D. Bird, M. Bernards, Jr., S. Hobel, B. Van Valkenburgh, A. Toga & D. Fudge (2014). Coiling and maturation of a high-performance fibre in hagfish slime gland thread cells. *Nature Communications* **5**, 3534.
281. Gordon, N.K. (2015). Hagfish Slime. http://embryogenesisexplained.org/2015/12/08/hagfish-slime/

282. Gardiner, J., P. McGee, R. Overall & J. Marc (2008). Are histones, tubulin, and actin derived from a common ancestral protein? *Protoplasma* **233**(1-2), 1-5.
283. Loose, M. & T.J. Mitchison (2014). The bacterial cell division proteins FtsA and FtsZ self-organize into dynamic cytoskeletal patterns. *Nature Cell Biology* **16**(1), 38-46.
284. RayChaudhuri, D. (1999). ZipA is a MAP-Tau homolog and is essential for structural integrity of the cytokinetic FtsZ ring during bacterial cell division. *EMBO Journal* **18**(9), 2372-2383.
285. Ohashi, T., C.A. Hale, P.A.J. de Boer & H.P. Erickson (2002). Structural evidence that the P/Q domain of ZipA is an unstructured, flexible tether between the membrane and the C-terminal FtsZ-binding domain. *Journal of Bacteriology* **184**(15), 4313-4315.
286. Szwedziak, P., Q. Wang, S.M.V. Freund & J. Löwe (2012). FtsA forms actin-like protofilaments. *EMBO Journal* **31**(10), 2249-2260.
287. Wickstead, B. & K. Gull (2011). The evolution of the cytoskeleton. *Journal of Cell Biology* **194**(4), 513-525.
288. Ward, R.J., K.D. Clements, J.H. Choat & E.R. Angert (2009). Cytology of terminally differentiated *Epulopiscium* mother cells. *DNA and Cell Biology* **28**(2), 57-64.
289. Cavalier-Smith, T. (2006). Cell evolution and Earth history: stasis and revolution. *Philosophical Transactions of the Royal Society (London) B Biological Sciences* **361**(1470), 969-1006.
290. Vogel, S.K., I. Raabe, A. Dereli, N. Maghelli & I. Tolić-Nørrelykke (2007). Interphase microtubules determine the initial alignment of the mitotic spindle. *Current Biology* **17**(5), 438-444.
291. Loughlin, R. (2009). Modeling microtubule-dependent length-scaling mechanisms in *Xenopus* spindles. In: *Computational Cell Biology, March 24 - 27, 2009, Cold Spring Harbor Laboratory*.
292. Douglas, M.E. & M. Mishima (2010). Still entangled: Assembly of the central spindle by multiple microtubule modulators. *Seminars in Cell & Developmental Biology* **21**(9), 899-908.
293. Masoud, K., E. Herzog, M.-E. Chabouté & A.-C. Schmit (2013). Microtubule nucleation and establishment of the mitotic spindle in vascular plant cells. *Plant Journal* **75**(2), 245-257.
294. Almonacid, M., M.-E. Terret & M.-H. Verlhac (2014). Actin-based spindle positioning: new insights from female gametes. *Journal of Cell Science* **127**(3), 477-483.
295. Sandquist, J.C., A.M. Kita & W.M. Bement (2011). And the dead shall rise: Actin and myosin return to the spindle. *Developmental Cell* **21**(3), 410-419.
296. Helmke, K.J., R. Heald & J.D. Wilbur (2013). Interplay between spindle architecture and function. *International Review of Cell and Molecular Biology* **306**, 83-125.
297. Siller, K.H. & C.Q. Doe (2009). Spindle orientation during asymmetric cell division. *Nature Cell Biology* **11**(4), 365-374.
298. Werts, A.D. & B. Goldstein (2011). How signaling between cells can orient a mitotic spindle. *Seminars in Cell & Developmental Biology* **22**(8), 842-849.
299. Peyre, E. & X. Morin (2012). An oblique view on the role of spindle orientation in vertebrate neurogenesis. *Development Growth & Differentiation* **54**(3), 287-305.

300. Stephens, A.D., R.A. Haggerty, P.A. Vasquez, L. Vicci, C.E. Snider, F. Shi, C. Quammen, C. Mullins, J. Haase, R.M. Taylor, J.S. Verdaasdonk, M.R. Falvo, Y. Jin, M.G. Forest & K. Bloom (2013). Pericentric chromatin loops function as a nonlinear spring in mitotic force balance. *Journal of Cell Biology* **200**(6), 757-772.
301. McIntosh, J.R., E. O'Toole, K. Zhudenkov, M. Morphew, C. Schwartz, F.I. Ataullakhanov & E.L. Grishchuk (2013). Conserved and divergent features of kinetochores and spindle microtubule ends from five species. *Journal of Cell Biology* **200**(4), 459-474.
302. Wikipedia (2014). Yeast. http://en.wikipedia.org/wiki/Yeast
303. Soll, D.R. (1986). The regulation of cellular differentiation in the dimorphic yeast *Candida albicans*. *BioEssays* **5**(1), 5-11.
304. Wikipedia (2014). *Schizosaccharomyces pombe*. http://en.wikipedia.org/wiki/Fission_yeast
305. Ribas, J.C. & J.C. Cortes (2014). Fission yeast morphogenesis: cytokinesis and cell polarity, Instituto de Biología Funcional y Genómica, Consejo Superior de Investigaciones Científicas (CSIC), Salamanca, Spain. http://jcgarciacortes.wix.com/ribas-pombe-lab
306. Blackwell, M., C.P. Kurtzman, M.-A. Lachance & S.-O. Suh (2009). The Tree of Life Web Project: Saccharomycotina, Saccharomycetales. http://tolweb.org/Saccharomycetales/29043
307. Witkin, K.L., J.M. Friederichs, O. Cohen-Fix & S.L. Jaspersen (2010). Changes in the nuclear envelope environment affect spindle pole body duplication in *Saccharomyces cerevisiae*. *Genetics* **186**(3), 867-883.
308. Firat-Karalar, E.N. & T. Stearns (2014). The centriole duplication cycle. *Philosophical Transactions of the Royal Society B-Biological Sciences* **369**(1650), #20130460.
309. Tulu, U.S., N.M. Rusan & P. Wadsworth (2003). Peripheral, non-centrosome-associated microtubules contribute to spindle formation in centrosome-containing cells. *Current Biology* **13**(21), 1894-1899.
310. Taylor II, R.M., A. Stephens, K. Bloom, L. Vicci, J. Verdaasdonk, S. Nedrud, M. Larson, M. Falvo, K. McKenzie & C. Holderman (2014). Mitotic Spindle. http://cismm.cs.unc.edu/research-collaborations/cell-mechanics-cluster/mitotic-spindle/
311. Johnson, B., R.M. Taylor II, K. Bloom, A. Stephens, L. Vicci, M. Falvo, J. Verdaasdonk, S. Nedrud, M. Larson, J. Lawrimore, J. Aicher, L. Jiang, K. McKenzie & C. Holderman (2013). Yeast Mitotic Spindle Model. https://www.youtube.com/watch?v=4E0Vh4kvyFQ
312. Smeets, M.F. & M. Segal (2002). Spindle polarity in *S. cerevisiae*: MEN can tell. *Cell Cycle* **1**(5), 308-311.
313. Wikipedia (2014). Endocytosis. http://en.wikipedia.org/wiki/Endocytosis
314. Kamasaki, T., R. Arai, M. Osumi & I. Mabuchi (2005). Directionality of F-actin cables changes during the fission yeast cell cycle. *Nature Cell Biology* **7**(9), 916-917.
315. Jones, L.A. & P.E. Sudbery (2010). Spitzenkörper, exocyst, and polarisome components in *Candida albicans* hyphae show different patterns of localization and have distinct dynamic properties. *Eukaryotic Cell* **9**(10), 1455-1465.

316. Amberg, D.C. (1998). Three-dimensional imaging of the yeast actin-cytoskeleton through the budding cell cycle. *Molecular Biology of the Cell* **9**(12), 3259-3262.
317. Tang, H., D. Laporte & D. Vavylonis (2014). Actin cable distribution and dynamics arising from cross-linking, motor pulling, and filament turnover. *Molecular Biology of the Cell* **25**(19), 3006-3016.
318. Chernyakov, I., F. Santiago-Tirado & A. Bretscher (2013). Active segregation of yeast mitochondria by Myo2 is essential and mediated by Mmrl and Ypt11. *Current Biology* **23**(18), 1818-1824.
319. Yin, H.W., D. Pruyne, T.C. Huffaker & A. Bretscher (2000). Myosin V orientates the mitotic spindle in yeast. *Nature* **406**(6799), 1013-1015.
320. Lichius, A., A. Berepiki & N.D. Read (2011). Form follows function - The versatile fungal cytoskeleton. *Fungal Biology* **115**(6), 518-540.
321. Mishra, M., J.Q. Huang & M.K. Balasubramanian (2014). The yeast actin cytoskeleton. *FEMS Microbiology Reviews* **38**(2), 213-227.
322. Bement, W.M. & G. von Dassow (2014). Single cell pattern formation and transient cytoskeletal arrays. *Current Opinion in Cell Biology* **26**(0), 51-59.
323. Martin, S.G. & R.A. Arkowitz (2014). Cell polarization in budding and fission yeasts. *FEMS Microbiology Reviews* **38**(2), 228-253.
324. Thomas, C.L., M.J. Blacketer, N.P. Edgington & A.M. Myers (2003). Assembly interdependence among the *S. cerevisiae* bud neck ring proteins Elm1p, Hsl1p and Cdc12p. *Yeast* **20**(9), 813-826.
325. Silván, U., C. Boiteux, R. Sütterlin, U. Schroeder, H.G. Mannherz, B.M. Jockusch, S. Bernèche, U. Aebi & C.-A. Schoenenberger (2012). An antiparallel actin dimer is associated with the endocytic pathway in mammalian cells. *Journal of Structural Biology* **177**(1), 70-80.
326. Schappi, J.M., A. Krbanjevic & M.M. Rasenick (2014). Tubulin, actin and heterotrimeric G proteins: Coordination of signaling and structure. *Biochimica et Biophysica Acta-Biomembranes* **1838**(2), 674-681.
327. Bidone, T.C., H. Tang & D. Vavylonis (2014). Dynamic network morphology and tension buildup in a 3D model of cytokinetic ring assembly. *Biophysical Journal* **107**(11), 2618-2628.
328. Jung, Y.-W. & M. Mascagni (2014). Constriction model of actomyosin ring for cytokinesis by fission yeast using a two-state sliding filament mechanism. *Journal of Chemical Physics* **141**(12), #125101.
329. Danielli, J.F. & E.N. Harvey (1935). The tension at the surface of mackerel egg oil, with remarks on the nature of the cell surface. *Journal of Cellular and Comparative Physiology* **5**(4), 483-494.
330. Harvey, E.N. (1936). The properties of elastic membranes with special reference to the cell surface. *Journal of Cellular and Comparative Physiology* **8**(2), 251-260.
331. Pollard, T.D. (2014). The value of mechanistic biophysical information for systems-level understanding of complex biological processes such as cytokinesis. *Biophysical Journal* **107**(11), 2499-2507.
332. Pollard, T.D. (2010). Mechanics of cytokinesis in eukaryotes. *Current Opinion in Cell Biology* **22**(1), 50-56.
333. Schwob, E. & R.P. Martin (1992). New yeast actin-like gene required late in the cell cycle. *Nature* **355**(6356), 179-182.

334. Clark, S.W. & D.I. Meyer (1994). *ACT3*: A putative centractin homolog in *Saccharomyces cerevisiae* is required for proper orientation of the mitotic spindle. *Journal of Cell Biology* **127**(1), 129-138.
335. Huang, M.E., J.L. Souciet, J.C. Chuat & F. Galibert (1996). Identification of *ACT4*, a novel essential actin-related gene in the yeast *Saccharomyces cerevisiae*. *Yeast* **12**(9), 839-848.
336. Yan, C., N. Leibowitz & R. Mélèse (1997). A role for the divergent actin gene, *ACT2*, in nuclear pore structure and function. *EMBO Journal* **16**(12), 3572-3586.
337. Subramanian, D., J. Huang, M. Sevugan, R.C. Robinson, M.K. Balasubramanian & X. Tang (2013). Insight into actin organization and function in cytokinesis from analysis of fission yeast mutants. *Genetics* **194**(2), 435-446.
338. Stroka, K.M. & H. Aranda-Espinoza (2010). A biophysical view of the interplay between mechanical forces and signaling pathways during transendothelial cell migration. *FEBS Journal* **277**(5), 1145-1158.
339. ImageJ (2006). Endothelial cells under the microscope. http://en.wikipedia.org/wiki/File:FluorescentCells.jpg
340. Dixit, R. (2012). Putting a bifunctional motor to work: insights into the role of plant KCH kinesins. *New Phytologist* **193**(3), 543-545.
341. Klotz, J. & P. Nick (2012). A novel actin-microtubule cross-linking kinesin, NtKCH, functions in cell expansion and division. *New Phytologist* **193**(3), 576-589.
342. Rodriguez, O.C., A.W. Schaefer, C.A. Mandato, P. Forscher, W.M. Bement & C.M. Waterman-Storer (2003). Conserved microtubule-actin interactions in cell movement and morphogenesis. *Nature Cell Biology* **5**(7), 599-609.
343. Waterman-Storer, C.M., J. Gregory, S.F. Parsons & E.D. Salmon (1995). Membrane/microtubule tip attachment complexes (TACs) allow the assembly dynamics of plus ends to push and pull membranes into tubulovesicular networks in interphase *Xenopus* egg extracts. *Journal of Cell Biology* **130**(5), 1161-1169.
344. Maninová, M., M.P. Iwanicki & T. Vomastek (2014). Emerging role for nuclear rotation and orientation in cell migration. *Cell Adhesion & Migration* **8**(1), 42-48.
345. Nelson, C.M. (2012). Symmetry breaking during morphogenesis in the embryo and in engineered tissues. *AIChE Journal* **58**(12), 3608-3613.
346. Naganathan, S.R., S. Fürthauer, M. Nishikawa, F. Jülicher & S.W. Grill (2014). Active torque generation by the actomyosin cell cortex drives left-right symmetry breaking. *Elife* **3**, e04165.
347. Gordon, R. (2015). Walking the tightrope: the dilemmas of hierarchical instabilities in Turing's morphogenesis [invited]. In: *The Once and Future Turing: Computing the World*. Eds.: S.B. Cooper & A. Hodges. Cambridge, Cambridge University Press: 150-164.
348. Franke, W.W., P.C. Rathke, E. Seib, M.F. Trendelenburg, M. Osborn & K. Weber (1976). Distribution and mode of arrangement of microfilamentous structures and actin in the cortex of the amphibian oocyte. *Cytobiologie* **14**(1), 111-130.

349. Sun, Q.Y. & H. Schatten (2006). Regulation of dynamic events by microfilaments during oocyte maturation and fertilization. *Reproduction* **131**(2), 193-205.

350. Lane, J. & V. Allan (1998). Microtubule-based membrane movement. *Biochimica et Biophysica Acta-Reviews on Biomembranes* **1376**(1), 27-55.

351. Chun, J.T., N. Limatola, F. Vasilev & L. Santella (2014). Early events of fertilization in sea urchin eggs are sensitive to actin-binding organic molecules. *Biochemical and Biophysical Research Communications* **450**(3), 1166-1174.

352. Elinson, R.P., M.L. King & C. Forristall (1993). Isolated vegetal cortex from *Xenopus* oocytes selectively retains localized mRNAs. *Developmental Biology* **160**(2), 554-562.

353. Iwao, Y., K. Yasumitsu, M. Narihira, J.Q. Jiang & Y. Nagahama (1997). Changes in microtubule structures during the first cell cycle of physiologically polyspermic newt eggs. *Molecular Reproduction and Development* **47**(2), 210-221.

354. Bohnsack, M.T., T. Stüven, C. Kuhn, V.C. Cordes & D. Görlich (2006). A selective block of nuclear actin export stabilizes the giant nuclei of *Xenopus* oocytes. *Nature Cell Biology* **8**(3), 257-263.

355. Imoh, H. & Y. Miyazaki (1984). Distribution of the germinal vesicle material during progesterone- induced oocyte maturation in *Xenopus* and in *Cynops*. *Development, Growth & Differentiation* **26**(2), 157-165.

356. Santacroce, M., F. Orsini, C. Perego, C. Lenardi, M. Castagna, S.A. Mari, V.F. Sacchi & G. Poletti (2006). Atomic force microscopy imaging of actin cortical cytoskeleton of *Xenopus laevis* oocyte. *Journal of Microscopy-Oxford* **223**, 57-65.

357. Gard, D.L. (2002). Confocal fluorescence microscopy of the cytoskeleton of amphibian oocytes and embryos. *Methods in Cell Biology* **70**, 379-416.

358. Gillette, R. (1944). Cell number and cell size in the ectoderm during neurulation (*Ambystoma maculatum*). *jJournal of Experimental Zoology* **96**, 201-222.

359. Desnitskii, A.G. (1975). Cell cycles in the early development of axolotl. *Soviet Journal of Developmental Biology* **5**(3), 248-250.

360. Desnitski, A.G. (1992). Changes in the cell cycles during early embryogenesis of the Mexican axolotl (*Ambystoma mexicanum*). *Axolotl Newsletter (Indiana University Axolotl Colony, Bloomington, Indiana)*(21), 20-22.

361. Inui, M., M. Montagner & S. Piccolo (2012). miRNAs and morphogen gradients. *Current Opinion in Cell Biology* **24**(2), 194-201.

362. Stappert, L., B. Roese-Koerner & O. Brüstle (2015). The role of microRNAs in human neural stem cells, neuronal differentiation and subtype specification. *Cell and Tissue Research* **359**(1), 47-64.

363. Sindelka, R., J. Jonák, R. Hands, S.A. Bustin & M. Kubista (2008). Intracellular expression profiles measured by real-time PCR tomography in the *Xenopus laevis* oocyte. *Nucleic Acids Research* **36**(2), 387-392.

364. Lipshitz, H.D. (2009). Follow the mRNA: a new model for Bicoid gradient formation. *Nature Reviews Molecular Cell Biology* **10**(8), 509-512.

365. Sindelka, R., M. Sidova, D. Svec & M. Kubista (2010). Spatial expression profiles in the *Xenopus laevis* oocytes measured with qPCR tomography. *Methods* **51**(1), 87-91.

366. Browder, L.W. (1996). Advanced Developmental Biology: Oocyte Polarization and mRNA Localization (*Xenopus*). http://people.ucalgary.ca/~browder/Polariz_Xenopus.html

367. Flint, R.W., R. Gordon, C.C. Martin & G.W. Brodland (1989). Simulation of the inversion of amphibian eggs in a gravitational field using hollow glass spheres. In: *Microgravity as a Tool in Developmental Biology, Selected papers from a special ESA Symposium held during the 11th International Congress of the International Society of Developmental Biologists, August 20-25, 1989, Utrecht*. Eds.: G.A. Ubbels, H. Oser & T.D. Guyenne. Paris Cedex, European Space Agency: 81-83.

368. Nouri, C., R. Luppes, A.E.P. Veldman, J.A. Tuszynski & R. Gordon (2008). Rayleigh instability of the inverted one-cell amphibian embryo. *Physical Biology* **5**(1), #015006.

369. Neff, A.W., M. Wakahara, A. Jurand & G.M. Malacinski (1984). Experimental analyses of cytoplasmic rearrangements which follow fertilization and accompany symmetrization of inverted *Xenopus* eggs. *Journal of Embryology and Experimental Morphology* **80**, 197-224.

370. Schatten, G. (1994). The centrosome and its mode of inheritance: The reduction of the centrosome during gametogenesis and its restoration during fertilization. *Developmental Biology* **165**(2), 299-335.

371. Terada, Y., G. Schatten, H. Hasegawa & N. Yaegashi (2010). Essential roles of the sperm centrosome in human fertilization: Developing the therapy for fertilization failure due to sperm centrosomal dysfunction. *Tohoku Journal of Experimental Medicine* **220**(4), 247-258.

372. Staveley, B.E. (2015). Molecular & Developmental Biology (BIOL3530): Vertebrate Development II: Axes and Germ Layers: *Xenopus* animal/vegetal axis: localized mRNAs. http://www.mun.ca/biology/desmid/brian/BIOL3530/DEVO_04/devo_04.html

373. Schroeder, K.E., N.L. Condic, L.M. Eisenberg & H.J. Yost (1999). Spatially regulated translation in embryos: Asymmetric expression of maternal Wnt-11 along the dorsal-ventral axis in *Xenopus*. *Developmental Biology* **214**(2), 288-297.

374. Kloc, M., C. Larabell & L.D. Etkin (1996). Confocal micrograph of a live stage 1 *Xenopus laevis* oocyte injected with Texas red-labeled *Xlsirt* mRNA (red) and counterstained for mitochondria (green), 36 hours post-injection. *Xlsirt* mRNA is localized in the disc at the vegetal pole (bottom). http://people.ucalgary.ca/~browder/Polariz_Xenopus.html

375. Iwao, Y. (2000). Mechanisms of egg activation and polyspermy block in amphibians and comparative aspects with fertilization in other vertebrates. *Zoological Science* **17**(6), 699-709.

376. Eisthen, H.L. (1989). Courtship and mating behavior in the axolotl. *Axolotl Newsletter*(18), 18-19.

377. Björklund, N.K. (1993). Small is beautiful: economical axolotl colony maintenance with natural spawnings as if axolotls mattered. In: *Handbook on Practical Methods*. Eds.: G.M. Malacinski & S.T. Duhon. Bloomington, Department of Biology, Indiana University: 38-47.

378. Mah, K. & Y.M. Binik (2001). The nature of human orgasm: A critical review of major trends. *Clinical Psychology Review* **21**(6), 823-856.

379. Roulin, A., P.L. Auer, M. Libault, J. Schlueter, A. Farmer, G. May, G. Stacey, R.W. Doerge & S.A. Jackson (2012). The fate of duplicated genes in a polyploid plant genome. *Plant Journal* **73**(1), 143-153.

380. Soltis, D.E., C.J. Visger & P.S. Soltis (2014). The polyploidy revolution then…and now: Stebbins revisited. *American Journal of Botany* **101**(7), 1057-1078.

381. Humphrey, R.R. & G. Fankhauser (1949). Three generations of polyploids in ambystomid salamanders. *Journal of Heredity* **40**, 7-12.

382. Fankhauser, G. (1945). Maintenance of normal structure in heteroploid salamander larvae, through compensation of changes in cell size by adjustment of cell number and cell shape. *Journal of Experimental Zoology* **100**(3), 445-455.

383. Luppes, R., B. Düz, H.J.L. van der Heiden, P. van der Plas & A.E.P. Veldman (2012). Numerical simulations of two-phase flow with ComFLOW: past and recent developments. In: *Proceedings of the 6th European Congress on Computational Methods in Applied Sciences and Engineering (ECCOMAS 2012), September 10-14, 2012, Vienna, Austria [CD-ROM]*. Eds.: J. Eberhardsteiner, H.J. Boehm & F.G. Rammerstorfer. Austria, Vienna University of Technology: http://www.math.rug.nl/~veldman/preprints/LuppesEtal-ECCOMAS2012.pdf.

384. Humphrey, R.R. (1963). Polyploidy in the Mexican axolotl (*Ambystoma mexicanum*) resulting from multinucleate ova. *Proceedings of the National Academy of Sciences USA* **50**(6), 1122-1127.

385. Fankhauser, G., J.A. Vernon, W.H. Frank & W.V. Slack (1955). Effect of size and number of brain cells on learning in larvae of the salamander, *Triturus viridescens*. *Science* **122**(3172), 692-693.

386. Englert, Y., F. Puissant, M. Camus, M. Degueldre & F. Leroy (1986). Factors leading to tripronucleate eggs during human in-vitro fertilization. *Human Reproduction* **1**(2), 117-119.

387. Kang, H.-J. & Z. Rosenwaks (2008). Triploidy - the breakdown of monogamy between sperm and egg. *International Journal of Developmental Biology* **52**(5-6), 449-454.

388. Fankhauser, G. (1945). The effect of changes in chromosome number on amphibian development. *Quarterly Review of Biology* **20**(1), 20-78.

389. Dale, B. & L. DeFelice (2011). Polyspermy prevention: facts and artifacts? *Journal of Assisted Reproduction and Genetics* **28**(3), 199-207.

390. Mio, Y., K. Iwata, K. Yumoto, Y. Kai, H.C. Sargant, C. Mizoguchi, M. Ueda, Y. Tsuchie, A. Imajo, Y. Iba & K. Nishikori (2012). Possible mechanism of polyspermy block in human oocytes observed by time-lapse cinematography. *Journal of Assisted Reproduction and Genetics* **29**(9), 951-956.

391. Dale, B. (2014). Is the idea of a fast block to polyspermy based on artifact? *Biochemical and Biophysical Research Communications* **450**(3), 1159-1165.

392. Vacquier, V.D. (2012). The quest for the sea urchin egg receptor for sperm. *Biochemical and Biophysical Research Communications* **425**(3), 583-587.

393. Kozlovsky, P. & A. Gefen (2013). Sperm penetration to the zona pellucida of an oocyte: a computational model incorporating acrosome reaction. *Computer Methods in Biomechanics and Biomedical Engineering* **16**(10), 1106-1111.

394. Hirohashi, N., N. Kamei, H. Kubo, H. Sawada, M. Matsumoto & M. Hoshi (2008). Egg and sperm recognition systems during fertilization. *Development Growth & Differentiation* **50**(Suppl 1), S221-S238.
395. McCulloh, D.H. & E.L. Chambers (1992). Fusion of membranes during fertilization. Increases of the sea urchin egg's membrane capacitance and membrane conductance at the site of contact with the sperm. *Journal of General Physiology* **99**(2), 137-175.
396. Ramalho-Santos, J., G. Schatten & R.D. Moreno (2002). Control of membrane fusion during spermiogenesis and the acrosome reaction. *Biology of Reproduction* **67**(4), 1043-1051.
397. Puppo, A., J.T. Chun, G. Gragnaniello, E. Garante & L. Santella (2008). Alteration of the cortical actin cytoskeleton deregulates Ca^{2+} signaling, monospermic fertilization, and sperm entry [with Nikon 5th Small World in Motion movie: http://journals.plos.org/plosone/article/asset?unique&id=info:doi/10.1371/journal.pone.0003588.s001]. *PLoS One* **3**(10), e3588.
398. Takeichi, T. & H.Y. Kubota (1984). Structural basis of the activation wave in the egg of *Xenopus laevis*. *Journal of Embryology and Experimental Morphology* **81**(JUN), 1-16.
399. Busa, W.B. & R. Nuccitelli (1985). An elevated free cytosolic Ca^{2+} wave follows fertilization in eggs of the frog, *Xenopus laevis*. *Journal of Cell Biology* **100**(4), 1325-1329.
400. Just, E.E. (1939). *The Biology of the Cell Surface*. Philadelphia, Blakiston.
401. Byrnes, W.M. & S.A. Newman (2014). Ernest Everett Just: Egg and embryo as excitable systems. *Journal of Experimental Zoology Part B: Molecular and Developmental Evolution* **322**(4), 191-201.
402. Gilbert, S.F. (1988). Cellular politics: Ernest Everett Just, Richard B. Goldschmidt, and the attempt to reconcile embryology and genetics. In: *The American Development of Biology; Friday Harbor Conference, Seattle, Washington, USA, September 1986*. Eds.: R. Rainger, K.R. Benson & J. Maienschein. Philadelphia, University of Pennsylvania Press: 311-346.
403. Byrnes, W.M. (2009). Ernest Everett Just, Johannes Holtfreter, and the origin of certain concepts in embryo morphogenesis. *Molecular Reproduction and Development* **76**(10), 912-921.
404. Iwao, Y., T. Murakawa, J. Yamaguchi & M. Yamashita (2002). Localization of γ-tubulin and cyclin B during early cleavage in physiologically polyspermic newt eggs. *Development Growth & Differentiation* **44**(6), 489-499.
405. Lim, D., E. Ercolano, K. Kyozuka, G.A. Nusco, F. Moccia, K. Lange & L. Santella (2003). The M-phase-promoting factor modulates the sensitivity of the Ca^{2+} stores to inositol 1,4,5-trisphosphate via the actin cytoskeleton. *Journal of Biological Chemistry* **278**(43), 42505-42514.
406. Santella, L., D. Lim & F. Moccia (2004). Calcium and fertilization: the beginning of life. *Trends in Biochemical Sciences* **29**(8), 400-407.
407. Fontanilla, R.A. & R. Nuccitelli (1998). Characterization of the sperm-induced calcium wave in *Xenopus* eggs using confocal microscopy. *Biophysical Journal* **75**(4), 2079-2087.
408. Stricker, S.A. (1999). Comparative biology of calcium signaling during fertilization and egg activation in animals. *Developmental Biology* **211**(2), 157-176.

409. Takahashi, T., T. Kikuchi, Y. Kidokoro & H. Shirakawa (2013). Ca^{2+} influx-dependent refilling of intracellular Ca^{2+} stores determines the frequency of Ca^{2+} oscillations in fertilized mouse eggs. *Biochemical and Biophysical Research Communications* **430**(1), 60-65.
410. Chun, J.T. & L. Santella (2009). Roles of the actin-binding proteins in intracellular Ca^{2+} signalling. *Acta Physiologica (Oxford)* **195**(1), 61-70.
411. Stricker, S.A. (1996). Repetitive calcium waves induced by fertilization in the nemertean worm *Cerebratulus lacteus*. *Developmental Biology* **176**(2), 243-263.
412. Bukowska, D., B. Kempisty, H. Piotrowska, P. Sosinska, M. Wozna, S. Ciesiolka, P. Antosik, J.M. Jaskowski, K.P. Bruessow & M. Nowicki (2013). The structure and role of mammalian sperm RNA: a review. *Veterinarni Medicina* **58**(2), 57-64.
413. Schatten, G. & H. Schatten (1981). Effects of motility inhibitors during sea urchin fertilization: microfilament inhibitors prevent sperm incorporation and restructuring of fertilized egg cortex, whereas microtubule inhibitors prevent pronuclear migrations. *Experimental Cell Research* **135**(2), 311-330.
414. Swope, R.E. & D.L. Kropf (1993). Pronuclear positioning and migration during fertilization in *Pelvetia*. *Developmental Biology* **157**(1), 269-276.
415. Nguyen, P.A., K. Ishihara, M. Wuhr & T.J. Mitchison (2012). Pronuclear migration: No attachment? No union, but a futile cycle! *Current Biology* **22**(10), R409-R411.
416. Reinsch, S. & E. Karsenti (1997). Movement of nuclei along microtubules in *Xenopus* egg extracts. *Current Biology* **7**(3), 211-214.
417. Whitaker, M. (1996). Control of meiotic arrest. *Reviews of Reproduction* **1**(2), 127-135.
418. Jeff, H. (2002). Dynamics of Development: Amphibians: Index: Fertilization/Rotation: Rotation: Gray Crescent schematic: Subcortical Rotation and the "Gray Crescent" [movie: spotmove.MOV]. http://worms.zoology.wisc.edu/embryology_main.html; http://worms.zoology.wisc.edu/frogs/fert/subcortrot/grescres.html
419. Houliston, E. & R.P. Elinson (1991). Evidence for the involvement of microtubules, ER, and kinesin in the cortical rotation of fertilized frog eggs. *Journal of Cell Biology* **114**(5), 1017-1028.
420. Weaver, C., G.H. Farr, 3rd, W. Pan, B.A. Rowning, J. Wang, J. Mao, D. Wu, L. Li, C.A. Larabell & D. Kimelman (2003). GBP binds kinesin light chain and translocates during cortical rotation in *Xenopus* eggs. *Development* **130**(22), 5425-5436.
421. Nouri, C., J.A. Tuszynski, M. Wiebe & R. Gordon (2012). Simulation of the effects of microtubules in the cortical rotation of amphibian embryos in normal and zero gravity. *BioSystems* **109**(3, Special Issue on Biological Morphogenesis), 444-449.
422. Tuszynski, J.A. & R. Gordon (2012). A mean field Ising model for cortical rotation in amphibian one-cell stage embryos. *BioSystems* **109**(3, Special Issue on Biological Morphogenesis), 381-389.
423. Jarosch, R. (1964). Screw-mechanical basis of protoplasmic movement. In: *Primitive Motile Systems in Cell Biology*. Eds.: R.D. Allen & N. Kamiya. New York, Academic Press: 599-622.

424. Jarosch, R. (1972). The participation of rotating fibrils in biological movements. *Acta Protozoologica* **11**(2), 23-40.
425. Walter, J.C., M. Laleman, M. Baiesi & E. Carlon (2014). Rotational dynamics of entangled polymers. *European Physical Journal-Special Topics* **223**(14), 3201-3213.
426. Brunnbauer, M., R. Dombi, T.-H. Ho, M. Schliwa, M. Rief & Z. Oekten (2012). Torque generation of kinesin motors is governed by the stability of the neck domain. *Molecular Cell* **46**(2), 147-158.
427. Larabell, C.A., B.A. Rowning, J. Wells, M. Wu & J.C. Gerhart (1996). Confocal microscopy analysis of living *Xenopus* eggs and the mechanism of cortical rotation. *Development* **122**(4), 1281-1289.
428. Ussing, A.P., R. Gordon, L. Ector, K. Buczkó, A.G. Desnitskiy & S.L. VanLandingham (2005). *The Colonial Diatom "Bacillaria paradoxa": Chaotic Gliding Motility, Lindenmeyer Model of Colonial Morphogenesis, and Bibliography, with Translation of O.F. Müller (1783), "About a peculiar being in the beach-water"*. Ruggell, A.R.G. Ganter Verlag Kommanditgesellschaft. **5**.
429. Weaver, C. & D. Kimelman (2004). Move it or lose it: axis specification in *Xenopus*. *Development* **131**(15), 3491-3499, doi: 3410.1242/dev.01284.
430. Rowning, B.A. & J.C. Gerhart (1990). Microtubule mediated cortical rotation and axis specification in the *Xenopus* egg. In: *Developmental Biology*. Eds.: E.H. Davidson, J.V. Ruderman & J.W. Posakony, Wiley-Liss: 61-78.
431. Milsum, J.H., Ed. (1968). *Positive Feedback, A General Systems Approach to Positive/Negative Feedback and Mutual Causality*. Oxford, Pergamon Press.
432. Cha, B.J. & D.L. Gard (1999). XMAP230 is required for the organization of cortical microtubules and patterning of the dorsoventral axis in fertilized *Xenopus* eggs. *Developmental Biology* **205**(2), 275-286.
433. Houliston, E. & R.P. Elinson (1991). Patterns of microtubule polymerization relating to cortical rotation in *Xenopus laevis* eggs. *Development* **112**(1), 107-117.
434. Rowning, B.A., J. Wells, M. Wu, J.C. Gerhart, R.T. Moon & C.A. Larabell (1997). Microtubule-mediated transport of organelles and localization of β-catenin to the future dorsal side of *Xenopus* eggs. *Proceedings of the National Academy of Sciences USA* **94**(4), 1224-1229.
435. Wang, H.F., I. Brust-Mascher & J.M. Scholey (2014). Sliding filaments and mitotic spindle organization. *Nature Cell Biology* **16**(8), 737-739.
436. Milo, R., R. Phillips & N. Orme (2015). *Cell Biology by the Numbers*, Garland Science.
437. Milo, R., R. Phillips & N. Orme (2015). Cell Biology by the Numbers: How fast do molecular motors move on cytoskeletal filaments? , http://book.bionumbers.org/how-fast-do-molecular-motors-move-on-cytoskeletal-filaments/
438. Le Clainche, C. & M.-F. Carlier (2008). Regulation of actin assembly associated with protrusion and adhesion in cell migration. *Physiological Reviews* **88**(2), 489-513.
439. MBInfo (2015). Actin filament polymerization generates protrusive force. http://www.mechanobio.info/topics/cytoskeleton-dynamics/go-0005856/go-0005884 - 5_go-0005884

440. Gerhart, J.C., M. Danilchik, T. Doniach, S. Roberts, B. Rowning & R. Stewart (1989). Cortical rotation of the *Xenopus* egg: consequences for the anteroposterior pattern of embryonic dorsal development. *Development* **107**(Suppl.), 37-51.

441. Tee, Y.H., T. Shemesh, V. Thiagarajan, R.F. Hariadi, K.L. Anderson, C. Page, N. Volkmann, D. Hanein, S. Sivaramakrishnan, M.M. Kozlov & A.D. Bershadsky (2015). Cellular chirality arising from the self-organization of the actin cytoskeleton. *Nature Cell Biology* **17**(4), 445-457.

442. Clarke, E.J. & V.J. Allan (2003). Cytokeratin intermediate filament organisation and dynamics in the vegetal cortex of living *Xenopus laevis* oocytes and eggs. *Cell Motility and the Cytoskeleton* **56**(1), 13-26.

443. Danilchik, M.V., E.E. Brown & K. Riegert (2006). Intrinsic chiral properties of the *Xenopus* egg cortex: an early indicator of left-right asymmetry? *Development* **133**(22), 4517-4526.

444. Kimball, J.W. (2013). The Cytoskeleton: Microtubules. http://users.rcn.com/jkimball.ma.ultranet/BiologyPages/C/Cytoskeleton.html - microtubules

445. Gordon, R. (2009). Microtubules and Left/Right Asymmetry of Amphibian Embryos [International Embryo Physics Course, PowerPoint presentation, Second Life®, April 2, 2009]. http://embryogenesisexplained.com/files/presentations/Gordon2009Microtubules.pdf

446. Kuroda, K. (1964). Behavior of naked cytoplasmic drops isolated from plant cells. In: *Primitive Motile Systems in Cell Biology*. Eds.: R.D. Allen & N. Kamiya. New York, Academic Press: 31-41.

447. Kachar, B., R. Urrutia, M.N. Rovolta & M.A. McNiven (1993). Myosin-mediated vesicular transport in the extruded cytoplasm of characean algae cells. *Methods in Cell Biology* **39**, 179-190.

448. Crisona, N.J. & N.R. Cozzarelli (2006). Alteration of *Escherichia coli* topoisomerase IV conformation upon enzyme binding to positively supercoiled DNA. *Journal of Biological Chemistry* **281**(28), 18927-18932.

449. Tsutsui, K.M., K. Sano, O. Hosoya, T. Miyamoto & K. Tsutsui (2011). Nuclear protein LEDGF/p75 recognizes supercoiled DNA by a novel DNA-binding domain. *Nucleic Acids Research* **39**(12), 5067-5081.

450. Brázdová, M., L. Navrátilová, V. Tichý, K. Němcová, M. Lexa, R. Hrstka, P. Pečinka, M. Adámik, B. Vojtesek, E. Paleček, W. Deppert & M. Fojta (2013). Preferential binding of hot spot mutant p53 proteins to supercoiled DNA in vitro and in cells. *PLoS One* **8**(3), e59567.

451. D'Annessa, I., A. Coletta, T. Sutthibutpong, J. Mitchell, G. Chillemi, S. Harris & A. Desideri (2014). Simulations of DNA topoisomerase 1B bound to supercoiled DNA reveal changes in the flexibility pattern of the enzyme and a secondary protein-DNA binding site. *Nucleic Acids Research* **42**(14), 9304-9312.

452. Vandenberg, L.N., J.M. Lemire & M. Levin (2013). It's never too early to get it Right: A conserved role for the cytoskeleton in left-right asymmetry. *Communicative & Integrative Biology* **6**(6), e27155.

453. Vandenberg, L.N. & M. Levin (2013). A unified model for left-right asymmetry? Comparison and synthesis of molecular models of embryonic laterality. *Developmental Biology* **379**(1), 1-15.

454. Tyler, S.E.B. & S.J. Kimber (2006). The dynamic nature of mollusc egg surface architecture and its relation to the microtubule network. *International Journal of Developmental Biology* **50**(4), 405-412.
455. Vogt, W. (1929). Gestaltungsanalyse am Amphibienkeim mit Örtlicher Vitalfärbung II. Teil. Gastrulation und Mesodermbildung bei Urodelen und Anuren/Analysis of the organization of the amphibian embryo with local vital dyes. Part II: Gastrulation and mesoderm development in urodeles and anurans. *Wilhelm Roux Archiv für Entwicklungsmechanik der Organismen* **120**(1), 384-706.
456. Méhes, E. & T. Vicsek (2014). Collective motion of cells: from experiments to models. *Integrative Biology (Cambridge)* **6**(9), 831-854.
457. Holtfreter, J. (1933). Die totale Exogastrulation, eine Selbstablösung des Ektoderms vom Entomesoderm. Entwicklung und funktionelles Verhalten nervenloser Organe [Total exogastrulation, an auto-dissociation of ectoderm from endomesoderm. Development and functional behavior of nerveless organs]. *Wilhelm Roux Archiv für Entwicklungsmechanik der Organismen* **129**(4), 669-793.
458. Holtfreter, J. & V. Hamburger (1955). Embryogenesis: progressive differentiation. Amphibians. In: *Analysis of Development*. Eds.: B.H. Willier, P.A. Weiss & V. Hamburger. Philadelphia, W.B. Saunders Co.: 230-296.
459. Danilchik, M.V., E.E. Brown & K. Riegert (2006). Normal first cleavage in *Xenopus* is chiral and not mirror-image symmetric. http://www.xenbase.org/common/showImage.do?imageId=75368
460. WikiHow (2015). How to Make a Water Bottle Cap Pop off with Air Pressure. http://www.wikihow.com/Make-a-Water-Bottle-Cap-Pop-off-with-Air-Pressure
461. Paleček, J., G.A. Ubbels & K. Rzehak (1978). Changes of the external and internal pigment pattern upon fertilization in the egg of *Xenopus laevis*. *J Embryol Exp Morphol* **45**, 203-214.
462. Wühr, M., E.S. Tan, S.K. Parker, H.W. Detrich, III & T.J. Mitchison (2010). A model for cleavage plane determination in early amphibian and fish embryos. *Current Biology* **20**(22), 2040-2045.
463. Kubota, T. (1966). Studies of the leavage in the frog egg. I. On the temporal relation between furrow determination and nuclear division. *Journal of Experimental Biology* **44**(3), 545-552.
464. Kubota, T. (1969). Studies of the cleavage in the frog egg. II. On determination of the position of the furrow. *Journal of Embryology and Experimental Morphology* **21**, 119-129.
465. Stepanek, L. & G. Pigino (2016). Microtubule doublets are double-track railways for intraflagellar transport trains. *Science* **352**(6286), 721-724.
466. Szabó, A., R. Ünnep, E. Méhes, W.O. Twal, W.S. Argraves, Y. Cao & A. Czirók (2010). Collective cell motion in endothelial monolayers. *Physical Biology* **7**(4), #046007.
467. Bystricky, K., P. Heun, L. Gehlen, J. Langowski & S.M. Gasser (2004). Long-range compaction and flexibility of interphase chromatin in budding yeast analyzed by high-resolution imaging techniques. *Proceedings of the National Academy of Sciences of the USA* **101**(47), 16495-16500.

Chapter 6

The Cell State Splitter and Differentiation Waves

As soon as the fertilized egg completes the first cleavage, it is no longer a single cell. For simple organisms like the nematode, determination begins with that first cleavage. The two daughters of the divided fertilized nematode egg are each determined during division and they become different cell types. The division is uneven, one large cell and one small cell. The major difference between yeast budding and the embryology of the nematode is that instead of the bud moving off to form an entirely separate organism, the determined cells of the nematode remain attached and as cell division progresses, determination occurs with each asymmetric division to eventually form a complete multi-cellular organism. The nematode *C. elegans* has 959 cells in the adult hermaphrodite, 1031 in adult males[1]. All are unique except for 50 pairs of like cells, with each pair different from all the other pairs[2] (Figure 2.15). In the nematode, determination is absolutely linked to cell division. In embryology the nematode is referred to as a *mosaic* organism. The term mosaic in this context is entirely unrelated to the mosaicism of females due to X inactivation (Chapter 4). If an embryologist cuts an early mosaic embryo in half, each half fails to develop properly as a whole organism, and that defines mosaicism.

The axolotl embryo first becomes a multicellular organism via rapid proliferative cell division without determination. In fact the early cleavages go directly from anaphase right into S2 phase without pausing to synthesize new RNA or proteins. Only after there is a ball of cells, will determination of groups of cells to form new tissues start. Before determination begins, each cell is totipotent, meaning it can form a whole new organism. If the embryologist cuts the early ball of amphibian cells in half, development proceeds normally with two identical larvae, although they will be smaller in size. They catch up in size later. In embryological terms this type of embryo is called a *regulating* embryo.

We humans are regulating organisms as well. We are reminded of that fact every time we see identical twins. We will revisit this distinction between mosaic and regulating embryos in Chapter 7. It has many exceptions[3-8]. Many animals exhibit characteristics of both types of embryogenesis (§2.03 in *Hierarchical Genome*[9]), suggesting that there is a common explanation.

A two dimensional tissue containing multiple cells of the same differentiated type all connected together is called an epithelium. The formation of an epithelium represented a huge leap in evolution. It was the leap required before highly complex embryos growing to large adult organisms could be formed. It was an essential step in the creation of multicellularity beyond simple clonal organisms like the cyanobacterium *Anabaena* (Figure 1.28) and the small mosaic animals such as the nematode *C. elegans* that exist today. Epithelia date back at least to the sponges[10]. The innovation of epithelial cell sheets with different functions and identities required decoupling cell division from cell determination. And so individual epithelial cells have one cytoskeletal structure, the spindle apparatus, for cell division and then this structure (we presume) was duplicated and diverged so that a second apparatus took over the determination function of embryogenesis in regulating embryos. We call this second structure the *cell state splitter*.

Like the polarisome of the yeast cytoskeleton during bud formation (Figures 5.46 and 6.1), the cell state splitter is located at one end of each cell. Like the spindle apparatus, it is a transient cytoskeletal structure, that is not membrane enclosed. The same components, microtubules, microfilaments and intermediate filaments are present. It is one of the acts of the cytoskeletal dance troupe. The difference is that, unlike the polarisome, which is disassembled after mitosis and only functions during cell division, the cell state splitter is not tied to cell division. The activation of the cell state splitter in the epithelial sheet can change the cell's differon without requiring a matching round of cell division. In other words, it allows an amitotic cell differentiation (Figure 6.2).

We first discovered the cell state splitter in the ectoderm of the axolotl[11]. It has the three components: a microfilament ring[12,13]; a mat of microtubules at and parallel to the apical surface; and an intermediate filament ring (Figure 6.3). All three can be seen in electron micrographs of the apical ends of neuroepithelium and epidermal cells (Figure 6.4). As in all

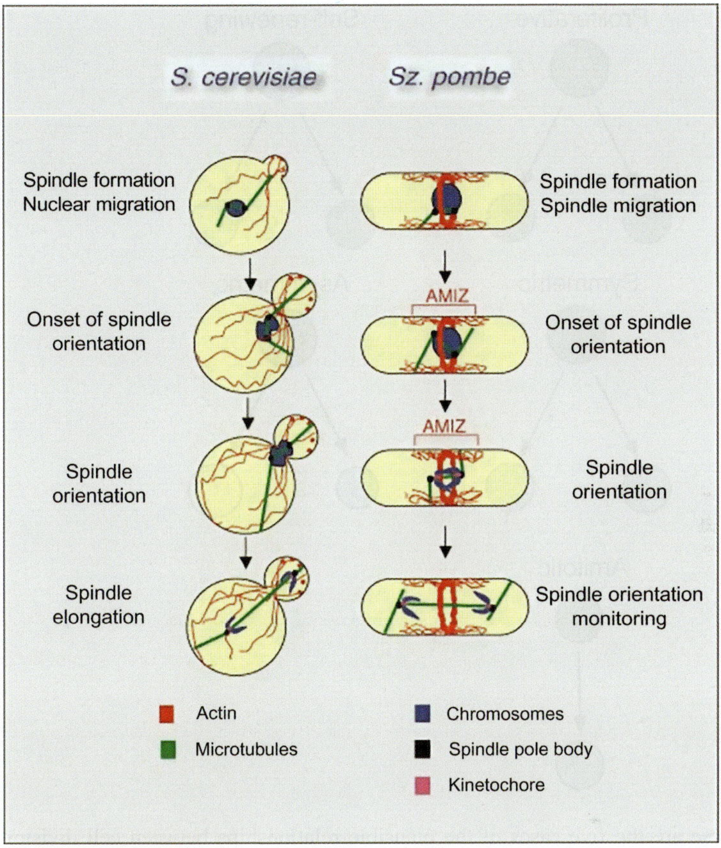

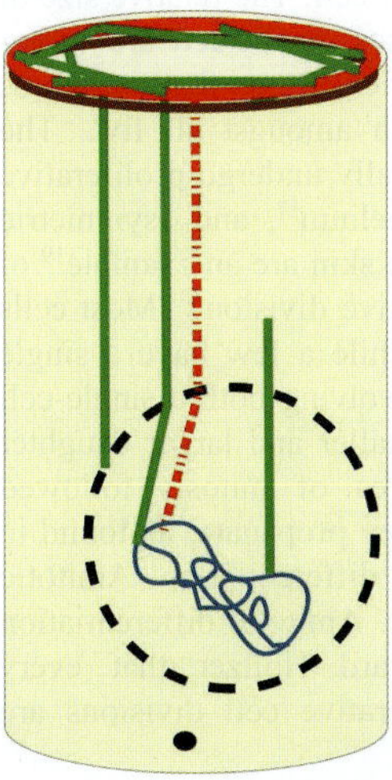

Figure 6.1. **Top**: Cytoskeletal elements of budding yeast (left, showing formation of the apparatus growing out from polarisomes = red dots) and fission yeast (right). AMIZ = astral microtubule interaction zone. From Gachet *et al.* (2006)[14] with permission of John Wiley and Sons. Yeast cell division can include a form of differentiation whereby the yeast bud will express different genes from the mother cell. **Bottom**: Cell sheet differentiation requires a cell state splitter on the apical (top) surface which is analogous in many ways. Both apparatuses have actin and microtubules acting in opposition. Both result in changes in gene expression. The cell state splitter has been disconnected from cell division allowing the cell to differentiate without requiring cell division. Color labels are the same. The intermediate filament component of the cell state splitter is in brown, the nuclear membrane is dashed in black.

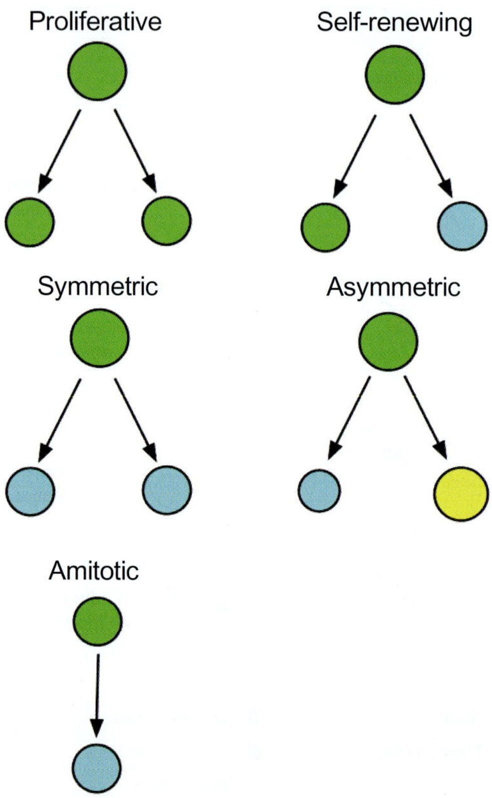

Figure 6.2. These are the five cases of the plausible relationships between cell division and cell differentiation. Each color represents a different kind of cell. The relative size of the mother cell is shown just before the event. The three cases on the left are "color symmetric", whereas the two on the right are "color asymmetric"[15,16], so that the words "symmetric" and "asymmetric" are insufficient to distinguish amongst all five. The asymmetric case is also size asymmetric. Epithelial cells generally undergo proliferative divisions if the daughter cells stay in the plane of the epithelium[17], and asymmetric divisions if one cell ends up below the sheet[18]. Keratinocytes in skin are an example[19] of the self-renewing case, though they can also undergo proliferative division[20]. Most cells in the nematode *C. elegans* undergo asymmetric divisions, while a few have a single proliferative division[21]. We refer to the asymmetric case as involving both a single-cell contraction wave and a single-cell expansion wave for the smaller and larger daughter cells, respectively (§4.06 in *Hierarchical Genome*[9]). Waves of mitosis followed immediately by differentiation of the tissues through which they propagate, as found in *Drosophila*[22,23] and *Xenopus*[24,25], are examples of symmetric differentiation. Amitotic differentiation may occur in multicellular differentiation waves. Amitotic differentiation is contrary to the "quantal mitosis" hypothesis of Howard Holtzer that every differentiation requires a preceding cell division[26-29]. Proliferative cell divisions are frequently referred to as "vegetative". Cf.[30,31].

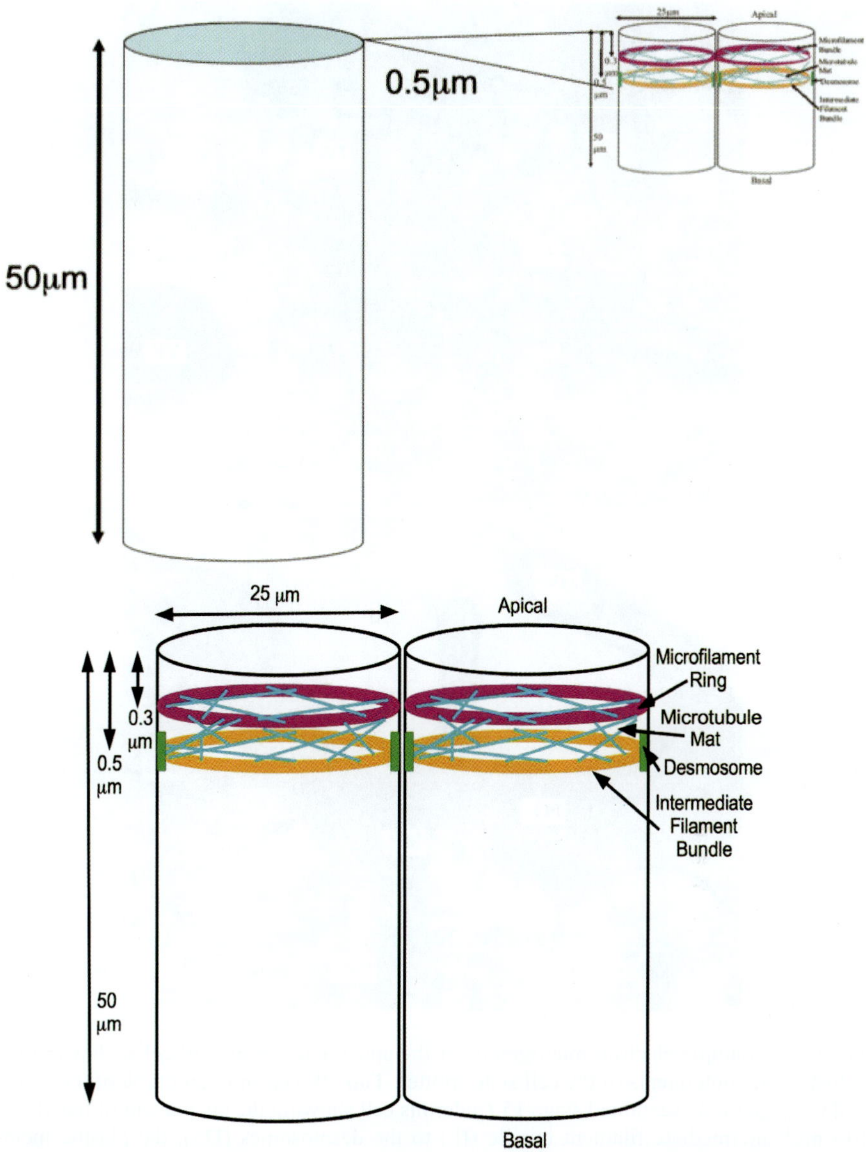

Figure 6.3. The cell state splitter is an organelle predicted[11] and then at first observed at the apical end of ectoderm cells in axolotl embryos at early gastrulation[32]. It consists of an upper microfilament ring with an intermediate filament ring below it, subtended by a mat of microtubules. As shown below, the cell state splitter occupies only 1% of the height of an ectoderm cell. (Below and in the inset, the vertical and horizontal scales are compressed, as 50 μm = 100x0.5 μm.)

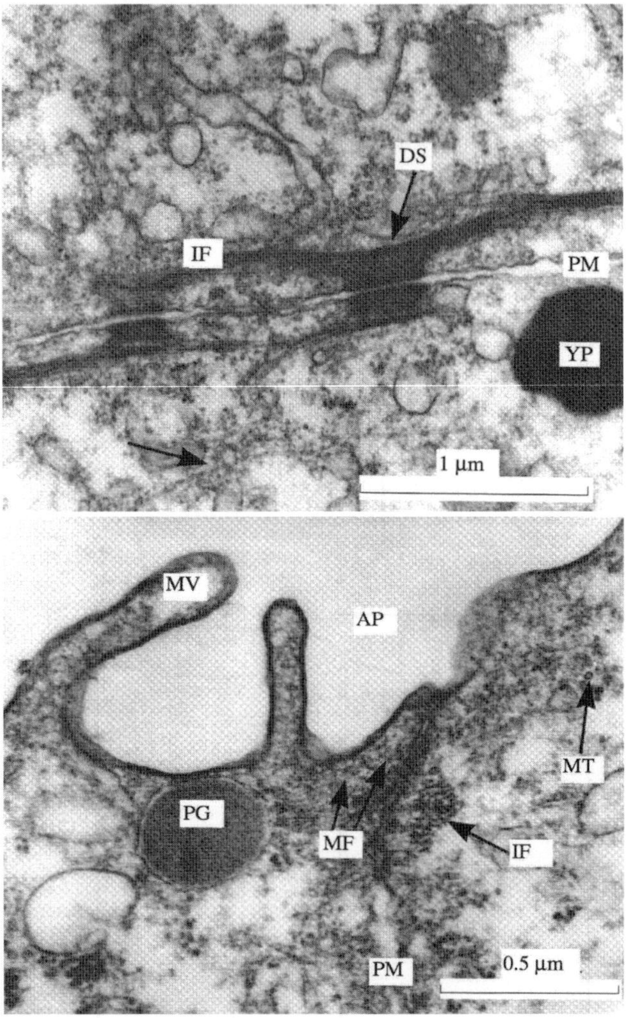

Figure 6.4. Sample electron micrographs of the apical ends of axolotl cells, showing the cytoskeletal components of the cell state splitter. **Top**: "Electron micrograph of the apical end of tangentially sectioned stage 15 epidermis cell showing the attachment of the 100 Å [0.1 nm] intermediate filament bundle (IF) to the desmosomes (DS), the plasma membrane (PM), a yolk platelet (YP) and one of many apical microtubules (unmarked arrow), some of which are bending around the membrane." A similar electron micrograph was taken by Beth Burnside[12]. **Bottom**: "Electron micrograph of the apical end of a radially sectioned stage 15 neural plate cell showing the apical surface (AP), microvilli (MV), a pigment granule (PG), the plasma membrane (PM), microtubule (MT), microfilaments (MF) and intermediate filament bundles (IF)." From Martin & Gordon (1997)[32] with kind permission of Sergei G. Vassetzky.

epithelia, the cells in the sheet are also connected by rivet-like structures near the apical end (called tight junctions and adherens junctions). The mat of microtubules may have first been observed by Beth Burnside in the California newt neuroepithelium and epidermis[12,33], when she was a graduate student with Antone G. Jacobson (Figures 6.5 and 6.6). Using transmission electron microscopy in an attempt to repeat and confirm Burnside's work, Chris Martin discovered that the cell state splitter includes the intermediate filament ring and he also showed that the microtubule mat is predominately along the periphery of the cell between the microfilament and intermediate filament rings[32], both at Stage 11 (Figure 6.7) and Stage 15 (Figure 6.8), in ectoderm cells before and after they become epidermal or neural plate cells. The structure is similar to the well documented microfilament and microtubule rings on the inner surface of the cell membrane of many nonmammalian, nucleated blood cells[34-36]. Mats of microtubules at the apical ends of cells, ringed by microfilaments, are found in various epithelia (Figure 6.9). There are models for apical constriction of epithelial cells (Figure 6.10).

Unlike the yeast microfilament ring, the microfilament ring in the cell state splitter does not remain the same thickness as it contracts. Instead the actin ring thickens (cf.[37]). We can infer then that it is not disassembled but instead becomes ever more tightly overlapping in a thickening and strengthening actin cable formation driven[38] by myosin 2[11]. Because the actin ring thickens this way, it is able to exert an increasing amount of force as it contracts (Figure 6.11). Conversely, the force will decrease if the cell's top (apical) diameter becomes larger and the ring is thinned.

The actin ring force is always directed inward, the contraction narrowing the apical end of the cell. Contraction of the microfilament ring can be triggered in two ways. One way is by release of tension across the ring. The second way is by a sharp abrupt tug on the ring which can cause contraction by inducing a smooth muscle type stretch activated contraction response[39,40]. Once the contraction is underway, if there are connections between cells, biochemical and mechanical signals relating to the smooth muscle contraction response cross these connections triggering a contraction in the next cell. This then propagates to the next cell, and the next and so on. We now have a wave of contraction propagating across the epithelial sheet of cells. Individual cells will participate in the contraction and then return to their original size. This

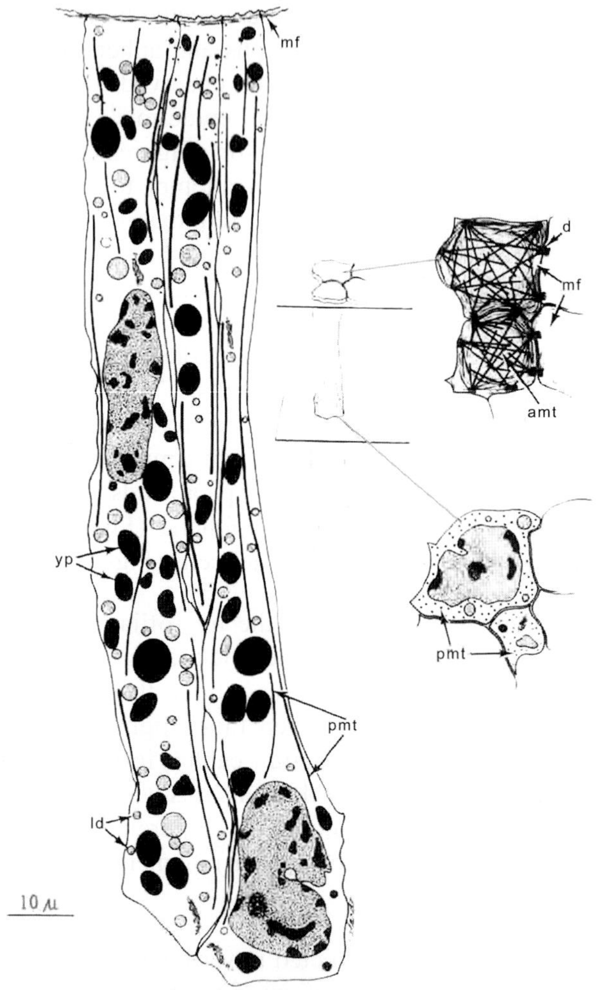

Figure 6.5. The original observation of neural plate apical microtubules by Beth Burnside in the newt *Taricha torosa*. "Schematic illustrations of the orientation of microfilaments and microtubules in elongating neural plate cells. Numerous microtubules are aligned parallel to the long axis of the cell (paraxial microtubules, *pmt*). These microtubules are seen in profile in longitudinal section and in cross section in transverse sections of the cell. Just beneath the free surface, microtubules are aligned parallel to the free surface in a layer across the cell apex (apical microtubules, *amt*). In the same plane with the apical layer of microtubules, 50-70 Å microfilaments (*mf*) are arranged into a circumferential bundle which encircles the cell apex in pursestring fashion. *yp*, yolk platelet; *ld*, lipid droplet; *d* desmosome." From Burnside (1971)[12] with permission of Elsevier. She discussed the roles of the microfilaments and paraxial microtubules, but assigned no role to the apical microtubules.

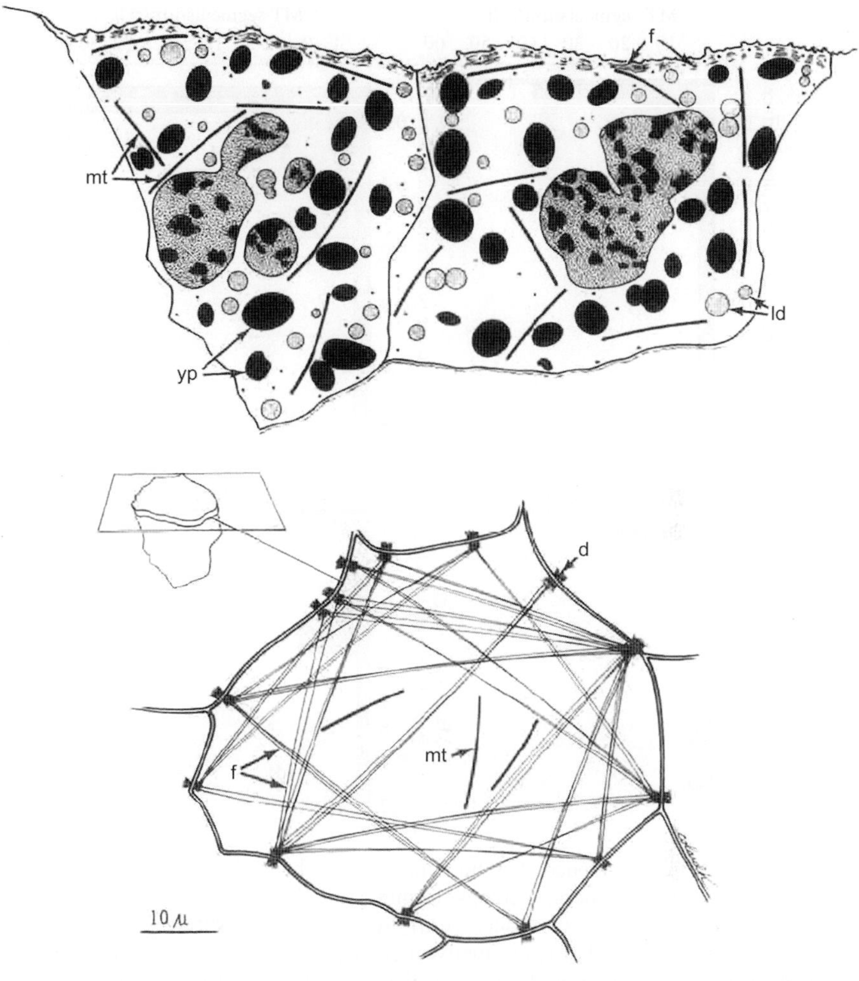

Figure 6.6. The original observation of epidermis apical microtubules by Beth Burnside in the newt *Taricha torosa*. "Schematic drawing of the orientation of microtubules and microfilaments in a flattening epidermal cell. Beneath the free surface, thicker filaments (*f*) (70-100 Å in diameter) are arranged into discrete bundles reminiscent of 'tonofibrils.' These bundles of filaments are often seen in continuity with desmosomal filaments, and are thought to span the cell from desmosome to desmosome (*d*). Microtubules (*mt*) in presumptive epidermal cells appear to be randomly oriented. *yp*, yolk platelet; *ld*, lipid droplet." From Burnside (1971)[12] with permission of Elsevier.

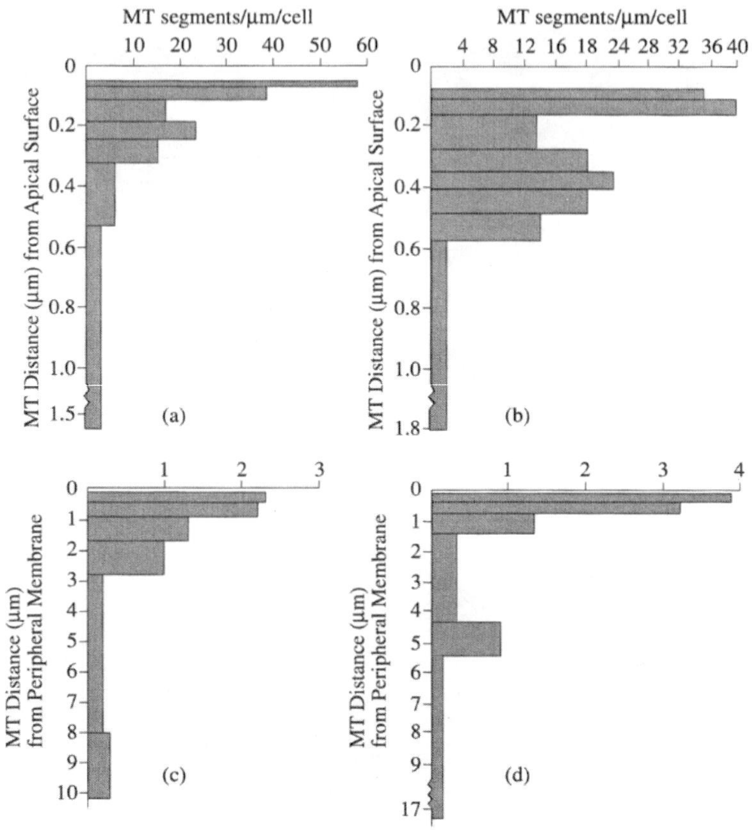

Figure 6.7. Statistical evidence that microtubules in the cell state splitter at axolotl Stage 11 are at its periphery, perhaps forming an annular mat. "Concentration of microtubule segments versus distance (μm) from the apical surface, (a) in stage 11 presumptive epidermis and (b) presumptive neural plate. Concentration of microtubule segments versus distance (μm) from the cell membrane to cell center, (c) in stage 11 presumptive epidermis and (d) presumptive neural plate." From Martin & Gordon (1997)[32] with kind permission of Sergei G. Vassetzky.

contraction is a transient event lasting about 10 minutes per cell in the axolotl. Cells that undergo transformation to bottle cells (Figures 2.60 and 2.61) contract even further but also take about ten minutes per cell. We presume the counterforce to the contraction which returns the cell to their original shape is the intermediate filament ring, acting like a circular spring that can be compressed or stretched and then returns to its resting diameter (Figure 6.12).

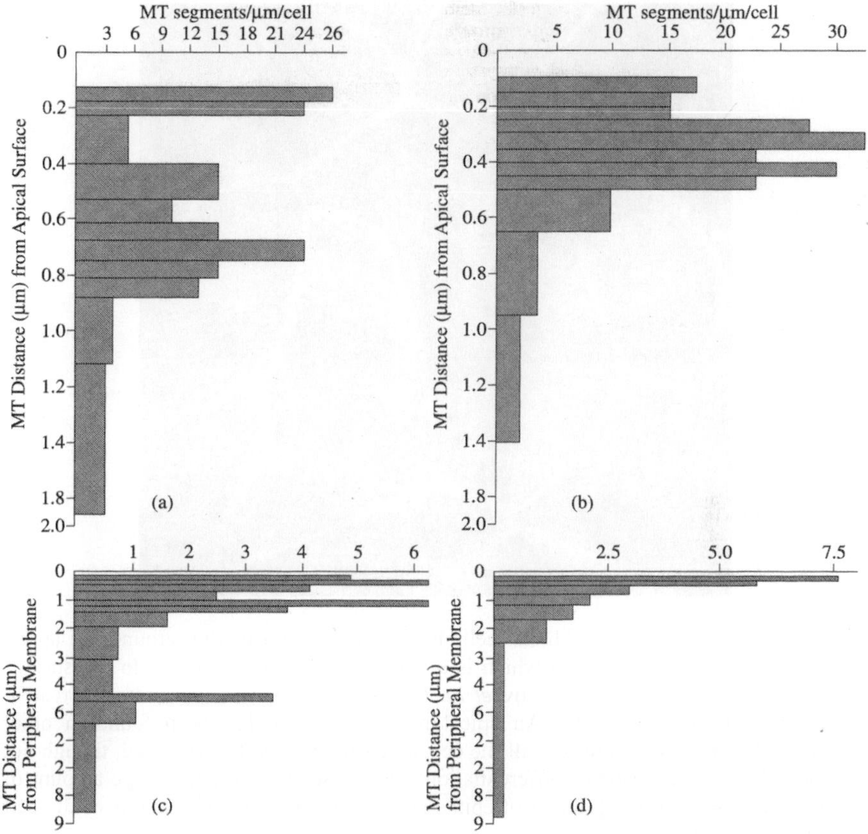

Figure 6.8. Statistical evidence that microtubules in the cell state splitter at axolotl Stage 15 are at its periphery, perhaps forming an annular mat. "Concentration of microtubule segments from the apical surface, distance (μm), (a) in stage 15 presumptive epidermis and (b) presumptive neural plate. Concentration of microtubule segments from the cell membrane to cell center, distance (μm), (c) in stage 15 presumptive epidermis and (d) presumptive neural plate." From Martin & Gordon (1997)[32] with kind permission of Sergei G. Vassetzky.

Working in the opposite direction from the microfilament ring is the microtubule mat. This set of microtubules goes across the microfilament ring at the edges. They also reach out and downwards to the lower intermediate filament ring (Figure 6.3). Altogether they form a annular mat which is open in the middle. The sum of the forces exerted by these individual microtubules results in a force that pushes outward (Figure 6.13). The force that microtubules exert is approximately

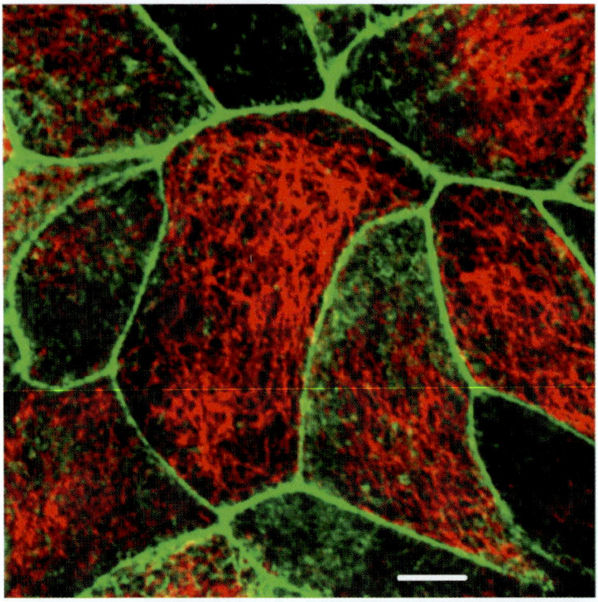

Figure 6.9. These cultured epithelial cells have a microfilament ring around the apical end (green) of each cell, within which is a mat of microtubules parallel to the surface (red). Scale bar 5 μm. From Ivanov *et al.* (2006)[41] with permission under a Creative Commons Attribution License. An apical microtubule mat has been found in other epithelia[42,43], so the structure we call the cell state splitter is widely depicted, though not interpreted as the organelle of differentiation[44], as we see it. There is a large amount of recent work on apical constriction of epithelia[45,46], what we call contractions of the cell state splitter involved in contraction waves.

constant as a function of cell radius[11]. Once the expansion due to microtubules occurs in one cell, a signal to expand propagates from cell to cell across a sheet in the same way that a contraction signal can propagate across a sheet. However, unlike with the contraction wave, the expanded cell remains expanded. We presume the slow steady pressure of the microtubules thins both the microfilament ring and alters the intermediate filament ring. The cell becomes flatter and squatter and it stays that way.

We can plot the predicted force balance equation of the microtubule mat versus the microfilament ring (Figure 6.14). When a sheet of cells is ready to differentiate, we hypothesize that each cell in it has a cell state splitter where the microtubule mat and the microfilament ring are in a

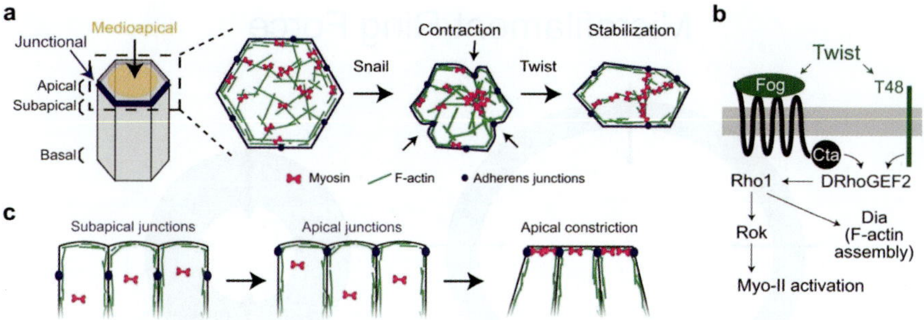

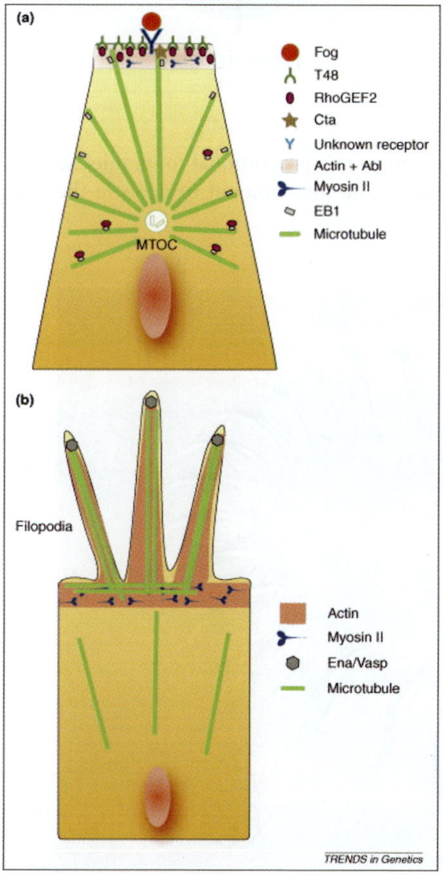

Figure 6.10. Apical constriction models during *Drosophila* gastrulation, illustrating roles of both apical microfilaments and apical microtubules, albeit without pointing out that cell differentiation is going on. **Top**: "(a) Diagram of ventral furrow cell and model of apical constriction. Snail [a protein] is required to initiate contraction, and Twist [another protein] stabilizes the contracted cell shape. During the stabilization, F-actin [microfilaments] and Myo-II are often remodelled from foci into fibres. (b) Schematic of the Twist pathway. Twist activates the expression of Fog and T48, leading to Rho1 activation, which is proposed to activate Myo-II through Rok and stimulate F-actin assembly through Dia. (c) Model of adherens junction remodelling in ventral furrow cells. Adherens junctions initially assemble subapically and then move apically into spot junctions when contraction initiates." From Mason *et al.* (2013)[45] with permission of Nature Publishing Group. See also Martin & Goldstein (2014)[38]. **Left**: "Microtubules in epidermal morphogenesis. The coordinated activity of both MTs and actin is crucial during *D. melanogaster* gastrulation **(a)** and the extension of filopodia in dorsal closure **(b)**. (a) MTs (green) can transport the Rho GTPase activator RhoGEF2 (pink ovals) when bound to the +tip protein EB1 (gray square) in tissue culture cells[47], in which case they grow from the microtubule organizing centre (MTOC). (b) Antiparallel microtubules (green) located mainly under the apical surface promote fast filopodia extension during dorsal closure[48]." From Quintin *et al.* (2008)[42] with permission of Elsevier.

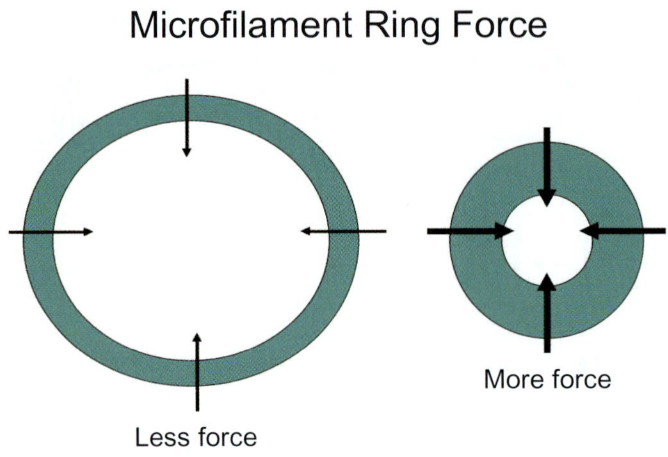

Figure 6.11. The force exerted by a microfilament ring is proportional to its thickness. If it gets thicker as it contracts, the force increases. The total amount of actin in the ring remains constant as it contracts.

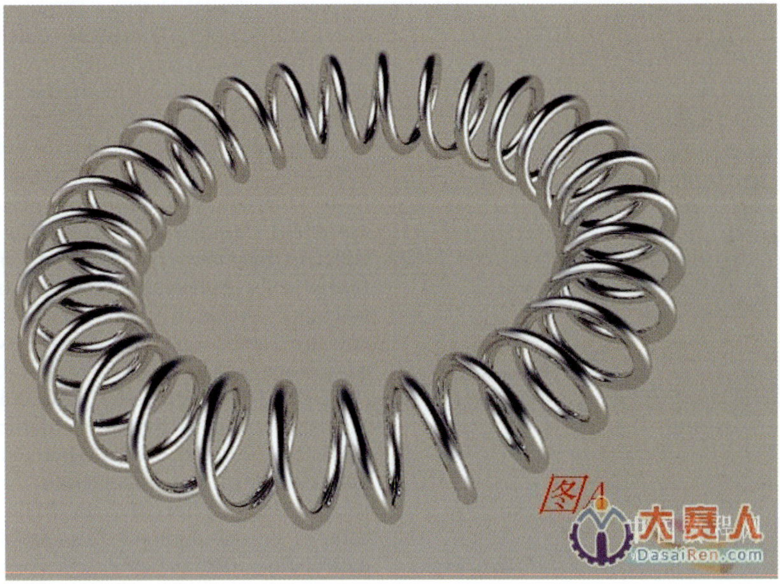

Figure 6.12. A circular spring provides a model for the mechanics of an intermediate filament ring as a component of the cell state splitter. Whether the ring is expanded or contracted, there is a restoring force to bring it back to its "resting" diameter. From iSee[49].

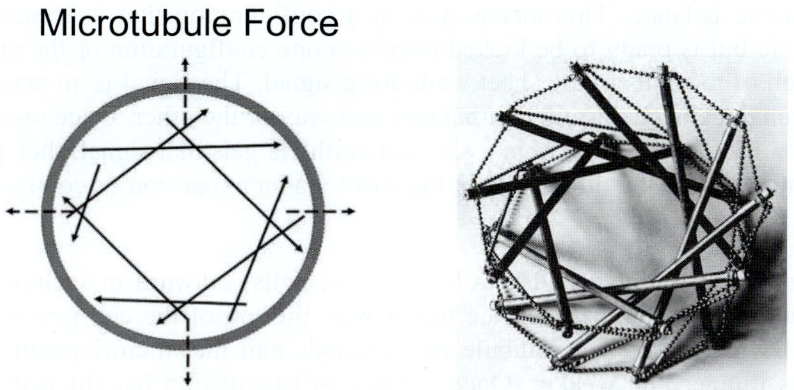

Figure 6.13. **Left**: The microtubule mat at the apical end of an ectoderm cell exerts an outward force via polymerization[50], which we have shown should be independent of cell radius[11]. Sliding of microtubules against one another (Chapter 5) could also generate an outward force. **Right**: A tensegrity structure analogous to the cell state splitter, by the artist Kenneth Snelson, with his kind permission[51]: Bead Chain Helix, 1959.

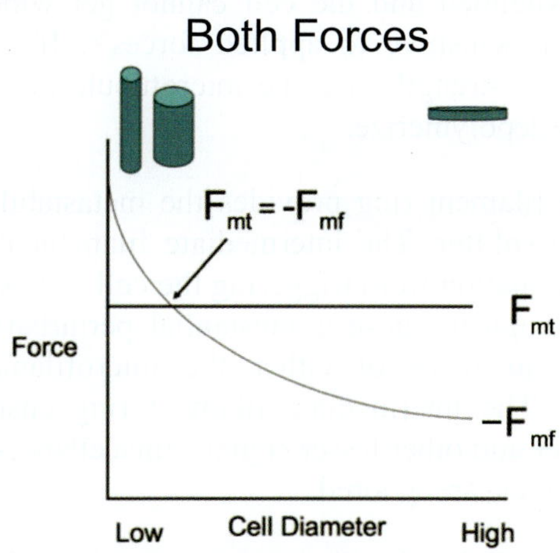

Figure 6.14. In a cell state splitter, since the force due to the apical mat of microtubules (mt) is independent of cell diameter, and the oppositely acting microfilament (mf) ring force falls of hyperbolically with cell diameter, there should be a cell diameter at which the forces are in mechanical balance, i.e.. $F_{mt} = -F_{mf}$. We assume that the cell state splitter is set up at this equilibrium point. However, this is an unstable equilibrium. At higher diameter $F_{mt} > -F_{mf}$ and at lower diameter $F_{mt} < -F_{mf}$. Thus, if the diameter goes higher than the equilibrium diameter, it will keep getting bigger, flattening the cell. If the diameter goes lower than the equilibrium diameter, it will keep getting smaller, turning the cell into a tall, narrow one. Adapted from Figure 10 in Gordon & Brodland (1987)[11].

bistable balance. This means it is in a configuration that is somewhat stable but is ready to be kicked over into one configuration or the other. Each of the cells in the sheet waits for a signal. The signal is an external mechanical signal to tip that balance one way or the other. Once one cell (or a group of cells) within a sheet of epithelia gets that signal, they pass it along to all the other cells in the sheet via an expansion or contraction wave.

If an external force pulls the first cell (or cells) outward in such a way that the microtubules win the tug-of-war, the top of the cell gets wider and wider as the microtubule mat expands and the microfilament ring gets thinner and weaker. Once the microfilament ring has thinned to a certain point there is no turning back. The cell top (apical end) continues to get wider and wider. The cell participates in a wave of expansion.

If the top of the cell gets narrower due to some external force, the microfilament ring thickens and the top becomes smaller and smaller. Once a certain degree of inward force is achieved, the microtubule mat is completely overwhelmed and the cell cannot get wider again. Microtubule assembly is sensitive to applied forces[52]. If the microfilament contracts with more strength than the microtubule mat can counter, the microtubules will depolymerize.

The intermediate filament ring provides the metastability aspect of the bistable cell state splitter. The intermediate filament ring prevents any minor random fluctuation from triggering the cell state splitter. The force must be large enough to cause a substantial perturbation and shift the balance strongly in favor of either the microfilament ring or the microtubule mat. The intermediate filament ring ensures the specific signal is the trigger and other lesser signals, miscellaneous fluctuations in the mechanical forces, are ignored.

If you watch the individual cells as they participate in the contraction wave you see the cell contract and then return to its original shape. However, the expansion wave behaves very differently. The expansion looks like a series of small jerks and then suddenly the apical end "gives" and expands rapidly. In the axolotl cells at least, the apical end spreads to double or triple the diameter and then stays that way. The intermediate filament ring "gets out of the way" in some fashion. We suspect an alpha/beta transition[53-55] occurs so that the intermediate filaments change from a more compact alpha configuration to an elongated beta (Chapter 5). Alpha/beta transitions are known to occur in response to a variety of signals such as the application of heat and the introduction of water and ions. A similar signalling event in the cell triggers the transition and allows the spread of the transition and the expansion wave.

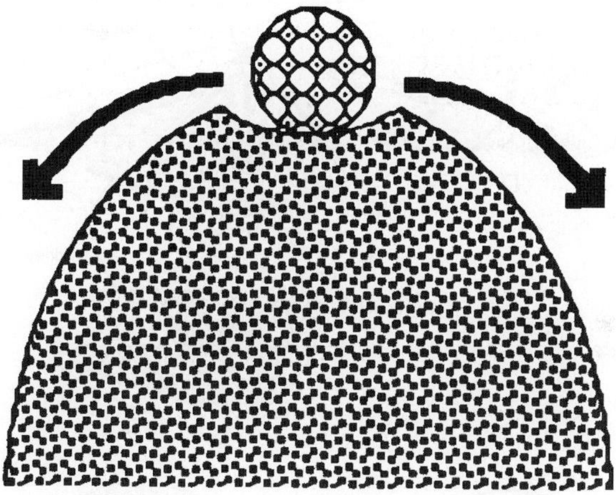

Figure 6.15. The bistability of the cell state splitter can be represented by a ball poised in a metastable situation at the top of a hill. It takes a finite push in one direction or the other to get it to run down the hill. In the case of the cell state splitter, there are only two ways to go: contraction of the microfilament ring (left) or expansion of the microtubule mat (right). The intermediate filament band, a component of the cell state splitter (Figure 6.3), may provide the small barrier via its elasticity that must be overcome. From Gordon & Brodland (1987)[11] with permission of Springer.

This may be represented abstractly by a potential diagram of the metastability (Figure 6.15). A change in the cytoskeleton with either rapid microfilament ring contraction or rapid microtubule mat expansion is more generally known as a cytoskeletal rearrangement. The activity of the cell state splitter is a specific form of cytoskeletal rearrangement required for determination of cells.

The cell state splitter may be regarded as a tensegrity structure (defined in Figure 5.37), as shown in Figure 6.13, but as an unusual nonlinear one (Proposition 57 in *Hierarchical Genome*[9], cf.[56-58]), in that it "snaps through" to one configuration or another from its metastable state. This contributes to the new field of biotensegrity[59,60].

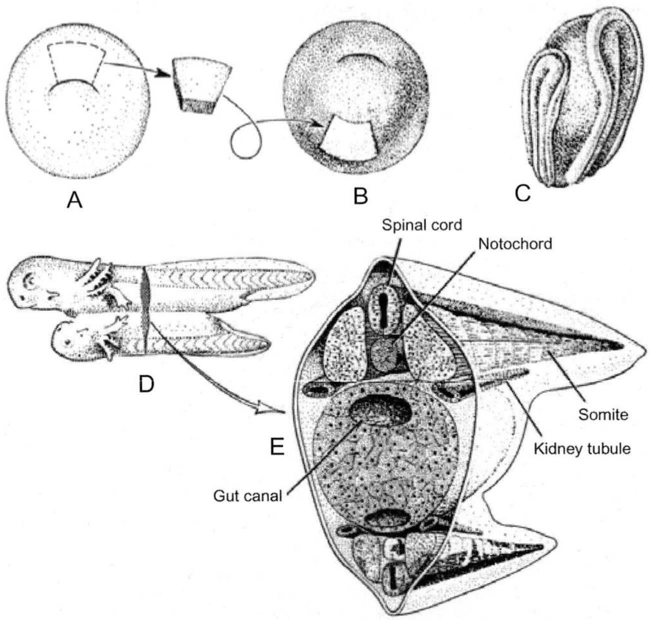
A
B
C
Spinal cord
Notochord
D
Somite
E
Kidney tubule
Gut canal

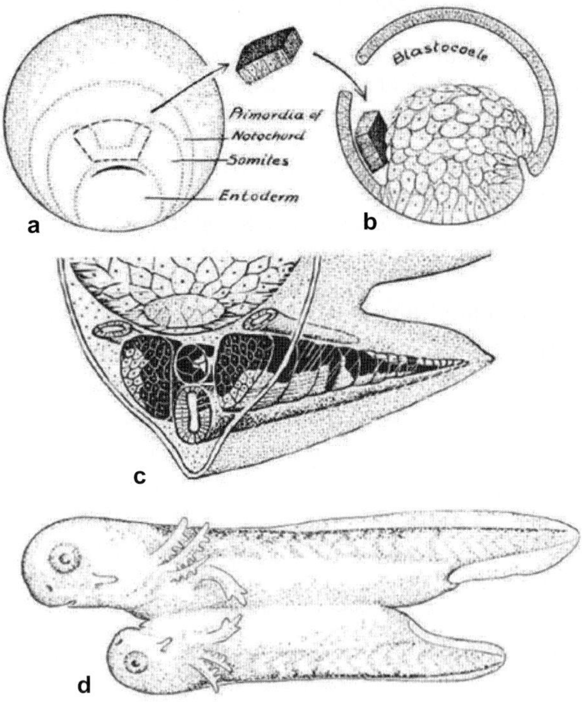
Blastocoele
Primordia of
Notochord
Somites
Entoderm
a
b
c
d

Figure 6.16. **Top**: Transplantation of a piece of ectoderm containing a portion of the dorsal lip of the blastopore (the "organizer") from embryo A to embryo B results in a second neural plate (C). The end result is a double embryo, i.e., conjoined twins. This shows that any portion of the ectoderm is capable of producing a neural plate and subsequent development. In an ordinary, single embryo the dorsal (top) part of the ectoderm produces the brain and spinal cord, while the lower (ventral) hemisphere produces the epidermis, which becomes skin. Next time you contemplate your navel, consider that the skin around it could have become a conjoined twin of you. This sketch, which depicts the original experiments by Hilde Mangold and Hans Spemann[63], is by Victor Twitty. From Twitty (1966)[64] with permission from Macmillan Education. Antone G. Jacobson, Dick's early collaborator[65-69], was trained as an embryologist by Twitty[70,71]. **Bottom**: The later, much easier Einsteck ("sticking in") experiment of Otto Mangold[72,73,61,74], in which the dorsal lip of the blastopore is inserted inside the blastocoel, giving the same result. From Holtfreter & Hamburger (1955)[75] with permission of Elsevier under the STM (International Association of Scientific, Technical & Medical Publishers) Permissions Guidelines. Note that Twitty's sketch (Top) is obviously a composite incorporating part of the earlier (Bottom) sketch. The perfection of the secondary neural plate shown was rarely achieved.

In *Pursued by the Differentiation Wave* in *Hierarchical Genome*[9] Dick wrote an informal history of our discovery of the ectoderm contraction wave. This story invokes some ideas that need explaining, and some that need correcting. The first is the problem of primary neural induction, for which Hans Spemann won the Nobel Prize in 1935[61,62]. The basic experiment, performed by his graduate student Hilde Mangold in 1921 and 1922[63,76], is illustrated in Figure 6.16. The dorsal lip of the blastopore is an arc of cells at the bottom of an amphibian embryo (Figure 2.33). Mangold cut this out and transferred it to a square hole cut in the ventral ectoderm in another embryo. To distinguish the transplant, which healed in, from the recipient embryo, two different species of the newt *Triton* were used, whose eggs have different degrees of pigmentation. The remarkable result was the appearance of a second neural plate (cf. Figure 2.55). The transplanted piece of tissue was called an "organizer" because most of the tissue in the new neural plate came from the host which was thus "organized" by the small piece that was patched in. The donor organizer somehow "induced" the surrounding ectoderm of the host to become a new neural plate, which could go on to to become part of a second embryo. It did not separate but rather became a conjoined twin. At that time it was thought that neural induction was the first "induction" (what we call determination) event. It was therefore called "primary neural induction".

Another point that came out of this work was the long standing embryological assumption that cells have a default fate they follow unless "induced" to do otherwise. So ectoderm will form skin, sweat glands and so forth unless induced to form neural tissue. We do not accept this idea of induction as a deviation from a default fate. Rather we see differentiation as a series of bifurcations where a divergence from one cell type to one of two possible descendant cell types occurs. Once the divergence occurs the original tissue ceases to exist. Thus ectoderm vanishes after the ectoderm waves have passed through it leaving neural epithelium in the wake of the portion that participates in a contraction wave and epidermal epithelium in the wake of the portion that participates in an expansion wave (Figure 6.17).

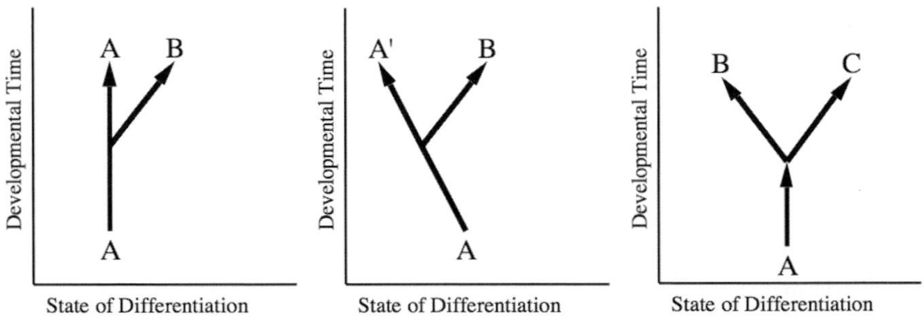

Figure 6.17. **Left**: Classical model of differentiation, in which tissue A continues as the same tissue, except for a portion which changes to tissue B when 'induced' by an external agent. **Middle**: Alternative classical model for differentiation, in which tissue type A is continually changing towards type A', except for a portion which changes to tissue B when 'induced' by an external agent[77,78]. **Right**: Our new view of differentiation, in which tissue A splits into two new tissues B and C, as contraction and expansion differentiation waves travel through complementary portions of tissue A. Tissue A ceases to exist. From *Hierarchical Genome*[9], Figures 5-7.

Recall the anatomical steps of early embryogenesis from Chapter 2. Ectoderm starts as a relatively simple primary tissue type. It subdivides into two tissue types. The first type is presumptive neural tissue. The ectoderm on the top half of the embryo in late gastrulation goes from rather squat cells to tall narrow cells. These tall narrow cells form a plate and then the edges rise and seal into the neural tube, which sinks down inside the embryo. The lower half of the ectoderm forms epithelial ectoderm (epidermis), which then goes on to form skin, sweat glands, and all related external tissue types as development proceeds. The epidermis expands until it covers the whole embryo. We studied primary neural induction out of a desire to understand the human birth defect, spina bifida. We discovered a propagating wave of contraction that went through the neural ectoderm at the time of neural ectoderm determination. We also found a propagating wave of expansion in the epithelial ectoderm.

Nature surprised us. The actual trajectory of what we came to call the ectoderm contraction wave (Figure 6.18) did not match the first simple mechanical model our group came up with to divide a sphere of ectoderm into two hemispheres, the top half neural epithelium and the bottom half epidermal epithelium (Figure 6.19). That was a very nice model. Without it we would never have gone looking for the ectoderm contraction wave. But it was wrong. The ectoderm contraction wave begins right at the spot where primary neural induction occurs just above the original Spemann organizer. It travels as an expanding circle, then ellipse, until it breaks over the dorsal lip of the blastopore, then the remaining part travels as an arc. The edges of the arc move more quickly than the middle and so it eventually reforms into another ellipse. That ellipse then closes down and vanishes at a point. The wave travels well ahead of the expanding archenteron and so cannot be explained away as simply due to underlying tissue movement.

The ectoderm contraction wave follows this weird trajectory apparently because: 1) a portion of it hits a barrier at the dorsal lip of the blastopore (which turns out to be another contraction wave) and disappears; 2) the remaining arc of the wave is distorted by the involution movement of mesoderm and endoderm from the outside to the inside of the embryo (Figure 2.57), so that its curvature changes from convex to concave, and the wave then self-annihilates by running into itself. Changing from an expanding circle to an arc and a vanishing ellipse involves waves running into one another. Not surprisingly, a cell can only participate in one wave at a time. If it is already involved in contracting or expanding in one wave, it can't simultaneously receive and propagate a second wave.

We have pictured the ectoderm contraction wave in many ways. This was necessary, because when we started we had no idea how complex its trajectory in time and space would be. First, and most startling, this wave

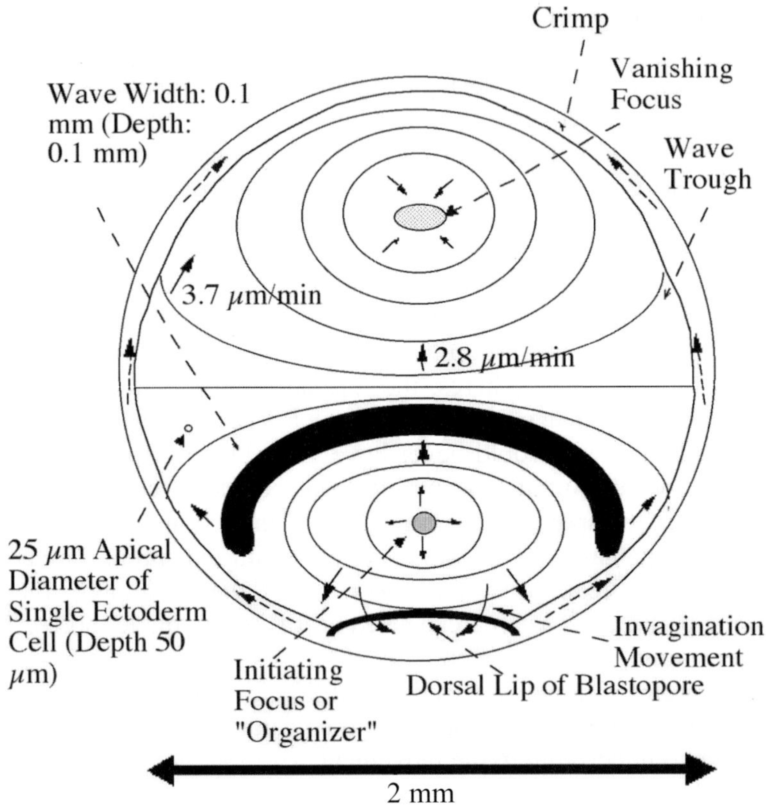

Figure 6.18. A dorsal view of the ectoderm contraction wave, which travels as a furrow. The bottom of the furrow is shown as a curved line at one hour intervals. The width of the furrow is indicated by the thick arc drawn at hour #4. After 11 hours the wave vanishes as it converges on itself. After the wave breaks into an arc on hitting the dorsal lip of the blastopore at hour #3, it switches from convex to concave at hour #6, by travelling faster near the equator of the embryo than along the midline. After arc formation at hour #3, the remaining wave travels away from the dorsal lip of the blastopore while tissue is invaginating in the opposite direction, going over the dorsal lip of the blastopore inside (curved arrows). For relative scale, the surface diameter of one cell is drawn. From Gordon *et al.* (1994)[79] with permission of Elsevier Science Publishers. Note that three mechanisms for "stopping" a wave are apparent here: 1) self-annihilation ("vanishing focus" at hour #11), 2) collision with another wave (with the dorsal lip of the blastopore, at hour #3), 3) shaping of the wave so that it is "stopped" from entering all of the tissue (bottom hemisphere of ectoderm) that could propagate it (perhaps due to tension generated by the invagination movement). Medical artist Jack Butler made a series of photos from a few angles of a clay model of the ectoderm contraction wave, which appeared in *Hierarchical Genome* as a flip movie[9] and on YouTube[80].

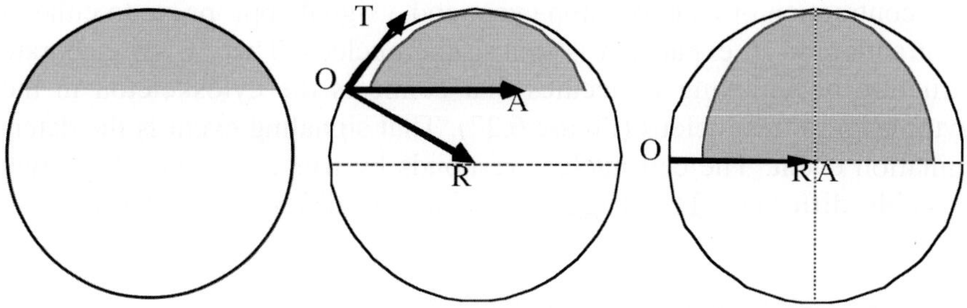

Figure 6.19. A thought experiment showing how a mechanical contraction wave can stop abruptly when it has traversed one hemisphere. **Left**: The grey spherical cap has contracted. **Middle**: Let's cut it off, so that it shrinks a bit. Now the force we would have to apply to pull the remaining spherical shell back to the cut off cap is directed along the vector OA. This vector can be broken into two perpendicular components, one towards the center of the sphere, OR, and one tangential to the surface of the sphere, OT. If the wave is transmitted as a stretch activated contraction, then it is the vector OT that causes the wave to keep propagating down over the surface. **Right**: However, when the wave reaches the equator of the sphere, OR = OA and the vector OT vanishes to nothing. Thus the upper hemisphere cannot pull on the lower hemisphere, and the wave halts when it reaches the equator. For a wave to keep propagating to the lower hemisphere, it must have a nonmechanical component. From *Hierarchical Genome*[9] Figures 11-13.

is a deep furrow on the surface of the embryo, its depth made visible with mirrors (Figures 6.20 and 6.21). We also were initially puzzled that in some embryos the wave spread out (Figure 6.22), or changed from convex to concave (Figure 6.23), or shrank to a point (Figure 6.24) and vanished. Using two time-lapse microscopes top and bottom, we discovered that the contraction wave was confined to the upper hemisphere (Figure 6.25). Putting it all together first on a tennis ball, then later by drawing on a clear hamster ball, we reconstructed the trajectory of the wave (Figure 6.26). A composite sketch in top view reveals that the wave starts at a point and ends at a point (Figure 6.18), as if it had gone from one focus to another.

Let's now return to the ectoderm cell sheet. We assume that each ectoderm cell has two possible sets of gene expression cascades that are ready to be triggered, i.e., two alternative differons. One differon can be triggered by the contraction of the microfilament ring. The other differon can be triggered by an expansion of the microtubule mat. The cell is ready for either event. It is within a cell sheet of similar cells all waiting for a signal. When the external signal comes, the cell either participates in a contraction or an expansion event and a signal appropriate to either a contraction or an expansion is sent to the nucleus. There is an elaborate structure of signaling molecules that connects the cytoskeleton in the cytoplasm to the nucleus (Figure 6.27). That signaling event is the determination event. The cell nucleus responds by triggering one of the two possible differons. That triggering of a new differon is determination

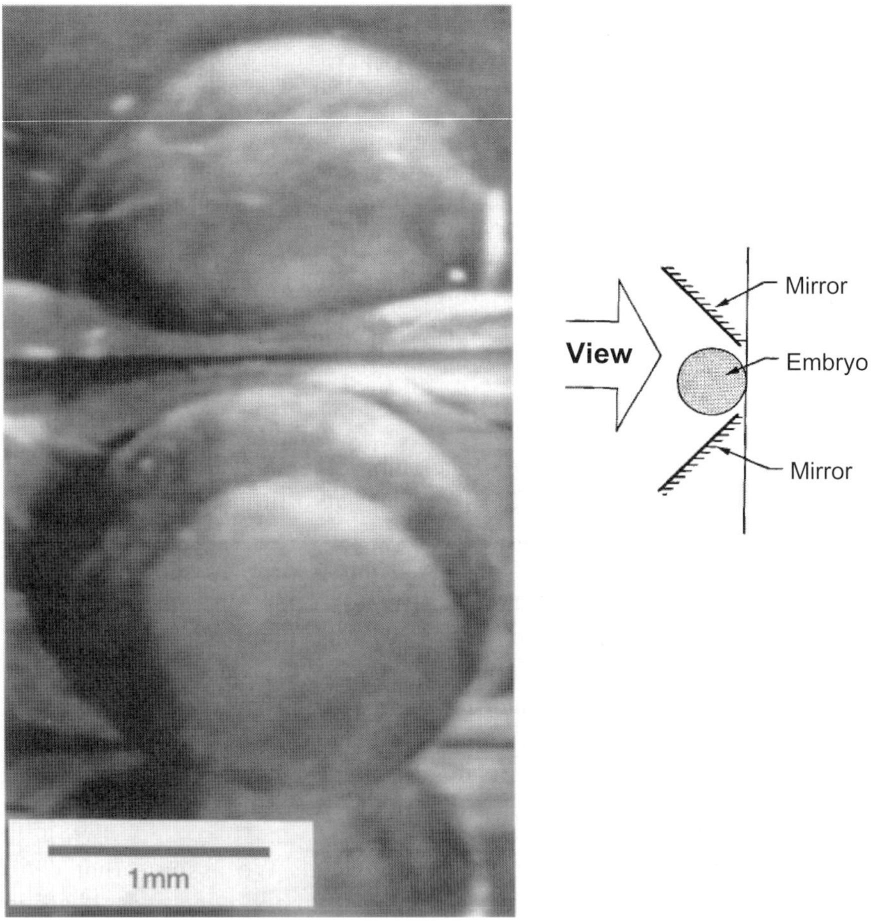

Figure 6.20. Looking down on an axolotl embryo with two mirrors near it angled at 45° and one underneath. The ectoderm contraction wave can be seen as a deep furrow on the surface. From Brodland *et al.* (1994)[81] with permission of John Wiley and Sons.

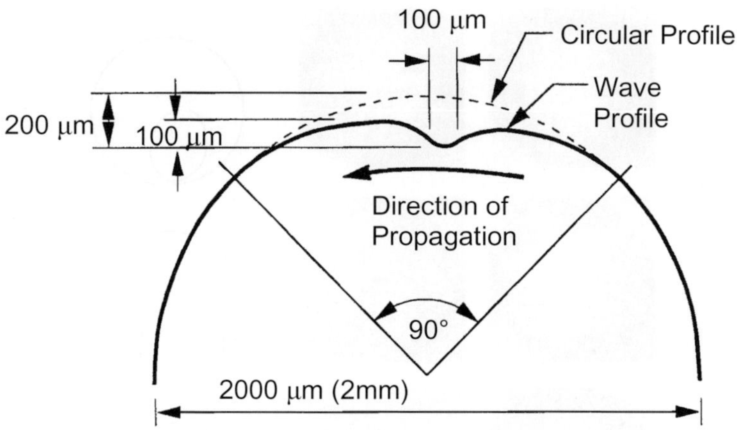

Figure 6.21. Left side view of the axolotl ectoderm contraction wave. "A region encompassing almost a 90° arc is indented moderately, producing an apparent flattening. Centered on this region is the furrow that is produced by cells that are intensely contracted at their apical (outside) surfaces. The ectoderm returns to the circular profile shown as the wave passes." From Brodland *et al.* (1994)[81] with permission of John Wiley and Sons. This figure was reproduced and discussed by the late Lionel Jaffe[82].

at the genetic level. Determination is followed by differentiation, as the genes in the triggered differon are expressed and the resulting RNAs and proteins do their things. The last step in the process is the setting up of a new cell state splitter (or resetting the old one) when it is time for the cells to become determined and to differentiate again. We will later devote the entire Chapters 8 and 9 to the specifics of the signaling between cell state splitter and the nucleus that changes gene expression.

This cell sheet differentiation process via cell state splitters may or may not involve cell division (mitosis) in the cell sheet. Commonly, there is a differentiation event, a round of cell division with proliferation and growth of the new cell type, followed by another new round of cell differentiation. These differentiation events and mitosis events often occur in tandem but, unlike single cell division, differentiation and mitosis of yeast, or the determination in mosaic embryos, the two do not have to happen together. They are independent. Differentiation can be amitotic (Figure 6.2).

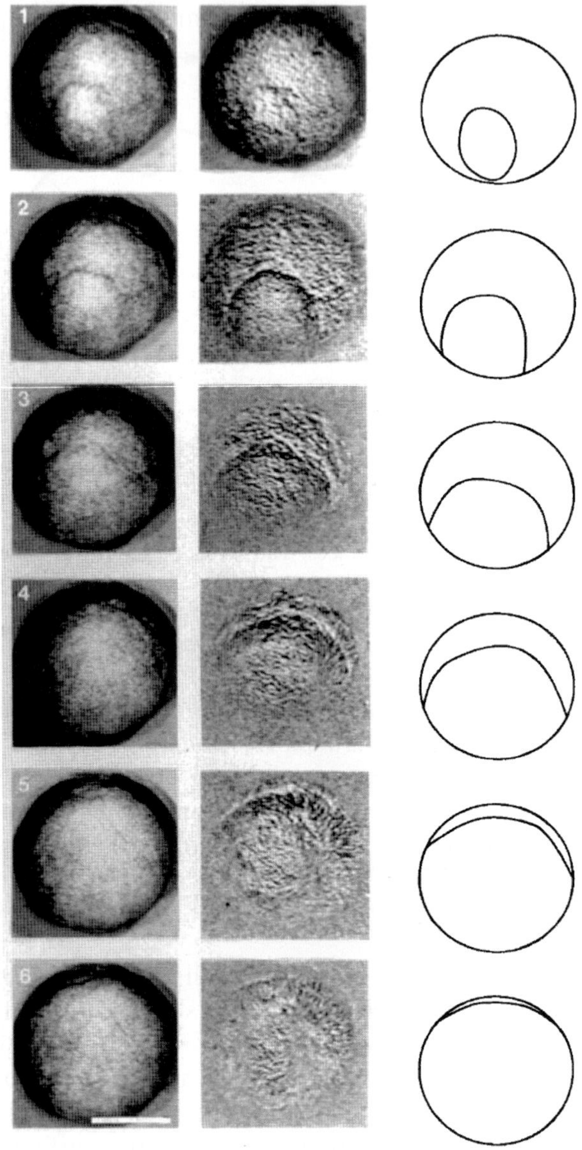

Figure 6.22. Time-lapse images of a single axolotl embryo are shown every hour, #1 starting at Stage 10.75 (Figure 2.33), ending at Stage 11.5. The pictures in the middle column are difference images, made by subtracting the left image from an image captured 5 minutes earlier. The sketch on the right is our interpretation of where the wave was. From Brodland *et al.* (1994)[81] with kind permission of John Wiley and Sons. Scale bar 1 millimeter.

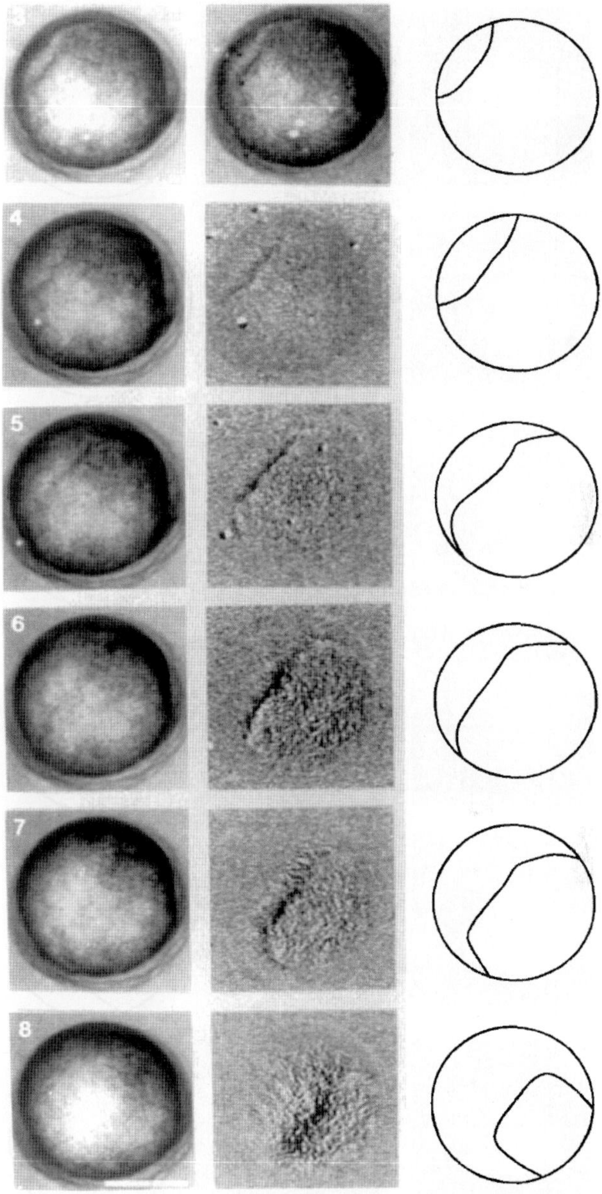

Figure 6.23. In this axolotl embryo, the ectoderm contraction wave is seen switching from convex to concave as it travels. From Brodland *et al.* (1994)[81] with kind permission of John Wiley and Sons. Scale bar 1 millimeter. Stages are approximately 11.5 through 12.

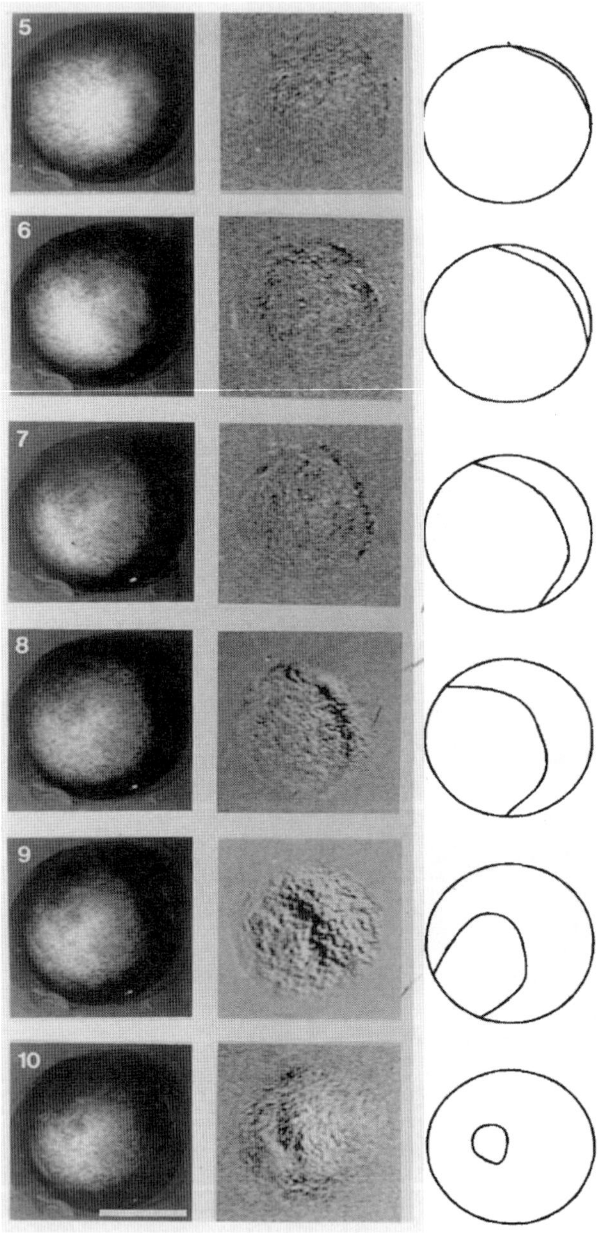

Figure 6.24. In this axolotl embryo the ectoderm contraction wave at Stage 11.5 is seen closing down to a small ellipse at Stage 12.5 to 13 (Figure 2.33). From Brodland *et al.* (1994)[81] with kind permission of John Wiley and Sons. Scale bar 1 millimeter.

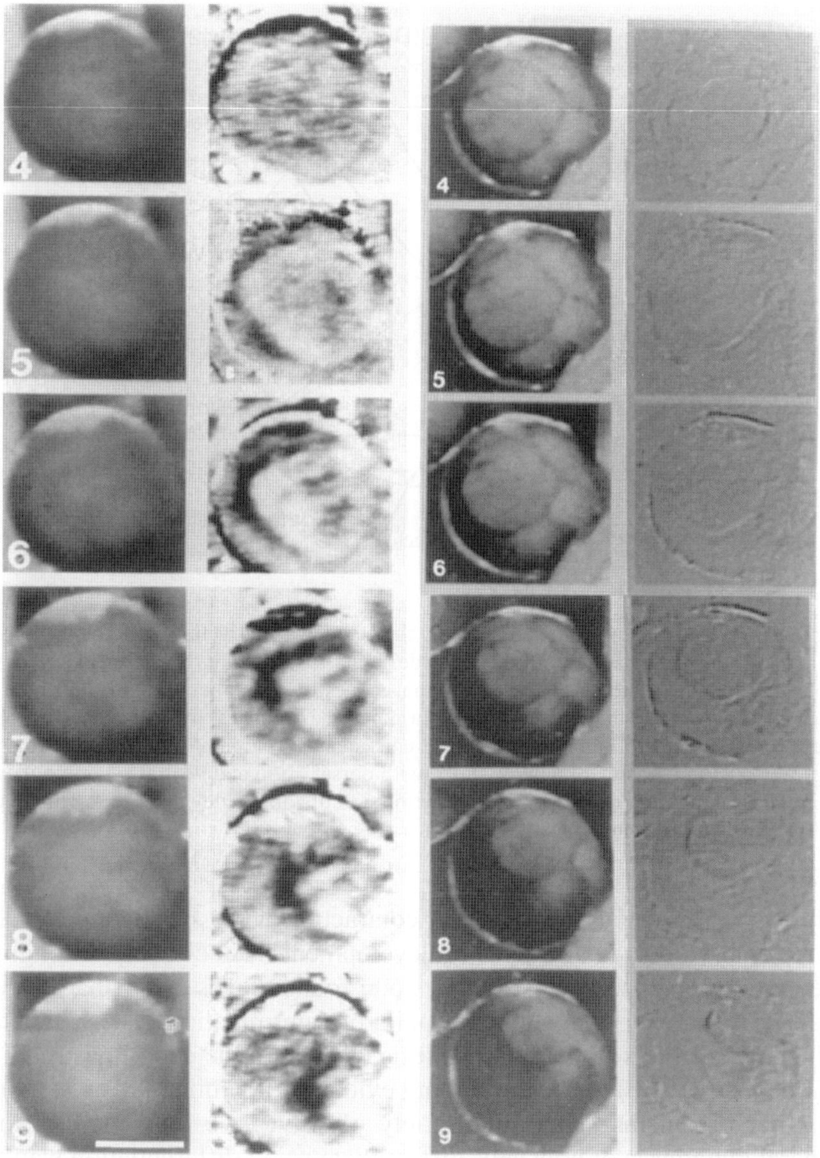

Figure 6.25. This axolotl embryo was imaged top (first column) and bottom (third column) simultaneously. The ectoderm contraction wave can be seen shrinking to a circle (as in Figure 6.24) in the second column, which shows difference images of the top, while only the dorsal lip of the blastopore shows up in the difference images of the bottom (fourth column). From Brodland *et al.* (1994)[81] with kind permission of John Wiley and Sons. Scale bar 1 millimeter. Stages are approximately 10.5 to 11.

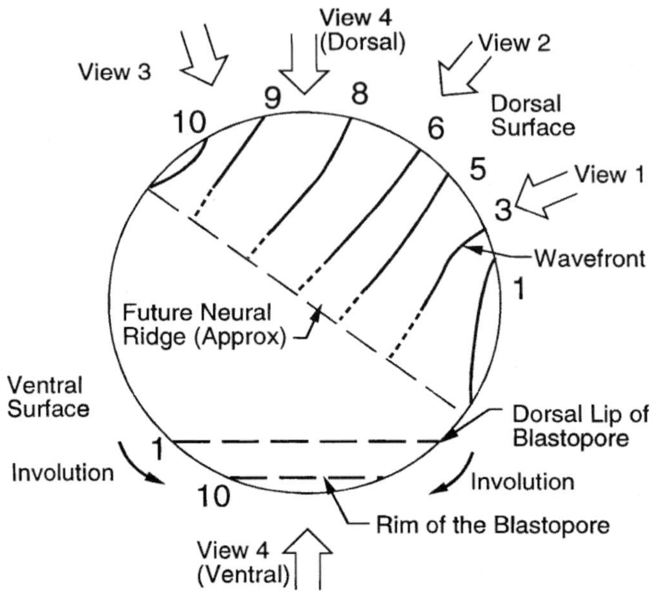

Figure 6.26. This diagram summarizes how the ectoderm contraction wave moves across an axolotl embryo, as seen from the left side. It starts at a point that spreads to a circle (hour #1), breaking into an arc (hour #3) that moves over one hemisphere, finishing off (hour #10) as a circle that comes down to a point and vanishes, over Stages 10.5 to 13. View 1 corresponds to Figure 6.22; View 2 corresponds to Figure 6.23; View 3 is that of Figure 6.24; View 4 is that of Figure 6.25, with dorsal = top and ventral = bottom views. From Brodland *et al.* (1994)[81] with kind permission of John Wiley and Sons. This figure is reproduced and discussed in Jaffe (1999)[82].

After we discovered the ectoderm contraction wave we started looking at the embryo during all stages of development from cleavage through to the end of neural tube closure. We soon found waves occurring all over the embryo at every known major point of determination. (We will detail all the waves of axolotl embryogenesis through to the end of gastrulation in Chapter 7.) We also began finding evidence of similar waves in a host of other organisms. The wide variety of waves we found led us to a general hypothesis of how cells differentiate using differentiation waves.

After examining case after case of differentiation waves and determination we found a pattern of cell divisions and differentiations, sometimes coupled and sometimes not. We can *a priori* expect the following cases of differentiation versus cell division to occur (Figure 6.2):

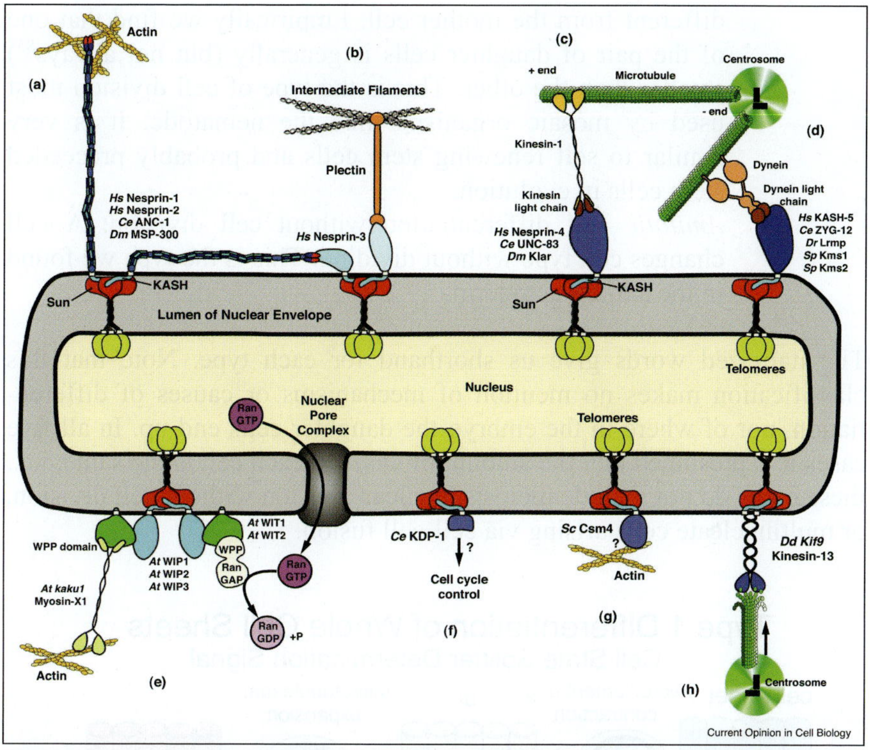

Figure 6.27. The nuclear envelope consists of two membranes that are spanned by a number of proteins with connections into the nucleus and out into the cytoplasm. Nuclear pores permit transport of molecules between nucleus and cytoplasm. From Luxton & Starr (2014)[83] with permission of Elsevier.

- *Proliferative* cell division without differentiation: Both daughter cells are the same type as the mother cell. This is often referred to as a "vegetative", "proliferative" or "symmetric"[84] cell division.
- *Self-renewing* stem cell division with differentiation: One cell stays of the same type as the mother cell, and the other changes to another kind.
- *Symmetric* cell division with symmetric differentiation: Both daughter cells are of the same kind, but different from the mother cell.
- *Asymmetric* cell division with asymmetric differentiation: Both daughter cells are different from one another, and also

different from the mother cell. Empirically we find that one of the pair of daughter cells is generally (but not always[85]) smaller than the other. This is the type of cell division most used by mosaic organisms like the nematode. It is very similar to self renewing stem cells and probably preceeded stem cells in evolution.

- *Amitotic* cell differentiation without cell division: A cell changes cell type without dividing. This is the type we found in the axolotl ectoderm.

The italicized words give us shorthand for each type. Note that this classification makes no mention of mechanisms or causes of differentiation, nor of where in the embryo the daughter cells end up. In all five cases it is presumed that the amount of DNA in each cell is the same, i.e., these cases do not include meiosis, nuclear division without cell division, or multinucleate cells arising via cell-cell fusion.

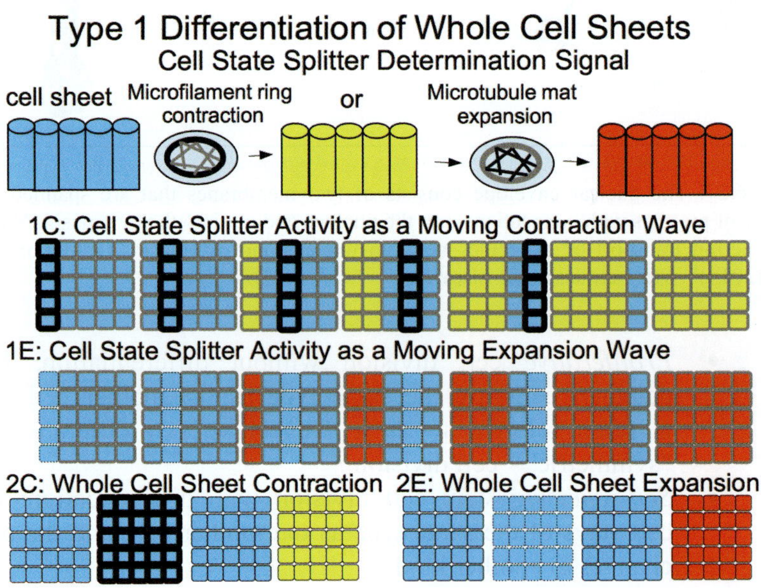

Figure 6.28. Type 1: *Whole Cell Sheet Differentiation*. **Top**: In Type 1 epithelial differentiation an entire sheet of cells (blue) undergoes either contraction (yellow) or expansion, via the microfilament rings or microtubule mats in its cell's cell state splitters. An example of the apical end of a cell is shown with the microfilament ring in black when it wins the tug-of-war in the cell state splitter, and another with the microtubule mat in

black when it wins the tug-of-war in the cell state splitter. The transient change in apical size may or may not persist, as differentiation proceeds, and so all cells are left the same size in this and subsequent figures. **1C**: The whole can change as result of a propagating contraction wave. A column of contracting cells is indicated by black-ringed cells (first frame), with the contraction propagating from left to right in this time sequence. The contraction relaxes, leaving a column of cells to the left (blue) that has been determined (second frame). These cells then differentiate (yellow column). The wave of contraction has in the meantime moved on to the right, until the whole tissue has been determined and the differentiated. **1E**: A propagating expansion wave, starting at the left with a column of expanded cells indicated by their thin outlines. Again the columns of cells to the left of the expansion wave is determined, and differentiates in the subsequent frame. The durations of contraction or expansion, the time between determination and discernible differentiation, and the wave speeds can all vary between tissues, between species, and with temperature. **2C**: All of the cells in a cell sheet may contract at once (second frame) and proceed synchronously through determination and differentiation (frames 3 and 4). **2E**: Same for expansion.

We will now focus on differentiation waves in epithelia. We have observed these waves either in the axolotl during embryogenesis or deduced them via careful examination of what is known about determination and differentiation in other organisms. Bearing in mind the kinds of differentiation and cell division (Figure 6.2) we concluded there are six types of differentiation waves in epithelial cell sheets. The six types of differentiation waves are as follows. The first three kinds of epithelia generation involve one step of differentiation per cell Figure 6.2, while the last three could be called "compound differentiation", as some cells undergo more than one step of differentiation.

Type 1. *Whole Cell Sheet Differentiation.* The Type 1 form of epithelial differentiation occurs when an entire sheet of cells differentiates together as a coordinated unit forming a new sheet of the next cell type (Figure 6.28). All the cells in the sheet become the new cell type together. All the cell state splitters do the same thing: they all contract or they all expand. We have observed two forms of the whole cell sheet differentiation waves. There is a moving contraction wave. This begins at one point, then spreads across the whole cell sheet or starts as a large circle that becomes smaller until it vanishes at a point. The wave launching domain is triggered by a mechanical signal generated outside the cells propagating the wave ("external"), but the propagation from cell to cell within a sheet is internal to the cell sheet. We have also observed contraction where all cells in an entire region simultaneously undergo a contraction or expansion together in response to an external mechanical event. These waves are purely amitotic in nature.

Type 2. *Alternating Pattern of Differentiation.* In Type 2, spatially alternating differentiation is the result. The surface area of the entire sheet remains the same but some cell state splitters contract and some cell state splitters expand resulting in an alternating pattern, often called a spacing pattern[86]. The force driving this alternating pattern has not yet been documented but it is likely that the individual cells in the sheet are relying more on mechanical signals than on biochemical signals. Cell sheets where an alternating pattern is found also have far fewer of the structural elements, such as gap junctions, that are required for cell-to-cell communication. Because of the poor cell-cell communication, a wave of contraction or expansion does not propagate from cell to cell. One cell contracts and in doing so, it pulls on the next cell so that the cell expands, which then pushes on the next cell causing it to contract and so on until the entire sheet has experienced alternating contraction or expansion. We have also seen evidence that sometimes it is not a single cell alternating pattern but rather alternating clusters of cells. The result is alternating cell types or alternating clusters of cell types. This can be a checkerboard pattern of individually different cells (Figure 1.40) or it can produce an alternating pattern of clusters of differentiating cells. If different regions of the sheet start independently of others this will lead to occasional irregularities in the spacing pattern. We have studied the spacing pattern in *Drosophila* eye imaginal discs (Figures 6.29, 6.30, 6.31, 6.32 and 6.33) but the pattern is also common in spacing of ciliated epidermal cells[87-89], feather buds[90] and plant trichomes[91-93]. We have also observed the alternating patterns of stomata in first leaves and/or cotyledons of seedling *Arabidopsis* plants (cf.[94-102]) This pattern is also amitotic in nature (Figure 6.34).

Type 3. *Two Cluster Differentiation.* In Type 3 differentiation, there are two clusters of two different kinds of cells formed. The cell state splitters in part of the sheet contract, the cell state splitters in the other part expand. Where expansion or contraction occurs it propagates from cell to cell affecting entire regions of the cell sheet. This is often accompanied by whole cell sheet movement

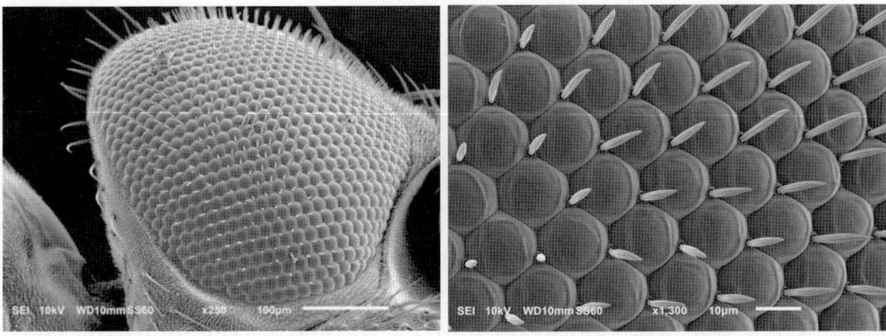

Figure 6.29. Adult *Drosophila* eye with closeup of the mature ommatidia by SEM (scanning electron microscopy). For the whole normal adult see Preface Figure 0.1. This one happens to be the mutant *wingless*. From a web page of Stanley A. Rice[103] with his kind permission.

and stress and/or strain caused by the movement, especially where bending occurs. These other mechanical forces limit the propagation of the signal to only one part of the cell sheet. This type occurs during axolotl gastrulation in endoderm, mesoderm, and ectoderm. We will describe this in detail in Chapter 7. This type is also amitotic (Figure 6.35).

Type 4. *Two-Cluster Differentiation with a Few Single Cells*. The Type 4 cell sheet is a form of epithelial differentiation which results in two horizontally adjacent cell sheets. This is similar to Type 3 except that Type 4 will also produce a few loose cells. High mechanical stresses at bend points break the gap junctions and cell adherens junctions and these result in some single cells leaving the sheet altogether while the rest of the cells in the sheet remain attached to each other. This typically occurs in regions of bending of the entire cell sheet. The boundary zone, where a few cells are removed, is the dividing line between the two new cell types. The cells that stay attached to the sheet differentiate into one of two new cell types depending on whether they have experienced a contraction or expansion (as in Type 3). The detached cells experience neither contraction nor expansion. When individual cells are removed from the cell sheet, the cell state splitter ceases to have any kind of function for them and the cell rounds up instead. The signal for these cells is that they are no longer in the cell sheet. These single cells will, in some situations, take on an amoeboid form and migrate to a new location before differentiating into a new cell type, or they may stay in place and become what are known as pleuripotential stem cells that may become involved in later regeneration. This is discussed further in Chapter 11. The precise cytoskeletal signal remains to be found but the answer may exist in the study of adherent cells in culture. A cultured cell that is chemically forced to release its adhesion to the culture dish undergoes dramatic cytoskeletal changes and even more dramatic changes to its nucleus. The nucleus changes shape and loses up to 50% of its volume[104] (Figure 6.36). These detached cells either migrate to a new location, or they rest near the cell sheet they once belonged to. An example of this type is the neural crest cells of the axolotl. The neural crest cells leave the cell sheet from the boundary zone between neural epithelial cells and future skin cells in the axolotl embryo. These individual ectoderm cells, freed from the cell sheet, give rise to the neural crest cells. Neural crest cells migrate to locations throughout the embryo and eventually form, for instance, the peripheral nerve cells that deliver signals between body and brain, and the pigment cells. This fourth type of wave is also amitotic but cells that fall out of the sheet will continue to have proliferative and/or self renewing divisions once free of the sheet (Figure 6.37).

Type 5. *Two Cluster Differentiation with Migrating Cells*. Type 5 of cell sheet differentiation is actually a variant of the Type 4. Part of the cell sheet forms two new cell types as in cell sheet Types 2 to 4. The differentiating cell sheet also has cells that leave the cell sheet, often at a point of extreme bending, but instead of just a few cells, many cells will leave the sheet. They join together and create a new cell sheet. This new cell sheet can be either beside the old one or they may migrate individually and join together to create a new cell sheet in another location. Type 5 cell sheet differentiation is found in early mammalian gastrulation when the mesodermal layer forms beneath differentiating ectoderm. Mesodermal intercalation in frogs can also be explained as being this form of differentiation. Though mostly amitotic, the cells that produce new migrating cells use other types of cell division such as proliferative, symmetric and asymmetric cell division to accomplish their final result in forming a second sheet of new tissue (Figure 6.38).

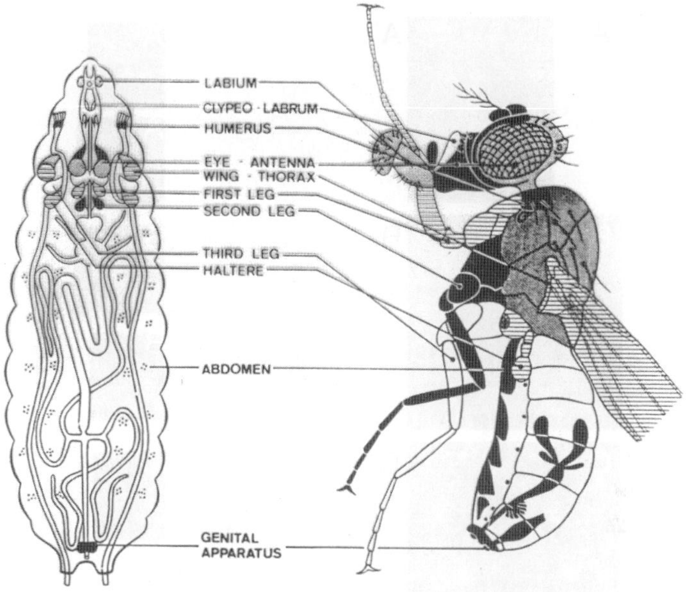

Figure 6.30. Imaginal discs in an insect larva (grub) are small tissues that differentiate and expand during metamorphosis forming most of the adult parts, except for the brain. Here we will focus on the eye imaginal disc. From Nöthiger (1972)[105] with permission of Springer.

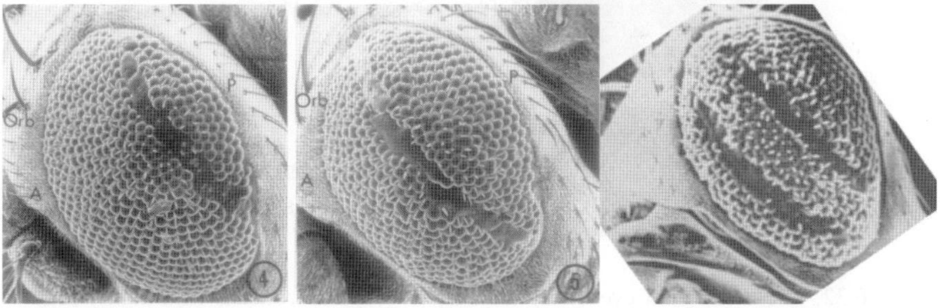

Figure 6.31. The temperature sensitive mutant *shibire*^{ts1} of the fruit fly *Drosophila melanogaster* may be used to record the motion of the differentiation wave called the "morphogenetic furrow" across the eye imaginal disc. **Left**: When the temperature is raised briefly from 22 to 29°C the wave keeps propagating, but the subsequent steps of differentiation of the cells into ommatidia fail. **Middle**: Here the temperature was raised briefly at a later time. The first two SEMs are from Poodry *et al.* (1973)[106] with permission of Elsevier. **Right**: When the temperature is raised briefly twice, two lines of ommatidia are missing. The third SEM is rotated to the same orientation of the first two and is from Suzuki (1974)[107] with permission of NRC Research Press.

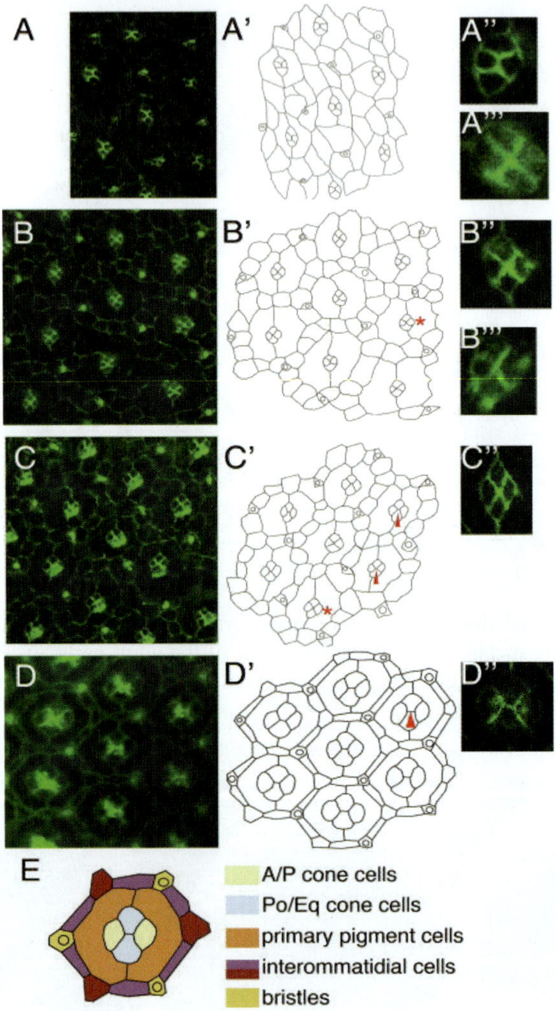

Figure 6.32. The compound eye of the fruit fly *Drosophila* consists of a highly regular array of ommatidia, each of which contains cells that focus light on retinal cells that then send nerve signals to the brain. The ommatidia start out in a spacing pattern that gradually comes to look almost crystalline. We have found earlier indications of this pattern (Figure 6.33). Details: "Confocal micrographs of cell junctions [cell-cell contacts] in fields of ommatidia, visualized with Arm-GFP, and accompanying tracings of 48 (A, A′), 60 (B, B′), 72 (C, C′), and 92 hour (D, D′) pupals eyes raised at 18°. Confocal micrographs of single ommatidia with distinctive configurations of cone cells (A″, A‴, B″, B‴, C″ and D″). (E) Color-coded key for identification of each cell type in ommatidium. Anterior is to the right for all panels." From Grillo-Hill & Wolff (2009)[108] with permission of John Wiley and Sons.

Type 6. *Stratification Differentiation.* Type 6 is a combination of the alternating pattern Type 2 and the single stem cell Type 4, with formation of a new cell sheet below the first. For details see Figure 6.39. Attachment points act as extrinsic cues allowing the spindle apparatus of individual cells to orient such that a new cell type is always dropped below the sheet[117]. One puzzling aspect of neural cell proliferation is that it is often followed by a round of cell death in an alternating pattern. This pattern of cell death can be explained by all the stem-like cells (yellow in Figure 6.39) in the sheet undergoing programmed cell death as their final differentiation once their role of creating the new cell layer below is completed[118]. This round of cell death could be triggered by a subsequent wave through the entire cell sheet (Figure 6.39).

The action of the cell state splitter is always involved in the determination step. First there is either a microfilament ring contraction or microtubule mat expansion. This creates the determination signal. The signal is either "contraction" or "expansion". Which signal is actually initiated in a given cell depends on the mechanics and physics of the embryo as a whole. Certain physical areas are mechanically affected such that the microfilament ring wins the tug-of-war and a contraction occurs at one region of cells. Alternatively, the microtubule mat is given the advantage by some mechanical event and the microtubule mat wins the tug-of-war and an expansion wave propagates over one region in the cell sheet.

Now let's take a step back and look at those mosaic organisms like nemotodes again. Most often their cell divisions are asymmetric and each produces one smaller cell and one larger cell. If we assume the division is linked to differentiation and the smaller cell undergoes the equivalent of a contraction and the larger cell undergoes the equivalent of an expansion, then we have a spindle apparatus that is both a cell state splitter and a mitotic spindle. Each is a mitotic single cell differentiation wave, the evolutionary precursor to the amitotic differentiation wave.

We have observed that there is an initial ten-minute contraction[79] during determination in the future nerve cells in axolotl ectoderm. This contraction relaxes and the ectoderm cells return temporarily to their previous shape. When the future nerve cells begin differentiating a permanent shape change into tall thin cells occurs (Figure 6.40). So

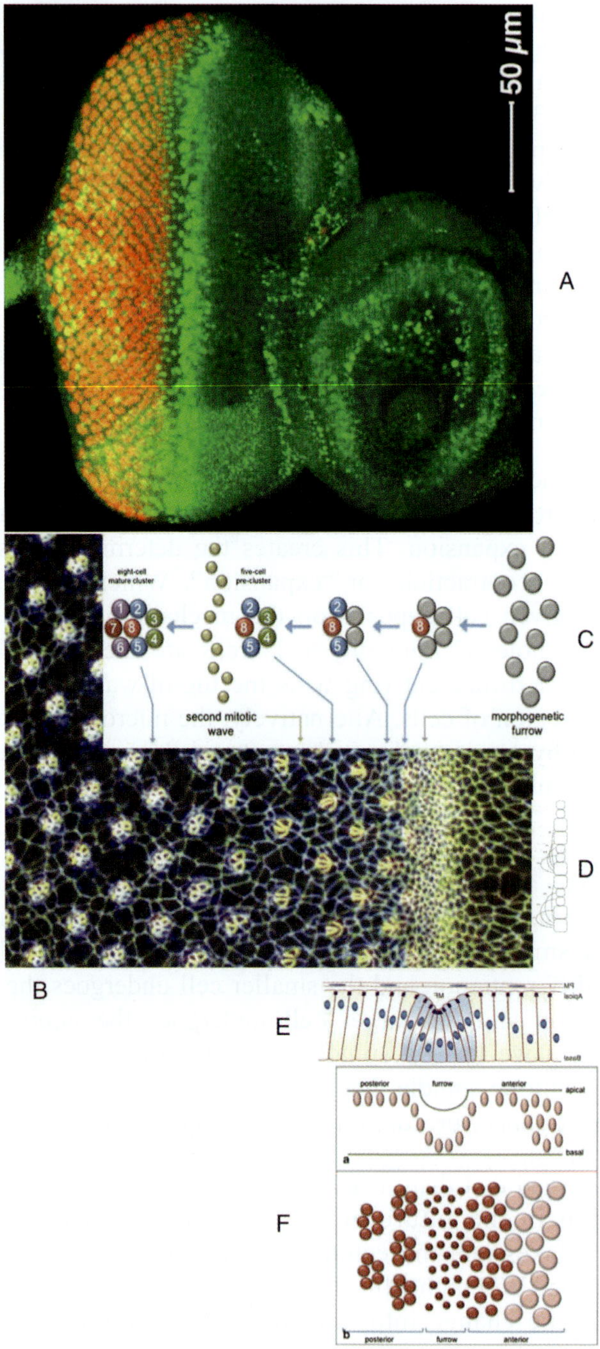
50 μm
A
eight-cell mature cluster
five-cell pre-cluster
second mitotic wave
morphogenetic furrow
C
D
B
E
posterior
furrow
anterior
apical
basal
a
F
b
posterior
furrow
anterior

Figure 6.33. **A**: The morphogenetic furrow is a differentiation wave in the *Drosophila* eye imaginal disc (left blob) oriented here propagating from left to right, posterior to anterior. Here the morphogenetic furrow has reached midway. As the furrow moves, some of the cells that have participated in the wave differentiate into ommatidia, which are the optical and neural components of the insect compound eye at this stage containing one each of 8 kinds of cells. At maturity a *Drosophila* ommatidium contains 23 cells. Most are recruited from the surrounding "undifferentiated" cells rather than by asymmetric cell divisions[109]. From Vrailas *et al.* (2006)[110] with kind permission of the Company of Biologists. **B**: Fluorescence image of the eye imaginal disc at sufficient magnification to show the apical ends of the individual cells, which are smallest for the cells in the furrow. From the cover image of *Genetics* by Konstantin Gaengel and Marek Mlodzik accompanying Weber *et al.* (2012)[111], with kind permission of Tracey DePellegrin Connelly of the Genetics Society of America. **C**: Sequence of development of the ommatidia after passing of the morphogenetic furrow through the cells. From Kumar (2013)[112] with permission of Springer. **D**: A detailed analysis of the steps of differentiation of the ommatidia[113] in terms of single cell differentiation waves was attempted in §9.12 *Differentiation Waves in Fly and Fish Eyes* of *Hierarchical Genome*[9], copied from Figure 52a. When we measured the sizes of cells along the bottom of the furrow from Wolff (1993)[114], we showed by variogram analysis that those cells exhibit an alternating pattern of two large cells followed by two small cells. This pattern precedes that previously described[115]. **E**: The apical ends of the cells contract, forming the furrow. "The columnar cells of the eye imaginal disc epithelium are apically constricted and shorter within the morphogenetic furrow. Dark blue lines indicate zonula adherens. A layer of squamous cells, the peripodial membrane (PM), overlies the columnar cells." From Schlichting & Dahmann (2008)[116] with permission of Elsevier, with the sketch left/right reversed to conform with the others. **F**: a) Interkinetic migration (vertical up/down migration of the cell nuclei). Before cells participate in the furrow (right), their nuclei are randomly distributed in the apical/basal direction. As they participate in the furrow, their nuclei migrate to the apical surface. While they are in the furrow, the nuclei plunge basally but ascend again as the furrow moves on. b) As cells participate in the furrow, their apical profiles constrict. After the furrow passes, their apical profiles expand. From Kumar (2013)[112] with permission of Springer.

apical contraction leads to tall skinny cells in ectoderm that become neural cells. Similarly, when the expansion wave goes through the lower half of the ectoderm, that expansion is then followed by a change to short squat cells. However, final shape change depends on subsequent differentiation events and not on the determination wave. We should not expect the final shape of all cell types will always be contraction wave = tall skinny cells and expansion wave = short squat cells.

Now let's compare waves to gradients. It is very common for embryologists and molecular biologists to speak of a "morphogen gradient" across developing tissue (Figure 6.41). Molecules presumed to be morphogens are the basis for the concept of positional information[119], which presumes that a cell can know its position by "reading" the concentration of the molecules of the gradients and then deciding what it is supposed to do, by "looking up" its coordinates in some sort of stored look up table[120] in its DNA. Despite a huge literature on morphogens,

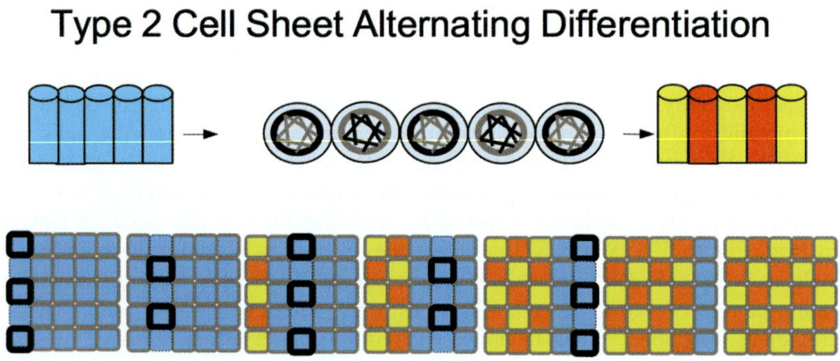

Figure 6.34. Type 2: *Alternating Pattern of Differentiation.* **Top**: All of the cells in a row (blue) alternate as to whether the microfilament ring or the microtubule mat (grey) wins the tug-of-war. The result on differentiation is alternating kinds of cells (yellow and red). **Bottom**: As for Type 1, a wave moves left to right in columns of alternating patterns, but in this case it is a wave of alternating contraction and expansion, followed by columns of alternating determination and differentiation.

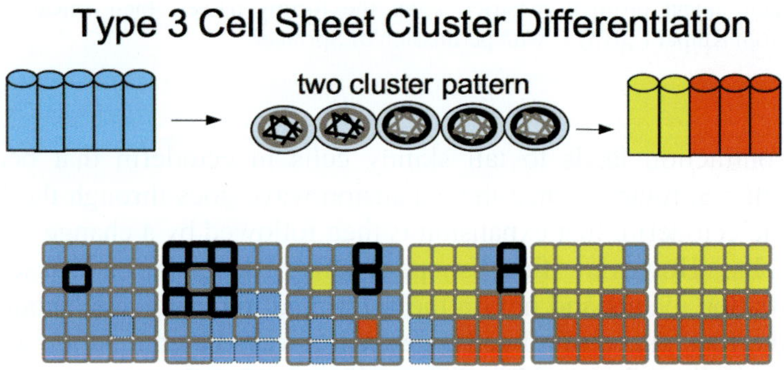

Figure 6.35. Type 3: *Two Cluster Differentiation.* All cells participate in a contraction or expansion in two clusters. A sheet of cells supports both a contraction wave and an expansion wave. There is a sharp boundary between their two dimensional trajectories. Mechanics of the cell sheet in the context of the whole embryo may determine the launching sites, propagation and termination of each wave.

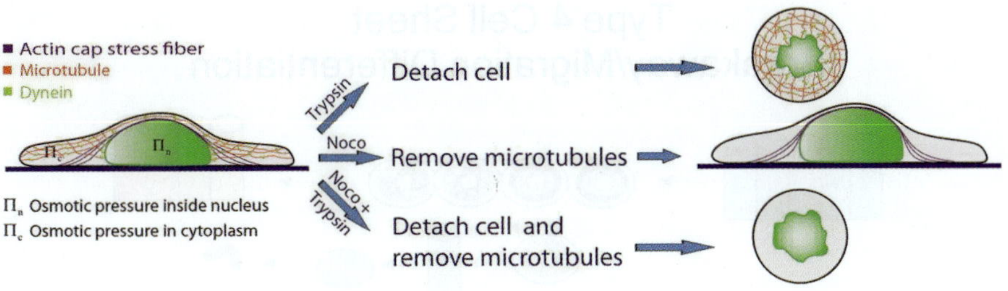

Figure 6.36. When cells are removed from a cell sheet at points of high stress during differentiation waves of contraction or expansion or at boundaries zones where two waves meet, individual cells can be forced out of the cell sheet. These cells round up and undergo differentiation appropriate to germ cells or stem cells. This can be followed by migration to a new location and cell proliferation to form a new cell layer, or quiescence, awaiting a future signal before proliferation and repair. The signal for such a cell is neither expansion nor contraction but rather a dramatic resetting of the genome. The mechanism may lie in the kind of changes observed in adherent cells when they are forced to leave the cell sheet, as shown here. From Kim *et al.* (2015)[104] with kind permission of the Company of Biologists. Noco = nocodazole.

even Lewis Wolpert, who coined the phrase in 1969[119], has expressed doubts about the reality of positional information[121,122]. This is not to deny the existence of gradients across cells in a tissue. Gradients of gene products absolutely do indeed exist and they can be measured. We have another explanation for gradients.

Once any individual cell has participated in an expansion wave or a contraction wave, determination is complete. Differentiation follows. If the cell sheet is examined as a whole during differentiation, there will be a gradient of new protein products that result from differentiation progressing across the sheet. The highest concentration of new proteins will be in the region where the wave began. The lowest concentration will be in the region where the wave ended. This gradient is merely a temporary epiphenomenon, a byproduct of differentiation wave activity (Figure 6.42). Once all differentiation is complete, the gradient will vanish and all cells of the same type will have the same level of gene expression over the entire sheet (Figure 6.42). There is one very good use for gradients of gene expression. Wherever they exist, one can expect to find a differentiation wave.

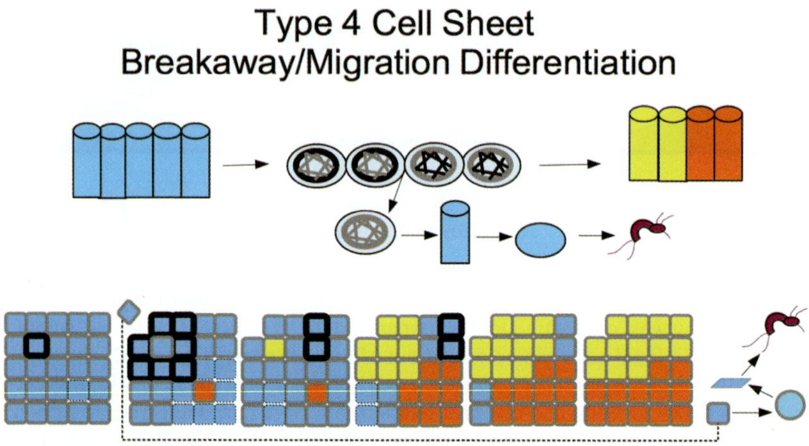

Figure 6.37. Type 4: *Two-Cluster Differentiation with a Few Single Cells.* All cells participate in a contraction or expansion in two clusters. A few cells leave the cell sheet and they round up. They then either migrate to a new location or remain with the sheet they originated from. If they remain with the cell sheet, they may form quiescent stem cells waiting for a signal to leave the quiescent state to take part in regeneration or healing. If they migrate, once they reach their final location they differentiate into their final form. As another step of differentiation is involved here, Type 4 is an example of compound epithelial differentiation. One example of this type is nerve cells (purple) which arise from neural crest cells released from the boundary between neural ectoderm and epithelial ectoderm at the beginning of neural tube closure.

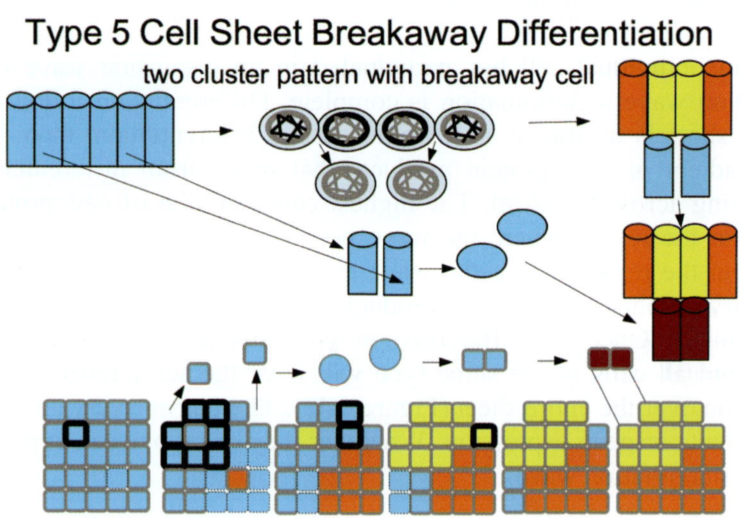

Figure 6.38. Type 5: *Two Cluster Differentiation with Migrating Cells.* All cells participate in a contraction or expansion in two clusters except for some cells which leave the sheet, migrate and then form another cell sheet near the original one. Type 5 is also an example of compound epithelial differentiation.

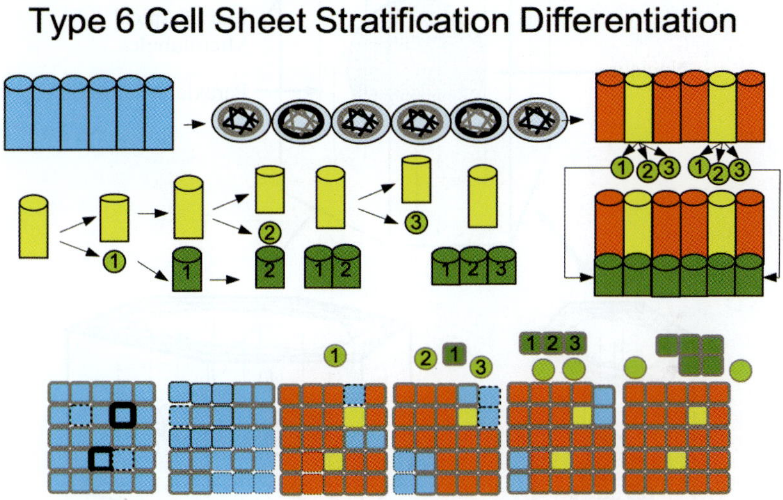

Figure 6.39. Type 6: *Stratification Differentiation.* This is another kind of compound epithelial differentiation. A sheet of cells (blue) first differentiates via the alternating pattern of Type 2 producing two new cell types in the same layer. One of the two new cell types (red) retains connections between individual cells and differentiate together and retain an epithelial character. Meanwhile the isolated cells of the other type within the alternating pattern (yellow) differentiate to become a stem-like cell. These stem-like cells then begin an asymmetric division that drops many new cells below the original cell sheet (green). These new cells, individual products of asymmetric cell divisions, then proliferate, interconnect and form a second cell sheet layer below the original one. Type 6 differentiation is important for the thickening and proliferation of the neural tube into the brain once neural tube closure is complete. The place of the stem-like cell (yellow) within the sheet of other cells keeps the formation of the new cell sheet properly oriented.

It is plausible that there are differences between early versus late developmental cell state splitter signal events. The nature of the signaling event between cells either in a cell sheet or between cells in asymmetric division is always a combination of a mechanical and a chemical event. Very early in embryogenesis, the mechanical aspects of the signal are much more important. Entire sections of the embryo contract and expand and enormous shape changes occur.

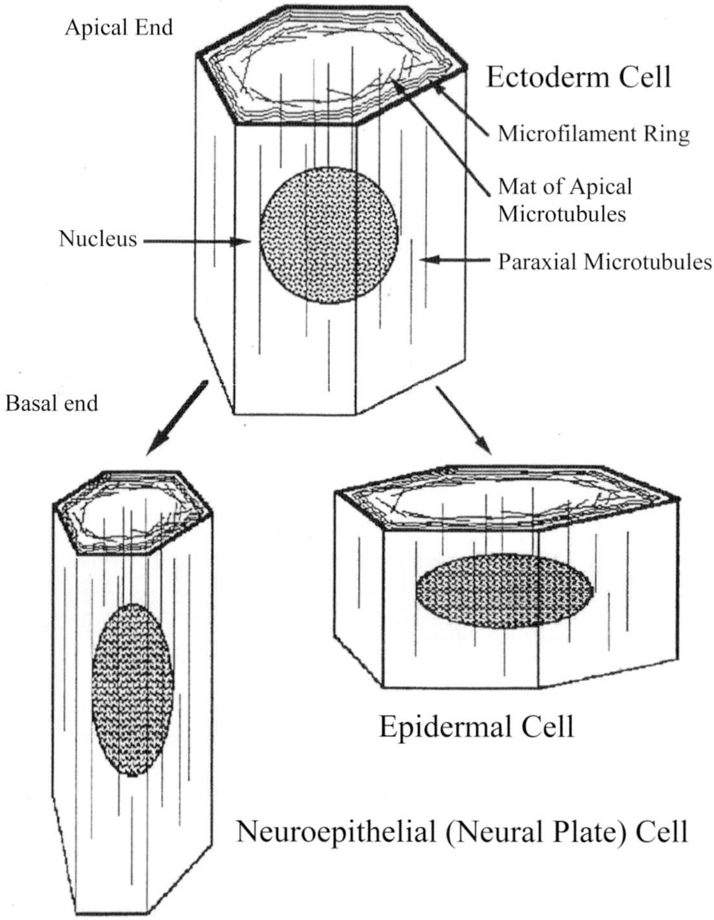

Figure 6.40. An embryonic ectoderm cell is a member of a sheet of cells forming an epithelium. At its apical end is a microfilament ring subtending a mat of apical microtubules. These are in a radial tug-of-war that results in either the microfilaments winning or the microtubules winning. If the microfilaments win there is a a prolonged contraction of the apical end (10 minutes in axolotl) and the cell changes differentiation state from ectoderm to neuroepithelium. If the apical microtubules win, the result is that the ectodermal cell becomes an epidermal cell. Paraxial microtubules have been shown to be oriented with plus ends nearer the basal end of epithelial cells and minus ends nearer the apical end[123]. It has been postulated that paraxial microtubules originate from a microtubule organizing center (MTOC) that spreads out at the apical end of an epithelial cell[124]. If this is the case, then the paraxial microtubules may be connected to the cell state splitter and also transport molecules and/or organelles to/from it. From Björklund & Gordon (1993)[125] with kind permission of Sergey Vassetzky. The nuclear shape changes are surmised.

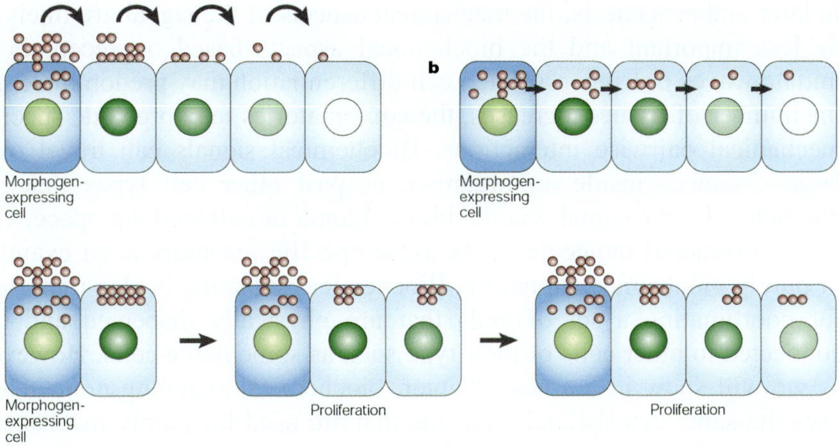

Figure 6.41. Three models for the generation of gradients of molecules presumed to be "morphogens", i.e., molecules that cells hypothetically use to decide, according to the concentration of the morphogen, whether or not to differentiate. "a) Diffusion through the extracellular space. b) Planar transcytosis. c) Displacement during growth". From Tabata (2011)[126] with permission of Nature Publishing Group.

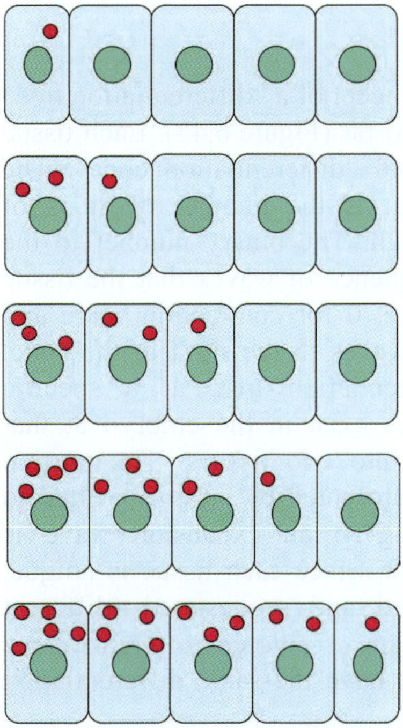

Figure 6.42. A contraction wave is depicted propagating from one cell to the next in an epithelial layer of cells. This initiates differentiation of the cell, which starts to produce cell type specific molecules, shown in red. If these molecules increase in number over time, as the wave propagates, a gradient of the molecules will develop across the epithelium, especially because differentiation waves propagate slowly, taking hours to cross it. The differentiation has already occurred, so the gradient is not causal of differentiation. In fact, the gradient may be called an "irrelevant epiphenomenon"[127].

In later embryogenesis, the mechanical aspects of the signal are likely to be less important and the biochemical aspects based on receptors to initiate waves or trigger isolated cell differentiation may predominate. As the number of tissues increases, the combinatorics may preclude specific mechanical pairwise interactions. Biochemical signals can travel over large distances inside an organism or past other cell types that are untouched by the signal, via the blood, lymph or extracellular space. The use of hormonal molecules with tissue specific receptors is an example of purely biochemical signaling. Biochemical signaling is also important for coordinating a set of cells that are physically dispersed and not connected to other cells of their type such as some nerve cells. However, as we will show in the next chapter, biochemical signaling nonetheless uses the same cytoskeletal elements that are used for purely mechanical signaling. The receptor is simply an adapted accessory to the cytoskeleton, an extension of it, allowing the cytoskeleton to respond to a purely biochemical determination signal for initiating differentiation. We will explore mechanical signaling events and their importance in early embryogenesis when we use the fate map of the axolotl embryo to map the alternation of contraction and expansion signals by cell state splitters during gastrulation and neurulation. We delve into the nature of the signaling to the nucleus in Chapters 8 and 9.

We end this chapter by introducing the concept of a "differentiation tree" as a way of summarizing what we have so far (Figure 6.43). Each tissue in an embryo is designated by a decimal "differentiation code". The integer to the left of the decimal point is the number of steps of differentiation the tissue has gone through. The binary number to the right of the decimal point shows the sequence of waves that the tissue went through to arrive at its current state: 0 for contraction wave and 1 for expansion wave, with the newest waves to the right in the code. Any horizontal line across the differentiation tree, at a specific developmental time, shows which tissues exist in the embryo at that moment. At each node a tissue splits into two tissues. The portion participating in a contraction wave is represented by an arrow tilted to the left, while the portion participating in an expansion wave is represented by an arrow tilted to the right. Since each tissue is unique, with different subsets of genes exposed and sequestered (different differons), each arrow representing a tissue is colored differently. We regard all of the cells in an embryo that have the same differentiation code as equivalent to one another, in the sense that they have the same subsets of genes exposed and sequestered, i.e., the same differons. Thus to distinguish our notion of a tissue from the less precise uses of the word, we will refer to the set of all cells with the same differentiation code as an "equivalence tissue". This language conforms to the notion of "equivalence class" in mathematics[128].

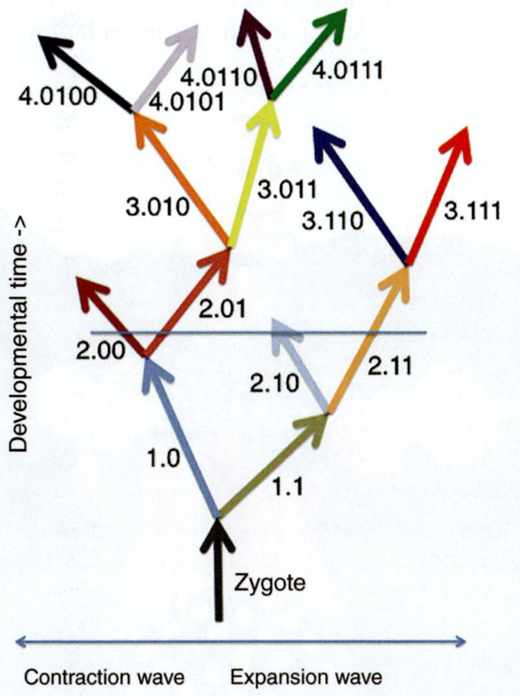

Figure 6.43. The general notion of a differentiation tree, first introduced theoretically in Gordon (1992)[129] and empirically, for the axolotl, in Gordon *et al.* (1994)[79]. Colored arrows represent distinct tissues, and are labeled by their differentiation codes. They are tilted left or right depending on whether a contraction or expansion wave was carried by the cells the arrow represents. All cells experiencing the same sequence of waves, designated by the sequence of 0's and 1's to the right of the decimal point in the differentiation code, are considered to define a tissue (whether the cells are in contact or not). The number to the left of the decimal point indicates how many waves brought the cells to that state of differentiation. If there is only one cell in a tissue[132], we may call this a "cellular differentiation tree". A horizontal line shows which tissues exist in the embryo at a given moment in developmental time, except for terminal cell types, whose arrows may either extend to adulthood or halt ealier due to cell death. This follows the convention of Sulston *et al.* (1983) for lineage trees[130], Figure 2.15.

If a tissue consists of a single cell, as it usually does in nematodes, for instance, then at each asymmetric division we refer to the larger daughter cell as having undergone a "single-cell expansion wave" and the smaller daughter cell as having undergone a "single-cell contraction wave" (Figure 6.44). In this case, we may call the organism's differentiation tree a "cellular differentiation tree", and the number of cells in any equivalence tissue is 1, i.e., the nematode consists primarily of one-cell tissues (Figure 2.15), since most of its cells have a distinct differentiation code. We are beginning to compare the differentiation trees of various

Figure 6.44. A balloon squeezed in the middle produces two spheres of equal size. This is a simple model for a "vegetative" cell division. A balloon squeezed off center results in a large sphere and a small sphere. This models asymmetric cell division, corresponding to a single cell expansion wave for the large cell and a single cell contraction wave for the small cell. Photograph by Mary Ann Marquez in 2012 of balloonist Kevin King = Floyd the Clown[131], with permission of both of them.

organisms across widely different phyla[132]. We expect to find some larger unifying structure of differentiation trees, perhaps illuminating the evolution of differentiation.

In the previous chapter, we introduced the marvelous cytoskeleton and everything it does in the fertilized egg of the axolotl. We described how the cytoskeleton sets up mechanical biases in the cell that will later be resolved. In this chapter, we defined and explained the cell state splitter, consisting of dynamic cytoskeletal components, and its role in determination and differentiation. We then elaborated on the assorted types of propagating waves the cell state splitter can produce and the concept of the differentiation tree. In the next chapter we will present the differentiation tree of the axolotl from cleavage to the end of gastrulation.

1. Wood, W.B. (1988). *The nematode Caenorhabditis elegans*. Cold Spring Harbor, New York, Cold Spring Harbor Laboratory.
2. Sulston, J.E. & H.R. Horvitz (1977). Post-embryonic cell lineages of the nematode *Caenorhabditis elegans*. *Developmental Biology* **56**, 110-156.
3. Vogt, W. (1928). Mosaikcharakter und Regulation in der Frühentwicklung des Amphibieneies/Mosaic character and regulation in the early development of the amphibian egg. *Verhandlungen der Deutschen Zoologischen Gesellschaft 32 Vers. München* **32**(Suppl.), 26-70.
4. Novikoff, A.B. (1939). Regulation in mosaic eggs. *Biological Bulletin* **77**(2), 308.
5. Yamana, K. & H. Kageura (1985). Mosaic and regulatory development of early amphibian embryos and their blastomeres. *Cell Structure and Function* **10**(4), 437.
6. McCain, E.R. & J.N. Cather (1989). Regulative and mosaic development of *Ilyanassa obsoleta* embryos lacking the A and C quadrants. *Invertebrate Reproduction & Development* **15**(3), 185-192.
7. Lawrence, P.A. & M. Levine (2006). Mosaic and regulative development: two faces of one coin. *Current Biology* **16**(7), R236-R239.
8. Callebaut, M., E. Van Nueten, F. Harrisson & H. Bortier (2007). Mosaic versus regulation development in avian blastoderms depends on the spatial distribution of Rauber's sickle material. *Journal of Morphology* **268**(7), 614-623.
9. Gordon, R. (1999). *The Hierarchical Genome and Differentiation Waves: Novel Unification of Development, Genetics and Evolution*. Singapore & London, World Scientific & Imperial College Press: [http://www.worldscientific.com/worldscibooks/10.1142/2755].
10. Leys, S.P. & A. Riesgo (2012). Epithelia, an evolutionary novelty of metazoans. *Journal of Experimental Zoology Part B-Molecular and Developmental Evolution* **318B**(6), 438-447.

11. Gordon, R. & G.W. Brodland (1987). The cytoskeletal mechanics of brain morphogenesis. Cell state splitters cause primary neural induction. *Cell Biophysics* **11**(1), 177-238.
12. Burnside, M.B. (1971). Microtubules and microfilaments in newt neurulation. *Developmental Biology* **26**, 416-441.
13. Baker, P.C. & T.E. Schroeder (1967). Cytoplasmic filaments and morphogenetic movements in the amphibian neural tube. *Developmental Biology* **15**, 432-450.
14. Gachet, Y., C. Reyes, S. Goldstone & S. Tournier (2006). The fission yeast spindle orientation checkpoint: a model that generates tension? *Yeast* **23**(13), 1015-1029.
15. Shubnikov, A.V. & N.V. Belov (1964). *Colored Symmetry*. New York, Pergamon Press.
16. Loeb, A.L. (1971). *Color and Symmetry*. New York, John Wiley & Sons.
17. Guillot, C. & T. Lecuit (2013). Adhesion disengagement uncouples intrinsic and extrinsic forces to drive cytokinesis in epithelial tissues. *Developmental Cell* **24**(3), 227-241.
18. Le Borgne, R., Y. Bellaïche & F. Schweisguth (2002). *Drosophila* E-cadherin regulates the orientation of asymmetric cell division in the sensory organ lineage. *Current Biology* **12**(2), 95-104.
19. Marella, N.V., B. Seifert, P. Nagarajan, S. Sinha & R. Berezney (2009). Chromosomal rearrangements during human epidermal keratinocyte differentiation. *Journal of Cellular Physiology* **221**(1), 139-146.
20. Houben, E., K. De Paepe & V. Rogiers (2007). A keratinocyte's course of life. *Skin Pharmacology and Physiology* **20**(3), 122-132.
21. Alicea, B., S. McGrew, R. Gordon, S. Larson & M. Watts (2014). From Differentiation Waves to "Wriggling": new views on *C. elegans* development. http://figshare.com/articles/From_Differentiation_Waves_to_Wriggling_new_views_on_C_elegans_development/1015760
22. Foe, V.E. (1989). Mitotic domains reveal early commitment of cells in *Drosophila* embryos. *Development* **107**(1), 1-22.
23. Foe, V.E. & G.M. Odell (1989). Mitotic domains partition fly embryos, reflecting early cell biological consequences of determination in progress. *American Zoologist* **29**(2), 617-652.
24. Maleyvar, R.P. & R. Lowery (1976). Correlation between the cell cycle in the neurectoderm and differentiation during the early development of *Xenopus laevis*. 1. BUdR sensitivity during gastrulation. *Cytobios* **17**(65), 21-30.
25. Maleyvar, R.P. & R. Lowery (1981). Correlation between the cell cycle in the neurectoderm and differentiation during the early development of *Xenopus laevis*. 2. Inhibition of DNA synthesis and mitosis during gastrulation. *Cytobios* **32**(126), 97-105.
26. Holtzer, H. (1970). Proliferative and quantal cell cycles in the differentiation of muscle, cartilage, and red blood cells. *Symposia of the International Society for Cell Biology* **9**, 69-88.
27. Holtzer, H., H. Weintraub, R. Mayne & B. Mochun (1972). The cell cycle, cell lineages, and cell differentiation. *Current Topics in Developmental Biology* **7**, 229-256.

28. Holtzer, H., N. Rubinstein, S. Fellini, G. Yeoh, J. Chi, J. Birnbaum & M. Okayama (1975). Lineages, quantal cell cycles, and the generation of cell diversity. *Quarterly Reviews of Biophysics* **8**(4), 523-557.
29. Holtzer, H., J. Biehl, P. Antin, S. Tokunaka, J. Sasse, M. Pacifici & S. Holtzer (1983). Quantal and proliferative cell cycles: how lineages generate cell diversity and maintain fidelity. *Progress in Clinical and Biological Research* **134**, 213-227.
30. Turo, K.A. & J.R. Florini (1982). Hormonal stimulation of myoblast differentiation in the absence of DNA synthesis. *American Journal of Physiology* **243**(5), C278-C284.
31. Grounds, M.D. & J.K. McGeachie (1987). A model of myogenesis in vivo, derived from detailed autoradiographic studies of regenerating skeletal-muscle, challenges the concept of quantal mitosis. *Cell and Tissue Research* **250**(3), 563-569.
32. Martin, C.C. & R. Gordon (1997). Ultrastructural analysis of the cell state splitter in ectoderm cells differentiating to neural plate and epidermis during gastrulation in embryos of the axolotl *Ambystoma mexicanum*. *Russian Journal of Developmental Biology* **28**(2), 71-80.
33. Burnside, M.B. (1973). Microtubules and microfilaments in amphibian neurulation. *American Zoologist* **13**, 989-1006.
34. Sinclair, G.D. & K. Brasch (1975). The nucleated erythrocyte: a model of cell differentiation. *Revue Canadienne de Biologie* **34**(4), 287-303.
35. Cohen, W.D., D. Bartelt, R. Jaeger, G. Langford & I. Nemhauser (1982). The cytoskeletal system of nucleated erythrocytes. I. Composition and function of major elements. *Journal of Cell Biology* **93**(3), 828.
36. Birgbauer, E. & F. Solomon (1989). A marginal band-associated protein has properties of both microtubule- and microfilament-associated proteins. *Journal of Cell Biology* **109**(4 Pt 1), 1609-1620.
37. Tucker, J.B. (1971). Microtubules and a contractile ring of microfilaments associated with a cleavage furrow. *Journal of Cell Science* **8**(2), 557-571.
38. Martin, A.C. & B. Goldstein (2014). Apical constriction: themes and variations on a cellular mechanism driving morphogenesis. *Development* **141**(10), 1987-1998.
39. Bülbring, E. (1955). Correlation between membrane potential, spike discharge and tension in smooth muscle. *Journal of Physiology-London* **128**(1), 200-221.
40. Kirber, M.T., J.V. Walsh & J.J. Singer (1988). Stretch-activated ion channels in smooth-muscle: A mechanism for the initiation of stretch-induced contraction. *Pflügers Archiv-European Journal of Physiology* **412**(4), 339-345.
41. Ivanov, A.I., I.C. McCall, B. Babbin, S.N. Samarin, A. Nusrat & C.A. Parkos (2006). Microtubules regulate disassembly of epithelial apical junctions. *BMC Cell Biology* **7**, #12.
42. Quintin, S., C. Gally & M. Labouesse (2008). Epithelial morphogenesis in embryos: asymmetries, motors and brakes. *Trends in Genetics* **24**(5), 221-230.
43. Sidhaye, V.K., E. Chau, V. Srivastava, S. Sirimalle, C. Balabhadrapatruni, N.R. Aggarwal, F.R. D'Alessio, D.N. Robinson & L.S. King (2012). A novel role for aquaporin-5 in enhancing microtubule organization and stability. *PLoS One* **7**(6), e38717.

44. Gordon, N.K. & R. Gordon (2016). The organelle of differentiation in embryos: the cell state splitter [invited review]. *Theoretical Biology and Medical Modelling* **13**(Special issue: Biophysical Models of Cell Behavior, Guest Editor: Jack A. Tuszynski), #11.
45. Mason, F.M., M. Tworoger & A.C. Martin (2013). Apical domain polarization localizes actin-myosin activity to drive ratchet-like apical constriction. *Nature Cell Biology* **15**(8), 926-936.
46. Ossipova, O., I. Chuykin, C.-W. Chu & S.Y. Sokol (2015). Vangl2 cooperates with Rab11 and Myosin V to regulate apical constriction during vertebrate gastrulation. *Development* **142**(1), 99-107.
47. Rogers, S.L., U. Wiedemann, U. Häcker, C. Turck & R.D. Vale (2004). *Drosophila* RhoGEF2 associates with microtubule plus ends in an EB1-dependent manner. *Current Biology* **14**(20), 1827-1833.
48. Jankovics, F. & D. Brunner (2006). Transiently reorganized microtubules are essential for zippering during dorsal closure in *Drosophila melanogaster*. *Developmental Cell* **11**(3), 375-385.
49. iSee (2008). Three AutoCAD 2008 Tutorial: circular arc springs and several spring painting. http://www.dasairen.com/AutoCAD/2305311391.html
50. Hill, T.L. (1981). Microfilament or microtubule assembly or disassembly against a force. *Proceedings of the National Academy of Sciences USA* **78**, 5613-5617.
51. Snelson, K. & E. Heartney (2013). *Kenneth Snelson; Art and Ideas*. New York, NY, USA, Marlborough Gallery.
52. Putnam, A.J., K. Schultz & D.J. Mooney (2001). Control of microtubule assembly by extracellular matrix and externally applied strain. *American Journal of Physiology-Cell Physiology* **280**(3), C556-C564.
53. Fudge, D.S., K.H. Gardner, V.T. Forsyth, C. Riekel & J.M. Gosline (2003). The mechanical properties of hydrated intermediate filaments: Insights from hagfish slime threads. *Biophysical Journal* **85**(3), 2015-2027.
54. Winegard, T., J. Herr, C. Mena, B. Lee, I. Dinov, D. Bird, M. Bernards, Jr., S. Hobel, B. Van Valkenburgh, A. Toga & D. Fudge (2014). Coiling and maturation of a high-performance fibre in hagfish slime gland thread cells. *Nature Communications* **5**, 3534.
55. Gordon, N.K. (2015). Hagfish Slime. http://embryogenesisexplained.org/2015/12/08/hagfish-slime/
56. Alippi, A., A. Bettucci, A. Biagioni, D. Conclusio, A. D'Orazio, M. Germano & D. Passeri (2012). Non linear behaviour of cell tensegrity models. In: *International Congress on Ultrasonics (Gdansk 2011), American Institute of Physics (AIP) Conference Proceedings*. **1433**: 329-332.
57. Burša, J., R. Lebis & J. Holata (2012). Tensegrity finite element models of mechanical tests of individual cells. *Technology and Health Care* **20**(2), 135-150.
58. Bansod, Y.D. & J. Burša (2014). Tensegrity principle based computational model of cytoskeleton. In: *6th World Conference on Structural Control and Monitoring (6WCSCM)*, 3317-3326.
59. Levin, S.M. (2014). *Biotensegrity, Bioarchitecture & Dynamic Anatomy: A Unified Model for Dynamical Anatomy, Stability, Mobility, Information Dissemination*. Second Life®, Embryo Physics Course.

60. Scarr, G. (2014). *Biotensegrity: The Architecture of Life*, Handspring Publishing.
61. Hamburger, V. (1988). *The Heritage of Experimental Embryology: Hans Spemann and the Organizer*. New York, Oxford University Press.
62. Fassler, P.E. (1996). Hans Spemann (1869-1941) and the Freiburg School of Embryology. *International Journal of Developmental Biology* **40**(1), 49-57.
63. Spemann, H. & H. Mangold (1924). Über Induktion von Embryonalanlagen durch Implantation artfremder 0rganisatoren [The induction of embryonic predispositions by implantation of organizers foreign to the species]. *Archiv für Mikroskopische Anatomie und Entwicklungsmechanik* **100**(3/4), 599-638.
64. Twitty, V.C. (1966). *Of Scientists and Salamanders*. San Francisco W. H. Freeman and Company
65. Jacobson, A.G. & R. Gordon (1976). Nature and origin of patterns of changes in cell shape in embryos. *Journal of Supramolecular Structure* **5**(4), 371-380.
66. Jacobson, A.G. & R. Gordon (1976). Changes in the shape of the developing vertebrate nervous system analyzed experimentally, mathematically and by computer simulation. *Journal of Experimental Zoology* **197**(2), 191-246.
67. Gordon, R. & A.G. Jacobson (1978). The shaping of tissues in embryos. *Scientific American* **238**(6), 106-113, 160.
68. Jacobson, A.G. (1980). Computer modeling of morphogenesis. *American Zoologist* **20**(4), 669-677.
69. Smith, S.L., E.L. Whitmore, A.G. Jacobson & R. Gordon (1987). Three dimensional reconstruction of neurulation. In: *Proceedings of the Microscopical Society of Canada, Fourteenth Annual Meeting, 17-19 June 1987, University of Manitoba, Winnipeg, Manitoba*. Toronto, Microscopical Society of Canada: 10-11.
70. Jacobson, A.G. (1955). The roles of the optic vesicle and other head tissues in lens induction. *Proceedings of the National Academy of Sciences USA* **41**(7), 522-525.
71. Wessells, N.K. (1998). Victor Chandler Twitty: November 5, 1901-March 22, 1967. *Biographical Memoirs. National Academy of Sciences (U.S.)* **74**, 333-346.
72. Mangold, O. (1926). Über formative Reize in der Entwicklung der Amphibien [About formative stimuli in the development of amphibians]. *Naturwissenschaften* **14**(50/51), 1169-1175.
73. Hamburger, V. (1960). *A Manual of Experimental Embryology*. Chicago, University of Chicago Press.
74. Slack, J.M.W. & D. Tannahill (1992). Mechanism of anteroposterior axis specification in vertebrates. Lessons from the amphibians. *Development* **114**(2), 285-302.
75. Holtfreter, J. & V. Hamburger (1955). Embryogenesis: progressive differentiation. Amphibians. In: *Analysis of Development*. Eds.: B.H. Willier, P.A. Weiss & V. Hamburger. Philadelphia, W.B. Saunders Co.: 230-296.
76. Spemann, H. & H. Mangold (2001). Induction of embryonic primordia by implantation of organizers from a different species (Reprinted from *Archiv Mikroskopische Anatomie Entwicklungsmechanik*, vol 100, pg 599-638, 1924) *International Journal of Developmental Biology* **45**(1 Spec No), 13-38.

77. Jacobson, A.G. (1966). Inductive processes in embryonic development:Most organs form as a result of gradual cumulative effects of interactions among embryonic tissues. *Science* **152**(3718), 25-34.
78. Gilbert, S.F. (2001). Continuity and change: paradigm shifts in neural induction. *International Journal of Developmental Biology* **45**(1 Spec No), 155-164.
79. Gordon, R., N.K. Björklund & P.D. Nieuwkoop (1994). Dialogue on embryonic induction and differentiation waves. *International Review of Cytology* **150**, 373-420.
80. Gordon, R. & J. Butler (1999). Clay model of axolotl ectoderm contraction wave. http://youtu.be/rfiYtwDMr7o
81. Brodland, G.W., R. Gordon, M.J. Scott, N.K. Björklund, K.B. Luchka, C.C. Martin, C. Matuga, M. Globus, S. Vethamany-Globus & D. Shu (1994). Furrowing surface contraction wave coincident with primary neural induction in amphibian embryos. *Journal of Morphology* **219**(2), 131-142.
82. Jaffe, L.F. (1999). Organization of early development by calcium patterns. *BioEssays* **21**(8), 657-667.
83. Luxton, G.W.G. & D.A. Starr (2014). KASHing up with the nucleus: novel functional roles of KASH proteins at the cytoplasmic surface of the nucleus. *Current Opinion in Cell Biology* **28**, 69-75.
84. Quinn, L.S., H. Holtzer & M. Nameroff (1985). Generation of chick skeletal muscle cells in groups of 16 from stem cells. *Nature* **313**(6004), 692-694.
85. Schulze, J. & E. Schierenberg (2009). Embryogenesis of *Romanomermis culicivorax*: An alternative way to construct a nematode. *Developmental Biology* **334**(1), 10-21.
86. Bonner, J.T. & M.E. Hoffman (1963). Evidence for a substance responsible for the spacing pattern of aggregation and fruiting in the cellular slime molds. *Journal of Embryology and Experimental Morphology* **11**, 571-589.
87. Deblandre, G.A., D.A. Wettstein, N. Koyano-Nakagawa & C. Kintner (1999). A two-step mechanism generates the spacing pattern of the ciliated cells in the skin of *Xenopus* embryos. *Development* **126**(21), 4715-4728.
88. Marnellos, G., G.A. Deblandre, E. Mjolsness & C. Kintner (2000). Delta-Notch lateral inhibitory patterning in the emergence of ciliated cells in *Xenopus*: experimental observations and a gene network model. *Pacific Symposium on Biocomputing 2004* **5**, 326-337.
89. Stubbs, J.L., L. Davidson, R. Keller & C. Kintner (2006). Radial intercalation of ciliated cells during *Xenopus* skin development. *Development* **133**(13), 2507-2515.
90. Kim, J.Y., S.W. Cho, W.C. Song, M.J. Lee, J.L. Cai, S.H. Ohk, H.K. Song, A. Degan & H.S. Jung (2005). Formation of spacing pattern and morphogenesis of chick feather buds is regulated by cytoskeletal structures. *Differentiation* **73**(5), 240-248.
91. Schnittger, A., U. Folkers, B. Schwab, G. Jurgens & M. Hülskamp (1999). Generation of a spacing pattern: The role of TRIPTYCHON in trichome patterning in *Arabidopsis*. *Plant Cell* **11**(6), 1105-1116.
92. Greese, B., K. Wester, R. Bensch, O. Ronneberger, J. Timmer, M. Hulskamp & C. Fleck (2012). Influence of cell-to-cell variability on spatial pattern formation. *IET Systems Biology* **6**(4), 143-153.

93. Yang, C.X. & Z.B. Ye (2013). Trichomes as models for studying plant cell differentiation. *Cellular and Molecular Life Sciences* **70**(11), 1937-1948.
94. Sachs, T. (1979). Cellular interactions in the development of stomatal patterns in *Vinca major* L. *Annals of Botany* **43**(6), 693-700.
95. Serna, L., J. Torres-Contreras & C. Fenoll (2002). Clonal analysis of stomatal development and patterning in *Arabidopsis* leaves. *Developmental Biology* **241**(1), 24-33.
96. Larkin, J.C., M.L. Brown & J. Schiefelbein (2003). How do cells know what they want to be when they grow up? Lessons from epidermal patterning in *Arabidopsis*. *Annual Review of Plant Biology* **54**, 403-430.
97. Galatis, B. & P. Apostolakos (2004). The role of the cytoskeleton in the morphogenesis and function of stomatal complexes. *New Phytologist* **161**(3), 613-639.
98. Lucas, J.R., J.A. Nadeau & F.D. Sack (2006). Microtubule arrays and *Arabidopsis* stomatal development. *Journal of Experimental Botany* **57**(1), 71-79.
99. Barlow, P. & J. Lück (2010). Transformations of cellular pattern: progress in the analysis of stomatal cellular complexes using L-systems. *Progress in Botany* **71**, 61-99.
100. Grebe, M. (2012). The patterning of epidermal hairs in *Arabidopsis* - updated. *Current Opinion in Plant Biology* **15**(1), 31-37.
101. Pillitteri, L.J. & K.U. Torii (2012). Mechanisms of stomatal development. *Annual Review of Plant Biology* **63**, 591-614.
102. Rudall, P.J. & E.V.W. Knowles (2013). Ultrastructure of stomatal development in early-divergent angiosperms reveals contrasting patterning and pre-patterning. *Annals of Botany* **112**(6), 1031-1043.
103. Rice, S.A. (2013). Electron Microscopy at the University of Tampa. http://utweb.ut.edu/hosted/faculty/srice/Elect.Micros.Page.htm
104. Kim, D.H., B. Li, F. Si, J.M. Phillip, D. Wirtz & S.X. Sun (2015). Volume regulation and shape bifurcation in the cell nucleus. *Journal of Cell Science* **128**(18), 3375-3385.
105. Nöthiger, R. (1972). The larval development of imaginal disks. In: *Results and Problems in Cell Differentiation, Vol. 5. The Biology of Imaginal Disk*. Eds.: H. Ursprung & R. Nöthiger. New York, Springer-Verlag: 1-34.
106. Poodry, C.A., L. Hall & D.T. Suzuki (1973). Temperature-sensitive mutations in *Drosophila melanogaster*. Part XV. Developmental properties of *shibire*^{ts1}: a pleiotropic mutation affecting larval and adult locomotion and development. *Developmental Biology* **32**(2), 373-386.
107. Suzuki, D.T. (1974). Behavior in *Drosophila melanogaster*: a geneticist's view. *Canadian Journal of Genetics and Cytology* **16**(4), 713-735.
108. Grillo-Hill, B.K. & T. Wolff (2009). Dynamic cell shapes and contacts in the developing *Drosophila* retina are regulated by the Ig cell adhesion protein Hibris. *Developmental Dynamics* **238**(9), 2223-2234.
109. Freeman, M. (1997). Cell determination strategies in the *Drosophila* eye. *Development* **124**(2), 261-270.
110. Vrailas, A.D., D.R. Marenda, S.E. Cook, M.A. Powers, J.A. Lorenzen, L.A. Perkins & K. Moses (2006). *smoothened* and *thickveins* regulate Moleskin/Importin 7-mediated MAP kinase signaling in the developing *Drosophila* eye. *Development* **133**(8), 1485-1494, doi: 1410.1242/dev.02334.

111. Weber, U., W.J. Gault, P. Olguin, E. Serysheva & M. Mlodzik (2012). Novel regulators of planar cell polarity: a genetic analysis in *Drosophila. Genetics* **191**(1), 145-162.
112. Kumar, J.P. (2013). Catching the next wave: Patterning of the *Drosophila* eye by the morphogenetic furrow. In: *Molecular Genetics of Axial Patterning, Growth and Disease in the Drosophila Eye*. Eds.: A. Singh & M. Kango-Singh, Springer New York: 75-97.
113. Wolff, T. & D.F. Ready (1993). Pattern formation in the *Drosophila* retina. In: *The Development of Drosophila melanogaster, Vols. 1 and 2*. Eds.: M. Bate & A. Martinez Arias. Plainview, New York, USA, Cold Spring Harbor Laboratory Press: 1277-1325.
114. Wolff, T. (1993). *Drosophila* third instar eye disc. In: *The Development of Drosophila melanogaster*. Eds.: M. Bate & A. Martinez Arias. Plainview, New York, Cold Spring Harbor Laboratory Press: poster in pocket.
115. Ready, D.F., T.E. Hanson & S. Benzer (1976). Development of the *Drosophila* retina, a neurocrystalline lattice. *Developmental Biology* **53**, 217-240.
116. Schlichting, K. & C. Dahmann (2008). Hedgehog and Dpp signaling induce cadherin Cad86C expression in the morphogenetic furrow during *Drosophila* eye development. *Mechanisms of Development* **125**(8), 712-728.
117. Siegrist, S.E. & C.Q. Doe (2005). Microtubule-induced Pins/Gαi cortical polarity in *Drosophila* neuroblasts. *Cell* **123**(7), 1323-1335.
118. Aaku-Saraste, E., B. Oback, A. Hellwig & W.B. Huttner (1997). Neuroepithelial cells downregulate their plasma membrane polarity prior to neural tube closure and neurogenesis. *Mechanisms of Development* **69**(1-2), 71-81.
119. Wolpert, L. (1969). Positional information and the spatial pattern of cellular differentiation. *Journal of Theoretical Biology* **25**(1), 1-47.
120. Wikipedia (2014). Lookup table. https://en.wikipedia.org/wiki/Lookup_table
121. Kerszberg, M. & L. Wolpert (2007). Specifying positional information in the embryo: looking beyond morphogens. *Cell* **130**(2), 205-209.
122. Wolpert, L. (2011). Positional information and patterning revisited. *Journal of Theoretical Biology* **269**(1), 359-365.
123. Akhtar, N. & C.H. Streuli (2013). An integrin–ILK–microtubule network orients cell polarity and lumen formation in glandular epithelium. *Nature Cell Biology* **15**(1), 17-27.
124. Bré, M.H., R. Pepperkok, A.M. Hill, N. Levilliers, W. Ansorge, E.H. Stelzer & E. Karsenti (1990). Regulation of microtubule dynamics and nucleation during polarization in MDCK II cells. *Journal of Cell Biology* **111**(6 Pt 2), 3013-3021.
125. Björklund, N.K. & R. Gordon (1993). Nuclear state splitting: a working model for the mechanochemical coupling of differentiation waves to master genes. *Russian Journal of Developmental Biology* **24**(2), 79-95.
126. Tabata, T. (2001). Genetics of morphogen gradients. *Nature Reviews Genetics* **2**(8), 620-630.
127. Gordon, R. (2006). Mechanics in embryogenesis and embryonics: prime mover or epiphenomenon? *International Journal of Developmental Biology* **50**(2/3), 245-253.
128. Wikipedia (2014). Equivalence class. http://en.wikipedia.org/wiki/Equivalence_class

129. Gordon, R. (1992). The fractal physics of biological evolution. In: *Dynamical Phenomena at Interfaces, Surfaces and Membranes*. Eds.: D. Beysens, N. Boccara & G. Forgacs. Commack, N.Y., NOVA Science Publishers: 99-111.

130. Sulston, J.E., E. Schierenberg, J.G. White & J.N. Thomson (1983). The embryonic cell lineage of the nematode *Caenorhabditis elegans*. *Developmental Biology* **100**(1), 64-119.

131. King, K. & B. Stewart (2010). *Hey Son! Guess What? Featuring Floyd "The Clown"*. Winnipeg Kevin King.

132. Alicea, B. & R. Gordon (2016). Quantifying mosaic development: Towards an Evo-Devo Postmodern Synthesis of the evolution of development via differentiation trees of embryos [invited]. *Biology (Basel)* **33**(Special Issue: Beyond the Modern Evolutionary Synthesis- what have we missed?, Ed. John S. Torday), doi:10.3390/biology5030033.

Chapter 7

The Differentiation Tree and the Fate Map of the Axolotl

In Chapter 2 we presented the anatomy of early embryogenesis in the salamander, *Ambystoma mexicanum*, whose common name is the axolotl (Figure 2.1). In Chapter 5 we explained the role of the cytoskeleton in a variety of cell functions. We introduced the cell state splitter, a specialized cytoskeletal organelle located at the apical end of the cells in a cell sheet (Figure 6.3), and we showed the types of expansion and contraction waves that the cell state splitter is capable of generating (Figures 6.28, 6.34, 6.35, 6.37, 6.38 and 6.39). We will now give a detailed case study of our own observations of differentiation waves on the axolotl. The bifurcating sequence of waves on an embryo is what we refer to as its differentiation tree (Figure 6.43). Here we present the empirical differentiation tree of the axolotl (Figure 7.1).

We introduced fate maps based on vital dye marking in Chapter 2 (Figures 2.68 and 2.69). Fate maps are drawn as if one point on an embryo in early development becomes a single point later in the embryo. The concept corresponds to one to one mapping in mathematics[1]. However, because of cell division and the detachment or shearing of daughter cells from one another, resulting in "changes in neighbors"[1], intercalation[2] or "cell mixing"[3,4] the concept of a one to one mapping fails. Mathematically these are one-to-many mappings (multivalued functions[5]).

Recall that embryos can be broadly divided into mosaic or regulating. We will stick with these classic designations instead of their more recent equivalents, determinate and indeterminate[6], respectively. A regulating embryo is one where up to some stage of development it is possible to split the embryo in two (or more) and still get complete individuals formed later. Humans are regulating since we have identical twins (and triplets, quadruplets and quintuplets). Regulating embryos are generally

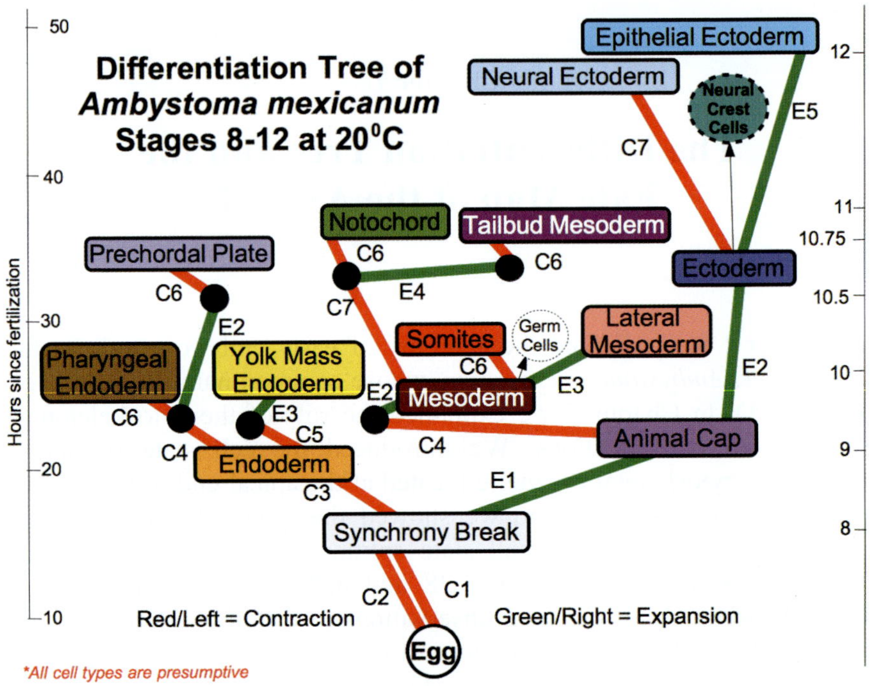

Figure 7.1. This axolotl differentiation tree indicates tissues determined from cleavage to the end of gastrulation. All tissues types should be considered "presumptive". Red, shift to left and the letter C indicate a contraction wave and green, shift to right, and the letter E an expansion wave. Stage numbers and hours since fertilization on the right and left, respectively, are from Bordzilovskaya *et al.* (1989)[7] and indicate the duration of each wave. Note each wave may pass through one tissue and continue into an adjacent tissue and so, for example, E3 begins in endodermal tissue and then continues into mesodermal tissue. See Björklund & Gordon (1994) and Gordon *et al.* (1994)[8,9] for the first, less complete versions of this differentiation tree, before we understood the animal/vegetal differentiation waves.

found in the clade known as deuterostomes in which the anus forms where the dorsal lip of the blastopore originated. Mosaic embryos are those where the first cell division is unequal and that division simultaneously determines cell fate. Mosaic embryos are generally found in the class of animals called proteostome where the mouth later forms at the blastopore and includes insects, mollusks and nematodes. At no point in development can a mosaic embryo be split and left to develop normally and produce identical whole individuals. (This type of mosaic organism has nothing to do with the mosaicism produced by

X inactivation, Figures 4.4, 4.12, or the fusion of two fertilized eggs to produce an embryo with 4 parents: see Figure 2.21 and Mintz, 1965[10], or even of two early embryos, as in Figure 2.12, all unfortunately also called "mosaic".) However, there are exceptions and mosaic versus regulating embryos should be considered only as a very broad and general classification.

All regulating embryos form and then develop first from a single cell and then undergo rounds of cell division. At a point where there are enough cells, a first round of determination occurs which separates the embryo into two primary tissues: endoderm, and ectoderm. Mesoderm forms the third primary layer and it also appears at this time as either its own third layer or from cells at the boundary between endoderm and ectoderm (Figure 7.2). From this point onward, if the embryo is split, the embryo develops abnormally. We have followed a decades old convention and colored these three classical tissues of first determination with the following primary color scheme for identification. Yellow is for endoderm, red is for mesoderm and blue is for ectoderm (Figure 2.53). We then expanded the original color scheme. We used different shades of blue for tissues that arise from ectoderm. Similarly we have used shades of red for tissues that arise from mesoderm and shades of yellow for tissues that arise from endoderm. By classical convention, notochord arises from mesoderm but is colored green. We have followed this convention in our illustrations because it is familiar, it makes embryogenesis easier to understand and it is prettier to look at.

The second convention is that the colored regions of the embryo are named according to what those cells will become rather than what they are at the time for which that fate map is drawn. They should technically be referred to as "presumptive ectoderm", "presumptive mesoderm", and "presumptive endoderm". In fact, all the tissue types in the fate map should properly have the term "presumptive" in front until they have actually reached the cell type for which they are named. This will occur much later in development for some tissues. For example, we have somitic mesoderm in our scheme. But somites will not actually form until much later, well after neural tube closure. What we call somitic mesoderm is actually "presumptive somitic mesoderm". It gets rather cumbersome to say and to read "presumptive" over and over again even though it is more accurate. The term "presumptive" has been deliberately dropped in this account for readability.

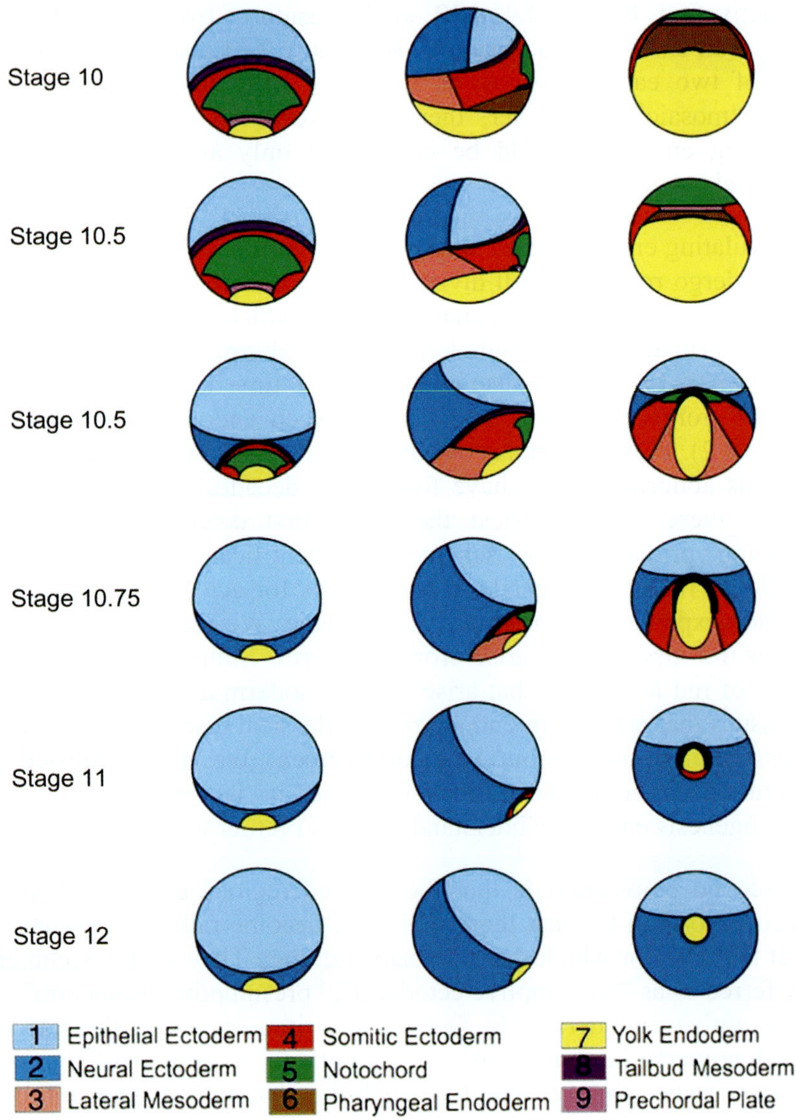

Figure 7.2. Invagination of the axolotl fate map during normal development, redrawn from Björklund & Gordon (1994)[8]. **Left column**: side view with the blastopore at the bottom. **Middle column**: left side. **Right column**: blastopore view. The three primary tissues are colored using the three primary colors, yellow for endoderm, red for mesoderm and blue for ectoderm. For Stage 8 see Figure 7.3, which also follows this convention. The standard staging sketches for this range of embryogenesis are in Figure 2.33. The numbered tissues correspond to Figure 7.3.

Recall from Chapter 2 how gastrulation is the period during which the embryo has a blastopore and all mesodermal and endodermal tissue is moved from the outside of the embryo into the interior. The embryo changes from a simple three zone organism to a three layered organism with endoderm on the inside, ectoderm on the outside and mesoderm in the middle (Figures 2.33 (Stages 10-1) and 2.55). In the axolotl, gastrulation takes about 40 hours depending on temperature[7]. The blastopore begins as a thin, short line of invaginating cells that contract immensely while moving from the outside to the interior producing the lip shape. The dorsal lip of the blastopore appears first and is visible in axolotl embryos as a dark line. The pigment granules in the apical ends of the cells get concentrated in a smaller, temporarily contracted apical area as each cell goes through a bottle shape (Figure 2.60). The blastopore elongates to a crescent shape, further changes to a half circle, then becomes nearly a full circle and finally forms a complete ring. The ring starts as a circle and as it decreases in size it changes to an oval, then a slit, and finally disappears. Once the blastopore closes to zero radius and invagination ceases, the discontinuity in the ectodermally derived tissue converts to a continuous external epithelial layer, with no trace of the hole left[11].

We have found that there are nine presumptive tissues that are determined during gastrulation. Two of them are ectodermal: the neural ectoderm, which will later form the neural plate, nerves, brain and spinal cord; and the epithelial ectoderm, which will form the future skin, sweat glands, hair and all the related external structures (scales, feathers, etc., depending on species). There are four mesodermal tissues determined during gastrulation. These are the somitic and lateral mesoderm, the notochord, and the tailbud mesoderm. There are three endodermal tissues, pharyngeal endoderm, prechordal plate, and the yolk mass endoderm. If we look at the fine details of the Vogt axolotl fate map (Figure 2.69), there are a lot of future tissue types that we don't mention. For example, we don't explain the eye, lens, sucker and ear regions that arise from ectoderm and that Vogt was able to locate using his fate map technique. This is because finer subdivisions of embryogenesis occur later in development well after gastrulation is finished. These tissues are determined from ectoderm but only after gastrulation. So this is how we ended up with a somewhat simplified version of the Vogt fate map with only nine presumptive tissue types determined by the end of gastrulation (Figure 7.3).

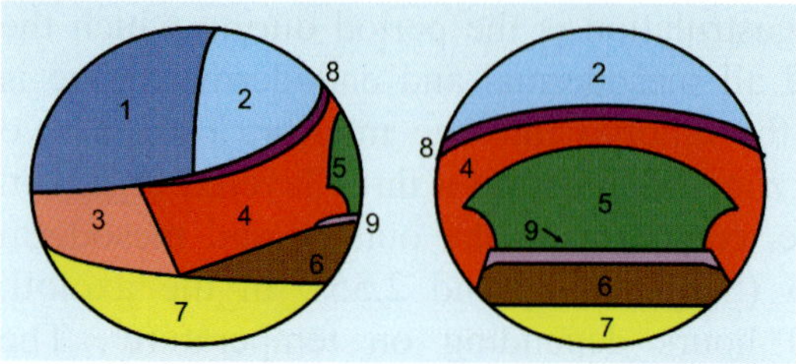

Figure 7.3. A simplified fate map of the key embryonic tissue types of the axolotl, *Ambystoma mexicanum* at gastrulation Stage 10, which we have adapted primarily from Vogt[12,13], with a few corrections and updates based on later work[14-22], mapped onto Stage 8 (Figure 2.33). Note that all tissue types are "presumptive" at this stage even though not labeled as such, meaning that they these areas will become those tissues later if development proceeds normally. **Left**: left side view. **Right**: posterior view. Top is gravitationally up, due to the yolk endoderm having the highest mass density. The full version includes many fine points of tissues that develop later including eye placodes and limb buds.

1. Epithelial ectoderm
2. Neural ectoderm
3. Lateral mesoderm
4. Somitic ectoderm
5. Notochord
6. Pharyngeal endoderm
7. Yolk endoderm
8. Tailbud mesoderm
9. Prechordal plate

In terms of actual methodology, we used only natural spawnings[23] from adult animals from our own colony (derived in 1989 from the colony of the late John B. Armstrong, University of Ottawa[24]). We removed the embryos' external jelly coats manually and staged them according to a standard chart of axolotl development[7] (Figures 2.31, 2.33 and 2.55). Embryos were videotaped either from the top and bottom simultaneously (a method pioneered by Koki Hara[25]) or from two sides using high resolution black/white VHS video cameras (before inexpensive digital cameras were available) mounted on two standard stereomicroscopes. Axolotls develop normally at temperatures between 8°C and 22°C, the warmer the temperature, the faster the development (Figure 7.4). They develop abnormally outside this range (Figure 7.5). We wanted the temperature to be as close to 8°C as possible so that development was slower, making it more likely that we would see potentially rapid events slowed down. We set up the entire apparatus in our basement lab during the colder months of fall and winter in Winnipeg, Canada with the thermostat set lower than is generally considered comfortable for most scientists. We had to wear our Canadian spring and fall clothes to keep warm. We also did a few "runs" at higher temperatures by way of comparison and to ensure we weren't missing anything. In this way, we were able to get many full videos of the entire course of gastrulation in many normal embryos. After videotaping, we allowed the embryos to develop until they were observed to eat normally. We also ensured that control embryos from the same spawning had hatched and were also eating normally. We only digitized videotapes of normal embryos from normal spawnings in our work. If the embryos or their controls were abnormal or failed to develop, we discarded the information.

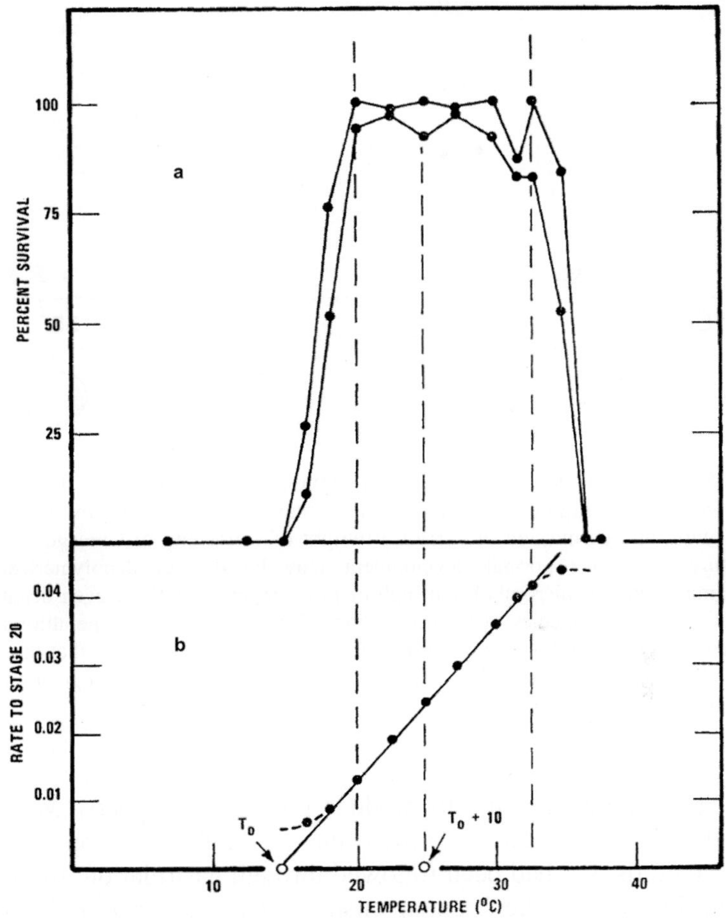

Figure 7.4. The speed of development varies with temperature for all amphibians (b) over their normal temperature range of development (a). These curves are for the frog *Bufo valliceps*. Note that over the whole of the normal temperature range the survival is nearly 100%. From Bachmann (1969)[26], drawn from data in Volpe (1957)[27], with kind permission of Perry Cartwright of the University of Chicago Press. For more such curves, showing that the shape of the survival curve is the same, but its temperature range varies with species, see Duellman & Trueb (1986)[28].

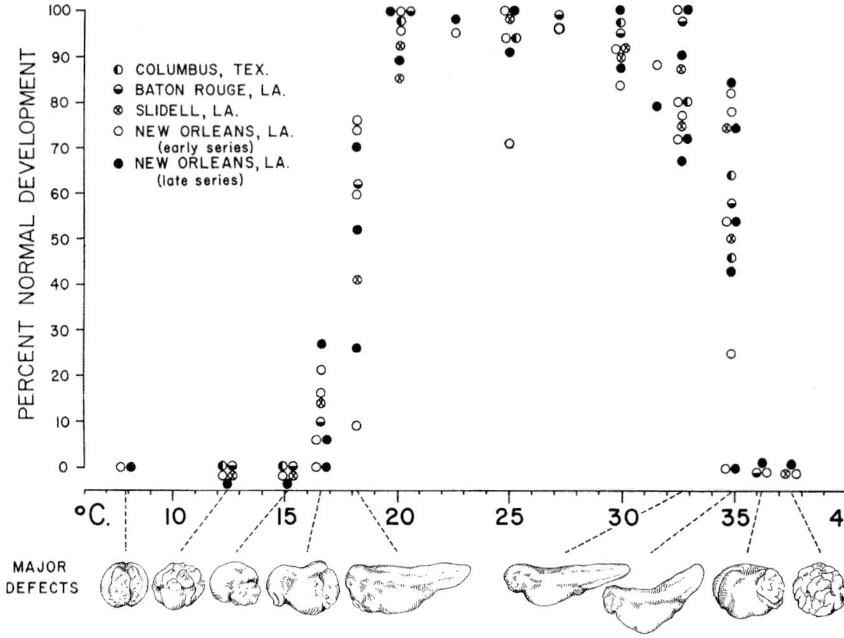

Figure 7.5. Outside the normal range of temperature, amphibian embryos develop major birth defects. As in Figure 7.4, this data is for the frog *Bufo valliceps*. From Volpe (1957)[27] with kind permission of Perry Cartwright of University of Chicago Press. The temperature limits for normal development may be due to depolymerization of microfilaments and/or microtubules at high or low temperature[32-34] or even denaturation of tubulin or actin[35]. Species differences (Figure 7.4) may reflect temperature stability differences in tubulin and/or actin[36]. Another contributing factor may be that the relative speeds of different differentiation waves vary with temperature (Proposition 82 in *Hierarchical Genome*[37]). Cf.[38].

We used computer imaging to produce our own fate maps. We superimposed an image of the simplified nine-tissue fate map on images of Stage 8 axolotls that had been followed through gastrulation. In this way we could follow how the fate map changes shape, at least on the outside of the embryo, as invagination through the blastopore proceeds during gastrulation.

Axolotls have variable pigmentation on the outer top hemisphere of the embryos, making it possible to follow small clusters of cells even when we couldn't see individual cells (Figure 2.5). We used[29] "NIH Image", the predecessor[30] to ImageJ[31], a scientists' image analysis software package, free open source computer program provided by the U.S. National Institutes of Health (NIH), to digitize our video tapes. We created a series of time lapse movies from the digitized images. At that time (early 1990s) computer disk storage was precious (we had 300 MB total for the whole laboratory), so even though the video tapes stored 30 frames per second (fps), we had to select frames to digitize. The time intervals we normally worked with were 15 or 20 minutes per digitized frame but in a few instances, in order to be able to figure out all the details we reduced that to 5 to 10 minutes between frames. We used the same NIH computer program to highlight the variable pigmentation spots in the time lapse movie, occasionally supplementing with pseudocolor[39] to aid in tracking pigmentation (Figure 7.17). We could also trace the movements of these spots on the time lapse movie as gastrulation occurred instead of waiting until a later stage and fixing and sectioning embryos. By matching pigment spots to the fate map on the time lapse movies, we could then show how the fate map changes as it invaginates (Figure 7.2). Thus, we were able to do the digital equivalent of putting vital dye on the surface of the embryo.

Another advantage of our method is that the embryos were never subjected to any surgical procedures beyond removing the jellycoat. A great deal of work in classical embryology is based on experiments that involve slicing out, transplanting and replacing parts (such as in Figure 6.16). It is possible to control for the effects of the operation itself by techniques such as matching sham surgeries. Nonetheless, to us it seemed counterproductive to subject embryos to physical forces caused by surgical experiments when attempting to study their normal mechanics. Furthermore, there is the possibility that cutting embryonic tissues generates the very kinds of waves we are trying to study (§9.22 *Uninvited Waves* in *Hierarchical Genome*[37]), and so we recommend that all such future experiments be imaged and followed up immediately with time lapse, to watch the healing process for waves, and not just measure its final consequences.

By using time lapse videos, we could also see and track the trajectory of waves of contraction and expansion from fertilization to the end of gastrulation. Because we were looking at spherical embryos with only two cameras we had to examine different embryos placed at different angles in order to visualize the entire wave for waves with complex trajectories like the ectoderm contraction wave. We also had to complete

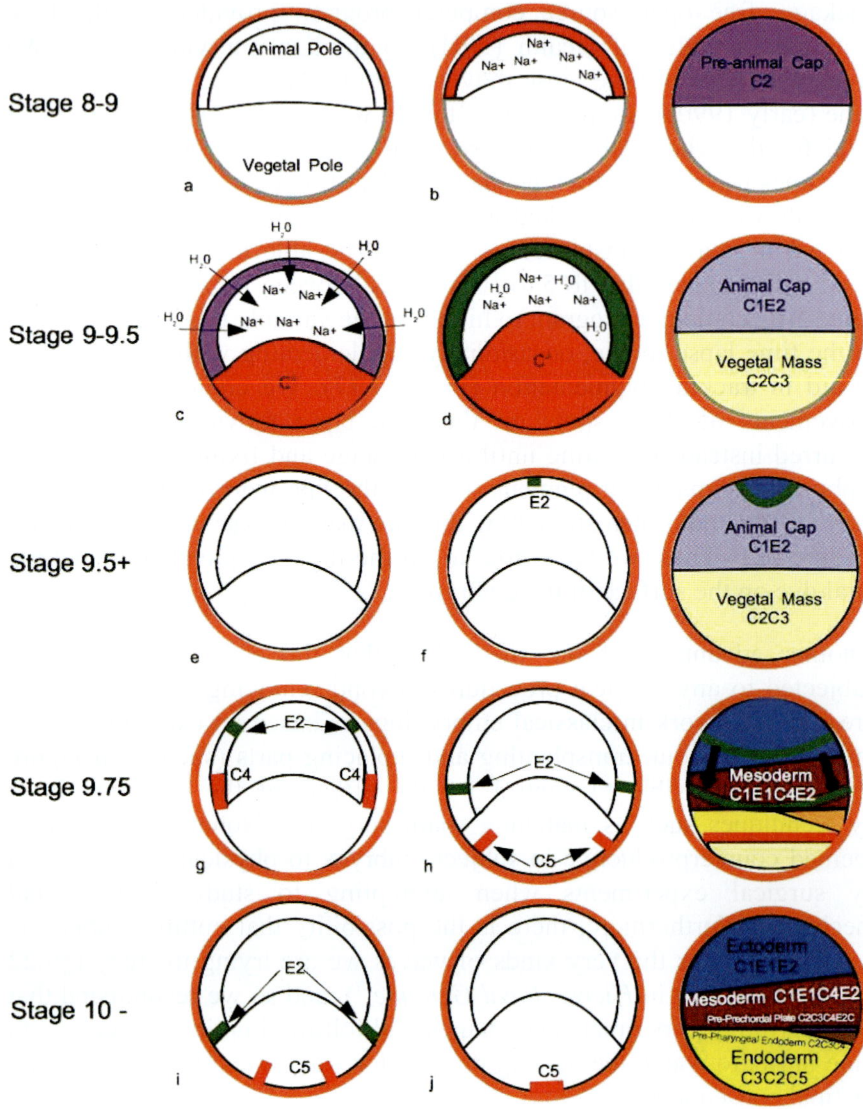

Figure 7.6. See text. Cross sectional images of the transition from a half moon to crescent shape of the blastocoel are available[40-42]. Columns 1 and 2 represent the sequence of events in cross section ordered a-j. The third column illustrates the developing fate map as tissues of the embryo are determined.

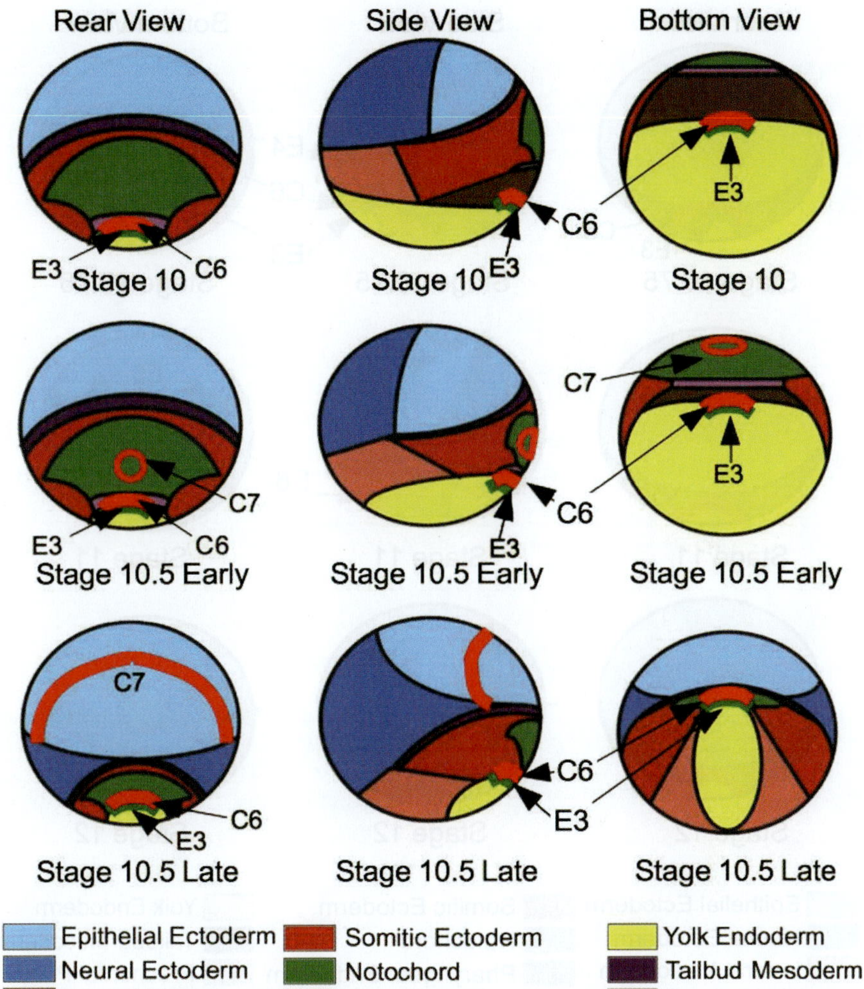

Figure 7.7. Details of the invagination of the fate map with the differentiation waves superimposed, with red arcs for contraction waves and green arcs for expansion waves. Numbers for the waves refer to their time sequence, the details of which are available in Figures 7.9 and 7.11. Three views of each stage are shown. **Left column**: side view with the blastopore at the bottom. **Middle column**: left side. **Right column**: blastopore view. Early in gastrulation the blastopore faces downward. As gastrulation proceeds and the endoderm and mesoderm move inside, the center of gravity of the embryo shifts and the embryo rolls so that the blastopore ends up horizontally at the equator. The third column is adjusted so that the view is always on the blastopore.

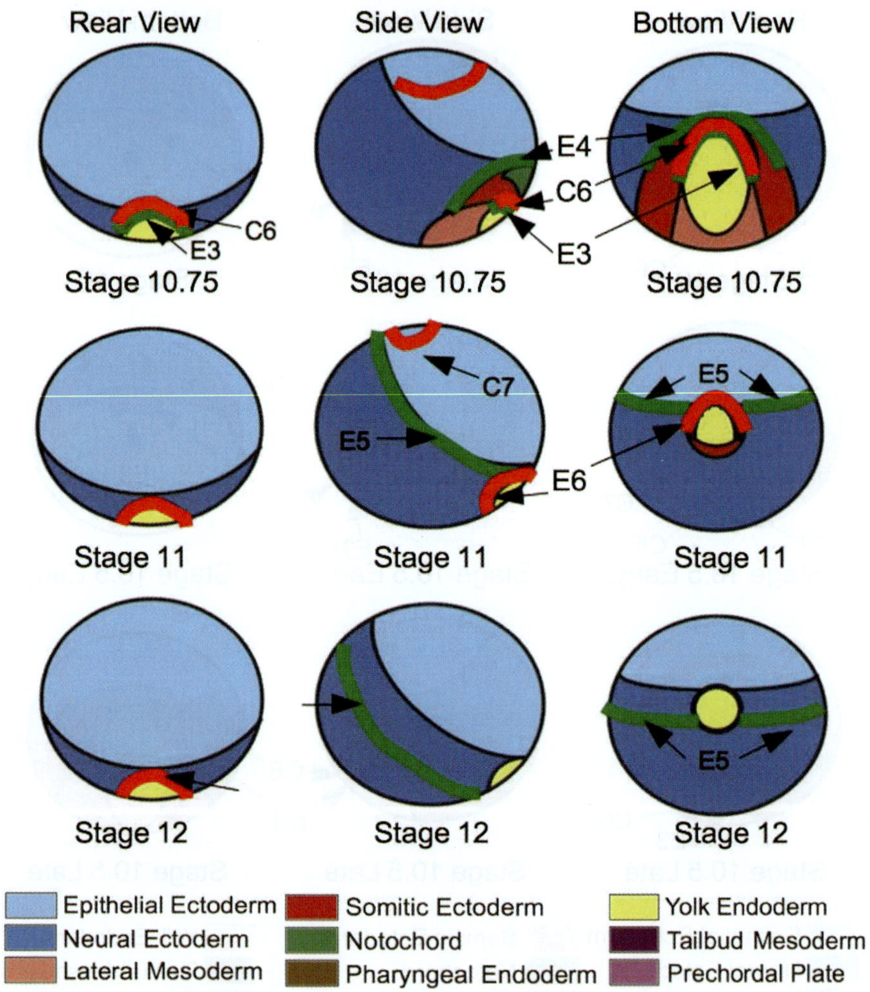

Figure 7.8. Continuation of the fate map through Stage 12. Details as in Figure 7.7.

several "runs" viewing the embryos from the side and not just from top and bottom. We also had a methodological rule that a wave had to be seen on at least three different embryos from at least three different batches or runs, where all the embryos went on to hatch and eat normally.

From this work, we were able to determine the sequence of contraction and expansion waves that traveled over the embryo during gastrulation and the relationship of these waves to the invaginating fate map as shown, for instance, in Figures 7.6, 7.7 and 7.8. In these figures we were able to construct a moving fate map that also included expansion and contraction waves, and derive a differentiation tree to match (Figure 7.1). There are 7 contraction waves and 5 expansion waves that occur between the time of loss of synchrony of cell division (midblastula transition, about Stage 8 to 9) and the end of gastrulation. Some of these waves pass through more than one tissue type.

As we presented in Chapter 2, the freshly fertilized egg goes through a period of rapid cell divisions. The first two cell divisions go from top to bottom creating four matched cells. The third cleavage goes horizontally but above rather than through the equator. The result is four larger, mostly yolky cells on the bottom and four smaller mostly pigmented cells on top (#4 in Figure 2.31). With the third cleavage a small space is also formed at the top where all three cleavage planes intersect, which is the beginning of the blastocoel (Figure 2.46). These rounds of cell division continue in a specific sequence with timing coordinated by waves of contraction that start at the top of the embryo and move downward[43,44]. These cleavage waves are much faster than differentiation waves, synchronizing and coordinating the cleavage of individual cells. The embryo soon becomes a loosely packed ball of cells (Figure 2.31: Stages 5 and 6). The bottom is full of heavier yolky material in densely packed larger cells (Figure 2.40). The embryo is now called a blastula.

The first differentiation wave in the axolotl embryo occurs at the end of the rapid cleavage of cells when the embryo is in the blastula stage. We published our first fate map paper in 1994. At that time we could not account for the difference between animal cap and vegetal cap at the beginning of gastrulation so we simply accepted it was there and carried on[8]. However, we knew, based on the binary code (Figure 7.9), that such a difference must exist. Further study of blastulation revealed where the animal/vegetal difference arises and so we describe the waves that cause it here for the first time.

The first breaking of cleavage synchrony in axolotls begins during Stage 8. The space that appeared with the third cleavage (Figure 2.46) has grown and is now called the blastocoel. The cells become quite small

Tissue #	Tissue name	Predecessor Waves	Differentiation code	Numerical differentiation code
1	Epithelial ectoderm	$C_1E_1E_2E_5$	CEEE	4.0111
2	Neural ectoderm	$C_1E_1E_2C_7$	CEEC	4.0110
3	Lateral mesoderm	$C_1E_1C_4E_2E_3$	CECEE	5.01011
4	Somitic ectoderm	$C_1E_1C_4E_2C_6$	CECEC	5.01010
5	Notochord	$C_1E_1C_4E_2C_7C_6$	CECECC	6.010100
6	Pharyngeal endoderm	$C_2C_3C_4C_6$	CCCC	4.0000
7	Yolk mass endoderm	$C_2\, C_3\, C_5\, E_3$	CCCE	4.0001
8	Tailbud mesoderm	$C_1E_1C_4E_2C_7E_4C_6$	CECECEC	7.0101010
9	Prechordal plate	$C_2C_3C_4E_2C_6$	CCCEC	5.00010

Figure 7.9. Table of wave sequence for each tissue type on the fate map determined during gastrulation. Specific waves, identified in Figure 7.11, are given subscripts. The differentiation code ignores the particular waves, assuming that from a cell's point of view all contraction waves or all expansion waves are equivalent. Therefore a wave traveling through more than one tissue type can trigger more than one type of differentiation depending on what wave sequence has occurred previously. The differentiation code can be expressed either as a sequence of letters "C" for contraction wave or "E" for expansion wave, in time sequence left to right, or numerically as a binary number. We use binary fraction notation, placing 0=C and 1=E after the "decimal" point. However, trailing zeros beyond the number of waves involved here don't have any meaning, and thus are not shown. We use a mixed notation, with a decimal count of the waves to the left of the decimal point. The numerical differentiation code is the decimal number of waves N leading to a given tissue, followed by a decimal point, followed by the N waves represented as "0" for contraction wave and "1" for expansion wave, in time sequence left to right. The right side is thus a binary fraction. While N may be redundant in this notation, it helps visually and serves as a check on consistent notation. We suggest that all cells with the same differentiation code are the same kind of cell, i.e., form the same tissue, whether contiguous or not. This also predicts that all cells with the same differentiation code have the same pleuropotency, i.e., have the potential to further differentiate along any of the subsequent branching pathways in the differentiation tree.

on the top (Figure 2.31: Stage 5, Figure 2.33: Stage 8), and as a result, the covering layer of cells that form the roof over the cap has become much thinner. Cells are still large enough that they can be individually watched using a stereomicroscope. The blastula surface appears rough and bubbly on top, though less obviously so as the cells get smaller. All the cells on the former animal cap suddenly sag noticeably pulling away from the vitelline membrane (Figure 7.10). These cells then undergo a change. They begin pumping sodium ions into the blastocoel[45-49].

Figure 7.10. An axolotl embryo imaged[50] from the side at Stage 7 to 8. The jelly sac encircles the embryo. The halo just above the pigmented top of the embryo is the vitelline membrane. In this image it is possible to see how the top of the embryo has begun to pull away from the vitelline membrane and droop loosely over the ever enlarging blastocoel inside.

Here, then, is the solution to the animal/vegetal difference. We designate the sagging of these cells of what is classically called the animal cap as a Type 1 contraction wave (all cells contract together and are determined to the same new cell type: Figure 6.27). This is the first contraction wave. The wave launches as the cells over the surface get thinner and thinner while the blastocoel grows in size below until the animal cap finally falls inward triggering a contraction wave. The animal cap contraction wave occurs at a time corresponding to the midblastula transition.

The result of this contraction wave is a determination signal to begin differentiation of the cells of the animal cap. They change into animal cap cells. They also begin to produce active ion pump membrane proteins. We have therefore named this contraction wave the Ion Pump Contraction Wave since it precedes the appearance of the ion pumps in the animal cap (Figure 7.1: C_1). The presence of ion pumps causes sodium ions to be pumped into the blastocoel to a higher concentration inside the blastocoel compared to the outside[51]. This causes water to move by osmosis into this hollow space. The result is an internal pressure in the blastocoel. Morphologically, the embryo loses its rough ball of cells look and puffs up so tautly that the surface becomes smooth against the vitelline membrane. The cells have also continued to divide so the cells are now too small to be seen by low power microscopy. The roof of the blastocoel is even thinner. When the embryo reaches Stage 9, the surface is so smooth that it is almost impossible to tell the difference between an embryo at this stage and a freshly fertilized egg by visual inspection under the light microscope. We designate this taut expansion of Stage 9 as the Animal Cap Expansion wave (Figure 7.1: E_1) and it is also a Type 1 wave. We now have all the tissue in the animal cap determined by two waves we designate as the sequence C_1E_1 (Figures 7.6: a,b and 7.11).

The tension on the animal cap increases until it is resolved by the bottom cells of the vegetal pole contracting and rising up into the blastocoel as a unit, lifting the floor of the blastocoel part way up into the space (Figure 7.6: Stage 9-9.5). We designate this as the Blastocoel Floor Contraction Wave (Figure 7.1: C_2) and it is also a Type 1 wave. By the contraction/expansion model, these cells have experienced a contraction wave. They should now behave in the same way as those of the roof of the blastocoel and begin differentiating as the same cell type. They should begin making sodium pumps and pumping out ions. They have no chance to do so.

The first contraction wave C_2 is immediately followed by a further contraction wave of the endoderm C_3 as the animal cap expands and elongates (Figure 7.6: a,b). The cross sectional shape of the blastocoel changes from a half moon (Figure 7.6: a) to an elongated crescent (Figure 7.6: c,d). We designate this second contraction as the Vegetal Contraction Wave (C_3) (Figures 7.1: C_3 and 7.11) and it is also a Type 1 wave. The vegetal half of the embryo is now partly tucked up inside the expanded animal cap. The visible border on the outside between pigmented and unpigmented regions drops (compare Figure 7.6: a with f). In the binary code of expansion and contraction, this means

Name of wave	#	Tissues for which this is the last wave	Wave Sequence	Stage wave launches	Stage wave ends
Ion Pump Contraction Wave	C_1	Pre-animal cap	C_1	8	8
Blastocoel Floor Contraction Wave	C_2	Pre-endoderm	C_2	9-	9-
Animal Cap Expansion wave	E_1	Animal cap	C_1E_1	9	9
Vegetal Contraction Wave	C_3	Primary Endoderm	C_2C_3	9	9
Blastula Expansion Wave	E_2	Ectoderm Mesoderm	$C_1E_1E_2$ $C_1E_1C_4E_2$	9-	10-
Central Constriction Band Contraction Wave	C_4	Premesoderm Frontal endoderm	$C_1E_1C_4$ $C_2C_3C_4$	9.5	9.5
Endoderm Contraction Wave	C_5	Endoderm	$C_2C_3C_5$	10-	10
Blastopore Lip Contraction Wave	C_6	Somites #4 Tailbud mesoderm #8 Pharyngeal endoderm #6 Notochord #5 Prechordal plate #9	$C_1E_1C_4E_2C_6$ $C_1E_1C_4E_2C_7E_4C_6$ $C_2C_3C_4C_6$ $C_1E_1C_4E_2C_7C_6$ $C_2C_3C_4E_2C_6$	10	12
Blastopore Expansion Wave (subduction wave)	E_3	Lateral mesoderm #3 Yolk mass endoderm #7	$C_1E_1C_4E_2E_3$ $C_2C_3C_5E_3$	10	12
Ectoderm Contraction Wave	C_7	Neural ectoderm #2 Prenotochord	$C_1E_1E_2C_7$ $C_1E_1C_4E_2C_7$	10.75	13
Tailbud Mesoderm Expansion Wave	E_4	Pretailbud mesoderm	$C_1E_1C_4E_2C_7E_4$	10.5	12
Ectoderm Expansion Wave	E_5	Epithelial ectoderm #1	$C_1E_1E_2E_5$	11	12+ to 13

Figure 7.11. Table of waves by name and the tissues for which they are the final wave determining the tissue. They are listed in order of launching and stop at various times (Figure 7.1). Some waves keep propagating through multiple tissues, thereby determining the next step for more than one kind of cell. The time a wave enters or launches and exits or terminates in a particular tissue can be read from Figure 7.1. The wave sequence experienced by a cell may not correspond to the launching sequence of the waves that encounter it. Tissues with #s are listed in Figure 7.9.

that the code for the vegetal half is C_2C_3 in contrast to the C_1E_1 of the animal cap (Figure 7.1: C_1E_1 versus C_2C_3 and Figure 7.11). These two sets of Type 1 contraction and expansion waves account for the preexisting difference in the animal and vegetal hemispheres when gastrulation begins that we could not explain in 1994 (Figure 7.6: a-d).

Let's take an aside from axolotls by looking at mammals. In mammals during compaction there is a wave of cortical microfilament contraction on the outside of the mammalian morula[52] driven by apical constriction of the microfilament rings (these are called "circumferential contractile networks"[53]). We postulated this is in fact a contraction differentiation wave. When it is finished, the cells on the interior, called the inner cell mass, give rise to the embryo proper. The rest of the cells on the outside form the placenta and trophoblast (Figure 7.12), and undergo a secondary expansion wave (Proposition 61 and §9.03 *The Compaction Wave* in *Hierarchical Genome*[37]). This primary sequence, CE and CC, should be exceptionally well preserved because it occurs so early in development. We expect to see a dual CC for vegetal and supportive tissue and CE for ectodermal or inner cell mass over and over again across phyla. That first contraction wave (or equivalent movement) should also always coincide with the midblastula transition or its equivalent.

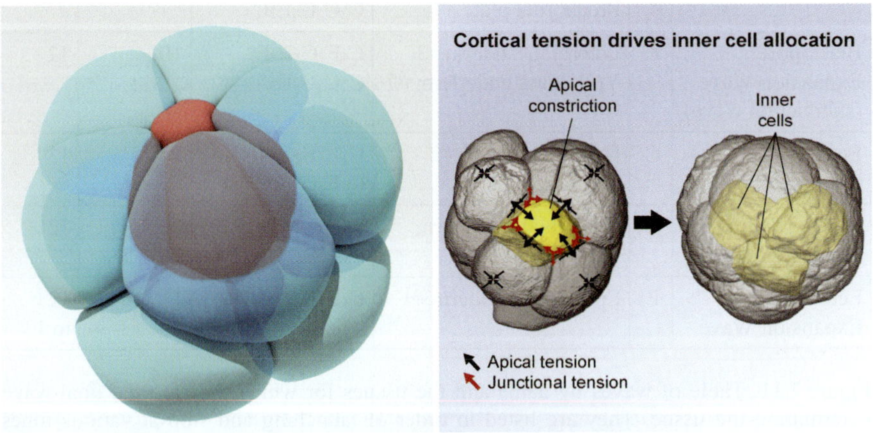

Figure 7.12. **Left**: "Cell Internalization at the 12-cell stage. A single cell (red) among the others (blue) at the 12-cell stage of the mouse embryo internalizes and is the first cell to contribute to the inner cell mass, which gives rise to the fetus." From Goldstein & Kiehart (2015)[54] with permission of Elsevier. **Right**: An asymmetric cell division, with

the larger cell going inside, results in the first cell that produces the inner cell mass which then goes on to become the mammalian embryo, while the outer cells form the placenta. Compaction, somewhat smoothing the outer surface of the embryo, is a contraction wave. From Samarage *et al.* (2015)[53] with permission of Elsevier. While in mouse a single cell starts off the inner cell mass, in other mammalian embryos it may be two[55] or more cells.

The next wave in the axolotl embryo is the Blastula Expansion Wave (E_2). This is also a Type 1 wave, but it is a wave that begins at one point and travels outward in an enlarging circle. The E_2 expansion wave begins from the place where the clear zone of the second polar body used to be. This is likely the starting point because, with little underlying actin, cells in this region are more susceptible to undergoing an expansion due to microtubules. The start point of the wave is set by the mechanical weakness at the top, a weakness that is left over from extrusion of the second polar body. Once started, the expansion wave of the microtubule mats propagates over the surface from cell to cell (Figures 7.6: f and 7.1: E_2).

The appearance of the specific mRNA DG42 in *Xenopus* (a frog that shares a lot of genes with the axolotl) is a candidate example of genes activated via the first traveling expansion wave (E_2). DG42 is an enzyme that is involved in the synthesis of hyaluronic acid. Hyaluronic acid is secreted to the outer surface of the cell membrane. This enzyme's mRNA appears first at the top of the embryo and then moves downward over the entire surface of the animal cap in a moving genetic expression wave at a time in the development of *Xenopus* that corresponds in timing to shortly after the second expansion wave (Blastula Expansion Wave E_2) of the axolotl[56,57].

As the expansion wave E_2 moves downward, the expanding top of the embryo places pressure on and triggers a sudden contraction of the midsection of the axolotl embryo (Figure 5.52: d). The result is a contraction that resembles a constriction band around the middle of the embryo. In the videos, the embryo visibly changes its shape from round to elliptical and the change lasts for fifteen minutes. The constriction band is wider on the grey crescent side and narrower on the successful sperm's entry point side of the original egg. Presumably, this difference in the width of the constriction band is a result of the mechanical bias left from the grey crescent formation (Figure 5.52). On the wider side of the band, the constriction also drops down into endoderm so that a small strip of the endoderm also participates in the contraction. This thin area will eventually form the endodermal derivative tissue called prechordal plate although it is not determined as such until later. The constriction band is labeled C_4 (Figure 7.11). It is also a Type 1 wave (Figure 7.1, C_4). We designate it with the name Central Constriction Band Contraction Wave (Figure 7.1, C_4).

The Blastula Expansion Wave (E_2) continues over the midsection of the embryo after the Central Constriction Band Contraction Wave (C_4) relaxes and it continues downward (Figure 5.52: g). By the end of blastulation and the beginning of gastrulation, the former animal cap of the embryo is neatly determined into two sections, ectoderm $C_1E_1E_2$, and mesoderm $C_1E_1C_4E_2$. Because of the constriction band, the endoderm has two sections, a frontal section of endoderm that has experienced $C_2C_3C_4$ and the lower and back endoderm which has experienced only C_2C_3. The embryo is now ready to begin gastrulation.

After the Central Constriction Band Contraction Wave (C_4) has relaxed, the mechanical stress created by the Blastula Expansion Wave (E_2) begins to show at the boundary between the animal cap and the endoderm. These 'come-and-go' dimplings we call problastopores because they appear prior to the appearance of the dorsal lip of the blastopore in a circle at the same level where the blastopore will finally appear (Figure 7.13). It is as if the embryo is testing the circle to find the weakest point. The next contraction wave is launched from the same region as a moving circle and that circle travels through the endoderm as a closing circle coming down to a point. This wave is called the Endoderm Contraction Wave (C_5) (Figures 7.6: h-j and 7.1: C_5) and it is a Type 1 wave traveling through all the endoderm. As the blastula expansion wave (E_2) reaches the boundary between mesoderm and endoderm (Figure 7.13), the embryo appears to 'give' in the region of the problastopores. The dorsal lip of the blastopore forms on the boundary between the yolk mass endoderm and the pharyngeal endoderm and invagination begins. We presume the embryo buckles inward at that point because of the preexisting stiffness bias created by grey crescent formation[9,37]). With the first appearance of the dorsal lip of the blastopore, gastrulation is underway. The embryo is now called a gastrula.

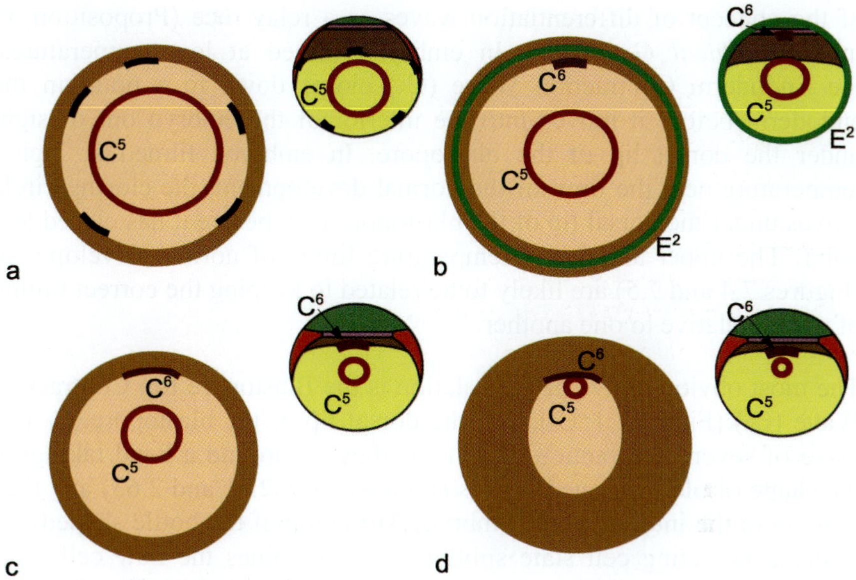

Figure 7.13. **a**: As the Blastula Expansion Wave (E_2 in green) approaches the boundary between the pigmented zone animal cap (brown) and the unpigmented vegetal zone (tan), the endoderm contraction wave (C_5) is launched as a circle which then closes down on itself. After this wave is launched, indentations resembling inward puckerings, which we call problastopores, form and vanish all around the circular boundary zone (black arcs). We interpret this as a mechanical strain due to the action of E_2 and C_5 on the embryo as a whole. The smaller figures show the embryo's approximate matching fate map. **b**: As the Blastula Expansion Wave (E_2 in green) reaches the boundary, the problastopores resolve to just one region and the dorsal lip of the blastopore appears (single black arc) launching the Blastopore Lip Contraction Wave C_6. Due to the mechanical bias created during fertilization, the former grey crescent region gives first and that is why in amphibians the subsequent blastopore C_6 forms approximately opposite to the sperm entry point. c,d: We also observed that when the temperature is at the colder end of the range of normal development, the final closure of the Endoderm Contraction Wave (C_5) is visible. In contrast, at the higher temperature of the normal range, the wave vanishes under the dorsal lip of the blastopore before completing closure. The upper and lower temperature limits of normal development are likely to be related to keeping the correct timing of waves relative to one another. And this timing is dependent on temperature limits for the functioning of microtubules, and not microfilaments.

The Endoderm Contraction Wave (C_5) continues to close down to a point independently of the formation of the dorsal lip of the blastopore (Figure 7.13: d). The dorsal lip of the blastopore always waits for the arrival of the Blastula Expansion Wave (E_2) before forming, an example of the concept of differentiation waves as a relay race (Proposition 82 in *Hierarchical Genome*[37]). In embryos filmed at low temperatures, the Endoderm Contraction Wave (C_5) closes down to a point in the endoderm before it moves into the interior of the embryo out of sight under the dorsal lip of the blastopore. In embryos filmed at higher temperature near the limit of the normal development, the closing circle moves under the dorsal lip of the blastopore well before it has closed to a point. The upper and lower temperature limits of normal development (Figures 7.4 and 7.5) are likely to be related to keeping the correct timing of waves relative to one another.

The most obvious wave of gastrulation is the Blastopore Lip Contraction Wave (C_6) (Figure 7.1: C_6), i.e., the dorsal lip of the blastopore. C_6 is a wave of severe contraction of cells as they go around a bend taking on the shape of old fashioned bottles (Figures 2.60, 2.61 and 2.63) and then moving to the interior of the embryo. At the top of the bottle shaped cell is the constricting cell state splitter that determines the new cell type. Unlike most contraction waves, the tissue travels over the lip while the visible zone of contraction remains stationary relative to the embryo as a whole. In all other differentiation waves that we have observed, the wave moves across a relatively stationary tissue. Which cells are in the dorsal lip of the blastopore is constantly changing as they slide over the edge, turning 180 degrees and curving back into the interior of the embryo.

There is a fundamental difference in the cells that make up the dorsal lip of the blastopore as compared to the other contraction waves we have studied. In other contraction waves, the microtubule mat of the cell state splitter remains even though the microtubules themselves shorten dramatically. In the bottle cells, the extreme contraction results in tiny wrinkles of membrane. These become vesicles. Microtubule tracks are used to carry off the vesicles essentially removing the microtubule mat (and probably the intermediate ring) altogether. This reduction in membrane coupled with the extreme contraction results in the bottle shape[59]. The pressure exerted by the expanding animal cap forces the extreme contraction response initially and it is maintained as the cells turn around the corner and move inside as gastrulation continues. Mesoderm in other species such as mammals and sea urchins is often formed entirely by cells breaking away from the boundary zone of ectoderm and endoderm. In frogs, this layer has cells that "intercalate" to form a two layer mesoderm. In the axolotl a specific gene that is turned

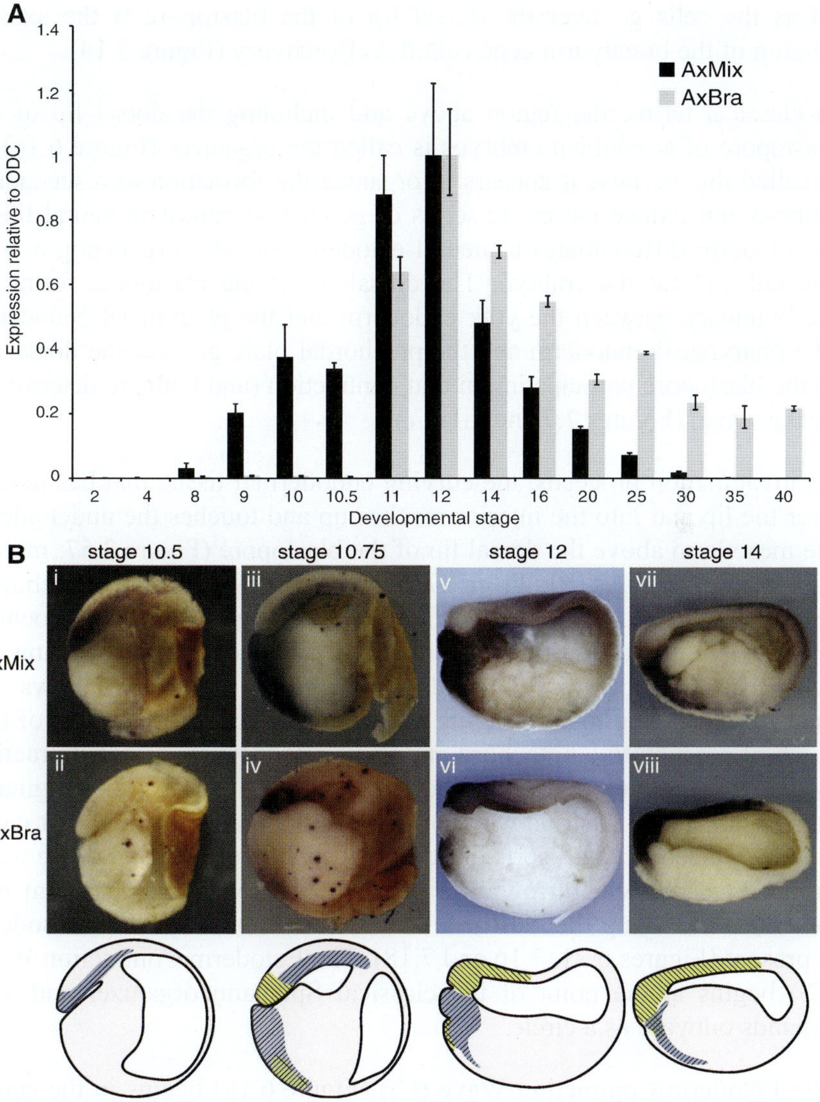

Figure 7.14. Examples of genes turned on due to a step of differentiation as cells participate in the waves forming the lip of the blastopore. "Analysis of AxMix and AxBrachyury expression during axolotl early development. (**A**) qPCR of AxMix and AxBra, normalised to ODC [i.e., ornithine decarboxylase was used as a standardized measure of enzyme activity] and then stage 12. (**B**) In situ hybridization on hemi-sectioned embryos. Stage 10.5, 10.75, and 12 images are the same embryo: dorsal = top, vegetal = left. The sketches below represent the combined expression of AxMix (blue) and AxBra (yellow)." From Swiers *et al.* (2010)[58] with permission of Elsevier.

on as the cells go over the dorsal lip of the blastopore is the axolotl version of the brachyuria gene called AxBrachyury (Figure 7.14).

In classical terms, the region above and including the dorsal lip of the blastopore of amphibian embryos is called the organizer (Figure 6.16). It is called this because it appears to organize the formation of a secondary embryo and induce the entire series of genes that appear in neural tissue as ectoderm differentiates to neural ectoderm, thereby organizing a head and tail end for the embryo. The dorsal lip of the blastopore forms on the boundary between the yolk endoderm and the pharyngeal endoderm. The pharyngeal endoderm and the prechordal plate go over the dorsal lip of the blastopore participating in that contraction (and both are determined by the wave) (Figures 7.7 and 7.1: C_6).

As invagination proceeds, the curving endodermal tissue that has moved over the lip and into the interior reaches up and touches the underside of the mesoderm above the dorsal lip of the blastopore (Figure 2.57: middle panes; also: Figures 10, 11 in Gordon *et al.*, 1994[9]). This attachment (Figure 7.15) reduces the tension on the outside presumptive notochord (Figure 7.7) due to the pressure inside the blastocoel by supporting the cells from underneath. The sudden reduction in tension allows the microfilament ring to win the tug of war in the cell state splitters of this patch of presumptive notochord and this starts the Ectoderm Contraction Wave (C_7), a Type 3 wave (Figures 7.1: C_7 and 6.35). (We originally named this as the gastrula wave or G wave but we abandoned that name when we discovered all sorts of other waves on the gastrula[60].) The wave produces a transient furrow as it propagates through the ectoderm into the tissue overlying the blastocoel even where no underlying mesoderm is present (Figures 6.21, 7.16 and 7.18). The Ectoderm Contraction Wave (C_7) begins at the point of the classical Spemann organizer and then expands outward as a circle.

The Ectoderm Contraction Wave (C_7) (Figure 6.18) begins in the center of the presumptive notochord (green on the simplified fate map, Figure 7.3) and it covers that tissue first. The Ectoderm Contraction Wave (C_7), passes through all of the notochord and the tailbud mesoderm in an expanding circle. This circle breaks over the dorsal lip of the blastopore. By the time the Ectoderm Contraction Wave (C_7) reaches the dorsal lip of the blastopore, the pharyngeal endoderm and prechordal plate are already inside. The Ectoderm Contraction Wave (C_7) vanishes where it meets the dorsal lip of the blastopore and it becomes an arc (Figure 7.7). The arc then travels over the top half of the embryo propagating from cell to cell. The ends move faster than the middle due to the rapid involution of the central portion of the embryo. As the Ectoderm Contraction Wave (C_7) reaches the opposite side of the embryo the wave ends meet, reform as an ellipse and then the ellipse closes down and vanishes (Figure 6.26).

There are three expansion waves that occur during gastrulation. The first expansion wave we originally named the subduction wave because anatomically it looks very much like a subducting geological ocean plate tectonically moving under a continental plate. All mesoderm and

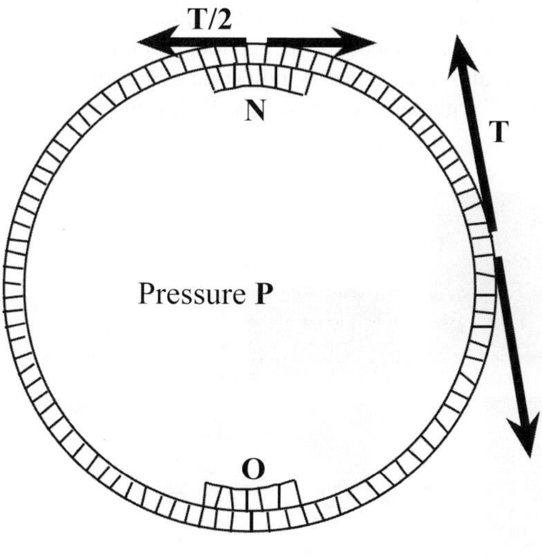

Figure 7.15. "A pressure **P** inside an embryo leads to a global tension **T** that pulls tangentially on each cell in a monolayer epithelium, such as the ectoderm of urodele amphibians prior to primary neural induction. Because the cell state splitters are attached via desmosomes and other junctions, this tension is borne by them. This simplified diagram, treating the embryo at early gastrulation as a spherical shell consisting of a monolayer of cells, may be taken as a transverse section[61] perpendicular to what will later be the axis or backbone of the adult. The pharyngeal endoderm **N** is the natural inducing tissue. Its adhesion to the ectoderm [presumptive notochord] doubles the tissue thickness and thus roughly halves the tension for the overlying ectodermal cells (Appendix I in Gordon & Brodland, 1987[62]). This leads to a force imbalance that causes the cell state splitters of these attached ectodermal cells to contract, launching a contraction wave. If another piece of tissue or an artificial substance, such as an 'organizer' **O**, is placed in the embryo ('Einsteck' = literally 'into stick' method[63-65]), and it adheres to the ectoderm [Figure 6.16], it likewise launches another contraction wave [Figure 7.1: C_7]. The only requirement for **O** is that it have a surface adhesive for the basal ends of [the invaginated and further differentiated] ectoderm cells. In mathematical terms, the metastable mechanical equilibrium is represented by $F_{MT} + T = -F_{MF}$ for unsupported ectoderm cells, while $F_{MT} + T/2 < -F_{MF}$ for ectoderm cells attached to an 'organizer', natural or artificial." The equations of Figure 6.14 are extended here to include T. From Figure 14 in *Hierarchical Genome*[37].

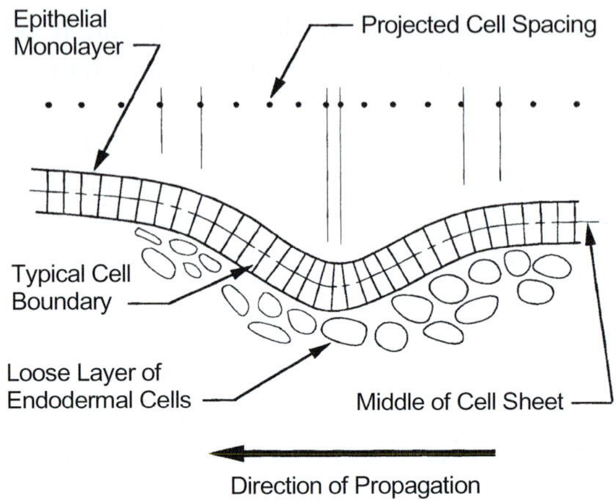

Figure 7.16. "Coupling of apical contraction and furrowing. When the apical end of a cell constricts, its enclosed cytoplasm forces it simultaneously to get longer or to become keystone-shaped[66]. The row of dots shown above the sheet is indicative of cell density as seen in plan." From Brodland *et al.* (1994)[29] with permission of John Wiley and Sons.

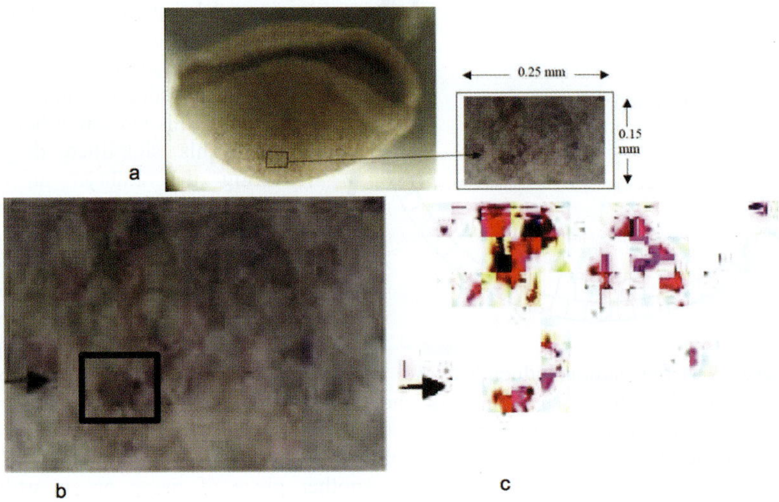

Figure 7.17. Example of using color enhancements to track variegated pigmentation of axolotl embryo cells. Beginning with Figure 2.5 (a and b), the image is color adjusted until it is oversaturated and a highly pigmented area (such as outlined in b) is clearly seen against a blank background (c). In the lower resolution videotapes of normal development that we made in the early 1990s we used a similar method with NIH Image pseudocolor to ensure we were tracking specific pigment cells at intervals and thereby mapped morphogenetic movements of cell sheets.

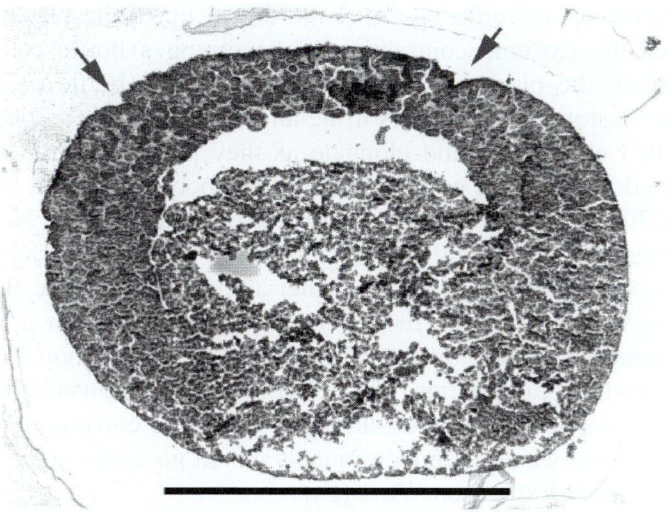

Figure 7.18. "A nearly sagittal section of a Stage 10 1/2 axolotl embryo, showing the ectoderm contraction wave cut at two points (arrows), prepared by microwave fixation. Note there is no underlying mesodermal tissue. Scale bar 1 mm. Thanks to Marc Del Bigio, Pierre Williot and Natalie K. Björklund." Figure 57 from *Hierarchical Genome*[37]. For a further explanation of the serial section views see Figure 7.19.

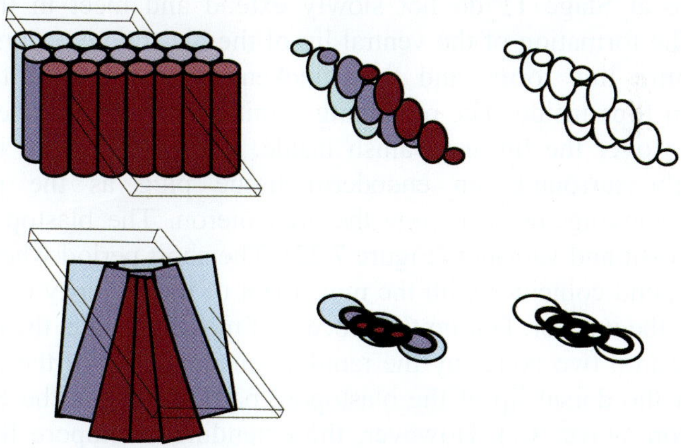

Figure 7.19. We show the results of a near sagital section through a cell sheet made up of uniform tall columnar (top) or bottled shaped cells (bottom). This figure shows a stylized cell sheet with a section through it and the resulting image of multiple overlapping circles and ellipses. Tracking of single cells through the entire set of registered serial sections confirms the single layer columnar shape in the roof of the blastocoel.

endoderm tissue will either go over the dorsal lip of the blastopore and experience the extreme contraction of becoming a bottle cell, or will slide beneath the blastopore lip without forming a bottle cell. Sliding under the blastopore lip appears to be almost as extreme as becoming a bottle cell. Cells stretch and elongate as they pass under the dorsal lip and so we designate this process as the Blastopore Expansion Wave (E_3) (Figure 7.20). Classically, this wave corresponds to the Nieuwkoop center (Figure 7.21). Many of the changes in gene expression that are known to occur in the Nieuwkoop center are triggered by the Blastopore Expansion Wave (E_3) in our model. The Spemann organizer is the result of determination caused by the Blastopore Lip Contraction Wave (C_6) and the Ectoderm Contraction Wave (C_7) which are intrinsic to the dorsal lip of the blastopore. The Nieuwkoop center is the corresponding result of the expansion wave in endoderm cells which go under and participate in the Blastopore Expansion Wave (E_3).

The lip of the blastopore does not extend ventrally in a smooth uniform fashion. There is an extended time period where the dorsal lip of the blastopore is a thin line. This is followed by a sudden very rapid extension to an arc. (Figure 7.22). All of this takes less than an hour and then there is another long period where the extensions forming the arc thicken and move inward to form a squat inverted U shape. The lateral lips of the blastopore at Stage 11 do not slowly extend and meet in the ventral region. The formation of the ventral lip of the blastopore occurs abruptly as an entire line forms and then thickens as the last of the lateral mesoderm flips inside. The remaining somitic mesoderm is the very last part to go over the lip and vanish inside. The now circular blastopore temporarily surrounds an endoderm tissue plug as the mesoderm continues moving inside to line the archenteron. The blastopore closes down to a slit and vanishes (Figure 7.22). The short period where the lips rapidly extend coincides with the movement of the majority of the mesoderm into the interior. During the process of moving inside, the mesoderm is divided into two zones by the rapid extensions. Part of the mesoderm rolls over the dorsal lip of the blastopore participating in the Blastopore Contraction Wave (C_6). However, the extending blastopore lips do not enclose all the mesoderm. They cut through it so that the rear portion slides under the blastopore lips and participates in the Blastopore Expansion Wave (E_3) instead (Figure 7.1: E_3). In this fashion mesoderm becomes neatly divided between somitic mesoderm and lateral mesoderm.

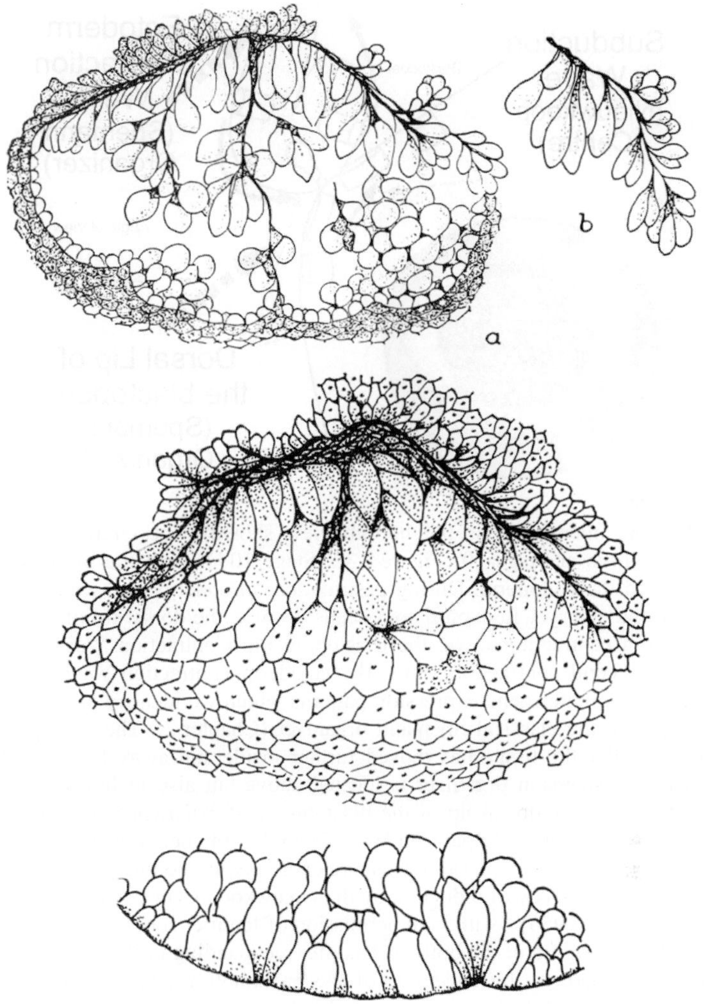

Figure 7.20. Cells stretch and elongate as they pass under the dorsal lip and so we designate this as the Blastopore Expansion Wave (E_3) This classical drawing by Johannes Holtfreter illustrates the expansion. **Top**: "*Triturus torosus*, gastrula. (a) Surface view of the blastoporal region after part of the surface layer has been removed, (b) Isolated bunch of flask [bottle] cells from the lateral blastopore lip, containing larger endodermal and smaller mesodermal cells. **Middle**: *Ambystoma punctatum*, gastrula. Surface view of the blastopore showing the accumulation of pigment and the stretching of the cells toward the lines of invagination. **Bottom**: *A. punctatum*, blastula. Section of the prospective blastoporal region. Note the black streaks of condensed coat material, and the flask cells attached to it"[67]. From Holtfreter (1943)[68] by way of Hill (2013)[67] with permission of John Wiley and Sons.

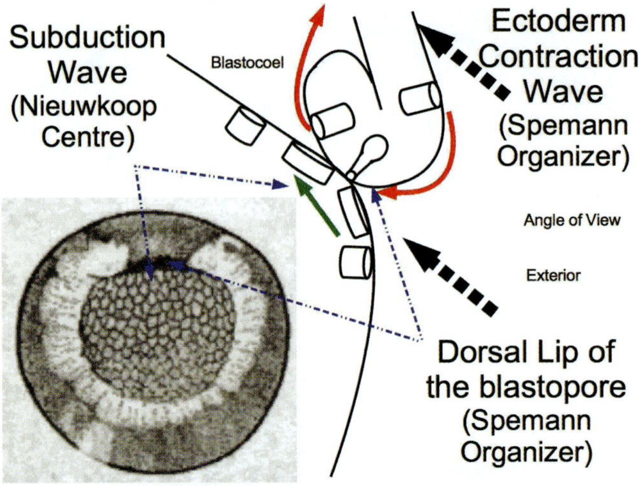

Figure 7.21. The micrograph is the bottom view showing the circular blastopore of the *Xenopus* embryo at Stage 10 (Figure 2.32: Stage 10+) when the blastopore lip has entirely enclosed the endoderm. During gastrulation, mesoderm and endoderm move to the interior of the embryo. In axolotl some cells go over the dorsal lip of the blastopore and experience a contraction wave, the Blastopore Lip Contraction Wave (C_6). This is followed by a second contraction wave, the ectoderm contraction wave (C_7). The C_7 contraction wave initiates gene expression patterns which we identify as those described as emanating from the classical Spemann organizer (Chapter 2). The gene products are detectable only after the tissue has moved further inside and away from the dorsal lip which is why the Spemann organizer is located above but also includes the dorsal lip. Other cells go under the dorsal lip of the blastopore and experience an expansion wave which we call the Blastopore Expansion Wave (also the subduction wave, E_3). The genes that are activated by the E_3 expansion wave are expressed in the lining of the archenteron and attributed to the activity of induction by the Nieuwkoop Center. Halftone image from Nieuwkoop & Faber (1956)[69] with kind permission of Jennifer Jones of Elsevier. See also Xenbase[70]. The leading tip of the bottle cells, after reexpanding in the blastocoel, goes on (red arrows) to adhere to the underside of the presumptive neural plate (Figure 2.57: middle panes; Figures 10, 11 in Gordon *et al.*, 1994[9]).

The second expansion wave of gastrulation (Tailbud Mesoderm Expansion Wave, E_4) is a transient short Type 1 wave that occurs during the same time that the lateral mesoderm is being enclosed by the blastopore lips. The region that will become tailbud mesoderm on the fate map is approaching the dorsal lip of the blastopore. It has just participated in the ectoderm contraction wave (C_7). The movement of tissue over the dorsal lip of the blastopore pauses. As the mesoderm is being rapidly divided by the extending lateral lips of the blastopore, the dorsal lip is paused and the tissue above is visibly elongated, appearing to be strained. We think that the tissue, having just participated in the ectoderm contraction wave, is paused because it is unable to immediately

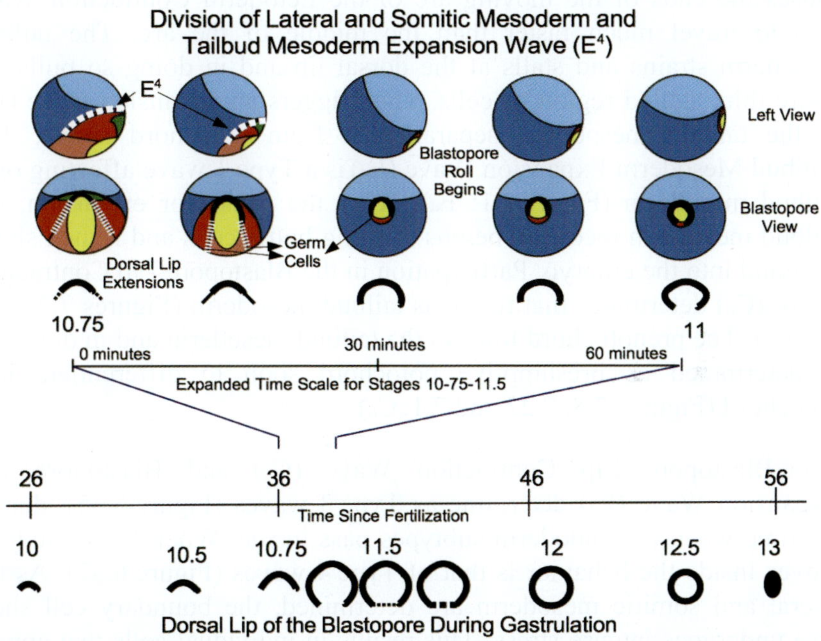

Figure 7.22. The blastopore begins as a line, then forms an arc. However the movement is not a smooth one. Time lapse reveals that between Stage 10.5 and 11 the dorsal lip has brief strong extensions of contraction that form ahead of the line of bottle cells (indicated by the finely dashed white lines). These extensions then go on to form bottle cells but because they go out straight initially, the line they create bisects the mesoderm. Some of the mesoderm then goes over the dorsal lip of the blastopore, participating in that contraction wave (C_6), and some goes under, participating in a expansion wave (E_3). During this time the future tailbud mesoderm first experiences the ectoderm contraction wave (C_7) and so is unable to form bottle cells as that wave reaches the dorsal lip of the blastopore. It undergoes a strong pulling that results in a transient expansion wave (E_4) in that region indicated by the coarsely dashed white line. The effect can be seen visually as a brief pause when the dorsal lip of the blastopore develops a break of uncontracted cells on the top which lasts about 10 minutes. After experiencing this expansion (E_4) it then recovers and contracts forming bottle cells that go over the dorsal lip of the blastopore. The prenotochord follows the tailbud mesoderm and in doing so, is determined as presumptive notochord soon to differentiate into notochord. The ventral lip of the blastopore is formed after a pause by the line of contracting cells joining the lateral lips that begin to contract (dashed black line on the bottom), and then the blastopore becomes a full circle.

accept the signal to begin forming bottle cells and go over the blastopore lip. This pause at the dorsal blastopore changes the mechanical properties of the embryo as a whole and triggers the sudden extension of the blastopore lips and the division of the mesoderm. This same tension also causes the ends of the moving arc of the Ectoderm Contraction Wave (C_7) to travel much faster than the middle of the arc. The tailbud mesoderm strains and stalls at the dorsal lip and in doing so pulls the entire thin, stalled region of cells. This triggers an expansion wave (E_4) in the tailbud mesoderm, separating it from notochord above. The Tailbud Mesoderm Expansion Wave (E_4) is a Type 1 wave affecting only tailbud mesoderm (Figure 7.1: E_4). After the pause for expansion, the tailbud mesoderm recovers, begins forming bottle cells and it then slides over and into the embryo. Participation in the Blastopore Lip Contraction Wave (C_6) determines that tissue as tailbud mesoderm (Figures 7.22 and 7.1: C_6). The prenotochord follows the tailbud mesoderm and in doing so, is determined as presumptive notochord soon to differentiate into notochord (Figures 7.8, 7.22 and 7.1: C_6).

The Blastopore Lip Contraction Wave (C_6) and Blastopore Lip Expansion Wave (E_3) determine as Type 3 waves (Figure 6.35), during the time when the endoderm subtypes pass inside. When the mesoderm moves inside the behavior is that of Type 4 waves (Figure 6.37). As the lateral and somitic mesoderm are determined, the boundary cell sheet area undergoes intense stress. This results in individual cells that appear to contract until they vanish altogether from this zone, particularly during the abrupt formation of the bottom of the blastopore lip. As this region of the embryo is also known to be the region in which future germ cells are determined[71,72], we consider these rapid extensions of the blastopore to be a Type 4 wave (Figure 6.37). Most of the tissue is determined as a cell sheet but some cells break away and become rounded free cells that then migrate to a new location. In this case, it is the germ cell line being established. This illustrates how one wave can pass through more than one tissue. It can behave differently in terms of wave type as it passes through each tissue (Figure 7.11). The wave behavior changes in response to mechanical strain and interaction with other waves. In the case of mesoderm, we see the first of the more complex wave types.

The distribution of the mass of the embryo shifts inwards during this time. This changes the center of gravity of the embryo. The result is that the embryo rolls about 90° so that the neural ectoderm is now on the top and the blastopore is on the side instead of the bottom[73]. The remaining ectoderm covers the lower half of the embryo and the blastopore is now located almost on the new equator where the eventual tail end will be (Figure 7.22). At this point the final wave of gastrulation begins. The neural ectoderm on the top part is undergoing rapid cell division and the cells become taller and horizontally thinner forming the neural plate, as part of carrying out their differentiation. As they do so, an expansion wave (E_5) begins on the boundary zone between neural ectoderm and the rest of the ectoderm. The Ectoderm Expansion Wave (E_5) travels downward from this boundary. This expansion, classically referred to as epiboly of the ectoderm, determines the remaining ectoderm to become epithelial ectoderm (Figures 7.8 and 7.1: E_5). The Ectoderm Expansion Wave (E_5) and the Ectoderm Contraction Wave (E_7) wave are also a Type 4 wave pair. Cells break away from the boundary zone between the neural ectoderm and the epithelial ectoderm and determine the neural crest cells. These presumptive neural crest cells will eventually migrate out and form the neural system connections to the brain and the pigment cells. They also differentiate into smooth muscle, teeth, and craniofacial cartilage and bone[74].

As gastrulation ends, every one of the nine tissues of the simplified fate map (Figure 7.3) is now determined by a unique series of expansion and contraction waves which the cells participated in. As this last expansion wave (E_5) travels downward the embryo now passes from the gastrula stage into the neurula stage.

The final set of waves we observed, the neural plate waves, were at the limit of our resolution with the methods we had. We will present the observations here as anecdotal as they did not meet our own standards for reaching definitive conclusions. These waves occurred in the open neural plate and we watched them as the edges of the neural plate rose, met and sealed to form the neural tube. The first of these was a contraction in which the entire neural plate flattened. This may not have been a contraction. It may have been due to differentiation triggered by the Ectoderm Contraction Wave (C_7).

The lower half of the plate then distinctly elongated and narrowed. The top half of the neural plate expanded. This results in two distinct portions, one future brain and the other future spinal cord. More transient contractions occurred in the spinal area in rapid succession. Meanwhile, the open brain section was also further subdivided. We noted an expansion in the future forebrain region and the region around it contracting. We then saw distinct contractions of groups of cells in the regions that will eventually become placodes. A placode is a flat round or oval disk that will then undergo further subdivisions, foldings, and reshaping to form complex structures like eyes and ears[75]. We saw these sudden dark patches indicating a region of contraction where future eye, nasal and otic (ear) placodes formed. These waves were of short duration, perhaps because they did not have far to go or they represented Type 1 contraction waves (Figure 6.28). The entire region would contract in a single image at our resolution of one image every ten minutes. We reduced the intervals of images to five minutes and one minute. We could still see the contraction but the image remained fleeting. The cells were so small that it was impossible to see them individually in our stereo-microscope. The embryos also moved and rolled sideways, and fell out of the focal plane as the rising neural tube closes, obstructing the view. The hardest to see was the otic placode. We had only one embryo lying at an angle allowing us to see into the groove being formed by the rising neural plate. Since our rule was three views in three normal embryos in three different batches we have not included these neurulation waves of the open neural plate in our fate map and differentiation tree.

Dick began our research on amphibian embryogenesis by trying to explain the shaping of the neural plate[1,76-79], which is where we end this chapter's discourse. In the 1970s computers were 10,000 times slower than now, and most three dimensional simulations were out of the question. Thus we were very fortunate that the urodele (newt, salamander, axolotl) embryo neural plate went through a step of flattening, which we have now put in the context of differentiation waves. It was at the point of that flattening that we began our computer simulation. It was strictly mechanical in nature, showing how a spatial distribution of rates of neural plate cell elongation perpendicular to the cell sheet, at constant volume per cell, coupled with elongation of the notochord and the overlying notoplate (Figure 7.23), could account for the change from a flat disc to a keyhole shape (Figure 7.24). Once the differentiation waves of the neural plate are better understood, future embryologists should be in a position to work out the relationship between the mechanics of embryonic tissue movements and differentiation waves in relationship to tissue differentiation. They should then be able to simulate fate maps quantitatively.

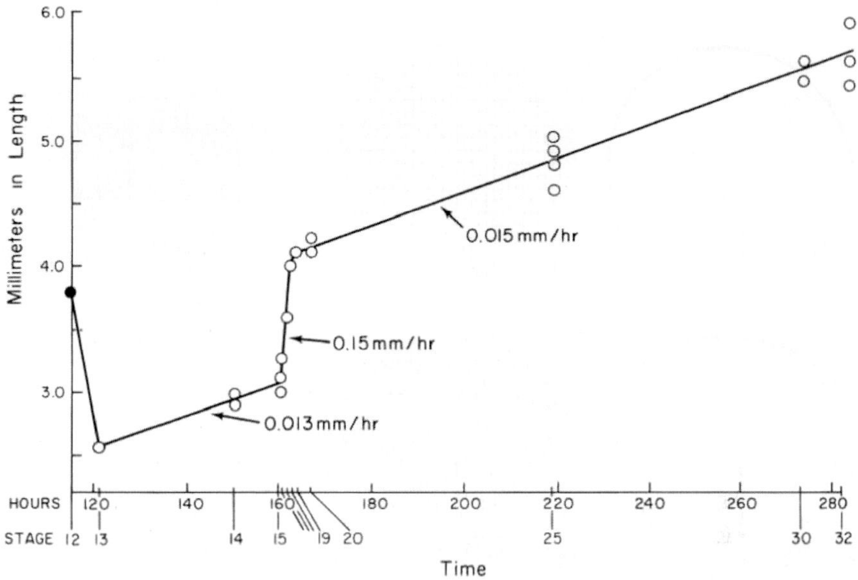

Figure 7.23. Length of the nervous system versus time in the California newt *Taricha torosa*. The initial decline is due to the flattening of the early neural plate from a hemisphere to a flat disc. Neural plate shaping is accompanied by a slow increase in length, followed by a more than ten times faster rate during neural tube closure. Afterwards the more leisurely rate of elongation continues. Note the correlation with total embryo volume versus time (Figure 2.50). The underlying notochord, and the portion of the neural plate attached to it, which we named the notoplate, are presumed to provide the force for elongation. From Jacobson & Gordon (1976)[1] with permission of John Wiley and Sons.

In summary we were able to construct the differentiation tree of the axolotl for the simplified Vogt's fate map. The strange shapes of the fate map regions, such as the wide flair that becomes the long narrow notochord suddenly made perfect sense. The embryo is self-organizing as it passes through gastrulation with waves triggering more waves, until all nine tissues have each experienced a unique differentiation code of expansion and contraction waves. We can now explain the shape of the fate map.

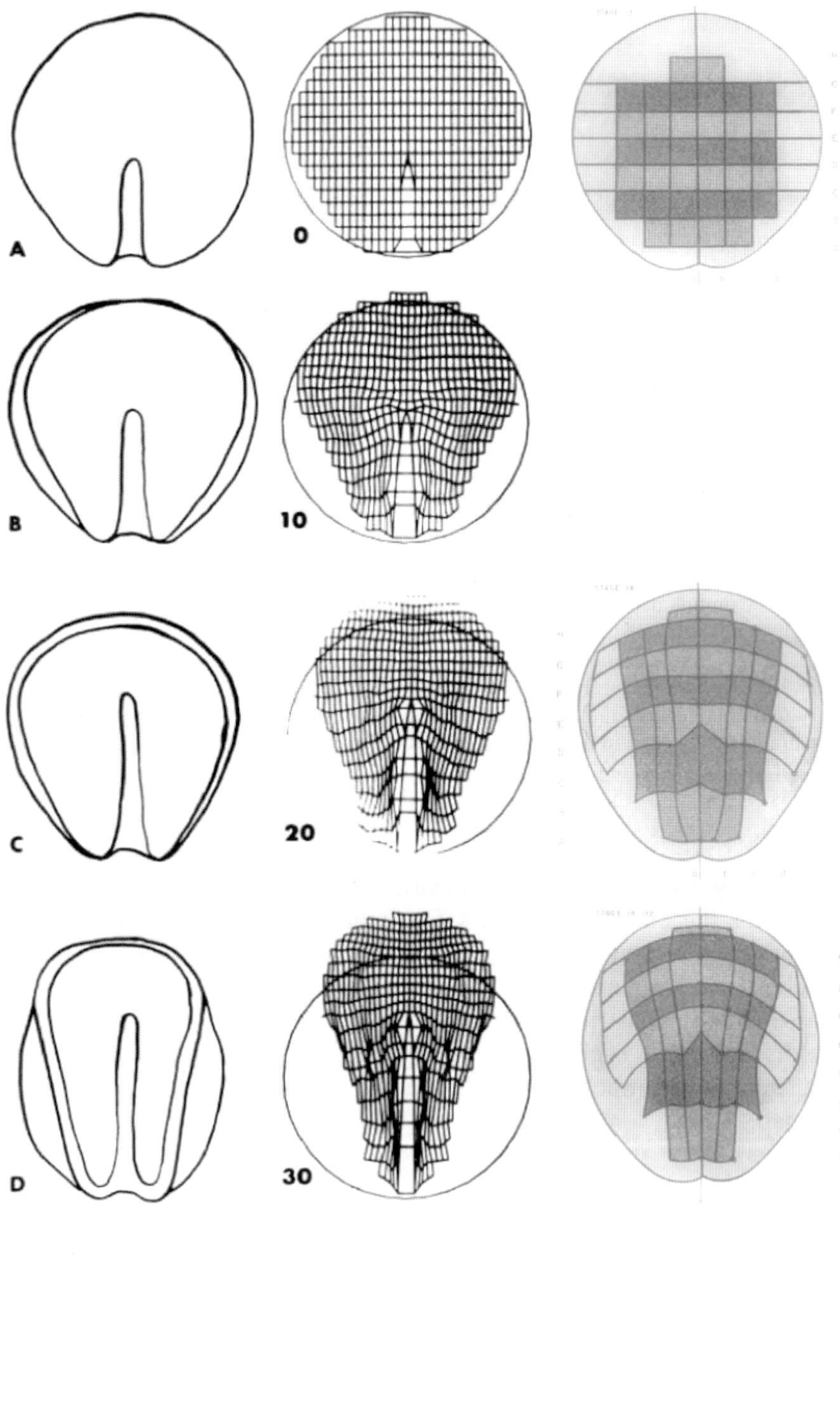
A
0
B
10
C
20
D
30

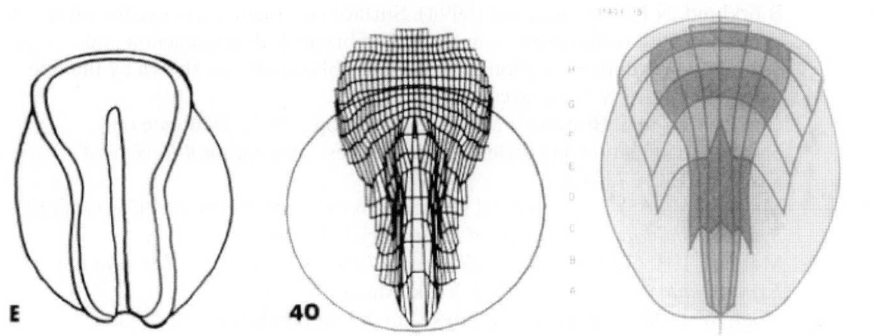

Figure 7.24. **Left column**: sketches of the neural plate changing to a keyhole shape over Stages 13 (A) to 15 (E). The elongation of the notochord and its overlying notoplate is indicated in the middle of the sketches. **Middle column**: Computer simulation of the change of shape of the neural plate. While flat, it needs to be imagined that it is wrapped onto the sphere of the embryo. From Jacobson & Gordon (1976)[1] with permission of John Wiley and Sons. **Right column**: A grid (à la D'Arcy Thompson[80,81]) was placed over the projected image of a film time-lapse movie at Stage 13, and clusters of pigmented cells were followed at each grid point to Stages 14, 14.5 and 15. From Burnside & Jacobson (1968)[82] with permission of Elsevier.

1. Jacobson, A.G. & R. Gordon (1976). Changes in the shape of the developing vertebrate nervous system analyzed experimentally, mathematically and by computer simulation. *Journal of Experimental Zoology* **197**(2), 191-246.
2. Carlson, M., W. Reeves & M. Veeman (2015). Stochasticity and stereotypy in the *Ciona* notochord. *Developmental Biology* **397**(2), 248-256.
3. Wetts, R. & S.E. Fraser (1989). Slow intermixing of cells during *Xenopus* embryogenesis contributes to the consistency of the blastomere fate map. *Development* **105**(1), 9-15.
4. Wilson, E.T., C.J. Cretekos & K.A. Helde (1995). Cell mixing during early epiboly in the zebrafish embryo. *Developmental Genetics* **17**(1), 6-15.
5. Wikipedia (2015). Multivalued function. https://en.wikipedia.org/wiki/Multivalued_function
6. Kimmel, C.B. & R.M. Warga (1987). Indeterminate cell lineage of the zebrafish embryo. *Developmental Biology* **124**(1), 269-280.
7. Bordzilovskaya, N.P., T.A. Dettlaff, S.T. Duhon & G.M. Malacinski (1989). Developmental-stage series of axolotl embryos [Erratum: Staging Table 19-1 is for 20°C, not 29°C]. In: *Developmental Biology of the Axolotl*. Eds.: J.B. Armstrong & G.M. Malacinski. New York, Oxford University Press: 201-219.

8. Björklund, N.K. & R. Gordon (1994). Surface contraction and expansion waves correlated with differentiation in axolotl embryos. I. Prolegomenon and differentiation during the plunge through the blastopore, as shown by the fate map. *Computers & Chemistry* **18**(3), 333-345.
9. Gordon, R., N.K. Björklund & P.D. Nieuwkoop (1994). Dialogue on embryonic induction and differentiation waves. *International Review of Cytology* **150**, 373-420.
10. Mintz, B. (1965). Experimental genetic mosaicism in the mouse. *Ciba Found Symp Preimplantation Stages of Pregnancy*, 194-216.
11. Morgan, T.H. (1897). *The Development of the Frog's Egg, An Introduction to Experimental Embryology*. New York, Macmillan Co.
12. Vogt, W. (1925). Gestaltungsanalyse am Amphibienkeim mit örtlicher Vitalfärbung. Vorwort über Wege und Ziele. I. Teil. Methodik und Wirkungsweise der örtlichen Vitalfärbung mit Agar als Farbträger/Analysis of the organization of the amphibian embryo with local vital dyes. Preface on methods and objectives. Part I. Methodology and method of operation of the local vital dye with agar as color carrier. *Wilhelm Roux' Archiv Entwicklungsmechanik Organismen* **106**, 542-610.
13. Vogt, W. (1929). Gestaltungsanalyse am Amphibienkeim mit Örtlicher Vitalfärbung II. Teil. Gastrulation und Mesodermbildung bei Urodelen und Anuren/Analysis of the organization of the amphibian embryo with local vital dyes. Part II: Gastrulation and mesoderm development in urodeles and anurans. *Wilhelm Roux Archiv für Entwicklungsmechanik der Organismen* **120**(1), 384-706.
14. Nakamura, O. (1938). Tail formation in the urodele, *Triturus pyrrhogaster. Zoological Magazine (Tokyo)* **50**(10), 442-446.
15. Nakamura, O. (1942). Die Entwicklung der hinteren Körperhälfte bei Urodelen [The development of the posterior half of the body of urodeles]. *Annotationes zoologicae Japonenses* **21**(4), 169-237.
16. Pasteels, J.J. (1942). New observations concerning the maps of presumptive areas of the young amphibian gastrula (*Amblystoma* and *Discoglossus*). *Journal of Experimental Zoology* **89**(2), 255-281.
17. Lovtrup, S. (1975). Fate maps and gastrulation in amphibia - a critique of current views. *Canadian Journal of Zoology-Revue Canadienne de Zoologie* **53**(4), 473-479.
18. Landström, U. & S. Løvtrup (1979). Fate maps and cell differentiation in the amphibian embryo - an experimental study. *Journal of Embryology and Experimental Morphology* **54**(DEC), 113-130.
19. Dale, L. & J.M.W. Slack (1987). Fate map of the 32-cell stage of *Xenopus laevis. Development* **99**(4), 527-551.
20. Delarue, M., S. Sanchez, K.E. Johnson, T. Darribère & J.C. Boucaut (1992). A fate map of superficial and deep circumblastoporal cells in the early gastrula of *Pleurodeles waltl. Development* **114**(1), 135-146.
21. Keller, R. & D. Shook (2008). Dynamic determinations: patterning the cell behaviours that close the amphibian blastopore. *Philosophical Transactions of the Royal Society B: Biological Sciences* **363**(1495), 1317-1332.

22. Kaneda, T. & J.-y.D. Motoki (2012). Gastrulation and pre-gastrulation morphogenesis, inductions, and gene expression: Similarities and dissimilarities between urodelean and anuran embryos. *Developmental Biology* **369**(1), 1-18.
23. Björklund, N.K. (1993). Small is beautiful: economical axolotl colony maintenance with natural spawnings as if axolotls mattered. In: *Handbook on Practical Methods*. Eds.: G.M. Malacinski & S.T. Duhon. Bloomington, Department of Biology, Indiana University: 38-47.
24. Armstrong, J.B., S.T. Duhon & G.M. Malacinski (1989). Raising the axolotl in captivity. In: *Developmental Biology of the Axolotl*. Eds.: J.B. Armstrong & G.M. Malacinski. New York, Oxford University Press: 220-227.
25. Hara, K. (1970). "Double camera" time-lapse micro-cinematography; simultaneous filming of both poles of the amphibian egg. *Mikroskopie* **26**(5-6), 181-184.
26. Bachmann, K. (1969). Temperature adaptations of amphibian embryos. *American Naturalist* **103**(930), 115-130.
27. Volpe, E.P. (1957). Embryonic temperature tolerance and rate of development in *Bufo valliceps*. *Physiological Zoology* **30**(2), 165-176.
28. Duellman, W.E. & L. Trueb (1986). *Biology of Amphibians*. New York, McGraw-Hill Book Co.
29. Brodland, G.W., R. Gordon, M.J. Scott, N.K. Björklund, K.B. Luchka, C.C. Martin, C. Matuga, M. Globus, S. Vethamany-Globus & D. Shu (1994). Furrowing surface contraction wave coincident with primary neural induction in amphibian embryos. *Journal of Morphology* **219**(2), 131-142.
30. Wikipedia (2015). ImageJ. http://en.wikipedia.org/wiki/ImageJ
31. Ferreira, T. & W. Rasband (2012). *ImageJ User Guide*. Bethesda, U.S. National Institutes of Health, IJ 1.46r.
32. Williams, R.C., J.J. Correia & A.L. Devries (1985). Formation of microtubules at low temperature by tubulin from Antarctic fish. *Biochemistry* **24**(11), 2790-2798.
33. Coakley, W.T. (1987). Hyperthermia effects on the cytoskeleton and on cell morphology. *Symposia of the Society for Experimental Biology* **41**, 187-211.
34. Coss, R.A. & W.A.M. Linnemans (1996). The effects of hyperthermia on the cytoskeleton: A review. *International Journal of Hyperthermia* **12**(2), 173-196.
35. Gulevsky, A.K. & L.I. Relina (2013). Molecular and genetic aspects of protein cold denaturation. *Cryoletters* **34**(1), 62-82.
36. Detrich III, H.W. (1997). Microtubule assembly in cold-adapted organisms: Functional properties and structural adaptations of tubulins from Antarctic fishes. *Comparative Biochemistry and Physiology A-Physiology* **118**(3), 501-513.
37. Gordon, R. (1999). *The Hierarchical Genome and Differentiation Waves: Novel Unification of Development, Genetics and Evolution*. Singapore & London, World Scientific & Imperial College Press: [http://www.worldscientific.com/worldscibooks/10.1142/2755].
38. Gordon, R. & R. Stone (2016). Cybernetic embryo. In: *Biocommunication*. Eds.: R. Gordon & J. Seckbach. London, World Scientific Publishing: Chapter 5, 111-163..

39. Sheppard Jr., J.J., R.H. Stratton & C. Gazley Jr. (1969). Pseudocolor as a means of image enhancement. *American Journal of Optometry and Archives of American Academy of Optometry* **46**(10), 735-754.
40. Fried, G. & G.J. Hademenos (1999). *Schaum's Outline of Theory and Problems of Biology*, McGraw-Hill.
41. Chegg Study (2016). Definition of Cleavage Patterns and Morphogenesis. https://www.chegg.com/homework-help/definitions/cleavage-patterns-and-morphogenesis-14
42. Uzwiak, D. (2016). Frog Blastula, Frog crescent blastopore. http://bio.rutgers.edu/~gb102/lab_7/702cm-blastula.html http://bio.rutgers.edu/~gb102/lab_7/702dm-cres_blastopore.html
43. Boterenbrood, E.C. & J.M. Narraway (1986). The direction of cleavage waves and the regional variation in the duration of cleavage cycles on the dorsal side of the *Xenopus laevis* blastula. *Roux's Archives of Developmental Biology* **195**, 484-488.
44. Boterenbrood, E.C. & J.M. Narraway (1990). Epiboly connected with cleavage in morula and early blastula stages of *Xenopus laevis*, a study using time-lapse photography. *Roux's Archives of Developmental Biology* **198**(5), 303-307.
45. Barth, L.J. (1964). *Development: Selected Topics*. Reading, Massachusetts, Addison-Wesley Publishing Co.
46. Barth, L.G. & L.J. Barth (1968). The role of sodium chloride in the process of induction by lithium chloride in cells of the *Rana pipiens* gastrula. *Journal of Embryology and Experimental Morphology* **19**, 387-396.
47. Barth, L.G. & L.J. Barth (1969). The sodium dependence of embryonic induction. *Developmental Biology* **20**, 236-262.
48. Barth, L.G. & L.J. Barth (1972). ^{22}Sodium and ^{45}calcium uptake during embryonic induction in *Rana pipiens*. *Developmental Biology* **28**, 18-34.
49. Barth, L.G. & L.J. Barth (1974). Ionic regulation of embryonic induction and cell differentiation in *Rana pipiens*. *Developmental Biology* **39**, 1-22.
50. Daniel (2007). Axolotl egg - cell differentiation: Stage 7 to 8, early blastula - 128 cells. http://www.caudata.org/forum/f1173-advanced-newt-salamander-topics/f24-photo-gallery-video-gallery-technique-discussion/f25-photo-video-gallery/51099-axolotl-egg-cell-differentiation.html
51. Barth, L.G. & L.J. Barth (1954). *The Energetics of Development: A Study of Metabolism in the Frog Egg*. New York, Columbia University Press.
52. Maître, J.L., R. Niwayama, H. Turlier, F. Nédélec & T. Hiiragi (2015). Pulsatile cell-autonomous contractility drives compaction in the mouse embryo. *Nature Cell Biology* **17**, 849-855.
53. Samarage, C.R., M.D. White, Y.D. Álvarez, J.C. Fierro-González, Y. Henon, E.C. Jesudason, S. Bissiere, A. Fouras & N. Plachta (2015). Cortical tension allocates the first inner cells of the mammalian embryo. *Developmental Cell* **34**(4), 435-447.
54. Goldstein, B. & Daniel P. Kiehart (2015). Moving inward: Establishing the mammalian inner cell mass. *Developmental Cell* **34**(4), 385-386.
55. Rivera, R.M., C.R. Youngs & S.P. Ford (1996). A comparison of the number of inner cell mass and trophectoderm cells of preimplantation Meishan and Yorkshire pig embryos at similar developmental stages. *Journal of Reproduction and Fertility* **106**(1), 111-116.

56. Rosa, F., T.D. Sargent, M.L. Rebbert, G.S. Michaels, M.M. Jamrich, H. Grunz, E. Jonas, J.A. Winkles & I.B. Dawid (1988). Accumulation and decay of DG42 gene products follow a gradient pattern during *Xenopus* embryogenesis. *Developmental Biology* **129**(1), 114-123.
57. Semino, C.E., C.A. Specht, A. Raimondi & P.W. Robbins (1996). Homologs of the *Xenopus* developmental gene *DG42* are present in zebrafish and mouse and are involved in the synthesis of Nod-like chitin oligosaccharides during early embryogenesis. *Proceedings of the National Academy of Sciences USA* **93**(10), 4548-4553.
58. Swiers, G., Y.H. Chen, A.D. Johnson & M. Loose (2010). A conserved mechanism for vertebrate mesoderm specification in urodele amphibians and mammals. *Developmental Biology* **343**(1-2), 138-152.
59. Lee, J.Y. & R.M. Harland (2010). Endocytosis is required for efficient apical constriction during *Xenopus* gastrulation. *Current Biology* **20**(3), 253-258.
60. Martin, C.C. & R. Gordon (1997). Ultrastructural analysis of the cell state splitter in ectoderm cells differentiating to neural plate and epidermis during gastrulation in embryos of the axolotl *Ambystoma mexicanum. Russian Journal of Developmental Biology* **28**(2), 71-80.
61. Ballard, W.W. (1964). *Comparative Anatomy and Embryology*. New York, Ronald Press.
62. Gordon, R. & G.W. Brodland (1987). The cytoskeletal mechanics of brain morphogenesis. Cell state splitters cause primary neural induction. *Cell Biophysics* **11**(1), 177-238.
63. Hamburger, V. (1960). *A Manual of Experimental Embryology*. Chicago, University of Chicago Press.
64. Hamburger, V. (1988). *The Heritage of Experimental Embryology: Hans Spemann and the Organizer*. New York, Oxford University Press.
65. Slack, J.M.W. & H.V. Isaacs (1994). The Einsteck-method: Position and structure of projections formed by implants of a ventral character. *Developmental Biology* **161**(1), 313-317.
66. Brodland, G.W. (1990). Large-strain kinematics of deforming cell sheets. In: *Biomechanics of Active Movement and Deformation of Cells, NATO ASI Series*. Ed.: N. Akkas. Berlin, Springer-Verlag. **H42**: 505-510.
67. Hill, M.A. (2013). Gastrula: File:Rugh 068.jpg [must paste in URL]. https://embryology.med.unsw.edu.au/embryology/index.php/File:Rugh_068.jpg
68. Holtfreter, J. (1943). A study of the mechanics of gastrulation, Part I. *Journal of Experimental Zoology* **94**(3), 261-318.
69. Nieuwkoop, P.D. & J. Faber (1956). *Normal table of Xenopus laevis Daudin. A Systematical and Chronological Survey of the Development from the Fertilized Egg Till the End of Metamorphosis*. Amsterdam, North Holland Publishing Co.
70. Karpinka, J.B., J.D. Fortriede, K.A. Burns, C. James-Zorn, V.G. Ponferrada, J. Lee, K. Karimi, A.M. Zorn & P.D. Vize (2015). Xenbase, the *Xenopus* model organism database; new virtualized system, data types and genomes. *Nucleic Acids Research* **43**(D1), D756-D763, http://www.xenbase.org/anatomy/alldev.do.
71. Nieuwkoop, P.D. & L.A. Sutasurya (1981). *Primordial Germ Cells in the Chordates: Embryogenesis and Phylogenesis*. Cambridge, Cambridge University Press.

72. Wakahara, M. (1996). Primordial germ cell development: is the urodele pattern closer to mammals than to anurans? *International Journal of Developmental Biology* **40**(4), 653-659.
73. Kellicott, W.E. (1913). *Outline of Chordate Development [Fig. 31: https://embryology.med.unsw.edu.au/embryology/index.php/Book_-_Text-Book_of_Embryology_5]*. New York, Henry Holt & Co.
74. Wikipedia (2015). Neural crest. https://en.wikipedia.org/wiki/Neural_crest
75. Schlosser, G. (2010). Making senses: Development of vertebrate cranial placodes. *International Review of Cell and Molecular Biology* **283**, 129-234.
76. Jacobson, A.G. & R. Gordon (1976). Nature and origin of patterns of changes in cell shape in embryos. *Journal of Supramolecular Structure* **5**(4), 371-380.
77. Jacobson, A.G. & R. Gordon (1977). Nature and origin of patterns of changes in cell shape in embryos. *Progress in Clinical Biological Research* **17**, 323-332.
78. Gordon, R. & A.G. Jacobson (1978). The shaping of tissues in embryos. *Scientific American* **238**(6), 106-113, 160.
79. Jacobson, A.G. (1980). Computer modeling of morphogenesis. *American Zoologist* **20**(4), 669-677.
80. Thompson, D.W. (1917). *On Growth and Form*. Cambridge, University Press.
81. Thompson, D.W. (1942). *On Growth and Form [2 vols.]*. Cambridge, Cambridge University Press, 2nd.
82. Burnside, M.B. & A.G. Jacobson (1968). Analysis of morphogenetic movements in the neural plate of the newt *Taricha torosa*. *Developmental Biology* **18**, 537-552.

Chapter 8

Signal Transduction from the Cell State Splitter to the Nuclear State Splitter

In May of 1990 we got our first hints of the presence of a physical wave in the ectoderm. By the end of 1991 we had the entire trajectory of that first wave well documented. There were also long stretches of time between waiting for spawnings. We had classes and teaching duties and grants to write at the laboratory. At home, we had children who needed things from us like school lunches, stories before bed, hugs and clean clothing. Still, the differentiation waves occupied our thoughts in every spare moment. A wave of contraction at the surface of the cell results in a signal to the nucleus. The nucleus then uses the signal to turn off one set of genes and turn on another. What exactly was that signal? We referred to the cytoskeletal apparatus on the top of the cell as the cell state splitter. It was inevitable that in our own discussions we would begin referring to the process of turning the signal from the cell state splitter into a specific subset of gene expression patterns as a "nuclear state splitter". Corresponding to the two outcomes for a cell's cell state splitter, contraction or expansion, we hypothesized two corresponding, readied states of the nucleus, only one of which would happen. In this conception, the nucleus too was in a specific bifurcation point, where it would go one way or the other: two choices, no more, no less. One choice would correspond to contraction of the cell state splitter, the other to expansion of the cell state splitter.

In our early discussions of the nuclear state splitter, we had two models. In one model, a purely biochemical signal would travel from the cell state splitter to the nucleus. This purely biochemical signal would be triggered by the contraction of microfilaments or by microtubule polymerization. In the second model, which we called the marionette model, we hypothesized that then unknown cytoskeletal elements in the nucleus would have physical attachments to the cytoskeletal elements at the top of the cell and the DNA would be strung onto that structure. The contraction or expansion of the cell state splitter would directly and physically alter the cytoskeleton in the nucleus and change gene expression by pulling or pushing on the DNA like strings on a puppet. As it turns out both models were each partly correct.

We scoured the literature over several months collecting every related paper we could find. Google did not exist in those days and there were no online journals to download articles in a PDF form. Finding papers meant hours of searching physical indexes or the limited, mostly keyword, computer searches that existed in those days. PubMed was a brand new tool. Using it was like being the proverbial kid let loose in the candy store. Once a reference was located via PubMed, we had to walk to the library and pull out physical copies of journals, carry them to a photocopier, and make a paper copy to work from. We would read that paper copy carefully, underlining or highlighting critical components. Each paper had multiple references to follow up on which meant more trips to the library. If the journal was not available in our library, and it often wasn't, we could try ordering it through interlibrary loan and it would arrive after a few weeks or months. (We would often find ourselves wondering why we ordered a particular paper once it finally came.) We would also contact the author and ask for a reprint. Most scientists were using email by that time and so our requests were acknowledged in a day or two with a promise to drop a reprint in the mail. Some of the scientists, especially those in key papers by senior members of the field, had to be petitioned in formal, politely worded paper letters. Each workday, one of us would run to check the mail to see what eagerly awaited gems had arrived by "snail mail". We would also trek to the library to see what precious items may have arrived via interlibrary loans. Our desks were soon piled high with towers of papers covered with notes. After six months of hard work, we found enough clues from the literature to create a plausible pathway between microfilament contraction and changes in gene expression. Molecular biology of eukaryotic cells was in its infancy back then. Like genetics, the field was exploding. We published the collected ideas as a working model in our paper, "Nuclear state splitting: a working model for the mechanochemical coupling of differentiation waves to master genes", in 1993 (Figure 8.1).

The basic premise was this. Contraction of microfilaments requires a calcium flux. It also requires adenosine triphosphate (Figures 3.15 and 3.16) in enormous quantities. ATP is converted to cyclic adenosine

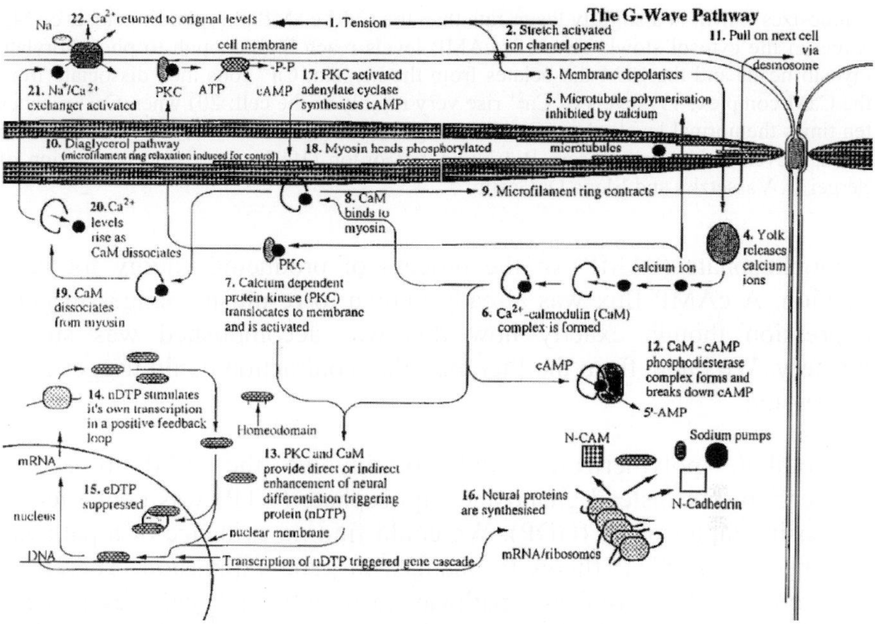

Figure 8.1. In the following original caption, G-wave = ectoderm contraction wave. "Nuclear state splitting. 1) Tension across the membrane surface is generated initially by a decrease in pressure which allows the microfilament ring to dominate the cell state splitter's instability and contract at the G-wave initiation site. The contraction results in tension across adjacent cells; 2) stretch-activated ion channels in the membrane are opened in response to tension. Ions are allowed to enter or leave the cell (or possibly both); 3) ions traveling through the stretch-activated ion channels cause the membrane to depolarize; 4) depolarization triggers a release of internally bound Ca^{2+} ions into cytosol. The Ca^{2+} is likely bound in the yolk or may be from the endoplasmic reticulum. How depolarization achieves this release is unclear but it is likely by the same mechanism that occurs in smooth muscle or via inositol triphosphate; 5) microtubule polymerization is inhibited by rising Ca^{2+} levels in the cytosol; 6) Ca^{2+} ions also bind to calmodulin forming an activated Ca^{2+}-calmodulin complex (CaM); 7) Ca^{2+} ions trigger the translocation of Calcium Dependent Protein Kinase (PKC) to the membrane and it is activated; 8) CaM binds to Myosin; 9) microfilament ring contracts in response to CaM binding; 10) PKC is probably activated via a diaglycerol pathway. This pathway simultaneously induces relaxation that allows smooth control of the microfilament ring contraction; 11) contraction triggers the opening of stretch-activated ion channels in the adjacent cell, propagating the G-wave; 12) CaM binds to a CaM-dependent cAMP phosphodiesterase. cAMP levels are lowered as a result; 13) the activation of PKC and CaM directly or indirectly enhance the neural differentiation triggering protein nDTP (a master gene regulatory protein); 14) the enhanced nDTP stimulates its own transcription, creating a positive feedback loop; 15) epithelial differentiation triggering protein, eDTP, is suppressed; 16) synthesis of neural proteins begins as nDTP triggers a neural gene cascade; 17) translocated PKC activates an adenylate cyclase which

synthesizes cAMP at a slightly faster rate than the CaM cAMP phosphodiesterase. cAMP levels in the cytosol slowly rise; 18) cAMP levels reach high enough to phosphorylate myosin heads and the CaM dissociates from the myosin. Ca^{2+} ions then dissociate from the CaM complex; 19) levels of Ca^{2+} rise very rapidly in the cell; 20) when Ca^{2+} reaches ten times the normal level a $Na+/Ca^{2+}$ exchanger in the cell membrane returns the Ca^{2+} to normal levels in the cell." From Björklund & Gordon (1993)[2] with kind permission of Sergei G. Vassetzky and Elena D. Gasilo.

monophosphate (cAMP) in the process of producing energy for contraction. A cAMP flux was already known to result in changes in gene expression though exactly how this was accomplished was still a mystery. We knew from the literature that contraction could trigger gene expression.

Microtubule polymerization did not involve the ATP to cAMP conversion, but instead guanosine triphosphate (GTP) was converted to guanosine diphosphate (GDP). We could find no evidence of a pathway directly linking GTP fluxes to changes in gene expression. Therefore, we were unable to find any pathway to directly link changes in gene expression to microtubule polymerization. We proposed that such a pathway must exist. The paper was almost purely speculative.

While theoretical papers are easy to publish in fields like physics, and it is quite respectable to do so, in biology theoretical papers are generally viewed with disdain. More than one colleague advised us to not publish the idea until we had more data. We knew we would have a very difficult time finding any standard journal to publish a mere idea. We did find a welcoming colleague in Russia, Lev Beloussov, who has a long history of investigating the physics of amphibian embryos[3-6]. We therefore published our idea, in Russian first[7], in the journal *Ontogenez*. Our English version appeared in their "translation" version of the journal a few months later, the *Russian Journal of Developmental Biology*, though of course that was the original and the Russian version was the translation[2]. While the English translation was in the process of being published, we had observed that there were more waves than the first wave we had seen and rather prematurely named as THE Gastrula wave. We took the opportunity of the delay to prepare an Addendum to the English version[8] correcting our error.

An important component of our model was the concept of bistability. The system can go one of two ways but once it has been triggered, the triggered state corresponds to the signal sent. But some kind of feedback from one response of the nucleus should inhibit the apparatus from triggering a response in the opposite direction. For the contraction signal, the inhibition of the opposing response was fairly obvious. Microfilament contraction requires release of calcium. Calcium inhibits microtubule polymerization. We put this in the first working model. We had no idea what would inhibit microfilament contraction that would also be present during microtubule polymerization. This left our original nuclear state splitter model with a lot of unlabeled arrows.

As the years passed, and biochemical, molecular biological and genetic knowledge grew in great leaps and bounds, more new pathways and interactions and proteins were collected and catalogued. It was not unusual to have a student spend their entire PhD characterizing a single protein within a complex pathway (Figure 8.2). Once the knowledge of the proteins was combined with the DNA sequence producing the proteins, families of biochemical components were discovered and their evolution and relationships across species were explored. Not surprisingly, the protein carefully studied in one organism often turned up in another organism in a closely related form. All too frequently this homologous protein would have an entirely different name or function ascribed to it by some other PhD student or postdoctoral fellow and his or her supervisor. Since the early days of the field, the general amount of knowledge of scientists studying these processes has doubled about every five years. We have tried to follow all of these developments as they came out and, while we found a lot of new detail, we never found anything contradicting the general layout of our first nuclear state splitter model. In fact, the more the scientific community learned, the more correct our original working model appeared to be. There was very little interest from anyone else in our early model. It is so easy to not see the forest for the trees, especially when you are trained to focus on leaves.

You'll need an introduction to basic biochemistry at this point. Biochemical pathways are both complex and simple, as are most things that work well. Today, all of this huge mass of data about how biochemical pathways work can be divided into two general classes, *metabolic pathways* and *cellular and molecular processes*. Metabolic pathways are those biochemical pathways that exist to either break down molecules in order to produce the raw components for making new molecules or to provide energy and make it available to the cell.

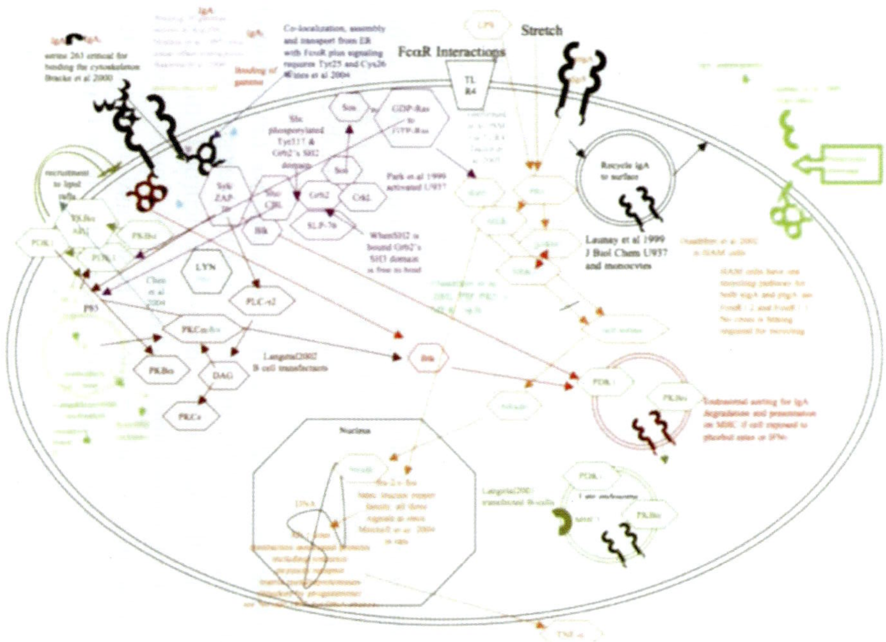

Figure 8.2. Natalie prepared this diagram in 2008 during a period she was reviewing the activity of the FcαR receptor. This receptor binds immunoglobulin type A and is involved in immune response. She was trying to understand the role of this receptor as a trigger for infection related preterm labor. Prior to starting labor the uterus undergoes a differentiation process. One remarkable aspect of this in depth review was how so many parts of the pathway for the 1994 nuclear state splitter, which has previously been simply proposed components denoted by unlabeled arrows, were now fully elucidated in the literature. The second remarkable fact was that many of the same players involved in uterine smooth muscle differentiation associated with initiation of labor were the same as the pathway originally proposed for axolotl ectoderm differentiation. Nature reuses what works.

Metabolic pathways can involve catabolism (breaking down) or anabolism (building up) of biochemical substances. Catabolism and anabolism together equal metabolism. Metabolism is the basis of life and it is undertaken by a multitude of enzymes that catalyze biochemical reactions. Here we inject our own views on use of the words "regulate" and "mediate". These words have their origins in human affairs, and are rather loosely applied to molecules. Regulate means "control by rules" and mediate means "interpose". We eschew "mediate" entirely in this book, except when quoting others. We use "regulate" only when there is a clear control of one molecule by another, extended sometimes to control of whole gene cascades. But we avoid it in broader contexts, such as "Wnt regulates noses". There is too much happening between the molecular activity of Wnt and nose morphogenesis levels to make such claims.

Understanding the metabolic pathways begins with one of the three major pathways that occur in some form in all living organisms. The first metabolic pathway is the Citrate Cycle (also called the Krebs cycle or the tricarboxylic acid cycle). This cycle takes amino acids and converts them to carbon dioxide and water while creating reduced nicotinamide adenine dinucleotide (NAD)[9]. Anaerobes do not technically have a citrate cycle because they do not use oxygen but the citrate cycle is thought to have evolved from the one they do use[10]. NAD is a critical component for energy generation in the second of the three metabolic pathways. This second metabolic pathway is called Glycolysis (also known as the Embden-Meyerhof Pathway). Glycolysis produces energy for *de novo* biosynthesis of molecules or, when reversed, glucose is created from noncarbohydrate sources[11].

The common denominator energy delivery package produced by glycolysis and used by cells is adenosine triphosphate (ATP). Chemical energy is produced by the breaking of the phosphate bond of ATP to produce adenosine diphosphate (ADP) (Figure 3.15). Adenosine diphosphate can also be converted to adenosine monophosphate (AMP) and the cyclic form of adenosine monophosphate (cAMP). cAMP is, as we saw in the old Nuclear State Splitter Model, an important signaling molecule used to signal "Hey, we need more ATP!", but it has other signaling functions as well. Conversion of cAMP, AMP and ADP back to ATP is therefore critical to cellular functioning and is the major purpose for glycolysis. Passing of those phosphate groups from one molecule to another is called *phosphorylation* and it is easily the single most important signaling reaction of all biochemical pathways, not just the metabolic ones. Phosphorylation reactions are used constantly all over the cell and we will detail this more below.

The third metabolic pathway is the Pentose Phosphate Cycle. The Pentose Phosphate Cycle uses glucose to produce nicotinamide adenine dinucleotide phosphate (NADP)[12]. This is similar to NAD[9] but is phosphorylated (has the phosphate group added) and is required for specific chemical reductions that create carbon rings. Carbon rings must be formed during nucleotide and amino acid synthesis. Since we need nucleotides for making DNA and amino acids for making proteins, it becomes obvious that this third metabolic pathway is an extremely important cycle to cells, especially cells of embryos that are growing and making many new cells.

Plants and some microbes also have photosynthesis, a mechanism for producing sugars from light, water and carbon dioxide. Photosynthesis is also considered part of the first metabolic pathway, since it uses the energy of light to run the Krebs Cycle backwards and thereby harvests energy of light to produce more ATP, NAD and sugars[13]. Chemosynthesis is the use of chemical energy instead of light (most commonly hydrogen sulfide), by the same general metabolic pathway. Another closely related metabolic pathway is the one used by those organisms, like yeast, that can produce energy by fermentation. So metabolic pathways come in three general forms with special extra adaptations used for photosynthesis, chemosynthesis, and fermentation by a subset of all living organisms. Now here is where you have an advantage over the biochemistry student. You have enough information about the three metabolic pathways to understand morphogenesis but you do not have to memorize and regurgitate (with correct spelling of all enzymes and accounting for every hydrogen, carbon and phosphorus atom) every step in the three metabolic pathways for your introductory biochemistry exam. Now some of our readers will find this stuff fascinating and if you are among them, you can follow up and learn more beginning with the references we provide. A fascinating overview of all the interrelated biochemistry of the cell has been created by the biotechnology company Roche. Their interactive "poster", which has gotten so complicated that it is now also available online as a teaching tool, is regularly updated and expanded[14]. We show a small portion in Figure 5.22.

All the rest of biochemistry and molecular biology is cellular and molecular processes. That means all the aspects of genes and gene products, the regulation of gene expression, and the function and regulation of the proteins that genes code for. Large, often overlapping, databases of information exist for these enzymes, substrates and products, and their genes, relating all the bits and pieces into one huge interlocking system. We have learned so much this system can no longer be expressed in any one diagram. The data has to be stored in giant databases.

The need for collections of accessible databases has resulted in the entirely new fields of bioinformatics and computational biology (cf.[140]). These two new biological fields essentially were born out of advances in computer science that have been driven by the need to analyze the growing masses of data biologists have been producing. Applying computer techniques to biological data is also responsible for wonderful new discoveries made by data mining existing information. An example of this kind of data mining would be comparing the sequence of a newly discovered gene to the sequence of all other similar genes in all of the species known and then deducing the function or evolution of the newly discovered gene. Comparisons of even larger subsections of the aggregated data can be done. An example of this is comparing whole pathways to determine the evolutionary history of these pathways and finding a common ancestor. All of this is also becoming easier with the new software tools bioinformaticians and computational biologists are constantly creating.

As this entire system of biochemical reactions has become far too complex to describe in any single diagram or individual conceptual understanding, by necessity, scientists and practical biochemists, like clinical metabolic geneticists, tend to deal only with those small subsections of this whole system that are directly relevant to their individual work. One of the more important ways to subdivide this whole enormous interconnected system is through the concept of the *"signal transduction pathway"*. We define a signal transduction pathway as the set of cascading chemical reactions in a cell that occur in response to an external signal such as a hormone or a physical force, and that ends in a predictable response from a targeted part of the cell, often (although not always) accompanied by a change in gene expression. For example, we can think of the arrival of lactose as triggering the signal transduction pathway that results in *E. coli* activating the *lac operon* we mentioned in previous chapters. This is one of the simplest of signal transduction pathways.

The various signal transduction pathways in the cell are becoming very well understood. The signal transduction pathway of one organism can be compared to and used as a basis for understanding the pathway of another organism. Since all of these pathways can eventually lead to control of gene expression, understanding the biochemical steps controlling gene expression in one area can lead to exciting breakthroughs in other fields such as medical science or bioenergy production. It is a bit artificial to think just in terms of signal transduction pathways, however. Most pathways have multiple branches and feedback loops to adjacent signal transduction pathways and this results in multiple subtle modulations of response. Still, defining things in terms of signal transduction pathways does allow us to work with understandable chunks of the entire complex biological system operating in a cell.

The first signal transduction pathways discovered by biologists began with surface receptors on a cell. Many biochemists still define a signal transduction pathway only in terms of a membrane receptor starting point. In 1970, the very first membrane surface receptor was characterized by Martin Rodbell[15]. That surface receptor would be activated when a specific molecule bound to it. These membrane receptor signal transduction pathways are the best understood of all signal transduction pathways because they are relatively easy to study. Add the signaling molecule and it will bind the receptor and something happens in a predictable pattern of response that can be measured. The measurement might be technically difficult but it is easily repeatable and therefore easy to understand. An experiment can be repeated as many times as required until each step along the pathway can be worked out. It is therefore no coincidence that most of what we know about signal transduction comes from studying surface receptor based signal transduction pathways. We now know that surface receptors are not the only type of signal transduction pathway. Receptors of varied types have been discovered to exist at all levels of cellular functioning from the outer surface all the way into the nucleus itself. There are also other forms of external and internal signal transduction systems based on nonbiochemical signals such as temperature, ion flux across membranes (voltage), or mechanical forces like shear and tension.

An important concept to bear in mind is that the signal transduction pathway often triggers the assembly of complex protein structures. Most of the parts of complex multiprotein structures are present already in the cell but do not assemble until signaled to do so. The assembly reaction is generally stalled by a critical component that is not available (Figure 8.3). As soon as the critical component is available, provided by a signal transduction pathway cascade, then the entire complex protein assembly cascade can rapidly assemble and begin functioning. The critical component is normally that component associated with the signal transduction pathway that triggers the assembly. The signal can be locked up in some fashion until it is needed or the signal can be waiting for an external molecule to come into the cell from some external source. The reverse is also true. A signal transduction pathway can also trigger the disassembly of a complex multiprotein structure.

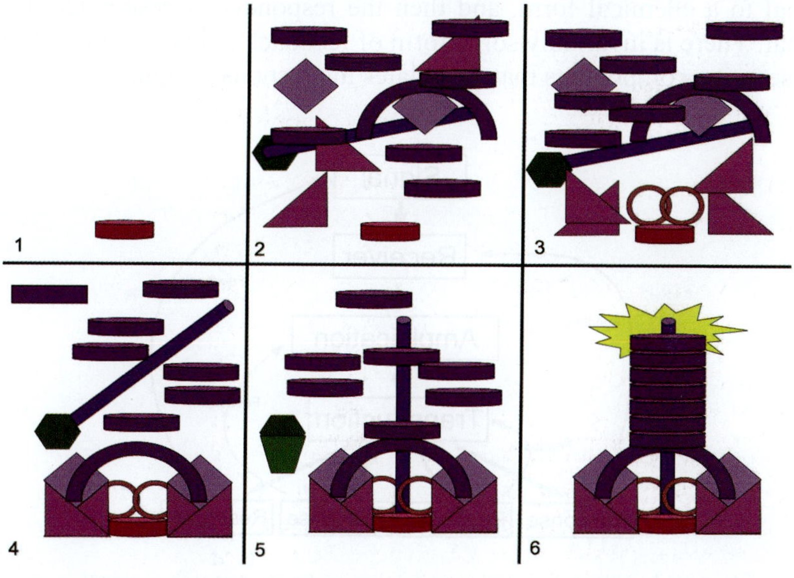

Figure 8.3. Protein Complex Assembly Cascade: This figure is a schematic representation of a complex protein assembly cascade. 1) A base protein is embedded in the membrane. No other proteins are present. No structure is present. 2) Differentiation occurs and the required proteins parts are manufactured and present except the first protein. No assembly takes place. 3) The missing protein is present, assembly begins. 4) One of the proteins in the cascade is present but capped by another protein. Partial assembly takes place but stalls at the point where the capped protein is required. 5) Another enzyme removes the capped protein by binding with the cap. Assembly

continues. 6) Assembly, which had been stalled now continues until the protein complex is fully assembled and it becomes functional (yellow). In the case of this hypothetical protein complex there are three control points. The first control point is the expression of the proteins. The second control point is the addition of the starting protein. The third control point is the uncapping of one protein required for completion of the structure. If the uncapping protein (shown as lighter green) must be freed by being disassembled from another structure in order to become active, the second structure will not activate until the first structure is gone. These types of regulation of assembly of cytoskeletal complexes are common and can work in the opposite direction resulting in disassembly. The control points to assembly cause "check point pauses" during critical processes in cell functioning such as regulation of mitosis. Building of nucleating centers for microtubules require this kind of cooperative building and is subject to these types of controls. The system can also work in reverse for disassembling structures.

In principal, signal transduction requires five components. These are the signal itself (often the missing component of a complex structure), the receiver of the signal, the amplifier of the signal, the transducers of the signal to a chemical form, and then the responder or responders to the signal. There is invariably some form of feedback to one or more of these consecutive components that modulates the response (Figure 8.4).

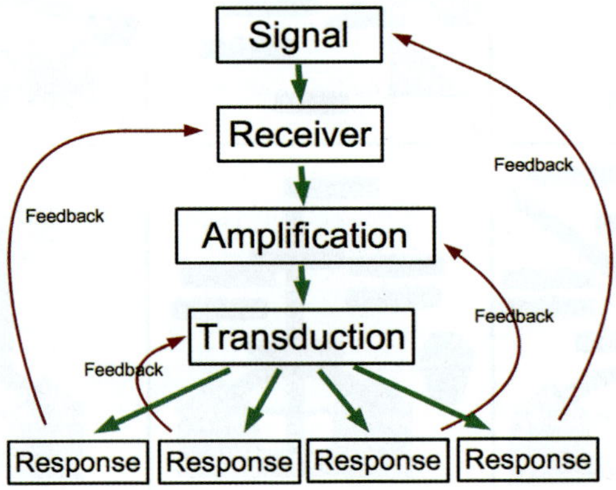

Figure 8.4. All biochemical signal transduction pathways contain a signal, a receiver, amplification of the original signal, transduction of the signal to biochemical form where the signal was originally something else such as a mechanical stress, and a response or responses to the signal. There is also feedback into the signal transduction pathway that modulates the response. While thinking of biochemical reactions in terms of signal transduction pathways is useful, it is important to recall that at each step there may be branch pathways, cycling and terminations of the signaling via feedback signaling from outside of the original pathway. Signal transductions pathways do not exist in isolation from what is going on in the rest of cell.

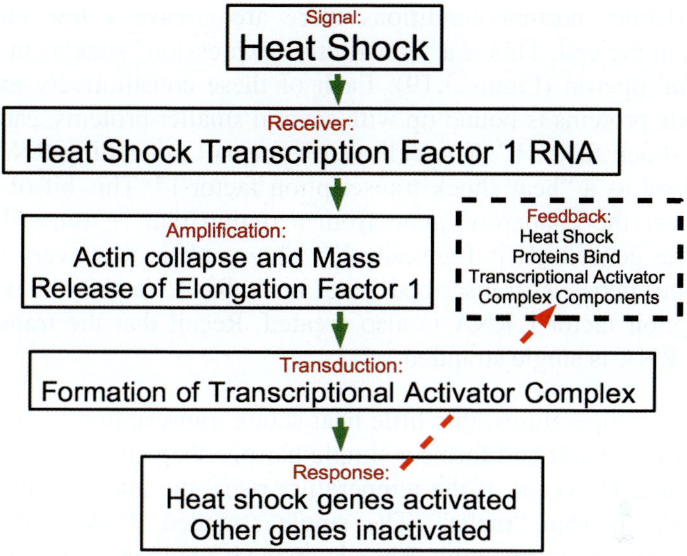

Figure 8.5. The general principal components of signal transduction pathways applied to the heat shock response, described in detail in text.

Recall heat shock proteins introduced in Chapter 3 on genetics. Heat shock proteins are actually chaperone proteins that are required for normal folding of many proteins (Figure 3.14). During a heat shock (or other stressful event) the number of these chaperone proteins required is dramatically increased. The heat shock has resulted in many proteins becoming unfolded then spontaneously misfolding. These misfolded proteins far outnumber any new proteins being made by regular protein synthesis. The cell responds to this by upregulating the expression of the gene that makes the chaperones. Many more mRNA copies are made than usually would be. These mRNAs are translated into new chaperone proteins. When all the misfolded proteins are repaired, and the cell has returned to a normal state, the extra expression of the heat shock proteins drops down to the normal amount required for regular protein folding. This says a lot, but of course there is a lot more detail to this process (Figure 8.5).

In fact the details of this particular signal transduction pathway are absolutely fascinating[16] and understanding this one as an example can provide important insights into all signal transduction pathways. So here we go. Under normal conditions there are always a few chaperone proteins in the cell. This is a "constitutive expression" state as in the case of the *lac* operon (Figure 3.19). Each of these constitutively expressed chaperone proteins is bound up with several smaller proteins, each called a "heat shock factor 1". The cell also has small noncoding RNAs, each categorized as a "heat shock transcription factor 1". This bit of RNA is coded near the chaperone gene from a region that is transcribed first before the gene itself is transcribed[17]. The result is that every time the chaperone gene is transcribed into an mRNA, a new heat shock transcription factor 1 RNA is also created. Recall that the transcription factor 1 RNA is single stranded.

At normal temperatures, this little heat shock transcription factor 1 RNA folds back on itself and forms a simple hairpin shape that does very little of anything. However, if the temperature rises, the internal binding that creates the hairpin "melts". The single stranded RNA works like a temperature sensor: once this RNA is opened it is in an active form. The active RNA binds to heat shock factor 1 proteins that are normally bound onto the chaperone. Heat shock factor 1 proteins change their conformation when the little single stranded RNA binds. Both the little heat shock transcription factor 1 RNA and the heat shock factor 1 proteins are released from the chaperone.

The heat shock factor 1 proteins don't just float about lost. Once they are released from their preferred position on the chaperone protein, they bind to each other in groups of three with one of the open temperature sensor RNAs stuck on. These three heat shock factor 1 proteins and a single heat shock transcription factor 1 RNA form a *transcriptional activator complex* which, as the name suggests, is a complex of proteins and RNA that can start DNA transcription. Once transported into the nucleus the transcriptional activator complex will bind to specific DNA sequences upstream of the sequences for chaperone proteins. When the transcriptional activator complex binds to the DNA, it displaces a repressor protein. (We saw this same displacing of a repressor protein in the *lac* operon: Figure 3.20) With the repressor protein removed, polymerases can now bind the DNA and begin making multiple mRNA copies of the chaperone gene. The mRNAs are then transported out of the nucleus and used to make many more chaperone proteins. These transcriptional activating complexes formed by heat shock factor 1 and the temperature sensing RNA appear in the nucleus within minutes of a heat shock. So how do those transcriptional activating complexes get down into the nucleus so fast? This requires another protein, one that is released by a *cytoskeletal rearrangement.* So we go from a signal transduction pathway that is part of the cellular biochemistry to the cytoskeleton. Surprise, surprise!

We know that normally the actin in the cell exists partly in free form but much of it is part of stable cables of microfilaments. The elongation factor 1 protein is required for both forms of actin. The first role of elongation factor 1 to be discovered was performing the delivery of loaded tRNAs (transfer RNAs each with an attached amino acid: Figure 3.12) to the ribosome where it adds to and elongates the amino acid chain that makes actin. If elongation factor 1 is not present, no tRNA is transferred into position and the elongation of the amino acid chain that is required for transcribing the mRNA for actin, fails. Because of the amount of actin being made all the time, any free elongation factor 1 floating about (and available to be isolated by a curious scientist studying the cell) is busy delivering the tRNAs to the ribosome. That was why the protein was called elongation factor 1. For a long time that was the only function we knew about. That is not the only thing this protein can do. Elongation factor 1 also binds microfilament polymers of actin into cables and other stable actin complexes. Unfortunately, in molecular biology, molecules are often named after their first discovered or presumed function, and the name sticks, no matter what else they do. It would be like naming a baby "Squally" after their first notable event. It's probably a better naming system than some others but such naming systems can leave us biased into not noticing other functions for proteins.

The better name for "elongation factor 1" really should be "micro-filament cable production protein". While some free elongation factor 1 is always busy moving amino acids to make actin, most of it is actually bound up in the making of stable microfilament cables. When a heat shock happens, the microfilament cytoskeleton collapses. With the collapse, there is a sudden and enormous change in the amount of free elongation factor 1. This newly released elongation factor 1 binds to the threesome of heat shock factor 1 proteins and the little RNA temperature sensors that form the transcriptional activating complex. The elongation factor 1 transports the transcriptional activating complex into the nucleus. While the mechanism is not precisely worked out, binding of elongation factor 1 to the transcriptional activating complex causes a change in conformation of the nuclear pores that exposes specific binding sites. This results in the transportation of the transcriptional activating complex to the inside of the nucleus[18]. Once in the nucleus, the transcriptional activating complex, with the attached elongation factor 1 from the actin, are then coated with multiple phosphate groups. (Coating with phosphate groups is the all important phosphorylation process we mentioned before.) The heat shock factor 1 proteins also have a short signal peptide added to signal where they should be directed to in the DNA. (A peptide is a very short protein, i.e., a very short amino acid chain.) With the phosphate groups and the signal peptide attached, the complex can bind to the DNA and displace the repressor. Polymerases move in to replace the repressor and make mRNA from the genes coding for chaperones. More temperature sensing RNA bits are made at the same time. Those mRNA copies of the chaperone code are moved out of the nucleus. The RNA temperature sensor travels with the mRNA outside the nucleus. The chaperone mRNAs are translated into more chaperone proteins.

The net result of a heat shock then is that all the cellular machinery making and maintaining the microfilament cytoskeleton and a variety of other proteins is abruptly repurposed for making heat shock chaperones and all the other proteins required to get all the existing misfolded proteins rapidly repaired and the cell back to its normal state. During a heat shock, the cell must respond quickly. So the collapse of the microfilament cytoskeleton is a "shortcut" speedway used to activate the whole heat shock system very rapidly, not unlike the way a red alert on a ship results in a whole lot of scurrying about while every sailor drops their usual task and rushes to battle stations[19].

So to summarize: a heat shock causes little sensor RNA strands to go from an inactive hairpin shape to an active open shape. The active open shape binds to heat shock factor 1 which is itself stuck on a chaperone protein. Heat shock factor 1 lets go of the chaperone and joins two other free heat shock factor 1 proteins and the little sensor RNA to form a transcriptional activator complex. Meanwhile, the cytoskeleton has collapsed and releases many elongation factor 1 proteins along with a small signal peptide. The transcriptional activator complex binds with elongation factor 1 and the signal peptide is added and the whole cluster gets covered with phosphate groups. All of them travel to the DNA and displace a repressor protein blocking RNA polymerases. The RNA polymerases can now get at the DNA and they make many copies of the mRNA for chaperone proteins. These are then carried out of the nucleus and translated into many chaperone proteins. In this signal transduction pathway, the original signal is the heat causing the little sensor mRNA to go from a hairloop to a single strand. The result is a cascade of reactions that causes many more copies of the chaperone protein to be made. But what exactly triggers the microfilaments to collapse? As it turns out, that little sensor RNA is not the only part of the cell that reacts directly to the heat shock. The cell's lipid membrane also reacts to a heat shock. And that reaction signals the cytoskeleton to collapse. It works in this fashion.

At any given time in the cell there are free actin molecules and there are actin molecules polymerized as microfilaments. In plants, the amount of free actin typically represents about 95% of all actin[20]. Only 5% of all plant actin is bound up in the polymerized form. The opposite is true in certain human brain cells where up to 95% of actin is in microfilament form[21]. However, both are always present. The reason is that *in vitro* experiments using pure microfilaments show that they will fall apart by themselves to free actin at a rate of 0.044 to 1.14 μm/min. This indicates that there must be binding proteins and enzymes in the cell to keep actin in its polymerized, microfilament form. The amount kept polymerized depends on the individual cell. However, in living cells of the frog *Xenopus*, for example, microfilaments depolymerize at up to 9 μm/min or eight times faster than spontaneously. Obviously, there are enzymes that can speed the natural tendency of microfilaments to depolymerize and there are other enzymes putting actin together into microfilaments and keeping it that way[22-24].

Eukaryotic cells have a specific family of microfilament depolymerizing enzymes. All of them are structurally similar and appear to have arisen from a single original sequence[25]. They can be loosely divided into two broad classes, gelosins and cofilins. There are tissue and species specific subclasses of these activating factors, each highly specialized for specific situations, but nonetheless all related. Gelosins are activated in the presence of calcium ions. Gelosin has strong microfilament capping ability (binding to the end) and it can cut into microfilaments when calcium ions are present. Cofilins seem to be more important and closer to the original microfilament depolymerizing enzyme. All eukaryotes are thought to have at least one copy of a cofilin protein gene in their genome[26].

Gelosins are inhibited from their microfilament capping by a set of signaling molecules called polyphosphoinositides. Polyphosphoinositides are proteins with a lipid tail attached to keep them firmly embedded in the underside of the membrane. They represent only a small percentage of membrane proteins and their function is not well understood but they are intimately tied to cytoskeletal rearrangements. They "autophosphorylate" and then can, in response to certain signals, either pass on their phosphate and phosphorylate nearby proteins or they can be phosphorylated and release small phosphoinositol molecules which act as so-called second messengers for phosphorylation based signaling. One of the best known is 1D-myo-inositol 1,4,5-triphosphate or IP_3. IP_3 is able to bind to receptors on vesicles filled with bound calcium and cause a calcium release with a resulting flux of free calcium ions, Ca^{2+}. IP_3 can also cross membranes so it is a good candidate molecule for cell-to-cell signaling required for propagating contraction waves over a membrane (Figure 8.6).

Nonphosphorylated cofilins are active and can sever microfilaments. When phosphorylated, they are inactive. This is not the entire story. Under correct conditions, active cofilin can also cause the creation of additional microfilament structures. This is because when cofilin severs a microfilament, the now exposed ends where the snipping occurred are suddenly available to make a two pronged, branched microfilament[27] (Figure 5.31). Active cofilin can therefore not only break down microfilaments to free actin but it can also participate in changing microfilaments to larger, multiple branched microfilament structures. Branched structures are much more stable and up to ten times stronger than any single microfilament. This branch creating activity of cofilin is critically important in amoeboid like cell movement. Cofilin activity results in the branching pattern of microfilaments needed at the front end of cells for exploratory ruffles as the cell moves. Note the branching apparent in Figure 5.33. Cofilin is also needed for contraction of the microfilaments at the rear end of an amoeboid cell as it moves and pulls up the back "skirting" of the cell membrane behind it. Phosphorylated (inactive) cofilin is found generally distributed in a moving cell. In the front ruffling end of the cell, where rapid branching and extension of microfilaments is required, cofilin is dephosphorylated and in its active form and working in the company of other proteins that quickly add actin monomers that extend branch points. Cofilin is also dephosphorylated in the rear zone where contraction and dismantling of microfilaments is

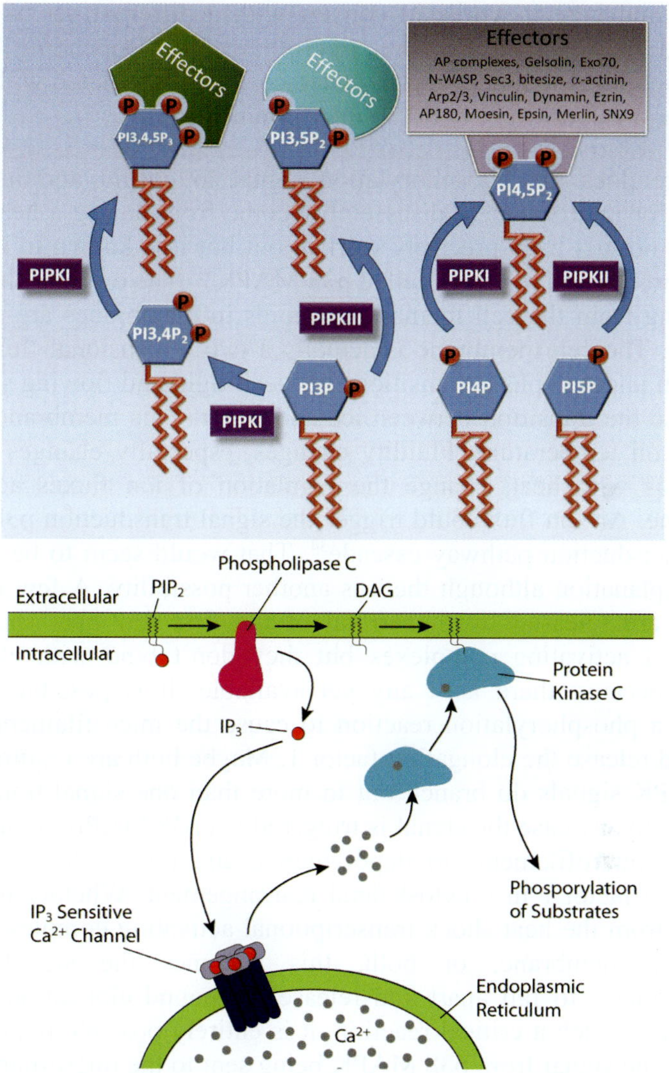

Figure 8.6. **Top**: "Specific phosphoinositide messengers regulate protein activities. Polyphosphoinositides can be generated from PtdIns3P (PI3P), PtdIns4P (PI4P) and PtdIns5P (PI5P) by enzymes of the PIPK family. Importantly, each of these polyphosphoinositides possesses a specific group of protein effectors within the cell, and the binding of these lipids to their protein partners acts to modulate protein activity. $PI3,4,5P_3$, $PtdIns(3,4,5)P_3$; $PI3,5P_2$, $PtdIns(3,5)P_2$; $PI3,4P_2$, $PtdIns(3,4)P_2$; $PI4,5P_2$, $PtdIns(4,5)P_2$." From Schill & Anderson (2009)[28] with permission under STM Permissions Guidelines. **Bottom**: "PIP_2 cleavage by PLC to release IP_3 and DAG". From Korte (2007)[29] with permission under a GNU Free Documentation License.

required and here it works in concert with a different set of proteins dedicated to dismantling the microfilaments. Depending on the enzymatic milieu that dephosphorylated cofilin is associated with, cofilin can either break down or build up microfilaments[30].

So where does the phosphorylation signal to cofilin and its related proteins for microfilament disassembly come from during a heat shock? That has not yet been precisely worked out but it is known to involve a signal transduction pathway called p38 MAPK[31]. The original signal may be coming from the cell membrane. Lipids in membranes are also heat sensitive. The cell membrane is actually a two dimensional fluid whose lipids can undergo phase transitions between rigid and flowing states[32,33], much like the transition between ice and water. This membrane fluidity depends on temperature. Fluidity changes, especially changes to more fluidity as with heat, change the regulation of ion fluxes across the membrane. An ion flux could trigger the signal transduction p38 MAPK signal transduction pathway cascade[34]. That would seem to be the most likely explanation although there is another possibility. A few repressor proteins are released by the arrival of the first few heat shock transcriptional activating complexes but they don't bind with elongation factor 1 because there isn't any yet available. It is possible that this initiates a phosphorylation reaction to cause the microfilaments to fall apart and release the elongation factor 1. Maybe both are required as the p38 MAPK signals do branch out to more than one signal transduction pathway. In any case the signal is triggered via p38 MAPK signaling and it causes microfilaments to depolymerize and release the rest of the elongation factor 1 in a cytoskeletal rearrangement. Whether the signal goes up from the heat shock transcriptional activating complex or down from the membrane, or both, this becomes the signal to the microfilaments to fall apart and release the bound elongation factor 1. Since this is such a critical reaction, it is entirely possible there is more than the one signal from p38 MAPK being sent to the microfilaments via the phosphorylation/dephosphorylation mechanism. There are differences in the reaction to heat shock in different cell types that make generalizations difficult[35,36].

So what does this tell us about cytoskeletal rearrangements? The amount of actin that is rapidly polymerized into microfilaments and just as rapidly taken apart in specifically localized regions of the cell is activity controlled by phosphorylation and dephosphorylation of cofilin and related proteins. We have also seen from the elongation factor 1 story that microfilaments can serve as a scaffold on which to bind a variety of proteins and in effect to store them in an inactive form. Rapid depolymerizing of the microfilaments results in an abrupt release of these proteins, creating a rapid activity flux. While microfilament depolymerization activates elongation factor 1, other enzymes are known to work in the opposite fashion. They become active only when bound to free actin. There is a constant build, take apart, build activity going on with the cytoskeleton and the result is that the activity is tightly controlled and confined to specific areas of the cell by various signal transduction pathways and mechanisms related to them.

Since microfilaments are also in the nucleus (with cofilin bound on), the same kind of rapid flux can occur in the nucleus itself. Just as changes can be localized to the leading edge of the moving cell, these fluxes can also be localized to specific areas of the nucleus. Changes in the microfilament portion of the cytoskeleton not only change shape, movement or contractility of a cell, which are the usual functions we think of for microfilaments. Microfilament rearrangement is also about rapid changes in gene expression in response to phosphorylation signaling by the kinases of signal transduction pathways[37]. There are so many actin or microfilament binding proteins, with more being discovered all the time, that scientists have created an online actin binding protein encyclopedia[38].

Microtubules, like microfilaments, can also act as phosphorylation controlled reservoirs for proteins. For example[39] NF-κB is another transcription modulator that activates a variety of genes, among them a set of genes that produce proteins that are triggered by an interruption of mitosis. These proteins cause the cell to commit suicide via what is known as *programmed cell death* or *apoptosis*. Programmed cell death is a very handy method for removing cells with huge mutations at the chromosome level due to gross failures of cell mitosis. IκBα is the enzyme that phosphorylates and activates NF-κB transcription factor. IκBα remains bound to microtubules in an inactive form. Rapid depolymerization of microtubules leads to release of the IκBα and then activation of the NF-κB transcription factor[41]. There are entire classes of microtubule binding proteins just as in the case of microfilaments. The influence of microtubules is accomplished by direct binding and/or release of specific proteins. This occurs via motor intermediates that use the microtubules to travel along them or by serving as a site for two molecules to bind in close proximity so that they can interact (Figure 8.7).

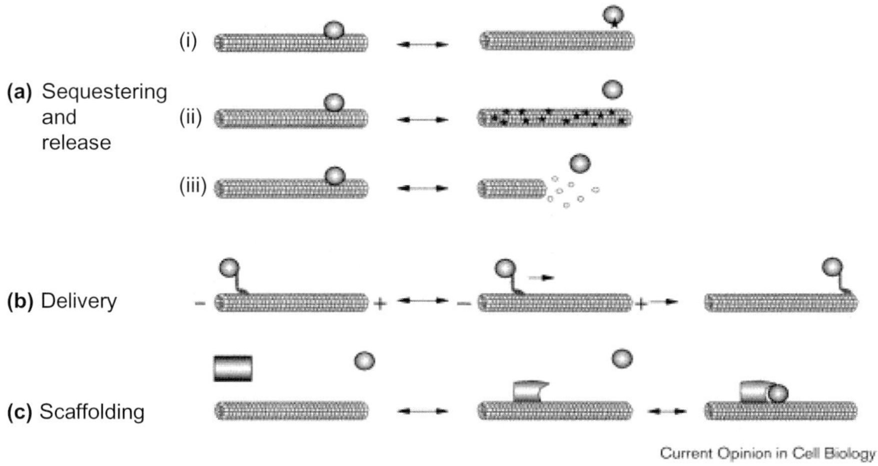

Figure 8.7. "Three basic mechanisms for MT-mediated signal transduction are shown. The signaling factor, represented by a sphere, **(a)** is shown to interact directly with the MT, **(b)** via a motor protein or **(c)** via a putative scaffolding factor. (a) For sequestering and release, three possible release mechanisms are envisioned: (i) signal induced modification (represented by a ★) of the factor itself, (ii) of the MT or (iii), enhanced MT depolymerization. (c) The scaffolding factor (represented by a cylinder) is shown to undergo a conformation change upon binding to the MT and this generates a binding site for the signaling factor. Other mechanisms (e.g. post-translational modification) may activate the binding of the scaffolding and signaling factors." From Gundersen & Cook (1999)[40] with permission of Elsevier Science Ltd. MT = microtubule.

The changes in gene expression that we considered when examining heat shock, elongation factor 1 and microfilaments, or NF-κB transcription factors, IκBα and microtubules, are changes related to normal cell function. The cells involved are not undergoing differentiation (although it could be argued programmed cell death is a form of terminal differentiation). These changes represent the set of genes accessible to the cell type they reside in. Those accessible gene sets, or using the term we previously defined, *regulons*, can be turned off or on at need. Whether a regulon is turned on or off depends on what proteins are bound to the cytoskeleton or not bound. Control of transcriptional activator complexes is also dependent on whether the proteins that control their formation are activated or inactivated when bound or unbound to cytoskeleton. Whether these proteins are bound or unbound depends on their phosphorylation state. Signal transduction pathways are involved in this phosphorylation or dephosphorylation, changing the state in response to a signal. So let's take a closer look at phosphorylation.

Phosphorylation involves the addition of a negatively charged molecule (i.e., an ion) made of a single phosphorus atom bound with four oxygen atoms in a group (PO_4^{-3}). This process is used throughout cellular biochemistry and normally involves transferring the third phosphate group from ATP to another molecule, releasing energy stored in the third bond and resulting in ADP plus the phosphorylated molecule in an active configuration (Figure 3.15). Enzymes can also transfer the phosphate group to another protein which then passes the phosphate on to a third protein and so on. It is not unlike runners in a relay race, racing down the biochemical pathways, with the phosphate group as the baton. Most enzymes, like transcription factors, are commonly only activated or deactivated via phosphorylation by other enzymes. These phosphorylating enzymes are collectively called protein kinases. The action of phosphorylation and dephosphorylation is so important to cellular functioning that 2% of all our own human genes code for protein kinases. We have over 500 genes for protein kinases that add or subtract phosphate groups to activate or deactivate important proteins. Individual protein kinases normally can bind more than one type of protein and what they bind to and act on depends on assorted signals they receive along signal transduction pathways[42].

Today, all the assorted signal transduction pathways are reasonable well understood. For example, in humans, all kinases that are known can be classified into seven different subgroups depending on the sequence of the gene producing the kinase and what the kinase interacts with. The first of these groups are the AGC kinases, including protein kinase A, protein kinase G and protein kinase C. Each of these kinases responds to levels of specific signaling molecules. Protein kinase A responds to the level of cyclic AMP and is critical for control of ATP producing pathways. Protein kinase C responds to increases in free calcium such as the calcium flux that occurs during fertilization as well as to the lipid based signal molecules, diaglycerols. The conversion of GTP to GDP during microtubule polymerization (and other reactions) can produce a cyclic GMP signaling molecule analogous to cyclic AMP (Figure 3.15). Not surprisingly, cyclic GMP has a matching set of kinases known as protein kinases G[43].

The second broad class of kinase is also associated with calcium fluxes but only when calcium is bound to a protein called calmodulin. Typically calmodulin acts by binding to proteins that have no calcium binding site of their own. By having a calmodulin bound to act as an adapter, there can be activation or deactivation of a protein that lacks a calcium binding site in response to a calcium flux. Calmodulin-type proteins are therefore a class of kinases called Ca^{2+}/calmodulin-dependent protein kinases[44].

The third type of kinase is the casein kinase 1 group that interacts with the protein casein much as in the case of calmodulin. This kinase is associated with changes in circadian rhythms but it also pops up as an important kinase for pathways involved in steps of embryogenesis and in transport of specific elongation factors (including the elongation factor 1 released from microfilaments in a heat shock). This kinase also binds microtubules, affects the phosphorylation of microtubule associated proteins, and has a role in the control of mitosis[45].

The fourth group of kinases is the CMGC group. (CMGC is alphabet soup speak for cyclin-dependent kinase, mitogen-activated protein kinase. The source of the last C evades us.) MAPK (the same protein as in the MAPK pathway required for microfilament collapse), glycogen synthase kinase, and CDC-like kinases of cell cycle regulation[46] are all CMGC kinases. This broad class of kinases are evolutionarily related and while there are multiple forms in any given species, they are also well conserved between species. This set of kinases has been extensively explored in our efforts to understand cancer because a failure in the regulation of one of these kinases is associated with most cancers. Of these, CDC and MAPK are the best understood. CDC is involved in the control of the cell cycle, especially when and if a cell goes into mitosis, and MAPK is required for so called "cell fate decisions" including differentiation. The MAPK pathway also has several kinases activated by receptors external to the cell. These receptors respond to signaling from the outside environment. Each of these kinases typically works in a complex with an average of eleven other proteins. Recall how when part of one group of proteins, cofilin, contracts and disassembles microfilaments. While cofilin is part of another group of proteins it causes branching and building of more elaborate microfilament structures[47]. The combinatorics of potential complexes with an average of eleven possible proteins per kinase, with each of these other kinases in different phosphorylation and allosteric states, means the signaling possibilities are nearly endless.

The fifth group of kinases consists of the tyrosine kinases that use ATP at the beginning of a signal transduction pathway by taking the phosphate of ATP and using it to phosphorylate another protein. There is a clinically relevant example of what can go wrong during embryogenesis when a tyrosine kinase is mutated. Fibroblast growth factor receptor 2 is a transmembrane protein with tyrosine kinase activity. In humans, this particular fibroblast growth factor receptor 2 is normally only present in a subset of early pharyngeal/mesodermal derived tissue (Figure 7.3) known as branchial arch 1. The tissue affected exists in both the fingers and the skull during early embryogenesis. The gene involved is found on Chromosome 10. Normally, the tyrosine kinase activity of this receptor is only initiated in the presence of specific "growth factors". (A growth factor is simply any naturally occurring substance that if present appears capable of somehow stimulating some form of cell proliferation or differentiation.) During normal development, some of the tissue in the hands, feet, and in the skull receives a signal to die instead of differentiate. The signal to die requires both an inactive tyrosine kinase and no bound growth factor. Mutations of the gene cause the tyrosine kinase protein to be active even when a stimulating growth factor is not present. The tissue made of these cells does not die when it is supposed to. The error results in Apert Syndrome.

People affected by Apert Syndrome have a constellation of specific malformations including a short and oddly angled thumb, and fused fingers and toes. They also have craniofacial abnormalities because the sutures of the skull fuse prematurely. This results in internal pressure on the developing brain and this pressure affects the formation of the face. Affected people have a long, narrow face that is sunken in the middle with widely spaced eyes. Their eyelids often don't quite close properly. There are also specific abnormalities of the teeth and ears due to malformations of the jaw. Most people with Apert Syndrome are intellectually normal and the malformations can be partly or completely repaired surgically. All of this is caused by a set of embryonic tissues that should have gone down the apoptosis route failing to do so and remaining present.

This failure of cells to die is usually due to a paternal age related mutation. Interestingly, this specific mutation is almost always inherited from a father who is entirely normal. When this particular mutation occurs in rapidly dividing presperm cells in the testis, the cells will divide more quickly and thereby produce far more sperm than cells without the mutation. The older the father is, the more likely that this mutation will have taken place in some of the presperm cells and therefore the higher the percentage of sperm that carry the mutation. There is a large class of such paternal age related genetic diseases[48].

Tyrosine kinases typically contain their own inhibitor group so their activity is short lived. They can therefore be considered an "on or off" signal switch. When they are stuck on due to some fault, they are often involved in the uncontrolled growth of cancer. Tyrosine kinases are therefore also a common target of cancer drugs. While many tyrosine kinases reside in the cytoplasmic membrane region, another entire class (including the tyrosine kinase "Lyn", to be discussed shortly, and SUN proteins) resides in the nuclear membrane and acts as a biochemical link between cytoplasmic signal transduction and nuclear signal transduction (Figure 5.14). When they reside in the nuclear membrane and act as linkers between signal transduction systems inside and outside the nucleus, tyrosine kinases act as biochemical rheostats, something that controls by continuously varying the signal strength rather than a simple on/off switch[49]. Another closely related class of kinase is the tyrosine-like kinases, which are lumped in with the tyrosine kinases in the six class set of kinases. Interesting enough, this kinase class is not found in fungi. Plants and animals have both tyrosine kinases and tyrosine-like kinases but in animals, the tyrosine kinases predominate while in plants the tyrosine-like kinases are the more common ones[50,51].

The last group of kinases, the sixth type, consists of the STR kinases. These are kinases originally found by studying yeast that were unable to bud or divide and were therefore sterile. The first mutated gene for the kinase was called "sterile" (yielding STeRile) and so this group of kinases got its name. The different types were referred to as yeast STR 7, STR 11, and STR 20 kinases. Not surprisingly, given their critically important role in one eukaryote, subsequently homologs were discovered in many other species. Humans, for example, have at least 28 versions of yeast STR 20. STR kinases are active during cell division and are important for a cell that enters apoptosis (programmed cell death). Because programmed cell death is avoided in cancer cells, many of these STR kinases are referred to as "tumor suppressors". When the STR kinases are working properly, damaged cells will go into apoptosis instead of forming tumors and so tumors do not appear[52]. There is a lot of overlap in activity with and interactions of STR kinases with the CMGC group of kinases.

So why do we have so many copies of all these kinases? We know that different subsets of the genome are opened and closed using specific control sequences. Once a gene is duplicated and copied and pasted into a different regulon, its sequence can drift and change with evolutionary pressures[53]. This evolutionary process would become ever more important as organisms developed multiple cell types. Each differon ends up requiring its own set of kinases. Different versions of a kinase will often have different affinities for what they would bind to. This means that in one cell type, there might be a fast and immediate response to a specific signal via its kinase while in another cell type, the same signal might elicit little or no responses. Different forms of the same kinase in different cell types can therefore result in specific reactions according to cell type. So, for example, Lyn kinases appear only in cells of the immune system even though almost identical non Lyn kinases are expressed in other cells types[54]. An interactive display of the many human protein kinases is available[55].

In the early days, when all of these signal transduction pathways and their assorted kinases were being worked out, we would attend seminars with PowerPoint "cartoons"[56] showing how all the various proteins interacted. More than once, we would return from the lecture hall, our heads spinning with the latest update on the latest newly discovered protein and what that protein bound to or did not bind to we would find ourselves humming the children's nursery rhyme *Dem Bones*[57]. We would sometimes relieve tension by spontaneously leaping up and singing the old song substituting the latest signal transduction pathway we had just been subjected to:

Ezkiel cried dem dry bones.
Ezkiel cried dem dry bones.
Ezkiel cried dem dry bones.
Oh hear the word of the Lord!
The growth factor is connected to the RTK.
The RTK is connected to the GRB2.
The GRB2 is connected to the Ras protein.
The Ras is connected to the MEK.
The MEK is connected to the MAPK.
The MAPK is connected to the ERK.
The ERK is connected to the FOS protein.
The FOS is connected to the DNA.
Now hear the word of the Lord!

The students generally loved it and more than once we had a funny spontaneous round of dry bones in the lab. Singing *Dem Bones* did bring us a few long looks from our more reserved colleagues, especially those who felt they were being mocked. If we did it in front of our children, one would invariably cry: "But that's not how the song goes!" We also developed such a strong aversion to alphabet soup and Alphabits cereal that both were practically banned from our kitchen. To get even with us for that, our children forbid the use of terms from molecular biology from all family Scrabble games. Is it any wonder none of our children became scientists?

Scientists now have enough knowledge of what is going on to see that in the whole picture there are really only two broad classes of biochemically based signal transduction pathways using all these kinases. These are the extracellular receptors and the intracellular receptors. The best understood of these two classes is the extracellular receptor class of the G protein coupled receptors. These are proteins that sit in the membrane and respond to external signals and trigger G protein activity. G proteins remove a phosphate from a guanosine triphosphate (GTP) changing it to guanosine diphosphate (GDP). The exchange of the phosphate underlies a G protein's ability to send an external signal down a signal transduction pathway. Another example of extracellular signaling is the tyrosine kinases that are embedded in the membrane. When a signaling molecule binds on the portion of such a protein that is outside the cell, a conformation change results in a change in phosphorylation on the inner side of the membrane. The change in phosphorylation can then continue down a signal transduction pathway to the nucleus and result in a change in gene expression. These reactions represent the extracellular receptors.

Intracellular receptors are the second class of biochemical signal transduction pathways. They are less well understood. There are receptors found throughout the cell including within the nucleus. Among the earliest documented are receptors that bind and respond to the presence of small molecules that have diffused passively across the cell membrane such as testosterone and other steroids. Intracellular signal transduction pathways also include the *second messengers*, so called because the first of such pathways discovered were triggered in response to the signaling from an extracellular signal transduction pathway. They were originally thought to be a secondary message system triggered only by an extracellular signal.

We now know these "second messengers" don't always need the extracellular "first" messenger to react but the name has unfortunately stuck. Second messengers typically involve signaling between the membrane proteins and cytoplasm. Calcium signaling is the release of the calcium ion, Ca^{2+}, moving from bound to free form. Calcium is normally sequestered in specific areas of the cell in bound and inactive form within membrane folds such as in the endoplasmic reticulum. With the correct stimulation of specific receptors in the membrane, sequestered calcium can be released as an ion flux into the cytoplasm and this in turn activates other signal transduction pathways. Calcium ion fluxes cause microtubules to depolymerize and simultaneously cause formation and contraction of microfilament structures. This flux is a good example of the class of second messenger biochemical signals. Second messengers are always small ions or molecules. cAMP, and nitric oxide (NO) are other examples of second messengers. IP_3 is an example of a second messenger of lipophilic molecules that are normally bound in the plasma membrane. Diaglycerol (which is produced by one of the polyphosphoinositides) is another example.

Mechanical stresses are also known to change gene expression. The endothelial cells lining our veins and arteries are subjected to an assortment of powerful mechanical stresses including laminar shear stress, turbulent sheer stress, and circumferential stretch. Each stress activates a specific regulon[58-60] (Figure 8.8). Most of our knowledge of mechanotransduction comes from studying diseases of the cardiovascular system because the forces there are so obvious and cardiovascular disease is such an important cause of morbidity and mortality. Mechanoreceptors are often surface molecules partially or entirely embedded in the membrane that can sense and respond to mechanical forces. These mechanoreceptors transduce those forces into biochemical signals that are then sent along the standard collection of signal transduction pathways.

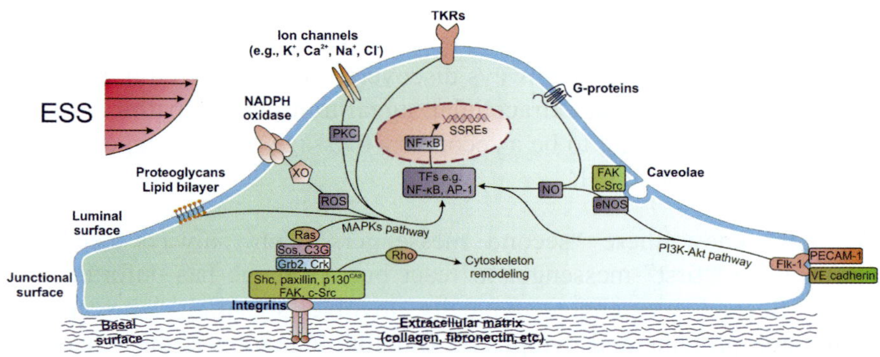

Figure 8.8. "Endothelial Mechanotransduction of ESS: Local endothelial shear stress (ESS) is sensed by luminal endothelial mechanoreceptors, such as ion channels (K^+, Ca^{2+}, Na^+, Cl^-), G-proteins, caveolae, tyrosine kinase receptors (TKRs), nicotinamide adenine dinucleotide phosphate (NADPH) oxidase and xanthine oxidase (XO), plasma membrane lipid bilayer, and heparin sulfate proteoglycans. Also, ESS signals are transmitted through the cytoskeleton to the basal or junctional endothelial surface, where certain integrins or a mechanosensory complex consisting of platelet endothelial cell adhesion molecule-1 (PECAM-1) and Flk-1 are activated, respectively, and initiate a downstream signaling cascade. Activated integrins phosphorylate and activate a multiple complex of non-receptor tyrosine kinases (FAK, c-Src, Shc, paxillin, and $p130^{CAS}$), adaptor proteins (Grb2, Crk), and guanine nucleotide exchange factors (Sos, C3G), thereby activating Ras family GTPase. Active Ras plays a pivotal role in intracellular transduction of ESS signals as it triggers various parallel downstream cascades of serine kinases; each of these kinases phosphorylates and hence activates the next one downstream, ultimately activating mitogen-activated protein kinases (MAPKs). Besides integrin-mediated mechanotransduction, ESS activates a number of other downstream signaling pathways initiated by luminal or junctional mechanoreceptors. These pathways include the production of reactive oxygen species (ROS) from NADPH oxidase and XO, activation of protein kinase C (PKC), activation of Rho family small GTPases (which mediate the remodeling cytoskeleton resulting in temporary or permanent structural changes of ECs), release of endothelial nitric oxide synthase (eNOS) and other signaling molecules from

caveolae, and activation of phosphoinositide-3 kinase (PI3K)-Akt cascade. Ultimately, all of these signaling pathways lead to phosphorylation of several transcription factors (TFs), such as nuclear factor-kappa β (NF-κB) and activator protein-1 (AP-1). These TF proteins bind positive or negative shear stress responsive elements (SSREs) at promoters of mechanosensitive genes inducing or suppressing their expression, thereby modulating cellular function and morphology." From Chatzizisis *et al.* (2007)[60] with permission of Elsevier.

These mechanochemical signals are much harder to study than chemical receptors because it is difficult to precisely replicate something like shear force in a petri dish. Multiple mechanical signals can be received by a cell, and all of these signals can be transduced to biochemical ones, and many will change gene expression. However, the signal transduction pathway typically activated by mechanical stresses is far less specific and less isolated than a purely chemical signal. With mechanochemical signal transduction there is a lot more cross talk and feedback between the various signal transduction pathways and their kinases. The cytoskeleton is also much more important for mechanotransduction. This makes characterizing mechanochemical signal transduction challenging and our knowledge has understandably lagged behind that of purely chemical signaling[61].

All of the signal transduction pathways, whether biochemical or mechanical, or some combination of both, eventually converge to the level of the DNA. The signal eventually activates or deactivates a specific set of proteins that bind to the DNA at specific regulatory sequences. All of this leads to the transcriptional activation or repression of specific regulons. Having now explained the concepts of metabolic pathways, signal transduction pathways, and their kinase components, and mechanical and chemical signaling, and how all of it interacts with the cytoskeleton, we are now ready to look at how this works in differentiation.

Integrin linked mechanotransduction that occurs during early differentiation of mesoderm is a good example of a signaling pathway used during differentiation. This signal transduction pathway is triggered in both zebrafish embryos and in the fruit fly (*Drosophila*) embryo meaning this pathway goes back at least to the point where the two species diverged on the evolutionary tree[62]. It is not a coincidence that this divergence of the two species occurred 570 million years ago, but after the time when mesoderm first appeared[63-67,62]. Both embryos have an early form of epiboly where presumptive ectoderm expands and in doing so compresses and triggers a contraction in the presumptive mesoderm. Nor is it a coincidence that this physical action corresponds to stage 9.5+ to stage 9.75 in the axolotl where our own work has demonstrated the Central Constriction Band (C_4) wave. The expression pattern matches the differentiation code $C_1E_1C_4$ in the axolotl (Figure 7.11: premesoderm). The result of triggering this integrin signal pathway in both the fruit fly and the zebrafish is that the transcription factor β-catenin translocates to the nucleus where it affects gene expression by binding to specific upstream sequences of exposed DNA[68].

A mesoderm specific gene brachyury (the same version of the protein is known as notail (ntl) in zebrafish and Twist in *Drosophila*) is turned on by integrin linked mechanotransduction. A pair of genes in axolotl called AxNodal-1 and -2 is also expressed in early presumptive mesoderm[69] coincident in time and place with the Central Constriction Band (C_4). We further illustrated in Chapter 7 how AxBrachyury, the axolotl version of brachyury, begins being expressed in bottle cells immediately after they go over the dorsal lip of the blastopore with the code $C_1E_1C_4E_2C_6$ (Figure 7.14). The signal transduction pathway for all of these early presumptive mesoderm differentiations requires some form of the brachyury protein whether the animal involved is a fruit fly, a zebrafish or an axolotl. The code involved in axolotl is CEC and then, in what appears to be a duplication and specialization, CECEC, (as CEC CEC to CECEC) for secondary early presumptive mesoderm differentiation. The signal sensor in all these cases (zebrafish, fruit flies and axolotls) is integrin linked mechanotransduction. The integrin triggers a common signal transduction pathway that uses the transcription factor β-catenin. β-catenin translocates to the nucleus and this determines a new pattern of gene expression. New gene expression triggers the early presumptive mesoderm differentiation.

Many of our colleagues call genes like brachyury "inducers" or "specifiers" or "organizers" and attribute to them the ability to cause specific determinations in the right cell, when it is in the right place, at the right time while failing to explain precisely how this is done beyond motherhood statements like canalized development or hierarchies of development. We propose that during early development the cell state splitter works entirely on *juxtacrine* based mechanochemical signaling.

Juxtacrine signaling is signaling that requires physical contact between cells and starts with mechanical signals that are then changed to chemical for signal transduction. In other words, the cell state splitter reacts to local mechanical signals by contracting or expanding. Contraction or expansion triggers the signal that the genome responds to. It is therefore purely mechanical forces that "induce" and "specify" mesoderm determination. Axolotl cells that participate in the $C_1E_1C_4$ code express AxNodal-1 and -2 proteins. Axolotl cells that later participate in the $C_1E_1C_4E_2C_6$ code express AxBrachyury proteins.

Juxtacrine signaling is signaling that requires physical contact and starts with mechanical signals that are then changed to chemical signaling for signal transduction. These are all that are required for early development. Later events of embryonic determination can include mechanochemical signal transduction combined with purely chemical signaling or determination can work via pure chemical signaling such as hormonal signaling in late development in complex organisms. But chemical signaling by itself is simply not a sufficient explanation for creating the right cell, in the right place, at the right time, during the early and most critical stages of embryogenesis. We know that axolotls can produce polyploid embryos that have fewer and larger cells. Yet these large celled axolotls still develop normally. That is because development is not linked to individual cell size but is linked to the mechanics of the entire cell sheet. It is a purely mechanical signal and it is therefore robust.

Thus, according to our model of the cell state splitter, either the entire tissue contracts as a unit (as in Central Constriction Band Contraction Wave), or more commonly, where pressure is exerted in only one region of a tissue, a contraction wave is initiated (as in the Ectoderm Contraction Wave). Once the contraction wave is initiated, the signal becomes the mechanical tugging of adjacent cells as they contract. The release of β-catenin from the plasma membrane and its translocation to the nucleus is the initial mechanochemical signal transduction. The mechanical force on the e-cadherin also results in the phosphorylation of specific proteins and this signal is then amplified by the triggering of a second messenger, a calcium flux. The calcium release causes a powerful reaction that stimulates contraction of the microfilament ring. Microtubules are disassembled due to the calcium flux as the microfilament contraction occurs. This changes the balance of free and bound cytoskeletal proteins. Amplification of the initial signal continues with release of more β-catenin and other transcription factors (probably previously bound to microtubules) that move to the nucleus. The contraction of the microfilament ring takes ten minutes. During that ten minutes, there are massive changes in the nucleus resulting in whole segments of DNA being sequestered and other segments are exposed. A new differon is exposed. As the microfilament ring relaxes, microtubules are polymerized and the related protein cofactors that were released are moved back onto the microtubules. The signal ends.

However, profound changes have occurred in the cell because it is now accessing a new differon. This results in changes in gene expression and changes in the types of proteins the cell produces. Changes in proteins synthesized for the membrane, for example, will change the properties of the membrane. Expression of different receptor proteins change how and what kind of signal the cell can now respond to. The final process of differentiation involves the setting up of a new cell state splitter (or resetting of an old) to await the next mechanical signal. Differons of embryogenesis are turned off or on in much the same way as differons for responses like heat shock, and by using the same signal transduction pathways that are used for changing regulons. The difference is the strength and length of the cytoskeletal rearrangement and the rearrangement of the sequestration state of the DNA. This differs sharply from the traditional model of the cytoskeleton as nothing more than a support structure for the cell. In classic embryological terms this is also the "activation-transformation model"[70].

Our model is not the traditional one. The traditional model of embryology is that specific cells, located in an "organizing center", secrete molecules that diffuse passively into the space around the nearby cells and bind to some form of surface receptors (Figure 6.41). Because of diffusion, regions close to the signaling cell receive high concentrations while cells far away receive low concentrations of signaling molecule. Cells decide their fate by reading the concentration they are in locally and cells respond with the correct gene expression for that level of morphogen. This is the gradient version of the positional information model of embryogenesis. This is the model currently taught in classrooms worldwide and that is presented in all the textbooks and in Wikipedia articles as factually correct and certain.

There are numerous problems with this model:

1. In order to maintain a gradient at steady state, besides a steady[71], spatially defined[72] source for the diffusing molecules, there has to be a sink. This means there must be a way in which diffusing molecules are destroyed or removed along the way and/or at some boundaries[73]. Most people who invoke gradients don't bother with analyzing and solving the partial differential equations for diffusion of molecules to show how the gradients work. Sinks are rarely, if ever, even considered when the gradient model is invoked.
2. A common supposition is that the molecules diffuse outside cells instead of through them[74], but diffusion must occur in a confined space, if a gradient is to be established. Otherwise the molecules will just diffuse away. If the diffusion is extracellular, then the diffusion is critically dependent on the existence of such confining boundaries. Most of the early development of the axolotl occurs in the outside layer of cells, and is normal whether or not the jelly layers and vitelline membrane are present. So in this case no such confined space exists.
3. The speed of development may not permit steady state to be reached[75-78]. This is sometimes considered an advantage in cases where the steady state could not possibly lead to the correct morphology. So we have a steady state invoked except when we don't want it[79,80]. In any case, the rate of development varies substantially with temperature over a species' temperature range for normal development (Figure 7.4). This would have to be matched to the temperature dependence of diffusion of the molecule[81] which is itself dependent on the temperature variation of the viscosity of the medium through which the molecule diffuses[82]. We will leave it as an exercise for the so inclined reader to calculate how robust this situation is.
4. Ordinary diffusion gradients do not scale well[83,84]. The consequence for embryos is that for embryos of different sizes there should be widely different proportions of parts but we know that is not the case[85]. There is a limit on the "range" of a morphogen gradient[86]. Such range limits also limits their potential role in growing tissues[87,88].
5. Diffusion gradients follow the superposition principle. This means that a gradient of one substance in, say the x-direction, and a gradient of the same substance in the y-direction, result in

a single one-dimensional gradient in the diagonal direction, not a two dimensional gradient. Yet biologists frequently invoke a two dimensional gradient to get the gradient model to fit their data. If you want a two dimensional gradient system you have to have two morphogen gradients with two different sources and sinks placed approximately perpendicular to one another.

6. The fundamental principal of gradients is that cells in high concentrations will respond in one way, while those at low concentrations respond in a different way while those in the middle respond in yet another way. Fluctuations in gradients always occur, especially if the number of diffusing molecules is low. Fluctuations of purported morphogen concentrations make response to particular concentration thresholds problematic[89-91].
7. Each cell has to be able to "read" the morphogen concentration accurately[92-97], lest boundaries between tissues become ragged[98]. Gradients are frequently invoked without any explanation how a cell measures a concentration.

Many doubts about the functioning[99,100] or existence[101] of these so-called "morphogen" gradients have been raised, with alternatives[102,103] and elaborations[104-113], and transport mechanisms other than diffusion[114-117] being proposed. While this literature makes assorted attempts to address these problems with gradients, they honestly remind us of the use of epicycles to retain the presumed perfectly circular motion of the planets: each planet rode around the Earth on a rotating circle borne by a larger rotating circle[118]. Like epicycles, multiple, overlapping gradients are sometimes required[119]. We won't review the numerous molecules that have been proposed to be morphogens, but as new biologically active molecules are discovered, they tend to be added to the list[120] and then later sometimes removed[121]. While gradients such as bicoid in *Drosophila* one-cell embryos are indeed found[122], as we pointed out in the Preface, the existence of polyembryonic wasps, reasonably similar to *Drosophila* in adult appearance (Preface Figure 0.1), raises doubts as to the importance of these gradients for embryogenesis.

Also, the gradient model ignores the entire mechanochemical signal transduction we now know exists, especially the entire cell-cell contact juxtacrine signaling via mechanosensitive protein connections like integrin. It also does not explain why the contraction and expansion waves occur. Why does the embryo go through the particular gyrations of differentiation wave activity, with contraction and expansion in such a well defined manner, if it all runs on gradients? Surely the juxtacrine signaling that is possible via waves is an obviously superior method to do cell fate signaling, especially in early embryogenesis.

There are signaling molecules used by cells to signal between one cell and another. They are commonly divided into autocrine, paracrine and endocrine signals. Autocrine signaling, where the cell releases molecules that bind to its own receptors and thereby signals itself, are common in development and in immune response. A cell releases a signaling molecule that signals the cell to grow and then it receives its own signal and grows. Cancer is a common result of autocrine signaling gone wrong. Paracrine signaling, where a cell uses release of molecules to signal adjacent cells, is also well established. Endocrine signaling, long distance signaling utilized for things like coordinating hormone activity between such distant cell types as the pituitary gland and ovaries during ovulation and menstruation, is also well documented. External membrane-attached receptors that bind signaling molecules certainly exist.

We therefore propose a two type signaling system. In early embryogenesis, cell-cell communication is based entirely on juxtacrine signaling through the use of the cell state splitter (Figure 8.9). As development proceeds, autocrine, paracrine and endocrine signaling appear and become more important as complexity increases (Figure 8.10). Autocrine, paracrine and endocrine signaling are secondary to juxtacrine signaling because for these signaling modalities to function, there must be specific cell and tissue receptors expressed in one cell type and not in others. This is the only way to keep the response specific. The signal has to be specific enough to affect only the intended target and not everything else in the vicinity. When the organism becomes so complex that mechanics alone can't do it, receptors and signaling molecules are required.

There is an even more fundamental issue to keep in mind. A cell receives a specific mechanical signal via waves. That is a determination event (or activation). After that signal, the cell produces a receptor and a signaling molecule in response. That receptor and signaling molecule and all subsequent activity are not the determination events. They are differentiation events that are secondary to determination (the transformation).

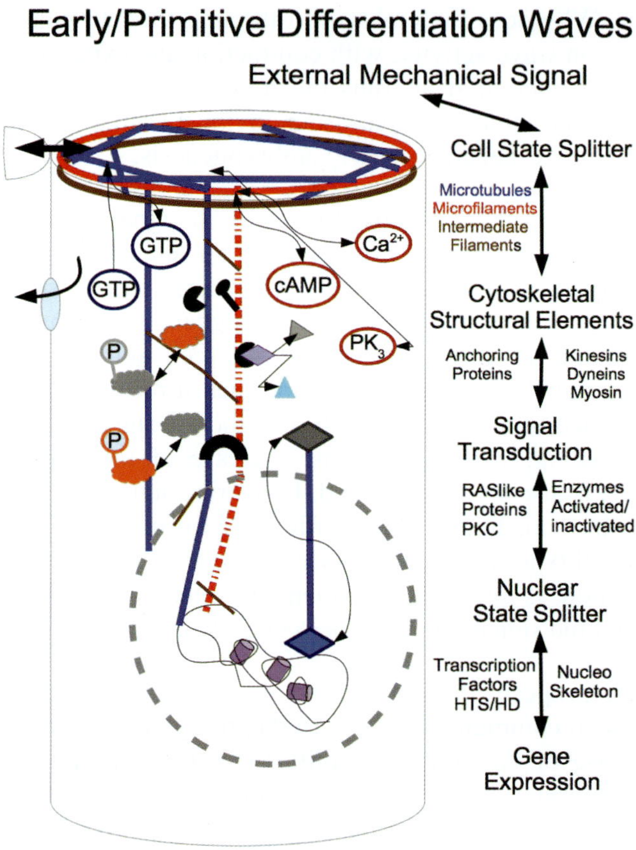

Figure 8.9. In early embryogenesis, cell-cell communication is based entirely on signaling from adjacent cells (juxtacrine signaling) through the use of the cell state splitter.

A great deal of the literature on receptor and molecular signaling during embryogenesis describes what are actually secondary differentiation/transformation events, not the initial determination/activation event. Let us now look at a specific textbook example of a typical "gradient" signaling morphogen molecule, Wnt.

All metazoans (all multicellular animals but not plants or fungi) have some form of gene for Wnt proteins. Humans carry 19 genes for members of the Wnt family[123], although it is likely many more Wnt proteins are formed because of splice variants. The Wnt protein is known

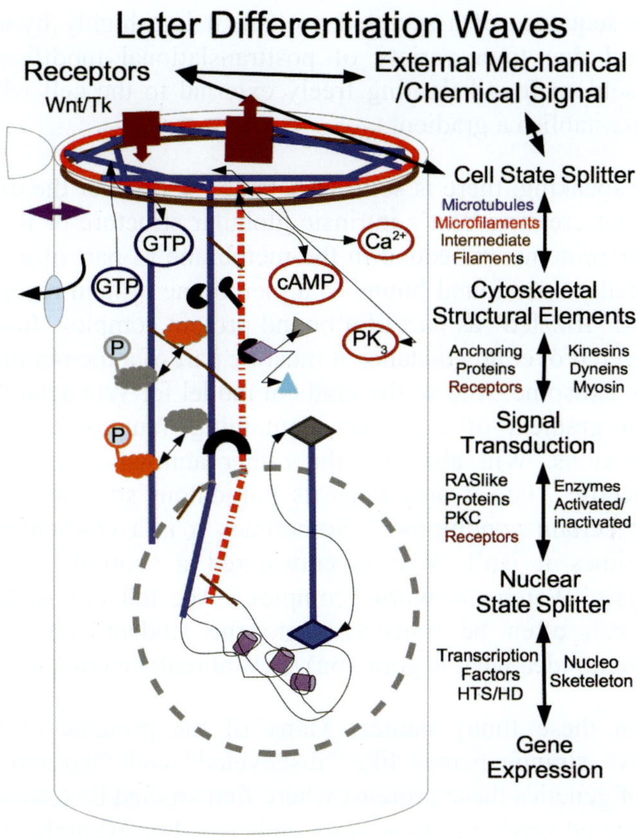

Figure 8.10. This is differentiation in later development that requires an external mechanical or chemical signal using a receptor to activate the cell state splitter. Autocrine, paracrine and endocrine signaling are secondary to juxtacrine signaling (Figure 8.9) and appear later in development.

to be a secreted protein with a globular structure. The transcript for Wnt contains the signal sequence for processing as a secretion type membrane protein. The Wnt family is therefore considered a secreted protein based on its genetic signal sequence. Wnt also undergoes acylation (the addition of fatty acid residues) during translation. There is a nearly invariant stretch of 22 copies of the amino acid cysteine used for disulphide protein bridges. Wnt can also be found in both the basal and apical ends of epithelial cells in cell layers where signals must be passed through the layer[124]. Although Wnt can be found in a free, secreted form and has the sequence of a secreted protein, it is a highly hydrophobic molecule and due to a variety of posttranslational modifications, it actually is seldom found floating freely external to the cell where it is presumed to establish a gradient.

Structurally speaking, there is some debate about whether the disulphide bridges are for creating Wnt's intrinsic globular structure or for binding Wnt to other proteins embedded in the membrane as part of a complex. Wnt is normally only found bound to a membrane protein receptor such as one called "frizzled" as part of a bound protein complex. In order for it to be delivered over any distance it must be part of a membrane vesicle known as an exosome. And so the gradient model for Wnt actually seems to require a gradient of exosomes containing Wnt, given how Wnt actually functions. Wnt also has the rather annoying habit of being "context specific", i.e., it only does its miraculous stuff in certain cell types under certain conditions[125]. Sometimes it is a gradient molecule and other times it isn't. Wnt is considered a "complex player" in embryogenesis. Watch the word "complex". We tell our students that "complex" can often be translated into the student speak "WTF?" (i.e., we have no idea what is going on). Is Wnt really a complex player?

An aside on these funny names: Many of the proteins of the Wnt pathway have strange names like "disheveled" and "frizzled". In the early days of genetics these proteins where first studied by examining the effect of mutated genes in *Drosophila* embryos. For example, a specific mutation on the gene for one protein produced a fruit fly that had a peculiar set of deformities that made it look "disheveled". Yet another poor fruit fly looked "frizzled" and so the gene was named. This habit led to fanciful names like "tinman" for a gene that when mutated produced an embryo with no heart, based on the tin man character in *The Wizard of Oz*[126]. It also indicates that a few graduate students had a bit too much fun and so there is a gene called "daughters against decapentaplegic" and "sonic" named for effects on "hedgehog" which was named for the popular video game of the era. This may or may not be a better system than naming a protein for the first thing it was discovered to do.

Frizzled is a class of membrane bound receptors that have one end on the outside of the membrane and are normally associated with G proteins. G proteins, you may recall, are kinases that are part of the assorted signal transduction networks that send signals along by an initial GTP to GDP phosphorylation reaction that is then passed along as phosphorylation signal transduction to other proteins[127]. The major role of Wnt is that when bound to Frizzled, various signal transduction pathways are triggered. Whether or not Wnt is bound to a transmembrane protein affects its conformation. Sometimes Wnt joins two of the transmembrane proteins, binding them to each other and changing the conformation of both. The tail ends of these transmembrane proteins form an underside complex. They bind to each other on the inner side of the cell membrane. Wnt related signal transduction pathways can signal the nucleus to change gene expression. Wnt itself therefore behaves more as an autocrine signalling protein than as a paracrine or endocrine one in spite of its secretion signal sequence. When Wnt was first studied, a specific "canonical" pathway was found to be triggered by Wnt binding. It is now known as the canonical β-catenin pathway because it includes that transcription factor translocating to the nucleus. Over time, other signal transduction pathways that included Wnt were discovered. These include the Wnt/Ca^{2+} pathway, the noncanonical Wnt pathway and some alternate pathways (Figure 8.11). If we go back to our conceptual assembly cascade (Figure 8.3), Wnt can be thought of as being just one of the purple-labeled components.

Just as there are multiple copies of Wnt per genome, so too there are often multiple copies of all the other proteins in the Wnt signal transduction complex. The signal pathway that is activated depends on the cellular context (which proteins are present in the complex and which are not) and the signal received. The signalling is accomplished via the ubiquitous phosphorylations and dephosphorylations of signal transduction pathway elements. The complex of proteins within which Wnt functions is localized to the apical end of the epithelial cells exactly where the cell state splitter also operates. Depending on the context, the Wnt signal transduction pathways can signal movement of different specific transcription factors. Several important transcription factors that can be activated or deactivated via a Wnt signal transduction pathway are β-catenin, JNK, Groucho, TCF/LEF and an entire class of "calcium mediated transcription factors".

What all Wnt signal transduction pathways have in common is that they end with a flux of transcription factors that move into the nucleus and this changes regulons in a particular cell type in response to a signal

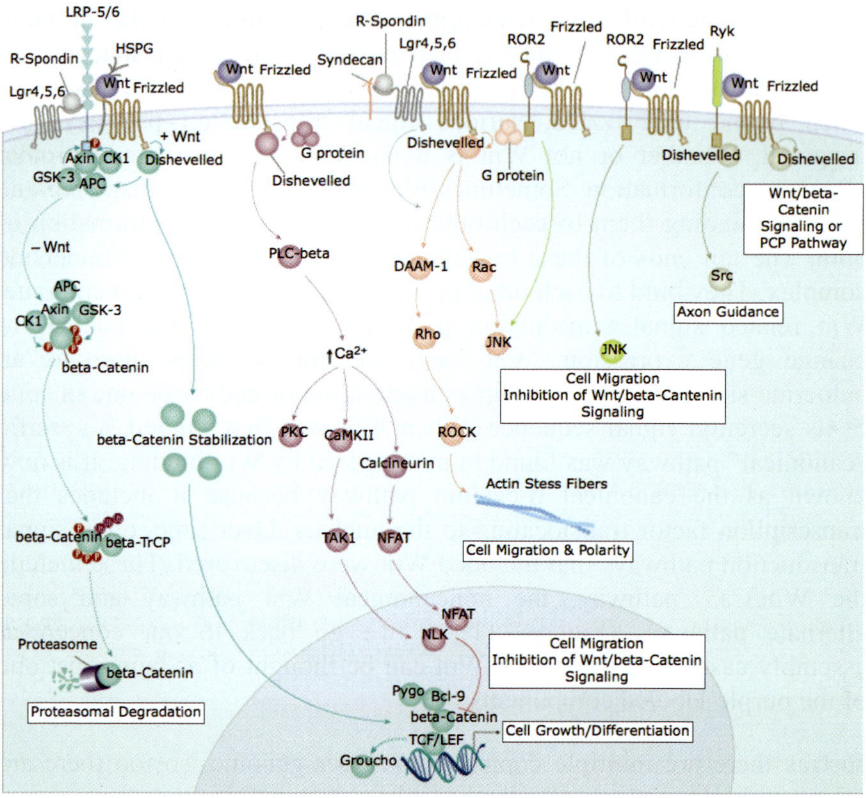

Figure 8.11. "Overview of Wnt Signaling. Wnt signaling has a central role in embryonic development, differentiation, cell motility, cell proliferation, and adult tissue homeostasis. Wnt ligands are a large family of secreted glycoproteins that signal through a canonical beta-Catenin-dependent pathway, and at least two well-characterized beta-Catenin-independent pathways, the Wnt/Ca^{2+} pathway and the planar cell polarity (PCP) pathway. The canonical beta-Catenin-dependent pathway is initiated by association of Dishevelled with the activated Wnt receptor. Dishevelled recruits the Axin protein complex (Axin, APC, CK1, GSK-3), leading to phosphorylation of the LRP-5/6 co-receptor by Glycogen Synthase Kinase-3 (GSK-3) and Casein Kinase 1 (CK1), and inhibiting phosphorylation of beta-Catenin by these kinases. Unphosphorylated beta-Catenin accumulates and subsequently translocates to the nucleus, where it associates with TCF/LEF family transcription factors and co-activators, such as Bcl-9 and Pygopus (Pygo), to induce the expression of Wnt target genes. In the absence of Wnt, cytoplasmic beta-Catenin is phosphorylated by CK1 and GSK-3, which creates a docking site on beta-Catenin for beta-Trcp, an E3 ubiquitin ligase that promotes its ubiquitination and proteasomal degradation.

"In addition to the beta-Catenin-dependent signaling pathway, there are at least two beta-Catenin-independent Wnt signaling pathways. The Wnt-Ca^{2+} pathway activates

Calcium/Calmodulin-dependent Protein Kinase type II (CaMKII) and Protein Kinase C (PKC). This occurs via a classical G protein-coupled signaling pathway that includes upstream activation of Phospholipase C (PLC) and its cleavage of PIP_2 into DAG and IP_3. DAG activates Protein Kinase C (PKC), while IP_3 promotes Ca^{2+} mobilization from intracellular stores. This pathway ultimately induces cell migration and inhibits the beta-Catenin-dependent pathway. The other major Wnt signaling pathway, the PCP pathway, regulates cell motility and tissue polarity. It involves activation of the Rho and Rac GTPases, Rho-Kinase (ROCK), and c-Jun N-terminal kinase (JNK). Like the Wnt/Ca^{2+} pathway, the PCP pathway also inhibits the beta-Catenin-dependent Wnt signaling pathway. Both the PCP pathway and the canonical beta-Catenin-dependent pathway are enhanced by R-Spondins binding to the leucine-rich repeat-containing, G protein-coupled receptors (Lgr), Lgr4, Lgr5, and Lgr6. R-Spondin/Lgr-enhanced Wnt/beta-Catenin signaling may require ligand-receptor internalization, as has also been suggested for the Wnt/Frizzled/LRP complex.

"Wnt ligands have also been shown to signal through a number of different co-receptors, including ROR2 and Ryk, which may or may not require a Frizzled receptor. Wnt-5a/ROR2 signaling activates JNK to regulate convergent extension in *Xenopus*, inhibits the beta-Catenin-dependent pathway, and may play a role in regulating the PCP pathway. In contrast, Wnt/Ryk signaling activates the beta-Catenin-dependent or PCP pathways and promotes axon guidance through Src activation. Deregulation of Wnt signaling is associated with a number of developmental disorders and human diseases, including bone and renal diseases, type II diabetes, and tumorigenesis. As a result, extracellular and intracellular modulators of Wnt signaling are of great interest." From R&D Systems[130] with kind permission of Laurie Mousel.

or the collection of regulons that form a differon in response to a determination signal from the cell state splitter. When canonical Wnt signaling is active, a transcriptional inhibitor called Groucho that is normally bound to specific signal sequences is displaced from the DNA and the transcriptional activator called TCF/LEF takes its place and genes previously inhibited are now activated[128]. Signals from the calcium pathway also cause the same effect, displacement of Groucho and activation of TCF/LEF. The calcium pathway for Wnt signal transduction is the same pathway that is triggered with microfilament contraction and involves a calcium flux.

The noncanonical pathway, (also called the PCP pathway) for signal transduction via Wnt signaling[129], causes phosphorylation of JNK and activates JNK while inhibiting TCF/LEF transcriptional activation. Wnt signal transduction is in one case inhibitory and the other stimulatory with respect to the genes controlled by TCF/LEF transcriptional activation. This makes the various Wnt signal transduction pathways ideal candidates to signal the nucleus that either an expansion or a contraction wave has occurred. The Wnt signal transduction pathway can sense what type of wave has occurred and send the signal so that the correct matching transcription factors are activated.

There are two additional noncanonical pathways. One also activates JNK. The other (at least in nerve cells) activates a tyrosine kinase called Src. Src is critical for nerve axon growth. Src in its phosphorylated form is active in the interaction between microfilaments and microtubules during axon formation. Src cannot function without the presence of microtubules, probably due to another protein that is bound to microtubules. And so the noncanonical pathways of Wnt signal transduction also leads to the cytoskeleton, both microfilaments and microtubules, but it requires polymerized microtubules.

The standard textbook way of looking at this is depicted in Figure 8.12. In this figure, stimulation of the canonical pathway leads to changes in gene expression via β-catenin pathways and it also causes microtubule stabilization with no particular reason given. The noncanonical Wnt pathway can lead to microtubule stabilization and to microfilament polymerization and JNK related changes in gene expression. The Wnt-Ca^{2+} pathway leads to microfilament polymerization and calcium related changes in gene transcription. There is no particular reason given for the microfilament polymerization to happen either. It just does. In this diagram at least, the changes in the cytoskeleton are a kind of a peculiar side epiphenomena of no import or interest. It's just cytoskeleton after all.

The cell state splitter model turns this on its head (Figure 8.13). If we assume that the microfilament contraction triggers one change in gene expression and microtubule polymerization triggers another, we can connect the Wnt pathway in a new way and create a new updated working model of the nuclear state splitter. We now have Wnt and the associated proteins sitting in the upper membrane ready to send signals from the cell state splitter. For the cell that starts the wave, the initial signal is purely mechanical. Integrin mechanotransduction is the initial signal, either expansion or contraction for propagating the wave.

That expansion or contraction signal is amplified by the Wnt receptor complex embedded within all the associated structures. With microfilament contraction, we have a calcium flux causing gene transcription

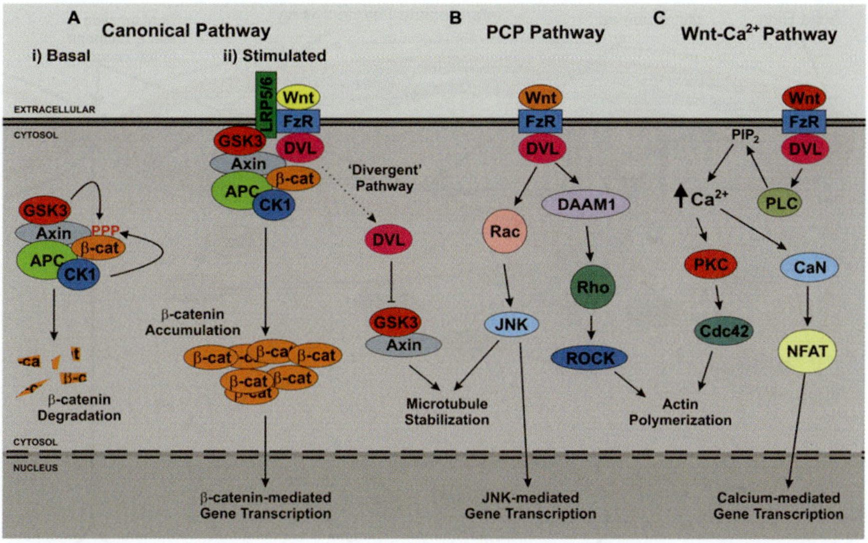

Figure 8.12. This version of the pathways has cytoskeletal rearrangements as unexplained dead ends. "Overview of Wnt signaling pathways. The three major branches of Wnt signaling – (A) the canonical, (B) planar cell polarity (PCP), (C) and Wnt-Ca^{2+} – pathways are illustrated. Note that in growing neurites a further branch has been reported, the so-called divergent canonical pathway, which impacts upon microtubule stability. APC, adenomatous polyposis coli; CaN, calcineurin; CK1, casein kinase 1; DAAM, dishevelled-associated activator of morphogenesis; DVL, dishevelled; FzR, frizzled receptor; GSK3, glycogen synthase kinase 3; JNK, c-Jun n-terminal kinase; LRP5/6, low-density lipoprotein receptor-related protein 5/6; NFAT, nuclear factor of activated T cells; PKC, protein kinase C; PLC, phospholipase C; ROCK, Rho-associated protein kinase." From Berwick & Harvey (2013)[131] with permission under a Creative Commons Attribution License.

via NFAT and β-catenin. With microtubule polymerization and cell expansion, we have a JNK related gene transcription with β-catenin being degraded instead of being sent to the nucleus. The Wnt signal transduction complex serves as a feedback modulator for the functioning cell state splitter apparatus (including a signal that acts to trigger the intermediate filament ring alpha to beta transition[132-134]). The Wnt complex signaling is changed via the inhibitory affect on GSK3 and on the AXIN/DVN binding. When microfilaments contract the JNK pathway is inhibited by the action of calcium and direct ROCK inhibition of JNK and because GSK3 is not inhibited, the AXIN/DVN binding is changed to activating DAAM1 and PLC. β-catenin is degraded instead of

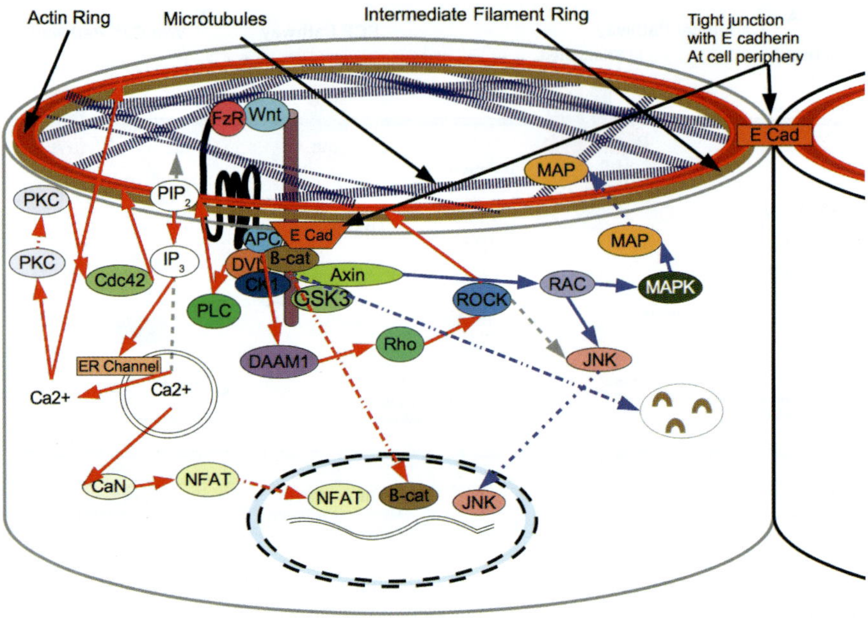

Figure 8.13. This signal transduction model is general. There will be replacement or alternate proteins for some of the proteins in our model in some tissues or organisms.

Nuclear State Splitter Participants: ***Wnt*** – Name is derived from "Wingless" and formerly considered "The Morphogen", Wnt acts primarily as an autocrine protein that binds one or more transmembrane proteins and affects their conformation and phosphorylation and thereby affects signal transduction pathways, often amplifying other signals. ***FzR-Frizzled*** transmembrane protein. ***Cdc42*** – a cell cycle protein that takes part in signal transduction pathways and can signal the cell to enter mitosis. It is a member of the small Rho-kinase family that is known to affect microfilament contraction and also can stimulate pathways promoting cell division. A round of determination is often followed by cell proliferation as part of differentiation and this could be via Rho-kinase. ***PKC*** – Protein Kinase C is a protein normally found in the cytosol of the cell in an inactive form but when phosphorylated by other signaling molecules, in particular calcium ions and diaglycerols, translocates to the cell membrane where it interacts with other kinases, especially RACK (receptor for activated C Kinase). PKC continues to signal long after the calcium flux has faded and the diaglycerol signal has ended and so can be considered a signal amplifier. ***CaN*** – Calcineurin, which is known to dephosphorylate the transcription factor **NFAT**. This causes a conformation change which exposes the NFAT nuclear import signal allowing it move to the nucleus and bind to specific DNA sequences and change gene expression. ***β-catenin*** – Cadherin associated protein beta one, a dual function protein that is active in cell-to-cell adhesion as well as gene expression. ***JNK-c-*Jun** terminal kinases first found because they bind and phosphorylate c-jun forming a transcriptional activator domain on DNA. JNK is a form of mitogen activated protein. ***PLC*** – An enzyme which cleaves PIP_2 to form two products, inositol 1,4,5-

triphosphate (IP_3) and diaglycerol, both of which are second messengers that commonly affect opening and closing of membrane channels. Diaglycerol and calcium combined can trigger Protein Kinase C translocation. ***PIP₂*** – Phosphatidylinositol 4,5-biphosphate, a phospholipid that is the major constituent of membranes that is cleaved by PLC. Such cleavage can activate PKC. ***Dvl*** – Dishevelled, cytoplasmic phosphoprotein that is required for canonical and noncanonical Wnt pathway signal transduction and Wnt signaling. ***CK1*** – Casein Kinase 1 involved in phosphorylation of Dishevelled. ***Axin*** – A dual domain cytoplasmic protein, with one domain binding the disheveled receptor and the other binding G proteins. ***APC*** – Adenomatous polyposis coli, so called because when one version is mutated in humans it is connected to development of colon cancer. It is a negative regulator which reduces β-catenin response and is also directly connected to e-cadherin. ***GSK3*** – glycogen synthase kinase 3, phosphorylates β-catenin (thereby signaling for it to be degraded) in response to signaling from Axin. ***DAAM1*** – Disheveled-associated activator of morphogenesis 1, a protein associated with microfilament polymerization, possibly by acting as a scaffold protein. It is activated by Rho. ***Rho*** – A member of the Ras homolog gene family, Rho is a kinase that is activated during microfilament contraction by directly stimulating microfilament polymerization undertaken by the formins. Formins recruit free actin monomers which are then used to elongate microfilaments. (They also capture and stabilize free ends of microfilaments required for ruffling in forward movements of cells.) ***ROCK*** – Rho associated protein kinase, this protein works downstream to Rho and can either trigger stabilization or destabilization of microfilaments. It can also trigger contraction by activating specific myosins. ***Rac*** – a serine/threonine-protein kinase that interacts with multiple other kinases. ***MAPK*** – Mitogen activated protein type K. ***MAP*** – microtubule associated protein, phosphorylated by MAPK, binds to microtubules and can either stabilize or destabilize microtubules depending on phosphorylation signals from other proteins. When signalled by Wnt signal transduction they generally stabilize and elongate microtubules.

joining JNK in the nucleus. Thus the cell state splitter at the top of the cell sends a signal, contraction or expansion, and that signal transduction is amplified and passed into the nucleus via the Wnt complex. Gene expression is changed in response. A new differon is accessed.

The Wnt signal transduction pathway is not the only pathway that exists for use by the cell state splitter. Another similar pathway is the Indian hedgehog signal transduction pathway. Indian hedgehog is also a small secreted protein which binds to a transmembrane receptor called *patch*. The Indian hedgehog protein is also part of a complex that leads to release of transcription factors, especially those bound to microtubules. Many of the same kinases that are phosphorylated or dephosphorylated by signaling in the Wnt signal transduction pathway are also used by the Indian hedgehog (and other hedgehog) pathways. There is even new evidence that Indian hedgehog also has a mechanotransduction component[135].

There must be some sort of signal transduction for the case where a cell is torn from a sheet and it responds by taking on stem cell or stem cell-like properties (including migration), as in the case of neural crest and germ cells. One possible explanation may be that when all the integrin connections are lost, the entire Wnt complex falls apart and spreads from the apical surface over the whole cell membrane. Normally cells can have entirely different Wnt on their apical and basal ends. But without integrins all of the Wnt signal transduction potential is lost. Vang (vangogh) and Pk (prickle), two proteins that localize to the basal end opposite to the cell state splitter may move upwards and interactions between all these components that normally don't mix in the membrane may provide a "stem cell" specific signal. Mixing types of Wnts all over the cell membrane might well result in the cell giving up on receiving a contraction or expansion signal and returning to a simpler state without organization and therefore acting like an undifferentiated stem cell[124].

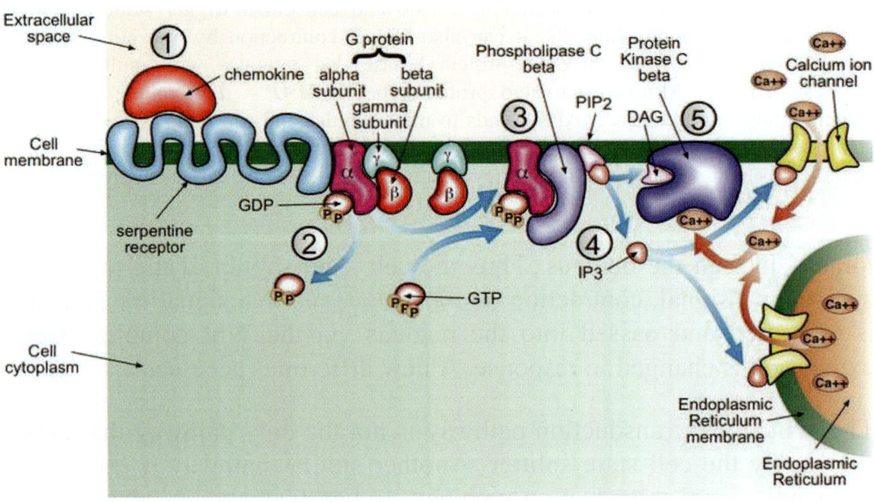

Figure 8.14. "G-protein linked - chemokines - protein kinase. (1) Binding of the serpentine receptor by a chemokine causes a conformational change in the GDP-bound alpha subunit of the G-protein. (2) GDP is exchanged for GTP and this causes the alpha subunit to dissociate from the beta and gamma subunits. (3) The alpha subunit binds to phospholipase C beta and activates the enzymatic function that converts membrane phospholipids (PIP_2) to IP_3 and DAG. (4) IP_3 activates calcium ion channels to release calcium into the cytoplasm. (5) Calcium ions (and DAG) activate protein kinase C." From Immunopaedia.org (2010)[137] with kind permission of Carina Kriel at Immunopaedia http://www.immunopaedia.org.za.

There is a huge family of assorted Wnt-like G protein-coupled receptors. Their more fanciful name is "serpentine receptors" because their main feature is a seven segment or loop transmembrane domain[136] (Figure 8.14). The same general structure of the serpentine receptors, like frizzled and a related protein called "smoothened", exists across the entire family of living organisms from bacteria to plants to insects to humans. In all cases, the serpentine proteins work by connecting signals such as light, mechanical forces, presence of proteins, ions, or lipids, and a multitude of other signals a cell must sense and respond to via some GTP to GDP change. So what clues do we have about the origin of serpentine receptors?

The original amino acid sequence for the first of these receptors has been modified so often over evolutionary time that it is nearly impossible to discern the common ancestor. However, by examining the sequence of specific portions of the protein and matching small segments to regions of importance for binding to other proteins (a technique called iterative BLAST (Basic Local Alignment Search Tool)[138]) it is possible to work out the relationship and evolutionary history of all of these nuclear state splitter signal transduction pathways.

Returning to our Chapter 1 and the first protolife, the first signal transduction pathway was used to initiate mitosis. The second one was the mitosis pathway copied and pasted to use for movement. After that event the copy and paste produced pathways for "move forward" and "move away". From there the pathways were copied, pasted and specialized, producing the first differentiated cell types, spores. This set of pathways includes a lifting and pulling of the nucleus to cause reversible nuclear deformation. Reversible nuclear deformation provided the basis for evolution of the changes in the structure of the nucleus required for changing gene expression (Figure 8.15). From the simplest of differentiations, over time, the pathway was copied and pasted and modified and specialized over and over and over again until we have the multiple forms with many cell types present today. It is reasonable to assume therefore, that embryogenesis arose from the need to perceive and respond to what is happening around the cell. The pathway used for cell division was copied and modified to use for sensing. It was again copied and specialized for differentiation. The same pathway has been replicated, and then altered again and again and again to produce the astonishingly wide variety of signal transductions systems that exist in all life forms on earth today[139].

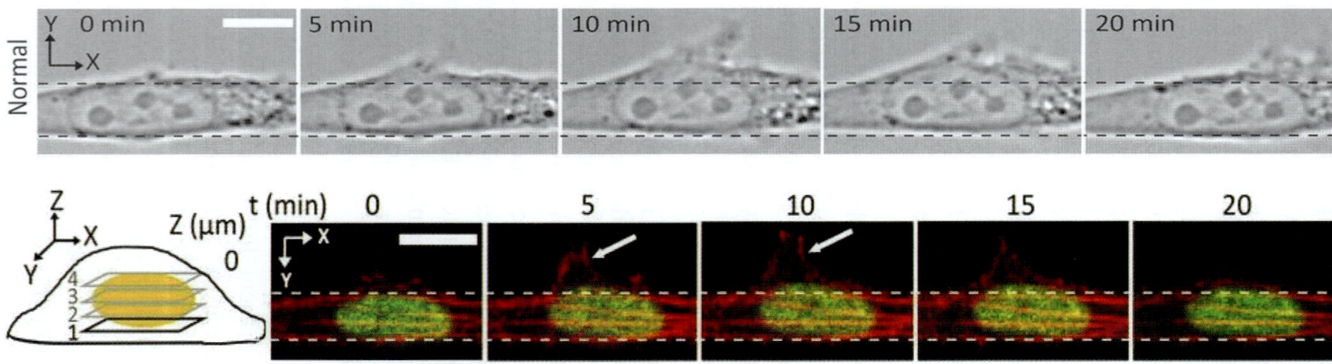

Figure 8.15. **Top**: These time lapse images show the deformation of the nucleus that occurs during normal forward movement of a fibroblast cell in the X direction. There is a lateral extension of cell material in the Y direction resulting in deformation of the nucleus, made clearer by its extension past the dashed lines. **Bottom**: These confocal microscopy images show the actin stained in red with the nucleus in green, with a lateral actin protrusion (arrows). In this case each image is of the slice of the cell just above the flat microscope slide. From Alam *et al.* (2015)[1] with kind permission of the Company of Biologists.

Our updated working model makes very specific predictions that can serve as the basis for further experimentation on the cell state splitter and the nuclear state splitter, especially when those studies examine the earliest stages of embryogenesis where the initial signaling is purely mechanical. The cytoskeleton is consistently treated in the scientific literature as merely a structural protein complex whose rearrangements are side effects to induction that are of no relevance in embryogenesis. We have moved the cytoskeleton from irrelevance to being the single most important component. The cytoskeleton is the starting point that allows the individual cell to become the right kind of cell, in the right place, at the right time.

You may have noticed that there is one rather critical point in all this we have glossed over. The transcription factors move into the nucleus and the DNA is rearranged so that there are changes in which portions of the DNA are exposed and which are sequestered. This rearranges the portion of the information that can be extracted from the DNA. So why and how does that happen? We explore this question in the next chapter.

1. Alam, S.G., D. Lovett, D.I. Kim, K.J. Roux, R.B. Dickinson & T.P. Lele (2015). The nucleus is an intracellular propagator of tensile forces in NIH 3T3 fibroblasts. *Journal of Cell Science* **128**(10), 1901-1911.
2. Björklund, N.K. & R. Gordon (1993). Nuclear state splitting: a working model for the mechanochemical coupling of differentiation waves to master genes. *Russian Journal of Developmental Biology* **24**(2), 79-95.
3. Beloussov, L.V. (1987). *Biological Morphogenesis*. Moscow, Moscow State University.
4. Beloussov, L.V. (1998). *The Dynamic Architecture of a Developing Organism: An Interdisciplinary Approach to the Development of Organisms*. Dordrecht, Kluwer Academic Publishers.
5. Beloussov, L.V. (2003). Developmental biology and physics of today. In: *Integrative Biophysics: Biophotonics*. Eds.: F.A. Popp & L.V. Beloussov, Springer: 149-178.
6. Beloussov, L.V. & A. Lipchinsky (2015). *Morphomechanics of Development*. Cham, Springer.
7. Björklund, N.K. & R. Gordon (1993). [Nuclear state splitting: a working model for the mechanochemical coupling of differentiation "waves" with the controlling genes (master genes)] [Russian]. *Ontogenez* **24**(2), 5-23.
8. Björklund, N.K. & R. Gordon (1993). Addendum to: Nuclear state splitting: a working model for the mechanochemical coupling of differentiation waves to master genes. *Ontogenez* **24**(2), 94-95.

9. Wikipedia (2015). Nicotinamide adenine dinucleotide. https://en.wikipedia.org/wiki/Nicotinamide_adenine_dinucleotide
10. Gest, H. (1987). Evolutionary roots of the citric acid cycle in prokaryotes. *Biochemical Society symposium* **54**, 3-16.
11. Wikipedia (2015). Glycolysis. https://en.wikipedia.org/wiki/Glycolysis
12. Wikipedia (2015). Nicotinamide adenine dinucleotide phosphate. https://en.wikipedia.org/wiki/Nicotinamide_adenine_dinucleotide_phosphate
13. Buchanan, B.B. & D.I. Arnon (1990). A reverse Krebs cycle in photosynthesis: Consensus at last. *Photosynthesis Research* **24**(1), 47-53.
14. Michal, G. (2014). Biochemical Pathways. http://www.roche.com/responsibility/supporting_communities/science_education/pathways.htm
15. Rodbell, M. (1995). Signal transduction: Evolution of an idea. *Bioscience Reports* **15**(3), 117-133.
16. Tomanek, L. & G.N. Somero (2002). Interspecific- and acclimation-induced variation in levels of heat-shock proteins 70 (hsp70) and 90 (hsp90) and heat-shock transcription factor-1 (HSF1) in congeneric marine snails (genus *Tegula*): implications for regulation of *hsp* gene expression. *Journal of Experimental Biology* **205**(5), 677-685.
17. Shamovsky, I., M. Ivannikov, E.S. Kandel, D. Gershon & E. Nudler (2006). RNA-mediated response to heat shock in mammalian cells. *Nature* **440**(7083), 556-560.
18. Speil, J., E. Baumgart, J.-P. Siebrasse, R. Veith, U. Vinkemeier & U. Kubitscheck (2011). Activated STAT1 transcription factors conduct distinct saltatory movements in the cell nucleus. *Biophysical Journal* **101**(11), 2592-2600.
19. Shamovsky, I. & E. Nudler (2008). New insights into the mechanism of heat shock response activation. *Cellular and Molecular Life Sciences* **65**(6), 855-861.
20. Smertenko, A.P., M.J. Deeks & P.J. Hussey (2010). Strategies of actin reorganisation in plant cells. *Journal of Cell Science* **123**(17), 3019-3029.
21. Farwell, A.P. & J.L. Leonard (1992). Dissociation of actin polymerization and enzyme inactivation in the hormonal regulation of type II iodothyronine 5'-deiodinase activity in astrocytes. *Endocrinology* **131**(2), 721-728.
22. Theriot, J.A. & T.J. Mitchison (1991). Actin microfilament dynamics in locomoting cells. *Nature* **352**(6331), 126-131.
23. Theriot, J.A., T.J. Mitchison, L.G. Tilney & D.A. Portnoy (1992). The rate of actin-based motility of intracellular *Listeria monocytogenes* equals the rate of actin polymerization. *Nature* **357**(6375), 257-260.
24. Rosenblatt, J., B.J. Agnew, H. Abe, J.R. Bamburg & T.J. Mitchison (1997). *Xenopus* actin depolymerizing factor/cofilin (XAC) is responsible for the turnover of actin filaments in *Listeria monocytogenes* tails. *Journal of Cell Biology* **136**(6), 1323-1332.
25. Hatanaka, H., K. Ogura, K. Moriyama, S. Ichikawa, I. Yahara & F. Inagaki (1996). Tertiary structure of destrin and structural similarity between two actin-regulating protein families. *Cell* **85**(7), 1047-1055.
26. Moon, A. & D.G. Drubin (1995). The ADF/cofilin proteins: stimulus-responsive modulators of actin dynamics. *Molecular Biology of the Cell* **6**(11), 1423-1431.

27. Pollard, T.D. & G.G. Borisy (2003). Cellular motility driven by assembly and disassembly of actin filaments. *Cell* **112**(4), 453-465.
28. Schill, N.J. & R.A. Anderson (2009). Out, in and back again: PtdIns(4,5)P_2 regulates cadherin trafficking in epithelial morphogenesis. *Biochemical Journal* **418**, 247-260.
29. Korte, E. (2007). PIP2 cleavage by PLC to release IP3 and DAG. https://commons.wikimedia.org/wiki/File:PIP2_cleavage_to_IP3_and_DAG.jpg
30. Bravo-Cordero, J.J., M.A.O. Magalhaes, R.J. Eddy, L. Hodgson & J. Condeelis (2013). Functions of cofilin in cell locomotion and invasion. *Nature Reviews Molecular Cell Biology* **14**, 405-415.
31. Meng, G., Y. Sun, W. Fu, Z. Guo & L. Xu (2011). Microcystin-LR induces cytoskeleton system reorganization through hyperphosphorylation of tau and HSP27 via PP2A inhibition and subsequent activation of the p38 MAPK signaling pathway in neuroendocrine (PC12) cells. *Toxicology* **290**(2-3), 218-229.
32. Marguet, D., P.-F. Lenne, H. Rigneault & H.-T. He (2006). Dynamics in the plasma membrane: how to combine fluidity and order. *EMBO Journal* **25**(15), 3446-3457.
33. Elson, E.L., E. Fried, J.E. Dolbow & G.M. Genin (2010). Phase separation in biological membranes: integration of theory and experiment. *Annual Review of Biophysics* **39**, 207-226.
34. Verghese, J., J. Abrams, Y. Wang & K.A. Morano (2012). Biology of the heat shock response and protein chaperones: Budding yeast (*Saccharomyces cerevisiae*) as a model system. *Microbiology and Molecular Biology Reviews* **76**(2), 115-158.
35. Gavrilova, L.P., I.I. Korpacheva, S.G. Semushina & V.A. Iashin (2012). Heat shock induces simultaneous rearrangements of all known cytoskeletal filaments in normal interphase fibroblasts [Russian]. *Tsitologiia* **54**(11), 837-846.
36. Luo, Q.B., X.Y. Song, C.L. Ji, X.Q. Zhang & D.X. Zhang (2014). Exploring the molecular mechanism of acute heat stress exposure in broiler chickens using gene expression profiling. *Gene* **546**(2), 200-205.
37. Nobusue, H., N. Onishi, T. Shimizu, E. Sugihara, Y. Oki, Y. Sumikawa, T. Chiyoda, K. Akashi, H. Saya & K. Kano (2014). Regulation of MKL1 via actin cytoskeleton dynamics drives adipocyte differentiation. *Nature Communications* **5**.
38. Maciver, S. (2015). The Encyclopaedia of Actin-Binding Proteins (and Drugs) http://www.bms.ed.ac.uk/research/others/smaciver/Encyclop/encycloABP.htm
39. Rosette, C. & M. Karin (1995). Cytoskeletal control of gene expression: depolymerization of microtubules activates NF-κB. *Journal of Cell Biology* **128**(6), 1111-1119.
40. Gundersen, G.G. & T.A. Cook (1999). Microtubules and signal transduction. *Current Opinion in Cell Biology* **11**(1), 81-94.
41. Mistry, P., K. Deacon, S. Mistry, J. Blank & R. Patel (2004). NF-κB promotes survival during mitotic cell cycle arrest. *Journal of Biological Chemistry* **279**(2), 1482-1490.
42. Wikipedia (2015). Signal Transduction. http://en.wikipedia.org/wiki/Signal_transduction
43. Wikipedia (2015). Protein kinase. http://en.wikipedia.org/wiki/Protein_kinase

44. Swulius, M.T. & M.N. Waxham (2008). Ca^{2+}/calmodulin-dependent protein kinases. *Cellular and Molecular Life Sciences* **65**(17), 2637-2657.
45. Behrend, L., M. Stoter, M. Kurth, G. Rutter, J. Heukeshoven, M. Deppert & U. Knippschild (2000). Interaction of casein kinase 1 delta (CK1δ) with post-Golgi structures, microtubules and the spindle apparatus. *European Journal of Cell Biology* **79**(4), 240-251.
46. Sanchez-Casalongue, M.E., J. Lee, A. Diamond, S. Shuldiner, R.D. Moir & I.M. Willis (2015). Differential phosphorylation of a regulatory subunit of protein kinase CK2 by target of rapamycin complex 1 signaling and the Cdc-like kinase Kns1. *Journal of Biological Chemistry* **290**(11), 7221-7233.
47. Varjosalo, M., S. Keskitalo, A. Van Drogen, H. Nurkkala, A. Vichalkovski, R. Aebersold & M. Gstaiger (2013). The protein interaction landscape of the human CMGC kinase group. *Cell Reports* **3**(4), 1306-1320.
48. Goriely, A., G.A.T. McVean, M. Röjmyr, B. Ingemarsson & A.O.M. Wilkie (2003). Evidence for selective advantage of pathogenic FGFR2 mutations in the male germ line. *Science* **301**(5633), 643-646.
49. Wikipedia (2015). LYN. http://en.wikipedia.org/wiki/LYN
50. Martin, D.M., D. Miranda-Saavedra & G.J. Barton (2009). The TKL group. http://www.compbio.dundee.ac.uk/kinomer/families/TKL.html
51. Martin, D.M., D. Miranda-Saavedra & G.J. Barton (2009). Kinomer v. 1.0: a database of systematically classified eukaryotic protein kinases. *Nucleic Acids Research* **37**(Database issue), D244-250.
52. Huang, C.Y.F., Y.M. Wu, C.Y. Hsu, W.S. Lee, M.D. Lai, T.J. Lu, C.L. Huang, T.H. Leu, H.M. Shih, H.I. Fang, D.R. Robinson, H.J. Kung & C.J. Yuan (2002). Caspase activation of mammalian Sterile 20-like kinase 3 (Mst3): Nuclear translocation and induction of apoptosis. *Journal of Biological Chemistry* **277**(37), 34367-34374.
53. Caffrey, D.R., L.A.J. O'Neill & D.C. Shields (1999). The evolution of the MAP kinase pathways: Coduplication of interacting proteins leads to new signaling cascades. *Journal of Molecular Evolution* **49**(5), 567-582.
54. Bimbó, A., Y. Jia, S.L. Poh, R.K. Karuturi, N. den Elzen, X. Peng, L. Zheng, M. O'Connell, E.T. Liu, M.K. Balasubramanian & J. Liu (2005). Systematic deletion analysis of fission yeast protein kinases. *Eukaryotic Cell* **4**(4), 799-813.
55. Cell Signaling Technology (2015). Protein Kinases: Interactive Human Kinome. http://www.cellsignal.com/common/content/content.jsp?id=kinases-human-kinome
56. Phillips, R. & J. Kondev (2004). Physical Biology of the Cell: Chapter 2, Cartoons as Models in Molecular Biology. http://www.rpgroup.caltech.edu/courses/aph161/Handouts/chap2.pdf
57. Page, A.L. & J.A. Shafferman (2000). *Favorite Sacred Songs for Children . . . Bible Stories & Songs of Praise*, Alfred Music Publishing.
58. Knudsen, H.L. & J.A. Frangos (1997). Role of cytoskeleton in shear stress-induced endothelial nitric oxide production. *American Journal of Physiology-Heart and Circulatory Physiology* **273**(1), H347-H355.
59. Ohura, N., K. Yamamoto, S. Ichioka, T. Sokabe, H. Nakatsuka, A. Baba, M. Shibata, T. Nakatsuka, K. Harii, Y. Wada, T. Kohro, T. Kodama & J. Ando (2003). Global analysis of shear stress-responsive genes in vascular endothelial cells. *Journal of Atherosclerosis and Thrombosis* **10**(5), 304-313.

60. Chatzizisis, Y.S., A.U. Coskun, M. Jonas, E.R. Edelman, C.L. Feldman & P.H. Stone (2007). Role of endothelial shear stress in the natural history of coronary atherosclerosis and vascular remodeling - Molecular, cellular, and vascular behavior. *Journal of the American College of Cardiology* **49**(25), 2379-2393.
61. Miller, C.J. & L.A. Davidson (2013). The interplay between cell signalling and mechanics in developmental processes. *Nature Reviews Genetics* **14**(10), 733-744.
62. Brunet, T., A. Bouclet, P. Ahmadi, D. Mitrossilis, B. Driquez, A.-C. Brunet, L. Henry, F. Serman, G. Béalle, C. Ménager, F. Dumas-Bouchiat, D. Givord, C. Yanicostas, D. Le-Roy, N.M. Dempsey, A. Plessis & E. Farge (2013). Evolutionary conservation of early mesoderm specification by mechanotransduction in Bilateria. *Nature Communications* **4**, 2821.
63. Rieger, R.M. & P. Ladurner (2003). The significance of muscle cells for the origin of mesoderm in bilateria. *Integrative and Comparative Biology* **43**(1), 47-54.
64. Rieger, R.M. & G. Purschke (2005). The coelom and the origin of the annelid body plan. *Hydrobiologia* **535**, 127-137.
65. Burke, A.C. (2007). Development and evolution of the vertebrate mesoderm. *Developmental Dynamics* **236**(9), 2369-2370.
66. Burton, P.M. (2008). Insights from diploblasts; the evolution of mesoderm and muscle. *Journal of Experimental Zoology Part B-Molecular and Developmental Evolution* **310B**(1), 5-14.
67. Davidson, B. (2008). Movers and shakers: Evolution and development of the mesoderm. *Journal of Experimental Zoology Part B-Molecular and Developmental Evolution* **310B**(1), 1-4.
68. Schepis, A., D. Sepich & W.J. Nelson (2012). αE-catenin regulates cell-cell adhesion and membrane blebbing during zebrafish epiboly. *Development* **139**(3), 537-546.
69. Swiers, G., Y.H. Chen, A.D. Johnson & M. Loose (2010). A conserved mechanism for vertebrate mesoderm specification in urodele amphibians and mammals. *Developmental Biology* **343**(1-2), 138-152.
70. Green, D., A.E. Whitener, S. Mohanty & A.C. Lekven (2015). Vertebrate nervous system posteriorization: Grading the function of Wnt signaling. *Developmental Dynamics* **244**(3), 507-512.
71. England, J.L. & J. Cardy (2005). Morphogen gradient from a noisy source. *Physical Review Letters* **94**(7), #078101.
72. Lander, A.D., Q. Nie & F.Y.M. Wan (2005). Spatially distributed morphogen production and morphogen gradient formation. *Mathematical Biosciences and Engineering* **2**(2), 239-262.
73. Dalessi, S., A. Neves & S. Bergmann (2012). Modeling morphogen gradient formation from arbitrary realistically shaped sources. *Journal of Theoretical Biology* **294**, 130-138.
74. Christian, J.L. (2012). Morphogen gradients in development: from form to function. *Wiley Interdisciplinary Reviews-Developmental Biology* **1**(1), 3-15.
75. Bergmann, S., O. Sandler, H. Sberro, S. Shnider, E. Schejter, B.Z. Shilo & N. Barkai (2007). Pre-steady-state decoding of the bicoid morphogen gradient. *PLoS Biology* **5**(2), 232-242.
76. de Lachapelle, A.M. & S. Bergmann (2010). Pre-steady and stable morphogen gradients: can they coexist? *Molecular Systems Biology* **6**, #428.

77. Berezhkovskii, A.M., C. Sample & S.Y. Shvartsman (2011). Formation of morphogen gradients: Local accumulation time. *Physical Review E* **83**(5), #051906.
78. Yin, H.W., X.Q. Wen & T.S. Zhou (2013). Local accumulation time for the formation of morphogen gradients from a Lévy diffusion process. *Physical Biology* **10**(5), #056012.
79. Bergmann, S., Z. Tamari, E. Schejter, B.Z. Shilo & N. Barkai (2008). Re-examining the stability of the bicoid morphogen gradient. *Cell* **132**(1), 15-17.
80. Saunders, T. & M. Howard (2009). When it pays to rush: interpreting morphogen gradients prior to steady-state. *Physical Biology* **6**(4), #046020.
81. Wikipedia (2015). Mass diffusivity. https://en.wikipedia.org/wiki/Mass_diffusivity - Liquids
82. Wikipedia (2015). Temperature dependence of liquid viscosity. https://en.wikipedia.org/wiki/Temperature_dependence_of_liquid_viscosity
83. McHale, P., W.J. Rappel & H. Levine (2006). Embryonic pattern scaling achieved by oppositely directed morphogen gradients. *Physical Biology* **3**(2), 107-120.
84. Barkai, N. & B.-Z. Shilo (2009). Robust generation and decoding of morphogen gradients. *Cold Spring Harbor Perspectives in Biology* **1**(5), #a001990.
85. de Lachapelle, A.M. & S. Bergmann (2010). Precision and scaling in morphogen gradient read-out. *Molecular Systems Biology* **6**, #351.
86. Kanodia, J.S., Y. Kim, R. Tomer, Z. Khan, K. Chung, J.D. Storey, H. Lu, P.J. Keller & S.Y. Shvartsman (2011). A computational statistics approach for estimating the spatial range of morphogen gradients. *Development* **138**(22), 4867-4874.
87. Hamaratoglu, F., K. Basler & M. Affolter (2009). Confronting morphogen gradients: How important are they for growth? *Science Signaling* **2**(94), pe67.
88. Yin, H.W., X.Q. Wen & T.S. Zhou (2012). Effects of tissue growth on robust formation of morphogen gradient. *International Journal of Modern Physics B* **26**(4), #1250015.
89. Eldar, A., R. Dorfman, D. Weiss, H. Ashe, B.Z. Shilo & N. Barkai (2002). Robustness of the BMP morphogen gradient in *Drosophila* embryonic patterning. *Nature* **419**(6904), 304-308.
90. Wu, Y.F., E. Myasnikova & J. Reinitz (2007). Master equation simulation analysis of immunostained Bicoid morphogen gradient. *BMC Systems Biology* **1**, #52.
91. Morishita, Y. & Y. Iwasa (2009). Accuracy of positional information provided by multiple morphogen gradients with correlated noise. *Physical Review E* **79**(6), #061905.
92. MacWilliams, H.K. & S. Papageorgiou (1978). A model of gradient interpretation based on morphogen binding. *Journal of Theoretical Biology* **72**(3), 385-411.
93. Kerszberg, M. (1996). Accurate reading of morphogen concentrations by nuclear receptors: a formal model of complex transduction pathways. *Journal of Theoretical Biology* **183**(1), 95-104.
94. Kerszberg, M. (1999). Morphogen propagation and action: towards molecular models. *Seminars in Cell and Developmental Biology* **10**(3), 297-302.

95. Gurdon, J.B. & P.-Y. Bourillot (2001). Morphogen gradient interpretation. *Nature* **413**(6858), 797-803.
96. Bothma, J.P., M. Levine & A. Boettiger (2010). Morphogen gradients: limits to signaling or limits to measurement? *Current Biology* **20**(5), R232-R234.
97. Tamari, Z. & N. Barkai (2012). Improved readout precision of the Bicoid morphogen gradient by early decoding. *Journal of Biological Physics* **38**(2), 317-329.
98. Emberly, E. (2008). Optimizing the readout of morphogen gradients. *Physical Review E* **77**(4), #041903.
99. Lander, A.D., Q. Nie & F.Y. Wan (2002). Do morphogen gradients arise by diffusion? *Developmental Cell* **2**(6), 785-796.
100. Gregor, T., E.F. Wieschaus, A.P. McGregor, W. Bialek & D.W. Tank (2007). Stability and nuclear dynamics of the bicoid morphogen gradient. *Cell* **130**(1), 141-152.
101. Nahmad, M. & A.D. Lander (2011). Spatiotemporal mechanisms of morphogen gradient interpretation. *Current Opinion in Genetics & Development* **21**(6), 726-731.
102. Kay, R.R. & C.R.L. Thompson (2009). Forming patterns in development without morphogen gradients: Scattered differentiation and sorting out. *Cold Spring Harbor Perspectives in Biology* **1**(6), a001503.
103. Payne, S., B. Li, Y. Cao, D. Schaeffer, M.D. Ryser & L. You (2013). Temporal control of self-organized pattern formation without morphogen gradients in bacteria. *Molecular Systems Biology* **9**, Unsp697.
104. Kerszberg, M. & L. Wolpert (1998). Mechanisms for positional signalling by morphogen transport: a theoretical study. *Journal of Theoretical Biology* **191**(1), 103-114.
105. Pagès, F. & S. Kerridge (2000). Morphogen gradients. A question of time or concentration? *Trends in Genetics* **16**(1), 40-44.
106. Baker, R.E. & P.K. Maini (2007). Travelling gradients in interacting morphogen systems. *Mathematical Biosciences* **209**(1), 30-50.
107. Ibanes, M. & J.C.I. Belmonte (2008). Theoretical and experimental approaches to understand morphogen gradients. *Molecular Systems Biology* **4**, #176.
108. Fosslien, E. (2009). The hormetic morphogen theory of curvature and the morphogenesis and pathology of tubular and other curved structures. *Dose-Response* **7**(4), 307-331.
109. Glimm, T., J. Zhang & Y.-Q. Shen (2009). Interaction of Turing patterns with an external linear morphogen gradient. *Nonlinearity* **22**(10), 2541-2560.
110. Lei, J.Z. & Y. Song (2010). Mathematical model of the formation of morphogen gradients through membrane-associated non-receptors. *Bulletin of Mathematical Biology* **72**(4), 805-829.
111. Glimm, T., J. Zhang, Y.-Q. Shen & S.A. Newman (2012). Reaction-diffusion systems and external morphogen gradients: the two-dimensional case, with an application to skeletal pattern formation. *Bulletin of Mathematical Biology* **74**(3), 666-687.
112. Yin, H.W., X.Y. Xiao, X.Q. Wen & T.S. Zhou (2013). Stability of regulatory protein gradients induced by morphogen dpp in *Drosophila* wing disc. *International Journal of Bifurcation and Chaos* **23**(8), #1350138.

113. Yin, H.W., X.Y. Xiao, X.Q. Wen & T.S. Zhou (2014). Mathematical analysis on a multidimensional model of morphogen transport with receptor synthesis. *International Journal of Biomathematics* **7**(5), #1450051.
114. Entchev, E.V. & M.A. González-Gaitán (2002). Morphogen gradient formation and vesicular trafficking. *Traffic* **3**(2), 98-109.
115. Esser, A.T., K.C. Smith, J.C. Weaver & M. Levin (2006). Mathematical model of morphogen electrophoresis through gap junctions. *Developmental Dynamics* **235**(8), 2144-2159.
116. Zhang, Y. & M. Levin (2009). Particle tracking model of electrophoretic morphogen movement reveals stochastic dynamics of embryonic gradient. *Developmental Dynamics* **238**(8), 1923-1935.
117. Plouhinec, J.-L., L. Zakin & E.M. De Robertis (2011). Systems control of BMP morphogen flow in vertebrate embryos. *Current Opinion in Genetics & Development* **21**(6), 696-703.
118. Wikipedia (2015). Deferent and epicycle. https://en.wikipedia.org/wiki/Deferent_and_epicycle
119. Struhl, G. (1989). Morphogen gradients and the control of body pattern in insect embryos. *Ciba Foundation Symposium* **144**, 65-91.
120. Inui, M., M. Montagner & S. Piccolo (2012). miRNAs and morphogen gradients. *Current Opinion in Cell Biology* **24**(2), 194-201.
121. Franceschi, R.T. (1992). Retinoic acid: morphogen or more mysteries? *Nutrition Reviews* **50**(1), 19-20.
122. Ephrussi, A. & D. St Johnston (2004). Seeing is believing: the bicoid morphogen gradient matures. *Cell* **116**(2), 143-152.
123. Nusse, R. (2015). The Wnt homepage. http://web.stanford.edu/group/nusselab/cgi-bin/wnt/main
124. Yamamoto, H., C. Awada, H. Hanaki, H. Sakane, I. Tsujimoto, Y. Takahashi, T. Takao & A. Kikuchi (2013). The apical and basolateral secretion of Wnt11 and Wnt3a in polarized epithelial cells is regulated by different mechanisms. *Journal of Cell Science* **126**(13), 2931-2943.
125. Gross, J.C. & M. Boutros (2013). Secretion and extracellular space travel of Wnt proteins. *Current Opinion in Genetics & Development* **23**(4), 385-390.
126. Wikipedia (2015). The Wonderful Wizard of Oz. https://en.wikipedia.org/wiki/The_Wonderful_Wizard_of_Oz
127. Willert, K. & R. Nusse (2012). Wnt proteins. *Cold Spring Harbor Perspectives in Biology* **4**(9), a007864.
128. Wikipedia (2014). TCF/LEF family. https://en.wikipedia.org/wiki/TCF/LEF_family
129. Wikipedia (2015). Paracrine signalling. https://en.wikipedia.org/wiki/Paracrine_signalling - The_noncanonical_Wnt_signaling_pathways
130. R&D Systems (2015). Wnt Signaling Pathways. https://www.rndsystems.com/pathways/wnt-signaling-pathways [cobweb site]
131. Berwick, D.C. & K. Harvey (2013). LRRK2: an éminence grise of Wnt-mediated neurogenesis? *Frontiers in Cellular Neuroscience* **7**, #82.
132. Fudge, D.S., K.H. Gardner, V.T. Forsyth, C. Riekel & J.M. Gosline (2003). The mechanical properties of hydrated intermediate filaments: Insights from hagfish slime threads. *Biophysical Journal* **85**(3), 2015-2027.

133. Winegard, T., J. Herr, C. Mena, B. Lee, I. Dinov, D. Bird, M. Bernards, Jr., S. Hobel, B. Van Valkenburgh, A. Toga & D. Fudge (2014). Coiling and maturation of a high-performance fibre in hagfish slime gland thread cells. *Nature Communications* **5**, 3534.
134. Gordon, N.K. (2015). Hagfish Slime. http://embryogenesisexplained.org/2015/12/08/hagfish-slime/
135. Jahan, E., A. Matsumoto, A.M. Rafiq, R. Hashimoto, T. Inoue, J. Udagawa, J. Sekine & H. Otani (2014). Fetal jaw movement affects Ihh signaling in mandibular condylar cartilage development: The possible role of Ihh as mechanotransduction mediator. *Archives of Oral Biology* **59**(10), 1108-1118.
136. Schaffer, J. (2006). 7-Spanner Serpentine Chemokine Receptor. http://jacquelineschaffermedicalillustration.com/work.asp
137. Immunopaedia.org (2010). G protein-associated receptors. http://www.immunopaedia.org/index.php?id=427&L=0
138. Wikipedia (2016). BLAST. https://en.wikipedia.org/wiki/BLAST
139. Graul, R.C. & W. Sadée (2001). Evolutionary relationships among G protein-coupled receptors using a clustered database approach. *AAPS Pharmsci* **3**(2), #12.
140. Gordon, R. (1983). Computational embryology of the vertebrate nervous system. In: *Computing in Biological Science*. Ed.: M. Geisow & A. Barrett. Amsterdam, Elsevier/North-Holland: 23-70.

Chapter 9

The Nuclear State Splitter

In the previous chapter we introduced the biochemistry of signal transduction. We showed our model for the signal transduction pathway from the cell state splitter to the nucleus. In Chapters 3 and 4 we looked at the control of individual genes or portions of the genome. After the passage of a wave caused by the cytoskeletal rearrangements, the signal transduction to the nucleus has taken place. The nucleus ends up in one of two new states, expressing one of two new possible differons. We call what happens inside the nucleus the "nuclear state splitter". In order to explain the nuclear state splitter, we first to need to look at what is known about the control of gene expression on the level of the entire genome.

Unfortunately, our understanding of the control of chromosome structure, let alone the whole genome, is in its infancy. When we began looking at signal transduction from the cell state splitter to the nucleus we had a few disparate pieces which we were able to put together into a working hypothesis[1]. There were a lot of blanks and unlabeled arrows in that model (Figure 8.1). Many of those blanks and unlabeled arrows have since been identified in our model because of more than two decades of advances in molecular biology. Nevertheless, there are still a lot of unlabeled arrows and disparate parts to our notion of what is going on in the physical genome when the signal from the cell state splitter arrives. We have no doubt that the future work of many graduate students and postdocs and their supervisors in many labs will eventually fill in these blanks too. And now some background.

One of the key kinds of proteins involved in controlling gene expression is the *transcription factor*. The concept of a transcription factor is actually rather muddy. It began by scientists finding a protein in a cell. The appearance of the protein, a change to the gene that coded for the protein, or removal of the protein altogether by blocking it, were associated with altered levels of proteins coded for by other genes. The alteration of protein expression could be either up or down. Any protein that was associated with changes in gene transcription levels of another protein was then labeled a transcription factor. It would not be unlike having a truck and removing some part at random. Remove the brake pads and the truck doesn't stop properly. Therefore, brake pads are truck transcription factors. Remove a spark plug and the truck doesn't go as fast and the engine misfires. Therefore spark plugs are also truck transcription factors. No key? The truck won't even start, so keys are obviously some major truck regulatory transcription factor. We are not trivializing this work. Biochemists often work with 'black boxes' and being able to make such a statement about the truck key could well represent years of painstaking work.

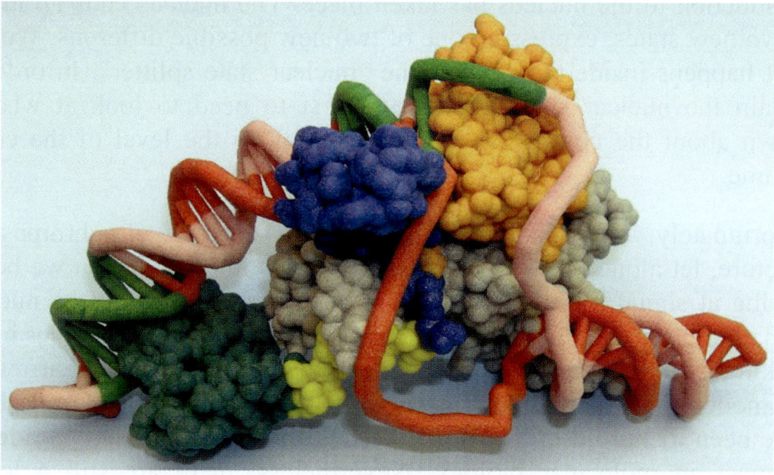

Figure 9.1. Transcription of a given gene generally requires binding of a number or proteins, plus an RNA polymerase, to a specific region of the DNA called the promoter, which is generally close to the gene along the DNA. This is a detailed model of the 3D arrangement of "Sigma factor (promoter recognition subunit) complexed with promoter DNA. Colors on sigma show various domains. Grey is part of core RNAP [RNA polymerase]. Green on DNA are promoter regions." Red on DNA = template strand. Pink on DNA = nontemplate strand. Left green on DNA = –35 element of the promoter. Right green on DNA = –10 element of the promoter. Orange protein = domain 2. Blue protein = domain 3. Yellow protein = region 3.2. Dark green protein = domain 4. The model shows how the sigma 3.2 loop extends into the active-site channel of RNAP during initiation. The red strand of DNA is the transcribed strand that extends into the active site of RNAP. Distance is measured from the first codon of the transcribed DNA. From Hata & Herman (2010)[2] with kind permission of Tommie S. Hata and Seth A. Darst, who provided the coordinates that allowed construction of the model[3]. DNA diameter is 2 nm.

Today, a transcription factor is more precisely defined as any protein that forms a structure that binds (or is part of a complex that binds) to a specific sequence of DNA and by that binding, changes the rate of transcription of a gene (Figure 9.1). Many scientists think a transcription factor should only be called a transcription factor when it directly binds DNA and all the other proteins involved in the complex should have alternative names. While we agree that such a definition is more precise, the history of the discovery of transcription factors has firmly fixed the title of "transcription factor" onto so many proteins that the distinction is almost meaningless. Nevertheless, we will try to adhere to it ourselves.

As we showed in Chapter 8, NFAT, β-catenin and JNK are three examples of transcription factors whose activity is directly linked to the expansion or contraction of the cell state splitter by the signal transduction pathway networks. Let's look at these three examples in more detail, especially their roles in the nucleus.

NFAT transcription factors are one example of a whole family of transcription factors found in every known eukaryote cell that has been tested for them. This family is collectively known as Rel or NF-kappaB (NF-κB) proteins. The general structure of the NFAT enzyme is like that of most transcription factors. There are two binding domains with a linker region in between (Figure 9.2). The N terminal region of the transcription factors selectively binds to specific DNA sequences in the 'enhancer region' of certain genes (Figure 9.3). If the enhancer does not have the correct sequence in it, NFAT doesn't bind. The C terminal region of NFAT binds nonspecifically (at any part) to the phosphate backbone of the DNA. NFAT is thereby stabilized in its adhesion if the N terminal region is also bound. NFAT is a true transcriptional factor in the purest sense because it binds directly to the DNA

NFAT is usually not found near the nucleus, but rather is localized in the cytoplasm. If the cell participates in a contraction wave, the cell state splitter triggers Ca^{2+}/calmodulin complex formation. This phosphorylates an enzyme in the Wnt pathway. The phosphorylated enzyme causes a conformation change in NFAT. This conformation change exposes an NFAT nuclear import signal, a part of its amino acid sequence that is normally folded inside of the protein, out of contact. With the import signal exposed NFAT will be transported into the nucleus. Other enzymes would normally prevent NFAT from being imported even with

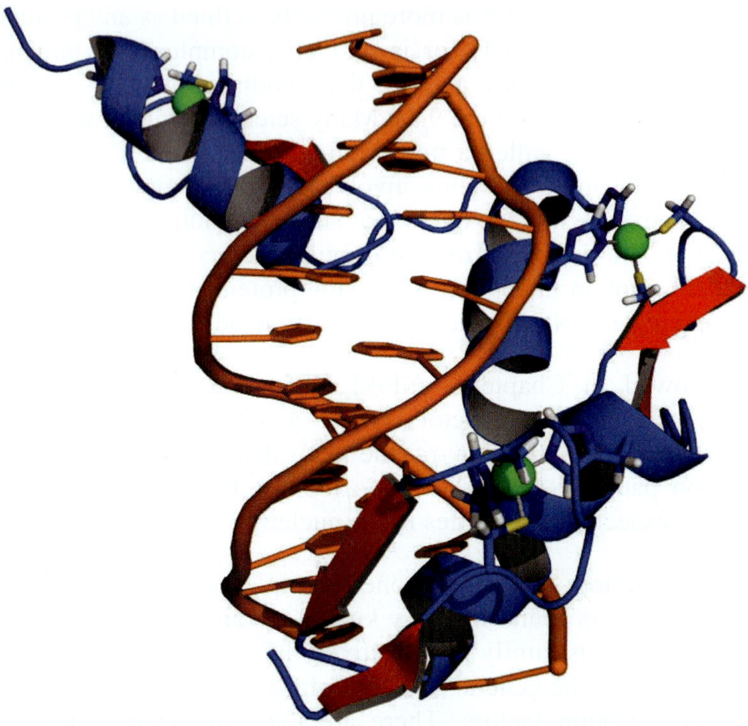

Figure 9.2. The transcription factor has a common motif, two DNA binding domains and a linker region. "Cartoon representation of a complex between DNA and the ZIF268 protein, containing 3 zinc finger motifs. The coordinating residues of the middle zinc finger are highlighted. Based on the x-ray structure of PDB 1A1L. Color coding: ZIF268: blue; DNA: orange; Zinc ions: green." From Splettstösser (2006)[4] with permission under the terms of the GNU Free Documentation License, Version 1.2.

the nuclear import signal exposed as part of a regulatory loop. However, these other enzymes are all inactivated by other phosphorylation events from the same signal transduction pathway. Even once in the nucleus there are proteins that bind and change the conformation of NFAT exposing an export signal sending it right back out unless they are also deactivated. So translocation of NFAT requires both a conformational change in NFAT exposing an import signal and deactivating of otherwise inhibitory proteins in the cytoplasm and the nucleus. There is also a feedback loop on the transcription of the gene for NFAT so NFAT is not only regulated by other proteins, it is also self-regulating at the level of transcription (Figure 9.4).

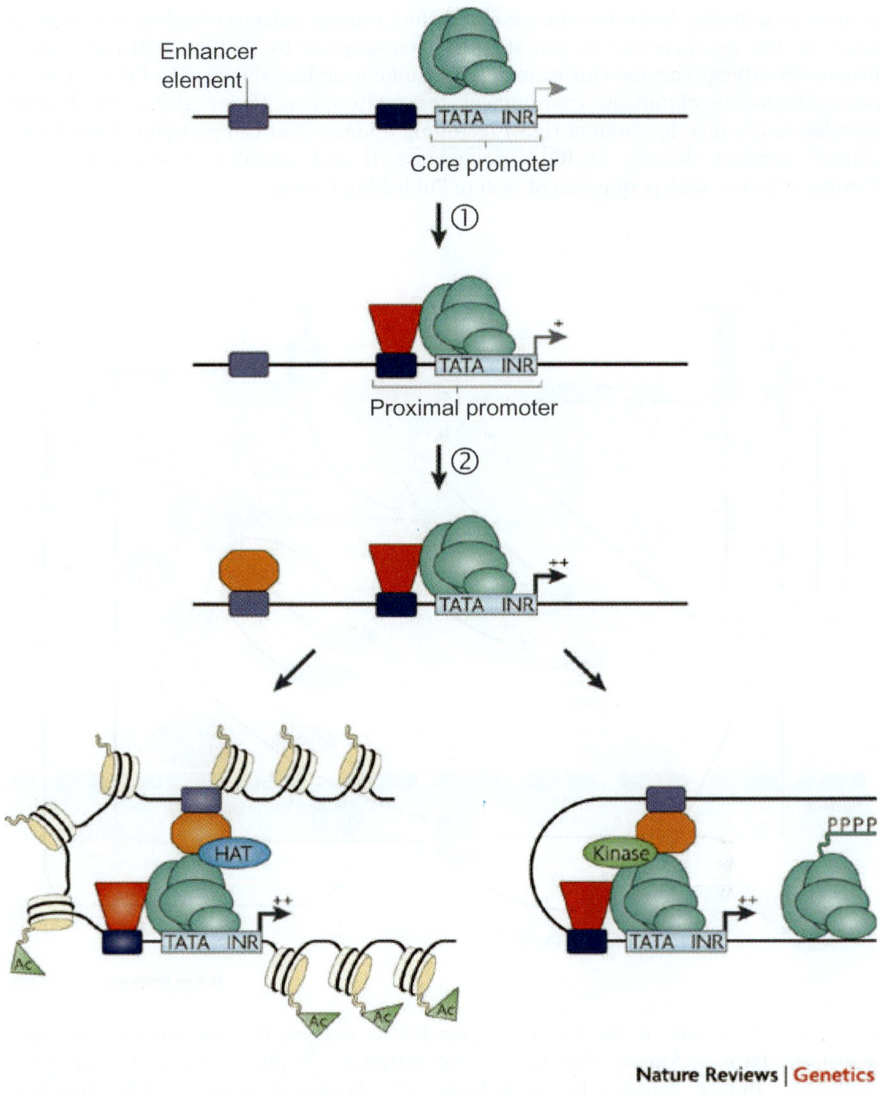

Figure 9.3. "General transcription factors (green ovals) bind to core promoter regions through recognition of common elements such as TATA boxes and initiators (INR). However, these elements on their own provide very low levels of transcriptional activity owing to unstable interactions of the general factors with the promoter region. Promoter activity can be increased (represented by +) by site-specific DNA-binding factors (red trapezoid) interacting with *cis* elements (dark blue box) in the proximal promoter region and stabilizing the recruitment of the transcriptional machinery through direct interaction of the site-specific factor and the general factors (step 1). Promoter activity can be further

stimulated to higher levels by site-specific factors (orange octagon) binding to enhancers (step 2). The enhancer factors can stimulate transcription by (bottom left) recruiting a histone-modifying enzyme (for example, a histone acetyltransferase (HAT)) to create a more favourable chromatin environment for transcription (for example, by histone acetylation (Ac)) or by (bottom right) recruiting a kinase that can phosphorylate (P) the carboxy-terminal domain of RNA polymerase II and stimulate elongation." From Farnham (2009)[5] with permission of Nature Publishing Group.

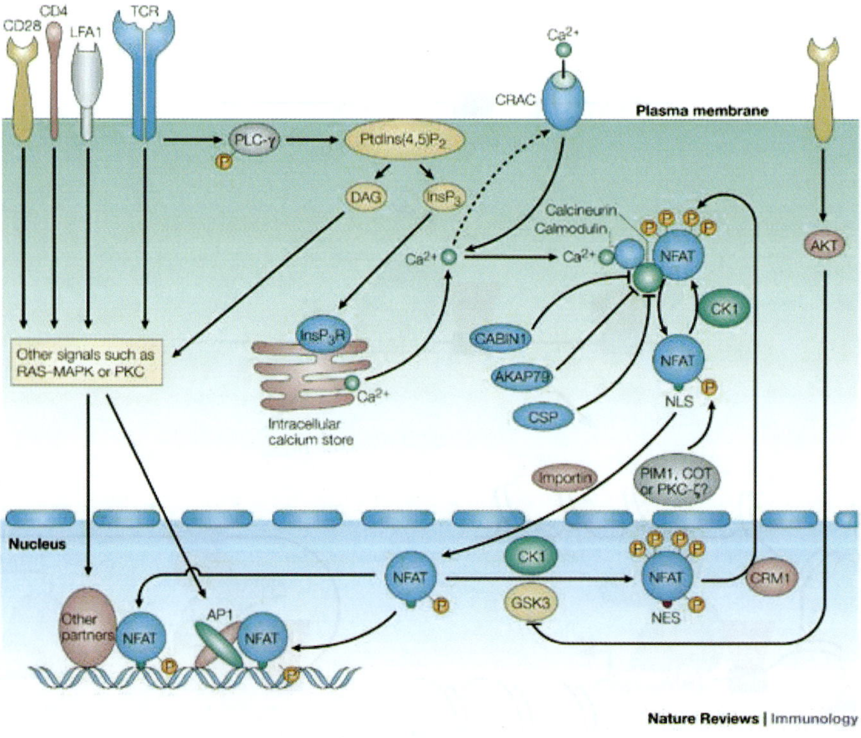

Figure 9.4. "Ligation of the T-cell receptor (TCR) triggers the activation of receptor-associated tyrosine kinases that lead to the activation of phospholipase C-γ (PLC-γ). Activated PLC-γ causes the hydrolysis of phosphatidylinositol-4,5-bisphosphate ($PtdIns(4,5)P_2$), which generates inositol-1,4,5-trisphosphate ($InsP_3$) and diacylglycerol (DAG). $InsP_3$ binds to its receptor ($InsP_3R$) and induces an increase in intracellular calcium levels that is caused by the depletion of intracellular stores. This increase triggers, through poorly characterized mechanisms, the opening of calcium-release-activated calcium (CRAC) channels in the plasma membrane, which leads to a sustained increase in intracellular calcium levels. Calcium binds calmodulin and activates calcineurin. Activated calcineurin dephosphorylates nuclear factor of activated T cells (NFAT) proteins, which exposes their nuclear-localization signal (NLS) and induces their nuclear translocation. Other kinases, such as protein kinase C-ζ (PKC-ζ), PIM1 (proviral

integration site 1) or COT (Cancer Osaka thyroid), might also phosphorylate NFAT and contribute to its activation…. The activity of calcineurin is also negatively regulated by calcineurin-binding protein 1 (CABIN1), calcipressins (CSPs) and A-kinase anchor protein 79 (AKAP79). After it has entered the nucleus, NFAT interacts with activator protein 1 (AP1) and other transcriptional partners to promote gene transcription. The activation of these partners during T-cell stimulation might be elicited by signals that are transmitted through different signalling pathways (such as the RAS–mitogen-activated protein kinase (MAPK) pathway or the PKC pathway), which are activated by the engagement of the TCR and/or other receptors. The activity of NFAT is also regulated by kinases, such as casein kinase 1 (CK1) and glycogen-synthase kinase 3 (GSK3), which help to maintain NFAT in a phosphorylated state in the cytosol (maintenance kinases) or induce the rephosphorylation of nuclear NFAT to expose a nuclear-export signal (NES) and translocate NFAT back to the cytosol (export kinases). GSK3 is negatively regulated by the kinase AKT, the activation of which is coupled to CD28 engagement. CRM1, chromosomal region maintenance 1; LFA1, lymphocyte function-associated antigen 1." From Macian (2005)[8] with permission of Nature Publishing Group. Note that the contraction of the microfilament ring of the cell state splitter can trigger NFAT translocation to the nucleus.

NFAT is also affected by noncoding RNA. When activated, NFAT moves into the nucleus and weakly binds to specific DNA sequences in regions called *enhancers*. Because the binding is weak, NFAT cannot change the transcription rates for a gene by itself. Other transcription factors bind to other sequences of DNA in the enhancer regions along with NFAT and in doing so change the NFAT binding and act as *transcriptional partners*. All the proteins bound together cooperatively[6,7] increase binding strength and form what is called an *enhanceosome* (Figure 9.5).

An enhanceosome is a complex of several transcription factors bound with other proteins that do not directly bind to the DNA. Altogether this complex binds the DNA containing the gene of interest to the RNA polymerase complex. The gene can then, and only then, be transcribed (Figure 9.6). More on this below.

There are some other transcription factors, such as *β-catenin*, which move to the nucleus and bind to other proteins and form the outer parts of the enhanceosome complex (Figure 9.7). This type of transcription factor is more properly called a transcriptional modulator because it does not bind directly to the DNA. JNK, the "mitogen activating protein", acts by its phosphorylation activity rather than by DNA binding. JNK moves into the nucleus and it phosphorylates several different transcription factors.

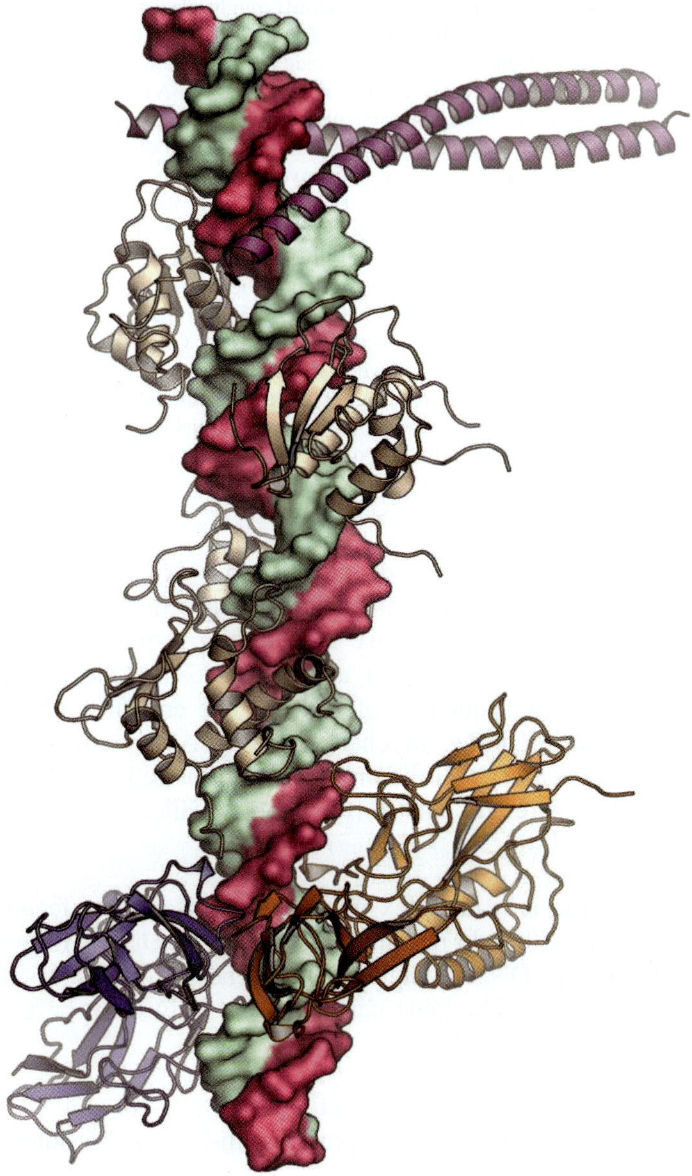

Figure 9.5. Model of the enhanceosome protein assembly on double-stranded DNA. "DNA: green/pink strands ATF-2/c-Jun coiled-coil DNA-binding domain: purple Interferon response factors: beige NF-kappa-B p105 subunit: orange, NF-kappa-B p65 subunit: blue." From Wikipedia (2015)[9] with permission under a Creative Commons Attribution-Share Alike 3.0 Unported License.

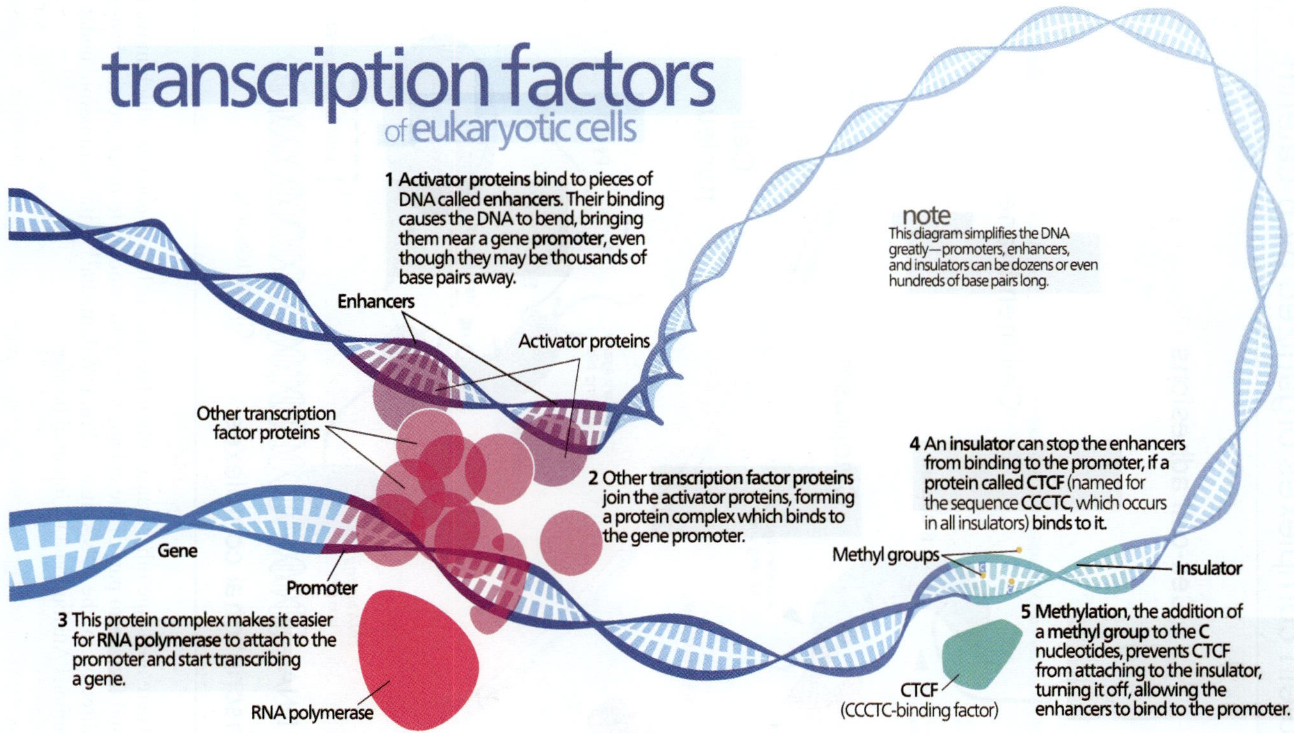

Figure 9.6. "Diagram of gene transcription factors." From Song (2012)[10] with permission under a Creative Commons Attribution 3.0 Unported license.

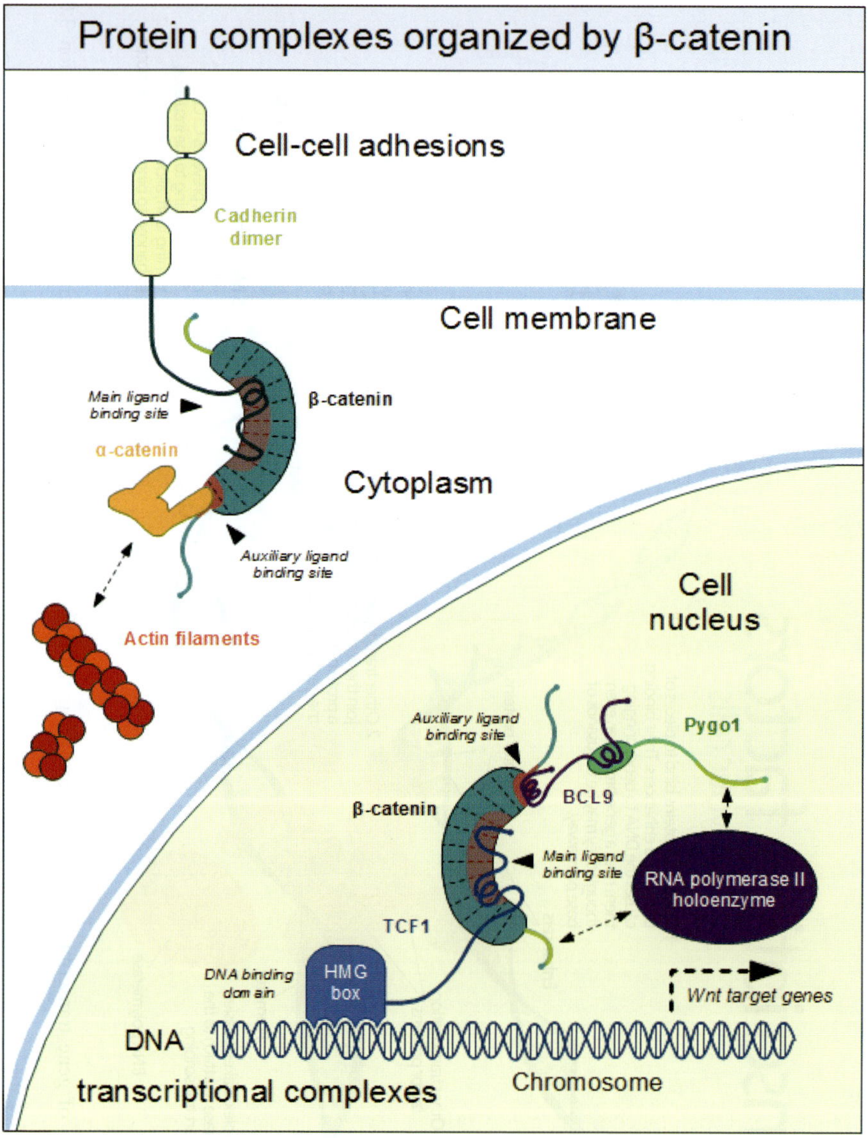

Figure 9.7. There are some other transcription factors, such as *β-catenin*, which move to the nucleus and then form part of the enhanceosome by binding to other proteins instead of binding directly to the DNA itself. From Wikimedia[11] with permission under a Creative Commons Attribution 3.0 Unported license.

For some transcriptional factors, this results in more gene transcription. For others, the JNK phosphorylation results in less gene transcription. Transcription factors, like JNK, can down regulate because of "silencer" regions.

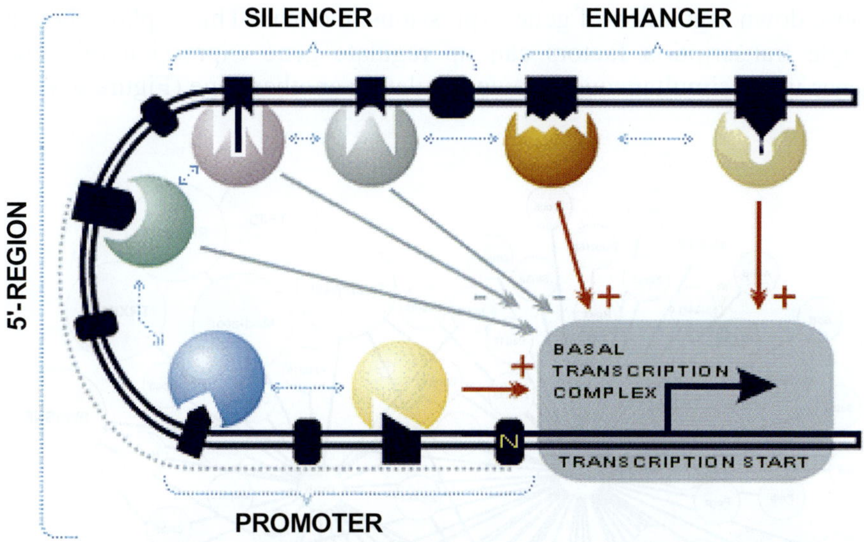

Figure 9.8. "TRRD [Transcription Regulatory Regions Database] is designed for accumulation of experimental data on extended transcription regulatory regions of eukaryotic genes. Eukaryotic gene transcription regulation is provided by complex interactions between transcription factors binding specifically to DNA and basal transcription complexes formed near the transcription start site." From Kolchanov *et al.* (2006)[12] with kind permission of Nikolay A. Kolchanov. "Glossary. Regulatory unit: extended DNA sequence (promoter region, enhancer, silencer, etc.), that alter the levels of gene transcription and can determine spatial patterns of gene expression. Promoter region: a specific region usually upstream of the transcription start site that binds and directs RNA polymerase to the correct transcriptional start site and contains binding sites for transcription factors that regulate the rate of transcription of the adjacent gene. Enhancer: a cis-acting sequence that increases the utilization of (some) eukaryotic promoters and can function in either orientation and in any location (upstream or downstream) relative to the promoter. Silencer: a cis-regulatory sequence that can reduce the levels of transcription from an adjacent promoter. Silencers can be 5' or 3' to the promoter they regulate. Transcription factor binding site: a short DNA sequence that is recognized by the transcription factor and binds them in a specific manner"[13]. See also Kolchanov *et al.* (2002)[14].

Silencer regions are sections of DNA with specific sequences that bind transcription factors. When transcription factors are bound, enhancer regions and promoters are blocked to the transcriptional machinery and gene expression is suppressed or silenced. Transcription factors can therefore bind to specific sites and cause up regulation of gene expression in some cases and simultaneously bind to silencer regions to cause down regulation of gene expressions in others. This explains how a single transcription factors can up regulate gene expression of some genes while simultaneously down regulating another gene (Figure 9.8).

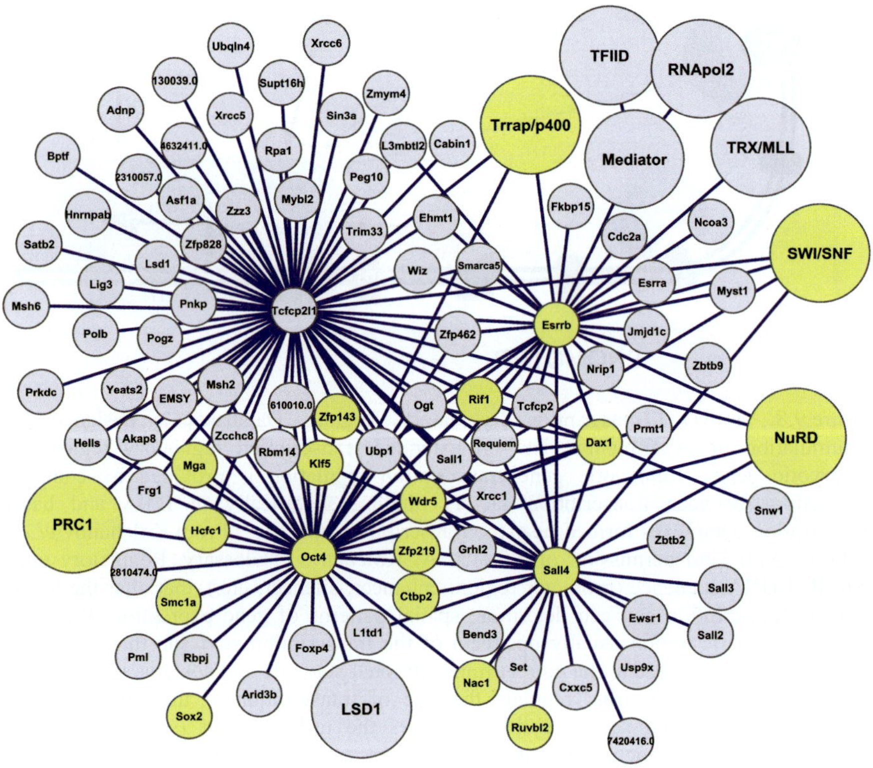

Figure 9.9. "Protein Interaction Network of Oct4 and Its Associated Proteins Sall4, Dax1, Tcfcp2l1, and Esrrb.... Complexes are shown as larger circles. Yellow coloring indicates importance for ESC [embryonic stem cell] self-renewal capacity." From van den Berg *et al.* (2010)[15] with permission under a Creative Commons Attribution License (CC BY).

The exact mechanism used by many other transcription factors is still unknown. The activity of all these transcription factors is often bewildering in both mechanism and interactions (Figure 9.9). In one cellular context you get this response. In another cellular context you get that response, in yet a third cellular context you get no response at all. An effort to sort out all this information is a project called ENCODE. ENCODE stands for ENCyclopedia Of DNA Elements (ENCODE) project[16]. It is a collaborative and multidisciplinary effort to collect and categorize all the functional elements of the human genome. Several principals of transcription factor function have come out of this and related work on noncoding DNA elements.

If we were to stretch our double stranded DNA in one diploid human cell out it would be about 2 meters long (2 chromosome sets x 3×10^9 base pairs per double stranded DNA x 0.34 nm per base pair)[17]. Yet it is typically packed into a nucleus of diameter about 10 μm, 1/200,000th the width. The smallest packing unit is a nucleosome. A nucleosome is a segment of DNA of 147 base pairs wrapped two and three quarters times around eight proteins called histones (two each of H3, H4, H2A, and H2B) (Figure 3.24). There is also a protein known as histone 1 which can bind onto the nucleosome and when it is bound, the DNA and histones together can be even more tightly condensed. As the nucleosome has proven to be a dynamic structure, we will refer to the 8 histones clumped together as a *histone octamer* and the histone octamer with DNA bound to it as a *nucleosome*.

Each histone in a nucleosome has "tails" sticking out at the N-terminal end amino acid sequences. These tails can be selectively phosphorylated, or dephosphorylated. The histone tails can also be subject to acetylation, methylation, hydrolyzing the imine group of arginines in the tail and attaching a keto group instead, ubiquitination, glycosylation, and ADP ribosylation (Figure 3.25). The strength of such modifications can also be varied. For example, there can be one, two or three methyl groups attached to the histone tail.

These changes to the histone tails serve to change how tightly the DNA is bound to the nucleosomes and how closely the nucleosomes are packed relative to each other. Acetylation loosens the binding of DNA by decreasing the interaction of the N termini of the histones with the negatively charged ions on phosphate groups of the DNA backbone.

Eukaryotic DNA is broadly divided into tightly bound *heterochromatin* that has heavily methylated histone tails and is sequestered, and *euchromatin* with histones that are acetylated instead of methylated. How strongly the DNA is wound up in the nucleosome affects the ability of other transcription factors to bind directly to the DNA and thus alters binding of both RNA polymerases and DNA polymerases. Euchromatin contains genes that are exposed and can be transcribed (Figure 9.10). For example, H4 is acetylated during preparation for mitosis when DNA is being replicated. One can infer then, that H4 acetylation loosens the binding of DNA to a sufficient extent that DNA polymerase can access the DNA for copying[18]. Changing the strength of DNA binding to histones by any changes to the tail of the histone can also change gene transcription rates. In addition to changing the binding strength of DNA to histones, modifications on the tails also serve as a code in and of

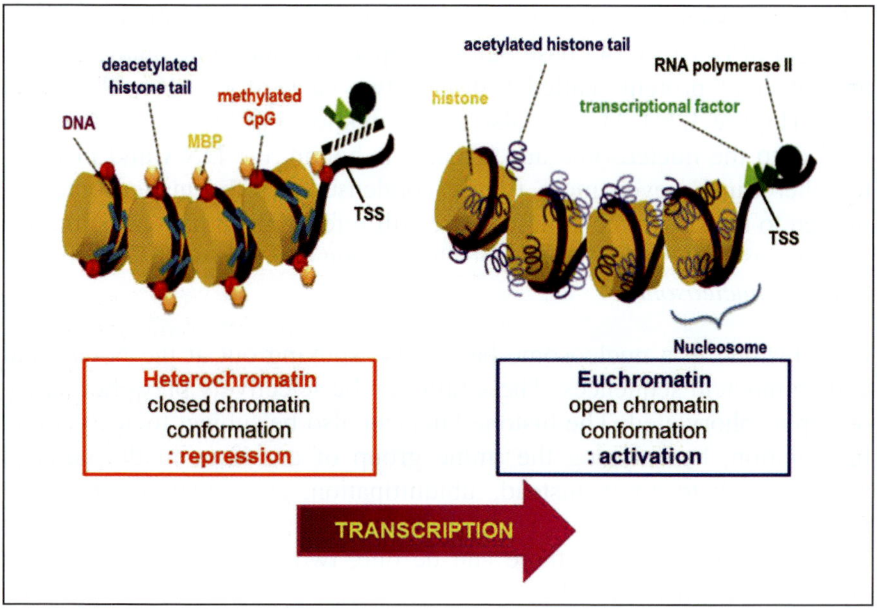

Figure 9.10. Our DNA is divided into tightly bound *heterochromatin* that has heavily methylated histone tails and *euchromatin* with histones that are acetylated instead of methylated and which contains genes that can be transcribed. These correspond to our use of sequestered and exposed, respectively. From Kim (2014)[19] with kind permission of Heon Yoo of the Korean Neuropsychiatric Association, Brain Tumor Research and Treatment, Neuro-Oncology Clinic, National Cancer Center.

themselves for proteins to detect[20,21]. And so histones with specific patterns of methylation or acetylation will attract and bind transcription factors/transcriptional modulators or repel them.

Along the length of the DNA, both the DNA around each nucleosome and the DNA between the nucleosomes, are sequences that bind specific transcription factors (Figure 9.11). One example is the CTCF proteins[22]. The CTCF binding site DNA sequences are normally invariant, i.e., they are all the same across the genome. A CTFC transcription factor encoded in a CTCF gene binds to both the CTCF DNA sequence and the nucleoskeleton (Figure 5.14), acting as an anchor point for loops of DNA. CTCF proteins are classified as transcription factors because they change the architecture of DNA loops and thus alter gene transcription. The first CTCF protein was discovered because it shuts off transcription of the c-*myc* gene in chickens when it is present[23]. It does this by binding

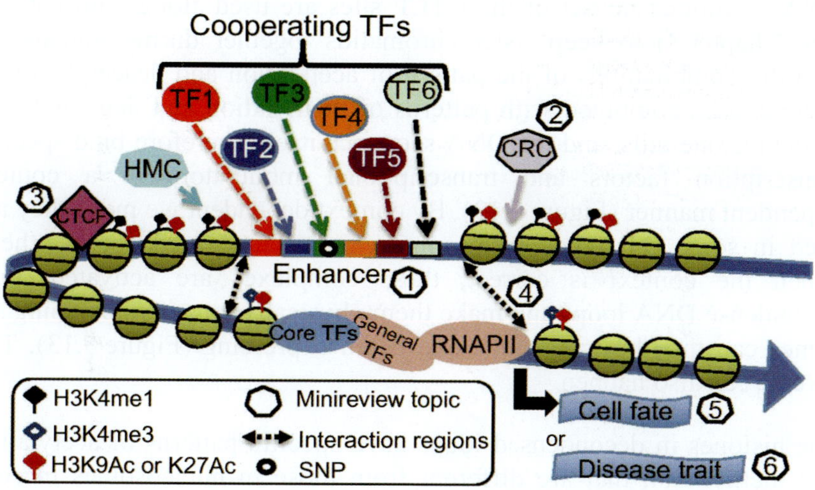

Figure 9.11. "Genome-wide characterizations of regulatory regions. Recent genome-wide ChIP-seq studies have revealed 100,00–200,000 regions of open chromatin per cell type, tens of thousands of which are marked by specifically modified histones (e.g. H3K4me1 and H3K27ac) and bound by many different site-specific transcription factors. Creation of these open chromatin regions requires the actions of histone-modifying complexes (HMC), chromatin-remodeling complexes (CRC), and boundary proteins such as CTCF. Distal regulatory regions are thought to function by interaction of bound site-specific factors with other transcription factors bound to core promoters via looping of the intervening DNA." From Farnham (2012)[24] with permission under a CC-BY license.

the DNA containing the gene into a conformation such that the RNA polymerase cannot access the gene. CTCF sequences flank the regions of DNA where genes are being expressed and regions where genes are not being expressed. CTCF also binds selectively to regions with different methylation patterns between the paternally and maternally inherited chromosomes and so is involved in imprinting effects. CTCF binds together strands of DNA by binding to two specific sites along the DNA, This creates chromatin loops and anchors the DNA to the inner lamin lining of the nuclear membranes or other protein structures in the nucleus (Figure 9.12).

Some CTCF DNA sequences are bound to CTCF proteins only in specific cell types. In other cell types, the same CTCF sequence is not used to bind the CTCF protein. This indicates that particular CTCF sequences are sequestered in some cell types and the same ones are exposed in others. This leads to cell type specific looping patterns in the DNA[25]. Another subset of the CTCF sites are used along with cohesin (see Chapter 3) to keep sister chromatids together during mitosis[28,29]. Specific combinations of the pattern of acetylation and deacetylation of histone tails, combined with patterns of methylation and demethylation of the histone tails, and the DNA sequence itself, therefore bind specific transcription factors and transcriptional modulators in a context dependent manner (Figure 9.13). By context dependent we mean they are used in some cell types or in some circumstances and not in others. When the context is correct, these complexes are activated, they decondense DNA loops and make them physically longer when doing so. Genes can then be transcribed and begin expressing (Figure 5.13). The reverse can also happen.

The histones in decondensed loops have specific patterns of acetylation and methylation that are different from those in nucleosomes in condensed DNA. The decondensed loops also have further attachment points known as "scaffold" attachment regions or equivalently "matrix" associated regions known as "S/Mars". One of the big puzzles about S/Mars has been that they are also context dependent, i.e., important in some cell types but irrelevant in others. Context dependency occurs because the S/Mars loops are only functional for DNA that is already decondensed and thus exposed. In condensed (sequestered) DNA the S/Mars sites are hidden and inaccessible. Therefore, we can surmise that DNA is organized in loops wrapped around nucleosomes which are

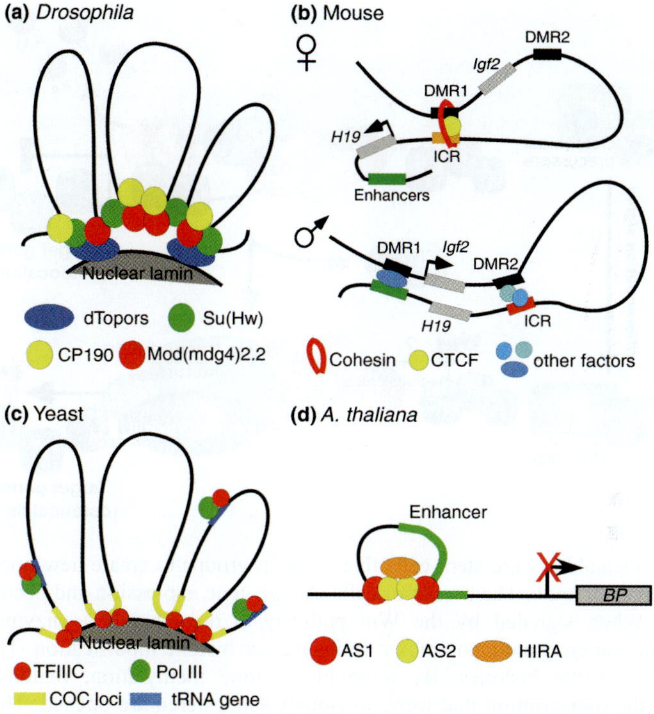

Figure 9.12. "Loop formation through intra- and inter-chromosomal interactions is a common strategy for genome organization and insulation in different organisms. **(a)** In *Drosophila*, the Su(Hw) protein binds to specific DNA elements and recruits the CP190 protein and Mod(mdg4)2.2 proteins. Interaction among these proteins results in the formation of chromatin loops. Mod(mdg4)2.2 attaches the chromatin to the nuclear periphery through its interaction with topoisomerase I-interacting RS protein (dTopors). **(b)** Monoallelic expression at the *Igf2-H19* locus is regulated by binding of CTCF to the imprinted control region (ICR). On the maternal allele, CTCF mediates interactions between ICR and DNA methylated region 1 (DMR1) that also involve joining of the DNA strands by cohesin, insulating *Igf2* from the influence of downstream enhancers. Methylated ICR sequences prevent CTCF from binding to the ICR on the paternal allele, allowing downstream enhancers to switch on *Igf2* transcription. **(c)** In *Schizosaccharomyces pombe*, TFIIIC binds to RNA polymerase (Pol) III at tRNA genes and acts as a barrier against the spreading of heterochromatin. It is also hypothesized to organize the chromatin into distinct loops by clustering various chromosome-organizing clamp (*COC*) loci to the nuclear periphery. **(d)** In *Arabidopsis thaliana*, binding of the ASYMMETRIC LEAVES1 (AS1)-AS2 complex at two specific DNA sites flanking the enhancer is required to silence the expression of the *BP* gene. Recruitment of the histone chaperone HIRA is necessary for this process, and it probably acts by facilitating looping of the enhancer element." From Ong & Corces (2009)[26] with permission of BioMed Central. Insulators are DNA binding proteins that keep transcriptional factors for different genes from interacting.

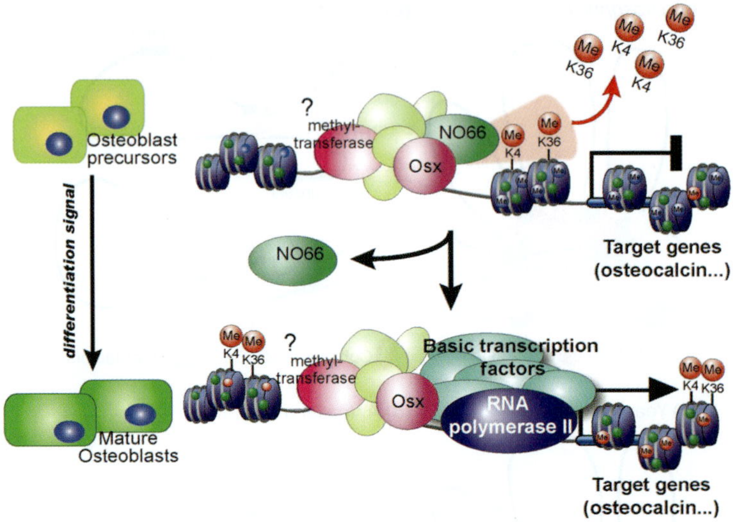

Figure 9.13. Osteoblasts are stem cells that work in groups to create new bone either in growth or repair of bone. The precursor cells change gene expression and become mature osteoblasts. When signaled by the Wnt pathway, a demethylating enzyme N066 is removed from the enzymatic complex responsible for histone modification. The result is demethylation of the histones. By removing histone methylation, regions of DNA upstream of the transcription that were previously sequestered are opened. The complex of histone modifying enzymes and transcriptional factors moves along the DNA. The histones behind the enzymatic machinery, and then transcriptional factors, are methylated. The precise mechanism remains to be discovered. However, the two prevalent models are opening of the loops around histones or sliding of loosened DNA along the histones, either of which can occur only with histone demethylation. "Histone demethylase is a negative regulator for Osterix (Osx)-mediated gene regulation in osteoblastogenesis. A histone H3K4 and K36 demethylase, nucleolar protein 66 (NO66), serves as a transcriptional co-repressor for Osx, a critical inducer of osteoblastogenesis. Through demethylation of active histone H3K4 and H3K36 methyl groups, surrounding chromatin areas are likely inactivated as a transcriptional control." From Kato *et al.* (2010)[27] with permission of Nature Publishing Group.

condensed (heterochromatin) or decondensed (euchromatin). Decondensed DNA is then further organized into smaller loops based on sequences like S/Mars. These smaller loops have regions with nucleosomes (sequestered) and regions that have no nucleosomes (exposed) (Figure 9.14). Note that we are extending the concepts of exposed and sequestered *genes*[30] in Chapter 3 to any portion of DNA in various levels of hierarchical organization. This organization is encoded in the DNA itself in the specific sequence of noncoding elements.

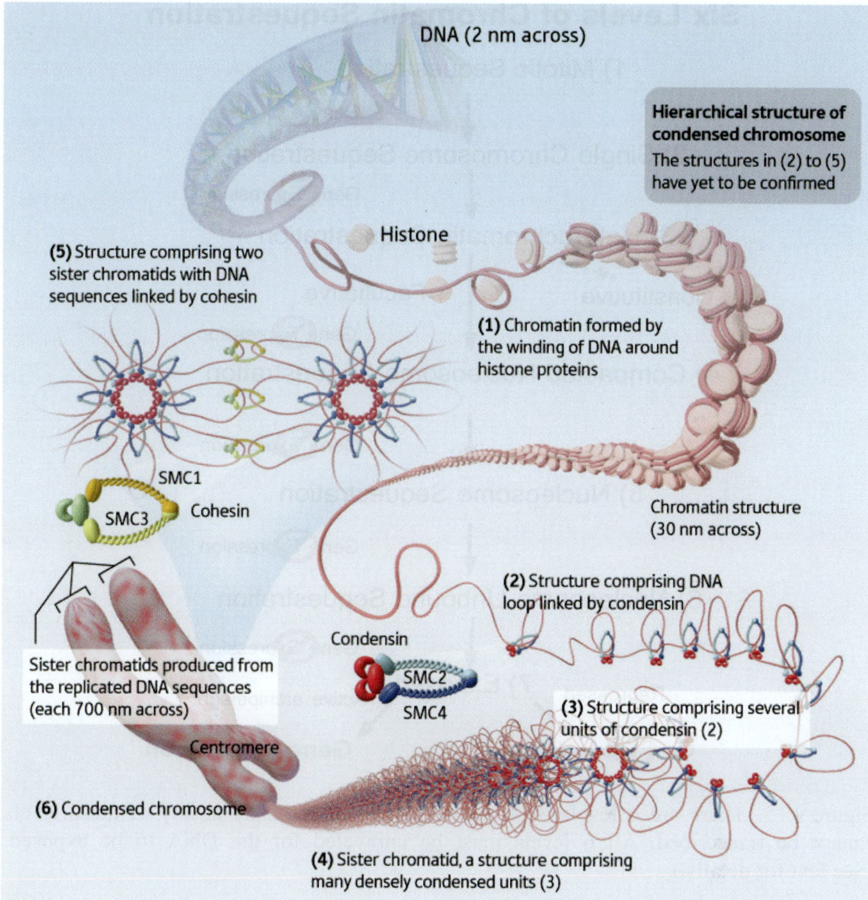

Figure 9.14. This figure shows DNA at 6 different levels of condensation. Each level of condensation requires the binding of specific proteins, in this case condensins and cohesins. These proteins do not bind at random to the DNA. Rather their binding is to specific sequences along the DNA itself. These sequences also bind the DNA around histones to form nucleosomes. Binding of certain enzymes to these sequences furthermore alters the histones tails. The modified tails can attract or repel specific proteins required for higher order organization. Thus, the strands of DNA encode their three dimensional structural hierarchy created using loops within loops within loops of organization. The loops of the lowest order of organization are not shown in this diagram but are illustrated in Figure 9.17. From Hirano (2010)[31] with permission of RIKEN. Whether the 30 nm coiling of nucleosomes is a correct structure or they are compacted in some other arrangement is still being debated, with experimental evidence pro and con[32-39].

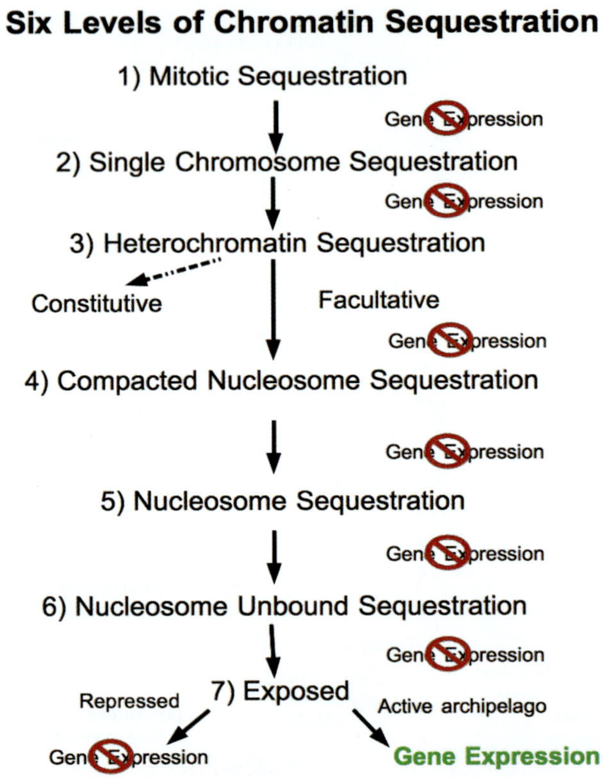

Figure 9.15. There are 6 levels of sequestration of DNA. DNA at any of these levels cannot be transcribed. All 6 levels must be unraveled for the DNA to be exposed. (See text for details.)

There are (at least) 6 levels of sequestration of DNA (Figure 9.15). DNA sequestered at any of these levels cannot be transcribed. All 6 levels must be unraveled for the DNA to be exposed. Unraveling sequestered DNA is commonly called decondensing, usually without specification of the level of sequestration of the DNA under consideration. Using the current state of the art, we identify these levels. The top 6 are states of sequestration, and the last level is the single state of exposure:

1. *Mitotic sequestration*: the whole genome is condensed in preparation for mitosis. This includes crosslinking of sister chromatids by cohesins. There is little to no gene expression that can occur.

2. *Single chromosome sequestration*: a whole chromosome remains sequestered after mitosis by cohesins. X inactivation in mammals is an example where this level persists throughout the cell cycle for the X chromosome. Again, a few genes from the whole chromosome may remain exposed. This kind of sequestration may also include a layer of proteins that cover the chromosome.
3. *Heterochromatin sequestration*: regions of each chromosome are sequestered with clustered loop-linked condensins. Other regions, referred to as euchromatin, are not. Heterochromatin has been divided into two categories. Constitutive heterochromatin consists of stretches of DNA that contain no genes, but may have structural functions, such as telomeres and centromeres. Whether this is coated with noncoding RNA/protein complexes in any way or produces its own noncoding RNA is unknown. Facultative heterochromatin contains genes which are sequestered by deacetylated histones and/or noncoding RNA.
4. *Compacted nucleosome sequestration*: facultative heterochromatin whose histones are at least partially acetylated and are not coated with noncoding RNA/protein complexes nonetheless remain sequestered. The 3D arrangement of the nucleosomes at this level is currently under dispute (Figure 9.14).
5. *Nucleosome sequestration*: nucleosomes are not compacted against each other, but the DNA is still tightly wound around the nucleosome, and cannot have its genes transcribed. The loosening of the nucleosomes from one another has been associated with further histone acetylation and other histone modifications.
6. *Nucleosome unbound sequestration*: a segment of DNA is separated from nucleosomes but nevertheless sequestered due to bound protein and/or RNAs. Two mechanisms have been proposed, both of which may be used: the nucleosomes slide along the DNA out of the way, or the DNA loops out from the histone octamer it is attached to. The latter is called the "seatbelt model" because you can visualize it by imagining pulling the shoulder strap of your seatbelt away from your chest.
7. *Exposed*: a segment of DNA is available for transcription. This may also require removal of repressors and the formation of a complex of transcriptions factors, transcriptional regulators and the RNA polymerase machinery.

RNA polymerase, contrary to how it is often pictured, is not floating about randomly in the nucleus looking for an exposed gene to jump on and transcribe. Rather RNA polymerase is part of a complex of proteins that forms the transcriptional machinery. The machinery is anchored to the nucleoskeleton in areas separate from chromosome territories (Figures 3.7 and 9.16). Decondensed loops of DNA are also in this area, moved in by the same enzymatic protein machinery that decondenses and thereby loosens up the loop. Only exposed DNA can be accessed by the RNA polymerase and transcribed.

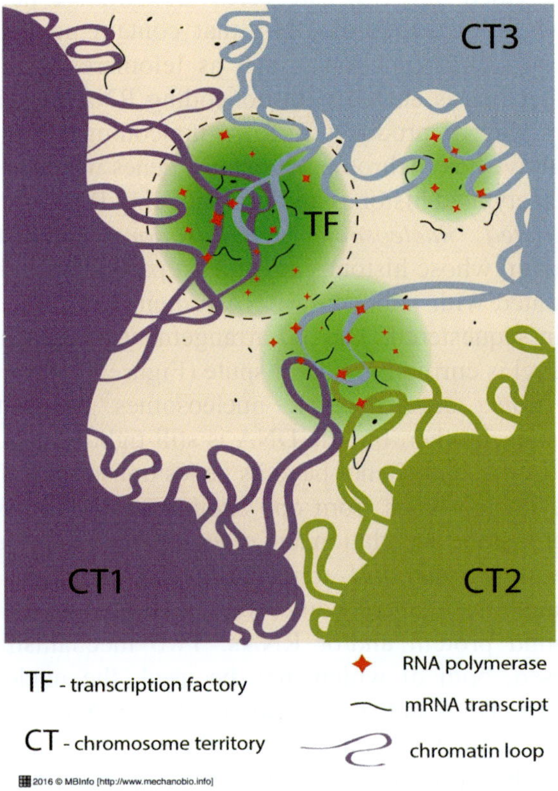

Figure 9.16. "Chromatin loops from the same or from different chromosome territories often share transcription factories (TFs). TFs are enriched in RNA polymerase II complexes that produce nascent RNA transcripts, using co-transcribed genes from chromatin loops as their templates." From MBInfo Wiki[40] with kind permission of Steven Wolf. Image provided courtesy of the Mechanobiology Institute, National University of Singapore.

The attachment points of exposed DNA can create loops that bring together parts of a gene or more than one gene into juxtaposition with the RNA polymerase complex for transcription. This collection of the smallest organizational loops of DNA, which are bound to the polymerase and held by the enhanceosome complex of transcription factors/modulators, is called a *regulatory archipelago*. The stretch of DNA that is noncoding between the enhancer DNA sequence and the gene that it enhances is a "gene desert". Like most deserts, we are finding there is actually a lot of regulatory life in there even if it isn't in standard gene form (Figure 9.17).

DNA in the genome also contains many specific kinds of enhancer binding sites (Figures 9.6 and 9.8). Like the term 'transcription factor', 'enhancer sequence' began as a rather nondescript term. If you changed the sequence of the DNA and decreased transcription or you found a transcription factor that could bind to that sequence and in doing so cause an increase in transcription, then that DNA sequence is an enhancer. There are literally tens of thousands of these DNA enhancer regions that can potentially attract transcription factors, though only when the DNA is decondensed. Enhancers are spread all over the genome in a nonrandom fashion. About one third of enhancers are located near the promoter region of the gene they enhance, as would be expected. However 2/3 of enhancers are located on entirely different regions of the DNA. There are even enhancers from one chromosome that affect transcription of a gene on another chromosome (Figure 9.3).

We now know that several loops of DNA from different parts of the genome can move out of the condensed chromosome territories, be decondensed and move into an area where RNA polymerase complexes are functioning. A transcription factor can bind both an exposed enhancer region on one loop of DNA and the gene promoter on another loop containing the gene. Both loops move to the transcription machinery where the gene is transcribed. And so enhancers that appear to be far away from the gene they control in the linear DNA sequence can be brought close together by the higher order archipelago looping structures of the genome. Splice variants of genes can also be made by changing the loops used (Figure 9.17). And so the three dimensional state of the DNA as well as the enhancer becomes important for gene expression. The 3D structure depends on the transcription factors/ transcriptional modulators that are bound to the DNA.

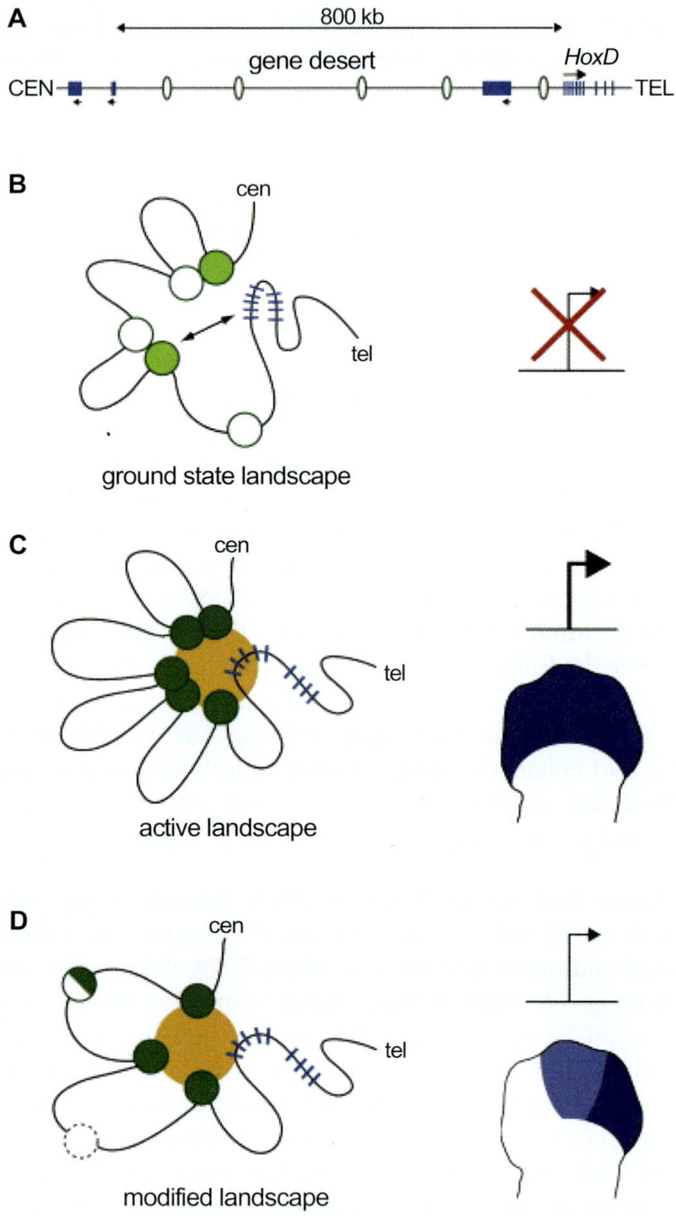

Figure 9.17. "A Regulatory Archipelago Controls Hoxd Gene Expression in Digits: (**A**) Map of the regulatory landscape, covering approximately 800 kb upstream the *HoxD* cluster, including the *Atp5g3-Lnp* gene desert. Multiple regulatory islands (green ovals) are required for *Hoxd* gene activation in digits. These islands may either have

enhancer activity or serve as anchor points. (**B–D**) Various conformations of the landscape (left), with schematic representations of the transcriptional outputs and distribution of *Hoxd* gene products in the autopod domain (right). (**B**) When *Hoxd* genes are not expressed, the landscape is in a spatial ground state involving interactions between a subset of regulatory islands, some of which also contact the *Hoxd13* locus. Some islands may display chromatin signatures typical of enhancers (light green). These contacts are, however, insufficient to activate transcription. (**C**) In digits, additional contacts are formed, leading to a fully active conformation paralleled by further changes in histone marks covering the islands (dark green) and the recruitment of RNAPII to active chromatin loops. This architecture ensures a strong expression of *Hoxd13* throughout the whole autopod region. (**D**) Various alterations in the structure of the archipelago, for example in the sequence of the islands, their spacing or relative order, could modify both the amount and spatial distribution of *Hoxd* gene products, thus providing a basis for morphological or pathological variations to occur. Although this model suggests simultaneous contacts, a more dynamic process is also possible. Islands are depicted schematically and do not reflect the behavior of individual elements…." From Spang *et al.* (2015)[41] with permission of Nature Publishing Group. Note that the sketches are projections to a plane of what would actually be a three dimensional structure.

Some transcription factors and transcriptional modulators are quite general and found in all cell types. These are "housekeeping" transcription factors and modulators, and would thus be part of "housekeeping regulons" (Chapter 3). If a gene they affect is exposed and conditions are right, they will work. If the DNA is sequestered, the gene won't be affected because the transcription factor can't bind to it. Each cell type will also have some transcription factors and transcriptional modulators that will be found only in related tissues such as in all mesoderm derived cells (Figure 9.18). Other transcription factors and transcriptional modulators will be tissue type specific, appearing in only a single tissue and only after that tissue is determined. It is this combination of common (housekeeping), less common and specific transcription factors/transcriptional regulators that result in a unique set of transcriptional control for each cell type. Each cell type will also have its own unique compliment of exposed and sequestered enhancer regions and condensed or decondensed genes.

Noncoding RNA can also bind to bound to DNA, transcription factors, and transcriptional modulators and affect transcription rates[43]. We are really only just beginning to appreciate the importance of noncoding RNA. In fact it may be that these noncoding RNAs, paired with transcription factors or transcriptional modulators, are of critical

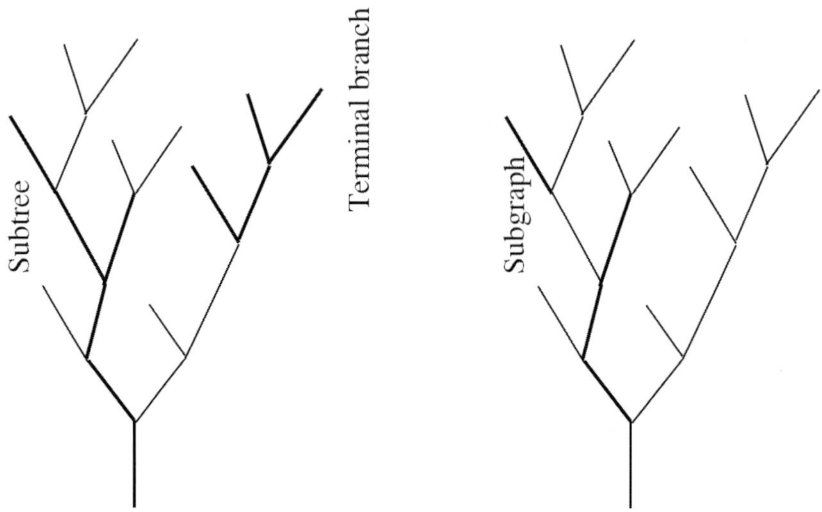

Figure 9.18. A differentiation tree showing a terminal branch and a subtree that is not a terminal branch. A terminal branch is a subtree ending in all terminal edges. Gene exposure is likely to follow subtrees, because a gene may be exposed in one cell type as part of its differon, and sequestered in various later, derived cells, at different later steps of differentiation. Gene expression itself will follow a subgraph of a subtree. The gene expression subgraph need not be a connected subtree itself. From Figure 37 in *Hierarchical Genome*[42]. The investigation of subgraphs and subtrees may then allow us to uncover general rules about how the differon changes when cells change types as they go through steps of determination and differentiation.

importance to all steps of development. These noncoding RNAs may work by guiding transcription factors or transcriptional modulators to their places[44]. This would dramatically increase the context specific binding of transcription factors and transcriptional modulators that are otherwise nonspecific.

There is a set of transcription factors/transcriptional modulators which are proteins that are often called *chromatin remodelers* because of their activity. These chromatin remodelers form large complexes of proteins such as the SWI/SNF Chromatin Remodeling Complex[46-49] (Figure 9.19). The SWI/SNF complex is so named because it was first discovered in the fission yeast, *Saccharomyces cerevisiae*. The complex is required for yeast mating type switching (SWI) and sucrose nonfermenting (SNF) genes. The SWI/SNF complex "recognizes" the pattern of histone acetylation unique to those histones in decondensed DNA that are also

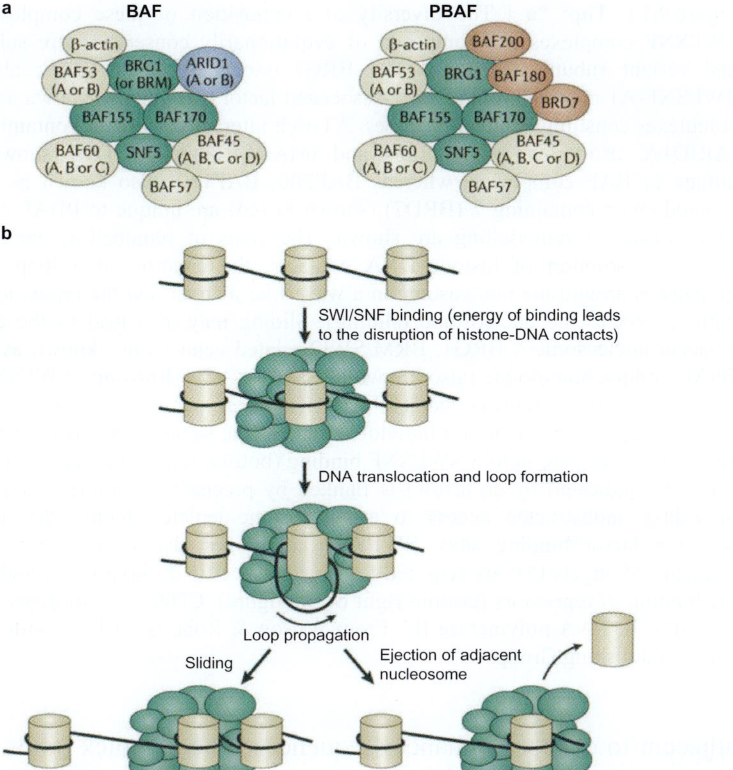

Nature Reviews | Cancer

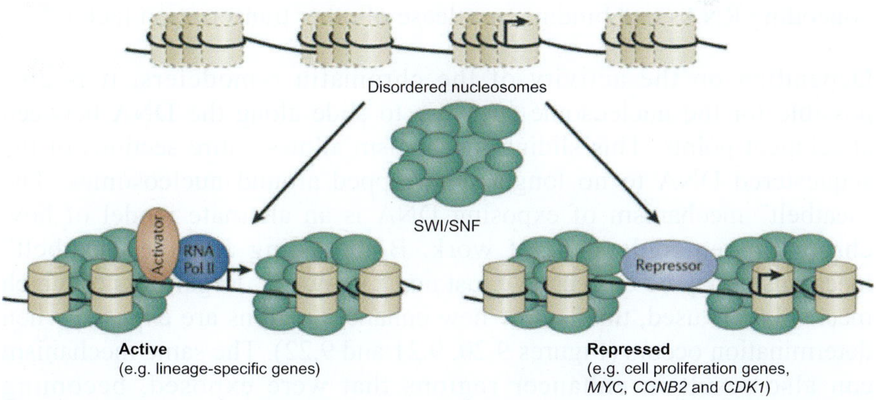

Nature Reviews | Cancer

Figure 9.19. **Top**: "**a** | The diversity of composition of these complexes is shown. SWI/SNF complexes are composed of evolutionarily conserved core subunits (green) and variant subunits (yellow). The BRG1-associated factor (BAF; also known as SWI/SNF-A) and polybromo BRG1-associated factor (PBAF; also known as SWI/SNF-B) complexes constitute major subclasses. AT-rich interactive domain-containing protein 1A (ARID1A; also known as BAF250A and SMARCF1) and ARID1B (shown in blue) are unique to BAF complexes, whereas BAF200, BAF180 (also known as PBRM1) and bromodomain-containing 7 (BRD7) (shown in red) are unique to PBAF complexes. **b** | Mechanisms of remodelling are shown. The steps of remodelling include SWI/SNF binding, disruption of histone–DNA contacts, the creation of a loop of DNA that propagates around the nucleosome in a wave-like manner and the repositioning of DNA with respect to the nucleosome (sliding). Sliding may also lead to the ejection of an adjacent nucleosome[25]. BRG1, BRM/SWI2-related gene 1 (also known as SMARCA4); BRM, brahma homologue (also known as SMARCA2)." **Bottom**: "SWI/SNF complexes are markedly under-represented at silent genes where nucleosomes are poorly positioned and are highly variable across individual cells of the same lineage (top of the figure). At active genes that are rich in SWI/SNF binding (bottom left of the figure) the transcription start site (indicated by an arrow) is flanked by precisely positioned nucleosomes, thus providing unobstructed access to a nucleosome-depleted region that contains transcription factor binding sites. SWI/SNF complexes also contribute to the dynamic silencing of targets that are required for lineage-specific differentiation and that facilitate the binding of repressors (bottom right of the figure). CDK1, cyclin-dependent kinase 1; RNA Pol II, RNA polymerase II." From Wilson & Roberts (2011)[45] with permission of Nature Publishing Group.

adjacent to exposed enhancer sequences. The complex binds to both the histones and the transcription factors on the exposed enhancers and stabilizes the loops of associated DNA. The functioning of the SWI/SNF complex is known to be changed by the binding of specific small noncoding RNAs and binding or release of other transcription factors[49].

Depending on the activity of the chromatin remodelers, it is also possible for the nucleosome structure to slide along the DNA between attachment points. This sliding mechanism allows entire sections of the sequestered DNA to no longer be wrapped around nucleosomes. The "seatbelt" mechanism of exposing DNA is an alternate model of how chromatin remodelers might work. Both sliding and the "seatbelt" mechanism may be correct. We just don't know yet. Regardless of which mechanism is used, this is how new enhancer regions are exposed when determination occurs (Figures 9.20, 9.21 and 9.22). The same mechanism can also result in enhancer regions that were exposed, becoming sequestered. Chromatin remodelers work by exposing or sequestering binding sequence regions of the DNA on the nucleosome[50] (Figure 9.23).

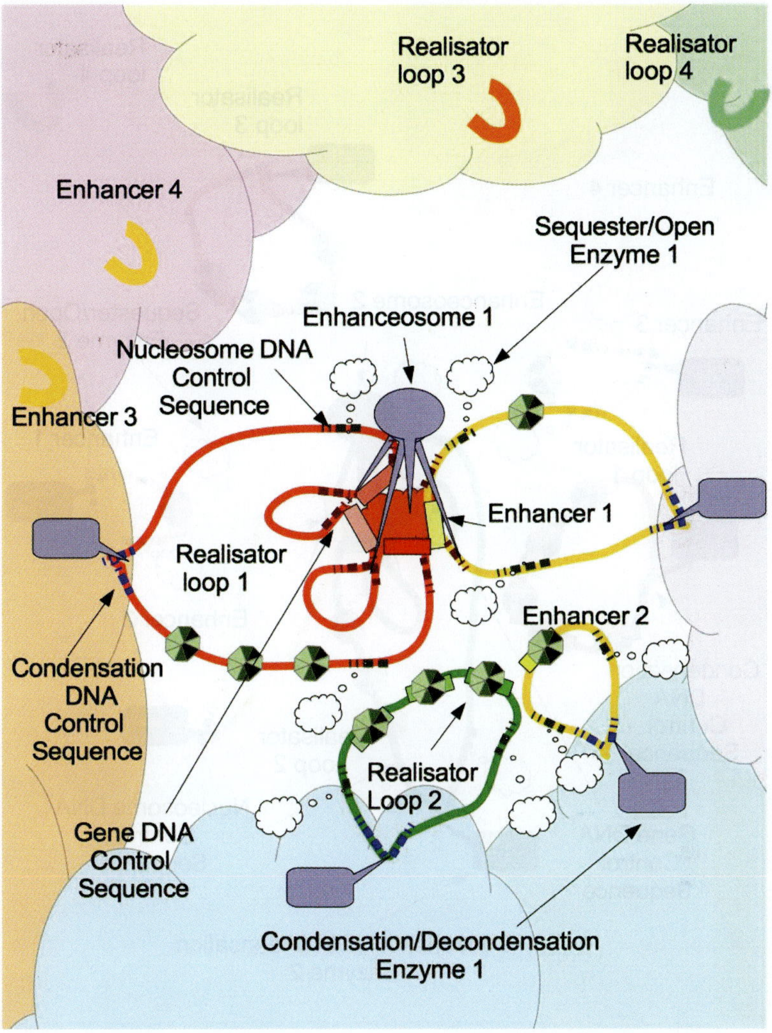

Figure 9.20. The nucleus has areas of chromosome territories (large colored clouds) and regions that are open containing RNA polymerases (small central orange cloud). A specific set of transcription factors (small white clouds) creates an enhanceosome that binds to a specific enhancer. It also binds to a realisator gene that is on its own loop divided into introns and exons. If the gene and enhancer is open, as in realisator loop 1, the DNA can be bound to the enhanceosome if the correct transcription factors are present and form a regulatory archipelago. The gene is then transcribed. If the DNA is sequestered, as in Realisator Loops 2, the DNA cannot be bound to the enhanceosome even if the correct transcription factors are present. If the DNA is condensed (as in Realisator Loops 3 and 4 and their enhancers) it is completely inaccessible.

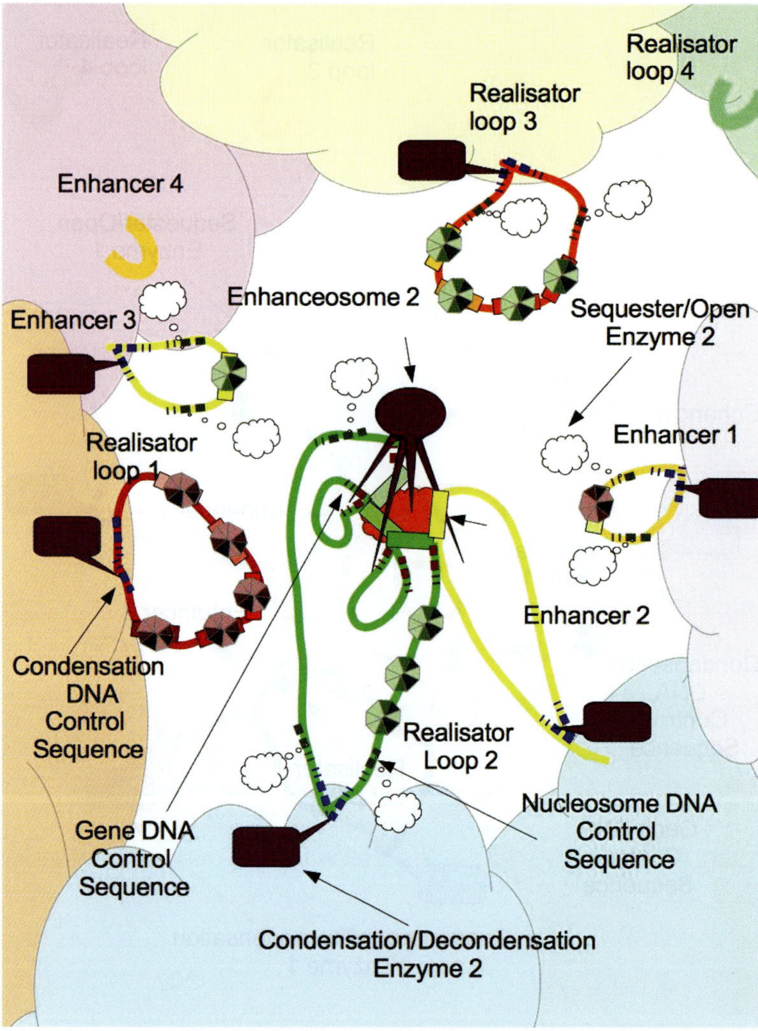

Figure 9.21. A signal from the cell state splitter causes a new set of transcription factors to be made. Nucleosomes in decondensed DNA will slide exposing new genes and old genes will be sequestered. A new enhanceosome is formed and it binds the freshly opened sequences. A new regulatory archipelago is formed. A new gene (Realisator Loop 2 and its enhancer) is transcribed. The portion of DNA required for the next round of determination is decondensed (Realisator Loop 2 and its enhancer). Meanwhile, Realisator Loop 1 is sequestered and histones are changed to mark them for condensation.

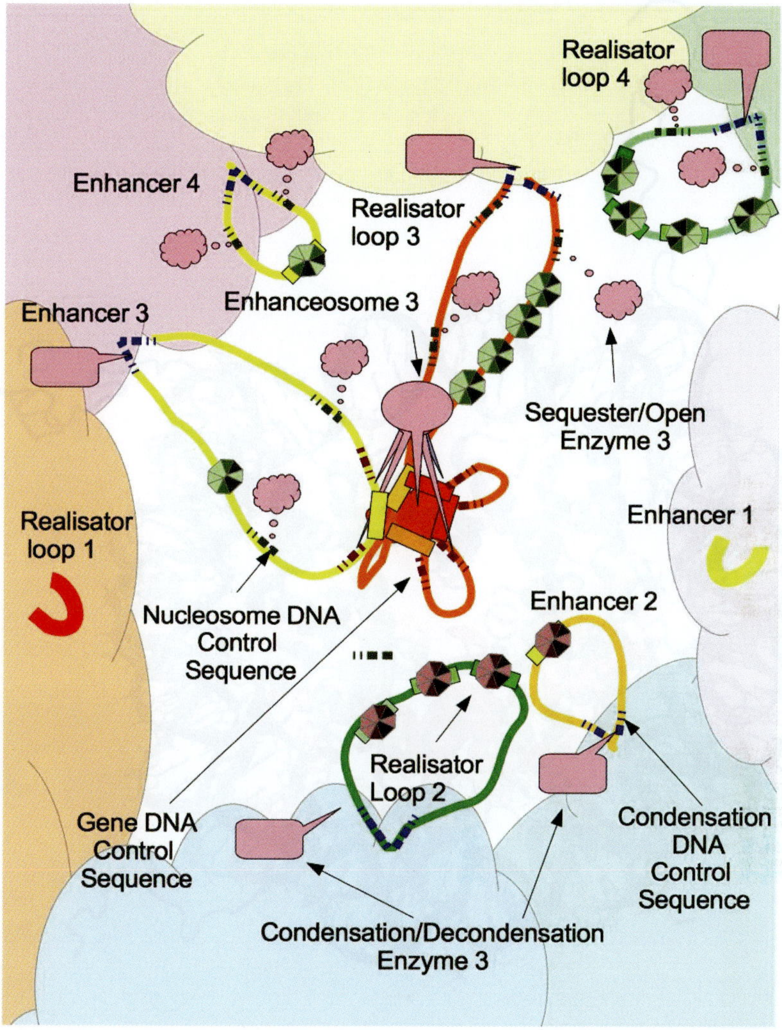

Figure 9.22. In the next step of determination, another signal from the restored cell state splitter causes a third set of transcription factors to be transcribed and translated. They effect the condensation of Realisator Loop 1 and Enhancer 1, sequestration of Realisator Loop 2 and Enhancer 2, opening of Realisator Loop 3 and Enhancer 4, and the decondensation of Realisator Loop 4 and Enhancer 4. We assume that sequestration, nucleosome sliding causing opening and nucleosome sliding causing sequestration, is a one way process. Therefore, the entire genome works as an irreversible ratcheting finite state machine (Figure 9.25).

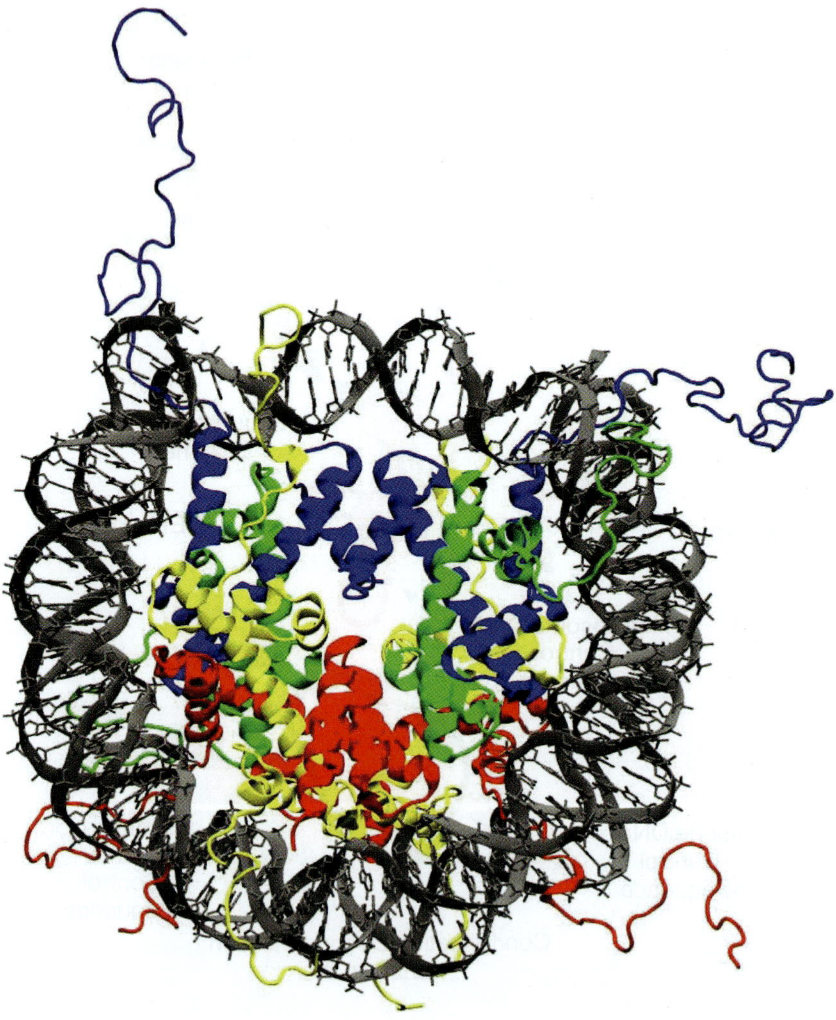

Figure 9.23. Nucleosome structure showing the 8 histone proteins with their tails, which extend past the DNA coiled around them. From Wheeler (2007)[51] with permission under a GNU Free Documentation License, with data from the Protein Data Bank[52] #1KX5 based on Davey *et al.* (2002)[53].

The SWI/SNF complex uses ATP to slide/twist the DNA on the nucleosomes. Once the slide/twist is complete the SWI/SNF complex stabilizes the new chromatin loop structure in its new configuration with the histone octamer bound to a new position. The SWI/SNF complex only binds to nucleosomes whose histones have a specific pattern of histone acetylation. The specific pattern of histone acetylation signals that this DNA has been decondensed and should be exposed. The SWI/SNF complex requires the presence of an enzyme called SAGA to change the pattern of histone acetylation before SWI/SNF binds. SAGA then is a higher order sequestration controller.

Other related chromatin remodeling enzymes include ISWI, INO80, and CHD whose function is primarily to center the DNA properly on the nucleosome guided by specific DNA sequences. The ISWI and CHD enzymes bind to both the DNA on the nucleosome and to adjacent exposed DNA that is not on the nucleosome. Thus, they sense and guide the sliding of the nucleosome along the DNA using ATP and a conformation change to slide the DNA from one DNA binding site to another. Imagine an inchworm with one set of feet on the nucleosome and the other reaching ahead to its binding sequence on the exposed DNA and then pulling the nucleosome along. The *engrailed* sequence is one of the sequences in DNA that chromatin modeling enzymes use to guide sliding. The gene for *engrailed* encodes a transcription factor that binds to any exposed *engrailed* DNA domain and interacts with chromatin modeling enzymes.

All of this information shows that while the genome's nucleotide sequence is one dimensional, it encodes its own 3D, and with the differentiation waves, its own 4D expression pattern (Figure 9.17). The cell-type specific enhancer sequences of exposed regions, the sequestered DNA on nucleosomes at various levels of sequestration along with their histone tail code all serve to organize the genome.

The study of the genetics of the fruit fly *Drosophila* have given us the general hierarchy of transcription factors required for development. The *Drosophila* embryo begins as an egg filled with maternal transcription factors and transcriptional modulators. There is rapid division of the egg's nucleus and then once multiple copies of the nucleus are present, due to mechanically transmitted waves of mitosis[54], these 6000 nuclei move to the exterior of the egg under the membrane. The membrane then comes down like a honeycomb curtain separating each nucleus into separate cells. Once the membrane has dropped the embryo reaches the *Drosophila* version of the midblastula transition and begins expressing its own genes. Gap genes are the genes that code for the first set of transcription factors and transcriptional modulators that are triggered in development. The Gap gene family proteins are present in the embryo's cells near their nuclear membranes. They are triggered during determination to move into the nucleus to the DNA. The trigger comes via the signal transduction pathways connected to the cell state splitter in each cell, or in the case of the *Drosophila* syncytium, each region of active cytoskeleton. Gap gene transcription factors are typically zinc fingers. Zinc fingers are small proteins with a fold in them containing one or more zinc ions stabilizing the fold. There are at least six types of zinc fingers and the Gap gene type zinc fingers in *Drosophila* are of the Cys_2His_2 class (Figure 9.2). The Gap genes act to trigger the "pair-rule" genes. We can therefore imagine Gap genes as initiating nucleosome sliding and exposing the pair-rule genes.

The pair-rule genes are genes like HOX. These are what we originally called the master genes in our earliest models for the nuclear state splitter[1]. All differons are controlled by a specific set of paired master genes which are transcribed and translated into transcription factors. These master genes work in a trans fashion, meaning they can affect many other genes all over the genome. When master genes are turned on, they produce transcripts for both noncoding RNAs and for protein transcription factors. These master genes control a cell type specific set of realisator genes[55-59]. Realisator genes will be organized in expression clusters that all have the same enhancer region encoded in the DNA. Some of these realisator genes are organized into quasi-operons with several genes for a single function (like a metabolic pathway) controlled by one DNA enhancer element. A realisator gene can also be a single gene by itself with the correct enhancer element[60].

The HOX genes are one set of master genes found throughout the genomes of all complex organisms on Earth from jellyfish to plants to humans. All HOX genes have a homeobox sequence (a DNA sequence for binding transcription factors) with an invariant 180 base pair homeodomain along with additional base pairs that are unique for each individual HOX gene. There are also other gene families (including the pair-rule genes of *Drosophila*) that have the homeobox-like domain although they are not considered true HOX genes. Once transcribed and translated, the HOX genes produce transcription factors that bind to and activate a nonrandom set of matching realisator genes.

The proteins produced by HOX genes are not the only DNA sequence elements controlling transcription. The homeobox domain is not specific enough by itself to precisely control all the genes in a differon. Rather, the HOX transcription factors form the core of a protein complex of other transcription factors that bind to tissue specific enhancers in DNA. These complexes work as a unit to activate specific realisator genes with the correct cell-type specific enhancers and only those genes. So the activation of a HOX gene results in a whole cascade of other genes being activated and that gene cascade is tissue specific and based on the DNA code. Of course, this is also contingent on whether or not the gene with the homeobox domain is exposed and therefore has an accessible enhancer.

HOX genes are organized in the genome in a fascinating fashion that provides insight into their functioning. HOX genes are found in clusters along the DNA. In vertebrates there are four clusters. Humans have four clusters of eleven, ten, nine, and nine HOX genes located on chromosomes 7, 17, 12 and 2 respectively. The HOX genes are organized in spatial collinearity meaning they are tandem to one another in a specific and extremely well conserved order along the DNA. What is even more fascinating is that they are accessed and expressed in a matching temporal collinearity and even a rough collinearity in the appearance and positioning of certain anatomical features[61-63]. Beginning from one end of the DNA, HOX genes are numbered HOX1, HOX2 and so on. HOX1 is expressed first during development, then HOX2 is expressed and then HOX3. There is also a quantitative collinearity in that where more than one HOX gene is expressed at the same time, the lower number genes are expressed more in the anterior (head) portion of the animal and the higher number of HOX genes are more expressed in the posterior (tail) regions. The more complex the organism the more tightly organized the HOX gene clusters are. While there are competing models that attempt to explain HOX genes[63-93], we will focus on the biophysical model of Spyros Papageorgiou that has HOX genes being pulled from a chromosome territory to where the transcriptional machinery is active[92]. The process of physically pulling the DNA allows HOX expression in an organized fashion down the gene cluster as it is pulled out of the chromatin domain (Figure 9.24). HOX genes are therefore accessed in an irreversible ratcheting fashion. We now combine this biophysical model with nucleosome sliding to imagine how that might work.

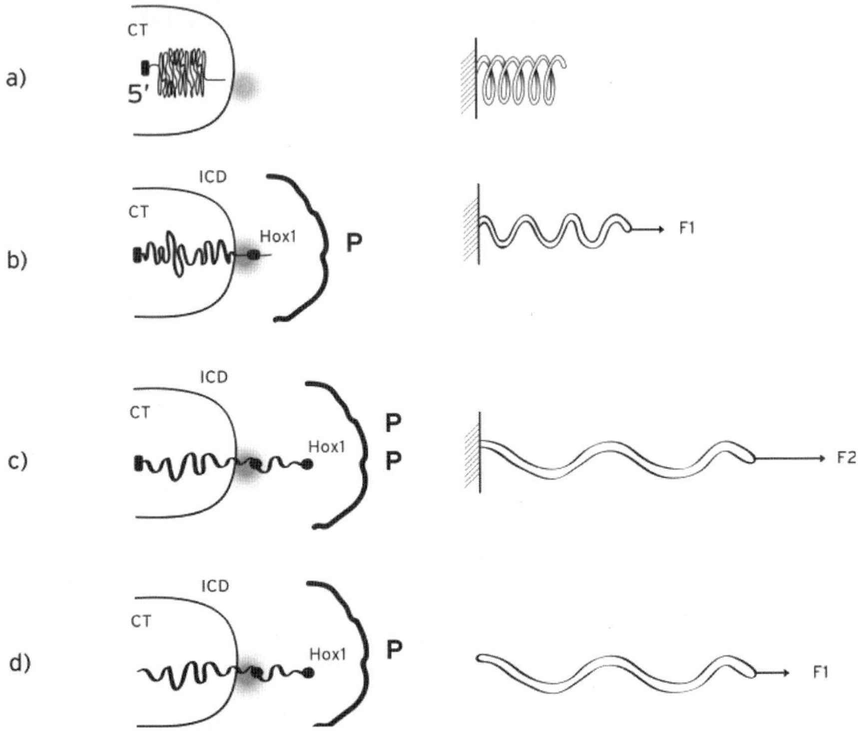

Figure 9.24. "Mechanical analogue of *Hox* cluster decondensation and extrusion. **a**) Left: before activation the *Hox* cluster is condensed inside the chromatin territory (CT) at the posterior 5' end. The transcription factory (TF) is represented by a grey domain. Right: mechanical analogue: an uncharged elastic spring is fixed at its left end. **b**) Left: the cluster is slightly decondensed and *Hox1* (black spot) is extruded in the interchromosome domain (ICD) in the area of the TF. The P-molecules are allocated opposite the cluster. Right: a small force **F1** is applied at the loose end of the spring and expands it slightly. **c**) Left: the cluster is further decondensed and the extruded *Hox2* is located in the transcription factory area. *Hox1* moves off the TF domain and its activation is reduced. Right: a bigger force (**F2** > **F1**) expands further the spring. **d**) Left: the posterior end of the cluster is cut-off. Right: the fixed end of the spring is removed and a smaller force (**F1**) expands and dislocates the spring as in **c** (right)." From Almirantis *et al.* (2013)[92] with permission under a Creative Commons Attribution License.

When the cell state splitter is activated, it undergoes a contraction or expansion. A set of triggered transcription factors and transcriptional modulators moves to the nucleus (Chapter 8). These expose one region of the DNA around a sequestered HOX gene by changing the normal sequestration pattern of histone methylations[94]. The signal to bind and expose the DNA exists on the DNA itself and in the pattern on the histones. Once exposed, one or more transcription factors cause the DNA on the nucleosomes to loosen and the histone octamers to slide along the DNA thereby exposing the correct HOX gene so the RNA polymerase can transcribe it. This produces the protein transcription factor from the HOX gene and associated noncoding RNAs. The transcription factors and noncoding RNAs then bind to their specific matching enhancer regions of DNA with the result that the regions controlled by HOX also become exposed. All the genes required by a specific cell type have their enhancers exposed by the HOX protein and RNAs. The enhancers are bound by HOX triggered transcription factors to form enhanceosomes. The genes are then transcribed.

We also predict the appearance of a specific group of HOX triggered transcription factors whose role is to sequester DNA enhancers from the previous cell type that are no longer required and to sequester the DNA and move it back into the chromosome territories. Differentiation follows as new proteins are transcribed and translated. The last part of differentiation is the preparation of a new set of transcription factors and transcriptional modulators that move outside the nucleus and will be ready to respond to the next cell state splitter event.

With more signaling from the cell state splitter, the next HOX gene is accessed and expressed and its matching set of realisator genes is turned on. In clusters where only one HOX gene is active per cell type, the previous one can be sequestered behind it by nucleosome sliding between HOX genes. The HOX genes are exposed for transcription in the same temporal collinearity as their spatial collinearity along the DNA. If more than one HOX gene is active, then there would be exposing, exposing again, and more exposing of sequestered DNA in an organized fashion without the HOX gene from the previous event being sequestered until some later point in development. Our general model derived from combining the biophysical model for HOX genes with nucleosome sliding is shown in Figures 9.20, 9.21 and 9.22.

When HOX genes (or their partners), are disrupted, normal development fails completely over entire segments of an organism. In the fruit fly, if the HOX genes are disrupted bizarre developmental changes happen. For example, if the HOX gene called *antennapedia* is mutated, the result is that the insect grows an antenna in place of the second leg. It is as if the chromosome sliding kept going past the defective or missing gene and the next HOX gene was activated instead. The wrong differon is then accessed and activated. HOX genes are master genes and must be expressed precisely in order, on the same loop of DNA. This explains why, over evolutionary time, HOX collinearity has been highly conserved. They are too high in the transcription factor hierarchy to be changed without causing a catastrophic and likely fatal outcome.

Nor are these the only ultraconserved sequences in DNA. In addition to enhancers, developmental genes such as *DACH*, *SOX9*, *DLX bigene* and *HOX-D* clusters also have ultraconserved regions (UCRs)[95]. These ultraconserved regions are so well conserved over evolutionary time that the regions for pufferfish match those for mice and humans. The ultraconserved regions are found in introns of genes, next to clusters of genes[96-99], or within "gene deserts". Gene deserts are long stretches of DNA we now know to be used to create the loops of regulatory archipelagos. These ultraconserved regions are associated with transcription factors, especially with transcription factors that bind to homeobox domains. Their possible functions include sites for chromatin rearrangements[98,99] and/or some sort of as yet unknown, noncoding RNA binding. And these transcription factors are those involved in development, differentiation and when things go wrong, they go horribly wrong and the embryo dies. This is why they are so well conserved.

Eggs need to have a selection of transcription factors and transcriptional modulators, and a specific region of decondensed DNA to start with. This is provided by maternal RNA (which can also be made by the egg itself during oogenesis) and the resetting of the epigenetic factors of the genome that occurs at fertilization. The sliding mechanism for nucleosomes, as in HOX genes, looks like it is generally irreversible because the sliding of nucleosomes is always unidirectional. Perhaps, then, the entire genome works as an irreversible bifurcating ratcheting finite state machine[100] (Figure 9.25) with the cell state splitter signaling when to move the ratchet (Figure 9.26). Which way each bifurcation goes is determined by the differentiation wave in which a cell participates, via its cell state splitter.

The genome as irreversible, bifurcating, ratcheting finite state machine gives new insight into stem cells[101,102]. Stem cells are cells set aside early in differentiation and left behind in a quiescent state. In other words, they are cells that have traversed a few steps of the differentiation tree, but then cease participating in differentiation waves, perhaps because they are located at the margins between the trajectories of differentiation waves[103,104]. One can think of each stem cell as if it were a vending machine with a few coins in it, but stuck until one or more additional coins are added (Figure 9.25). Meanwhile the rest of the tissue continues its forward bifurcating genome ratcheting. In the event of injury or a requirement to regenerate, the stem cells are activated and then they continue the forward ratcheting from their pause point, forming replacement cells. This is why stem cells for liver can regenerate liver cells but not brain cells. The path from ectoderm to brain cells was closed off when the cells became early liver cells. The germ cell line, the ultimate in a stem cell, is set aside very early in development. It can

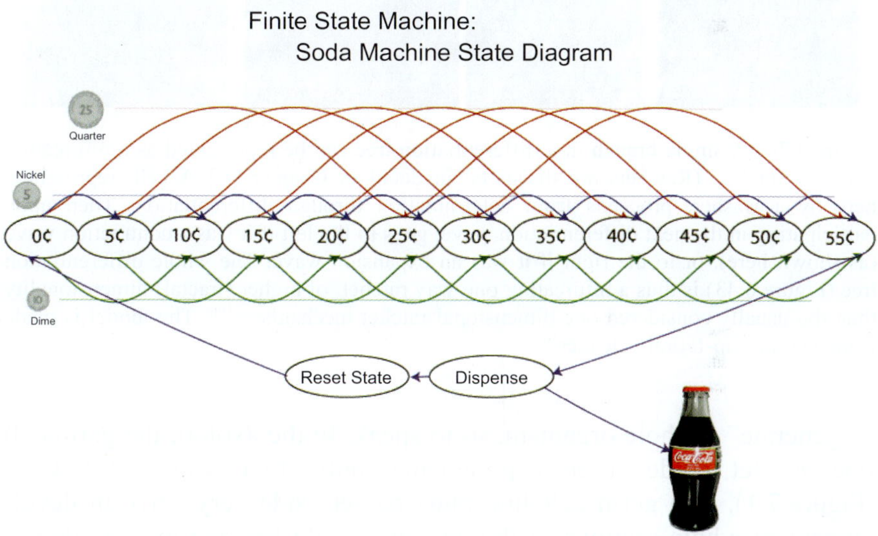

Figure 9.25. An example of the logic of a finite state machine. In this vending machine, there are 12 states. Each time a particular coin is added, a transition occurs to a higher state. Quarter = 25¢, nickel = 5¢, dime = 10¢. Except for the Reset State instruction, this is a one-way, irreversible finite state machine. (This one does not give change if you go over the price of 55¢.) In general: "A finite state machine is a mathematical abstraction used to design algorithms. The machine is in only one state at a time. It can change from one state to another when initiated by a triggering event or condition; this is called a transition. A particular FSM is defined by a list of its states, and the triggering condition for each transition." From Mancuso (2015)[105], widely circulated online and unattributed. If only a few coins are put in, a vending machine simply waits for more, like a stem cell.

Figure 9.26. A single branch in a differentiation tree can be represented as a bifurcating one-way ratchet. (Ratchets are discussed in detail in Chapter 11.) A cell, represented here by a washer, proceeds from determination through differentiation. Then upon participating in the next differentiation wave, goes to the left if it was a contraction wave (as shown here), or to the right if it was an expansion wave. The whole differentiation tree (Figure 6.43) is thus a bifurcating one-way ratchet, of higher (fractal) dimensionality than the usually considered one-dimensional ratchet mechanisms[106]. The model is made from a cutup Zip-It drain cleaner[107].

"regenerate" a whole organism, so to speak. In the axolotl, the germ cell line is set aside after experiencing only four waves $E_1C_2C_4E_2$ (Figure 7.1). The germ cell line must be set aside very early in development or whole sections of the genome would become inaccessible to future generations.

So what about nuclear transplant experiments[108-111]? It is possible to take cells whose genomes are far along in the ratcheting process and reset them by putting them inside an egg. Putting the nucleus inside an egg means the transcription factors present in the egg seek out and expose egg appropriate binding sequences and ignore whatever other sequences are exposed. In doing so, the transcription factors reset the genome of the transplanted nucleus. The resetting is not complete though. The methylation patterns are faulty because they are set for later in development and so cloned animals can have unexpected problems once development is completed.

This model also explains another of those puzzles of cell functioning. Different types of cells have nuclei with different shapes[112,113]. But the shape of the nucleus from one cell to the other of the same tissue type is the same[114-116]. As different loops are pulled in and out of the chromosome territories for transcription, the shape of the nucleus is altered to accommodate these loops.

But we have left something out. Recall the specific transcription factors and transcriptional modulators that respond to the cell state splitter. They translocate to the nucleus in response to signal transduction and they up regulate one set of genes while simultaneously down regulating another set of genes. Prior to each cell state splitter event, two sets of master genes have DNA sequence control elements for each determination event. These master gene control elements receive the activated transcriptional complexes that translocate in response to a contraction or expansion wave. When one of this pair of master genes is up regulated by the wave, the other is downregulated. The reverse occurs for the opposite wave. Thus we call these "differentiation coupled master genes". The complexity of the NFAT transcription factor becomes clearer. NFAT translocation at determination must up regulate one set of genes while simultaneously down regulating another set (Figure 9.4).

Although we present these control elements for coupled master genes as being separate elements we need to recall the lessons from genes like the one responsible for both Prader-Willi and Angelman syndromes (Figure 4.16). The control elements for coupled master genes may be a single DNA sequence that responds differently due to imprinting, splice variations, or other clever RNA linked devices. Also, while we have discussed each round of differentiation as having only one set of master genes, we predict that there will be more than one, and often multiple sets of master genes, working in concert to effect determination events.

With this we can say with reasonable confidence that we have a working hypothesis of differentiation that covers all aspects of embryogenesis. It begins with an egg. The genome is set at a starting point. The methylation patterns on the histones and DNA are all reset to a ground state with maternal and paternal DNA designated by imprinting where required. The egg undergoes multiple rounds of cell division until it reaches a state where mechanical differences can be used to trigger a cytoskeletal based expansion or contraction wave. In regulating organisms this will involve clusters of identical cells that have reached some equivalent of a blastula stage. In mosaic organisms this will occur with the first unequal cell division. (Mammals are a special exception in that they use unequal cell division to cause compaction, forming the trophoblast and inner cell mass, and then go on to form a secondary blastocyst stage in the inner cell mass to form the embryo proper.)

At each stage of differentiation, a bistable cell state splitter is triggered by mechanical signals to cause an expansion or a contraction wave. Which is triggered will depend on the physical properties of the embryo itself. In clusters of cells, the initial mechanical signal can be transferred through a cell sheet via a differentiation wave. A specific set of transcription factors in an individual cell is triggered in response to the kind of wave it just participated in. Paired sets of master genes are bound by these transcription factors and one set of master genes is activated and the other is suppressed, depending on whether the signal corresponds to an expansion or a contraction. Once specific master genes are activated, the DNA configuration is changed to expose a unique differon (unique complement of genes) for each cell type. At each step of differentiation, previous genes that are no longer required are sequestered. The genome is therefore exposed and sequestered in a ratcheting fashion with each wave that the cell or its descendants participate in.

Depending on the sequence of expansions and contractions, unique differons are accessed in cells of each tissue. Which cells experience which waves, is determined by the trajectories of the differentiation waves. This process is irreversible and so very early in development, germ cells must be set aside for the formation of future egg and sperm. Before new cell types arise, in preparation for the next wave, a cell state splitter and matching transcriptions factors, transcriptional modulators, and access to master genes are prepared. Where and when to express genes is the product of mechanical signaling via the cell state splitters and the differentiation waves they propagate. The series of bifurcating changes to the structure of the DNA results in a unique set of genes being exposed in each cell type in an adult organism. Which genes is inherent in the DNA code itself, written into its nucleotide sequence by elements such as enhancers and loop attachment points. This is how the organism gets the right cell, in the right place, at the right time. Embryogenesis is explained.

1. Björklund, N.K. & R. Gordon (1993). Nuclear state splitting: a working model for the mechanochemical coupling of differentiation waves to master genes. *Russian Journal of Developmental Biology* **24**(2), 79-95.
2. Hata, T.S. & T. Herman (2010). Pingry S.M.A.R.T. Team Website. http://www.pingrysmartteam.com
3. Murakami, K.S., S. Masuda, E.A. Campbell, O. Muzzin & S.A. Darst (2002). Structural basis of transcription initiation: An RNA polymerase holoenzyme-DNA complex. *Science* **296**(5571), 1285-1290.
4. Splettstösser, T. (2006). Zinc finger DNA complex. https://commons.wikimedia.org/wiki/File:Zinc_finger_DNA_complex.png
5. Farnham, P.J. (2009). Insights from genomic profiling of transcription factors. *Nature Reviews Genetics* **10**(9), 605-616.
6. Stefan, M.I. & N. Le Novère (2013). Cooperative binding. *PLoS Computational Biology* **9**(6), e1003106.
7. Wikipedia (2015). Cooperative binding. https://en.wikipedia.org/wiki/Cooperative_binding
8. Macian, F. (2005). NFAT proteins: key regulators of T-cell development and function. *Nature Reviews Immunology* **5**(6), 472-484.
9. Opabinia regalis (2015). Enhanceosome. https://en.wikipedia.org/wiki/Enhanceosome
10. Song, K. (2012). Transcription factors of eukaryotic cells. https://commons.wikimedia.org/wiki/File:Transcription_Factors.svg
11. Bubus12 (2013). The moonlighting of beta-catenin. https://commons.wikimedia.org/wiki/File:Beta-catenin-moonlighting.png
12. Kolchanov, N., E. Ignatieva, O. Podkolodnaya, E. Ananko, I. Stepanenko, T. Merkulova, T. Khlebodarova, V. Merkulov, N. Podkolodny, D. Grigorovich, A. Poplavsky & A. Romashchenko (2006). Transcription Regulatory Regions Database (TRRD). http://wwwmgs.bionet.nsc.ru/mgs/gnw/trrd/db_schema.shtml
13. Kolchanov, N., E. Ignatieva, O. Podkolodnaya, E. Ananko, I. Stepanenko, T. Merkulova, T. Khlebodarova, V. Merkulov, N. Podkolodny, D. Grigorovich, A. Poplavsky & A. Romashchenko (2006). Transcription Regulatory Regions Database (TRRD): A source of experimentally confirmed data on transcription regulatory regions of eukaryotic genes. In: *Bioinformatics of Genome Regulation and Structure II*. Eds.: N. Kolchanov, R. Hoffestaedt & L. Milanesi, Springer: 43-53.
14. Kolchanov, N.A., E.V. Ignatieva, E.A. Ananko, O.A. Podkolodnaya, I.L. Stepanenko, T.I. Merkulova, M.A. Pozdnyakov, N.L. Podkolodny, A.N.

Naumochkin & A.G. Romashchenko (2002). Transcription Regulatory Regions Database (TRRD): its status in 2002. *Nucleic Acids Research* **30**(1), 312-317.

15. van den Berg, D.L.C., T. Snoek, N.P. Mullin, A. Yates, K. Bezstarosti, J. Demmers, I. Chambers & R.A. Poot (2010). An Oct4-centered protein interaction network in embryonic stem cells. *Cell Stem Cell* **6**(4), 369-381.
16. Wikipedia (2014). ENCODE. http://en.wikipedia.org/wiki/ENCODE
17. Greulich, K.O. (2005). Single- molecule studies on DNA and RNA. *ChemPhysChem* **6**(12), 2458-2471.
18. Jasencakova, Z., A. Meister, J. Walter, B.M. Turner & I. Schubert (2000). Histone H4 acetylation of euchromatin and heterochromatin is cell cycle dependent and correlated with replication rather than with transcription. *Plant Cell* **12**(11), 2087-2100.
19. Kim, Y.Z. (2014). Altered histone modifications in gliomas. *Brain Tumor Research and Treatment* **2**(1), 7-21.
20. Rando, O.J. (2012). Combinatorial complexity in chromatin structure and function: revisiting the histone code. *Current Opinion in Genetics & Development* **22**(2), 148-155.
21. Kühn, S. (2014). *Organic Codes and their Identication: Is the Histone Code a True Organic Code? [Master's Thesis]*. Matieland, South Africa, Department of Biochemistry, University of Stellenbosch.
22. Kim, S., N.-K. Yu & B.-K. Kaang (2015). CTCF as a multifunctional protein in genome regulation and gene expression. *Experimental and Molecular Medicine* **47**, e166.
23. Klenova, E.M., R.H. Nicolas, H.F. Paterson, A.F. Carne, C.M. Heath, G.H. Goodwin, P.E. Neiman & V.V. Lobanenkov (1993). CTCF, a conserved nuclear factor required for optimal transcriptional activity of the chicken c-*myc* gene, is an 11-Zn-finger protein differentially expressed in multiple forms. *Molecular and Cellular Biology* **13**(12), 7612-7624.
24. Farnham, P.J. (2012). Thematic minireview series on results from the ENCODE project: integrative global analyses of regulatory regions in the human genome. *Journal of Biological Chemistry* **287**(37), 30885-30887.
25. Merkenschlager, M. & D.T. Odom (2013). CTCF and cohesin: linking gene regulatory elements with their targets. *Cell* **152**(6), 1285-1297.
26. Ong, C.-T. & V.G. Corces (2009). Insulators as mediators of intra- and inter-chromosomal interactions: a common evolutionary theme. *Journal of Biology (London)* **8**, #73.
27. Kato, S., K. Inoue & M.-Y. Youn (2010). Emergence of the osteo-epigenome in bone biology. *IBMS BoneKEy* **7**(9), 314-324.
28. Klein, F., P. Mahr, M. Galova, S.B. Buonomo, C. Michaelis, K. Nairz & K. Nasmyth (1999). A central role for cohesins in sister chromatid cohesion, formation of axial elements, and recombination during yeast meiosis. *Cell* **98**(1), 91-103.
29. Merkenschlager, M. (2010). Cohesin: a global player in chromosome biology with local ties to gene regulation. *Current Opinion in Genetics & Development* **20**(5), 555-561.
30. Puck, T.T., P. Webb & R. Johnson (1998). Genome exposure and regulation in mammalian cells. *Somatic Cell and Molecular Genetics* **24**(5), 291-301.
31. Hirano, T. (2010). Unlocking the chromosome. http://phys.org/news/2010-12-chromosome.html

32. Staynov, D.Z. (2008). The controversial 30 nm chromatin fibre. *BioEssays* **30**(10), 1003-1009.
33. Maeshima, K., S. Hihara & M. Eltsov (2010). Chromatin structure: does the 30-nm fibre exist in vivo? *Current Opinion in Cell Biology* **22**(3), 291-297.
34. Fussner, E., R.W. Ching & D.P. Bazett-Jones (2011). Living without 30 nm chromatin fibers. *Trends in Biochemical Sciences* **36**(1), 1-6.
35. Grigoryev, S.A. & C.L. Woodcock (2012). Chromatin organization - The 30 nm fiber. *Experimental Cell Research* **318**(12), 1448-1455.
36. Maeshima, K., R. Imai, T. Hikima & Y. Joti (2014). Chromatin structure revealed by X-ray scattering analysis and computational modeling. *Methods* **70**(2-3), 154-161.
37. Razin, S.V. & A.A. Gavrilov (2014). Chromatin without the 30-nm fiber Constrained disorder instead of hierarchical folding. *Epigenetics* **9**(5), 653-657.
38. Ausio, J. (2015). The shades of gray of the chromatin fiber: Recent literature provides new insights into the structure of chromatin. *BioEssays* **37**(1), 46-51.
39. Travers, A. & G. Muskhelishvili (2015). DNA structure and function. *FEBS Journal* **282**(12), 2279-2295.
40. MBInfo Wiki (2014). DNA-templated transcription. http://www.mechanobio.info/topics/synthesis/go-0006351
41. Spang, A., J.H. Saw, S.L. Jorgensen, K. Zaremba-Niedzwiedzka, J. Martijn, A.E. Lind, R. van Eijk, C. Schleper, L. Guy & T.J.G. Ettema (2015). Complex archaea that bridge the gap between prokaryotes and eukaryotes. *Nature* **521**(7551), 173-179.
42. Gordon, R. (1999). *The Hierarchical Genome and Differentiation Waves: Novel Unification of Development, Genetics and Evolution*. Singapore & London, World Scientific & Imperial College Press: [http://www.worldscientific.com/worldscibooks/10.1142/2755].
43. Herriges, M.J., D.T. Swarr, M.P. Morley, K.S. Rathi, T. Peng, K.M. Stewart & E.E. Morrisey (2014). Long noncoding RNAs are spatially correlated with transcription factors and regulate lung development. *Genes & Development* **28**(12), 1363-1379.
44. Lee, N. & J.A. Steitz (2015). Noncoding RNA-guided recruitment of transcription factors: A prevalent but undocumented mechanism? *BioEssays* **37**(9), 936-941.
45. Wilson, B.G. & C.W.M. Roberts (2011). SWI/SNF nucleosome remodellers and cancer. *Nature Reviews Cancer* **11**(7), 481-492.
46. Vignali, M., A.H. Hassan, K.E. Neely & J.L. Workman (2000). ATP-dependent chromatin-remodeling complexes. *Molecular and Cellular Biology* **20**(6), 1899-1910.
47. Roberts, C.W.M. & S.H. Orkin (2004). The SWI/SNF complex - Chromatin and cancer. *Nature Reviews Cancer* **4**(2), 133-142.
48. Biggar, K.K. & K.B. Storey (2009). Perspectives in cell cycle regulation: Lessons from an anoxic vertebrate. *Current Genomics* **10**(8), 573-584.
49. Liu, N., C.L. Peterson & J.J. Hayes (2011). SWI/SNF- and RSC-catalyzed nucleosome mobilization requires internal DNA loop translocation within nucleosomes. *Molecular and Cellular Biology* **31**(20), 4165-4175.
50. Kulić, I.M., R. Thaokar & H. Schiessel (2006). From sliding nucleosomes to twirling DNA motors. *Journal of Computational and Theoretical Nanoscience* **3**(6), 912-921.

51. Wheeler, R. (2007). Nucleosome 1KX5 colour coded. https://commons.wikimedia.org/wiki/File:Nucleosome_1KX5_colour_coded.png
52. Protein Data Bank (2015). An Information Portal to 112387 Biological Macromolecular Structures. http://www.rcsb.org/pdb/home/home.do
53. Davey, C.A., D.F. Sargent, K. Luger, A.W. Maeder & T.J. Richmond (2002). Solvent mediated interactions in the structure of the nucleosome core particle at 1.9 Å resolution. *Journal of Molecular Biology* **319**(5), 1097-1113.
54. Idema, T. & A.J. Liu (2014). Mechanical signaling via nonlinear wavefront propagation in a mechanically excitable medium. *Physical Review E* **89**(6), #062709.
55. Pradel, J. & R.A.H. White (1998). From selectors to realizators. *International Journal of Developmental Biology* **42**(3), 417-421.
56. Lovegrove, B., S. Simões, M.L. Rivas, S. SotilloS, K. Johnson, E. Knust, A. Jacinto & J. Castelli-Gair Hombría (2006). Coordinated control of cell adhesion, polarity, and cytoskeleton underlies Hox-induced organogenesis in Drosophila. *Current Biology* **16**(22), 2206-2216.
57. Foronda, D., L.F. de Navas, D.L. Garaulet & E. Sánchez-Herrero (2009). Function and specificity of *Hox* genes. *International Journal of Developmental Biology* **53**(8-10), 1409-1419.
58. Choo, S.W. & S. Russell (2011). Genomic approaches to understanding Hox gene function. *Advances in Genetics* **76**, 55-91.
59. Sotillos, S., M. Aguilar & J. Castelli-Gair Hombría (2013). Forces shaping a Hox morphogenetic gene network. *Proceedings of the National Academy of Sciences USA* **110**(11), 4303-4308.
60. Osbourn, A. (2010). Gene clusters for secondary metabolic pathways: An emerging theme in plant biology. *Plant Physiology* **154**(2), 531-535.
61. Dekker, E.J., M. Pannese, E. Houtzager, A. Timmermans, E. Boncinelli & A. Durston (1992). Xenopus Hox-2 genes are expressed sequentially after the onset of gastrulation and are differentially inducible by retinoic acid. *Development*(Suppl.), 195-202.
62. Krumlauf, R. (1992). Evolution of the vertebrate Hox homeobox genes. *BioEssays* **14**(4), 245-252.
63. Duboule, D. (1994). Temporal colinearity and the phylotypic progression: A basis for the stability of a vertebrate Bauplan and the evolution of morphologies through heterochrony. *Development (Cambridge)* (Suppl.), 135-142.
64. Gardiner, D.M. & S.V. Bryant (1996). Molecular mechanisms in the control of limb regeneration: the role of homeobox genes. *International Journal of Developmental Biology* **40**(4), 797-805.
65. Papageorgiou, S. & Y. Almirantis (1996). Gradient model describes the spatial-temporal expression pattern of *Hoxa* genes in the developing vertebrate limb. *Developmental Dynamics* **207**(4), 461-469.
66. van der Hoeven, F., P. Sordino, N. Fraudeau, J.C. Izpisua-Belmonte & D. Duboule (1996). Teleost *HoxD* and *HoxA* genes: comparison with tetrapods and functional evolution of the *HOXD* complex. *Mechanisms of Development* **54**(1), 9-21.
67. Wakahara, M. (1996). Heterochrony and neotenic salamanders: possible clues for understanding the animal development and evolution. *Zoological Science* **13**(6), 765-776.

68. Papageorgiou, S. (1998). Cooperating morphogens control hoxd gene expression in the developing vertebrate limb. *Journal of Theoretical Biology* **192**(1), 43-53.
69. Almirantis, Y. & S. Papageorgiou (1999). Modes of morphogen cooperation for limb formation in vertebrates and insects. *Journal of Theoretical Biology* **199**(3), 235-242.
70. Hérault, Y., J. Beckers, M. Gérard & D. Duboule (1999). *Hox* gene expression in limbs: colinearity by opposite regulatory controls. *Developmental Biology* **208**(1), 157-165.
71. Kmita, M., F. van Der Hoeven, J. Zákány, R. Krumlauf & D. Duboule (2000). Mechanisms of *Hox* gene colinearity: transposition of the anterior *Hoxb1* gene into the posterior *HoxD* complex. *Genes & Development* **14**(2), 198-211.
72. Papageorgiou, S. (2001). A physical force may expose *Hox* genes to express in a morphogenetic density gradient. *Bulletin of Mathematical Biology* **63**(1), 185-200.
73. Kmita, M., N. Fraudeau, Y. Hérault & D. Duboule (2002). Serial deletions and duplications suggest a mechanism for the collinearity of *Hoxd* genes in limbs. *Nature* **420**(6912), 145-150.
74. Reid, A.I. & S.J. Gaunt (2002). Colinearity and non-colinearity in the expression of Hox genes in developing chick skin. *International Journal of Developmental Biology* **46**(2), 209-215.
75. Kmita, M. & D. Duboule (2003). Organizing axes in time and space; 25 years of colinear tinkering. *Science* **301**(5631), 331-333.
76. Chambeyron, S. & W.A. Bickmore (2004). Chromatin decondensation and nuclear reorganization of the *HoxB* locus upon induction of transcription. *Genes & Development* **18**(10), 1119-1130.
77. Duboule, D. & J. Deschamps (2004). Colinearity loops out. *Developmental Cell* **6**(6), 738-740.
78. Papageorgiou, S. (2004). A cluster translocation model may explain the collinearity of *Hox* gene expressions. *BioEssays* **26**(2), 189-195.
79. Wacker, S.A., H.J. Jansen, C.L. McNulty, E. Houtzager & A.J. Durston (2004). Timed interactions between the Hox expressing non-organiser mesoderm and the Spemann organiser generate positional information during vertebrate gastrulation. *Developmental Biology* **268**(1), 207-219.
80. Chambeyron, S., N.R. Da Silva, K.A. Lawson & W.A. Bickmore (2005). Nuclear re-organisation of the *Hoxb* complex during mouse embryonic development. *Development* **132**(9), 2215-2223.
81. Papageorgiou, S. (2006). Pulling forces acting on Hox gene clusters cause expression collinearity. *International Journal of Developmental Biology* **50**(2/3), 301-308.
82. Tarchini, B. & D. Duboule (2006). Control of *Hoxd* genes' collinearity during early limb development. *Developmental Cell* **10**(1), 93-103.
83. Montavon, T., J.F. Le Garrec, M. Kerszberg & D. Duboule (2008). Modeling Hox gene regulation in digits: reverse collinearity and the molecular origin of thumbness. *Genes & Development* **22**(3), 346-359.
84. Morey, C., N.R. Da Silva, M. Kmita, D. Duboule & W.A. Bickmore (2008). Ectopic nuclear reorganisation driven by a *Hoxb1* transgene transposed into *Hoxd*. *Journal of Cell Science* **121**(5), 571-577.

85. Papageorgiou, S. (2009). A biophysical mechanism may control the collinearity of *Hoxd* genes during the early phase of limb development. *Human Genomics* **3**(3), 275-280.
86. Durston, A.J., H.J. Jansen & S.A. Wacker (2010). Review: Time-space translation regulates trunk axial patterning in the early vertebrate embryo. *Genomics* **95**(5), 250-255.
87. Mazza, M.E., K. Pang, A.M. Reitzel, M.Q. Martindale & J.R. Finnerty (2010). A conserved cluster of three PRD-class homeobox genes (*homeobrain*, *rx* and *orthopedia*) in the Cnidaria and Protostomia. *Evodevo* **1**, #3.
88. Papageorgiou, S. (2011). Physical forces may cause *Hox* gene collinearity in the primary and secondary axes of the developing vertebrates. *Development Growth & Differentiation* **53**(1), 1-8.
89. Papageorgiou, S. (2011). Evidence favoring the biophysical model for Hox gene collinearity. http://spyrospapage.blogspot.com/
90. Papageorgiou, S. (2012). Comparison of models for the collinearity of Hox genes in the developmental axes of vertebrates. *Current Genomics* **13**, 245-251.
91. Papageorgiou, S. (2012). *A biophysical model can explain the multiscale phenomenon of collinearity of clustered HOX genes [PowerPoint presentation]*. Silver Bog Research, Embryo Physics Course.
92. Almirantis, Y., A. Provata & S. Papageorgiou (2013). Evolutionary constraints favor a biophysical model explaining *Hox* gene collinearity. *Current Genomics* **14**(4), 279-288.
93. Nolte, C., T.B. Alexander & R. Krumlauf (2015). Mammalian embryo: *Hox* genes. In: *Encyclopedia of Life Sciences*. John Wiley & Sons, Ltd: doi: 10.1002/9780470015902.a9780470000740.pub9780470015903.
94. Mazzoni, E.O., S. Mahony, M. Peljto, T. Patel, S.R. Thornton, S. McCuine, C. Reeder, L.A. Boyer, R.A. Young, D.K. Gifford & H. Wichterle (2013). Saltatory remodeling of *Hox* chromatin in response to rostrocaudal patterning signals. *Nature Neuroscience* **16**(9), 1191-1198.
95. Sandelin, A., P. Bailey, S. Bruce, P.G. Engström, J.M. Klos, W.W. Wasserman, J. Ericson & B. Lenhard (2004). Arrays of ultraconserved non-coding regions span the loci of key developmental genes in vertebrate genomes. *BMC Genomics* **5**, #99.
96. Feldmann, R.J. (2001). *Algorithmic determination of connectrons [US Patent WO2001094542 A2]*. Washington, DC, USA, US Patent Office.
97. Feldmann, R.J. & J.V. Oberthaler (2005). *Flanking DNA sequences that control expression in the human genome [US Patent WO2005035715 A2]*. Washington, DC, USA, US Patent Office.
98. Harmston, N., A. Baresic & B. Lenhard (2013). The mystery of extreme non-coding conservation. *Philosophical Transactions of the Royal Society B-Biological Sciences* **368**(1632).
99. Harmston, N. & B. Lenhard (2013). Chromatin and epigenetic features of long-range gene regulation. *Nucleic Acids Research* **41**(15), 7185-7199.
100. Yeo, Z.X., S.T. Wong, S.N. Arjunan, V. Piras, M. Tomita, K. Selvarajoo, A. Giuliani & M. Tsuchiya (2007). Sequential logic model deciphers dynamic transcriptional control of gene expressions. *PLoS One* **2**(1), e776.
101. Lu, K., T. Cao & R. Gordon (2012). A cell state splitter and differentiation wave working-model for embryonic stem cell development and somatic cell

epigenetic reprogramming. *BioSystems* **109**(3, Special Issue on Biological Morphogenesis), 390-396.

102. Lu, K., R. Gordon & T. Cao (2013). Reverse engineering the mechanical and molecular pathways in stem cell morphogenesis. *Journal of Tissue Engineering and Regenerative Medicine*, doi: 10.1002/term.1672.

103. Gordon, R. (2006). Mechanics in embryogenesis and embryonics: prime mover or epiphenomenon? *International Journal of Developmental Biology* **50**(2/3), 245-253.

104. Gordon, R. (2011). Epilogue: the diseased breast lobe in the context of X-chromosome inactivation and differentiation waves. In: *Breast Cancer: A Lobar Disease*. Ed.: T. Tot. London, Springer: 205-210.

105. Mancuso, E. (2015). Tired of conditionals? State pattern could help. Tips, hints and reflexions about designing a State Machine. http://bits.citrusbyte.com/state-design-pattern-with-ruby/

106. Alicea, B. (2013). Ratchets in Nature: a short review. http://syntheticdaisies.blogspot.com/2013/02/ratchets-in-nature-short-review.html

107. Luoma, G. (2015). Zip-It. http://zipitclean.com

108. Gurdon, J.B. (1962). Adult frogs derived from the nuclei of single somatic cells. *Developmental Biology* **4**(2), 256-273.

109. Gordon, R. & G.W. Brodland (1989). Cell state splitters cause primary neural induction and differentiation [+ poster: Gordon vs Gurdon]. *Cell Differentiation and Development* **27**(Supplement 1: From DNA to Body Plan, Advances in the Field of Human, Animal and Plant Development: Abstracts of the 11th International Congress of the International Society of Developmental Biologists, Utrecht, The Netherlands, August 20-25, 1989), S69.

110. Johnson, M.H. & J. Cohen (2012). Reprogramming rewarded: the 2012 Nobel prize for Physiology or Medicine awarded to John Gurdon and Shinya Yamanaka. *Reproductive Biomedicine Online* **25**(6), 549-550.

111. Gurdon, J.B. (2013). The egg and the nucleus: A battle for supremacy (Nobel lecture). *Angewandte Chemie-International Edition* **52**(52), 13890-13899.

112. Gerner, C. & G. Sauermann (1999). Nuclear matrix proteins specific for subtypes of human hematopoietic cells. *Journal of Cellular Biochemistry* **72**(4), 470-482.

113. Collings, D.A., C.N. Carter, J.C. Rink, A.C. Scott, S.E. Wyatt & N.S. Allen (2000). Plant nuclei can contain extensive grooves and invaginations. *Plant Cell* **12**(12), 2425-2440.

114. Webster, M., K.L. Witkin & O. Cohen-Fix (2009). Sizing up the nucleus: nuclear shape, size and nuclear-envelope assembly. *Journal of Cell Science* **122**(10), 1477-1486.

115. Le Douarin, N.M. & F. Dieterlen-Lièvre (2013). How studies on the avian embryo have opened new avenues in the understanding of development: A view about the neural and hematopoietic systems. *Development, Growth & Differentiation* **55**(1), 1-14.

116. Wikipedia (2015). Nicole Marthe Le Douarin. https://en.wikipedia.org/wiki/Nicole_Marthe_Le_Douarin

Chapter 10

Irritable Protoplasm: Forerunners to Differentiation Waves

We live in the era of the ascendancy of molecular biology. In this book we are turning the idea that molecules control development on its head, and proposing that development controls the molecules of which we are made. Of course, both mechanics and molecules are involved, but we eschew the idea that molecules do it all by themselves. The idea that waves are involved in embryogenesis has a reasonably long history, which we shall now review to show that differentiation waves fit into a more general framework of embryo research.

An old fashioned criterion for life was "Does it wiggle if we poke it?" This used to be called "irritability", and the careful experimental study of muscle contraction in frogs dates to Albrecht von Haller, first published in 1753[1]. Luigi Galvani, who worked with his wife Lucia, demonstrated in 1781 that the concept of "Hallerian irritability" involves both electrical and mechanical phenomena[2] (Figure 10.1). Outside of studies of muscle contraction and nerve conduction[3] (Proposition 205 in *Hierarchical Genome*[4]), the concept of irritability was only rarely discussed[5,6]. It appears to have been introduced into embryology by Thomas Hunt Morgan by 1906[7-9], and indirectly in the work of Ralph S. Lillie[10].

Lillie gave a clear description linking irritability to waves of membrane permeability changes in 1911[11] (Figure 10.2). He ended citing earlier work on permeability changes in eggs, but did not explicitly link that to waves on the eggs.

The first tentative observations of a wave on an embryo started in 1877[17]. These were later reviewed by Ernest Just in 1919[16] (Figure 10.3). He pointed out[18] that in eggs of the sand dollar *Echinarachnius*: "...one can follow the wave in the ectoplasm [i.e., the cell cortex, not the ectoplasm of the movie *Ghostbusters*[19]] which begins at the point of sperm-entry

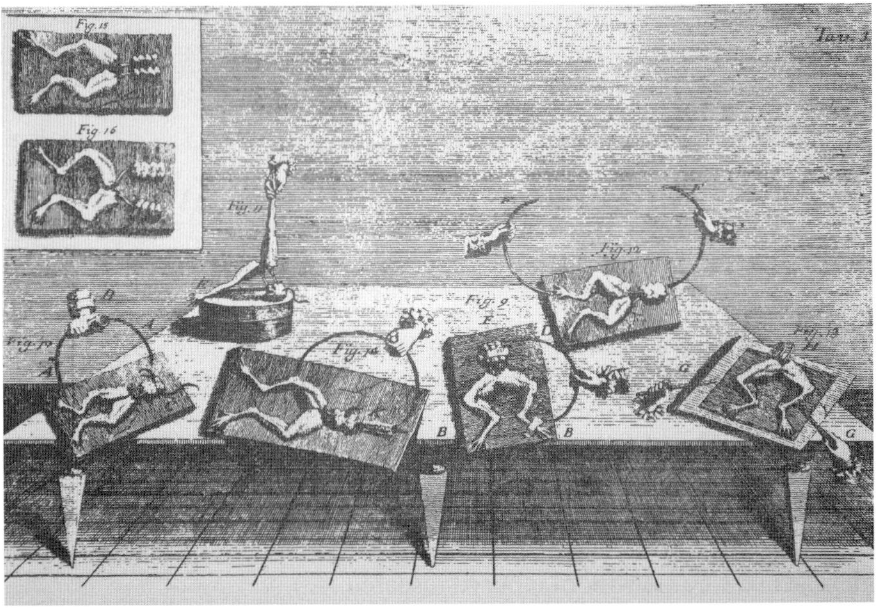

Figure 10.1. Luigi Galvani stimulated frog legs with an arc consisting of two different metals in contact within the grasping hand shown here. As a result the legs spread apart and twitched. The generation of current by two dissimilar metals in contact was later used by Alessandro Volta to invent the electric battery. From Galvani (1791)[12], cf. English translation[13].

and sweeps over the egg. As this wave progresses, it renders the egg immune to the entry of supernumerary attached spermatozoa." Toki-o Yamamoto used eggs of the lamprey[20] and an aquarium fish called medaka (*Oryzias latipes*), noting that the breakdown of the cortical alveoli just under the cell membrane of the egg takes place in a wave-like manner. This wave begins near the animal pole and ends at the vegetal pole. The same kind of wave was observed when the egg was pricked, but started from the site of pricking. The activation wave on a one-cell embryo traverses the whole sphere, and does so no matter where the activation begins, travelling to the opposite side of the egg, quitting there[21]. By centrifuging some of the alveoli away from the membrane, Yamamoto proved that fertilization involves "an invisible wave of some sort", i.e., not caused by the alveoli, and postulated that it is a "wave of excitation" involving calcium ions[22], later described as a propagating impulse[23,24]. Floyd Moser pricked unfertilized eggs of the sea urchin

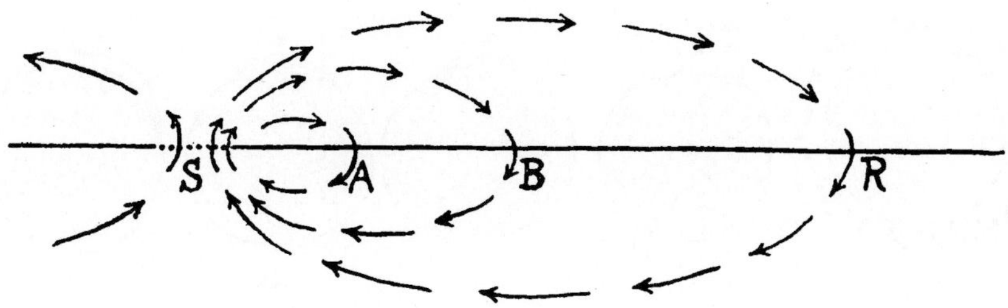

Figure 10.2. The straight line represents an irritable or excitable membrane. The polarization breaks down at patch S, allowing ions to flow across the membrane at that point. This generates an electric (ionic) current that is sufficiently intense and lasts long enough, out to point R, to change the permeability of the membrane at points between S and R. As a result, the process propagates from left to right. From Lillie (1915)[14] with kind permission of Martin Mould, American Physiological Society, and in public domain. While the process, as drawn, can equally propagate right to left, once the distance SR has been depolarized, if the depolarization lasts long enough, there will be no propagation from R towards S, i.e., in general the process proceeds in one direction. In Lillie's words: "…the effect of a local stimulus is never confined to the region directly stimulated but spreads to adjacent regions. In other words, the region already in a state of excitation excites the adjoining regions, so that an excitation wave travels from point to point along the irritable element at a more or less rapid rate. On the present theory the wave of excitation is accompanied by, or equivalent to, a wave of depolarization, i.e., of increased permeability."

Arbacia and he also observed a "wave of cortical granule breakdown" in 1939[25,26]. The calcium component of the wave was made visible by Lionel Jaffe in 1978[27,28].

Mechanical phenomena also accompany the fertilization wave. Edwin Conklin, in 1905, observed a rapid movement of the peripheral cytoplasm within 5 minutes after fertilization of a sea squirt (Figure 10.4). This egg has small "test cells" that apparently adhere to the outer surface of the egg. Their simultaneous downward movement suggests that the cell membrane actually contracts around the sperm entrance point, which must be compensated by membrane expansion of the upper hemisphere. This may correspond to what was later called the "activation contraction"[29] in frogs, depicted in Figure 10.5. There is perhaps a 50% increase in membrane surface area during activation. Undoubtedly much of this increase is because the membrane is initially wrinkled (since the egg remains spherical). It is possible that the exocytosed cortical granule membranes[30] are added to the egg membrane upon activation (Figure 10.4: bottom right). While the activation contraction wave in the

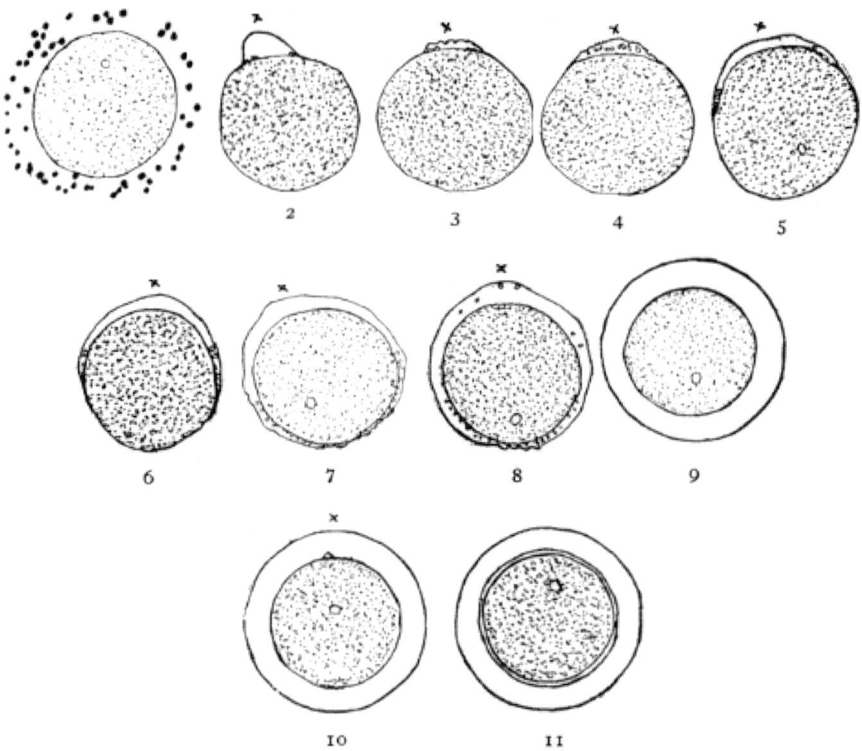

Figure 10.3. First published, by Ernest Just in 1919, this sequence depicts the propagation of the fertilization wave starting at the sperm entry point. The lifting of the vitelline membrane is preceded by a then invisible wave of change of the cell surface preventing polyspermy. The egg used here is the sand dollar *Echinarachnius parma* of diameter[15] 135 μm. "*Before the actual elevation of the membrane, some cortical change beginning at the point of sperm entry* [marked x] *sweeps over the egg immunizing it to other sperm; the direct opposite pole of the site of sperm entry is the last point affected*.... 1: Freshly shed egg. Its pigmented jelly is as yet unswollen by the sea water. 2: An egg ten seconds after the disappearance of the sperm head within the egg. 3: Same. 4: Four seconds after the membrane began lifting. Free vesicles beneath the membrane. 5: About two seconds later than #4. 6: Two to three seconds later than #5. Note vesicles forming. 7: About two seconds later than #6. Membrane incomplete at pole opposite entrance-point of the sperm. 8: Twelve seconds after beginning of membrane lifting. The membrane is fully formed but still wrinkled, especially at point at which it lifted last. 9: Two minutes after insemination. Membrane is fully rounded, equidistant at all points from the now spherical egg. 10: About same time as in #9. Cone is clearly visible. 11: Ten minutes after insemination. Egg with hyaline plasma layer formed." From Just (1919)[16] with kind permission of the Editors of the *Biological Bulletin* and in the public domain.

pigmented cortex has received the most attention by theoretical biologists[31,32], none have modeled the mechanical contraction of the cortex or the expansion of the membrane in the lower half (Figure 10.4).

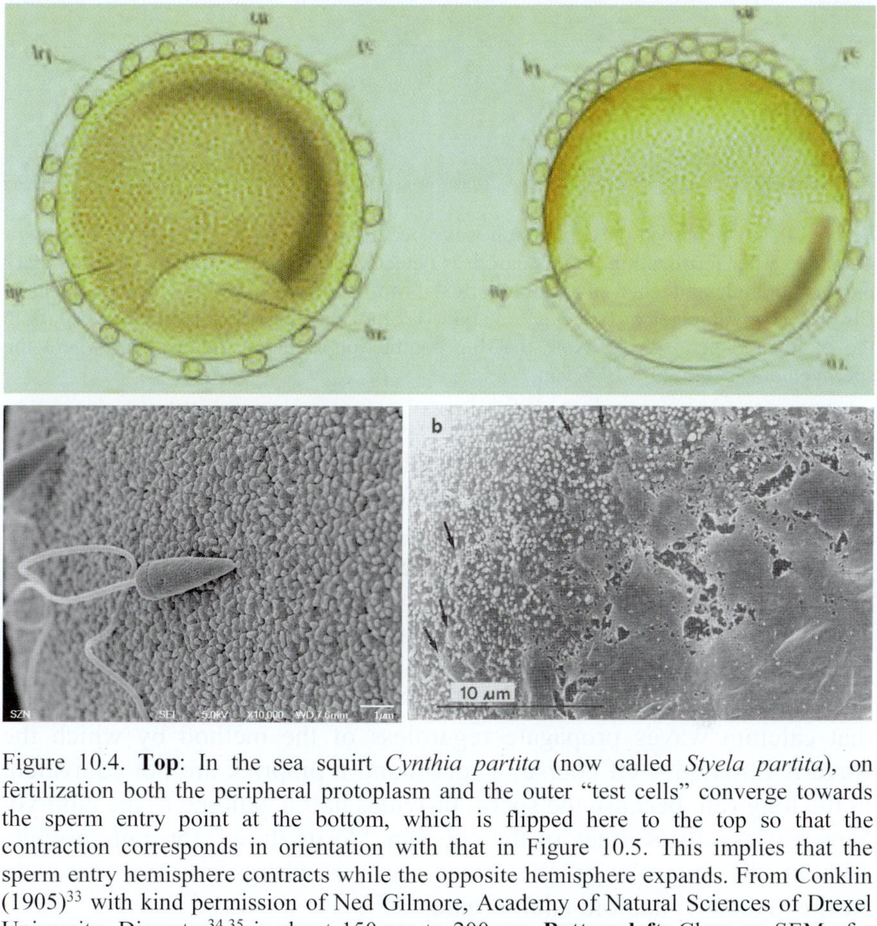

Figure 10.4. **Top**: In the sea squirt *Cynthia partita* (now called *Styela partita*), on fertilization both the peripheral protoplasm and the outer "test cells" converge towards the sperm entry point at the bottom, which is flipped here to the top so that the contraction corresponds in orientation with that in Figure 10.5. This implies that the sperm entry hemisphere contracts while the opposite hemisphere expands. From Conklin (1905)[33] with kind permission of Ned Gilmore, Academy of Natural Sciences of Drexel University. Diameter[34,35] is about 150 µm to 200 µm. **Bottom left**: Close-up SEM of a sea urchin egg just before fertilization by sperm, with kind permission of Luigia Santella. The egg surface area is greatly enlarged compared to a sphere by the membrane around its protruding microvilli. **Bottom right**: During propagation of the fertilization wave the microvilli flatten. The arrows point to possible membrane additions by exocytosed cortical granules. This flattening would expand part of the membrane. From Dale *et al.* (1989)[36] with permission of John Wiley & Sons.

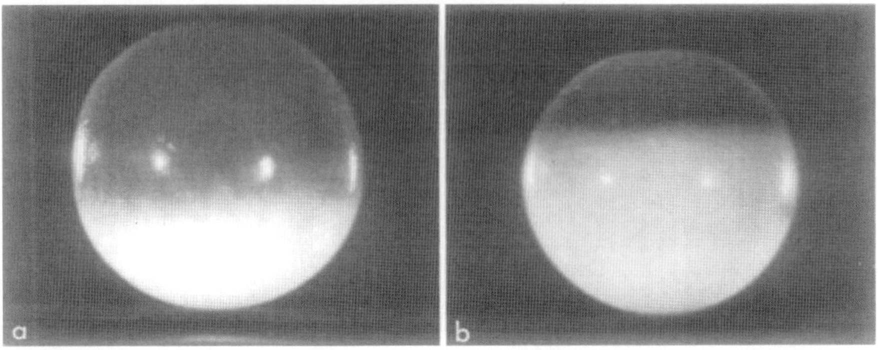

Figure 10.5. The activation contraction wave occurs in the egg of the frog *Rana pipiens* 4 minutes after fertilization. In this case it is detectable by the upward motion of the dark pigment granules attached inside the cortex. From Elinson (1975)[37] with permission of Elsevier. The diameter of this egg[38] is about 1.7 mm. In this case sperm entry occurs in the pigmented hemisphere. Note that the contraction is in the hemisphere where the sperm enters as in Figure 10.4 Top.

We thus see that irritability in one-cell embryos consists of a triggering event, followed by propagation of a calcium wave that is accompanied in the cases examined so far by a mechanical contraction wave of the cortex affecting the hemisphere in which sperm entry occurs. Simultaneously the opposite hemisphere undergoes a wave of expansion. Since both sperm and pinpricks can start some of these waves, the triggering and the propagation of a wave are independent phenomena. Ernest Just called the propagation "independent irritability"[18,39,40]. Mathematical models show that calcium waves propagate regardless of the method by which the wave is initiated[41]. Of course, a sperm and a pinprick are not equivalent in the long run, because the sperm provides half a genome to the embryo. Nevertheless, the surface of the egg can "do its thing" for quite a while without the sperm's genetic input, even resulting in some cases in parthenogenetic, haploid adult organisms[42].

Here we have focused on the first wave discovered, the activation or fertilization wave. Many other waves on amphibian embryos have been discovered, in large part due to the 1970s work of Koki Hara:

1. Associated with the fertilization or activation wave (Figure 5.60), are propagating "dark-light-dark zones" starting from the insemination site to the opposite side of the egg[43].

2. Contraction of the pigmented hemisphere occurs ("activation contraction")[43].
3. A wave of cortical rigidity propagates through the pigmented hemisphere[44].
4. Cortical rotation may be thought of as a wave[45,46].
5. Post-fertilization waves occur[47].
6. Cleavage associated surface contraction waves occur[48].

These waves are summarized in Figures 10.6 and 10.7 and have been reviewed in detail[30,49]. Some of them are related to internal events, such as calcium fluxes or growth of the sperm aster (microtubules emanating from a microtubule organizing center[44]. In general, except for the activation wave, the calcium waves observed in invertebrate one-cell embryos have yet to be correlated with these various optical and physical waves of vertebrate amphibians (Figures 5.60, 5.61, 5.62 and 10.3). Correlations of any of these waves with changes in expression of maternal RNAs[50] have yet to be made. Alterations in these one-cell waves are caused by the human reproductive technology ICSI (intracytoplasmic sperm injection) and may result in prevention of the cortical flash (Figure 5.62). The possible consequences, if any, to later development are under investigation[51].

These one-cell embryo waves are followed by a series of cleavage waves that result initially in synchronized cell divisions, with eventual loss of synchrony by gastrulation[52,53,50]. The cleavage waves alternate with surface expansion waves, in the top ("animal") hemisphere, and slight contractions of the surface in the bottom ("vegetal") hemisphere[56]. There are many other calcium waves in organisms[57]. These other calcium waves have a wide range of speeds that vary systematically with temperature[58,55,59,60]. The various waves in embryos[61] fall neatly onto these plots[55] (Figure 10.8).

These waves are now understood mathematically to be what are called in general waves in active/excitable media[62]. Waves on water are not excitable waves. Water waves do not occur unless some force is applied to the water, by wind, or by a boat making a wake, or by an earthquake generating a tsunami. The water is passive. A wave in an active medium gets its energy from the medium through which it is propagating. A simple example is a forest, grass or brush fire, which can propagate from plant to plant, obtaining its energy from their combustion (Figure 10.9).

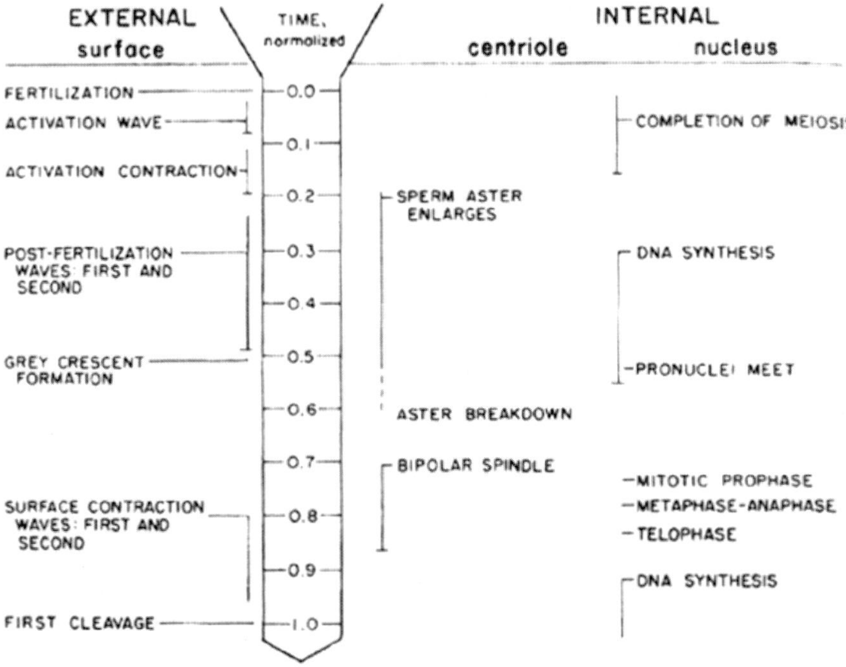

Figure 10.6. At least 7 waves over the surface of eggs have been reported from the time of fertilization to the first cleavage or cell division of the egg into two cells. From Kirschner & Gerhart (1981)[54] with permission of Oxford University Press. "Grey crescent formation" is the same as cortical rotation (Figure 2.28).

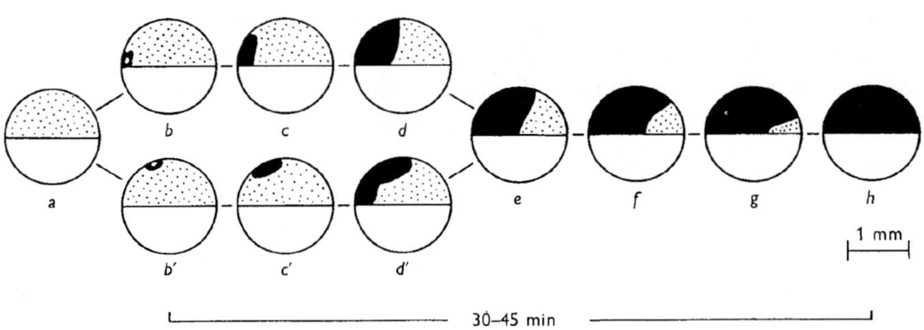

Figure 10.7. The change in rigidity of the pigmented hemisphere of the frog egg *Rana nigromaculata*, shown here, has not been called a wave, but it does propagate, and therefore qualifies as such. It's starting point, the sperm entry point, is variable, as indicated by the two cases, but it nevertheless ends up covering just the one hemisphere. From Kubota (1967)[44] with kind permission of the Company of Biologists.

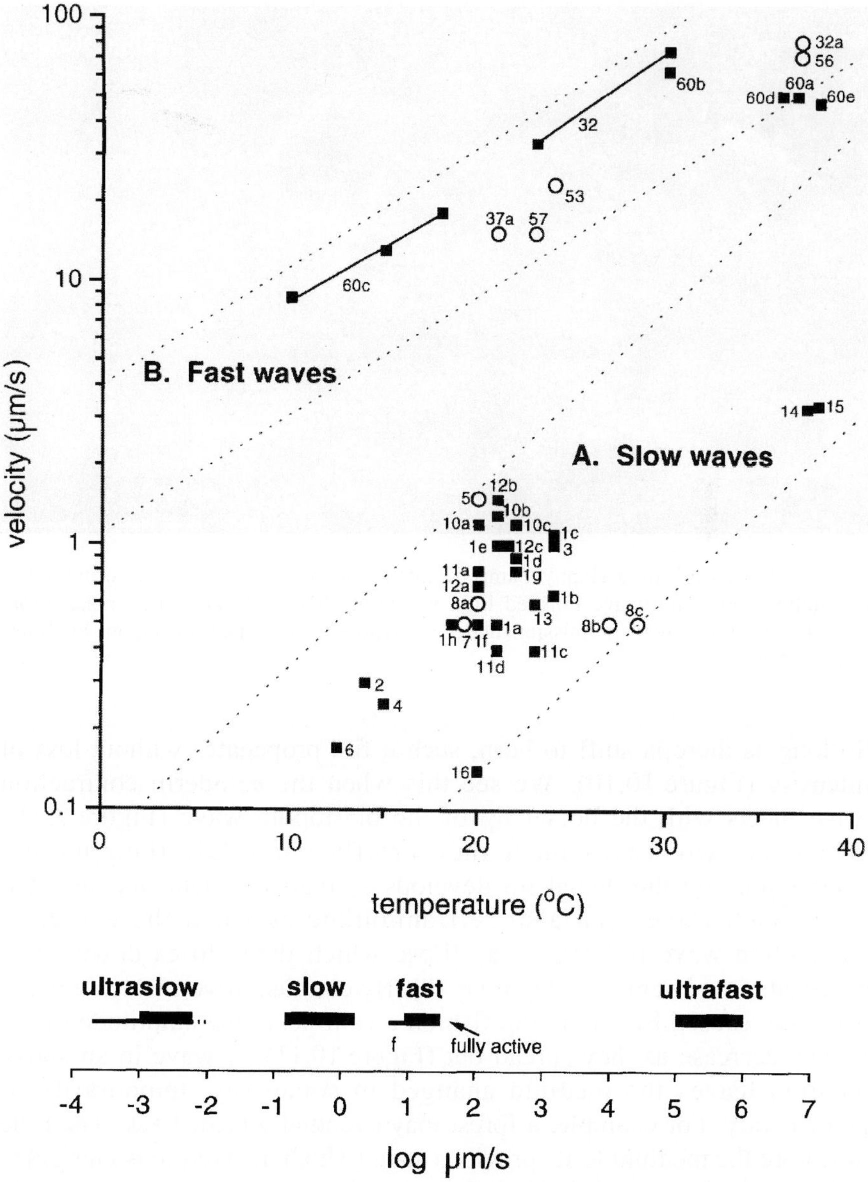

Figure 10.8. **Top**: Speeds of various "calcium" waves in an assortment of organisms and embryos versus temperature. **Bottom**: Calcium waves fall into four categories, with all data here at "room" temperature. The ectoderm contraction wave, #16 (Figure 6.18), is an ultraslow wave. From Jaffe & Créton (1998)[55] with permission of Elsevier.

Figure 10.9. A wildfire at Happy Camp, Klamath Falls National Park, California, taken by satellite using shortwave infrared light to see the flames through the smoke. From DigitalGlobe Inc.[63] with permission under a Creative Commons Attribution 4.0 International License.

So long as there is stuff to burn, such a fire propagates without loss of intensity (Figure 10.10). We see this when the ectoderm contraction wave meets with the dorsal lip of the blastopore wave (Figure 7.22). When the two waves meet they briefly annihilate their mutual contraction and the dorsal lip develops a small transient uncontracted zone. Such waves can also self annihilate as when the ectoderm contraction wave reforms as a ellipse which then closes down, self-annihilates, and vanishes (Figure 6.18). By contrast, if you watch a water wave say caused by a jumping fish on a calm lake, the amplitude of the ripples decrease as they spread out (Figure 10.11). A wave in an active medium leaves the medium changed in some way, temporarily or permanently. For example, a forest may eventually grow back. The time to restore the medium to its previous state (which also requires energy) is called its "refractory period". A forest's refractory period is measured in decades. For nerve fibers carrying the nerve "impulse", the refractory period can be around a millisecond[66]. When there is a sequence of different kinds of active waves going through a medium, such as the waves listed above that consecutively traverse the same one cell embryo,

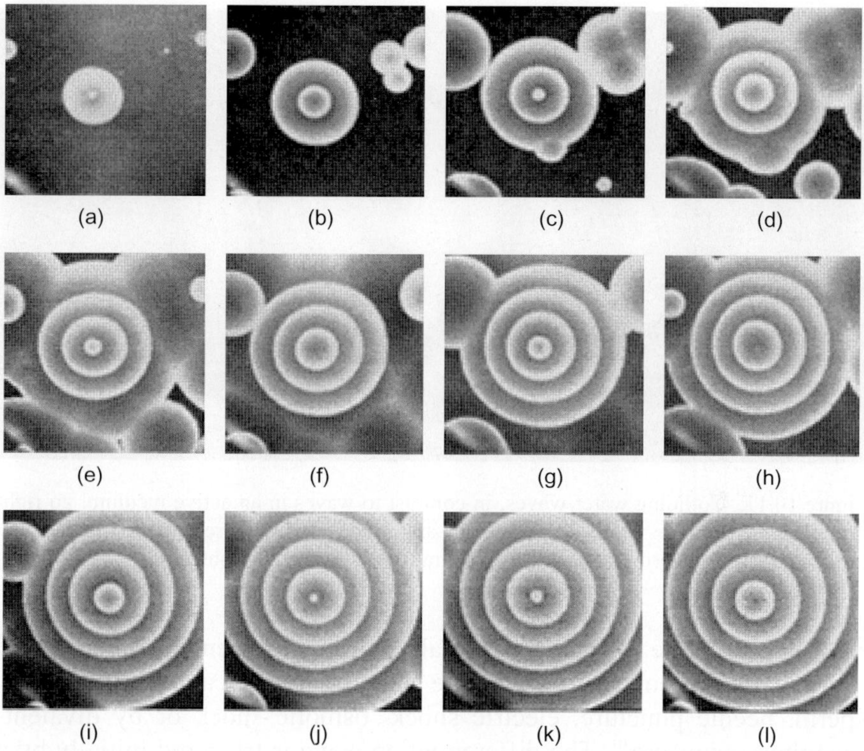

Figure 10.10. A wave in an excitable medium propagates without loss of amplitude. This one is triggered repeatedly, with a short refractory period, giving rings of waves that propagate outwards from the trigger point. Note in the upper left hand corner two waves annihilating when and where they meet, as apparently happens with differentiation waves. From Zhabotinsky (2007)[64] with kind permission of Scholarpedia. Rings in a comparable experiment are 0.5 cm apart[77].

the medium, the cell surface, never returns to a previous condition or state, so the changes are permanent on the time scale of the life of the organism. Alternatively, we may think of the refractory period for each wave as equal to the generation time for the species, for each wave recurs in the next generation.

Again, for all excitable media, we distinguish the mechanism of the trigger from that of the propagation of the wave. A forest fire can be triggered by lightning, volcanic eruption, a struck match, discarded cigarette, or arson. However triggered, the fire propagates itself. This

Figure 10.11. Colliding water waves, in contrast to waves in an active medium, go right through one another, while decreasing in amplitude as they spread. From Davidhazy (2013)[65] with kind permission of the photographer, Andrew D. Johnson.

distinction, while seemingly trivial, has often been ignored in the investigation of embryogenesis. The activation wave can be triggered by sperm, needle puncture, electric shock, osmotic shock or by divalent-ionophore exposure[30]. The differentiation wave is triggered initially by a mechanical event. It then moves from cell to cell across a sheet of cells. The trigger event and the propagation events are separate phenomena. Fire, seen as a wave in an active medium, stops where and when it runs out of fuel. The differentiation waves are limited by self-annihilation or upon reaching cells that do not have a cell state splitter ready. Adjacent tissues that do not have a cell state splitter ready will not propagate the wave. It will stop at the tissue boundary. If a differentiation wave runs into cells participating on its arrival in another differentiation wave, it also stops, because a cell state splitter can only participate in one wave at a time.

In summary, waves on embryo surfaces are a mix of mechanics, chemistry, and electrical currents that are involved in the embryo surface being an excitable medium.

The following observations by Koki Hara require confirmation:

1. The side where the sperm has entered the *Xenopus* egg becomes the ventral side of the embryo[21], suggesting, as we have seen in Figure 2.21, that the pigment in the pigmented hemisphere has no direct involvement in embryogenesis.
2. Nevertheless, no matter where the sperm entered, the activation contraction wave involves only the pigmented hemisphere (as in Figure 10.5).

The observations suggest that the activation wave of fertilization is not dominated by mechanics, whereas the contraction of the pigmented cortex (activation contraction) is dominated by mechanics. Perhaps the contraction of the pigmented cortex is caused by contraction of the actin microfilaments attached to the cell membrane[67] and this is normally used for moving the sperm inside the cortex (see Chapter 5, and Figure 5.60). Furthermore, triggering of the activation contraction wave is not directly caused by sperm entry[37]. The halting of the cortical rigidity wave once it has traversed the pigmented hemisphere fits no present model for wave propagation. We need a lot more observations and experiments on one-cell embryo waves. We also do not at this time have any more confirmation that there is an electric component to differentiation waves other than the speculation that they are calcium waves[68,61]. Tools are now available to distinguish biochemically versus mechanically propagated waves. Using such tools, it has been recently shown that the 12-13 waves of mitosis producing approximately 6000 nuclei in *Drosophila* during the early, syncytial phase[69,70] (before membranes separate the nuclei) are primarily mechanical[71]. This discovery supports our conclusion that early differentiation waves are primarily mechanical in nature, especially since pattern formation has taken place[72] and therefore the cellularized nuclei should be differentiated from one another.

We can then conclude that waves on one-cell embryos are at least both mechanical and electrical, in varying proportions, and undoubtedly involve other biochemical events. Some of the observed waves, such as the wave of breakdown of the alveoli, may be consequences of other waves, i.e., the coupling for some components can be one-way rather than reciprocal. In multicellular differentiation waves, the proportions of mechanical versus electrical components may vary in strength between the six types of differentiation waves (Figures 6.28, 6.34, 6.35, 6.37, 6.38 and 6.39). This is all ripe for careful investigation requiring simultaneous observations of all propagating components of waves on embryos, and the relationship between one wave and the next. It is all too descriptive right now. There are deep mathematical relationships between differentiation waves and Turing's reaction/diffusion equations, kink waves (§1.15 *Chemical or Mechanochemical Instabilities?* in *Hierarchical Genome*[4]), the theory of waves in excitable media, both chemical and mechanical[73-76], critical dynamics, self-organizing systems and chaotic phenomena, waiting to be explored and tested against observations and manipulations in living embryos.

1. Frixione, E. (2006). Albrecht von Haller (1708-1777). *Journal of Neurology* **253**(2), 265-266.
2. Cajavilca, C., J. Varon & G.L. Sternbach (2009). Luigi Galvani and the foundations of electrophysiology. *Resuscitation* **80**(2), 159-162.
3. Lillie, R.S. (1920). The recovery of transmissivity in passive iron wires as a model of recovery processes in irritable living systems. Part I. *Journal of General Physiology* **3**(1), 107-128.
4. Gordon, R. (1999). *The Hierarchical Genome and Differentiation Waves: Novel Unification of Development, Genetics and Evolution*. Singapore & London, World Scientific & Imperial College Press: [http://www.worldscientific.com/worldscibooks/10.1142/2755].
5. Peirce, G.J. & F.A. Randolph (1905). Studies of irritability in algae. *Botanical Gazette* **40**(5), 321-350.
6. Peirce, G.J. (1906). Studies of irritability in plants. *Annals of Botany* **20**(77-80), 449-465.
7. Morgan, T.H. (1906). The origin of the organ-forming materials in the frog's embryo. *Biological Bulletin* **11**(3), 124-136.
8. Morgan, T.H. (1909). The dynamic factor in regeneration. *Biological Bulletin* **16**(5), 265-276.
9. Morgan, T.H. (1912). The role of irritability and contractility as dynamic factors in development and regeneration. In: *Proceedings of the Seventh International Zoological Congress, Boston, 19-24 August, 1907*. Cambridge, Massachusetts, USA, University Press. **7**: 483-490.
10. Gerard, R.W. (1952). Ralph Stayner Lillie: 1875-1952. *Science* **116**(3019), 496-497.
11. Lillie, R.S. (1911). The relation of stimulation and conduction in irritable tissues to changes in the permeability of the limiting membranes. *American Journal of Physiology* **28**(4), 197-222.
12. Galvani, L. (1791). De viribus electricitatis in motu musculari commentarius. *De Bononiensi Scientiarum et Artium Instituto atque Academia Commentarii (Accademia delle scienze dell'Istituto di Bologna)* **7**, 363-418.
13. Galvani, L., G. Aldini & R.M. Green (1953). *Commentary on the Effect of Electricity on Muscular Motion. (A Translation of Luigi Galvani's De Viribus Electricitatis in Motu Musculari Commentarius.) By Robert Montraville Green.*

[Translated from the Edition Published in 1792 by Giovanni Aldini, Together with a Translation of Aldini's Introductory Dissertation and an Introduction by Giulio C. Pupilli.]. Cambridge, Mass., Licht.

14. Lillie, R.S. (1915). The conditions of conduction of excitation in irritable cells and tissues and especially in nerve. II. *American Journal of Physiology* **37**(2), 348-370.
15. Costello, D.P. & C. Henley (1971). *Methods for Obtaining and Handling Marine Eggs and Embryos [*http://darchive.mblwhoilibrary.org/handle/1912/295*]*. Woods Hole, Lancaster Press, 2nd.
16. Just, E.E. (1919). The fertilization reaction in *Echinarachnius parma*. I. Cortical response of the egg to insemination. *Biological Bulletin* **36**(1), 1-10.
17. Fol, H. (1877). Sur le commencement de l'hénogénie chez divers animaux. *Archives des Sciences Physiques et Naturelles Geneve* **6**, 145-169.
18. Just, E.E. (1939). *The Biology of the Cell Surface*. Philadelphia, Blakiston.
19. Wikia (2016). Ghostbusters Wiki: Ectoplasm. http://ghostbusters.wikia.com/wiki/Ectoplasm
20. Yamamoto, T. (1944). On the excitation-conduction gradient in the unfertilized egg of the lamprey, *Lampetra planeri*. *Proceedings of the Imperial Academy (Tokyo)* **20**(1), 30-35.
21. Hara, K., P. Tydeman & C.W.A. Gorlee (1979). A new method for local artificial insemination of *Xenopus laevis* eggs. *Mikroskopie (Wien)* **35**, 10-12.
22. Yamamoto, T. (1954). Physiological studies on fertilization and activation of fish eggs V. The role of calcium ions in activation of *Oryzias* eggs. *Experimental Cell Research* **6**(1), 56-68.
23. Yamamoto, T. (1949). Physiological studies on fertilization and activation of fish eggs IV. Fertilization and activation in narcotized eggs. *Cytologia* **15**(1-2), 1-7.
24. Yamamoto, T. (1949). Physiological studies on fertilization and activation of fish eggs III. The activation of the unfertilized egg with electric current. *Cytologia* **14**(3-4), 219-225.
25. Moser, F. (1939). Studies on a cortical layer response to stimulating agents in the *Arbacia* egg II. Response to chemical and physical agents. *Journal of Experimental Zoology* **80**(3), 447-471.
26. Kitching, J.A. & F. Moser (1940). Studies on a cortical layer response to stimulating agents in the *Arbacia* egg. IV. Response to chemical and physical agents in the absence of oxygen, and observations of the effects of low oxygen tensions and high hydrostatic pressures upon amoeboid eggs. *Biological Bulletin Wood's Hole* **78**, 80-91.
27. Gilkey, J.C., L.F. Jaffe, E.B. Ridgway & G.T. Reynolds (1978). A free calcium wave traverses the activating egg of the medaka, *Oryzias latipes*. *Journal of Cell Biology* **76**(2), 448-466.
28. Jaffe, L. (2006). The discovery of calcium waves. *Seminars in Cell & Developmental Biology* **17**(2), 229-229.
29. Stewart-Savage, J. & R.D. Grey (1982). The temporal and spatial relationships between cortical contraction, sperm trail formation, and pronuclear migration in fertilized *Xenopus* eggs. *Wilhelm Roux's Archives of Developmental Biology* **191**(4), 241-245.

30. Gerhart, J.C. (1980). Mechanisms regulating pattern formation in the amphibian egg and early embryo. In: *Biological Regulation and Development Vol. 2: Molecular Organization and Cell Function*. Ed.: R.F. Goldsberger, Plenum Press: 133-316.
31. Wagner, J., Y.X. Li, J. Pearson & J. Keizer (1998). Simulation of the fertilization Ca^{2+} wave in *Xenopus laevis* eggs. *Biophysical Journal* **75**(4), 2088-2097.
32. Fall, C.P., J.M. Wagner, L.M. Loew & R. Nuccitelli (2004). Cortically restricted production of IP_3 leads to propagation of the fertilization Ca^{2+} wave along the cell surface in a model of the *Xenopus* egg. *Journal of Theoretical Biology* **231**(4), 487-496.
33. Conklin, E.G. (1905). The organization and cell lineage of the ascidian egg. *Journal of the Academy of Natural Sciences of Philadelphia* **13 Ser. 2**, 1-119.
34. Conklin, E.G. (1931). The development of centrifuged eggs of ascidians. *Journal of Experimental Zoology* **60**(1), 1-119.
35. Saad, G.A., S.S. Hamed & E.H. Radwan (2015). Ultrastructural study of the neural gland-ganglionic complex and the ovarian development in the ascidian, *Styela partita* (Stimpson,1852). *Journal of Advances in Biology* **6**(3), 1005-1022.
36. Dale, B., B. Hagström & L. Santella (1989). Partially fertilized sea urchin eggs: An electrophysiological and morphological study. *Development Growth & Differentiation* **31**(2), 165-170.
37. Elinson, R.P. (1975). Site of sperm entry and a cortical contraction associated with egg activation in the frog *Rana pipiens*. *Developmental Biology* **47**, 257-268.
38. Shumway, W. (1940). Stages in the normal development of *Rana pipiens*. I. External form. *Anatomical Record* **78**(2), 139-148.
39. Byrnes, W.M. (2009). Ernest Everett Just, Johannes Holtfreter, and the origin of certain concepts in embryo morphogenesis. *Molecular Reproduction and Development* **76**(10), 912-921.
40. Newman, S.A. (2009). E.E. Just's "independent irritability" revisited: the activated egg as excitable soft matter. *Molecular Reproduction and Development* **76**(10), 966-974.
41. Atri, A., J. Amundson, D. Clapham & J. Sneyd (1993). A single-pool model for intracellular calcium oscillations and waves in the *Xenopus laevis* oocyte. *Biophysical Journal* **65**(4), 1727-1739.
42. Rougier, N. & Z. Werb (2001). Minireview: parthenogenesis in mammals. *Molecular Reproduction and Development* **59**(4), 468-474.
43. Hara, K. & P. Tydeman (1979). Cinematographic observation of an "activation wave" (AW) on the locally inseminated egg of *Xenopus laevis*. *Roux's Archives of Developmental Biology* **186**(1), 91-94.
44. Kubota, T. (1967). A regional change in the rigidity of the cortex of the egg of *Rana nigromaculata* following extrusion of the second polar body. *Journal of Embryology and Experimental Morphology* **17**(2), 331-340, http://dev.biologists.org/content/317/332/331.
45. Gerhart, J.C., M. Danilchik, T. Doniach, S. Roberts, B. Rowning & R. Stewart (1989). Cortical rotation of the *Xenopus* egg: consequences for the anteroposterior pattern of embryonic dorsal development. *Development* **107**(Suppl.), 37-51.

46. Tuszynski, J.A. & R. Gordon (2012). A mean field Ising model for cortical rotation in amphibian one-cell stage embryos. *BioSystems* **109**(3, Special Issue on Biological Morphogenesis), 381-389.
47. Hara, K., P. Tydeman & R.T.M. Hengst (1977). Cinematographic observation of 'post-fertilization waves' (PFW) on the zygotes of *Xenopus laevis*. *Roux's Archives of Developmental Biology* **181**, 189-192.
48. Hara, K. (1971). Cinematographic observation of "surface contraction waves" (SCW) during the early cleavage of axolotl eggs. *Wilhelm Roux' Archiv für Entwicklungsmechanik der Organismen* **167**, 183-186.
49. Ubbels, G.A., K. Hara, C. Koster & M. Kirschner (1983). Evidence for a functional role of the cytoskeleton in determination of the dorsoventral axis in *Xenopus laevis* eggs. *Journal of Embryology and Experimental Morphology* **77**(1), 15-37.
50. Pelczar, H., S. Caulet, C. Thibier, G. Aubet, R. Poulhe, I. Vallianou, M. Yamashita & Y. Andéol (2007). Characterization and expression of a maternal axolotl Cyclin B1 during oogenesis and early development. *Development Growth & Differentiation* **49**(5), 407-419.
51. Santella, L. & B. Dale (2015). Assisted yes, but where do we draw the line? *Reproductive BioMedicine Online* **31**(4), 476-478.
52. Boterenbrood, E.C., J.M. Narraway & K. Hara (1983). Duration of cleavage cycles and asymmetry in the direction of cleavage waves prior to gastrulation in *Xenopus laevis*. *Roux's Archives of Developmental Biology* **192**(5), 216-221.
53. Boterenbrood, E.C. & J.M. Narraway (1986). The direction of cleavage waves and the regional variation in the duration of cleavage cycles on the dorsal side of the *Xenopus laevis* blastula. *Roux's Archives of Developmental Biology* **195**, 484-488.
54. Kirschner, M.W. & J.C. Gerhart (1981). Spatial and temporal changes in the amphibian egg. *BioScience* **31**(5), 381-388.
55. Jaffe, L.F. & R. Créton (1998). On the conservation of calcium wave speeds. *Cell Calcium* **24**(1), 1-8.
56. Boterenbrood, E.C. & J.M. Narraway (1990). Epiboly connected with cleavage in morula and early blastula stages of *Xenopus laevis*, a study using time-lapse photography. *Roux's Archives of Developmental Biology* **198**(5), 303-307.
57. Jaffe, L.F. (2008). Calcium waves. *Philosophical Transactions of the Royal Society B-Biological Sciences* **363**(1495), 1311-1316.
58. Jaffe, L.F. (1993). Classes and mechanisms of calcium waves. *Cell Calcium* **14**(10), 736-745.
59. Lefresne, J., Y. Andéol & J. Signoret (1998). Evidence for introduction of a variable G1 phase at the midblastula transition during early development in axolotl. *Development, Growth & Differentiation* **40**(5), 497-508.
60. Jaffe, L. (2002). On the conservation of fast calcium wave speeds. *Cell Calcium* **32**(4), 217-229.
61. Jaffe, L.F. (1999). Organization of early development by calcium patterns. *BioEssays* **21**(8), 657-667.
62. Markin, V.S., V.F. Pastushenko & Y.A. Chizmadzhev (1987). *Theory of Excitable Media*. New York, John Wiley & Sons.
63. DigitalGlobe (2014). Revealing the Hidden World with Shortwave Infrared (SWIR) Imagery. http://www.digitalglobeblog.com/2014/09/03/revealing-the-hidden-world-with-shortwave-infrared-swir-imagery/

64. Zhabotinsky, A.M. (2007). Belousov-Zhabotinsky reaction. *Scholarpedia* **2**(9), 1435.
65. Davidhazy, A. (2013). Colliding water ripples #4565. http://people.rit.edu/andpph/exhibit-3.html
66. Kimura, J., T. Yamada & R.L. Rodnitzky (1978). Refractory period of human motor nerve fibres. *Journal of Neurology, Neurosurgery, and Psychiatry* **41**(9), 784-790.
67. Franke, W.W., P.C. Rathke, E. Seib, M.F. Trendelenburg, M. Osborn & K. Weber (1976). Distribution and mode of arrangement of microfilamentous structures and actin in the cortex of the amphibian oocyte. *Cytobiologie* **14**(1), 111-130.
68. Jaffe, L.F. (1995). Calcium waves and development. *Ciba Foundation Symposium* **188**, 4-17.
69. Mazumdar, A. & M. Mazumdar (2002). How one becomes many: blastoderm cellularization in *Drosophila melanogaster*. *BioEssays* **24**(11), 1012-1022.
70. Harris, T.J.C., J.K. Sawyer & M. Peifer (2009). How the cytoskeleton helps build the embryonic body plan: models of morphogenesis from *Drosophila*. *Current Topics in Developmental Biology* **89**, 55-85.
71. Idema, T., J.O. Dubuis, L. Kang, M.L. Manning, P.C. Nelson, T.C. Lubensky & A.J. Liu (2013). The syncytial *Drosophila* embryo as a mechanically excitable medium. *PLoS One* **8**(10), e77216.
72. Kimelman, D. & B.L. Martin (2012). Anterior-posterior patterning in early development: three strategies. *Wiley Interdisciplinary Reviews-Developmental Biology* **1**(2), 253-266.
73. Odell, G.M., G. Oster, P. Alberch & B. Burnside (1981). The mechanical basis of morphogenesis. I. Epithelial folding and invagination. *Developmental Biology* **85**(2), 446-462.
74. Brodland, G.W. & D.A. Clausi (1994). Embryonic tissue morphogenesis modeled by FEM. *Journal of Biomechanical Engineering* **116**(2), 146-155.
75. Forgacs, G. & S.A. Newman (2005). *Biological Physics of the Developing Embryo*. Cambridge, Cambridge University Press.
76. Idema, T. & A.J. Liu (2014). Mechanical signaling via nonlinear wavefront propagation in a mechanically excitable medium. *Physical Review E* **89**(6), #062709.
77. Field, R.J. & R.M. Noyes (1974). Oscillations in chemical systems. V. Quantitative explanation of band migration in the Belousov-Zhabotinskii reaction. *Journal of the American Chemical Society* **96**(7), 2001-2006.

Chapter 11

Why Evolution is Progressive

One of the outstanding recent scandals of biology has been the notion that evolution is not progressive[1,2], a concept that flaunts the evidence of our eyes[3]. Any reading of the fossil record shows simpler starts followed by increasing sophistication, along many lines. Even where morphology doesn't appear to change, competition between organisms can lead to an arms race, known as the Red Queen hypothesis[4-6] from the children's book *Through the Looking-Glass, and What Alice Found There*: "Now, here, you see, it takes all the running you can do, to keep in the same place"[7]. Overall, there is indeed an increase in biodiversity over evolutionary time (Figure 11.1), which we could deliberately maintain[8]. To be sure, there is some apparent backsliding, as when an independent organism, even a bacterium[9], becomes a parasite[10,11], or parts become vestigial[12] or discarded, or features disappear and recur in closely related species[13], or a comet or an asteroid wipes out 95% of living species (Figure 1.24). But on the whole, some species in each branch of the tree of life tend to march on with richer diversity of structure and perhaps behavior[13]. One caveat is that the total number of extant species, though perhaps not their complexity, may have peaked 530 million years ago. A drop in the total number of extant species since that time may be due to subsequent extinctions[14].

Those species that stay behind may not progress, or at least not appear to do so in morphology. The most blatant example of the latter is that we are descended from bacteria[15], yet there are still plenty of our brethren bacteria around. And while we can't be sure yet[16] that modern bacteria are any more complex than the earliest ones, it is clear that the number of niches they live in has increased, and the corresponding biochemical and genetic diversity amongst bacteria has evolved[17,18]. The diversity of minerals has increased with the evolution of life[19] (Figure 11.2), so for example we can expect a parallel increase in the number of environments on minerals for bacteria[20], let alone environments in and on other organisms, in the evolution of bacterial film communities[21], symbioses[22] and communities on and in eukaryotes that have been called symbiocosms[23]. The heterogeneity and geographic separation of places that bacteria can live contributes to their diversity[24], and many have been classified as extremophiles[25] or polyextremophiles[26] for the extreme environments they live in. Therefore progress may be occurring at every level.

Some work on progressive evolution has been couched in the language of "increase in complexity", and thus centers around the construction of measures of complexity and empirical data on whether complexity is increasing[27], as in Figure 1.3. For example, John Tyler Bonner counted the number of cell types in an organism as a measure of complexity and applied it to evolution[28]. In our work, because of the direct link between the number of cell types and the number of differentiation waves, this

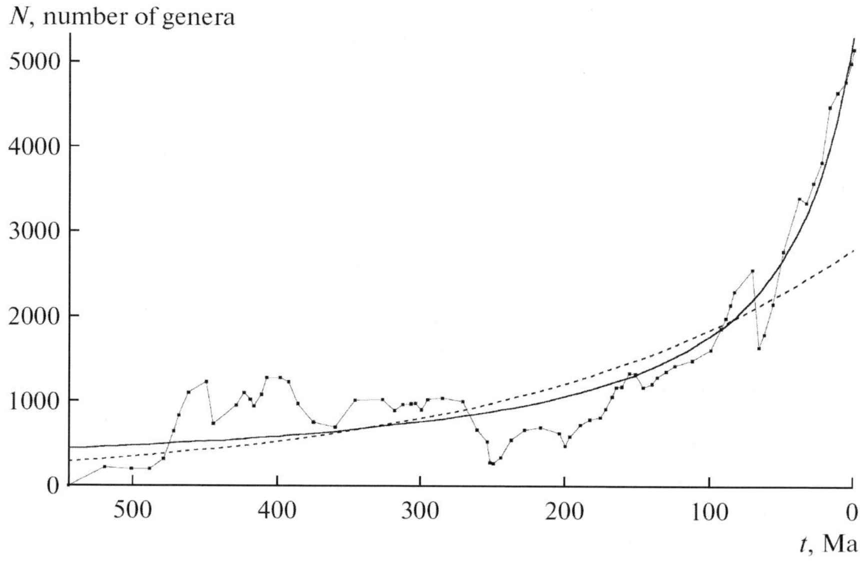

Figure 11.1. Biodiversity is measured here as the number of genera of marine animals versus time, measured in millions of years ago (Ma). The curves fit to the data suggest that biodiversity increases faster than exponential (dashed curve), as shown by the hyperbola (solid curve). Thus the rate of increase in biodiversity may itself be a function of the biodiversity. In any case, biodiversity does increase, and thus, at least in this sense, evolution is progressive. From Dmitriev (2011)[29] with permission of Springer and Pleiades Publishing, Ltd.

TABLE 1 THREE ERAS AND TEN STAGES OF EARTH'S MINERAL EVOLUTION

Era/Stage	Age (Ga)	Cumulative no. of species
Prenebular "Ur-Minerals"	>4.6	12
Era of Planetary Accretion (>4.55 Ga)		
1. Primary chondrite minerals	>4.56 Ga	60
2. Achondrite and planetesimal alteration	>4.56 to 4.55 Ga	250
Era of Crust and Mantle Reworking (4.55 to 2.5 Ga)		
3. Igneous rock evolution	4.55 to 4.0 Ga	350 to 500*
4. Granite and pegmatite formation	4.0 to 3.5 Ga	1000
5. Plate tectonics	>3.0 Ga	1500
Era of Biologically Mediated Mineralogy (>2.5 Ga to Present)		
6. Anoxic biological world	3.9 to 2.5 Ga	1500
7. Great Oxidation Event	2.5 to 1.9 Ga	>4000
8. Intermediate ocean	1.9 to 1.0 Ga	>4000
9. Snowball Earth events	1.0 to 0.542 Ga	>4000
10. Phanerozoic era of biomineralization	0.542 Ga to present	4400+

* Depending on the volatile content of the planet or moon

Figure 11.2. The increase of the number of mineral "species" on Earth roughly parallels the increasing complexity of life. From Hazen & Ferry (2010)[30] with kind permission of J. Alex Speer of the Mineralogical Society of America.

count is clearly a measure of both cell differentiation and morphogenesis, i.e., it operates at widely different size scales. On the other hand, the differentiation tree is not a linear, but rather a branched structure, so it is not subject to simple, scalar measures of complexity. The problem of estimating the complexity of differentiation trees in a way that allows comparison is a subset of the problem of complexity of networks, which is under active investigation[31,32]. Insofar as the differentiation tree is the genetic program of the organism, a bifurcating alternation of molecular events and physics (differentiation waves), it may be possible to someday calculate its algorithmic complexity, defined as the length of the shortest computer program[33] that successfully mimics a particular organism's embryogenesis.

"Progress" became a dirty word in biology because it smacks of "purpose", and the implication that the purpose of evolution was to produce us. The concept suggests that evolution indeed has a direction, which to many religious people (including many scientists[34,35]), means "God did it". Thus, for a biologist to speak of progress is said to play into the hands of the creationists and the Intelligent Designers (who distinguish themselves from other creationists by not uttering the word "God"). Rather than take this head in the sands approach, we will confront progress in evolution in this chapter.

The intricate motivations of many scientists, starting with Darwin's "bulldog" Thamas Huxley, to keep progress out of the profession of evolution, while promulgating it in popular accounts of evolution, can be read in the eye opening book *Monad to Man: The Concept of Progress in Evolutionary Biology* by philosopher of science Michael Ruse[36]. While Ruse does not pass judgment, this two-faced approach is probably part of the reason that the debate with creationists[37-39] is never settled.

On the question of the centuries old debate of the relationship between science and religion, we stand firmly in the belief that the universe is one[40], deserving one unified explanation that works, and do not opt for *nonoverlapping magisteria*, the purported lack of conflict between science and religion due to the lack of overlap between their respective domains[41]. That is an imaginary line in the sand. Attempts to draw it, or to claim the whole territory, from either side, are failing[42]. Rather than propose a God of the gaps[43-45] or science of the gaps[46,47], we prefer to push our investigations as far as we can with testable hypotheses, and leave the gaps of both kinds for later generations to work on. Our plan in this chapter is, then, to delve into the evolutionary consequences of differentiation waves, ending with a plausible demonstration that they may be responsible for progress in evolution.

Let's begin by reviewing the structure of a differentiation tree (Figure 6.43). The cell state splitter (Chapter 6) and the nuclear state splitter (Chapter 9) result in a cell changing type to one of two new kinds. We orient the differentiation tree in two ways: up is increasing developmental time. Left indicates a contraction wave, and right

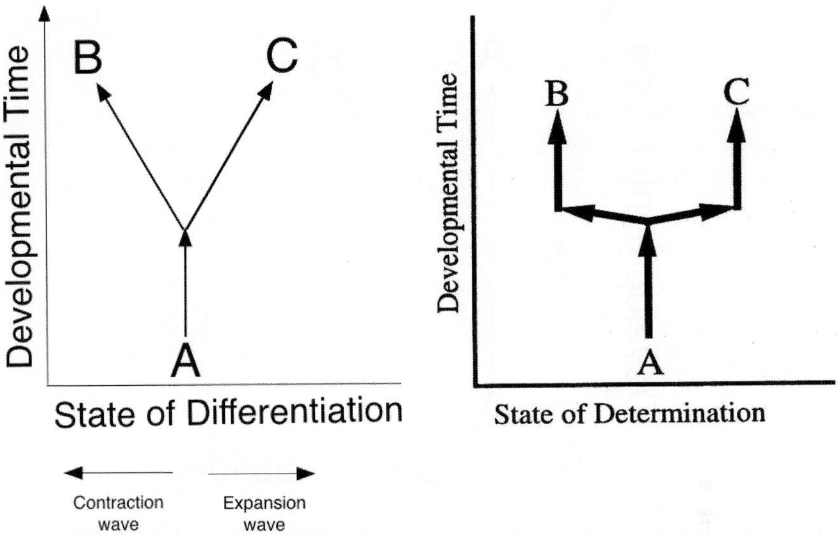

Figure 11.3. **Left**: The differentiation waves model of a single step of cell differentiation, shown as one branch of a differentiation tree. Cells of type A become type B if they participate in a contraction wave, and become type C if they participate in an expansion wave. Modified from Figure 7 in *Hierarchical Genome*[48]. **Right**: State of determination. If we distinguish the state of determination of a tissue from its state of differentiation, we see that the state of determination changes abruptly (nearly horizontal arrows) as a cell participates in a contraction (left arrow) or an expansion (right arrow) differentiation wave, followed by a longer period of differentiation (vertical arrows). Copy of Figure 8 in *Hierarchical Genome*[48].

indicates an expansion wave (Figure 11.3). Actually, this represents the choices before up to tens of thousands of cells in an embryo. For example, those ectoderm cells that participate in a contraction wave become neural plate cells, and those ectoderm cells that participate in an expansion wave become epidermal cells (Figure 6.40). The cell state splitter and the nuclear state splitter thus create a bifurcation, a branch point in the fate of a cell, depending on which kind of differentiation wave it participates in.

Previous models for cell differentiation in embryos were based on the notion of induction[49], a term that seems to have been introduced in 1894 by Hans Driesch, who distinguished external (i.e., environmental) from internal induction, regarding the latter as more reliable[50]. What induction came to mean in embryology was the notion that cells of, say, type A

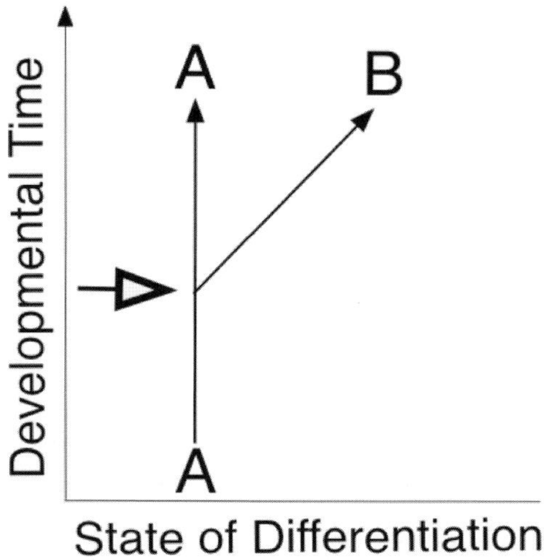

Figure 11.4. Standard model of differentiation, in which some A cells become B cells when "induced" to do so by another tissue, indicated by the large arrow. The unaffected A cells continue to be of type A. Modified from Figure 5 in *Hierarchical Genome*[48].

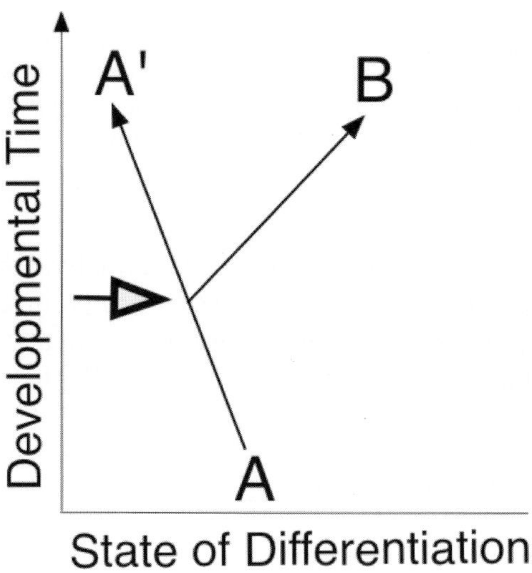

Figure 11.5. Modified standard model of cell differentiation in embryos. Cell type A slowly changes to cell type A', unless some of the A cells are induced by another tissue touching them to become type B cells. Modified from Figure 5 in *Hierarchical Genome*[48].

would remain type A until some of them made contact with some other tissue and were "induced" by it to change to cell type B. Those cells not involved in this initial induction would stay of type A forever, unless something else happened (Figure 11.4). The initial example was the induction of the lens of the vertebrate eye from epidermis by the forming eye[49,51]. A variant on this presumption was that cells of type A slowly matured to cells of type A', unless some of them were induced to become B cells (Figure 11.5).

In contrast, in the differentiation waves model the cells of type A actively become one of two new kinds of cells, B or C, when they participate in a contraction or an expansion wave, respectively (Figure 11.3) Those cells and their descendants are no longer A cells. Unless some A cells miss both waves altogether (which is possible: Figure 11.6), none remain in the embryo.

The notion of induction in embryology means that somehow one tissue causes another tissue to differentiate. The most famous case is that the dorsal lip of the blastopore transplanted to ventral ectoderm induced that tissue to make another whole embryo (Figures 2.65 and 6.16). This view of embryogenesis left each tissue passive unless it was induced to do something new. It still prevails. It led to two strange concepts, that would seem to be inconsistent: Self-induction and planar induction.

In some species, there was no other tissue found to be touching the "induced" tissue, so it was postulated that somehow the tissue induced itself. Thus we got the weird concept of self-induction (Proposition 74 in *Hierarchical Genome*[48]). This was called autoinduction when the triggering was done experimentally[52]. The belief that induction was essential to development even led to the rejection of a relative of the axolotl, *Ambystoma maculatum*, as an appropriate animal for studying neural development, because it used self-induction when it "should have" used regular induction by another tissue[53]! In other cases, such as eye induction, autoinduction was attributed to a second, independent inducer[54].

The other confusing concept was planar induction, i.e., somehow the effect of induction spread from the point of contact over a specific region of an epithelium such as ectoderm, from one cell to the next, where the cells were all of the *same* kind. So this also violated the definition of

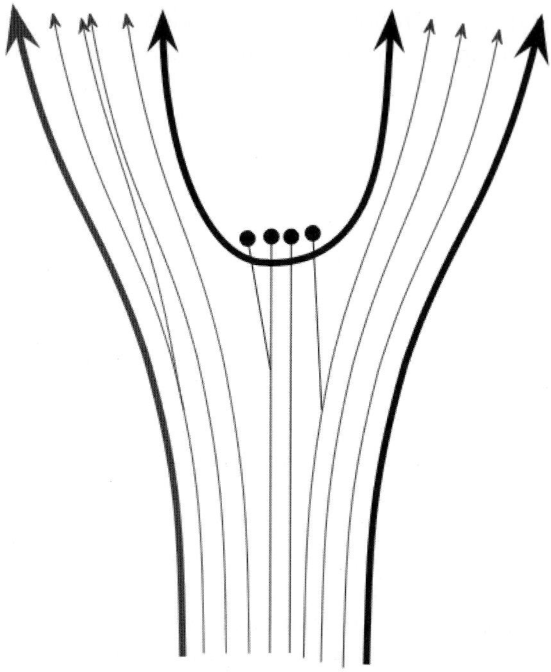

Figure 11.6. The thicker arrows outline a single branch of a differentiation tree (as in Figure 11.3 Left). The vertical direction defines increasing developmental time. Inside is a portion of the cell lineage tree with the thin lines indicating individual cells. Those cells on the left experience a contraction wave while those on the right experience an expansion wave, as in Figure 11.3 Left. Thus, cells at the earlier stage of development at the bottom, all of one type, become cells of two types. However, some cells may be missed by both waves and become stem cells (Figure 6.37), as indicated by the dots now off the differentiation tree (cf. a model in which adjacent regions at a given "growth phase" are defined by thresholds in antagonistic transcription factors and stem cells form in regions below both thresholds[55]). Cell divisions are indicated by branch points of the cell lineage tree, and do not necessarily coincide with differentiation waves. Both trees, followed backward in time, would converge at the one-cell stage of the embryo, namely the fertilized egg. From Gordon (2011)[56] with permission of Springer.

normal induction, which is supposed to occur between different kinds of cells. How planar induction stopped from going everywhere, or what its mechanism might be, were never figured out[57-60], though a "wave" of transient gene expression was observed[61].

We can now see through this contorted thought pattern. In retrospect, we can see that so-called induction is the launching of a differentiation wave, planar induction is its propagation, and the region that changes cell type is determined by the trajectory of a given differentiation wave and the mechanism(s) by which it stops propagating (Figures 6.18 and 6.26). Differentiation waves stop either by self-annihilation, by running into other waves, or by running into a tissue whose cells do not have cell state splitters readied. In some cases they don't stop at a tissue boundary but do enter an adjacent tissue, whose cell state splitters presumably are ready to be triggered. For the ectoderm contraction wave, the trigger mechanism may be the binding of pharyngeal endoderm (Figures 7.3 and 7.13) to a small region on the underside of the ectoderm resulting in a force imbalance (Figure 7.15). This mechanism also suggests the possibility that the choice of launching a contraction or an expansion wave depends on the embryo physics, as discussed for each wave in Chapter 7. We thus suspect that global mechanical effects[62] may be what launch differentiation waves from points, arcs, or full circles[63], perhaps by causing some cells to pass a threshold for triggering expansion or contraction of their cell state splitters. Self-induction is then launching of a wave by a mechanism that does not require direct cell-cell contact between two tissues, perhaps triggered by globally generated mechanical strains.

The generalization of the picture in Figure 11.5 is our concept of the *differentiation tree*. Our empirical tree for early axolotl development (Figure 7.1) is a differentiation tree. Each edge represents a new tissue formed during embryogenesis, and the nodes (points where edges meet) represent the process of splitting of a tissue into two new tissues. To make this clear, in Figure 6.43 we have given each edge of a hypothetical differentiation tree a different color.

The notion that each cell type in the course of embryogenesis can, at a step of determination, become only (at most) two cell types (except for stem cells: Figure 11.6), and not three or more, is an underlying, untested assumption (*Proposition 3: all cell state splitters are bistable*, in *Hierarchical Genome*[48]). It would seem to be contradicted by the hierarchy of cell types found in, for instance, the mouse, for which some of the 6900 identified cell types[64] branch to up to five new cell types (*Permissions and note added in proof* in *Hierarchical Genome*[48]). This apparent contradiction warrants further investigation. It may be due to cryptic cell types, i.e., cells that look the same by histological criteria that are nevertheless different kinds, so that some bifurcations were not counted. It may also be due to nonuniform criteria for defining cell types[65]. Our ability to classify cells into equivalence groups depends on the criteria we develop and our technical ability to measure differences between cells that we deem are important, especially in regards to whether or not the cells are of different differentiation types. Traditionally histology depended on morphology, tissue architecture[66] and chemical staining[67] of cells. New criteria include housekeeping genes[68], mechanical properties[69], infrared spectra[70] and DNA replication timing[71]. If the differentiation tree proves universal, it would form an objective basis for a binary hierarchy of cell types.

We may think of the differentiation tree drawn on a page as a flattened version of the development of the tissues of an organism. Of course, at a given stage the tissues each have a geometry in space, and they are intimately associated with one another. The actual differentiation tree is thus a 4D (four dimensional) diagram, with the three spatial dimensions at a given time being collapsed onto a horizontal line on the page. The branches are then rearranged so that tissues participating in contraction waves lean to the left and those participating in expansion waves lean to the right, and none of the line segments representing tissues overlap. The differentiation tree differs from a lineage tree, such as that of the nematode *C. elegans* (Figure 2.15), in the left/right ordering of each bifurcation. In the mathematical jargon of graph theory[72], a flattened differentiation tree, as a graph called a tree, is a directed graph embedded in a plane, or planar graph for short. It is a rooted tree, with the fertilized egg being its root[73]. Unflattened, embedded in 4D space (3D + time), with its root placed at time = 0, the 4D differentiation tree is still a planar graph in terms of its topology (Figures 11.7 and 11.8). One could consider the labeling of each tissue as forming from a contraction or an expansion wave as a 5th dimension to the differentiation tree, but since it is a just two-valued parameter, we choose to portray it via two colors for the corresponding edges of the graphs.

A tree canopy seen from above provides a rough analogy (Figure 11.9). The tops of the trees correspond to the adult tissues, each terminal branch taking up space, as tissues do. Looked at from the side, a deciduous tree shows how the branches contribute over time to the final form and to the canopy (Figure 11.10). If all the earlier branches are erased (Figure 11.11), then we have a side view representation of the canopy or adult organism. Most of the earlier, transient embryonic tissues no longer exist as such in

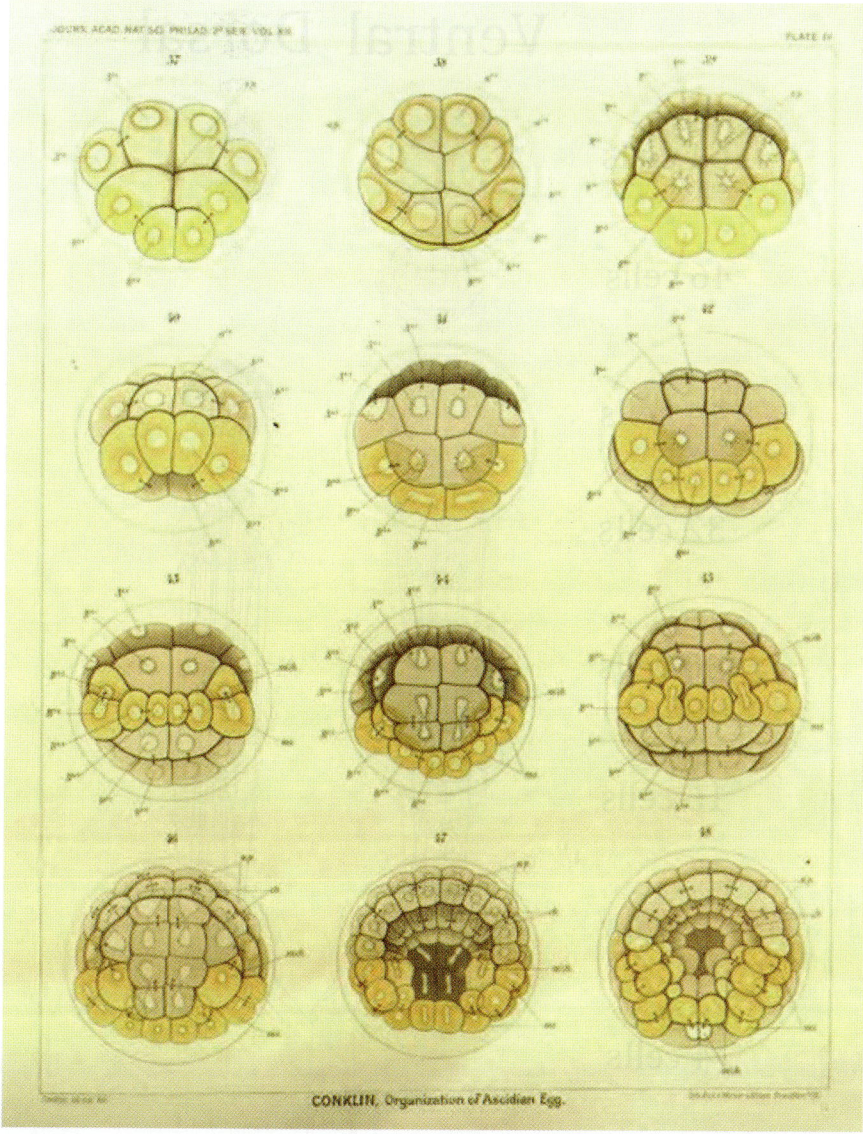

Figure 11.7. Edwin Conklin was able to name and follow individual cells and so determine their lineages in the ascidian *Cynthia partita*, in part using color differences[74]. Here embryos are shown from 16 cells to gastrulation. The small, double headed arrows indicate pairs of daughter cells. Each cell can be given a label indicating its history, and thus its lineage. The position of it and its line of ancestors could potentially then be traced back over time and space. From Conklin (1905)[75] with kind permission of Ned Gilmore, Academy of Natural Sciences of Drexel University.

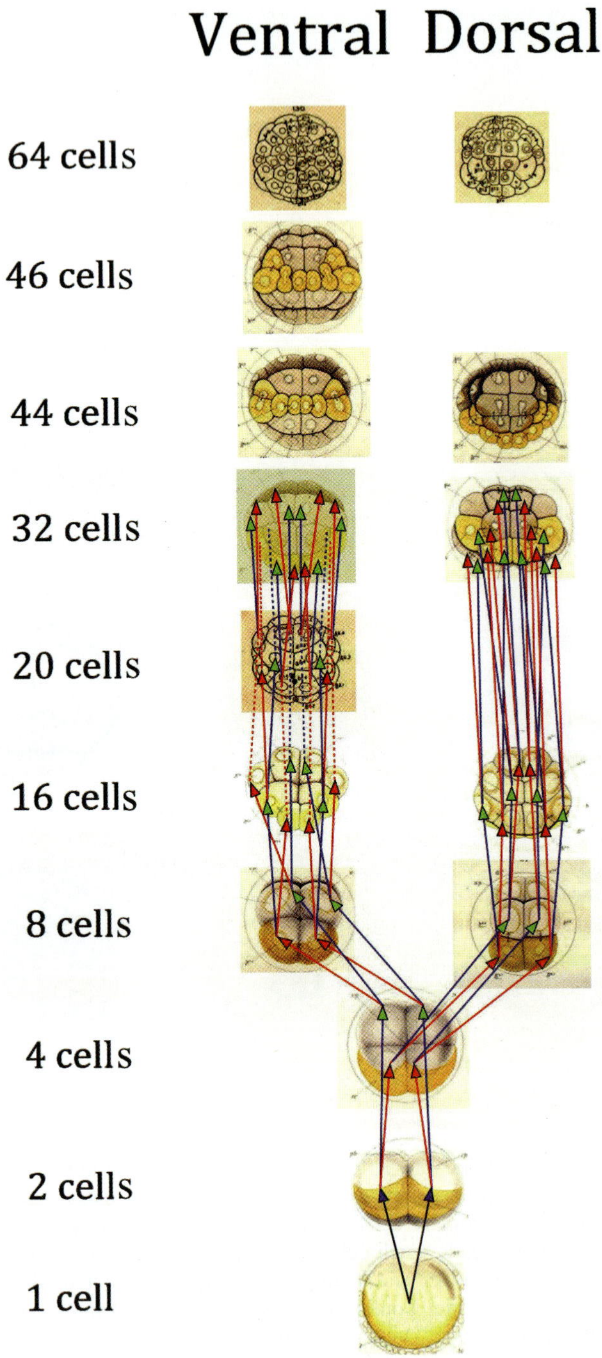
Ventral
Dorsal
64 cells
46 cells
44 cells
32 cells
20 cells
16 cells
8 cells
4 cells
2 cells
1 cell

Figure 11.8. Using sketches from Conklin (1905)[75] we have constructed a first guess at the early portion of the 4D-tree of the ascidian *Cionus partita*, now renamed *Styela partita*. The first cell division appears to be symmetric. After that we presume that all cell divisions are asymmetric, with a large cell corresponding to a single cell expansion wave (blue arrows with green tips) and a small cell corresponding to a single cell contraction wave (red arrows). Dashed lines indicate that the mother cell divides at the next stage above. For instance, there is a 20-cell stage because 4 cells divide prior to the 12 others, breaking the synchrony of cell division of earlier stages. The ventral and dorsal views could, of course, be combined into a single 3D structure versus time, which could be individually rotated with the arrows following. The 4D-tree shows the spatiotemporal flow of development. The reader is invited to continue the 4D-tree beyond the 32-cell stage. Relative cell sizes here have been eyeballed from Conklin's sketches, and may have to be reversed in some cases using modern data, so this 4D-tree is presented only as a draft of what could be a great 4D computer visualization tool.

Figure 11.9. The tops of trees are analogous to the adult tissues at the terminal branches of a differentiation tree. One no longer sees the connections and hierarchy of the branches below, many of which would represent transient embryonic tissues, as the vertical dimension here represents time. From Williams (2007)[76] with kind permission of John Williams.

the adult. This analogy may help us understand the spatial irregularity of growth, such as has been found amongst various species of piranha fish[77]: each bough in the canopy, corresponding to each tissue type, could be envisaged to have its own rate of growth[78].

The differentiation tree, at least for multicellular organisms, is what evolves. This can happen in two fundamentally different ways. Each line segment represents a single tissue type, either transient in existence during embryogenesis, or terminally differentiated, as in most adult tissues. As such, each edge represents a different subset of the genes that are exposed for transcription, represented by each edge being colored differently (Figure 6.43). Any of those genes can mutate. The result would be a topologically similar differentiation tree, but with the color of that edge changed to reflect the mutation. Another type of change that can occur is to the actual topology of the differentiation tree. In this case edges are added or removed by mutation.

The word macroevolution is a vague term generally used to refer to evolution on a grand scale, or over long periods of time[79]. Thus macroevolution refers to all that higher level structure in the tree of life[80] (Figures 11.12 and 11.13), i.e., big changes above the species level. Debates have raged since the late 1800s on how macroevolution came about, and whether it is directed in any sense. We would like to give it a new, precise definition that will lead us to a possible DNA basis for macroevolution: *macroevolution* can be said to have occurred whenever there has been a topological change in the differentiation tree. *Microevolution*, (evolutionary changes on the small scale, such as changes in gene frequencies within a population[77]), are all other changes to the differentiation tree that preserve its topology. Examples of microevolution would range from mutation in a single gene to heterochrony (Figure 11.14), in which the relative timing of the development of tissues shifts (Proposition 78 in *Hierarchical Genome*[48]).

Simple forms of microevolution occur in plant and animal breeding and in the laboratory. By our definition, experimental macroevolution has already been achieved, though it wasn't recognized as such, in a mutant of the nematode *C. elegans* (Figure 11.16) that altered the topology of its lineage tree, and thus of its differentiation tree[73]. With deliberate manipulation of the differentiation tree we might be able to produce new species, genera or even phyla.

Figure 11.10. The vertical axis of a deciduous tree corresponds roughly to developmental time. Those branches that are visible from the top contribute to the canopy, while, by analogy, lower branches of the differentiation tree support and lead to the upper branches.

Figure 11.11. To emphasize the loss of morphological and developmental history in the adult organism, we have here erased the understory of the tree in Figure 11.10. Only the canopy, corresponding to adult tissues, may be seen, albeit from the side, rather than from the top, as in Figure 11.9.

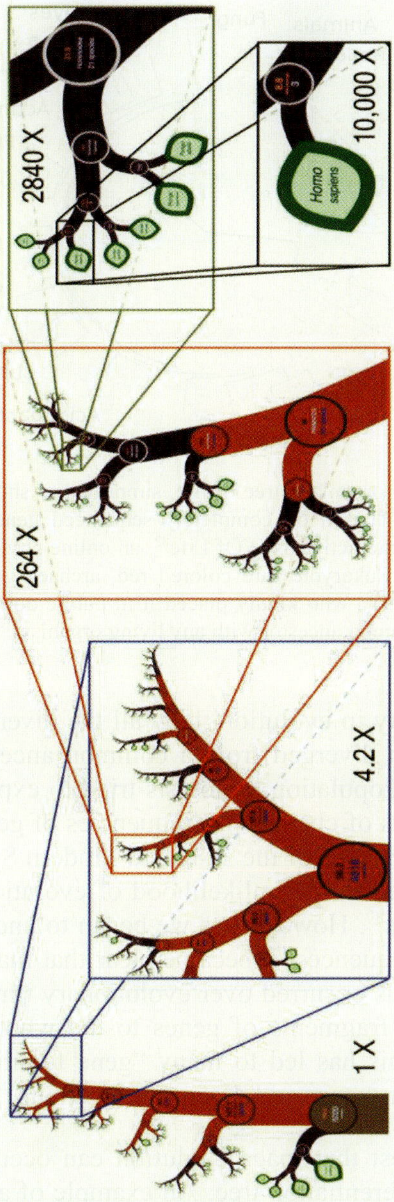

Figure 11.12. The concept of fractals allows interactive representation of the tree of life by dynamically expanding or contracting any portion of it, here starting with the branch containing the mammals, expanded in steps to show *Homo sapiens*. From Rosindell & Harmon (2012)[80] with permission under a Creative Commons Attribution License.

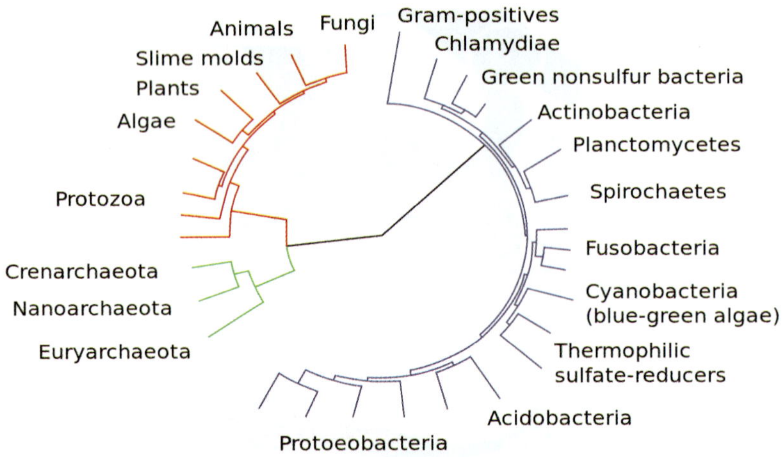

Figure 11.13. A radially organized tree of life, simplified to show just the major groups and their relationships, "based on completely sequenced genomes[81]. The image was generated using iTOL: Interactive Tree Of Life[82], an online phylogenetic tree viewer and Tree Of Life resource. Eukaryotes are colored red, archaea green and bacteria blue." From Tim Vickers (2009)[83] who kindly placed it in public domain. Another interactive tree of life finds our common ancestor with any living organism[14].

The origin of novelty in evolution, how all the diversity of creatures we are surrounded with diverged from a common ancestor, is basically an unsolved problem. Population biologists tried to explain the grand scale of evolution in terms of changes in frequencies of genes in a population. This became the core idea of the so-called Modern Synthesis[88]. Progress, if any, was attributed to the unlikelihood of evolution going backwards by retracing its steps[89]. However, as we began to understand the genome in terms of DNA sequences, it became clear that massive duplication of parts of the DNA has occurred over evolutionary time[90], with duplicated parts ranging from fragments of genes to the whole genome[91]. At the level of the gene this has led to many "gene families", i.e., genes and proteins of related structure and divergent functions[92] (Figure 11.16).

We therefore suggest that macroevolution can occur by duplication of branches of the differentiation tree. An example of a sequence of events that could produce such a duplication is shown in Figures 11.17 and 11.18. The basic idea is that the branching structure of the differentiation tree has to be represented somehow along the linear DNA molecule. Here we model the simplest arrangement, in which all of the genes

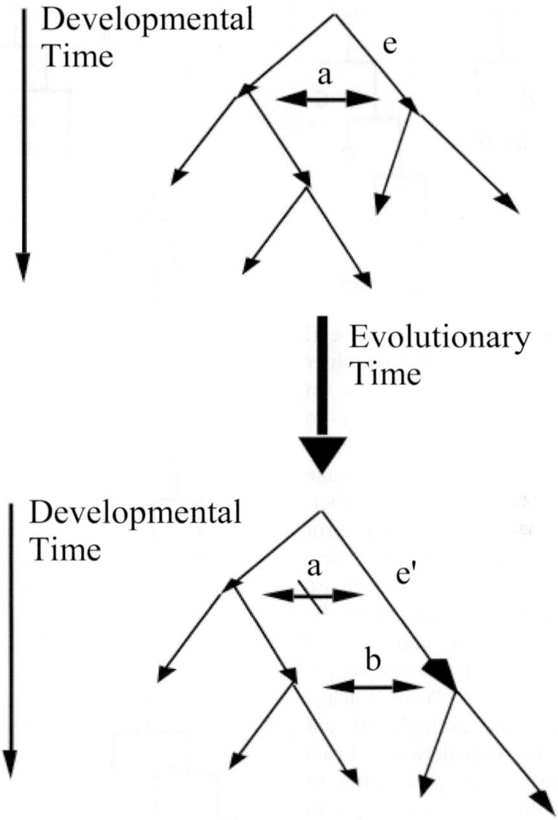

Figure 11.14. Heterochrony is a form of microevolution. It involves simple nonuniform stretching of the differentiation tree. Inductive relationships between tissues are opportunistic and therefore not fundamental to evodevo (evolution and development), an observation that has been made, for instance, of nematode evolution[84]. Copy of Figure 29 in *Hierarchical Genome*[48].

involved in a given cell type (its differon) are adjacent to one another. The connections between gene cascades are presumed to be made by molecules that are produced or released from the DNA of one cascade and trigger off the next cascade, including transcription factors. For example, molecule *ab* is released in response to a contraction wave and *ac* in response to an expansion wave. A portion of DNA extending from the production site of *ab* and *ac* incorporating part of cascade *A* and all of cascades *B* and *C* is copied (Figure 11.17) and then pasted to the end of *C*, in one possible mutation (Figure 11.18). Cf.[93].

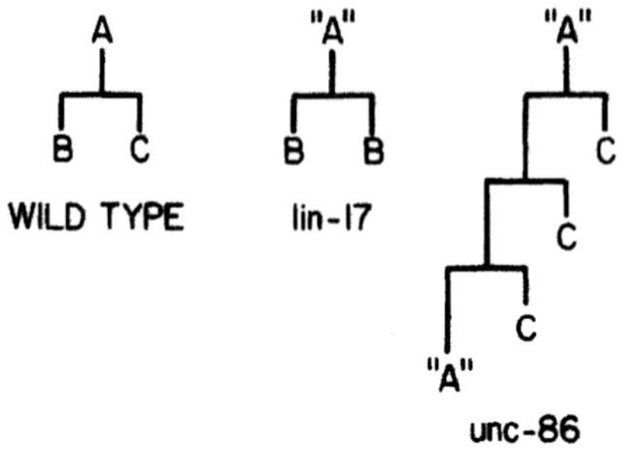

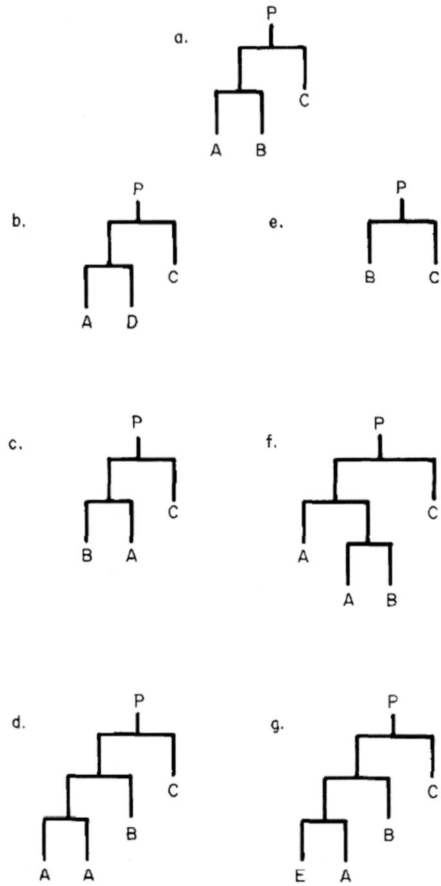

Figure 11.15. **Left**: "In a typical cell division in the wild-type, the two daughter cells (B and C) have different fates. In some cell divisions in *lin-17* and *unc-86* mutants, one daughter fails to generate its normal lineage but generates a lineage characteristic of its sister or its mother, respectively. In the *unc-86* case, the error is compounded at each division, giving the lineage indefinite length." The *unc-86* is a clear case of topology change of the lineage tree (Figure 2.15), and thus of the differentiation tree of *Caenorhabditis elegans*. Therefore this "cell lineage mutant" represents an example of our definition of macroevolution. From Hedgecock (1985)[85] with permission of Elsevier. Perhaps this is a case of incipient speciation[86], if reproductive isolation is involved. **Right**: Proposed cell lineage tree transformations thought to be involved in speciation of nematodes[87], all of which qualify as macroevolutionary. From Sternberg & Horvitz (1981)[87] with permission of Elsevier. Note that developmental time increases downwards here.

The result is a spaghetti-like mess (Figure 11.18: left). If the individual containing these extra gene cascades with all of the associated cross talk survives and reproduces, the altered genome has a chance of spreading in its population. While it does so, further smaller mutations can independently occur in all five cascades. We indicate this by B' and C' becoming distinct from B and C. At the same time, connecting molecules could also become distinct by mutation, unscrambling the spaghetti (Figure 11.18: right), an example of what has been called "post-adaptive adjustment"[95].

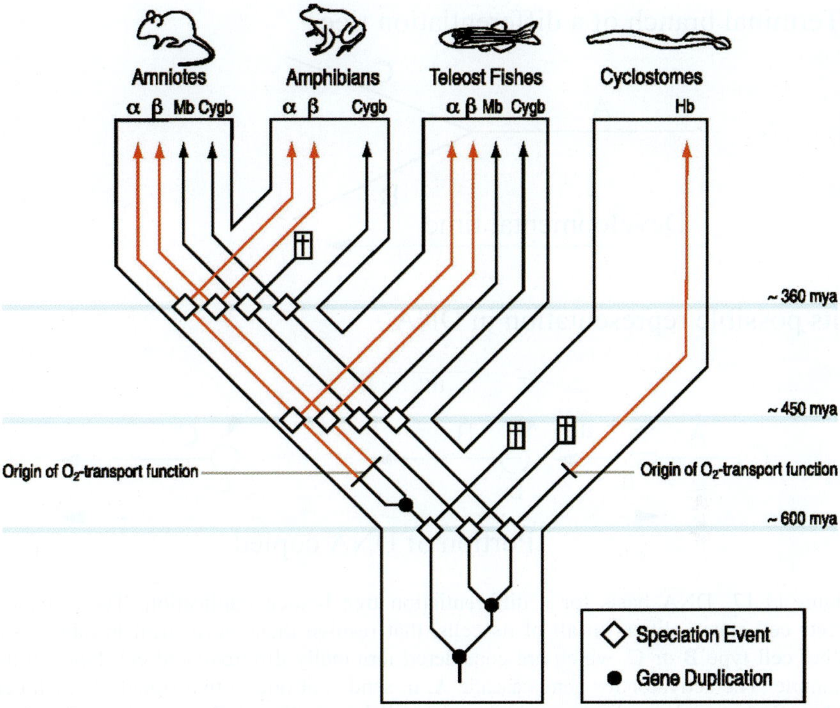

Figure 11.16. Example of a family of genes. Duplication of an ancestral globin gene in the common ancestor of the vertebrates (middle black dot) produced one descendant gene lineage that eventually gave rise to the α- and β-hemoglobins via a gene duplication (left black dot). A subsequent duplication (right black dot) gave rise to myoglobin (Mb) (which was lost in cyclostomes), Cygb, and cyclostome hemoglobin (Hb). Independent origins of oxygen (O_2) transport functions are denoted by orange lines, which involve different chemical mechanisms. From Hoffmann *et al.* (2010)[94] with kind permission of Diane Sullenberger, US National Academy of Sciences.

We assume that fitness increases in the process, especially because later generations now have two additional, new tissues with which to face the world. Except for the scale of the change in the DNA, duplication of a branch of a differentiation tree is substantially similar to the commonly observed duplication of single genes, followed occasionally by them acquiring different functions[96] (Figure 11.16).

Beyond duplications, we now have many instances of transfer of genes and even whole genomes between species. While these are generally documented at the single cell species level, it has been hypothesized

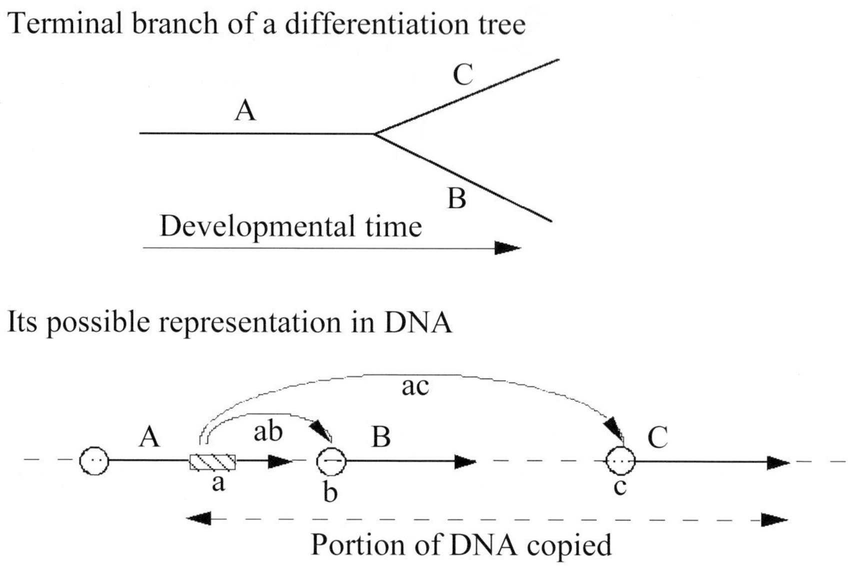

Figure 11.17. DNA basis for a differentiation tree branch duplication. Tissue type **A** erects cell state splitters in all of its cells, that resolve their mechanical instabilities as either cell type **B** or **C**, which are considered terminally differentiated cell types in this example. The activator for gene cascade **A**, **a**, sends out one of two signals in each cell: either **ab** or **ac**, depending on whether the cell has participated in a contraction or an expansion wave, respectively. These signals activate either gene cascade **B** or **C**, respectively. Copy of Figure 24 in *Hierarchical Genome*[48].

by Donald Williamson that organisms that undergo major changes in their life cycle (metamorphosis) are actually concatenations of two genomes[97,98].

We can put our imaginations to work in trying to guess at the topology of differentiation trees corresponding to different kinds of metamorphosis. The simplest case is the one where the larval and adult animal have no obvious relationship to one another at all. This is the case of the sea squirt, also known as a tunicate[99] or ascidian. The larva has a notochord, as we do during early development[100] (Figures 2.17: top and 11.19). On metamorphosis, all but one of the tissues degenerate. The remaining tissue appears to differentiate, and then forms the adult tunicate, which in no serious way looks like us (Figure 2.17: bottom). This process can be visualized by a differentiation tree such as that in Figure 11.20. This indeed looks like one tree grafted onto another. Five different kinds of

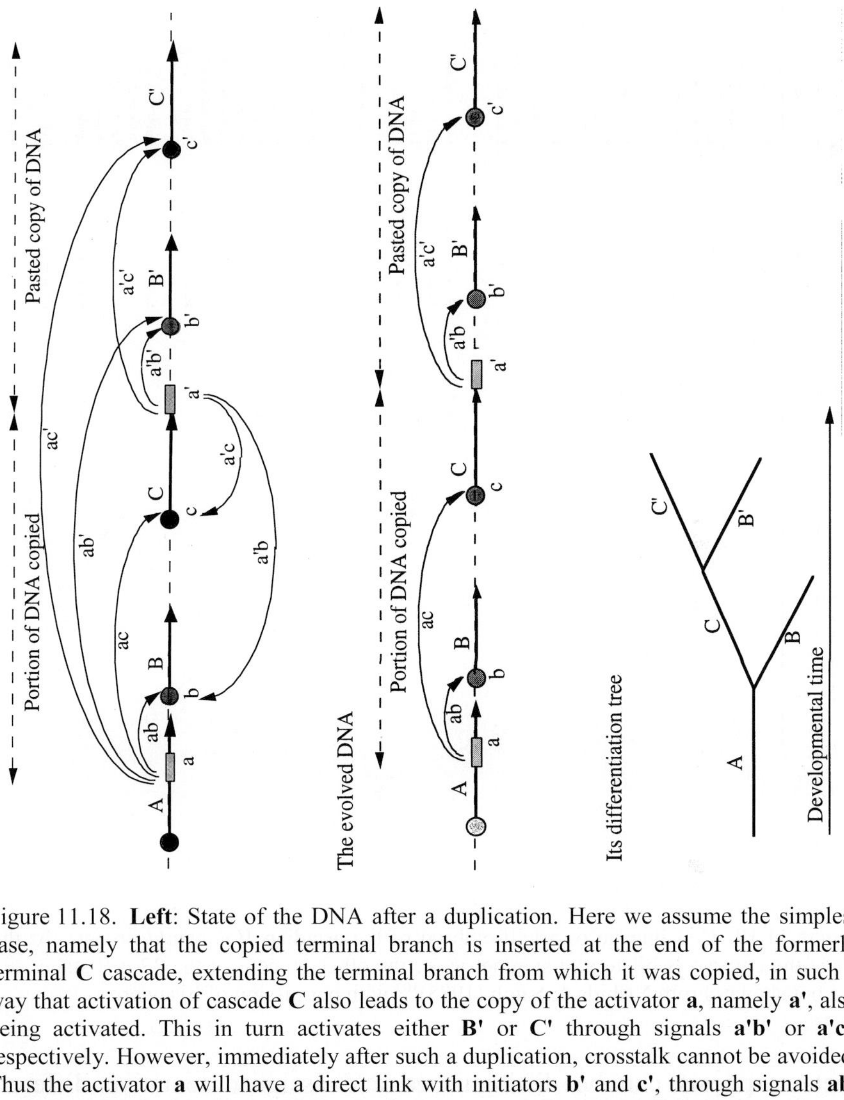

Figure 11.18. **Left**: State of the DNA after a duplication. Here we assume the simplest case, namely that the copied terminal branch is inserted at the end of the formerly terminal **C** cascade, extending the terminal branch from which it was copied, in such a way that activation of cascade **C** also leads to the copy of the activator **a**, namely **a'**, also being activated. This in turn activates either **B'** or **C'** through signals **a'b'** or **a'c'**, respectively. However, immediately after such a duplication, crosstalk cannot be avoided. Thus the activator **a** will have a direct link with initiators **b'** and **c'**, through signals **ab'** and **ac'**. Similarly, signal **a'** will also back react to cascades **B** and **C** through the signals **a'b** and **a'c**, respectively. Of course, in terms of molecular structure, **a=a'**, **b=b'**, **c=c'**, and **ab=a'b'=a'b**, **ac=a'c'=a'c**. **Middle**: Coevolution of DNA after a duplication. Activator **a'** is assumed to coevolve with initiators **b'** and **c'**, so that a new, cleanly separated terminal branch of the differentiation tree results (**Right**). In this diagram, the old cascades are shown unchanged, though they, too, could continue evolving. Thus we can end up with **a'≠a**, **b'≠c**, **c'≠c**, **a'b'≠ab**, **a'c'≠ac**, **B'≠B**, and **C'≠C**. Copies of Figures 25-26 from *Hierarchical Genome*[48]. Figure 11.15 Right f may be an example of duplication of a branch of a differentiation tree.

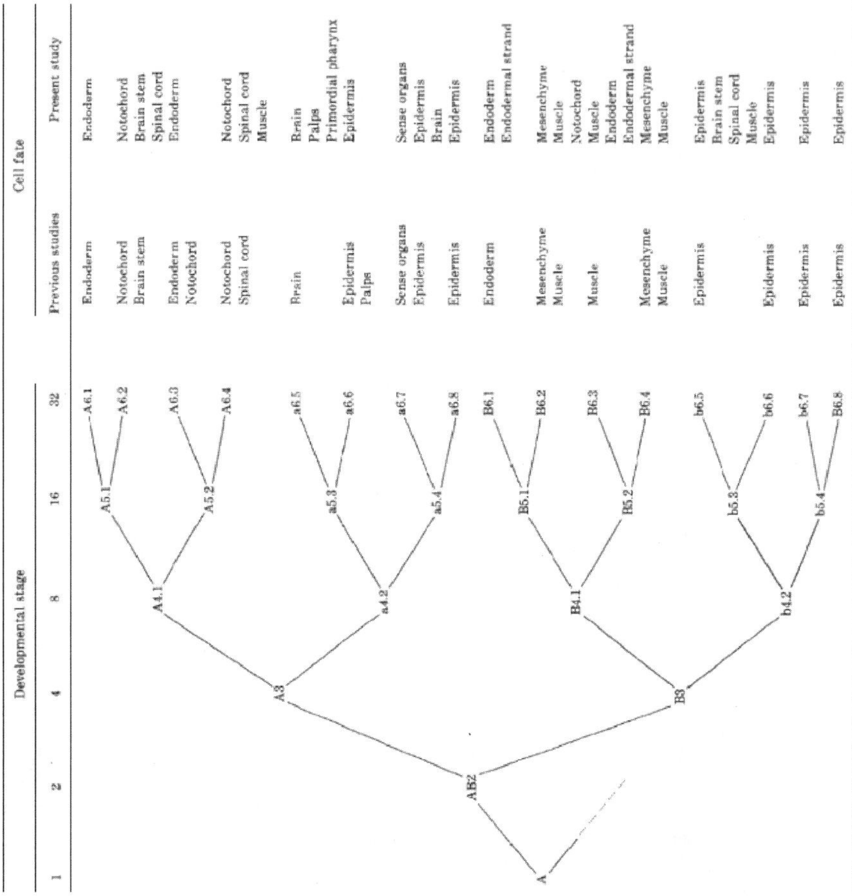

Figure 11.19. Cell lineage of a half-embryo of the ascidian *Halocynthia roretz*. At this stage the embryo is bilaterally symmetric, so the other 16 cells at the 32 cell stage, etc., are not shown. From Nishida & Satoh (1985)[103] with permission of Elsevier.

metamorphosis can be distinguished and assigned tentative differentiation trees (Proposition 224 in *Hierarchical Genome*[48]). We[48] and others[101] have lent credence to the idea of genome concatenation, perhaps too quickly. For insects, such as the caterpillar to butterfly metamorphosis, the idea of concatenation so far has not stood up to scrutiny[102]. Its value here is that it makes us think about the topology of differentiation trees, which could, of course, have evolved by more than one pathway.

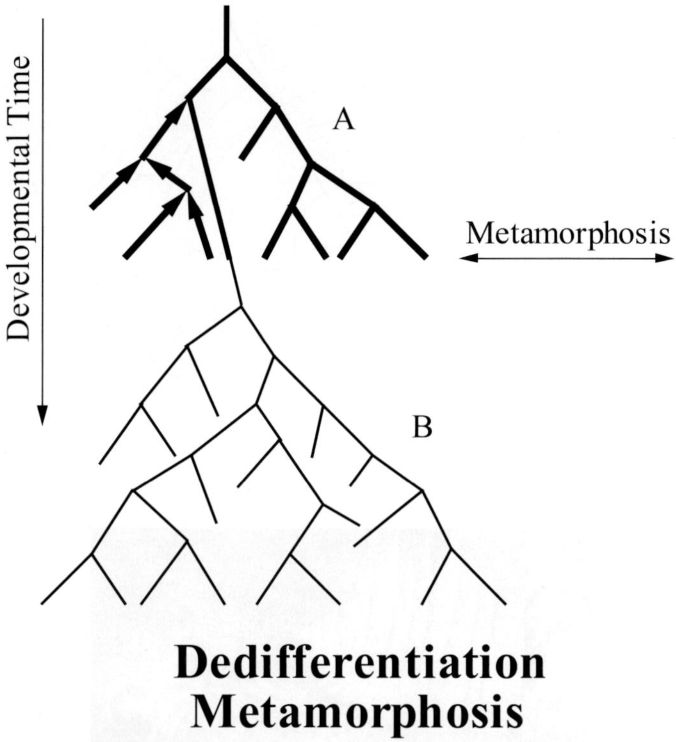

Figure 11.20. Two concatenated differentiation trees hypothesized to schematically represent metamorphosis in tunicates (also known as ascidians or sea squirts: Figure 2.17). This was called dedifferentiation metamorphosis in *Hierarchical Genome*[48]. In *dedifferentiation metamorphosis* a few or most of the larval tissues (differentiation tree A) dedifferentiate (reverse arrows) and then redifferentiate into the adult (differentiation tree B). The remaining tissues may contribute in a nutritive fashion. This is a possible case of the concatenation of two very different genomes[97]. Thus the ascidian larva may be close to our common ancestor, but the ascidian adult would not be. Copy of Figure 49 from *Hierarchical Genome*[48]. Note that developmental time increases downwards here.

To establish progress in macroevolution, we need make only one assumption: the probability of addition of branches to a differentiation tree exceeds the probability of pruning (Proposition 179 in *Hierarchical Genome*[48]). This is reasonable, because once a branch has been added, generating two new tissues, these functions generally diverge, perhaps in part avoiding a gene dose effect[104]. Once new functions evolve, loss of those functions may subsequently generally reduce fitness. Of course, there are exceptions, such as parasitism[105,106]. The net result is that

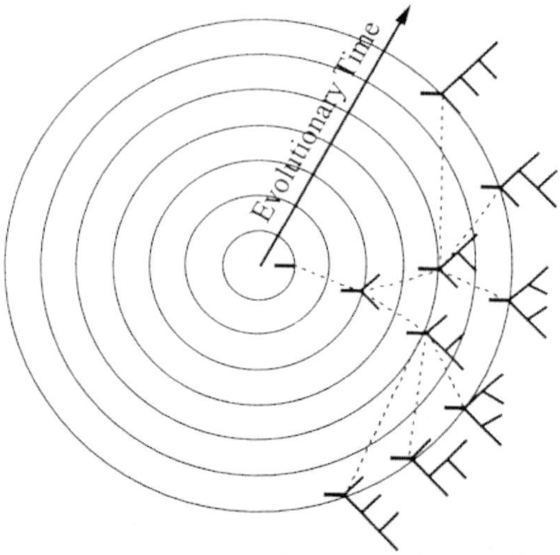

Figure 11.21. **Top**: The mystery of progressive evolution vanishes if the probability of growth of the differentiation tree exceeds that of pruning. Growth of the tree of life (top: dashed lines) is like frost on a window pane (**bottom**), and as inevitable. Evolution is depicted (top) as a radially organized phylogenetic tree (cf. Figure 11.13) whose nodes are differentiation trees (i.e., a tree of trees). Each macroevolutionary step involves a change in the topology of the differentiation tree. Note that two trees with the same topology can nevertheless be quite different in the origin of their branches, and the genes expressed in each branch. Taken as a whole, each differentiation tree also represents a different genome. Copy of Figure 60 from *Hierarchical Genome*[48]. **Bottom**: Frost on a window from Friedl (2014)[110] with kind permission of Jeffrey Friedl. Cf. Figure 11.12. See also Figure 7 in Alicea & Gordon (2014)[111].

progressive evolution is as inevitable (and as unpredictable in detail) as the growth of frost on a window pane (Figure 11.21). No special law of progress in evolution is needed[107], and progress in biological evolution is not a transfer of the notion of human cultural progress to biology[36].

Progressive evolution via growth of differentiation trees could be considered as the highest level version of what has been called *constructive neutral evolution*. In this notion, the retention of many duplications of parts of the genome leads to an increase in the number of parts of an organism[108,109]. The initial event of duplication is called neutral, because it in itself is not in any sense caused by natural selection.

Progress in evolution is thus seen as a nearly one-way ratchet mechanism: differentiation trees are more likely to ratchet to larger sizes than to smaller. The ratcheting mechanism is natural selection working on organisms, which are phenotypes whose genotype is represented by their differentiation trees. Each step forward is by growth of the differentiation tree.

Ratchets occur at all levels in biology, but to speak of them we need to understand human-made ratchets first. There are at least four kinds of ratchets involving gears, distinguished by whether or not the different components, the gear and/or the pawl, are symmetric (Figure 11.22). We can imagine the same options for an inchworm ratchet (Figure 11.23). Only ratchet mechanisms containing some sort of asymmetry can be one-way ratchets. Every motor molecule is a ratcheting mechanism[116,117] (Figure 5.7). Most models for kinesin, which walks on microtubules, and myosin, which walks on microfilaments, assume that the cytoskeletal molecule on which they walk is asymmetric, like a gear with asymmetric teeth. But the binding sites on the microtubules, for instance, for the misnamed "heads" of kinesin (they are "feet"), are symmetric, and the asymmetry appears to be in the ATP driven chemical cycle that creates mechanical potential energy in the kinesin, changing it from a symmetric to an asymmetric conformation. This potential energy is relieved as it takes a one-way 8 nm step and returns to a symmetric conformation. Then another ATP repeats the cycle and the kinesin moves on[118-121]. Bigger structures such as ribosome synthesizing proteins[122] and RNA polymerase[123] are also ratchets. In each case it will be necessary to determine which component(s) of the ratcheting system is/are asymmetric to achieve one-way ratcheting.

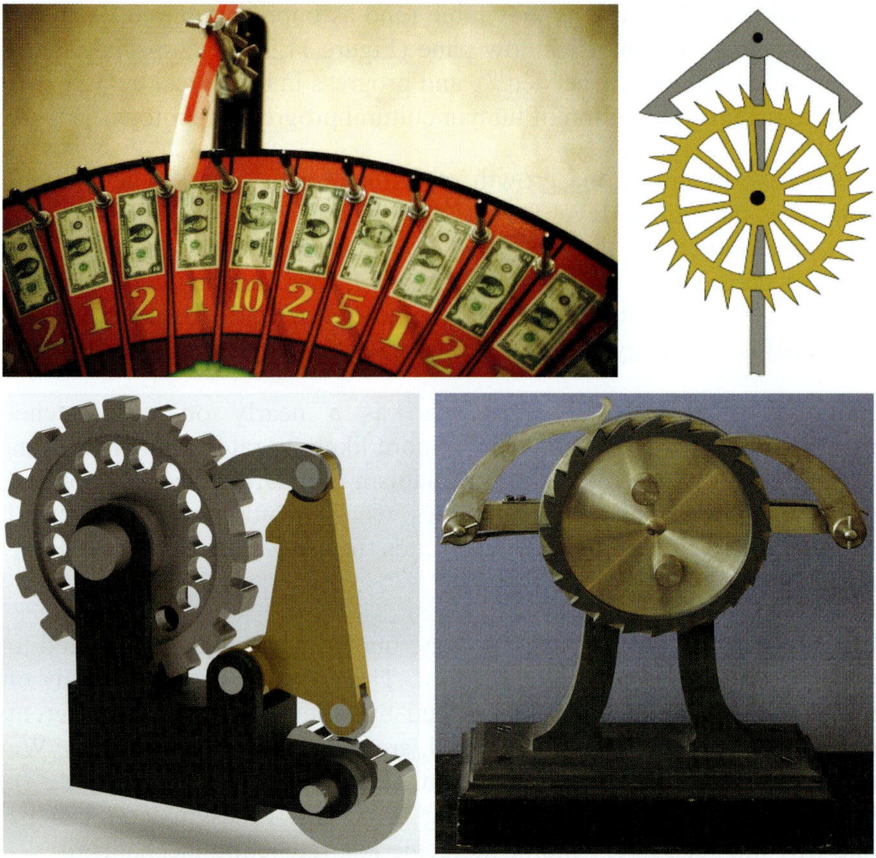

Figure 11.22. Historically a ratchet was defined as any device that turned smooth movement into small steps. **Top left**: The money wheel is a bidirectional ratchet, because it can be spun in either direction. Note that the stops and the pawl are both symmetric with respect to both directions. From Florida Casino Company[112]. **Top right**: A one-way ratchet with an asymmetric gear and a symmetric pawl. From Chetvorno[113] with permission under a Creative Commons CC0 1.0 Universal Public Domain Dedication, where an animation may be seen. **Bottom left**: A one-way ratchet driven by a rotating cam. The gear is symmetric and the pawl is asymmetric. From Trinityscsp[114], where a movie can be found. **Bottom right**: A one-way ratchet with an asymmetric gear and an asymmetric pawl. Original photo provided with kind permission by Francis Charles Moon of Cornell University. See Moon (2002)[115] for a movie. Note that in each case the gear is not the ratchet. The whole system is the rachet. But only one component need be asymmetric to achieve one-way ratcheting movement.

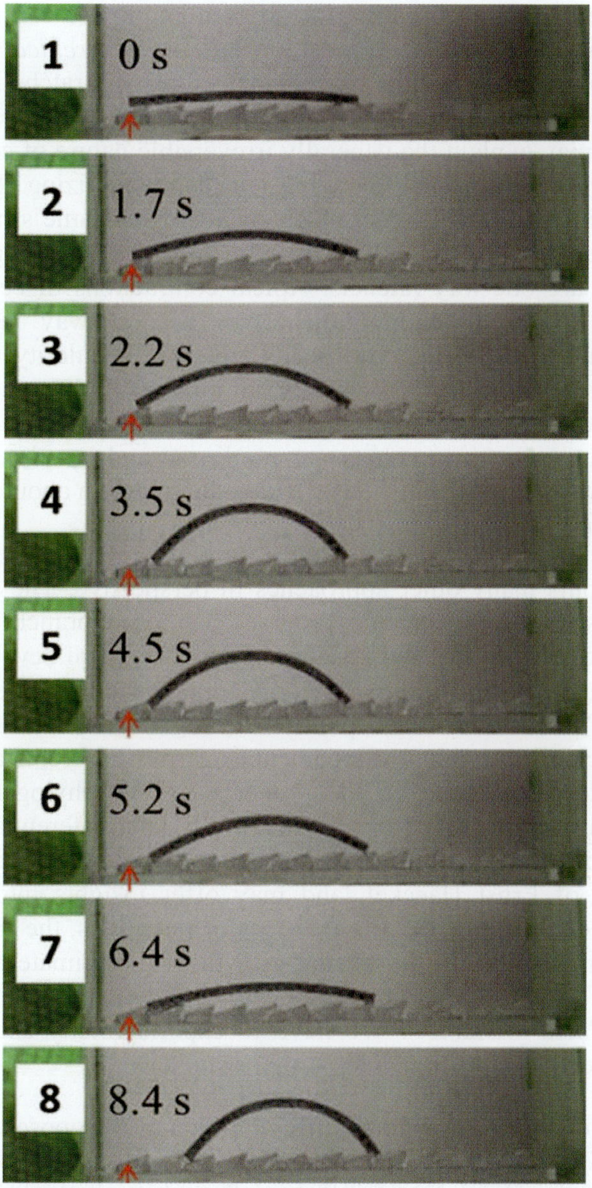

Figure 11.23. In this device an asymmetric linear "gear" (of infinite radius of curvature) interacts with a 2.4 cm pawl that is left/right symmetric, that moves from left to right in an inchworm-like movement. The pawl consists of a strip of a composite material that changes shape symmetrically via a changing magnetic field. From Kimura *et al.* (2012)[124] with permission of the Royal Society of Chemistry (Great Britain).

We have seen how each edge of a differentiation tree can be regarded as a one-way ratchet itself (Proposition 187 in *Hierarchical Genome*[48]), based on our generalization of the inchworm-like ratcheting seen in SWI/SNF remodelers and perhaps homeobox genes (Figures 9.23 and 9.25). This may explain the generally irreversible nature of cell differentiation[125]. Microtubules polymerizing and depolymerizing (Figure 5.23) are alternating one-way ratchet mechanisms of the same structure, both driven by energy from GTP. Cells replicating or changing neighbors (Chapter 7) are one-way ratchets. Apical contractions of cells are one-way ratchets[126,127]. Differentiation waves are one-way ratchets. The symbioses that led to eukaryotic cells and multicellularity may have a one-way ratcheting basis[128]. Basically, all biological phenomena that use energy to go from one state to another are one-way ratchets. In terms of their physics, one-way ratchets are nonlinear and operate away from thermodynamic equilibrium[129], as we have long known about life itself.

The differentiation tree as the central feature of evolution of multicellular organisms puts many other approaches in perspective. The logic of the genome, its so-called "genetic program", at each branch point is an alternation of the physics and genetics of embryological development. As such it forms the physical basis for Conrad Waddington's *epigenetic landscape*[130-132], and is the bauplan for an organism. The differentiation tree is Richard Dawkins' fundamental germ line replicator[142]. The differentiation tree could be the proper basis for phylogeny, i.e., for working out the details of the tree of life[73]. The differentiation tree is the basis for understanding the biogenetic law (ontogeny recapitulates phylogeny) of Ernst Haeckel, and may offer insights into stasis and punctuated equilibrium, Cope's Law, major radiations, the fractal nature of evolution[133], and the exponential growth of the primate brain. These predictions cannot be elaborated on now beyond the detailed expositions in *Hierarchical Genome*[48]. We have to experimentally ascertain a lot more differentiation trees of a lot more organisms.

In terms of evolution, a mutated differentiation tree proposes, and natural selection disposes (cf.[134]). Since the differentiation tree is both a representative of the hierarchy of differentiation waves, which are physical phenomena on the scale of the whole embryo, and of the corresponding hierarchical genome, the philosophical headaches of deciding whether physics or genetics prevails in morphogenesis and evolution[135-137] are resolved: the genetic program is a bifurcating alternation of genetic and mechanical events (Proposition 14 in *Hierarchical Genome*[48]).

In summary, not only have we explained embryogenesis, but we have also indicated how embryology can finally[138] be reunited with the grand sweep of evolution, after their "messy" intellectual divorce of a century ago[139,140]. In fact, it may be time to go a step beyond Theodosius Dobzhansky's 1973, oft repeated statement that "Nothing in biology makes sense except in the light of evolution"[141] and declare that furthermore evolution can only be understood in the light of embryogenesis[36].

1. Gould, S.J. (1988). Trends as changes in variance: a new slant on progress and directionality in evolution. *Journal of Paleontology* **62**(3), 319-329.
2. Gould, S.J. (1994). The evolution of life on the earth. *Scientific American* **271**(4), 84-91.
3. Huxley, J. (1923). *Essays of a Biologist*. New York, Alfred A. Knopf.
4. Van Valen, L.M. (1973). A new evolutionary law. *Evolutionary Theory* **1**, 1-30.
5. Ridley, M. (1993). *The Red Queen: Sex and the Evolution of Human Nature*. New York, Macmillan Publishing Co.
6. Voje, K.L., Ø.H. Nolen, L.H. Liow & N.C. Stenseth (2015). The role of biotic forces in driving macroevolution: beyond the Red Queen. *Proceedings of the Royal Society B-Biological Sciences* **282**(1808).
7. Carroll, L. (1875). *Through the Looking-Glass, and What Alice Found There*. New York, Macmillan & Co.
8. Smith, T.B., M.T. Kinnison, S.Y. Strauss, T.L. Fuller & S.P. Carroll (2014). Prescriptive evolution to conserve and manage biodiversity. *Annual Review of Ecology, Evolution, and Systematics* **45**, 1-22.
9. Mira, A., H. Ochman & N.A. Moran (2001). Deletional bias and the evolution of bacterial genomes. *Trends in Genetics* **17**(10), 589-596.
10. Poulin, R. (1995). Evolution of parasite life history traits: myths and reality. *Parasitology Today* **11**(9), 342-345.
11. Ebert, D. (1998). Evolution - Experimental evolution of parasites. *Science* **282**(5393), 1432-1435.
12. Fong, D.W., T.C. Kane & D.C. Culver (1995). Vestigialization and loss of nonfunctional characters. *Annual Review of Ecology, Evolution, and Systematics* **26**, 249-268.
13. West-Eberhard, M.J. (2002). *Developmental Plasticity and Evolution*. Oxford, Oxford University Press.
14. Eisenberg, L. (2008). The Tree of Life. https://www.evogeneao.com/learn/tree-of-life
15. Margulis, L. (1993). *Symbiosis in Cell Evolution: Microbial Communities in the Archean and Proterozoic Eons*. New York, W.H. Freeman and Co., 2nd.

16. Perez-Jimenez, R., A. Inglés-Prieto, Z.M. Zhao, I. Sanchez-Romero, J. Alegre-Cebollada, P. Kosuri, S. Garcia-Manyes, T.J. Kappock, M. Tanokura, A. Holmgren, J.M. Sanchez-Ruiz, E.A. Gaucher & J.M. Fernandez (2011). Single-molecule paleoenzymology probes the chemistry of resurrected enzymes. *Nature Structural & Molecular Biology* **18**(5), 592-596.
17. Kreimer, A., E. Borenstein, U. Gophna & E. Ruppin (2008). The evolution of modularity in bacterial metabolic networks. *Proceedings of the National Academy of Sciences USA* **105**(19), 6976-6981.
18. Gribaldo, S. & C. Brochier (2009). Phylogeny of prokaryotes: does it exist and why should we care? *Research in Microbiology* **160**(7), 513-521.
19. Hazen, R.M. (2010). Evolution of minerals. *Scientific American* **302**(3), 58-65.
20. Douglas, S. (2005). Mineralogical footprints of microbial life. *American Journal of Science* **305**(6-8), 503-525.
21. Caldwell, D.E. & J.W. Costerton (1996). Are bacterial biofilms constrained to Darwin's concept of evolution through natural selection? *Microbiologia* **12**(3), 347-358.
22. Mathieu, L.G. & S. Sonea (1996). Review of the unique mode of evolution of bacteria: An opinion. *Symbiosis* **21**(3), 199-207.
23. Reid, R.G.B. (2007). *Biological Emergences: Evolution by Natural Experiment*. Cambridge, MIT Press.
24. Papke, R.T. & D.M. Ward (2004). The importance of physical isolation to microbial diversification. *FEMS Microbiology Ecology* **48**(3), 293-303.
25. Pikuta, E.V., R.B. Hoover & J. Tang (2007). Microbial extremophiles at the limits of life. *Critical Reviews in Microbiology* **33**(3), 183-209.
26. Seckbach, J., A. Oren & H. Stan-Lotter, Eds. (2013). *Polyextremophiles: Life Under Multiple Forms of Stress*. Dordrecht, Springer Netherlands.
27. McShea, D.W. (1991). Complexity and evolution: what everybody knows. *Biology & Philosophy* **6**(3), 303-324.
28. Bonner, J.T. (1988). *The Evolution of Complexity by Means of Natural Selection*. Princeton, Princeton University Press.
29. Dmitriev, V.Y. (2011). Evolution of biodiversity: Hyperbola or exponent? *Paleontological Journal* **45**(6), 705-708.
30. Hazen, R.M. & J.M. Ferry (2010). Mineral evolution: mineralogy in the fourth dimension. *Elements* **6**(1), 9-12.
31. Holland, J.H. (2012). *Signals and Boundaries: Building Blocks for Complex Adaptive Systems*, MIT Press.
32. Standish, R.K. (2012). Complexity of networks (reprise). *Complexity* **17**(3), 50-61.
33. Mikhailovsky, G. & A. Levich (2015). Entropy, information and complexity or which aims the arrow of time? *Entropy* **17**(7), 4863-4890.
34. Long Jr., E.L. (1952). *Religious Beliefs of American Scientists*. Philadelphia, Westminster Press.
35. Ecklund, E.H. & J.Z. Park (2009). Conflict between religion and science among academic scientists? *Journal for the Scientific Study of Religion* **48**(2), 276-292.
36. Ruse, M. (1996). *Monad to Man: The Concept of Progress in Evolutionary Biology*. Cambridge, Massachusetts, Harvard University Press.
37. Dembski, W.A. & M. Ruse (2004). *Debating Design: From Darwin to DNA*. Cambridge, Cambridge University Press.

38. Seckbach, J. & R. Gordon, Eds. (2008). *Divine Action and Natural Selection: Science, Faith and Evolution*. Singapore, World Scientific.
39. Seckbach, J., Ed. (2016). *Theology and Science: Discussions about Faith and Facts [in preparation]*, World Scientific Publishing.
40. Gordon, R. (2008). Preface 3. Monotheism: the basis for unifying Abrahamic religion and science? In: *Divine Action and Natural Selection: Science, Faith and Evolution*. Eds.: J. Seckbach & R. Gordon. Singapore, World Scientific: xlv-lii.
41. Gould, S.J. (1997). Nonoverlapping magisteria. *Natural History* **106**(2), 16-24.
42. Kaye, H.L. (2008). Are science and religion complementary perspectives? *Society* **45**(2), 152-154.
43. Larmer, R. (2002). Is there anything wrong with "God of the gaps" reasoning? *International Journal for Philosophy of Religion* **52**(3), 129-142.
44. Berlinski, D. (2008). The God of the gaps. *Commentary* **125**(4), 34-40.
45. Shermer, M. (2013). Gods of the gaps: Arguments of divine intervention—alien or otherwise—start with ignorance. *Scientific American* **309**(1), 92.
46. Gordon, R. (2008). Over-confident anti-creationists versus over-confident creationists. In: *Divine Action and Natural Selection: Science, Faith and Evolution*. Eds.: R. Gordon & J. Seckbach. Singapore, World Scientific: 216-247.
47. Anderson, O. (2013). Kinds of gaps in knowledge: The conflict of appeals to God and methodological naturalism in developing explanations of the world. *Heythrop Journal-A Quarterly Review of Philosophy and Theology* **54**(4), 574-589.
48. Gordon, R. (1999). *The Hierarchical Genome and Differentiation Waves: Novel Unification of Development, Genetics and Evolution*. Singapore & London, World Scientific & Imperial College Press: [http://www.worldscientific.com/worldscibooks/10.1142/2755].
49. Spemann, H. (1938). *Embryonic Development and Induction*. New York, Hafner Publishing Co., 1962 reprint.
50. Driesch, H.A.E. (1894). *Analytische Theorie der Organischen Entwicklung/Analytic Theory of Organic Development*. Leipzig, Verlag Von Wilhelm Engelman.
51. Lopashov, G.V. (1961). The development of the vertebrate eye in relation to the problem of embryonic induction. *Annales Botanici Societatis Zoologicæ-Botanicæ Fennicæ "Vanamo"* **22**(6), 1-17.
52. Byrnes, W.M. (2009). Ernest Everett Just, Johannes Holtfreter, and the origin of certain concepts in embryo morphogenesis. *Molecular Reproduction and Development* **76**(10), 912-921.
53. Barth, L.G. & L.J. Barth (1954). *The Energetics of Development: A Study of Metabolism in the Frog Egg*. New York, Columbia University Press.
54. Jacobson, A.G. (1955). The roles of the optic vesicle and other head tissues in lens induction. *Proceedings of the National Academy of Sciences USA* **41**(7), 522-525.
55. Cummings, F.W. & K. Lu (2015). Self-organized patterning and morphogenesis in embryonic epithelia from antagonistic signals and fiber networks. http://dx.doi.org/10.1101/023069

56. Gordon, R. (2011). Epilogue: the diseased breast lobe in the context of X-chromosome inactivation and differentiation waves. In: *Breast Cancer: A Lobar Disease*. Ed.: T. Tot. London, Springer: 205-210.
57. Doniach, T., C.R. Phillips & J.C. Gerhart (1992). Planar induction of anteroposterior pattern in the developing central nervous system of *Xenopus laevis*. *Science* **257**(5069), 542-545.
58. Keller, R.E., J. Shih, A.K. Sater & C. Moreno (1992). Planar induction of convergence and extension of the neural plate by the organizer of *Xenopus*. *Developmental Dynamics* **193**(3), 218-234.
59. Nieuwkoop, P.D. (1993). Vertical and planar induction in early amphibian development during meso-endoderm formation and neural plate development. *Acta Biologiae Experimentalis Sinica* **26**(4), 307-315.
60. Nieuwkoop, P.D. & K. Koster (1995). Vertical versus planar induction in amphibian early development. *Development Growth & Differentiation* **37**(6), 653-668.
61. Blitz, I.L. & K.W.Y. Cho (1995). Anterior neurectoderm is progressively induced during gastrulation: the role of the *Xenopus* homeobox gene *orthodenticle*. *Development* **121**(4), 993-1004.
62. Fleury, V. & R. Gordon (2012). Coupling of growth, differentiation and morphogenesis: an integrated approach to design in embryogenesis. In: *Origin(s) of Design in Nature: A Fresh, Interdisciplinary Look at How Design Emerges in Complex Systems, Especially Life*. Eds.: L. Swan, R. Gordon & J. Seckbach. Dordrecht, Springer: 385-428.
63. Gordon, R., N.K. Björklund & P.D. Nieuwkoop (1994). Dialogue on embryonic induction and differentiation waves. *International Review of Cytology* **150**, 373-420.
64. Bard, J.B.L., M.H. Kaufman, C. Dubreuil, R.M. Brune, A. Burger, R.A. Baldock & D.R. Davidson (1998). An internet-accessible database of mouse developmental anatomy based on a systematic nomenclature. *Mechanisms of Development* **74**(1-2), 111-120.
65. Gordon, R. & J.B.L. Bard (1999). Establishing a HUGO Tissue Nomenclature Committee. *Human Genome News*, in press, then "rejected for political reasons".
66. Bracegirdle, B. (1977). The history of histology: A brief survey of sources. *History of Science* **15**(28), 77-101.
67. Conn, H.J. (1930). The history of staining: Anilin dyes in histology. *Stain Technology* **5**(1), 3-12.
68. Oyolu, C., F. Zakharia & J. Baker (2012). Distinguishing human cell types based on housekeeping gene signatures. *Stem Cells* **30**(3), 580-584.
69. Du, G., A. Ravetto, Q. Fang & J.M.J. den Toonder (2011). Cell types can be distinguished by measuring their viscoelastic recovery times using a micro-fluidic device. *Biomedical Microdevices* **13**(1), 29-40.
70. Martin, F.L., J.G. Kelly, V. Llabjani, P.L. Martin-Hirsch, I.I. Patel, J. Trevisan, N.J. Fullwood & M.J. Walsh (2010). Distinguishing cell types or populations based on the computational analysis of their infrared spectra. *Nature Protocols* **5**(11), 1748-1760.
71. Ryba, T., I. Hiratani, J. Lu, M. Itoh, M. Kulik, J. Zhang, T.C. Schulz, A.J. Robins, S. Dalton & D.M. Gilbert (2010). Evolutionarily conserved replication timing profiles predict long-range chromatin interactions and distinguish closely related cell types. *Genome Research* **20**(6), 761-770.

72. Diestel, R. (2010). *Graph Theory*. Heidelberg, Springer-Verlag, 4th.
73. Alicea, B. & R. Gordon (2016). Quantifying mosaic development: Towards an Evo-Devo Postmodern Synthesis of the evolution of development via differentiation trees of embryos [invited]. *Biology (Basel)* **33**(Special Issue: Beyond the Modern Evolutionary Synthesis- what have we missed?, Ed. John S. Torday), doi:10.3390/biology5030033.
74. Fankhauser, G. (1972). Memories of great embryologists. Reminiscences of F. Baltzer, H. Spemann, F. R. Lillie, R. G. Harrison, and E. G. Conklin. *American Scientist* **60**(1), 46-55.
75. Conklin, E.G. (1905). The organization and cell lineage of the ascidian egg. *Journal of the Academy of Natural Sciences of Philadelphia* **13 Ser. 2**, 1-119.
76. Williams, J. (2007). Kite Aerial Photography (KAP) / Discuss: Something more green? , http://www.flickr.com/groups/kiteaerialphotography/discuss/72157600025192212/
77. Zelditch, M.L., H.D. Sheets & W.L. Fink (2001). The spatial complexity and evolutionary dynamics of growth. In: *Beyond Heterochrony: The Evolution of Development*. Ed.: M.L. Zelditch. New York, Wiley-Liss: 145-194.
78. Agutter, P.S. & J.A. Tuszynski (2011). Analytic theories of allometric scaling. *Journal of Experimental Biology* **214**(7), 1055-1062.
79. Agnoli, G.L. & P. Rosa (2010). Glossary of Biological, Ecological, Evolutionary terms. http://www.chrysis.net/chrysis/resources/glossary_bio.htm
80. Rosindell, J. & L.J. Harmon (2012). OneZoom: A Fractal Explorer for the Tree of Life. *PLoS Biology* **10**(10), e1001406.
81. Ciccarelli, F.D., T. Doerks, C. von Mering, C.J. Creevey, B. Snel & P. Bork (2006). Toward automatic reconstruction of a highly resolved tree of life [+Erratum]. *Science* **311**(5765), 1283-1287.
82. Letunic, I. & P. Bork (2007). Interactive Tree Of Life (iTOL): an online tool for phylogenetic tree display and annotation. *Bioinformatics* **23**(1), 127-128.
83. Vickers, T. (2009). A highly resolved Tree of Life. https://commons.wikimedia.org/wiki/File:CollapsedtreeLabels-simplified.svg
84. Sommer, R.J. (2000). Evolution of nematode development. *Current Opinion in Genetics & Development* **10**(4), 443-448.
85. Hedgecock, E.M. (1985). Cell lineage mutants in the nematode *Caenorhabditis elegans*. *Trends in Neurosciences* **8**(6), 288-293.
86. Thomas, C.G., W. Wang, R. Jovelin, R. Ghosh, T. Lomasko, T. Quang, L. Kruglyak, L.D. Stein & A.D. Cutter (2015). Full-genome evolutionary histories of selfing, splitting, and selection in *Caenorhabditis*. *Genome Research* **25**(5), 667-678.
87. Sternberg, P.W. & H.R. Horvitz (1981). Gonadal cell lineages of the nematode *Panagrellus redivivus* and implications for evolution by the modification of cell lineage. *Developmental Biology* **88**(1), 147-166.
88. Huxley, J.S. (1942). *Evolution, The Modern Synthesis*. London, George Allen & Unwin.
89. Teotonio, H. & M.R. Rose (2001). Perspective: reverse evolution. *Evolution* **55**(4), 653-660.
90. Ohno, S. (1970). *Evolution by Gene Duplication*. New York, Springer-Verlag.
91. DePamphilis, M.L. & S.D. Bell (2010). *Genome Duplication*, Garland Publishing.

92. Henikoff, S., E.A. Greene, S. Pietrokovski, P. Bork, T.K. Attwood & L. Hood (1997). Gene families: the taxonomy of protein paralogs and chimeras. *Science* **278**(5338), 609-614.
93. Ramadas, R. & M. Thattai (2013). New organelles by gene duplication in a biophysical model of eukaryote endomembrane evolution. *Biophysical Journal* **104**(11), 2553-2563.
94. Hoffmann, F.G., J.C. Opazo & J.F. Storz (2010). Gene cooption and convergent evolution of oxygen transport hemoglobins in jawed and jawless vertebrates. *Proceedings of the National Academy of Sciences USA* **107**(32), 14274-14279.
95. Seaborg, D.M. (1999). Evolutionary feedback: a new mechanism for stasis and punctuated evolutionary change based on integration of the organism. *Journal of Theoretical Biology* **198**(1), 1-26.
96. Conant, G.C. & K.H. Wolfe (2008). Turning a hobby into a job: how duplicated genes find new functions. *Nature Reviews Genetics* **9**(12), 938-950.
97. Williamson, D.I. (1992). *Larvae and Evolution: Toward a New Zoology*. New York, Chapman & Hall.
98. Williamson, D.I. (2003). *The Origins of Larvae*, Kluwer Academic Publishers, 2nd.
99. Ferrebee, W.M. (2014). Tunicates. https://ferrebeekeeper.wordpress.com/2014/11/21/tunicates/
100. Get (2011). [Tunicate larva]. http://gwis2.circ.gwu.edu/~atkins/newwebpages/Embryo/Compembryo.html
101. Margulis, L. & D. Sagan (2003). *Acquiring Genomes: A Theory of the Origins of Species*. New York, Basic Books.
102. Willis, J.H. & D.L. Cox-Foster (2010). Insect metamorphosis via hybridogenesis: An evidentiary rebuttal. *Journal of Insect Physiology* **56**(4), 333-335.
103. Nishida, H. & N. Satoh (1985). Cell lineage analysis in ascidian embryos by intracellular injection of a tracer enzyme. II. The 16- and 32-cell stages. *Developmental Biology* **110**(2), 440-454.
104. Gordon, R. (1997). Digital Burgess Conference Reviews: Digital Digital Burgess Conference Follow-up: More on differentiation trees and how they relate to embryonic development, genetics, and evolution. http://www.biota.org/conf97/gordon.html
105. Keeling, P.J. & C.H. Slamovits (2005). Causes and effects of nuclear genome reduction. *Current Opinion in Genetics & Development* **15**(6), 601-608.
106. Olson, P.D., M. Zarowiecki, F. Kiss & K. Brehm (2012). Cestode genomics - progress and prospects for advancing basic and applied aspects of flatworm biology. *Parasite Immunology* **34**(2-3), 130-150.
107. McShea, D.W. & R.N. Brandon (2010). *Biology's First Law: The Tendency for Diversity and Complexity to Increase in Evolutionary Systems*, University of Chicago Press.
108. Stoltzfus, A. (1999). On the possibility of constructive neutral evolution. *Journal of Molecular Evolution* **49**(2), 169-181.
109. Lukeš, J., J.M. Archibald, P.J. Keeling, W.F. Doolittle & M.W. Gray (2011). How a neutral evolutionary ratchet can build cellular complexity. *IUBMB Life* **63**(7), 528-537.

110. Friedl, J. (2014). Pretty Frost Pattern on a Window. http://regex.info/blog/2014-01-09/2367
111. Alicea, B. & R. Gordon (2014). Toy models for macroevolutionary patterns and trends. *BioSystems* **122**(Special Issue: Patterns of Evolution), 25-37.
112. Florida Casino Company (2015). Money Wheel. http://www.floridacasinocompany.com/galleries/?galAlbum=11 - prettyPhoto
113. Chetvorno (2014). File:Anchor escapement animation 217x328px.gif. https://commons.wikimedia.org/wiki/File:Anchor_escapement_animation_217x328px.gif
114. Trinityscsp (2014). Cam-operated Ratchet Pawl. https://grabcad.com/library/cam-operated-ratchet-pawl-1
115. Moon, F.C. (2002). Model: N01 Ratchet and Pawl Mechanism. http://kmoddl.library.cornell.edu/model.php?m=247&movie=show
116. Duke, T. (2002). Modelling motor protein systems. In: *Physics of biomolecules and cells. Physique des biomolécules et des cellules*. Eds.: F. Flyvbjerg, F. Jülicher, P. Ormos & F. David, Springer Berlin/Heidelberg. **75**: 95-143.
117. Chowdhury, D. (2013). Stochastic mechano-chemical kinetics of molecular motors: A multidisciplinary enterprise from a physicist's perspective. *Physics Reports-Review Section of Physics Letters* **529**(1), 1-197.
118. Asbury, C.L., A.N. Fehr & S.M. Block (2003). Kinesin moves by an asymmetric hand-over-hand mechanism. *Science* **302**(5653), 2130-2134.
119. Bolterauer, H., J.A. Tuszynski & E. Unger (2005). Directed binding: A novel physical mechanism that describes the directional motion of two-headed kinesin motor proteins. *Cell Biochemistry and Biophysics* **42**(2), 95-119.
120. Bolterauer, H., J.A. Tuszynski & E. Unger (2006). A directed binding mechanism of processive motion for the kinesin motor protein families. *Acta Physica Polonica B* **37**(5), 1425-1443.
121. Bolterauer, H., J.A. Tuszynski & E. Unger (2006). *The Kinesin and Myosin walk, principles, models and calculations [slides]*. Les Houches, France, L'Ecole de Physique des Houches.
122. Wikipedia (2015). Ribosome. http://en.wikipedia.org/wiki/Ribosome
123. Bustamante, C., W. Cheng & Y.X. Meija (2011). Revisiting the central dogma one molecule at a time. *Cell* **144**(4), 480-497.
124. Kimura, T., Y. Umehara & F. Kimura (2012). Magnetic field responsive silicone elastomer loaded with short steel wires having orientation distribution. *Soft Matter* **8**(23), 6206.
125. Herrick, C.E. (1920). Irreversible differentiation and orthogenesis. *Science* **51**, 621-625.
126. Sawyer, J.M., J.R. Harrell, G. Shemer, J. Sullivan-Brown, M. Roh-Johnson & B. Goldstein (2010). Apical constriction: A cell shape change that can drive morphogenesis. *Developmental Biology* **341**(1), 5-19.
127. Mason, F.M., M. Tworoger & A.C. Martin (2013). Apical domain polarization localizes actin-myosin activity to drive ratchet-like apical constriction. *Nature Cell Biology* **15**(8), 926-936.
128. Libby, E. & W.C. Ratcliff (2014). Ratcheting the evolution of multicellularity. *Science* **346**(6208), 426-427.

129. Sokolov, I.M. (2001). Irreversible and reversible modes of operation of deterministic ratchets. *Physical Review E: Statistical, Nonlinear, and Soft Matter Physics* **63**(2 Pt 1), 021107.
130. Waddington, C.H. (1957). *The Strategy of the Genes, A Discussion of Some Aspects of Theoretical Biology*. London, George Allen and Unwin.
131. Waddington, C.H. (1966). *Principles of Development and Differentiation*. New York, Macmillan Co.
132. Waddington, C.H. (1977). Stabilization in systems: chreods and epigenetic landscapes. *Futures* **9**(2), 139-146.
133. Gordon, R. (1992). The fractal physics of biological evolution. In: *Dynamical Phenomena at Interfaces, Surfaces and Membranes*. Eds.: D. Beysens, N. Boccara & G. Forgacs. Commack, N.Y., NOVA Science Publishers: 99-111.
134. Batten, D., S. Salthe & F. Boschetti (2008). Visions of evolution: Self-organization proposes what natural selection disposes. *Biological Theory* **3**(1), 17-29.
135. Linde-Medina, M. (2010). Two "EvoDevos". *Biology Theory* **5**(1), 7-11.
136. Linde-Medina, M. (2010). Natural selection and self-organization: A deep dichotomy in the study of organic form. *Ludus Vitalis* **18**(34), 25-56.
137. Ramírez-Trejo, L., B. Demarest, J. Van Poucke & G. Van de Vijver (2011). Conflated epistemology or how to lose the organism (again). *Ludus Vitalis* **19**(36), 353-385.
138. Yoshida, Y. & H. Nakao (2015). EvoDevo as a motley aggregation: local integration and conflicting views of genes during the 1980s. *Biology Theory* **10**(2), 156-166.
139. Devillers, C., J. Chaline & B. Laurin (1990). Plaidoyer pour une embryologie évolutive [Plea for an evolutionary embryology] [French]. *Recherche* **21**(222), 802-809.
140. Williams, D.M. & P.L. Forey (2004). *Milestones in Systematics*, CRC Press.
141. Dobzhansky, T. (1973). Nothing in biology makes sense except in the light of evolution. *American Biology Teacher* **35**(3), 125-129.
142. Dawkins, R. (1982). *The Extended Phenotype, The Gene as the Unit of Selection*. Oxford, Oxford University Press.

Chapter 12

Wholeness and the Implicate Embryo: Embryogenesis as Self-Construction of the Observer

A remarkable book was published in 1926 by statesman Jan Christiaan Smuts under the title *Holism and Evolution*[1,2]. For such a busy and world-involved person, a founder of the League of Nations, the predecessor to the United Nations, and prime minister governing the large country of pre-Apartheid South Africa[3,4], to master the fundamentals of the then new fields of quantum mechanics and relativity well enough to write about them, is perhaps unprecedented. The word holism, which he coined[5], has since been hijacked for all sorts of wooly thinking. Just search the Internet for "12 holistic beauty tips to get beautiful skin", "holistic camping", and "holistic spirituality". On Amazon.com you can buy books on *Holistic Herbal Healing Medicine*, *Holistic Tarot*, *The Holistic Orchard*, *Holistic Aromatherapy for Animals*, *Career Counseling: A Holistic Approach*, with more than 5500 other such titles available. Our canine companions Fred and Trusty eat "Holistic Dog Food". Google Books lists over 200 volumes on holism, starting with *Holism and Ecology* and *Holism: A Wonderful Faith*. The database Web of Science lists over 12,000 publications with the words "holistic" or "holism" in the title. Going by Google Scholar, about 1700 academic publications cite Smuts' book. Smuts has had quite an impact on our popular and learned culture.

Holism is frequently contrasted with reductionism[6], as if these are two irreconcilable views of how the universe is constructed. We need to discuss this here because molecular developmental biology has been primarily reductionist in its approach[7], while embryogenesis would seem to be very much a holistic phenomenon[8,9]. As a result, many scientists will accept a molecular answer to a morphogenetic question, without being aware that the gap between the molecules and a full explanation of how an embryo builds itself (or a part of itself) is enormous[10]. The fundamental question here is the relationship between what we identify as wholes and parts. As we shall see, arguments about this extend to the relationship between the universe and its parts, and even the relationship of our universe to other hypothesized universes. Thus, we have to examine ideas at many levels, not just the in between one of embryogenesis.

Consider your own body and what you do with it. Let's just go down to the level of cells, of which an "average" adult[11] has at least 37.2 trillion (37.2×10^{12}). Is everything you do the result of the interactions of these 40 trillion cells, or do you exercise some top-down control? If we go down to the "inanimate" level of atoms[12], you are made of 7×10^{27} of them banging around. Are your cells (the parts), or the atoms (their 2×10^{15} per cell), or are you (the whole), commanding your actions? If you take the holistic point of view, you are in charge. If you take the reductionist point of view, your cells or atoms are in charge. Which of these three versions of you is reading this book? Is your whole indeed something more than the sum of your parts, or is that wishful thinking? In this chapter we are going to turn this question over and over, examining it top to bottom, so to speak. We won't come to any definitive answers, but will end up with a new way of looking at embryos that suggests new experimental and conceptual approaches to understanding how embryos build themselves. In writing this we found it a mind blowing, headache generating trip. You may too.

Reductionism has been around for thousands of years, starting at least from the ancient Indian and Greek atomists[13-15]. The Greeks in particular proposed that the universe, insofar as they comprehended it, is made of particles that are indivisible, indestructible and immutable, that bang around in the "void" and sometimes stick to one another, and which thereby form everything we know. The deduction from chemical experiments that atoms actually exist and combine with one another in fixed proportions, by John Dalton[16] in the early 1800s, clinched the atomic view, and gave reductionism what seemed a firm footing. Despite the success of theories based on the existence of atoms by Ludwig Boltzmann and others, there was widespread skepticism. Scientists of the day doubted that atoms existed and their doubt may have contributed to Boltzmann's suicide[17-19]. Ironically Boltzmann died just about when Albert Einstein came up with a proof of atoms. Einstein showed that

Brownian motion, the jittering of small particles seen by light microscopy, is due to their being banged at random by moving water molecules[20]. Dick can remember lifting his own skepticism half a century after Einstein's abstract proof, in 1967, when he saw atoms moving in a live image formed by a field ion emission microscope in the laboratory of Eugene S. Machlin[21] at Columbia University, and got to use the microscope himself. For Natalie atoms became real at a demonstration using an atomic force microscope[22] to see the atoms of carbon. Indeed, here seeing is believing.

In biology, the atom is generally the level at which we start our studies. The energies involved in splitting atoms or fusing atomic nuclei releases ionizing radiation which damages living organisms. So we think of organisms as made up of stable atoms, and usually do not have to trouble our thoughts with what is going on at lower, subatomic levels. Exceptions are when we have to think about the key role of natural background radiation in generating mutations, and thus in evolution[23]. This energy also keeps the ground warm in winter[24], permitting life to go on under the snow[25].

From atoms to molecules to cells to tissues to organisms is a familiar litany, and presumed in the reductionist point of view to explain everything. Some take it further, extending reductionism to social groups, populations, ecological systems, and all of biota. Our job as reductionist scientists is to but work out the details, explaining each higher order phenomenon in terms of the ones below. Indeed, Dick expressed this attitude in introducing a Festschrift for his Ph.D. mentor, the late Terrell Hill: "It is, after all, statistical mechanics that has shown us exactly how it is possible to intellectually bridge the gap between hierarchical levels"[26]. Statistical mechanics is the discipline of predicting temperature, pressure, etc., from the random motions of molecules[27]. Reductionism has indeed explained a lot, and its very success makes many doubt that holism is necessary, productive or in any way a proper model for the universe.

So let's get back to Smuts, to see what was bugging him about reductionism that led him to organize his thoughts about holism[28,29]. Smuts starts with the observation that "matter, life and mind still remain utterly disparate phenomena"[1]. This is still true today. We do not yet understand the relationships between these three facets of our universe.

Embryogenesis obviously bridges all three, since a living, developing embryo contains and takes in and ejects matter, is alive, and self-constructs itself into a thinking organism. It is generally assumed that life arose from matter. As we showed in Chapter 1, embryogenesis began shortly after the origin of life. So indeed, embryogenesis may be the key to understanding how matter, life and mind are intertwined.

These are the wholes Smuts observed in the world of 1926:

1. atoms
2. molecules
3. cells
4. organisms
5. personalities

The first two are now "explained" by quantum mechanics. Let's look at how this comes about. Quantum mechanics has proved very much to be the physics of wholeness. Perhaps it can help us understand wholeness at other "higher" levels, perhaps not. So let's take the plunge and try to understand what physicists have been up to for the past century.

In its simplest form, quantum wholeness can be seen in the very formulation of the Schrödinger equation[30]. This equation is the basic equation for quantum mechanics. In its nonrelativistic form it reads:

$$-\frac{h^2}{8\pi^2}\sum_{n=1}^{N}\frac{1}{m_n}\nabla_n^2\Psi_k(\mathbf{r}_1,\ldots,\mathbf{r}_N,t)+V(\mathbf{r}_1,\ldots,\mathbf{r}_N,t)\Psi_k(\mathbf{r}_1,\ldots,\mathbf{r}_N,t)=E_k\Psi_k(\mathbf{r}_1,\ldots,\mathbf{r}_N,t)$$

Let's consider what this equation means for a single molecule of water, H_2O, compared to the gases that can be (explosively) combined to make it, H_2 and O_2. A hydrogen atom (H) consists of an atomic nucleus containing one proton, with one electron in orbit about it. A molecule H_2 therefore has 4 particles, $n = 1,2,3,4$ in the summation ($\mathbf{\Sigma}$) in the equation. The $N = 4$ particles are all attracting and repelling one another, these interactions being summarized by the potential energy V. Their masses are given by m_n. The quantum wave function $\boldsymbol{\psi}_k$ (Greek letter psi), also known as the "state" of the system, with energy E_k, is therefore a function of the positions $\mathbf{r}_n$ of the 4 particles at a given moment in time t. (h is Planck's constant.) Each position requires 3 coordinates in space to specify it, i.e., $\mathbf{r}_n = (x_n,y_n,z_n)$, so $\boldsymbol{\psi}_k$ is therefore a time dependent function of 12 variables for molecular hydrogen. $\boldsymbol{\psi}_k$ represents a wave from which one can calculate the probability of finding the molecule in various states. The states are indexed by *k*. The "quantum" in quantum mechanics is reflected in the fact that the Schrödinger equation usually only has solutions for discrete values of the energy E_k, i.e., the energy levels are "quantized". One way to appreciate the quantization of energy levels is by watching and listening to the vibrations of a guitar string (Figure 12.1).

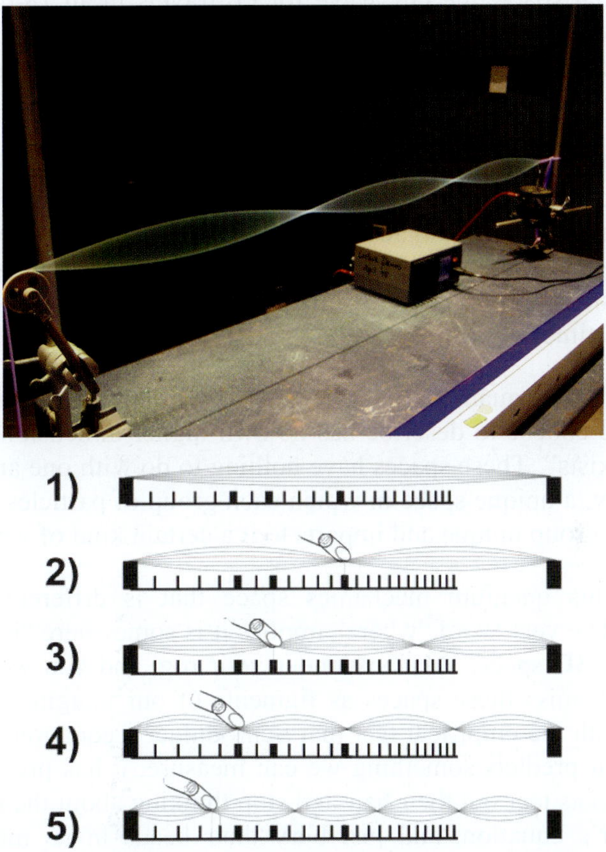

Figure 12.1. **Top**: A vibrating string held taut between two points has an integer (quantized) number of standing waves on it. In this case there are 3 standing waves. From Rueckner *et al.* (2014)[31] with kind permission of Allen Crockett. **Bottom**: The first 5 harmonics of a guitar string can be heard by placing the plectrum at the designated positions, which suppresses the other harmonics. From Frudua Guitar Works[32].

Now an oxygen atom has one nucleus of charge +8 (8 protons, stabilized by 8 uncharged neutrons) surrounded by 8 negatively charged electrons in various orbitals, so the molecule O_2 has $N = 18$ particles. The atomic nucleus is counted as one particle. A water molecule then has $N = 22$ particles. It would be nice if there were some relationship between ψ_{H_2}, ψ_{O_2} and ψ_{H_2O}. But there isn't. Each one is a "wave" in an N-dimensional space, where the spaces are entirely different for each molecule. Since N depends on how many particles are present then each molecule also has different values of N. The ψ_{H_2} wave for H_2 "travels" over time in a 4-dimensional space, the ψ_{O_2} wave for O_2 travels in an 18-dimensional space, and the ψ_{H_2O} wave for a water molecule travels in a 22-dimensional space. Furthermore, these spaces are not subspaces of one another: each is unique and different from the others.

It's hard for us to wrap our brains around any space of more than three dimensions. This mental gymnastics is usually avoided by merely making lists of the coordinates of these spaces. But there are mathematicians who delight in trying to visualize the generalization, say, from a two dimensional square, to a 3-dimensional cube, to a 4-D hypercube, etc.[33] Space-time is 4-dimensional.

The problem with quantum mechanics is worse than this. Every group of particles we choose to describe has its own higher dimensional space in which it "exists". These spaces have nothing to do with one another. It is this property, a unique space in which each group of particles exists, that makes each group unique and imparts to it a certain kind of wholeness.

Where is this quantum mechanics space that is different for every molecule? The very word "where" implies it is somewhere "in" our own, comfortable 3D space. But look under your rug, and you won't find it. We could dismiss these spaces as figments of our imagination, but we are faced with the empirical fact that Schrödinger's equation works, that every time it predicts something we can measure, it has proven correct. Most physicists tear out their hair and stop thinking about the meaning of Schrödinger's equation, and just bury their heads in its mathematical sand. Where are these spaces? Don't ask. In summary, every system that we can isolate that consists of atoms can be described by its own higher-dimensional quantum space and thus has its own "Schrödinger quantum wholeness". In this sense, Smuts was quite right that atoms and molecules are legitimate wholes.

But what most scientists who study "quantum wholeness" think about is not this simple "Schrödinger quantum wholeness". For them, quantum wholeness is about the more esoteric phenomenon of quantum entanglement[34]. Quantum entanglement was initially predicted by Albert Einstein, Boris Podolsky and Nathan Rosen a decade after Smuts' book. They proved what seemed to them to be an absurd consequence of quantum mechanics[35,36]. They concluded that quantum mechanics was incomplete. But the joke was on them, because the "absurdity" turned out to be a really observable, measurable and correct aspect of our universe.

To understand entanglement, let's go back to our example of the water molecule. Consider the chemical reaction:

$$2H_2 + O_2 \Leftrightarrow 2H_2O$$

This is a typical balanced chemistry equation, with the number and kinds of particles adding up to the same sums on both sides: 20 electrons, 4 H (hydrogen) atomic nuclei, 2 O (oxygen) atomic nuclei. Thus both sides could be considered as representing a single whole in a 26-dimensional space, and the chemical reaction a change in the corresponding *single* 26-D Schrödinger equation. Since this formulation does not depend on the distance between the electrons and nuclei in our "normal" 3D space, they have an instantaneous, nonlocal connection to one another, even if very far apart, and so are entangled. In other words, if the electrons and atomic nuclei of a pair of water molecules (right side of our chemistry equation) turn into 2 molecules of hydrogen and one of oxygen, those three molecules remain entangled no matter how far apart they may roam.

By way of example, let's start with something much simpler than water, that is entangled. We can experimentally generate two entangled particles with opposite spins that then fly apart. If you measure the spin of one particle, the spin of the other one is instantaneously determined no matter how far apart the particles have flown. According to Einstein's theory of relativity light is the fastest any "thing" can move. But the speed of this quantum entanglement is not limited by the speed of light. It may be instantaneous[37]. For this reason Einstein called quantum entanglement "spooky action at a distance"[38,39,34]. Quantum entanglement has been shown to be real[34] and applicable not only to pairs of subatomic particles as they move well over 100 kilometers apart[40,41], but also to "large" systems[42] of up to 10^{12} atoms[43]. Based on quantum entanglement, now quantum teleportation, the measurement and transmission of a quantum state without transport of matter, is a reality[44]. Quantum computing[45] may follow.

What this means is that quantum mechanics itself is holistic. Reductionism does not work. Reductionism is at best an approximation. And this conclusion applies to groups of particles of all sizes. David Bohm, in his book *Wholeness and the Implicate Order*[46], proposed that entanglement applies to all of the particles in the universe at once. The universe runs under one big (albeit relativistic) Schrödinger equation. Ultimately, the whole universe is but one "unbroken wholeness,... the *implicate* or *enfolded* order"[46]. Anything less is not a whole. What happens anywhere affects everything everywhere.

If this seems like Superman leaping from atoms to the whole universe in a single bound, you're not alone. Describing the universe by a single quantum wave function of a zillion dimensions would seem to be philosophically or religiously satisfying: "Hear, O Israel: the Lord our God, the Lord is one"[47], provided we equate God with the universe as a whole.

There are, however, a few caveats. We still do not have the long sought unified theory of gravity and quantum mechanics[48]. We speculate about what goes on inside black holes and outside our visible universe. And we've discovered the existence of dark matter[49] (26.8%) and dark energy[50] (68.3%). We do not yet understand what they are, but together they make up at least 95.1% of the universe. The matter that we seem to be made of is only 4.9% of all stuff. Even our matter is hypothesized to made up of tiny "strings" that we can't yet experimentally investigate[51]. There is a lot more to the universe than we knew in Bohm's days (1917-1992)[52].

How do we deal, then, with anything whose size is in between tiny strings and the whole universe? The common approach is to regard the whole universe as a nested set of wholes. While in reductionism we started from the bottom and tried to reason up, in holism we start from the top and try to reason down. Going from top down is paralleled by The Great Chain of Being, starting with God on down, a concept that can be considered as dating back to at least the ancient Greeks[53-55]. However, unlike neatly nested Russian matryoshka dolls, in a holistic quantum view drawing boundaries around any given group of particles is arbitrary. Bohm called such groupings "independent subwholes"[56], and Arthur Koestler named them Janus-faced holons[57-60]. Holons "looked up" to higher levels in the hierarchy of wholes, and "looked down" to lower levels. This was an attempt to have one's cake and eat it too. If the universe is fundamentally quantum and thereby holistic, then subwholes and holons are arbitrary mental constructs.

Indeed, evolutionary thought is rife with the concept of "levels" of selection and organization[61-65]. These are in effect subwholes or holons. Arguments occur whether each of these levels has its own laws of nature, and whether biology in general is explainable via physics (the reductionist question again) or is a law unto itself[66,67]. The field of biosemiotics[68,69] is an attempt to deal with symbols and their meaning in and between living organisms. For example, consider the transfer RNAs (Figure 3.12). These match up a sequence of three nucleotides at one end with a specific amino acid at the other end. We call the nucleotide triplet a codon precisely because it acts as a biological code for translating mRNA into protein. Biosemiotics makes a claim that evolution consists of multiple innovations, each of which involves creation of a new biological code. We can regard the differentiation code (Figure 6.43), the mapping between the sequence of differentiation waves cells experience and their resultant cell types, as one of these biological codes. Codes mean that objects (like tRNA) or processes (like differentiation waves) act as symbols for either other objects or processes. Thus codons and differentiation waves take on meanings. This "meaning" is not a parameter measured by physicists. The meanings are kinds of holons that "really" only exist in our heads, it would seem. From the point of view of physics, holons have no characteristics that are describable.

Despite all of its success, the field of quantum mechanics is still in flux. The square of the wave function $\boldsymbol{\psi}_k$ calculated from the Schrödinger equation (strictly speaking, its complex conjugate, if you are familiar with complex numbers) is usually interpreted as the probability of the particles being in a given state. Thus it does not describe the trajectories of individual particles. Some physicists would like to believe that particles move in some continuous fashion, and do not probabilistically appear in one place, then another, magically jumping around. For example, Louis de Broglie[70] proposed in 1927 that each particle is accompanied by a "pilot" wave[71] that guides it from place to place in a smooth fashion. This concept was not well accepted by his peers and he had doubts about his own math. He abandoned the idea[72]. The pilot wave idea was later rediscovered and further developed by Bohm[73]. De Broglie both claimed his priority[74] and nevertheless continued to doubt the veracity of the idea of pilot waves[75].

Now we have a beautiful experimental model for visualizing pilot waves, consisting of a drop of liquid that repeatedly bounces off waves that it itself generates[76] (best appreciated in movies[77-79]). The bouncing ends up in resonance with those waves[80]. These waves cause the bouncing drop to change position and/or direction, usually at every bounce (Figure 12.2), during which the drop generates yet another wave in the fluid off of which it bounces. The waves can superimpose, interfere, and reflect off

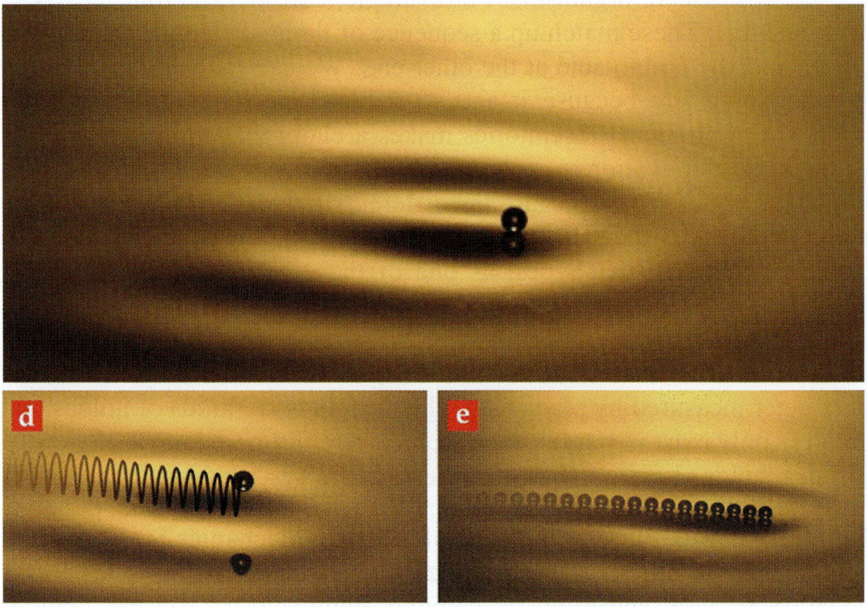

Figure 12.2. **Top**: A fluid drop bouncing on the pilot wave it generated. Due to the angle of impact, it bounces along from left to right. From Harris & Bush (2014)[81] with permission of Cambridge University Press. These drops are often referred to as "walkers". **Bottom**: "(d) The walking drop, whose reflection is also visible, bounces along the liquid surface, as shown by the oscillatory black line. (e) This strobed image of the walker illustrates the drop surfing on its own wave field." From Bush (2015)[80] with kind permission of John W.M. Bush.

boundaries, creating a quite complicated rippling of the whole surface, making it quite difficult to predict the trajectory of the bouncing drop. Nevertheless, this model system is macroscopic (the drops used are millimeters in size), and would seem to be deterministic, as we are dealing with plain old Newtonian mechanics. Yet via its deterministic chaos[82], when many "walking" particles are tracked, in sum they end up reproducing some of the particle interference patterns that so mystify people about quantum mechanics.

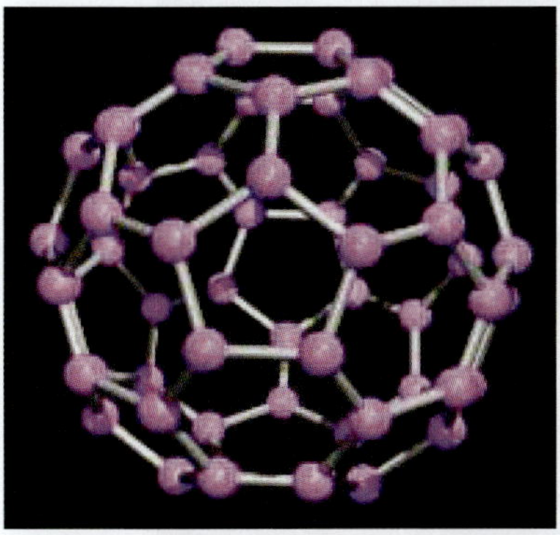

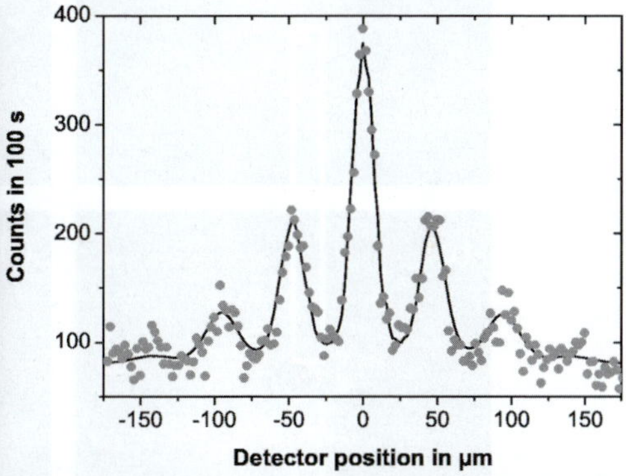

Figure 12.3. **Left**: A Buckyball is a molecule consisting of 60 carbon (C) atoms. From Ball (2000)[83] with permission of Nature Publishing Group. **Right**: The double slit interference pattern of Buckyballs, which were separated in space far enough so there would be no interactions between them as they travelled through the apparatus. The detector was 1.25 m away from the slits. Reproduced from Nairz *et al.* (2003)[84] with permission of American Association of Physics Teachers.

For example, take the infamous double slit experiment[85]. A Buckyball is a ball made of 60 carbon atoms (Figure 12.3). We throw the Buckyballs at a barrier with two openings called slits. Quantum mechanics presumably informs us that a Buckyball cannot be said to have passed through one slit and not the other. Somehow it nevertheless gets to the other side but quantum mechanics does not explain how. This experiment has a new interpretation. When quantum mechanics is taught at an elementary level, we are asked to get used to the idea that sometimes a photon of light acts like a particle, sometimes like a wave. How can a particle also be a wave? In this alternative interpretation, it isn't. It is indeed a particle but the wave properties occur because the particle is accompanied by a pilot wave of its own making[86]. This pilot wave is what diffracts around both slits, not the particle itself[87]. The particle does indeed go through one slit and not the other (Figure 12.4).

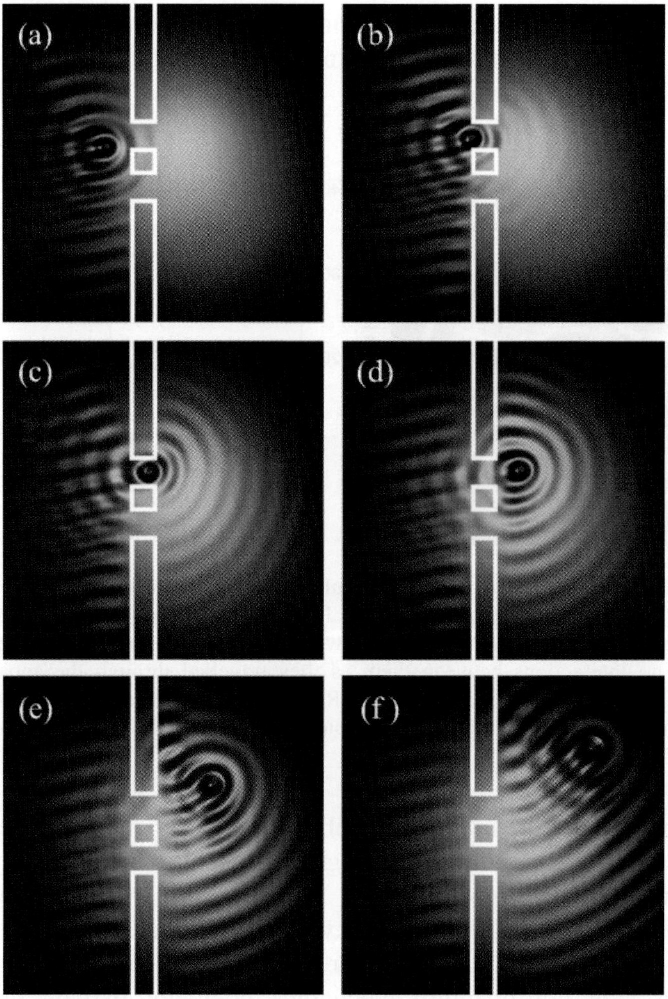

Figure 12.4. "Time series of a walking droplet and the associated wave in the double-slit experiment. The red dot highlights the position of the droplet, and the double slit is accentuated in white." In this case, while the bouncing drop goes through one specific slit of the two, while its pilot wave goes through both slits, there is no obvious interference pattern. From Andersen *et al.* (2015)[94] with permission of the American Physical Society. For a motion picture of a similar experiment, see Freeman (2011)[77].

Albert Einstein protested against quantum mechanics that "God does not throw dice"[88,89]. By this he meant that an individual particle indeed has a specific trajectory in space that is smooth and predictable if indeed we had a more complete theory[90,91]. We can excuse him for not having a convincing way to express his doubts about quantum mechanics. Einstein died before chaos theory developed into its own discipline and understanding how a pilot wave explains the double slit experiment in a smooth way requires chaos theory[92]. Chaos theory is the mathematics of systems that are extremely sensitive to initial conditions and tiny perturbations[92]. It took off as a field in 1963 when Edward Lorentz reported on the sensitivity of his computer simulations of weather to tiny changes in the input data. Formation of weather systems can be dramatically altered by tiny perturbations in initial conditions. This is the so called "butterfly effect" where the answer is "possibly yes" to: "Does the flap of a butterfly's wings in Brazil set off a tornado in Texas?"[93]. Weather is chaotic.

The bouncing drops create and are then confronted with a very complex set of pilot waves. Each bounce is quite sensitive to the exact superimposition of these waves on impact, and so the direction and height of each bounce, and thus its overall trajectory, is chaotic. Despite this, the motion of the drop is smooth in space and time: you can see and follow the drop in fine detail. The individual drop indeed goes through one slit and not the other. The idea of chaotic systems was not formalized in Einstein's time. Therefore he could not use that discipline and apply it to de Broglie's proposal of pilot waves and their potential role in quantum mechanics.

But as we go to press the story has gained a twist: the experiment of the bouncing drops has been shown to have inadequate statistics to conclude that the expected interference patterns (Figure 12.5) are being reproduced[94] (Figure 12.4). We have to wait for the experimenters to get better statistics before we can conclude that the droplet/pilot wave experiment actually helps us understand quantum mechanics, or not.

We turned to quantum mechanics as the example par excellence of wholeness and are left with some strange notions. Not only does holism prevail, but the parts seem to have no independence, even on the cosmic scale. Does the approximation of subwholes or holons help us at all? Bohm did take a stab at wholeness in embryogenesis. He suggested that

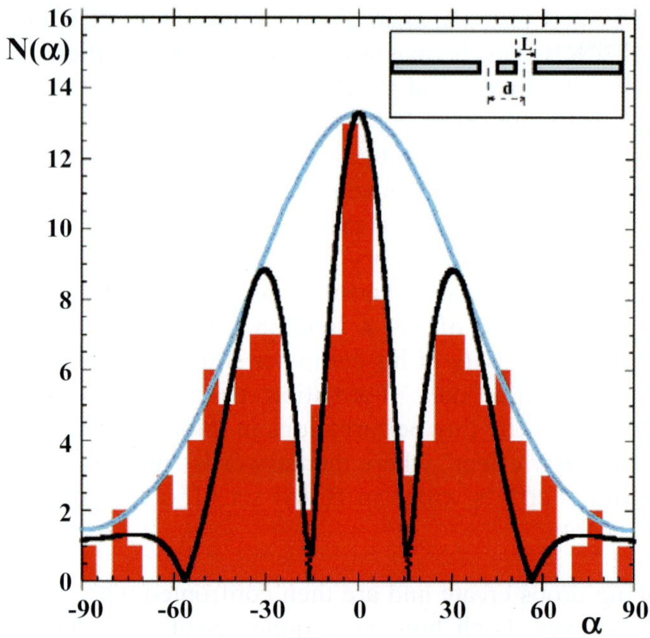

Figure 12.5. "Histogram for the deviation of 75 particles through two slits…", which while looking like a diffraction pattern, as it roughly follows the peaks in a diffraction (black) curve, may not meet statistical criteria for goodness of fit[94]. This of course suggests that more experiments, with variation of the many physical parameters, are needed, along with further theoretical analysis and computer simulation. From Couder & Fort (2006)[87] with permission of the American Physical Society.

embryos consist of parts that interact with each other and the whole embryo[46]. What he is doing here is drawing an analogy[56,95], not applying his quantum mechanics to the embryo. He makes an analogy of the role of DNA in (plant) embryos to entanglement[46], but does not discuss any physics of the embryo. We cannot conclude from his remarks that quantum entanglement has anything to do with embryogenesis. Attempts by Bohm to relate his implicate order to embryogenesis[7,95] thus leave much to be desired.

As we noted in the title of this chapter, embryogenesis is the self-construction of the observer. A further feature of quantum mechanists is that they are obsessed with the role of "the observer", generally conscious, during measurement of physical phenomena[96-100,72,101-107]. However, quantum mechanists never discuss how this observer comes to be. If the observer is required for quantum mechanics, then embryogenesis must also be part of the package. It is the means by which the universe acquires observers.

However, if we as observers are merely quantum states of a set of preexisting groups of particles, there would seem to be some circular reasoning here. This leads to weird behavior on the part of some physicists. These physicists concern themselves as to whether or not the quantum state of a cat in a box, that will be poisoned or not depending on a quantum event in the box, corresponds to it being dead or alive. In this mind game, the cat could have a wave function $\boldsymbol{\psi}_{\text{dead}}$ or $\boldsymbol{\psi}_{\text{alive}}$, or any mixture (superposition) of the two. But there is nothing in Schrödinger's equation that corresponds to "Schrödinger's cat" being dead or alive. As we all well know, people can be brain dead and donate their live hearts and other organs to others who need them, so the very definition of dead depends on context.

The all important observer in cosmology had its first appearance in the form of the anthropic principle[108,109]. The basic notion is that the ratios of certain measurable physical "constants" have to be reasonably close to what they are or else atoms would not exist[110,112,113]. If no atoms, then presumably no intelligent life. In other words, the supposition is that the fundamental parameters of *our* universe[111], like the speed of light[112] (Figure 12.6), must be what they are in order for life and thus observers (or at least physicists!) to exist. This amounts to saying that in "other" universes, where these parameters could be different, things like atoms and molecules might not exist, and thus presumably neither would conscious beings exist.

String theory is the notion that all matter consists of one-dimensional strings[51]. It has been argued that string theory may allow us to calculate values of the "fundamental" parameters[117]. This would make the anthropic principle irrelevant. All of these hypothesized universes would have the same constants. In that case observers, and thus embryogenesis, could exist in all universes. And other universes may have been observed[118], or perhaps not[119].

Such is the nature of modern physics, sort of an extension of medieval arguments about "how many Angels may fit upon the point of a Needle?" (Richard Baxter, 1667, in Wikipedia[120]). Like Einstein, it is

Quantity	Symbol	Value in our universe
Speed of light	c	$299792458\ \mathrm{m\,s^{-1}}$
Gravitational constant	G	$6.673 \times 10^{-11}\ \mathrm{m^3\,kg^{-1}\,s^{-2}}$
(Reduced) Planck constant	$\hbar$	$1.05457148 \times 10^{-34}\ \mathrm{m^2\,kg\,s^{-1}}$ [1]
Planck mass-energy	$m_{\mathrm{Pl}} = \sqrt{\hbar c/G}$	1.2209×10^{22} MeV
Mass of electron; proton; neutron	m_{e}; m_{p}; m_{n}	0.511; 938.3; 939.6 MeV
Mass of up; down; strange quark	m_{u}; m_{d}; m_{s}	(Approx.) 2.4; 4.8; 104 MeV
Ratio of electron to proton mass	β	$(1836.15)^{-1}$
Gravitational coupling constant	$\alpha_{\mathrm{G}} = m_{\mathrm{p}}^2/m_{\mathrm{Pl}}^2$	5.9×10^{-39}
Hypercharge coupling constant	α_1	1/98.4
Weak coupling constant	α_2	1/29.6
Strong force coupling constant	$\alpha_{\mathrm{s}} = \alpha_3$	0.1187
Fine-structure constant	$\alpha = \alpha_1\alpha_2/(\alpha_1 + \alpha_2)$	1/127.9 (1/137 at low energy)
Higgs vacuum expectation value	v	246.2 GeV
QCD scale	Λ_{QCD}	≈ 200 MeV
Yukawa couplings	$\Gamma_i = \sqrt{2}m_i/v$	Listed in Tegmark et al. (2006)
Hubble constant	H	$71\ \mathrm{km\,s^{-1}\,Mpc^{-1}}$ (today)
Cosmological constant (energy density)	$\Lambda(\rho_\Lambda)$	$\rho_\Lambda = (2.3 \times 10^{-3}\ \mathrm{eV})^4$ [4]
Amplitude of primordial fluctuations	Q	2×10^{-5}
Total matter mass per photon	ξ	≈ 4 eV
Baryonic mass per photon	ξ_{baryon}	≈ 0.61 eV

Using the definitions in Burgess & Moore (2006). Many of these quantities are listed in Tegmark et al. (2006), Burgess & Moore (2006, Table A.2) and Nakamura (2010). Unless otherwise noted, standard model coupling constants are evaluated at m_Z, the mass of the Z particle, and hereafter we will use Planck units: $G = \hbar = c = 1$, unless reintroduced for clarity. Note that often in the fine-tuning literature (e.g. Carr & Rees 1979; Barrow & Tipler 1986, p. 354), the low energy weak coupling constant is defined as $\alpha_w \equiv G_F m_e^2$, where $G_F = 1/\sqrt{2}v^2 = (292.8\ \mathrm{GeV})^{-2}$ is the Fermi constant. Using the definition of the Yukawa coupling above, we can write this as $\alpha_w = \Gamma_e^2/2\sqrt{2} \approx 3 \times 10^{12}$. This means that α_w is independent of α_2.

Figure 12.6. Fundamental and derived physical and cosmological parameters. From Barnes (2012)[112] with kind permission of Luke Barnes, CSIRO Publishing, and Cambridge University Press. Cited literature[113-115,111,116]. Superscripts in boxes are corrections by Luke Barnes.

hard for us to take some modern physicists seriously. Consider the concept of the vacuum energy, the idea that so-called empty space isn't empty at all. The notion is that fluctuations in this energy can produce pairs of particles that exist for a short time, annihilate because they are matter/antimatter pairs, and return to the vacuum. This has measurable predictions. However, the theoretical estimates[121] for the density of the vacuum energy vary from 10^{-9} to 10^{113} joules per cubic meter, so it is hardly a robust theory. Assuming the multiuniverse has infinite time and infinite space and that the vacuum fluctuations can make particles, then any structure made of matter could exist. Such fluctuations have been called "Boltzmann brains", named after Ludwig Boltzmann, one of the founders of statistical mechanics[19,122].

Out of all possible structures we can think of as being produced by vacuum fluctuations, physicists have chosen to focus on "Boltzmann brains". Boltzmann brains are hypothetical constructs that spring forth spontaneously from vacuum fluctuations in space that come complete with imaginations that they themselves are real. Therefore Boltzmann brains are observers. How can we tell them from us, the "real" observers? We can't, because we might be Boltzmann brains too. We can't tell if we are real or not. Moreover, the math of these infinities in our inflationarily expanding universe[123] shows the universe would eventually be dominated by Boltzmann brains[124]. However, it seems that man-eating "Boltzmann sharks" could be just as well imagined and would make the ecology of infinite multiverses a bit more interesting.

And so it goes. Some add the concept that the vacuum energy itself has its own quantum levels, and that we might live in a bubble of a higher quantum energy that is about to collapse. We discussed solar systems flung out of galaxies by encounters with pairs of black holes in Chapter 1. This has been proposed as a means of preserving us observers against the hypothesized decay of our local vacuum to a lower, (hypothesized again) matter disrupting state[125]. Despite the expansion of the universe isolating clusters of galaxies from one another, we observers may be able to hang on for the slingshot ride around the black holes[126,127]. In reaction to all of this (nonsense?), some physicists argue that just maybe the universe does not require us observers to exist[100,72,128].

In summary, the old view of physics was reductionist, in that it started with fundamental particles and/or fields (such as magnetic or electric) and described the universe from the bottom up. The new view is "holistic", in that the universe is regarded as a single whole, and the parts must always be described in that context, a top down view. Both are terribly unsatisfactory, if only because we don't have any sensible bottom to start up from, and the "whole" is flying apart faster than the speed of light, so we have no sensible top to start down from. In today's parlance the two "ends", bottom and top, string theory and multiverses, have no empirical data on which they rest. Going up we encounter so-called emergent phenomena[129], while coming down we arbitrarily create a hierarchy of intermediate wholes that may be nothing but mental constructs. The inability of physics to deal with such issues in any satisfactory way might be called the "poverty of physics"[130]. The robes of the Queen of Sciences, Physics, are in tatters, and even her King, Mathematics, is not in much better shape[131-136]. So now it's time to return to the Court Jester, Biology, in its most in your face performance, Embryogenesis.

In all of this, the "observer" pops up in the minds of physicists at all levels, starting with Einstein's theory of relativity[137-139,103]. Whatever the case, embryogenesis is the self-construction of the observer. A fertilized axolotl embryo will develop into an axolotl larva in distilled water at 1 atmosphere of pressure between 8 and 24°C. The embryo grows with no assist from the rest of its environment other than gas and water exchange. It hardly seems to need the help of the whole universe. Gravity[140], the Earth's magnetic field, the solar photon and neutrino flux, background radiation, cosmic rays, gravity waves, dark matter, dark energy, etc., would seem to be benign contexts, and thus irrelevant. But physics, cosmology and embryogenesis should be intimately tied together, if the above musings have any validity. Can we make any sense of this "union of the observer and observed"[7]? If the conscious observer is part of physics, then which is more fundamental? These are heady questions, so let's try to bring things back down to Earth.

In practice, or at least in their teaching, most physicists implicitly invoke wholeness. Let's consider a simple nonquantum example. A drop of water is said to round up "due to surface tension". But surface tension, and the boundary between that drop and the rest of the universe, are mental constructs. There is no force due to surface tension. The forces are between the water molecules, but we can conveniently assign a boundary representing the drop and talk as if there is a force called surface tension keeping it round. Yet we speak of surface tension as if it were a real force. It is responsible for the shape of liquid droplets and has multiple effects[141], even though it does not exist. Quantum mechanical wholeness of the drop is of no help here, because it also does not describe surface tension in any way. On the other hand, the surface tension gives the drop a kind of wholeness, beyond its quantum wholeness. The surface tension "controls" the positions of the water molecules inside the drop, at least enough to keep it round. We can see this kind of wholeness in the mathematics of what happens when we indent a ball (Figure 2.58) or a bacterium (Figure 12.7). We could call surface tension in a drop of water an emergent phenomenon.

Beyond Smuts' list, scientists now recognize many other forms of what we now call emergent phenomena[142,129], from pattern formation to populations to social networks and crowd behavior[143] in the living world, and from atoms to solar systems to galaxies to universes in the physical world[144]. What we wish to accomplish here is to consider to what extent the concept of wholeness helps us understand embryogenesis. In their extreme forms these "explanations" are labeled reductionist versus holistic, how versus why[142], materialistic versus spiritual, or rational versus religious. Both approaches have had remarkable successes in some domains, and thus are in a tug-of-war for our attention, as if we have to choose between them.

But first, let's try to get a clearer picture of Smuts' definition of holism. He sees it as a deep something about the universe that tends to create wholes. But what characterizes a whole? He says "it is itself an action factor like its parts". This means that the whole interacts with, and may thereby change, its parts. In a way, the whole has the same status as its parts, and indeed, this is an obvious conclusion from quantum mechanics, evidenced in the diffraction of Buckyballs. In present day jargon, the parts may be considered to be agents that interact with and to some extent are controlled by the whole. The bacterial colony becomes a whole when quorum sensing is going on, and each bacterium thus responds to the whole colony (Chapter 1). We respond to traffic reports and then change when, where and how we drive. The global whole alters the behavior of its parts[143]. In other words, if you are driving in heavy automobile traffic, listening to traffic reports, and change your route as a result, you are an "agent" participating in a whole/part interaction. This goes beyond the notion that everything in the universe is intertwined,

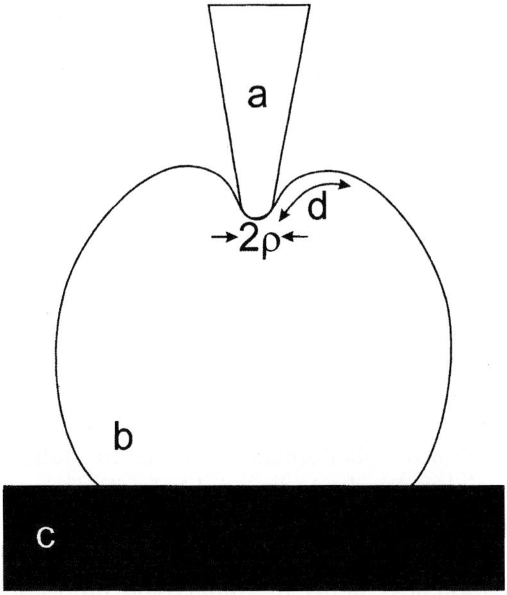

Figure 12.7. Poking a bacterium is an example of mathematical wholeness: the free energy F is an integral over the whole surface of the surface tension σ, stretching in area A with lateral stretching modulus λ, with the applied force f, less an integral over the whole volume V of the turgor pressure p:

$$F = \oint\!\!\oint \left[\sigma + \frac{\lambda}{2}\left(\frac{\delta A}{A}\right)^2 - (\mathbf{f}\cdot\mathbf{n})\psi\right] dA - p \iiint dV$$

From Boulbitch (2000)[145] with permission of Oxford University Press. Furrows formed by differentiation contraction waves would have an analogous global effect on the whole embryo (Figure 6.21) or tissue (Figure 6.33) through which they propagate, and thereby might be involved in altering or launching other differentiation waves. Apical constriction of epithelial cells may be sufficient to generate the forces involved[146,147].

because each whole in such an example has an immediate and direct influence on its parts. Smuts[1] understood this. In other words, in modern jargon, the whole is itself an agent.

Ever since the clockwork universe was bequeathed to us by the ancient atomists and formalized in Newtonian mechanics, the view has been that the universe just muddles along via laws of interaction of its parts. Quantum wholeness has upset Newton's apple cart. Perhaps Newton should have waited for a second apple to fall[148]?

Given all of this, we will not subject the reader to the far flung speculations of a holistic quantum mechanics on beauty, truth, religion, consciousness, free will and creativity, which abound in the popular, philosophical and scientific literature. What Smuts' and quantum wholeness does is give us license to take a new approach to embryogenesis that is neither reductionist nor presuming an act of God, and ask your indulgence to take us at least half seriously in this regard, whether your own predilection is bottom up or top down or uncommitted to either.

Robert Stone and Dick recently introduced a new way to look at embryogenesis, using concepts from the field of cybernetics[122]. Cybernetics started out as the study of control systems with feedbacks, initially for tracking and shooting down enemy aircraft in World War II. Biological applications were immediately apparent[149]. It came together as a discipline with publication[150] of *Cybernetics: Or Control and Communication in the Animal and the Machine* by Norbert Wiener[151] in 1948. In general, cybernetics deals with systems that have goals.

These goals need not be based on high principles. They are not lofty goals. They need not be conscious. A thermostat has a mundane goal, namely a temperature to which it attempts to drive its surroundings. If it is linked to a heater and/or cooler of the surroundings, its "purpose" can actually be realized. Such a goal or purpose is not part of physics, but then neither is surface tension. We apparently do have to depart from the strict reductionism of most physicists to make any headway. In our manmade world we surround ourselves with goal seeking machines. Elevators seek the correct floor. Sump pumps empty our basements of ground water. Toilets flush and refill to a given level. Computer programs of all sorts seek out information for us or do innumerable tasks. We now have vacuum cleaning robots that traverse our homes by themselves, and plug themselves in when their batteries are low. All of this happens nowadays without delving into the realm of science fiction. The internal logic by which such machines work has been beautifully described in a series of "vehicles" of increasing complexity[152]. While cybernetic ideas have been introduced to physicists[153], the author notes: "Feedback and its big brother, control theory, are such important concepts that it is odd that they usually find no formal place in the education of physicists". That is because cybernetics is not (yet?) a part of fundamental physics.

In the living world, we can doubt that a mosquito has an "intention" to bite us and take our blood, but it certainly is good at accomplishing that goal. Goals can be simple, even in living organisms. We started this book by postulating a goal of early cellular life to find food (Figure 1.13), or sporulate to survive without food (Figure 1.16) or escape harm (Figure 1.20). These are acts of perception[130] and action based upon them (Figure 1.14), analogous to the way a thermostat "perceives" the difference between the current temperature and its goal, and "acts" upon it by turning on or off the appropriate cooler or heater. Those cells who succeeded in having such goals survived, while those not having those goals died or failed to leave progeny. So in biology we can be comfortable with the concept of goal seeking, without necessarily invoking consciousness, even if such talk does not fit in physics.

We can also see that this kind of goal seeking requires a system that may be considered to be a kind of a whole. The thermostat, for example, does not do much in isolation. It requires an enclosed space on which to operate (analogous to a cell's membrane or one's skin), a means of communicating with the apparatus that can adjust the whole towards its goal (heater and/or cooler), besides its own set point (a form of memory), and the energy to run the heater and/or cooler. All of that taken together might be called a "whole", and it is a whole is Smuts' sense. One way of putting this is that a thermostat is one component of a "network". This then leads to systems biology and concepts of wholeness in networks[154]. While this leads to important insights, in general systems biology, "the [so-called] next wave in the evolution of the biosciences"[155], ignores spatial relationships and thus morphology, which is what embryogenesis is primarily about.

We are now ready to end our book with our concept of the cybernetic embryo[122]. The idea can be summarized recursively as follows:

1. Each tissue divides into two tissues via a pair of differentiation waves, one contraction wave and one expansion wave.
2. These waves, upon launching, each have a three-part goal: a) spread through a portion of the cells in the tissue, b) cause the cell(s) through which they pass to undergo a step of differentiation, according to the differentiation code and the binary choice (expansion/contraction) of differons (which is encoded in the DNA), c) set up the conditions for launching of the next

pair of differentiation waves in the tissue through which the differentiation wave has propagated.
3. For each wave in a pair of waves, go to #1.

The result is a bifurcating cybernetic system, which we represent as a differentiation tree (Figure 12.8). One nuance is that some waves go through more than one tissue, perhaps providing some global coordination[122]. The cell state splitters and nuclear state splitters triggered by these waves, representable by the differentiation code, result in the right kinds of cells in an organism in the right place, at the right time.

Differentiation waves are reasonably robust, in that they can accommodate variations in cell numbers, cell size, embryo size, and a wide range of temperature and other environmental fluctuations, including the hand of the experimenter, i.e., in multicell tissues they are traversing. By specifying "cell(s)" the above (#2) applies equally to mosaic and regulating organisms, and to tissues of both types in both, as most organisms partake of mosaicism or regulation in various degrees, times and places. As a kind of bifurcating relay race (Chapter 7), differentiation waves may explain the temperature independence of embryogenesis over a wide range of temperatures, despite the dependence of the overall speed of development on temperature (Figure 7.4). Perhaps it is the relay race itself that collapses outside the species-specific temperature range of normal development (Figure 7.5).

We noted the uncanny similarities between fate maps of widely different animals (Figure 2.52), some highly mosaic while others are highly regulating as embryos. If we take the extreme of mosaicism, such as in the nematode *C. elegans*, where essentially every cell is its own tissue, then the fate map is equivalent to its differentiation tree. We can then ask whether the differentiation tree of such an organism has any relationship to, say, the differentiation tree of the axolotl? If the answer (presently under investigation[165,166]) is yes, then what we call a multicellular tissue in a regulating embryo could be looked upon as a cellularized syncytium, analogous to the cellularization of the 6000 nuclei in an early syncytial *Drosophila* embryo (Figure 12.9).

Looked at this way, a regulating embryo is then basically equivalent to a mosaic embryo in terms of its tissues. Pairs of expansion and contraction waves traversing a tissue and dividing it into two new tissues then are but

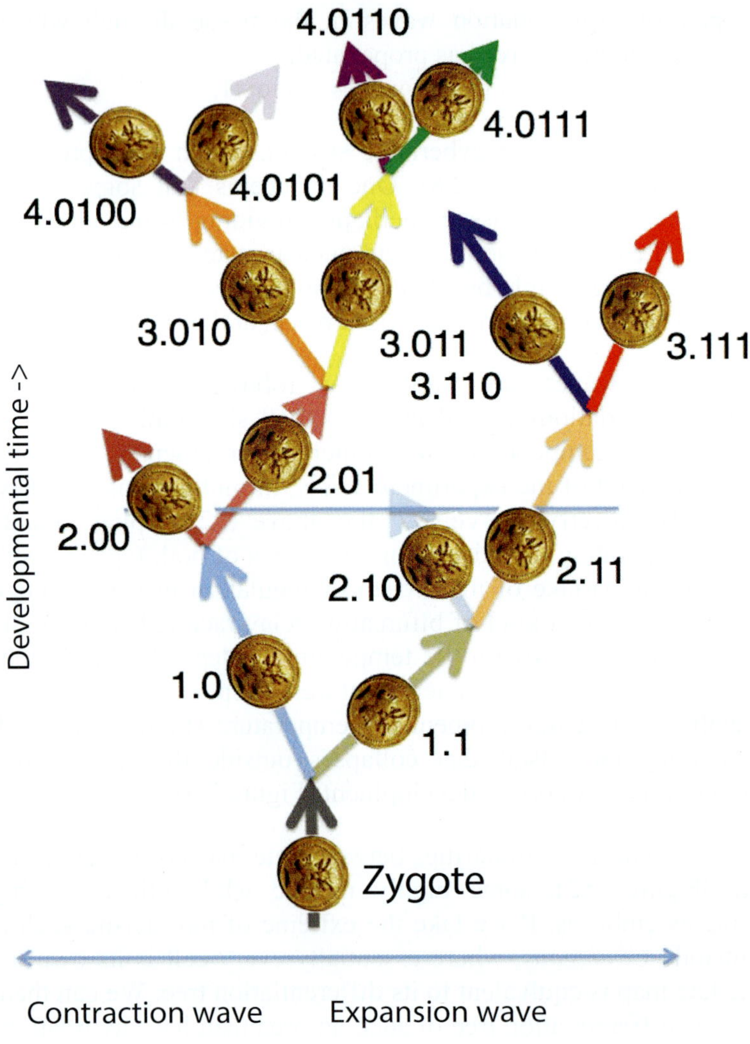

Figure 12.8. The differentiation tree is depicted with each edge, which represents a tissue with a distinct cell type, as being a Janus-faced cybernetic holon. It looks back to the cell type from which it arose, and forward to its goal of setting up the next pair of differentiation waves, one contraction wave and one expansion wave, resulting in two new tissues. The Roman coin is in the Capitoline Museums, Rome, Italy[156]. There are a few intellectual predecessors to the cybernetic embryo. Wilfred Agar suggested that we "...ascribe perception, and therefore final causation and purpose, not only to all independent living organisms but also to embryonic and tissue cells"[157,158], a view amplified by Edmund Sinnott[159,160]. The idea of a tissue hierarchy in development has been around for a while[161,162,57,163].

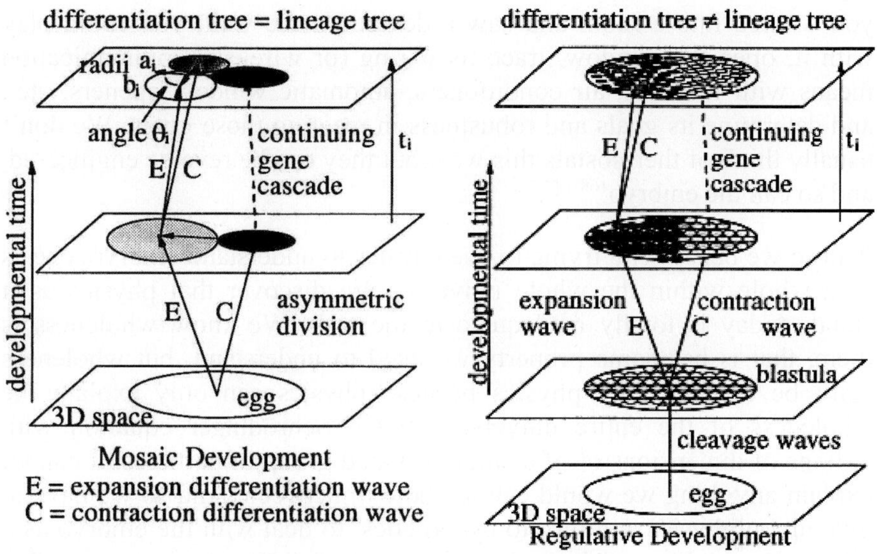

Figure 12.9. One way of looking at regulative development (**right**) is to regard each multicellular tissue as a cellularized version of a single celled tissue of a mosaic organism (**left**). In each case we've flattened the 3D organism to represent the differentiation tree versus time. In terms of multicellularization, the blastula of the regulative embryo maps to the egg of the mosaic embryo, and is when differentiation starts. Cf. Figure 11.8. From Figures 22, 23 in *Hierarchical Genome*[164]. If homologous tissues are formed by the similar bifurcating sequences of events in mosaic and regulating embryos, then there may be a "universal" differentiation tree (as suggested in Chapter 7).

extensions of the asymmetric cell division. We've then come full circle. Instead of regarding an asymmetric cell division as a pair of single-cell differentiation waves (Figure 6.44), we regard an "asymmetric tissue division" via a pair of expansion and contraction waves as if it were a big "tissue-cell" undergoing an asymmetric "tissue-cell" division. If the notion works, it will unify concepts of development across a vast array of phyla. Perhaps we and the worm are indeed but one in developmental mechanism: there is some degree of universality to differentiation trees, and maybe this is why our numbers of genes are similar (Figure 3.11).

With our discovery of differentiation waves, the cybernetic embryo has become a testable hypothesis. As is required of any good theory[167], we can observe and measure and decide if each differentiation wave meets criteria for a goal directed system, and what those goals are. After all, if you walked into a room and saw a device on the wall, you could play with it, open the window, trace its wiring (or wireless) communication means with furnaces, air conditioners, automatic window openers, etc., and determine its goals and robustness in meeting those goals. We don't usually think of thermostats this way, but they can be reverse engineered, and so can the embryo[168-182].

And so we began with trying to use physics to understand embryogenesis as a whole within the whole universe. We discover that physics as it stands today is totally inadequate to the task. We know wholeness is there, that is has some property we need to understand, but wholeness can't be explained by physics because physics can only explain the wholeness of the entire universe using a Schrödinger equation with zillions of dimensions or of a single isolated group of particles. It cannot explain any thing we would call a whole in between. And so we have to give up on physics and turn to cybernetics, to deal with the embryo as a whole. By thinking of the embryo in terms of differentiation waves that have goals in the cybernetic sense, we have a theory of how the embryo builds itself that is experimentally testable.

1. Smuts, J.C. (1926). *Holism and Evolution*. New York, Macmillan Company.
2. Smuts, J.C. (1927). *Holism and Evolution [*http://archive.org/details/holismandevoluti032439mbp*]*. London, Macmillan and Co., 2nd.
3. Wikipedia (2014). Apartheid. http://en.wikipedia.org/wiki/Apartheid
4. Wikipedia (2015). Jan Smuts. https://en.wikipedia.org/wiki/Jan_Smuts
5. Wikipedia (2015). Holism. https://en.wikipedia.org/wiki/Holism
6. Jones, R.H. (2000). *Reductionism: Analysis and the Fullness of Reality*. Lewisburg, Bucknell University Press.
7. Grene, M. (1969). Bohm's metaphysics and biology. In: *Towards a Theoretical Biology. 2. Sketches. An IUBS Symposium*. Ed.: C.H. Waddington. Chicago, Aldine Publishing Co.: 61-71.
8. Sinnott, E.W. (1961). Introduction. In: *Holism and Evolution*. By: J.C. Smuts. New York, Viking Press: ix-xvii.
9. Beloussov, L.V. (2003). Developmental biology and physics of today. In: *Integrative Biophysics: Biophotonics*. Eds.: F.A. Popp & L.V. Beloussov, Springer: 149-178.
10. Beloussov, L.V. & R. Gordon (2006). Preface. Developmental morphodynamics: bridging the gap between the genome and embryo physics. *International Journal of Developmental Biology* **50**(2-3), 79-80.

11. Bianconi, E., A. Piovesan, F. Facchin, A. Beraudi, R. Casadei, F. Frabetti, L. Vitale, M.C. Pelleri, S. Tassani, F. Piva, S. Perez-Amodio, P. Strippoli & S. Canaider (2013). An estimation of the number of cells in the human body. *Annals of Human Biology* **40**(6), 463-471.
12. Kross, B. (2015). How many atoms are in the human body? , http://education.jlab.org/qa/mathatom_04.html
13. Bruhacs, L. (2008). *Vasubandhu's Argument Against Atomism in the Twenty Verses [Master of Arts thesis]*. Frankfurt, Germany, Institut für Philosophie, Univesität-Bibliothek Frankfurt am Main.
14. May, L. (2010). Atomism before Dalton. In: *Atoms in Chemistry: From Dalton's Predecessors to Complex Atoms and Beyond.* Ed.: C.J. Giunta. Washington, DC, USA, American Chemical Society. **1044**: 21-33.
15. Wikipedia (2014). Atomism. http://en.wikipedia.org/wiki/Atomism
16. Wikipedia (2015). John Dalton. https://en.wikipedia.org/wiki/John_Dalton
17. Lindley, D. (2001). *Boltzmann's Atom: The Great Debate that Launched a Revolution in Physics*. New York, Free Press.
18. Cercignani, C. & R. Penrose (2006). *Ludwig Boltzmann: The Man Who Trusted Atoms*, OUP Oxford.
19. Wikipedia (2015). Ludwig Boltzmann. https://en.wikipedia.org/wiki/Ludwig_Boltzmann
20. Einstein, A. (1905). Über die von der molekularkinetischen Theorie der Wärme geforderte Bewegung von in ruhenden Flüssigkeiten suspendierten Teilchen [On the movement of small particles suspended in a stationary liquid demanded by the molecular-kinetic theory of heat]. *Annalen der Physik* **322**(8), 549-560.
21. Machlin, E.S., A. Freilich, D.C. Agrawal, J.J. Burton & C.L. Briant (1975). Field ion microscopy of biomolecules. *Journal of Microscopy* **104**(2), 127-168.
22. Wikipedia (2016). Atomic-force microscopy. https://en.wikipedia.org/wiki/Atomic-force_microscopy
23. Møller, A.P. & T.A. Mousseau (2013). The effects of natural variation in background radioactivity on humans, animals and other organisms. *Biological Reviews* **88**(1), 226-254.
24. Jellinek, A.M. & M.G. Jackson (2015). Connections between the bulk composition, geodynamics and habitability of Earth. *Nature Geoscience* **8**(8), 587-593.
25. Marchand, P.J. (2014). *Life in the Cold: An Introduction to Winter Ecology*. Hanover, University Press of New England, 4th.
26. Gordon, R. & Y. Chen (1987). From statistical mechanics to molecular biology: a Festschrift for Terrell L. Hill. *Cell Biophysics* **11**, i-iii.
27. Wikipedia (2015). Statistical mechanics. https://en.wikipedia.org/wiki/Statistical_mechanics
28. Brush, F.W. (1984). Jan Christian Smuts and his doctrine of holism. *Ultimate Reality and Meaning* **7**(4), 288-297.
29. Poynton, J.C. (1987). Smuts *Holism and Evolution* 60 years on. *Transactions of the Royal Society of South Africa* **46**(Part 3), 181-189.
30. Wikipedia (2015). Schrödinger equation. https://en.wikipedia.org/wiki/Schr%C3%B6dinger_equation
31. Rueckner, W., D. Goodale, D. Rosenberg & A. Crockett (2014). Harvard Natural Sciences Lecture Demonstrations: Standing Waves.

http://isites.harvard.edu/icb/icb.do?keyword=k16940&panel=icb.pagecontent343270%3Ar%241%3Fname%3Dindepth.html&pageid=icb.page80866&pageContentId=icb.pagecontent273326&view=view.do&viewParam_name=indepth.html

32. Frudua, G. (2010). Guitar Strings Vibration. http://www.frudua.com/guitar_strings_vibration.htm

33. Roberts, S. (2009). *King of Infinite Space: Donald Coxeter, the Man Who Saved Geometry*. Toronto, House of Anansi Press.

34. Wikipedia (2015). Quantum entanglement. https://en.wikipedia.org/wiki/Quantum_entanglement

35. Einstein, A., B. Podolsky & N. Rosen (1935). Can quantum mechanical description of physical reality be considered complete? *Physical Review* **47**(10), 777-780.

36. Wikipedia (2015). EPR paradox. https://en.wikipedia.org/wiki/EPR_paradox

37. Yin, J., Y. Cao, H.-L. Yong, J.-G. Ren, H. Liang, S.-K. Liao, F. Zhou, C. Liu, Y.-P. Wu, G.-S. Pan, Q. Zhang, C.-Z. Peng & J.-W. Pan (2013). Bounding the speed of 'spooky action at a distance'. http://arxiv.org/abs/1303.0614

38. Einstein, A., M. Born & H. Born (1971). *The Born-Einstein Letters: Correspondence Between Albert Einstein and Max and Hedwig Born from 1916-1955, with Commentaries by Max Born*, Macmillan.

39. Bell, J.S. (1987). *Speakable and Unspeakable in Quantum Theory*. Cambridge, Cambridge University Press.

40. Ma, X.-S., T. Herbst, T. Scheidl, D. Wang, S. Kropatschek, W. Naylor, B. Wittmann, A. Mech, J. Kofler, E. Anisimova, V. Makarov, T. Jennewein, R. Ursin & A. Zeilinger (2012). Quantum teleportation over 143 kilometres using active feed-forward. *Nature* **489**(7415), 269-273.

41. Yin, J., J.-G. Ren, H. Lu, Y. Cao, H.-L. Yong, Y.-P. Wu, C. Liu, S.-K. Liao, F. Zhou, Y. Jiang, X.-D. Cai, P. Xu, G.-S. Pan, J.-J. Jia, Y.-M. Huang, H. Yin, J.-Y. Wang, Y.-A. Chen, C.-Z. Peng & J.-W. Pan (2012). Quantum teleportation and entanglement distribution over 100-kilometre free-space channels. *Nature* **488**(7410), 185-188.

42. Beduini, F.A., J.A. Zielińska, V.G. Lucivero, Y.A. de Icaza Astiz & M.W. Mitchell (2015). Macroscopic quantum state analyzed particle by particle. *Physical Review Letters* **114**(12), 120402.

43. Vedral, V. (2011). Living in a quantum world. *Scientific American* **304**(6), 38-43.

44. Pirandola, S., J. Eisert, C. Weedbrook, A. Furusawa & S.L. Braunstein (2015). Advances in quantum teleportation. *Nature Photonics* **9**(10), 641-652.

45. Wikipedia (2015). Quantum computing. https://en.wikipedia.org/wiki/Quantum_computing

46. Bohm, D. (1980). *Wholeness and the Implicate Order*. London, Routledge & Kegan Paul.

47. Wikipedia (2015). Shema Yisrael. https://en.wikipedia.org/wiki/Shema_Yisrael

48. Wikipedia (2015). Unified field theory. https://en.wikipedia.org/wiki/Unified_field_theory

49. Wikipedia (2015). Dark matter. https://en.wikipedia.org/wiki/Dark_matter

50. Wikipedia (2015). Dark energy. https://en.wikipedia.org/wiki/Dark_energy

51. Wikipedia (2015). String theory. https://en.wikipedia.org/wiki/String_theory

52. Wikipedia (2015). David Bohm. https://en.wikipedia.org/wiki/David_Bohm

53. Lovejoy, A.O. (1936). *The Great Chain of Being: A Study of the History of an Idea*. Cambridge, Harvard University Press.
54. Kuntz, M.L. & P.G. Kuntz (1987). *Jacob's Ladder and the Tree of Life: Concepts of Hierarchy and the Great Chain of Being*. New York, Peter Lang.
55. Wikipedia (2015). Great chain of being. https://en.wikipedia.org/wiki/Great_chain_of_being
56. Bohm, D. (1985). Hidden variables and the implicate order. *Zygon* **20**(2), 111-133.
57. Koestler, A. (1967). *The Ghost in the Machine*. New York, Macmillan.
58. Koestler, A. (1969). Beyond atomism and holism—the concept of the holon. In: *Beyond Reductionism, New Perspectives in the Life Sciences: The Alpbach Symposium 1968*. Eds.: A. Koestler & J.R. Smythies. New York, Macmillan Company: 192-232.
59. Koestler, A. (1970). Beyond atomism and holism—the concept of the holon. *Perspectives in Biology and Medicine* **13**(2), 131-154.
60. Koestler, A. (1978). *Janus: A Summing Up*, Random House.
61. Hecht, M.K. (1965). The role of natural selection and evolutionary rates in the origin of higher levels of organization. *Systematic Zoology* **14**, 301-317.
62. Eldredge, N. (1983). Phenomenological levels and evolutionary rates. *Systematic Zoology* **31**, 338-347.
63. Keller, L. (1999). *Levels of Selection in Evolution*, Princeton University Press.
64. Calcott, B. & K. Sterelny, Eds. (2011). *The Major Transitions in Evolution Revisited*. Cambridge, MA USA, MIT Press.
65. Rainey, P.B. & B. Kerr (2011). Conflicts among levels of selection as fuel for the evolution of individuality. In: *The Major Transitions in Evolution Revisited*. Ed.: B. Calcott & K. Sterelny. Cambridge, MA USA, MIT Press: 141-162.
66. Mayr, E. (2004). The autonomy of biology. *Ludus Vitalis* **12**(21), 15-27.
67. Mayr, E. (2004). *What Makes Biology Unique? Considerations On the Autonomy of a Scientific Discipline*. Cambridge, Cambridge University Press.
68. Barbieri, M. (2001). *The Organic Codes: The Birth of Semantic Biology*. Ancona, Italy, Casa editrice Pequod.
69. Barbieri, M., Ed. (2008). *The Codes of Life: The Rules of Macroevolution*, Springer.
70. Wikipedia (2015). Louis de Broglie. https://en.wikipedia.org/wiki/Louis_de_Broglie
71. de Broglie, L. (1927). La mécanique ondulatoire et la structure atomique de la matière et du rayonnement [Wave mechanics and atomic structure of matter and radiation]. *Journal de Physique et le Radium* **8**(5), 225-241.
72. Goldstein, S. (1998). Quantum theory without observers - Part two. *Physics Today* **51**(4), 38-42.
73. Bohm, D. & B.J. Hiley (1993). *The Undivided Universe: An Ontological Interpretation of Quantum Theory*. London, Routledge.
74. de Broglie, L. (1951). Remarques sur la théorie de l'onde pilote. *Comptes rendus hebdomadaires des séances de l'Académie des sciences, Paris* **233**(12), 641-644.
75. de Broglie, L., P. Barut, G. Lochak & A. van der Merwe (2012). *Heisenberg's Uncertainties and the Probabilistic Interpretation of Wave Mechanics: with Critical Notes of the Author*, Springer Netherlands.

76. Bush, J.W.M. (2015). Pilot-wave hydrodynamics. *Annual Review of Fluid Mechanics* **47**, 269-292.
77. Freeman, M. (2011). Through the Wormhole - Wave/Particle - Silicon Droplets. https://www.youtube.com/watch?v=fnUBaBdl0Aw
78. Harris, D.M. & J.W.M. Bush (2012). The pilot-wave dynamics of walking droplets. https://www.youtube.com/watch?v=nmC0ygr08tE
79. Parry, B. (2015). Pilot wave dynamics of walking droplets; Cal Poly SLO Physics. https://www.youtube.com/watch?v=JUI_DtzXdw4
80. Bush, J.W.M. (2015). The new wave of pilot-wave theory. *Physics Today* **68**(8), 47-53.
81. Harris, D.M. & J.W.M. Bush (2014). Droplets walking in a rotating frame: from quantized orbits to multimodal statistics. *Journal of Fluid Mechanics* **739**, 444-464.
82. Perrard, S., M. Labousse, E. Fort & Y. Couder (2014). Chaos driven by interfering memory. *Physical Review Letters* **113**(10), #104101.
83. Ball, P. (2000). Buckyball superconductors hot up. http://www.nature.com/news/2000/001130/full/news001130-9.html
84. Nairz, O., M. Arndt & A. Zeilinger (2003). Quantum interference experiments with large molecules. *American Journal of Physics* **71**(4), 319-325, doi: 310.1119/1111.1531580.
85. Wikipedia (2015). Double-slit experiment. https://en.wikipedia.org/wiki/Double-slit_experiment
86. Grössing, G. (2014). Emergence of quantum mechanics from a sub-quantum statistical mechanics. *International Journal of Modern Physics B* **28**(26), #1450179.
87. Couder, Y. & E. Fort (2006). Single-particle diffraction and interference at a macroscopic scale. *Physical Review Letters* **97**(15), #154101.
88. Einstein, A., H. Born & M. Born (1969). *Albert Einstein, Hedwig und Max Born, Briefwechsel: 1916-1955*, Edition Erbrich, 1982 reprint.
89. Wikipedia (2015). Bohr–Einstein debates. https://en.wikipedia.org/wiki/Bohr-Einstein_debates
90. Beller, M. (1993). Einstein and Bohr's rhetoric of complementarity. *Science in Context* **6**(1), 241-255.
91. Fine, A. (1993). Einstein's interpretations of the quantum theory. *Science in Context* **6**(1), 257-273.
92. Wikipedia (2015). Chaos theory. https://en.wikipedia.org/wiki/Chaos_theory - History
93. Wikipedia (2015). Butterfly effect. https://en.wikipedia.org/wiki/Butterfly_effect
94. Andersen, A., J. Madsen, C. Reichelt, S.R. Ahl, B. Lautrup, C. Ellegaard, M.T. Levinsen & T. Bohr (2015). Double-slit experiment with single wave-driven particles and its relation to quantum mechanics. *Physical Review E* **92**(1), #013006.
95. Pylkkänen, P.T.I. (2006). *Mind, Matter and the Implicate Order*, Springer Berlin Heidelberg.
96. Bhandari, R. (1976). Observer in world-view of physics. *Current Science* **45**(23), 817-818.
97. Molchanov, Y.B. (1981). Problem of the subject (observer) in modern physics [Russian]. *Voprosy Filosofii [Problems of Philosophy]*(7), 52-63.

98. Dennett, D.C. & M. Kinsbourne (1992). Time and the observer: the where and when of consciousness in the brain. *Behavioral and Brain Sciences* **15**, 183-200.

99. Mitchell, M. & A. Goswami (1992). Quantum mechanics for observer systems. *Physics Essays* **5**, 526-530.

100. Goldstein, S. (1998). Quantum theory without observers - Part one. *Physics Today* **51**(3), 42-46.

101. Thaheld, F.H. (1998). Can we determine the role of the observer in quantum mechanics? *Physics Essays* **11**(2), 235-240.

102. Goldstein, S. & S. Teufel (2001). Quantum spacetime without observers: Ontological clarity and the conceptual foundations of quantum gravity. In: *Physics Meets Philosophy at the Planck Scale*. Ed.: C. Callender & N. Huggett. Cambridge, Cambridge University Press: 275-289.

103. Rowland, D.R. (2006). Noninertial observers in special relativity and clock synchronization debates. *Foundations of Physics Letters* **19**(2), 103-126.

104. Osoroma, D.S. (2010). Nature of the physical observer. In: *Search for Fundamental Theory*. Eds.: R.L. Amoroso, P. Rowlands & S. Jeffers. **1316**: 420-454.

105. Garrett, T. (2011). The observer class hypothesis. http://arxiv.org/abs/1101.2198

106. Sutner, K. (2012). Universality, Turing incompleteness and observers. In: *A Computable Universe: Understanding and Exploring Nature As Computation*. Eds.: A.G. Ladde, G.S. Ladde & H. Zenil. Singapore, World Scientific Publishing Company Incorporated.

107. Da Silva, M.A., R.M. Serra & L.C. Céleri (2013). Observer invariance of the collapse postulate of quantum mechanics. *International Journal of Modern Physics B* **27**(01n03), 1345013.

108. Breuer, R. (1990). *The Anthropic Principle, Man as the Focal Point of Nature*. Boston, Birkhäuser.

109. Wikipedia (2015). Anthropic principle. https://en.wikipedia.org/wiki/Anthropic_principle

110. Adams, F.C. (2016). Constraints on alternate universes: stars and habitable planets with different fundamental constants. *Journal of Cosmology and Astroparticle Physics*(2), doi:10.1088/1475-7516/2016/1002/1042.

111. Tegmark, M., A. Aguirre, M.J. Rees & F. Wilczek (2006). Dimensionless constants, cosmology, and other dark matters. *Physical Review D* **73**(2), #023505.

112. Barnes, L.A. (2012). The fine-tuning of the universe for intelligent life. *Publications of the Astronomical Society of Australia* **29**(4), 529-564.

113. Carr, B.J. & M.J. Rees (1979). The anthropic principle and the structure of the physical world. *Nature* **278**(5705), 605-612.

114. Barrow, J.D. & F.J. Tipler (1986). *The Anthropic Cosmological Principle*. Oxford, Clarendon Press.

115. Burgess, C. & G. Moore (2006). *The Standard Model: A Primer*. Cambridge, Cambridge University Press.

116. Nakamura, K., K. Hagiwara, K. Hikasa, H. Murayama, M. Tanabashi, T. Watari, C. Amsler, M. Antonelli, D.M. Asner, H. Baer, H.R. Band, R.M. Barnett, T. Basaglia, E. Bergren, J. Beringer, G. Bernardi, W. Bertl, H. Bichsel, O. Biebel, E. Blucher, S. Blusk, R.N. Cahn, M. Carena, A. Ceccucci, D.

Chakraborty, M.C. Chen, R.S. Chivukula, G. Cowan, O. Dahl, G. D'Ambrosio, T. Damour, D. de Florian, A. de Gouvea, T. DeGrand, G. Dissertori, B. Dobrescu, M. Doser, M. Drees, D.A. Edwards, S. Eidelman, J. Erler, V.V. Ezhela, W. Fetscher, B.D. Fields, B. Foster, T.K. Gaisser, L. Garren, H.J. Gerber, G. Gerbier, T. Gherghetta, C.F. Giudice, S. Golwala, M. Goodman, C. Grab, A.V. Gritsan, J.F. Grivaz, D.E. Groom, M. Grunewald, A. Gurtu, T. Gutsche, H.E. Haber, C. Hagmann, K.G. Hayes, M. Heffner, B. Heltsley, J.J. Hernandez-Rey, A. Hoecker, J. Holder, J. Huston, J.D. Jackson, K.F. Johnson, T. Junk, A. Karle, D. Karlen, B. Kayser, D. Kirkby, S.R. Klein, C. Kolda, R.V. Kowalewski, B. Krusche, Y.V. Kuyanov, Y. Kwon, O. Lahav, P. Langacker, A. Liddle, Z. Ligeti, C.J. Lin, T.M. Liss, L. Littenberg, K.S. Lugovsky, S.B. Lugovsky, J. Lys, H. Mahlke, T. Mannel, A.V. Manohar, W.J. Marciano, A.D. Martin, A. Masoni, D. Milstead, R. Miquel, K. Moenig, M. Narain, P. Nason, S. Navas, P. Nevski, Y. Nir, K.A. Olive, L. Pape, C. Patrignani, J.A. Peacock, S.T. Petcov, A. Piepke, G. Punzi, A. Quadt, S. Raby, G. Raffelt, B.N. Ratcliff, P. Richardson, S. Roesler, S. Rolli, A. Romaniouk, L.J. Rosenberg, J.L. Rosner, C.T. Sachrajda, Y. Sakai, G.P. Salam, S. Sarkar, F. Sauli, O. Schneider, K. Scholberg, D. Scott, W.G. Seligman, M.H. Shaevitz, M. Silari, T. Sjostrand, J.G. Smith, G.F. Smoot, S. Spanier, H. Spieler, A. Stahl, T. Stanev, S.L. Stone, T. Sumiyoshi, M.J. Syphers, J. Terning, M. Titov, N.P. Tkachenko, N.A. Tornqvist, D. Tovey, T.G. Trippe, G. Valencia, K. van Bibber, G. Venanzoni, M.G. Vincter, P. Vogel, A. Vogt, W. Walkowiak, C.W. Walter, D.R. Ward, B.R. Webber, G. Weiglein, E.J. Weinberg, J.D. Wells, A. Wheeler, L.R. Wiencke, C.G. Wohl, L. Wolfenstein, J. Womersley, C.L. Woody, R.L. Workman, A. Yamamoto, W.M. Yao, O.V. Zenin, J. Zhang, R.Y. Zhu, P.A. Zyla, G. Harper, V.S. Lugovsky & P. Schaffner (2010). Review of particle physics. *Journal of Physics G-Nuclear and Particle Physics* **37**(7A), 1-1422.

117. Kane, G.L., M.J. Perry & A.N. Zytkow (2002). The beginning of the end of the anthropic principle. *New Astronomy* **7**(1), 45-53.
118. MIT Technology Review (2010). Astronomers Find First Evidence of Other Universes.
119. Chary, R. (2016). Spectral variations of the sky: Constraints on alternate universes. *Astrophysical Journal* **817**(1), #33.
120. Wikipedia (2015). How many angels can dance on the head of a pin? , https://en.wikipedia.org/wiki/How_many_angels_can_dance_on_the_head_of_a_pin%3F
121. Wikipedia (2015). Vacuum energy. https://en.wikipedia.org/wiki/Vacuum_energy
122. Gordon, R. & R. Stone (2016). Cybernetic embryo. In: *Biocommunication.* Eds.: R. Gordon & J. Seckbach. London, World Scientific Publishing: Chapter 5, 111-163.
123. Tegmark, M. (2005). What does inflation really predict? *Journal of Cosmology and Astroparticle Physics*(4), #001.
124. De Simone, A., A.H. Guth, A. Linde, M. Noorbala, M.P. Salem & A. Vilenkin (2010). Boltzmann brains and the scale-factor cutoff measure of the multiverse. *Physical Review D* **82**(6), #063520.
125. Kashyap, S.P., S. Mondal, A. Sen & M. Verma (2015). Surviving in a metastable de Sitter space-time. *Journal of High Energy Physics*(9), #139.

126. Lemonick, M.D. (2014). You've Heard of Shooting Stars, but This is Ridiculous. http://time.com/3602047/hypervelocity-stars/
127. Woo, M. (2014). The Fastest Stars in the Universe May Approach Light Speed. http://www.wired.com/2014/12/superfast-stars-black-holes/
128. Tumulka, R. (2004). Understanding Bohmian mechanics: A dialogue. *American Journal of Physics* **72**(9), 1220-1226.
129. Morowitz, H.J. (2002). *The Emergence of Everything, How the World Became Complex*. Oxford, Oxford University Press.
130. Martin, C.C. & R. Gordon (2001). The evolution of perception. *Cybernetics & Systems* **32**(3-4), 393-409.
131. Dawson Jr., J.W. (1997). *Logical Dilemmas: The Life and Work of Kurt Godel*. Wellesley, Massachusetts, USA, A K Peters Ltd.
132. Calude, C.S. (2005). Algorithmic randomness, quantum physics, and incompleteness. *Lecture Notes in Computer Science* **3354**, 1-17.
133. Goldstein, R. (2005). *Incompleteness: The Proof and Paradox of Kurt Gödel*. New York, W.W. Norton, 1st.
134. Bailly, F. & G. Longo (2008). Phenomenology of incompleteness: From formal deductions to mathematics and physics. In: *Deduction, Computation, Experiment: Exploring the Effectiveness of Proof*. Ed.: R.C.G. Lupacchini, Springer: 243-271.
135. Sutner, K. (2013). Universality, Turing incompleteness and observers. In: *Computable Universe: Understanding and Exploring Nature as Computation*. Ed.: H. Zenil. Singapore, World Scientific Publishing: 435-449.
136. Wikipedia (2015). Gödel's incompleteness theorems. https://en.wikipedia.org/wiki/G%C3%B6del%27s_incompleteness_theorems
137. Einstein, A. (1905). Zur Elektrodynamik bewegter Körper [On the electrodynamics of moving bodies]. *Annalen der Physik* **17**(10), 891-921.
138. Einstein, A. (1907). Über das Relativitätsprinzip und die aus demselben gezogenen Folgerungen [On the relativity principle and the conclusions drawn from it]. *Jahrbuch der Radioaktivität and Elektronik* **4**, 411-462.
139. Einstein, A. (1936). Lens-like action of a star by the deviation of light in the gravitational field. *Science* **84**(2188), 506-507.
140. Crawford-Young, S. (2006). Effects of microgravity on cell cytoskeleton and embryogenesis. *International Journal of Developmental Biology* **50**(2/3), 183-191.
141. Wikipedia (2015). Surface tension. https://en.wikipedia.org/wiki/Surface_tension
142. Corning, P.A. (2002). The re-emergence of "emergence": a venerable concept in search of a theory. *Complexity* **7**(6), 18-30.
143. Friesen, M.R.P., R. Gordon & R.D. McLeod (2014). Exploring emergence within social systems with agent based models [invited]. In: *Interdisciplinary Applications of Agent-Based Social Simulation and Modeling*. Eds.: D.F. Adamatti, G.P. Dimuro & H. Coelho. Hershey, Pennsylvania, USA IGI Global: 52-71.
144. Wallace, D. (2012). *The Emergent Multiverse: Quantum Theory According to the Everett Interpretation*, OUP Oxford.
145. Boulbitch, A. (2000). Deformation of the envelope of a spherical Gram-negative bacterium during the atomic force measurements. *Journal of Electron Microscopy* **49**(3), 459-462.

146. Osterfield, M., X.X. Du, T. Schüpbach, E. Wieschaus & S.Y. Shvartsman (2013). Three-dimensional epithelial morphogenesis in the developing *Drosophila* egg. *Developmental Cell* **24**(4), 400-410.
147. Polyakov, O., B. He, M. Swan, J.W. Shaevitz, M. Kaschube & E. Wieschaus (2014). Passive mechanical forces control cell-shape change during *Drosophila* ventral furrow formation. *Biophysical Journal* **107**(4), 998-1010.
148. McKie, D. & G.R. de Beer (1951). Newton's apple. *Notes and Records of the Royal Society of London* **9**(1), 46-54.
149. Rosenblueth, A., N. Wiener & J. Bigelow (1943). Behavior, purpose and teleology. *Philosophy of Science* **10**(1), 18-24.
150. Wiener, N. (1948). *Cybernetics: Or Control and Communication in the Animal and the Machine*. New York, John Wiley & Sons.
151. Wiener, N. (1953). *Ex-prodigy: My Childhood and Youth*, Simon & Schuster.
152. Braitenberg, V. (1984). *Vehicles: Essays in Synthetic Psychology*. Cambridge, MIT Press.
153. Bechhoefer, J. (2005). Feedback for physicists: A tutorial essay on control. *Reviews of Modern Physics* **77**(3), 783-836.
154. Haruna, T. & Y.-P. Gunji (2009). Wholeness and information processing in biological networks: An algebraic study of network motifs. In: *Natural Computing, Proceedings*. Eds.: Y. Suzuki, M. Hagiya, H. Umeo & A. Adamatzky. **1**: 70-80.
155. Andrews, B., D. Figeys, M. Kaern, S. Kauffman, E. Emberly, M. Hallett, P. Hieter, B. Ingalls, J. Parkinson, T. Pawson, M. Surette, P. Swain & S. Wodak (2008). Integrative Systems Biology: The 21st Century Challenge to Biological, Biomedical and Biotechnological Research in Canada. Briefing Paper and Working Recommendations. https://www.sysbiosociety.ca/documents
156. Musei Capitolini (2015). Coin (half stater) from the "oath" series: double faced head of Janus and oathing swearing scene, 225-217 BC. http://en.museicapitolini.org/collezioni/percorsi_per_sale/medagliere_capitolino/medagliere_capitolino/moneta_mezzo_statere_della_serie_del_giuramento_testa_di_giano_bifronte_e_scena_di_giuramento
157. Agar, W.E. (1943). *A Contribution to the Theory of the Living Organism*. Melbourne, Melbourne University Press & Oxford University Press.
158. Agar, W.E. (1952). *A Contribution to the Theory of the Living Organism*, 2nd.
159. Sinnott, E.W. (1962). *The Biology of The Spirit*. New York The Viking Press.
160. Sinnott, E.W. (1963). *The Problem of Organic Form*. New Haven, Yale University Press.
161. Agar, W.E. (1938). The concept of purpose in biology. *Quarterly Review of Biology* **13**(3), 255-273.
162. Clayton, R.M. (1964). Differentiation. In: *Penguin Science Survey B 1964. Biology. Genetics. Agriculture. Medicine. Zoology. Botany*. Eds.: A. McLaren & S.A. Barnett. Harmondsworth, Middlesex, UK, Penguin Books Ltd.: 68-88.
163. Maynard Smith, J. (1983). Evolution and development. In: *Development and Evolution*. Eds.: B.C. Goodwin, N. Holder & C.C. Wylie. Cambridge, Cambridge University Press: 33-45.
164. Gordon, R. (1999). *The Hierarchical Genome and Differentiation Waves: Novel Unification of Development, Genetics and Evolution*. Singapore &

London, World Scientific & Imperial College Press: [http://www.worldscientific.com/worldscibooks/10.1142/2755].
165. Alicea, B. & R. Gordon (2015). The Network Architecture of Embryo Developmental Regulation [poster]. In: *Network Frontier Workshop, Northwestern University, December 6-7, 2015, Northwestern University, Evanston, Illinois, USA*. Evanston, Illinois, USA, Northwestern University.
166. Alicea, B. & R. Gordon (2016). Quantifying mosaic development: Towards an Evo-Devo Postmodern Synthesis of the evolution of development via differentiation trees of embryos [invited]. *Biology (Basel)* **33**(Special Issue: Beyond the Modern Evolutionary Synthesis- what have we missed?, Ed. John S. Torday), doi:10.3390/biology5030033.
167. Woodger, J.H. (1945). Review of: *A Contribution to the Theory of the Living Organism* by Wilfred Eade Agar. *Philosophy* **20**(77), 265-267.
168. Stein, M.B. & R. Gordon (1982). Epithelia as bubble rafts: a new method for measuring cell shape and intercellular adhesion in embryonic and other epithelia. *Journal of Theoretical Biology* **97**(4), 625-635.
169. Gordon, R. & C.A. Melvin (2003). Reverse engineering the embryo: a graduate course in developmental biology for engineering students at the University of Manitoba, Canada. *International Journal of Developmental Biology* **47**(2/3), 183-187.
170. Kumar, R., F. Stepanek & A. Mantalaris (2004). An oxygen transport model for human bone marrow microcirculation. *Food and Bioproducts Processing* **82**(C2), 105-116.
171. Brodland, G.W., V. Conte, P.G. Cranston, J. Veldhuis, S. Narasimhan, M.S. Hutson, A. Jacinto, F. Ulrich, B. Baum & M. Miodownik (2010). Video force microscopy reveals the mechanics of ventral furrow invagination in *Drosophila. Proceedings of the National Academy of Sciences USA* **107**(51), 22111-22116.
172. Cranston, P.G., J.H. Veldhuis, S. Narasimhan & G.W. Brodland (2010). Cinemechanometry (CMM): a method to determine the forces that drive morphogenetic movements from time-lapse images. *Annals of Biomedical Engineering* **38**(9), 2937-2947.
173. Chiou, K.K., L. Hufnagel & B.I. Shraiman (2012). Mechanical stress inference for two dimensional cell arrays. *PLoS Computational Biology* **8**(5), e1002512.
174. Conte, V., F. Ulrich, B. Baum, J. Muñoz, J. Veldhuis, W. Brodland & M. Miodownik (2012). A biomechanical analysis of ventral furrow formation in the *Drosophila melanogaster* embryo. *PLoS One* **7**(4), e34473.
175. Ishihara, S. & K. Sugimura (2012). Bayesian inference of force dynamics during morphogenesis. *Journal of Theoretical Biology* **313**, 201-211.
176. Lobo, D., W.S. Beane & M. Levin (2012). Modeling planarian regeneration: a primer for reverse-engineering the worm. *PLoS Computational Biology* **8**(4), e1002481.
177. Ishihara, S., K. Sugimura, S.J. Cox, I. Bonnet, Y. Bellaïche & F. Graner (2013). Comparative study of non-invasive force and stress inference methods in tissue. *European Physical Journal E* **36**(4), #45.
178. Lu, K., R. Gordon & T. Cao (2013). Reverse engineering the mechanical and molecular pathways in stem cell morphogenesis. *Journal of Tissue Engineering and Regenerative Medicine*, doi: 10.1002/term.1672.

179. Sugimura, K., Y. Bellaïche, F. Graner, P. Marcq & S. Ishihara (2013). Robustness of force and stress inference in an epithelial tissue. In: *2013 35th Annual International Conference of the IEEE Engineering in Medicine and Biology Society*. IEEE: 2712-2715.

180. Brodland, G.W., J.H. Veldhuis, S. Kim, M. Perrone, D. Mashburn & M.S. Hutson (2014). CellFIT: A cellular force-inference toolkit using curvilinear cell boundaries. *PLoS One* **9**(6), e99116.

181. Lobikin, M., D. Lobo, D.J. Blackiston, C.J. Martyniuk, E. Tkachenko & M. Levin (2015). Serotonergic regulation of melanocyte conversion: A bioelectrically regulated network for stochastic all-or-none hyperpigmentation. *Science Signaling* **8**(397), ra99.

182. Portegys, T., G. Pascualy, R. Gordon, S. McGrew & B. Alicea (2016). Morphozoic, cellular automata with nested neighborhoods as a metamorphic representation of morphogenesis [invited]. In: *Multi-Agent Based Simulations Applied to Biological and Environmental Systems*. Ed.: D.F. Adamatti, IGI Global: Submitted.

Index